Discrete Wavelet Transformations

Discrete Wavelet Transformations

An Elementary Approach with Applications

Patrick J. Van Fleet

University of St. Thomas
St. Paul, Minnesota

Second Edition

WILEY

This edition first published 2019

Edition History
Wiley-Interscience, (1e in English, 2008)

Registered Office
John Wiley & Sons, Inc., 111 River Street, Hoboken, NJ 07030, USA

Editorial Office
111 River Street, Hoboken, NJ 07030, USA

For details of our global editorial offices, customer services, and more information about Wiley products visit us at www.wiley.com.

Wiley also publishes its books in a variety of electronic formats and by print-on-demand. Some content that appears in standard print versions of this book may not be available in other formats.

Library of Congress Cataloging-in-Publication Data

Names: Van Fleet, Patrick J., 1962- author.
Title: Discrete wavelet transformations : an elementary approach with
applications / Patrick J. Van Fleet (University of St. Thomas).
Description: 2nd edition. | Hoboken, NJ : John Wiley & Sons, Inc., [2019] |
Includes bibliographical references and index. |
Identifiers: LCCN 2018046966 (print) | LCCN 2018055538 (ebook) | ISBN
9781118979327 (Adobe PDF) | ISBN 9781118979310 (ePub) | ISBN 9781118979273
(hardcover)
Subjects: LCSH: Wavelets (Mathematics) | Transformations (Mathematics) |
Digital images--Mathematics.
Classification: LCC QA403.3 (ebook) | LCC QA403.3 .V36 2019 (print) | DDC
515/.2433--dc23
LC record available at https://lccn.loc.gov/2018046966

Cover design: Wiley
Cover image: Courtesy of Patrick J. Van Fleet

Set in 10/12pt NimbusRomNo9L by SPi Global, Chennai, India

Printed in the United States of America

V10015037_102119

For Andy

CONTENTS

PREFACE TO THE FIRST EDITION

(Abridged and edited)

Why This Book?

How do you apply wavelets to images? This question was asked of me by a bright undergraduate student while I was a professor in the mid-1990s at Sam Houston State University. I was part of a research group there and we had written papers in the area of multiwavelets, obtained external funding to support our research, and hosted an international conference on multiwavelets. So I fancied myself as somewhat knowledgeable on the topic. But this student wanted to know how they were actually *used* in the applications mentioned in articles she had read. It was quite humbling to admit to her that I could not exactly answer her question. Like most mathematicians, I had a cursory understanding of the applications, but I had never written code that would apply a wavelet transformation to a digital image for the purposes of processing it in some way. Together, we worked out the details of applying a discrete Haar wavelet transformation to a digital image, learned how to use the output to identify the edges in the image (much like what is done in Section 4.4), and wrote software to implement our work.

My first year at the University of St. Thomas was 1998–1999 and I was scheduled to teach Applied Mathematical Modeling II during the spring semester. I wanted to select a current topic that students could immediately grasp and use in concrete applications. I kept returning to my positive experience working with the undergraduate student at Sam Houston State University on the edge detection problem. I was surprised by the number of concepts from calculus and linear algebra that we had reviewed in the process of coding and applying the Haar wavelet transformation. I was also impressed with the way the student embraced the coding portion of the work and connected to it ideas from linear algebra. In December 1998, I attended a wavelet workshop organized by Gilbert Strang and Truong Nguyen. They had just authored the book *Wavelets and Filter Banks* [88], and their presentation of the material focused a bit more on an engineering perspective than a mathematical one. As a result, they developed wavelet filters by using ideas from convolution theory and Fourier series.

I decided that the class I would prepare would adopt the approach of Strang and Nguyen and I planned accordingly. I would attempt to provide enough detail and background material to make the ideas accessible to undergraduates with backgrounds in calculus and linear algebra. I would concentrate only on the development of the *discrete* wavelet transformation. The course would take an "applications first" approach. With minimal background, students would be immersed in applications and provide detailed solutions. Moreover, the students would make heavy use of the computer by writing their own code to apply wavelet transformations to digital audio or image files. Only after the students had a good understanding of the basic ideas and uses of discrete wavelet transformations would we frame general filter development using classical ideas from Fourier series. Finally, wherever possible, I would provide a discussion of *how* and *why* a result was obtained versus a statement of the result followed by a concise proof and example. The first course was enough of a success to try again. To date, I have taught the course seven times and developed course materials (lecture notes, software, and computer labs) to the point where colleagues can use them to offer the course at their home institutions.

As is often the case, this book evolved out of several years' worth of lecture notes prepared for the course. The goal of the text is to present a topic that is useful in several of today's applications involving digital data in such a way that it is accessible to students who have taken calculus and linear algebra. The ideas are motivated through applications – students learn the ideas behind discrete wavelet transformations and their applications by using them in image compression, image edge detection, and signal denoising. I have done my best to provide many of the details for these applications that my SHSU student and I had to discover on our own. In so doing, I found that the material strongly reinforces ideas learned in calculus and linear algebra, provides a natural arena for an introduction of complex numbers, convolution, and Fourier series, offers motivation for student enrollment in higher-level undergraduate courses such as real analysis or complex analysis, and establishes the computer as a legitimate learning tool. The book also introduces students to late-twentieth century mathematics. Students who have grown up in the digital age learn how mathematics is utilized to solve problems they understand and to which they can easily relate. And although students who read this book may not be ready to perform high-level mathematical research, they will be at a point where they can identify problems and open questions studied by researchers today.

To the Student

Many of us have learned a foreign language. Do you remember how you formulated an answer when your instructor asked you a question in the foreign language? If you were like me, you mentally translated the question to English, formulated an answer to the question, and then translated the answer back to the foreign language. Ultimately, the goal is to omit the translation steps from the process, but the language analogy perfectly describes an important mathematical technique.

Mathematicians are often faced with a problem that is difficult to solve. In many instances, mathematicians will *transform* the problem to a different setting, solve the problem in the new setting, and then transform the answer back to the original domain. This is exactly the approach you use in calculus when you learn about u-substitutions or integration by parts. What you might not realize is that for applications involving discrete data (lists or tables of numbers), *matrix multiplication* is often used to transform the data to a setting more conducive to solving the problem. The key, of course, is to choose the correct matrix for the task.

In this book, you will learn about discrete wavelet transformations and their applications. For now, think of the transformation as a matrix that we multiply with a vector (audio) or another matrix (image). The resulting product is much better suited than the original image for performing tasks such as compression, denoising, or edge detection. A wavelet filter is simply a list of numbers that is used to construct the wavelet matrix. Of course, this matrix is very special, and as you might guess, some thought must go into its construction. What you will learn is that the ideas used to construct wavelet filters and wavelet transformation matrices draw largely from calculus and linear algebra.

The first wavelet filters will be easy to construct and the ad hoc approached can be mimicked, to a point, to construct other filters. But then you will need to learn more about convolution and Fourier series in order to systematically construct wavelet filters popular in many applications. The hope is that by this point, you will understand the applications sufficiently to be highly motivated to learn the theory. If your instructor covers the material in Chapters 8 and 9, work hard to master it. In all likelihood, it will be new mathematics for you but the skill you gain from this approach to filter construction will greatly enhance your problem-solving skills.

Questions you are often asked in mathematics courses start with phrases such as "solve this equation," "differentiate/integrate this function," or "invert this matrix." In this book, you will see *why* you need to perform these tasks since the questions you will be asked often start with "denoise this signal," "compress this image," or "build this filter." At first you might find it difficult to solve problems without clear-cut instructions, but understand that this is exactly the approach used to solve problems in mathematical research or industry.

Finally, if your instructor asks you to write software to implement wavelet transformations and their inverses, understand that learning to write good mathematical programs takes time. In many cases, you can simply translate the pseudocode from the book to the programming language you are using. Resist this temptation and take the extra time necessary to deeply understand how the algorithm works. You will develop good programming skills and you will be surprised at the amount of mathematics you can learn in the process.

To the Instructor

The technique of solving problems in the transform domain is common in applied mathematics as used in research and industry, but we do not devote as much time to it as we should in the undergraduate curriculum. It is my hope that faculty can use this book to create a course that can be offered early in the curriculum and fill this void.

I have found that it is entirely tractable to offer this course to students who have completed calculus I and II, a computer programming course, and sophomore linear algebra. I view the course as a post-sophomore capstone course that strengthens student knowledge in the prerequisite courses and provides some rationale and motivation for the mathematics they will see in courses such as real analysis.

The aim is to make the presentation as elementary as possible. Toward this end, explanations are not quite as terse as they could be, applications are showcased, rigor is sometimes sacrificed for ease of presentation, and problem sets and labs sometimes elicit subjective answers. It is difficult as a mathematician to minimize the attention given to rigor and detail (convergence of Fourier series, for example) and it is often irresistible to omit ancillary topics (the FFT and its uses, other filtering techniques, wavelet packets) that almost beg to be included. It is important to remind yourself that you are preparing students for more rigorous mathematics and that any additional topics you introduce takes time away from a schedule that is already quite full.

I have prepared the book and software so that instructors have several options when offering the course. When I teach the course, I typically ask students to write modules (subroutines) for constructing wavelet filters or computing wavelet transformations and their inverses. Once the modules are working, students use them for work on inquiry-type labs that are included in the text. For those faculty members who wish to offer a less programming-intensive course and concentrate on the material in the labs, the complete software package is available for download.

In lieu of a final exam, I ask my students to work on final projects. I usually allow four or five class periods at the end of the semester for student work on the projects. Project topics are either theoretical in design, address a topic in the book we did not cover, or introduce a topic or application involving the wavelet transformation that is not included in the book. In many cases, the projects are totally experimental in nature. Projects are designed for students to work either on their own or in small groups. To make time for the projects, I usually have to omit some chapters from the book. Faculty members who do not require the programming described above and do not assign final projects can probably complete a large portion of the book in a single semester. Detailed course offerings follow in the preface to the second edition.

What You *Won't* Find in This Book

There are two points to be made in this section. The first laments the fact that several pet topics were omitted from the text and the second provides some rationale for the presentation of the material.

For those readers with a background in wavelets, it might pain you to know that except for a short discussion in Chapter 1 and asides in Sections 4.1 and 9.4, there is no mention of scaling functions, wavelet functions, dilation equations, or multiresolution analyses. In other words, this presentation does not follow the classical approach where wavelet filters are constructed in $L^2(\mathbb{R})$. The approach here is entirely discrete – the point of the presentation is to draw as much as possible from courses early in a mathematics curriculum, augment it with the necessary amount of Fourier series, and then use applications to motivate more general filter constructions. For those students who might desire a more classical approach to the topic, I highly recommend the texts by Boggess and Narcowich [7], Frazier [42], Walnut [100] or even Ruch and Van Fleet [77].

Scientists who have used nonwavelet methods (Fourier or otherwise) to process signals and images will not find them discussed in this book. There are many methods for processing signals and images. Some work better than wavelet-based methods, some do not work as well. We should view wavelets as simply another tool that we can use in these applications. Of course, it is important to compare other methods used in signal and image processing to those presented in this book. If those methods were presented here, it would not be possible to cover all the material in a single semester.

Some mathematicians might be disappointed at the lack of rigor in several parts of the text. This is the conundrum that comes with writing a book with modest prerequisites. If we get too bogged down in the details, we lose sight of the applications. On the other hand, if we concern ourselves only with applications, then we never develop a deep understanding of how the mathematics works. I have skewed my presentation to favor more applications and fewer technical details. Where appropriate, I have tried to point out arguments that are incomplete and provided suggestions (or references) that can make the argument rigorous.

Finally, despite the best efforts of friends, colleagues, and students, you will not find an error-free presentation of the material in this book. For this, I am entirely to blame. Any corrections, suggestions, or criticisms will be greatly appreciated!

P. J. VAN FLEET

St. Paul, Minnesota USA
June 2007

PREFACE

What Has Changed?

Prior to the first edition appearing in January 2008, I had taught a version of this course eight times. Supported by National Science Foundation grants (DUE) I began offering workshops and mini-courses about the wavelets course in 2006. Several of the attendees at these events taught a wavelets course at their home institutions and fortunately for me, provided valuable feedback about the course and text. A couple of popular points of discussion were the necessity of a sophomore linear algebra class as a prerequisite and possibility of using non-Fourier methods for the ad hoc filter development. These two points were the primary motivating factors for the creation of the second edition of the text.

While I am not ready to claim that a sophomore linear algebra course is not needed as a prerequisite for a course in wavelets, I am comfortable saying that for a course constructed from particular topics in the book, the necessary background on vectors and matrices can be covered in Chapter 2. In such a case, a prerequisite linear algebra course serves to provide a student with a more mathematical appreciation for the material presented.

The major change in the second edition of the book is the clear delineation of filter development methods. The book starts largely unchanged through the first three chapters save for the inclusion of a section on convolution and filters in Chapter 2. The Haar wavelet transformation is introduced in Chapter 4 as a tool that efficiently concentrates the energy of a signal/image for applications such as compression. Daubechies filters and certain biorthogonal spline filter pairs are developed in Chapters 5 and 7 using conditions on the transformation matrix as well as requirements imposed on associated highpass filters. In particular, the concept of *annihilating filters* introduced in Section 2.4 takes the place of Fourier series in the derivation of wavelet filters. The ad hoc construction is limiting so complex numbers and Fourier series are introduced in Chapter 8 and these ideas are used in subsequent chapters to construct orthogonal filters and biorthogonal filter pairs in a more systematic way.

There have been chapter additions and subtractions as well. Gone is the chapter (previously Chapter 11) on algorithmic development of the biorthogonal wavelet transformation from the first edition although the material on symmetric biorthogonal filter pairs appears in Section 7.3 in this edition. Two new chapters have been added. Wavelet packets are introduced in Chapter 10. A main feature of this chapter is coverage of the FBI Wavelet Scalar Quantization Specification for compressing digital fingerprints. The LeGall wavelet transformation and lifting was discussed in Section 12.3 of the first edition of the text. This material now serves as an introduction for an entire chapter (Chapter 11) on the lifting method. This chapter makes heavy use of the Z-transformation and serves as a more mathematically rigorous topic than what typically appears in the text.

Finally, due to strong reader feedback, the clown image has been removed from the new edition!

To the Instructor

For those instructors who have taught a course from the first edition of the book, you will notice a significant reordering of some topics in addition to a couple of new chapters. The book now naturally separates into two parts. After some background review in matrix algebra and digital images, the first part of the text can be viewed as ad hoc wavelet filter development devoid of Fourier methods. Old applications such as image edge detection, image compression, and de-noising and still present. Applications of wavelet transformations to image segmentation and image pansharpening are new to this edition. When I last taught the course, I primarily covered Chapter 2–7. I also work through the labs (with a fair amount of in-class time devoted to them) that emphasize the software development of transformations as well as applications. For a group of students with a good background in linear algebra, I would do a cursory review of Chapters 3 and 4 and then cover Chapters 8 – 11 with Chapter 12 optional. I would not assign much lab work instead emphasizing the live examples so that students can get an appreciation for applications. Detailed course outlines are given later in the preface.

If you do decide to have your students work through the transformation development labs (Labs 4.1–4.3, 5.1–5.3, 7.1, and 7.2), you should have your students work Problems 4.8, 4.10, 5.12, 5.13, 5.22, 5.23, 7.13, 7.14, 7.32, and 7.33. These problems are similar in nature but give some insight into developing efficient algorithms for computing wavelet transformations and their inverses. Problem 2.23 is useful for the construction of two-dimensional transformations.

Problem Sets, Software Package, Labs, Live Examples

Problem sets and computers labs are given at the end of each section. There are 521 problems and 24 computer labs in the book. Some of the problems can be viewed as basic drill work or an opportunity to step through an algorithm "by hand." There are basic problem-solving exercises designed to increase student understanding of the material, problems where the student is asked to either provide or complete the proof of a result from the section, and problems that allow students to furnish details for ancillary topics that we do not completely cover in the text. To help students develop good problem-solving skills, many problems contain hints, and more difficult problems are often broken into several steps. The results of problems marked with a ★ are used later in the text.

With regards to software, there are some new changes. The `DiscreteWavelets` package has been replaced by the package `WaveletWare`. The new package is a complete rewrite of `DiscreteWavelets` and among the many new improvements and enhancements are new visualization routines as well as routines to compute wavelet packet transformations.

Labs have become more streamlined. The primary focus of most labs are the development of algorithms either to construct wavelet filter (pairs) or wavelet transformations. Some labs do investigate new ideas or extensions of ideas introduced in the parent section. I often require my students to complete final projects and in these instances, I assign selected "programming" labs so that the students will have developed filter or transformation modules for use on the computational aspects of the projects.

Most of the computations are now addressed in the form of *live examples*. There are 123 examples in the text and 91 of these can be reproduced via live example software. Moreover, live example software often contains further investigation suggestions or "things to try" problems. Those examples that are live examples conclude with a link to relevant software.

Text Web Site

The software package, computer labs and live examples are available in MATLAB and Mathematica and are available on the course web site

`stthomas.edu/wavelets.`

Text Topics

The book begins with a short chapter (essay) entitled *Why Wavelets?* The purpose of this chapter is to give the reader information about the mathematical history of wavelets, how they are applied, and why we would want to learn about them. In Chapter Chapter 2, we review basic ideas about vectors and matrices and introduce the notions of convolution and filters. Vectors and matrices are central tools in our development. We review vector inner products and norms, matrix algebra, and block matrix arithmetic. The material in Sections 2.1 and 2.2 is included to make the text self-contained and can be omitted if students have taken a sophomore linear algebra course. Students rarely see a detailed treatment of block matrices in lower level math courses, and because the text makes heavy use of block matrices, it is important to cover Section 2.3. The final section introduces the convolution product of bi-infinite sequences and filters. After connecting filters to convolution, we discuss finite impulse response, lowpass and highpass filters. A new concept to the second edition is that of filters that annihilate polynomial data. The ad hoc construction of wavelet filters in Chapters 5 and 7 makes use of annihilating filters.

Since most of our applications deal with digital images, we cover basic ideas about digital images in Chapter 3. We show how to write images as matrices, discuss some elementary digital processing tools, and define popular color spaces. The application of image compression is visited often in the text. The final step in an image compression algorithm is the encoding of data. There are many sophisticated coding methods but for the ease of presentation, we utilize *Huffman coding*. This method is presented in Section 3.3. Quantization is a typical step in image compression or audio denoising. Quantization functions specific to particular applications as needed in the text, but for generic quantization needs, we use *cumulative energy*. This function and its role in compression is presented in Section 3.4. *Entropy* and *peak signal-to-noise ratio* are tools used to measure the effectiveness of image compression methods. They are also introduced in Section 3.4.

We learn in Section 3.3 that encoding image data without preprocessing it leads to an inefficient image compression routine. The *Haar wavelet transformation* is introduced in Chapter 4 as one possible way to process data ahead of encoding it. The first three sections discuss the one-dimensional transformation, the iterated transformation, and the two-dimensional transformation. The chapter concludes with a section on applications of the transformation to image compression and image edge detection.

Chapters 5 and 7 consider ad hoc construction of filters needed to create wavelet transformation matrices. In Chapter 5, we construct filters from *Daubechies' family of orthogonal filters* [32]. The motivation for this construction is the inability for the Haar filter to correlate data – longer filter are better in this regard and can be constructed so that the wavelet transformation matrix is orthogonal. We learn in Chapter 7 that symmetric filters are best equipped to handle values at "beginning" or "end" of a data set. Unfortunately, Daubechies filters, other than the Haar filter, are not symmetric, so we continue our ad hoc construction by deriving *biorthogonal spline filter pairs*. Each filter is symmetric but the matrices they generate are not.

However, the wavelet matrices are related in that the inverse of one is the transpose of the other. Thus the construction preserves the basic structure of the wavelet transformation matrix. After deriving pairs of lengths 3 and 5 (Section 7.1) and 4 and 8 (Section 7.2), we discuss ways to exploit the symmetry of the filters in order to further improve the wavelet transformation's ability to process data ahead of encoding in image compression. Chapter 7 concludes with a section on applications of biorthogonal wavelet transformations to image compression and image panharpening. Sandwiched between Chapters 5 and 7 is a chapter that considers wavelet-based methods for denoising data. In particular, we introduce the VISUShrink [36] and SUREShrink [37] methods and illustrate how the latter can be employed in the application of image segmentation.

The ad hoc construction methods employed in Chapters 5 and 7 are limiting. In particular, we are unable to use these ideas to construct the entire family of biorthogonal spline filter pairs or an orthogonal family of filters called *Coiflets*. In order to perform more systematic construction of wavelet filters, we need ideas from complex analysis and Fourier series. These concepts are introduced in Chapter 8. The chapter concludes with a section that connects Fourier series to filter construction. The background material in Chapter 8 is implemented in Chapter 9 where Fourier methods are used to characterize filter construction (both lowpass and highpass) in the Fourier domain. The general results are given in Section 9.1 and Daubechies filters, Coiflets, biorthogonal spline filter pairs and the *CDF97 biorthogonal filter pair* are constructed in subsequent sections.

Chapters 10 and 11 are new to the second edition of the book. The *wavelet packet transformation* (Chapter 10) is a generalization of the wavelet transformation and allows facilitates a more application-dependent decomposition of data. In Section 10.1, the wavelet packet transformation is introduced for both one- and two-dimensional data and the *best basis algorithm* is outlined in Section 10.2 as a way of determining the most efficient way to transform the data, relative to a given cost function. Probably the most well-known application of the wavelet packet transformation is the *FBI Wavelet Scalar Quantization Specification* for compressing digital fingerprints and this method is presented in Section 10.3. The first edition of the book contained a section on the *LeGall filter* and its implementation via *lifting*. The immediate application of this filter and lifting is in the lossless compression component of the *JPEG2000 image compression standard*. In the lossless compression case, it is imperative that integer-valued data are mapped to integers so that the quantization step can be skipped. Computation of the wavelet transformation via lifting facilitates this step. Chapter 11 begins with lifting and the LeGall filter pair and then in Section 11.2 introduces the *Z-transform* and *Laurent polynomials*. These are the necessary tools to perform lifting for any given orthogonal filter or biorthogonal filter pair. The general lifting method is given in Section 11.3 with three example constructions give in the final section of the chapter.

The book concludes with a presentation (Chapter 12 of the JPEG2000 image compression standard. This standard makes use of lifting for lossless compression and the CDF97 filter pair in conjunction with the biorthogonal wavelet transformation for lossy compression.

Course Outlines

The book contains more material than can be covered in a one-semester course and this allows some flexibility when teaching the course.

I taught a version of the course using this manuscript during the Fall 2016 semester. The course was comprised of 40 class periods meeting three days a week for 65 minutes per meeting. The students were best suited for an applications-driven version of the course so the design included time for in-class lab work as well as work on final projects that occurred during the last seven meetings of the class. Here is a breakdown of the material covered.

1 meeting: Outline the ideas in Chapter 1.

2 meetings: Exams.

4 meetings: Chapter 2 with a main focus on the material in Section 2.4.

7 meetings: Chapter 3. We did not cover conversion to and from HSI space and spent roughly two days on in-class labs.

10 meetings: Probably the most important material for this group of students is in Chapter 4. Along with the chapter material, we worked on three in-class labs. The labs consisted of writing code to implement the one- and two-dimensional Haar wavelet transformation and then labs each on image compression and image edge detection.

4 meetings: Sections 5.1 and 5.2 with one in-class lab on the implementation of the wavelet transformation.

3 meetings: Sections 7.1 and 7.3 with some time allotted for an in-class lab on image compression.

2 meetings: Section 11.1 and an in-class lab on the associated lifting method

7 meetings: Final projects. Topic material was taken from Sections 5.3, 7.4, Chapter 10 and ideas in [1] on CAPTCHAs.

For a more mathematically mature audience, I might use the following outline.

1 meeting: Outline the ideas in Chapter 1.

3 meetings: Exams.

2 meetings: Section 2.4.

3 meetings: Chapter 3 and two labs on Huffman coding and the functions in Section 3.4.

5 meetings: Chapter 4 and labs on the applications in Section 4.4.

5 meetings: Chapter 8.

7 meetings: Chapter 9.

7 meetings: Chapter 11.

7 meetings: Final projects with material taken from Section 7.4 and Chapters 6, 10 and 12.

P. J. Van Fleet

St. Paul, Minnesota USA
May 2017

ACKNOWLEDGMENTS

There are the usual motivations for writing a second edition of a book – timely updates, error corrections, or basic reorganization of topics are likely reasons. For this book the driving motivation was the realization that it is possible to present the material to students with limited background in numerical analysis. I am indebted to David Ruch and Catherine Bénéteau for suggesting that I consider this notion. I taught the course a couple of times using their suggestions and was convinced to make the suggested changes. The idea is to develop some wavelet filters using non–Fourier methods. To do so meant a major rewrite of early chapters and a shift of Fourier topics to the end of the book. I am grateful for their suggestions.

I am also grateful to Bruce Atwood, Catherine Bénéteau, Caroline Haddad, Kristin Pfabe, David Ruch, and Roger Zarnowski for the careful proofreading of chapter drafts. I appreciate the work of two St. Thomas students on this project. Naomi Latt worked through many of the live examples and created a significant amount of Mathematica and MATLAB code. Emma Western worked through all the exercises in the first seven chapters of the book, checking for correctness of solutions and clarity.

Thanks in advance to all readers who suggest improvements in the presentation, create new computer labs, or develop new problems. Please contact me with your ideas!

P.V.F.

CHAPTER 1

INTRODUCTION: WHY WAVELETS?

Why wavelets? This is certainly a fair question for those interested in this book and also one that can be posed in multiple contexts. Students wonder why they need to learn about wavelets. Researchers ask theoretical questions about properties of wavelets with an eye towards tailoring them for applications. Scientists want to know why they should consider using wavelets in applications instead of other popular tools. Finally, there is curiosity about the word itself – why *wavelets*? What are wavelets? Where did they come from?

In this short introduction to the book, we answer these questions and in the process, provide some information about what to expect from the chapters that follow.

Image Compression

In keeping with the true spirit of the book, let's start with an application in image compression.

Suppose you wish to use the Internet to send a friend the digital image file plotted in Figure 1.1. Since the file is quite large, you first decide to *compress* it. For this application, it is permissible to sacrifice image resolution in order to minimize

Discrete Wavelet Transformations: An Elementary Approach With Applications, Second Edition.
Patrick J. Van Fleet.

transmission time. The dimensions of the image, in pixels,[1] are 512 rows by 512 columns so that the total number of elements that comprise the image is $512 \times 512 = 262{,}144$.

Figure 1.1 A digital image.

What does it even mean to compress the image? How do you suppose that we compress the image? How do we measure the effectiveness of the compression method?

As you will learn in Chapter 3, each pixel value is an integer from 0 (black) to 255 (white) and each of these integers is stored on a computer using eight *bits* (the value of a bit is either 0 or 1). Thus we need $262{,}144 \times 8 = 2{,}097{,}152$ bits to represent the image. To compress an image, we simply wish to reduce the number of bits needed to store it. Most compression methods follow the basic algorithm given in Figure 1.2.

The first step is to *transform* the image. The goal here is to map the integers that comprise the image to a new set of numbers. In this new setting we alter (*quantize*) some or all of the values so that we can write (*encode*) the modified values using fewer bits. In Figure 1.3, we have plotted one such transformation of the image in Figure 1.1. The figure also contains a plot of the quantized transformation. The

[1]You can think of a pixel as a small rectangle on your computer screen or paper that is rendered at some gray level between black and white.

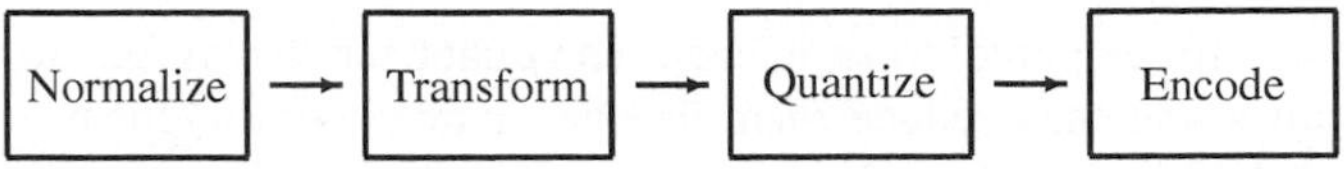

Figure 1.2 A basic algorithm for compressing a digital image.

transformation in Figure 1.3 is a *discrete wavelet transformation.* In fact, it is the same transformation as that used in the JPEG2000 image compression standard.[2]

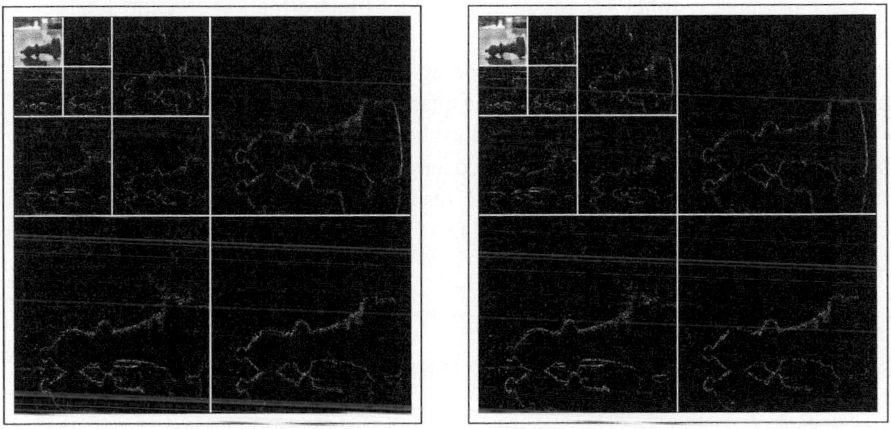

(a) Wavelet transformation (b) Quantized wavelet transformation

Figure 1.3 A transformation and the quantized transformation of the image in Figure 1.1. The white lines that appear in both images are superimposed to separate the different parts of the transformation.

In mathematical terms, if the image is represented by the square matrix A, then the transformation is performed by constructing special matrices W and $\widetilde{W}^T$ and next computing $WA\widetilde{W}^T$. Of course, the key to the process is understanding the construction of $W, \widetilde{W}$ and how $WA\widetilde{W}^T$ concentrates most of the nonzero values (nonblack pixels) in the upper left-hand corner of the transformation. You might already begin to see why the wavelet transformation is useful for reducing the number of bits. The transformation has created large regions that are black or nearly black (i.e., regions where most of values are either at or near 0). It is natural to believe that you would need less information to store these values than what is required for the original image.

The next step in the process is to quantize the transformation. It turns out that zero is the only integer value in the transformed output. Thus each value must be

[2]You will learn about JPEG2000 in Chapter 12.

converted (rounded, for example) to an integer. The quantizer also makes decisions about certain values and may reduce them in size or even convert them to zero. Values that are reduced or converted to zero are those that the quantizer believes will not adversely affect the resolution of the compressed image. After quantization, it is impossible to exactly retrieve the original image.[3] It may be hard to see in Figure 1.3, but nearly all the values of the quantized transformation are different from the corresponding values in the transformation.

The final step before transmission is to *encode* the quantized transformation. That is, instead of using eight bits to store each integer, we will try to group together like integers and possibly use a smaller number of bits to store those values that occur with greater frequency. Since the quantized wavelet transformation contains a large number of zeros (black pixels), we expect the encoded transformation to require fewer bits. Indeed, using a very naive encoding scheme (see Section 3.3), we find that only 543,613 bits are needed to store the quantized wavelet transformation. This is about 26% of the number of bits required to store the original!

To recover the compressed image, a decoder is applied to the encoded data and the inverse transformation is applied to the resulting data stored in matrix form.[4] The compressed image appears in Figure 1.4. It is very difficult to tell the images in Figures 1.1 and 1.4 apart!

Other Applications That Use Wavelet Transformations

Wavelets, or more precisely, wavelet transformations, abound in image-processing applications. The Federal Bureau of Investigation uses wavelets in their Wavelet Scalar Quantization Specification to compress digital images of fingerprints [11]. In this case there is an objective measure of the effectiveness of the compression method – the uncompressed fingerprint file must be uniquely matched with the correct person. In many instances it is desirable to identify edges of regions that appear in digital images. Wavelets have proved to be an effective way to perform edge detection in images [69, 67]. Other image-processing applications where wavelets are used are image morphing [54] and digital watermarking [103, 39]. Image morphing is a visualization technique that transforms one image into another. You have probably seen image morphing as part of a special effect in a movie. Digital watermarking is the process by which information is added (either visible or invisible) to an image for the purpose of establishing the authenticity of the image.

Wavelets are used in applications outside of image processing. For example, wavelet-based methods have proved effective in the area of signal denoising [35, 98, 102]. Cipra [22] recounts an interesting story of Yale University mathematician Ronald Coifman and his collaborators using wavelets to denoise an old cylinder

[3]For those readers with some background in linear algebra, $W^{-1}, \widetilde{W}^{-1}$ both exist so it is possible to completely recover A before the quantization step.

[4]In some image compression algorithms, a dequantization function might be applied before computing the inverse transformation of the decoded data.

Figure 1.4 The uncompressed digital image. Whereas each pixel in the original image is stored using eight bits, the pixels in this image require 2.07 bits storage on average.

recording (made of a radio broadcast) of Brahms playing his *First Hungarian Dance* [4]. It is impossible to identify the piano in the original but a restored version (using wavelets) clearly portrays the piano melody. Stanford University music professor and Coifman collaborator Jonathan Berger maintains a very interesting web site on this project [5]. Wavelet-based denoising (wavelet shrinkage) is also used in applications in finance and economics [45]. We study wavelet shrinkage in Chapter 6.

Wavelets have been used to examine electroencephalogram data in order to detect epileptic seizures [3], model distant galaxies [6], and analyze seismic data [56, 57]. Finally, in more mathematical applications, wavelets have been used to estimate densities and to model time series in statistics [98] and in numerical methods for solving partial differential equations [28].

Wavelet Transformations Are Local Transformations

We have by no means exhausted the list of applications that utilize wavelets. Understand as well that wavelets are not always the best tool to use for a particular application. There are many applications in audio and signal processing for which Fourier methods are superior and applications in image processing where filtering methods other than wavelet-based methods are preferred.

There are several reasons why wavelets work well in many applications. Probably the most important is the fact that a discrete wavelet transformation is a *local transformation*. To understand what is meant by this term, consider the 50-term signal in Figure 1.5(a). Four of the signal values are altered and the result is plotted in Figure 1.5(b).

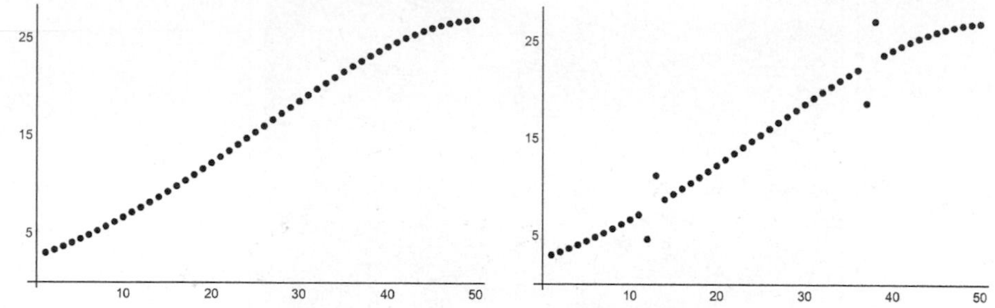

(a) Original signal

(b) Altered signal

Figure 1.5 A signal and alteration of the signal.

In Figure 1.6 we have plotted the discrete wavelet transformation of each signal from Figure 1.5. The wavelet transformation in Figure 1.6(a) is again simply a matrix we use to multiply with the input vector. The output consists of two parts. The first 25 components of the transformation serve as an approximation of the original signal. The last 25 elements basically describe the differences between the original signal and the approximation.[5] Since the original signal is relatively smooth, it is not surprising that the differences between the original signal and the approximation are quite small.

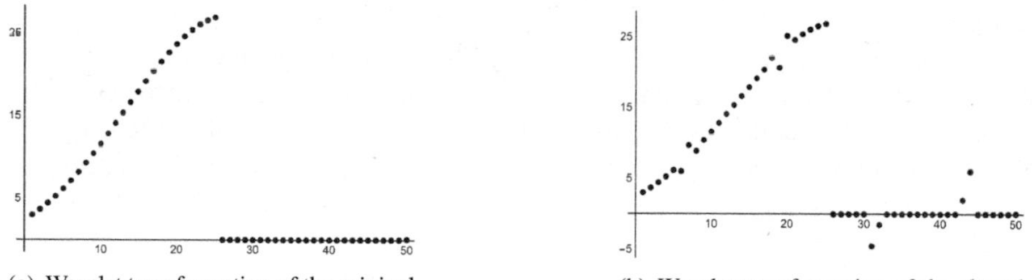

(a) Wavelet transformation of the original signal

(b) Wavelet transformation of the altered signal

Figure 1.6 Discrete wavelet transformations of each of the signals in Figure 1.5.

Recall that the altered signal shown in Figure 1.5(b) differed only slightly from the original signal plotted in Figure 1.5(a). The same holds true when we compare

[5]Due to the scale in Figure 1.5, many of the values in the differences portion of the transformation appear to be zero. This is not they case – see Figure 1.7(a).

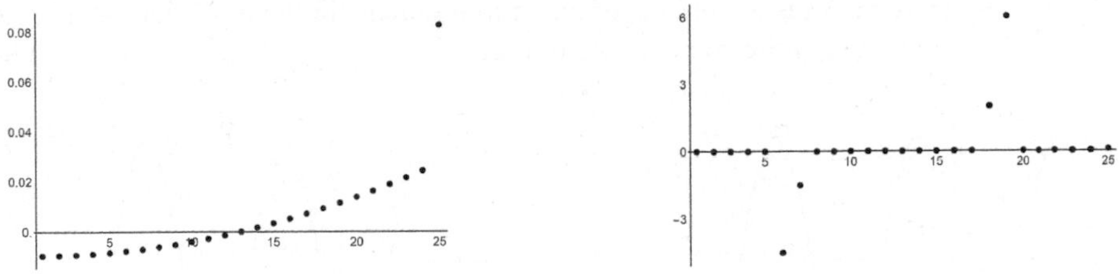

(a) Differences portion of the original signal's transformation

(b) Difference portion of the altered signal's transformation

Figure 1.7 The differences portions of the discrete wavelet transformations from Figure 1.6.

the differences portions (plotted in more detail in Figure 1.7) of the wavelet transformations of the signals. Thus, small changes in the input data result in small changes in the wavelet-transformed data.

One of the most popular and useful tools for signal-processing applications is the *discrete Fourier transformation* (DFT).[6] Its complex values are built using sampled values of cosine and sine functions. Although we will not delve into the specifics of the construction of the DFT and how it is used in applications, it is worthwhile to view the DFT of the signals in Figure 1.5. We have plotted the modulus (see Section 8.1) of these DFTs in Figure 1.8.

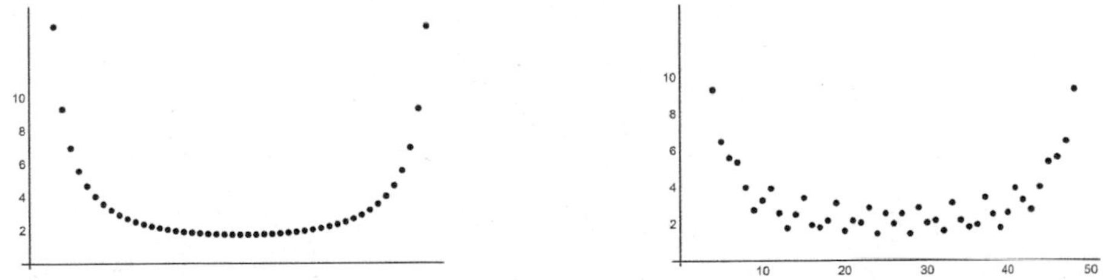

(a) |DFT| of original signal

(b) |DFT| of altered signal

Figure 1.8 The moduli of the discrete Fourier transformations of each of the signals in Figure 1.5.

Look at what happened to the DFT of the altered data. Although we changed only four values, the effect of these changes on the transformation is *global*.[7] The DFT's building block functions, sine and cosine, oscillate between ± 1 for all time

[6]A nice derivation of the fast Fourier transformation, a method for efficienty computing the DFT, appears in Kammler [62].

[7]After examining the images in Figure 1.8, an expert in Fourier analysis would undoubtedly be able to ascertain that a few of the values in the original signal had been altered but would have no way of knowing the locations of the altered values.

and never decay to zero. Thus the effects of any minor change in the input data will reverberate throughout the entire transformed output.

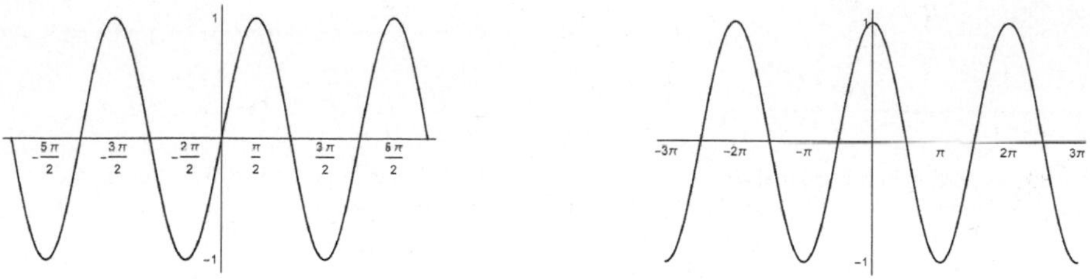

(a) $\sin t$ (b) $\cos t$

Figure 1.9 A few periods of sine and cosine. Both functions oscillate between ± 1 and never decay to zero.

Classical Wavelet Theory in a Nutshell

The classical approach to wavelet theory utilizes the setting of square integrable functions on the real line. The idea is to retain those attributes from transformations constructed from oscillatory functions (e.g. sine and cosine; Figure 1.9) and at the same time, use oscillatory functions that, unlike sine and cosine, are required to decay to zero. That is, the underlying functions are tiny waves or *wavelets.* A couple of popular wavelet functions are plotted in Figure 1.10.

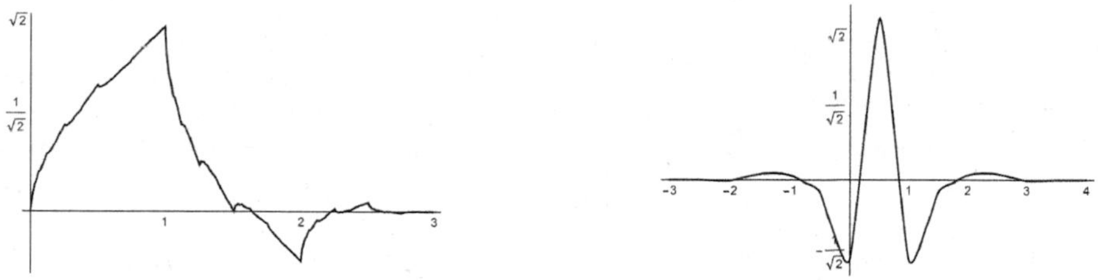

(a) Daubechies four–tap scaling function (b) CDF97 wavelet function

Figure 1.10 Some functions from classical wavelet theory.

Note that the functions in Figure 1.10 are oscillatory and have value zero outside a finite interval. The function in Figure 1.10(a) is probably one of the most famous in all of wavelet theory. It is Ingrid Daubechies' *four–tap scaling function*[8] [32].

[8] A scaling function is called a *father wavelet function* by some researchers.

It gives rise to the four numbers we use to build the wavelet transformation matrix in Section 5.1. The function in Figure 1.10(b) is a wavelet function constructed by Albert Cohen, Ingrid Daubechies, and Jean-Christophe Feauveau [23]. It provides the elements of the wavelet transformation matrix used in the JPEG2000 image compression standard.

If these functions really do lead to the discrete wavelet transformations used in so many applications, several questions naturally arise: Where do these functions come from? How are they constructed? Once we have built them, how do we extract the elements that are used in the discrete wavelet transformations?

Although the mathematics of wavelets can be traced back to the early twentieth century,[9] most researchers mark the beginning of modern work in wavelet theory by the 1984 work [50] of French physicists Jean Morlet and Alexander Grossmann. Yves Meyer [71] (in 1992) and Stéphane Mallat [68] (in 1989) produced foundational work on the so-called *multiresolution analysis* and using this construct, Ingrid Daubechies created wavelet functions [29, 30] (in 1988 and 1993, respectively) that give rise to essentially all the discrete wavelet transformations found in this book. Many of these ideas are discussed in the book [77] by David K. Ruch and Patrick Van Fleet.

To (over)simplify the central idea of Daubechies, we seek a function $\phi(t)$ (use the function in Figure 1.10(a) as a reference) that satisfies a number of properties. In particular, $\phi(t)$ should be zero outside a finite interval and $\phi(t)$ and its integer translates $\phi(t-k), k = 0, \pm1, \pm2, \pm3, \ldots,$ (see Figure 1.11) should form a *basis* for a particular space. The function $\phi(t)$ should be suitably smooth[10] and the functions should also satisfy

$$\int_{\mathbb{R}} \phi(t)\phi(t\quad k)\,\mathrm{d}t = 0, \qquad k = \pm1, \pm2, \ldots \tag{1.1}$$

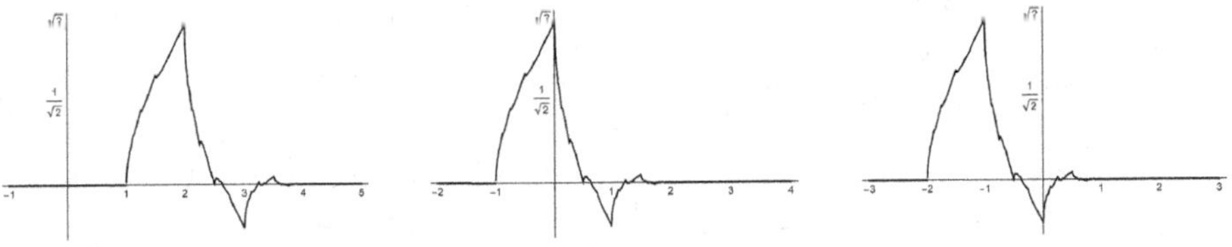

(a) $\phi(t-1)$ (b) $\phi(t+1)$ (c) $\phi(t+2)$

Figure 1.11 Some integer translates of the function $\phi(t)$ from Figure 1.10(a).

[9]Yves Meyer [71] gives a nice discussion of the origins of wavelet theory in the twentieth century. A wonderful exposition on the history of wavelet theory, its origins, and uses in applications can be found in a 1998 book by Barbara Burke Hubbard [58].

[10]Believe it or not, in a certain sense, the function in Figure 1.10(a) and its integer translates can be used to represent linear polynomials!

Finally, we should be able to write $\phi(t)$ as a combination of dilations (contractions, actually) and translations of itself. For example, the function $\phi(t)$ in Figure 1.10(a) satisfies the *dilation equation*

$$\phi(t) = h_0\phi(2t) + h_1\phi(2t-1) + h_2\phi(2t-2) + h_3\phi(2t-3). \tag{1.2}$$

The function $\phi(t)$ in Figure 1.10(a) is nonzero only on the interval $[0, 3]$. The function $\phi(2t)$ is a contraction – it is nonzero only on $[0, \frac{3}{2}]$. For $k = 1$, $\phi(2t-1) = \phi(2(t-\frac{1}{2}))$, so we see that $\phi(2t-1)$ is simply $\phi(2t)$ translated $1/2$ units right. In a similar way, we see that $\phi(2t-2)$ and $\phi(2t-3)$ are obtained by translating $\phi(2t)$ one and $3/2$ units right, respectively. The functions on the right side of (1.2) are plotted in Figure 1.12.

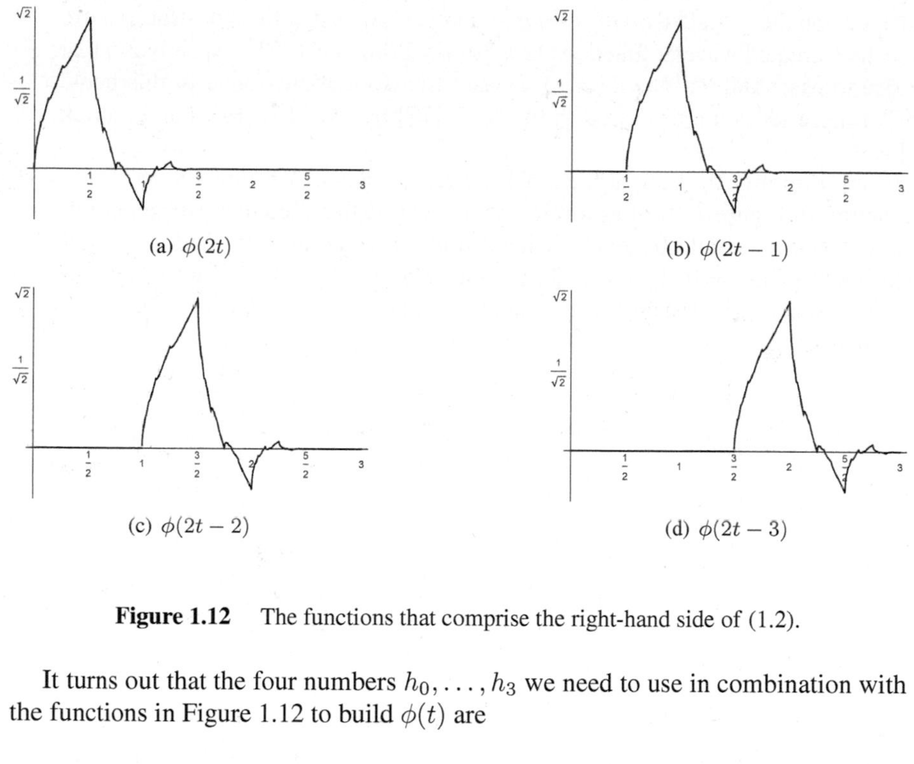

Figure 1.12 The functions that comprise the right-hand side of (1.2).

It turns out that the four numbers $h_0, \ldots, h_3$ we need to use in combination with the functions in Figure 1.12 to build $\phi(t)$ are

$$h_0 = \frac{1+\sqrt{3}}{4\sqrt{2}} \approx 0.482963 \qquad h_1 = \frac{3+\sqrt{3}}{4\sqrt{2}} \approx 0.836516$$

$$h_2 = \frac{3-\sqrt{3}}{4\sqrt{2}} \approx 0.224144 \qquad h_3 = \frac{1-\sqrt{3}}{4\sqrt{2}} \approx -0.129410.$$

We derive these numbers in Section 5.1. Each of the properties satisfied by $\phi(t)$ affects the discrete wavelet transformation. The integral condition (1.1) ensures that the discrete wavelet transformation matrix W is not only invertible, but *orthogonal*. As we will see in Section 2.2, orthogonal matrices satisfy the simple inverse formula $W^{-1} = W^T$. Insisting that $\phi(t)$ is suitably smooth allows us to form accurate approximations of smooth data. Requiring $\phi(t)$ to be zero outside a finite interval results in a finite list of numbers $h_0, \ldots, h_N$ that are used in the discrete wavelet transformation. Finally, the dilation equation allows us to build a transformation that can "zoom" in on complicated segments of a signal or image. This zoom-in property is not entirely obvious – we talk more about it in Chapters 4 and 5.

The Approach in This Book

The classical approach of Daubechies [29] to derive the numbers h_0, h_1, h_2, and h_3 for the function in Figure 1.10(a) is a journey through some beautiful mathematics, but it is difficult and requires more mathematical background than most undergraduate students possess.[11]

On the other hand, the matrix used to produce the output in Figure 1.6 is quite simple in structure. It is 50 rows by 50 columns and has the following form:

$$W = \left[\begin{array}{cccccccccccc}
\frac{3}{4} & \frac{1}{2} & -\frac{1}{4} & 0 & 0 & 0 & 0 & 0 & 0 & 0 & \cdots & 0 \\
-\frac{1}{8} & \frac{1}{4} & \frac{3}{4} & \frac{1}{4} & -\frac{1}{8} & 0 & 0 & 0 & 0 & 0 & \cdots & 0 \\
0 & 0 & -\frac{1}{8} & \frac{1}{4} & \frac{3}{4} & \frac{1}{4} & -\frac{1}{8} & 0 & 0 & 0 & \cdots & 0 \\
0 & 0 & 0 & 0 & -\frac{1}{8} & \frac{1}{4} & \frac{3}{4} & \frac{1}{4} & -\frac{1}{8} & 0 & \cdots & 0 \\
\vdots & & & & & & \ddots & & & & & \vdots \\
0 & 0 & 0 & 0 & \cdots & 0 & -\frac{1}{8} & \frac{1}{4} & \frac{3}{4} & \frac{1}{4} & -\frac{1}{8} & 0 \\
0 & 0 & 0 & 0 & \cdots & 0 & 0 & 0 & -\frac{1}{8} & \frac{1}{4} & \frac{5}{8} & \frac{1}{4} \\
\hline
-\frac{1}{2} & 1 & -\frac{1}{2} & 0 & 0 & 0 & 0 & 0 & 0 & 0 & \cdots & 0 \\
0 & 0 & -\frac{1}{2} & 1 & -\frac{1}{2} & 0 & 0 & 0 & 0 & 0 & \cdots & 0 \\
0 & 0 & 0 & 0 & -\frac{1}{2} & 1 & -\frac{1}{2} & 0 & 0 & 0 & \cdots & 0 \\
0 & 0 & 0 & 0 & 0 & -\frac{1}{2} & 1 & -\frac{1}{2} & 0 & 0 & \cdots & 0 \\
\vdots & & & & & & \ddots & & & & & \vdots \\
0 & 0 & 0 & 0 & 0 & 0 & \cdots & 0 & -\frac{1}{2} & 1 & -\frac{1}{2} & 0 \\
0 & 0 & 0 & 0 & 0 & 0 & \cdots & 0 & 0 & 0 & -1 & 1
\end{array}\right]. \qquad (1.3)$$

Note that we have placed a line between the first 25 rows and the last 25 rows. There are a couple of reasons for partitioning the matrix in this way. First, you might notice that rows 2 through 24 are built using the same sequence of five nonzero

[11] For those students interested in a more classical approach to wavelet theory, I strongly recommend the books by David Walnut [100], Albert Boggess and Francis Narcowich [7], or Michael Frazier [42].

values $(-\frac{1}{8}, \frac{1}{4}, \frac{3}{4}, \frac{1}{4}, -\frac{1}{8})$, and rows 26 through 49 are built using the same sequence of three nonzero values $(-\frac{1}{2}, 1, -\frac{1}{2})$.[12] Another reason for the separator is the fact that the product of W with an input signal consisting of 50 elements has to produce an approximation to the input (recall Figure 1.6) that consists of 25 elements. So the first 25 rows of W perform that task. In a similar manner, the last 25 rows of W produce the differences we need to combine with the approximation to recover the original signal.

There is additional structure evident in W. If we consider row two as a list of 50 numbers, then row three is obtained by cyclically shifting the elements in row two to the right two units. Row four is obtained from row three in a similar manner, and the process is continued to row 24. Row 27 can be constructed from row 26 by cyclically shifting the list $(-\frac{1}{2}, 1, -\frac{1}{2}, 0, 0, \ldots, 0, 0)$ two unit right. It is easy to see how to continue this process to construct rows 26 through 49. Rows 1, 25, and 50 are different from the other rows and you will have to wait until Section 7.3 to find out why.

Let's look closer at the list $(-\frac{1}{8}, \frac{1}{4}, \frac{3}{4}, \frac{1}{4}, -\frac{1}{8})$. When we compute the dot product of row two say with an input signal $(x_1, \ldots, x_{50})$, we obtain

$$-\frac{1}{8}x_1 + \frac{1}{4}x_2 + \frac{3}{4}x_3 + \frac{1}{4}x_4 - \frac{1}{8}x_5.$$

If we add the elements of our list, we obtain

$$-\frac{1}{8} + \frac{1}{4} + \frac{3}{4} + \frac{1}{4} - \frac{1}{8} = 1$$

and see that the list creates a weighted average of the values $(x_1, \ldots, x_5)$. The idea here is no different from what your instructor uses to compute your final course grade. For example, if three hourly exams are worth 100 points each and the final exam is worth 200 points, then your grade is computed as

$$\text{grade} = \frac{1}{5} \cdot \text{exam 1} + \frac{1}{5} \cdot \text{exam 2} + \frac{1}{5} \cdot \text{exam 3} + \frac{2}{5} \cdot \text{final}$$

Of course, most instructors do not use negative weights like we did in W. The point here is that by computing weighted averages, we are creating an approximation of the original input signal.

Taking the dot product of row 26 with the input signal produces

$$-\frac{1}{2}x_1 + 1 \cdot x_2 - \frac{1}{2}x_3.$$

This value will be near zero if x_1, x_2, and x_3 are close in size. If there is a large change between x_1 and x_2 or x_2 and x_3, then the output will be a relatively large

[12]We mentioned earlier in this chapter that we seek orthogonal matrices for use as wavelet transformations. The reader with a background in linear algebra will undoubtedly realize that W in (1.3) is not orthogonal. As we will see later in the book, it is desirable to relinquish orthogonality so that our wavelet matrices can be built with lists of numbers that are symmetric.

(absolute) value. For applications, we may want to convert small changes to zero (compression) or identify the large values (edge detection).

The analysis of W and its entries leads to several questions: How do we determine the structure of W? How do we build the lists of numbers that populate it? How are the rows at the top and bottom of the different portions of W formed? How do we design W for use in different applications? How do we make decisions about the output of the transformation? The matrix W consists of a large number of zeros; can we write a program that exploits this fact and computes the product of W and the input signal faster than by using conventional matrix and vector multiplication?

We answer all of these questions and many more in the chapters that follow. We will learn that our lists of numbers are called *filters* by electrical engineers and other scientists. We first learn about the structure of W by understanding how signals are processed by filters. We then learn how to create filters that perform tasks such as approximating signals or creating the differences portion of the transformation. These are the ideas that are covered first in Section 2.4 and later in Section 8.3.

The book is roughly divided into two parts. The first is an ad hoc development of wavelet transformations and an introduction to some applications that use them. As we ask more from our transformations, the methods we use to construct them increase in complexity. But at this point, you should have a good understanding of the discrete wavelet transformation, how to use it in applications, how to write software for its implementation, and what limitations we still face. With this understanding comes an appreciation of the need to step back and model our construct in a more general mathematical setting. In the second half of the book we take a more theoretical tack toward transformation design. The payoff is the ability to design wavelet transformations like that given in (1.3) and use them in applications such as JPEG2000 or the FBI Wavelet Scalar Quantization Specification.

Thus, in the pages that follow, you will discover the answers to the questions asked at the beginning of the chapter. It is my sincere hope after working through the material, problem sets, software, and computer labs that your response to these questions is: *Why wavelets, indeed!*

CHAPTER 2

VECTORS AND MATRICES

In this chapter we cover some basic concepts from linear algebra necessary to understand the ideas introduced later in the book. Signals and digital audio are typically represented as vectors and digital images are usually described with matrices. Therefore, it is imperative that before proceeding, you have a solid understanding of some elementary concepts associated with vectors and matrices.

If you have completed a sophomore linear algebra class, the first two sections of this chapter will undoubtedly be a review for you. Make sure though, that you take time to understand concepts such as orthogonal vectors and orthogonal matrices. We also discuss matrix multiplication as a way of "processing" another matrix or a vector. Although you may have mastered multiplication with matrices, it is worthwhile to think about the concepts related to matrix multiplication presented in Section 2.2.

The third section of the chapter deals with block (partitioned) matrix algebra. This material is typically not covered in detail in a sophomore linear algebra class but is very useful for analyzing and designing the wavelet transformations that appear in later chapters.

We conclude the chapter with a discussion of convolution and filters. The act of convolving two sequences is at the very heart of signal/image processing and typically one of the convolved sequences possesses special characteristics that influence

Discrete Wavelet Transformations: An Elementary Approach With Applications, Second Edition.
Patrick J. Van Fleet.

the convolution product. We will think of this fixed sequence as a *filter* and look at a couple of special filters upon which wavelet transformations are based.

2.1 Vectors, Inner Products, and Norms

We remember vectors from a linear algebra course. In $\mathbb{R}^2$, vectors take the form

$$\mathbf{v} = \begin{bmatrix} v_1 \\ v_2 \end{bmatrix}$$

where v_1 and v_2 are any real numbers. You may recall the idea of vectors in $\mathbb{R}^n$:

$$\mathbf{v} = \begin{bmatrix} v_1 \\ v_2 \\ \vdots \\ v_n \end{bmatrix}$$

where $v_1, \ldots, v_n$ are any real numbers. When we think of vectors we generally think of the vectors described above. It is worth mentioning, however, that in a general sense, a vector is best thought of as some element of a given vector space. For example, if the space we are working with is the space of all quadratic polynomials, then we consider $f(x) = x^2 + 3x + 5$ a vector in that space. If the space under consideration is all 2×2 matrices, then certainly

$$A = \begin{bmatrix} 3 & 2 \\ 5 & 1 \end{bmatrix}$$

is a vector in that space. For our purposes though, we will think of vectors as elements of $\mathbb{R}^n$.

Vector Arithmetic

You should be familiar with the addition and subtraction of vectors. Figure 2.1 is a plot of the sum and difference of $\mathbf{u} = \begin{bmatrix} 4 \\ 1 \end{bmatrix}$ and $\mathbf{v} = \begin{bmatrix} 2 \\ 3 \end{bmatrix}$. The sums and differences are computed element–wise:

$$\begin{bmatrix} 4 \\ 1 \end{bmatrix} + \begin{bmatrix} 2 \\ 3 \end{bmatrix} = \begin{bmatrix} 4+2 \\ 1+3 \end{bmatrix} = \begin{bmatrix} 6 \\ 4 \end{bmatrix} \quad \text{and} \quad \begin{bmatrix} 4 \\ 1 \end{bmatrix} - \begin{bmatrix} 2 \\ 3 \end{bmatrix} = \begin{bmatrix} 4-2 \\ 1-3 \end{bmatrix} = \begin{bmatrix} 2 \\ -2 \end{bmatrix}.$$

Vectors cannot be multiplied together, but we can multiply vectors by real numbers or *scalars*. The scalar multiplication of real number c and vector $\mathbf{v}$ is defined as

$$c \begin{bmatrix} v_1 \\ v_2 \\ \vdots \\ v_n \end{bmatrix} = \begin{bmatrix} cv_1 \\ cv_2 \\ \vdots \\ cv_n \end{bmatrix}.$$

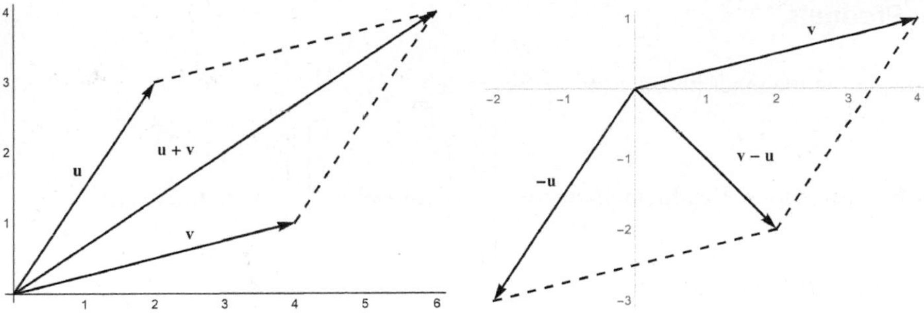

(a) Sum of **u** and **v** (b) Difference of **u** and **v**

Figure 2.1 Plots of the sum and difference of two vectors.

For $\mathbf{v} = \begin{bmatrix} 4 \\ 2 \end{bmatrix}$, the scalar multiples $(-1)\mathbf{v} = -\mathbf{v} = \begin{bmatrix} -4 \\ -2 \end{bmatrix}$, $\frac{1}{2}\mathbf{v} = \begin{bmatrix} 2 \\ 1 \end{bmatrix}$ and $2\mathbf{v} = \begin{bmatrix} 8 \\ 4 \end{bmatrix}$ are plotted in Figure 2.2.

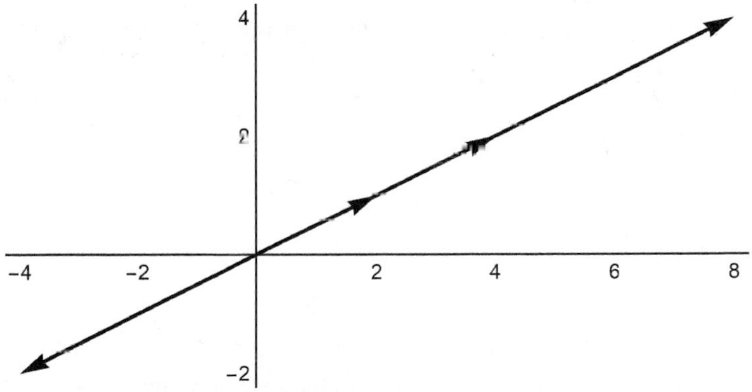

Figure 2.2 Scalar multiples of the vector $\mathbf{v} = [4, 2]^T$ with $c = -1, \frac{1}{2}, 2$.

Two important concepts associated with vectors are the *inner (dot) product* and *norm* (length).

Inner Products

We now define the inner product for vectors in $\mathbb{R}^n$.

Definition 2.1 (Inner Product). *Let* $\mathbf{v} = \begin{bmatrix} v_1 \\ v_2 \\ \vdots \\ v_n \end{bmatrix}$ *and* $\mathbf{w} = \begin{bmatrix} w_1 \\ w_2 \\ \vdots \\ w_n \end{bmatrix}$ *be two vectors in* $\mathbb{R}^n$. *We define the* inner product $\mathbf{v} \cdot \mathbf{w}$ *as*

$$\boxed{\mathbf{v} \cdot \mathbf{w} = \sum_{k=1}^{n} v_k w_k.} \tag{2.1}$$

□

You might recall the *transpose* of a matrix from your linear algebra class. We cover the matrix transpose in Section 2.2, but since we are discussing inner products, it is useful to talk about transposing a vector $\mathbf{v}$.

Definition 2.2 (Vector Transpose). *We define the* transpose *of vector* $\mathbf{v} \in \mathbb{R}^n$ *as*

$$\mathbf{v}^T = [v_1, v_2, \ldots, v_n].$$

□

We find it convenient to write vectors as $\mathbf{v} = [v_1, v_2, \ldots, v_n]^T$:

$$\mathbf{v}^T = \begin{bmatrix} 1 \\ 2 \\ 3 \end{bmatrix}^T = [1, 2, 3].$$

The transpose of a vector is often used to express the inner product of two vectors. Indeed, we write

$$\boxed{\mathbf{v} \cdot \mathbf{w} = \mathbf{v}^T \mathbf{w}.}$$

Next, we look at some examples of inner products.

Example 2.1 (Inner Products). *Compute the following inner products:*

(a) $\mathbf{v} = \begin{bmatrix} 1 \\ 4 \\ -1 \end{bmatrix}$ *and* $\mathbf{w} = \begin{bmatrix} 3 \\ 2 \\ -2 \end{bmatrix}$

(b) $\mathbf{v} = \begin{bmatrix} -1 \\ 3 \end{bmatrix}$ *and* $\mathbf{w} = \begin{bmatrix} 6 \\ 2 \end{bmatrix}$.

Solution.
We have

$$\begin{bmatrix}1\\4\\-1\end{bmatrix}\cdot\begin{bmatrix}3\\2\\-2\end{bmatrix}=[1,4,-1]\begin{bmatrix}3\\2\\-2\end{bmatrix}=1\cdot 3+4\cdot 2+(-1)\cdot(-2)=13$$

and

$$\begin{bmatrix}-1\\3\end{bmatrix}\cdot\begin{bmatrix}6\\2\end{bmatrix}=[-1,3]\begin{bmatrix}6\\2\end{bmatrix}=(-1)\cdot 6+3\cdot 2=0.$$

(Live example: Visit `stthomas.edu/wavelets` *and click on Live Examples.)*

□

The inner product value of 0 that we computed in part (b) of Example 2.1 has geometric significance. We plot these vectors in Figure 2.3.

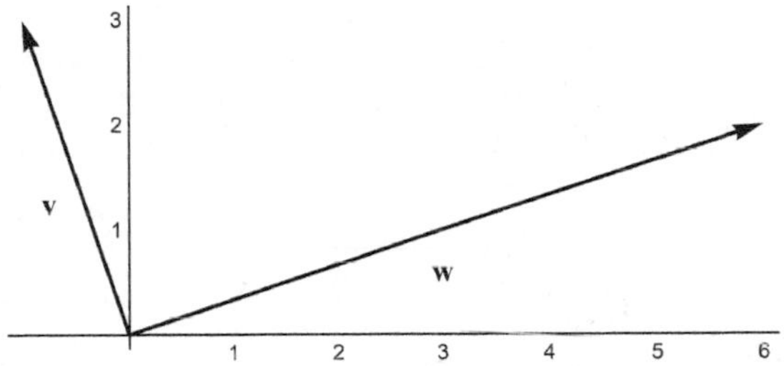

Figure 2.3 The vectors **v** and **w** from Example 2.1(b).

Notice that these vectors are perpendicular to each other. Using a term more common in linear algebra, we say that **v** and **w** are *orthogonal*. We can extend this notion to vectors of length n and make the following definition.

Definition 2.3 (Orthogonal Vectors). *Let* $\mathbf{v},\mathbf{w}\in\mathbb{R}^n$. *Then we say that* $\mathbf{v}$ *and* $\mathbf{w}$ *are* orthogonal *if*

$$\mathbf{v}\cdot\mathbf{w}=\mathbf{v}^T\mathbf{w}=0.$$

We say that a set of vectors is orthogonal or forms an orthogonal set *if all distinct vectors in the set are orthogonal to each other.* □

As we shall see, there are many advantages to working with orthogonal vectors.

Vector Norms

It is important to be able to measure the length or *norm* of a vector. Typically, the norm of a vector $\mathbf{v}$ is written as $\|\mathbf{v}\|$. To motivate our definition, let's assume that $\mathbf{v}\in\mathbb{R}^2$. Consider Figure 2.4:

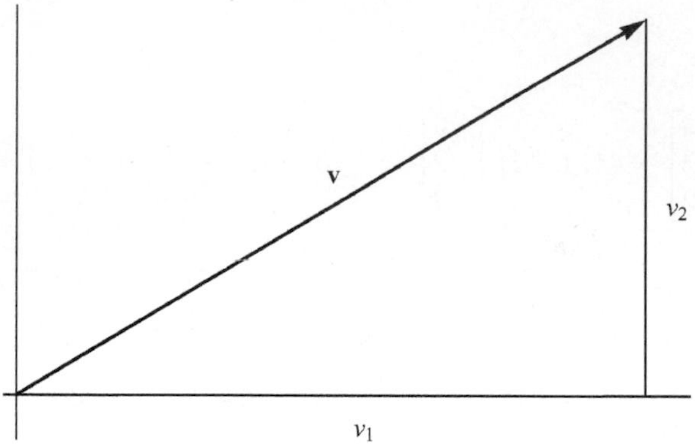

Figure 2.4 Vector $\mathbf{v} = [v_1, v_2]^T$.

In addition to graphing $\mathbf{v}$ in Figure 2.4, we have added the horizontal and vertical line segments with lengths v_1 and v_2, respectively. We see that the length of $\mathbf{v}$ might be measured by considering $\mathbf{v}$ as the hypotenuse of a right triangle and using the Pythagorean theorem to write the length as

$$\|\mathbf{v}\| = \sqrt{v_1^2 + v_2^2}.$$

This idea can be formalized for vectors of length n. We give the following definition for $\|\mathbf{v}\|$.

Definition 2.4 (Vector Norm). *Let* $\mathbf{v} \in \mathbb{R}^n$. *We define the* norm *of* $\mathbf{v}$ *by*

$$\boxed{\|\mathbf{v}\| = \sqrt{\sum_{k=1}^{n} v_k^2}.}$$

□

You should notice a connection between $\|\mathbf{v}\|$ and the definition of the inner product (2.1). In Problem 2.4 you will be asked more about this relationship.

Example 2.2 (Vector Norms). *Let* $\mathbf{v} = \begin{bmatrix} 2 \\ -5 \\ 4 \end{bmatrix}$ *and* $\mathbf{w} = \left[\frac{1}{2}, \frac{1}{2}, \frac{1}{2}, \frac{1}{2}\right]^T$. *Compute* $\|\mathbf{v}\|$ *and* $\|\mathbf{w}\|$.

Solution.

$$\|\mathbf{v}\| = \sqrt{2^2 + (-5)^2 + 4^2} = 3\sqrt{5} \quad \text{and} \quad \|\mathbf{w}\| = \sqrt{\frac{1}{4} + \frac{1}{4} + \frac{1}{4} + \frac{1}{4}} = 1.$$

(Live example: Visit `stthomas.edu/wavelets` *and click on Live Examples.)*

□

PROBLEMS

2.1 For each of the pairs $\mathbf{v}$ and $\mathbf{w}$ below, compute the inner product $\mathbf{v} \cdot \mathbf{w}$. *Hint:* To simplify the inner products in (c) – (e), you can use well-known summation formulas for $\sum_{k=1}^{n} k$ and $\sum_{k=1}^{n} k^2$. See, e.g., Stewart [84], Chap. 5.

(a) $\mathbf{v} = [2, 5, -3, 1]^T$ and $\mathbf{w} = [-1, 0, -1, 0]^T$

(b) $\mathbf{v} = [1, 0, 1]^T$ and $\mathbf{w} = [0, 50, 0]^T$

(c) $\mathbf{v} = [1, 2, 3, \ldots, n]^T$ and $\mathbf{w} = [1, 1, 1, \ldots, 1]^T$

(d) $\mathbf{v} = \left[1, 1, 2, 4, 3, 9, 4, 16, \ldots, n, n^2\right]^T$ and $\mathbf{w} = [1, 0, 1, 0, \ldots, 1, 0]^T$

(e) $\mathbf{v} = \left[1, 1, 2, 4, 3, 9, 4, 16, \ldots, n, n^2\right]^T$ and $\mathbf{w} = [0, 1, 0, 1, \ldots, 0, 1]^T$

2.2 Compute $\|\mathbf{v}\|$ and $\|\mathbf{w}\|$ for each of the vectors in Problem 2.1(a)-(e).

2.3 Find $\|\mathbf{v}\|$ when $\mathbf{v} = [v_1, \ldots, v_n]^T$ with

(a) $v_k = c$, $k = 1, \ldots, n$, where c is any real number

(b) $v_k = k$, $k = 1, \ldots, n$

(c) $v_k = \sqrt{k}$, $k = 1, \ldots, n$

Hint: The summation formulas referred to in Problem 2.1(c)-(e) will be useful for (b) and (c).

2.4 Let $\mathbf{v} \in \mathbb{R}^N$ and show that $\|\mathbf{v}\|^2 = \mathbf{v} \cdot \mathbf{v}$.

2.5 Show that for vectors $\mathbf{u}, \mathbf{v}, \mathbf{w} \in \mathbb{R}^n$ and $c \in \mathbb{R}$, we have

(a) $\mathbf{u} \cdot \mathbf{v} = \mathbf{v} \cdot \mathbf{u}$

(b) $(c\mathbf{u}) \cdot \mathbf{v} = c(\mathbf{u} \cdot \mathbf{v})$

(c) $(\mathbf{u} + c\mathbf{v}) \cdot \mathbf{w} = \mathbf{u} \cdot \mathbf{w} + c\mathbf{v} \cdot \mathbf{w}$

★**2.6** Let c be any real number. Show that for any vector $\mathbf{v} \in \mathbb{R}^n$, $\|c\mathbf{v}\| = |c| \cdot \|\mathbf{v}\|$.

2.7 Let $\mathbf{u}, \mathbf{v} \in \mathbb{R}^N$. Show that $\mathbf{u} \cdot \mathbf{v} = \frac{1}{4}\|\mathbf{u} + \mathbf{v}\|^2 - \frac{1}{4}\|\mathbf{u} - \mathbf{v}\|^2$.

2.8 Prove the *Cauchy-Schwarz inequality*. That is, show that for vectors $\mathbf{u}, \mathbf{v} \in \mathbb{R}^n$, we have

$$|\mathbf{u} \cdot \mathbf{v}|^2 \leq \|\mathbf{u}\|^2 \|\mathbf{v}\|^2. \tag{2.2}$$

Hint: Consider the nonnegative expression $0 \leq (\mathbf{u} - t\mathbf{v}) \cdot (\mathbf{u} - t\mathbf{v})$. Expand this expression and observe that it is a quadratic polynomial in t that is nonnegative. What do we know then, about the roots of this quadratic polynomial? Write this condition using the discriminant from the quadratic formula.

2.9 Show that in Problem 2.8, equality holds if and only if $\mathbf{u} = c\mathbf{v}$ for some $c \in \mathbb{R}$.

2.10 In this problem you will see how to construct a vector that is orthogonal to a given vector.

(a) Let $\mathbf{u} = [1, 2, 3, 4]^T$. Set $\mathbf{v} = [4, -3, 2, -1]^T$. Show that $\mathbf{u}$ and $\mathbf{v}$ are orthogonal and that $\|\mathbf{v}\| = \|\mathbf{u}\|$. Do you see how $\mathbf{v}$ was constructed from $\mathbf{u}$?

(b) If you have guessed what the construction is, try it on your own examples in $\mathbb{R}^n$. Make sure to choose vectors that are of even length and vectors that are of odd length. What did you find?

★**2.11** Let n be even and let $\mathbf{u} = [u_1, u_2, \ldots, u_n]^T$. Define $\mathbf{v} = [v_1, v_2, \ldots, v_n]^T$ by the rule

$$v_k = (-1)^k u_{n-k+1}, \qquad k = 1, 2, \ldots, n$$

Show that $\mathbf{u}$ and $\mathbf{v}$ are orthogonal and that $\|\mathbf{v}\| = \|\mathbf{u}\|$. The following steps will help you show that $\mathbf{u} \cdot \mathbf{v} = 0$.

(a) Show that

$$\mathbf{u} \cdot \mathbf{v} = \sum_{k=1}^{n} u_k (-1)^k u_{n-k+1}.$$

(b) Separate the even and odd terms in (a), showing that

$$\mathbf{u} \cdot \mathbf{v} = -\sum_{k \text{ odd}} u_k u_{n-k+1} + \sum_{k \text{ even}} u_k u_{n-k+1}.$$

(c) In the second sum above, make the index substitution $j = n - k + 1$ so that $k = n - j + 1$. If k runs over the values $2, 4, \ldots, n$, what are the values of j?

★**2.12** Often, it is desirable to work with vectors of length 1. Suppose that $\mathbf{v} \in \mathbb{R}^n$ with $\mathbf{v} \neq \mathbf{0}$. Let $\mathbf{u} = \frac{1}{\|\mathbf{v}\|}\mathbf{v}$. Show that $\|\mathbf{u}\| = 1$. This process is known as *normalizing* the vector $\mathbf{v}$. *Hint:* Problem 2.6 will be helpful.

2.13 Use Problem 2.12 to normalize the following vectors:

(a) $\mathbf{v} = [5, 0, 12]^T$

(b) $\mathbf{v} = [2, 6, 3, 1]^T$

(c) $\mathbf{v} \in \mathbb{R}^n$ with $v_k = -2$, $k = 1, 2, \ldots, n$

★**2.14** We measure the *distance between* vectors $\mathbf{v}$ and $\mathbf{w}$ by computing $\|\mathbf{v} - \mathbf{w}\|$. Let $\mathbf{u} = [1, 0, 0]$, $\mathbf{v} = \left[\frac{1}{3}, \frac{1}{3}, \frac{1}{3}\right]$, and $\mathbf{w} = \left[0, \frac{1}{2}, \frac{1}{2}\right]$.

(a) Compute $\|\mathbf{u} - \mathbf{w}\|$ and $\|\mathbf{v} - \mathbf{w}\|$.

(b) Show that $\frac{\|\mathbf{v}-\mathbf{w}\|}{\|\mathbf{u}-\mathbf{w}\|} = \frac{1}{3}$.

★**2.15** Suppose $\mathbf{v}, \mathbf{w} \in \mathbb{R}^n$ are two nonzero vectors with $0 \leq \theta \leq \pi$ the angle between them. Show that $\mathbf{v} \cdot \mathbf{w} = \|\mathbf{v}\|\|\mathbf{w}\| \cos\theta$. The following steps will help you organize your work.

(a) Let $\theta = 0$. Then $\cos\theta = 1$ and $\mathbf{w} = c\mathbf{v}$ where $c > 0$. Use Problem 2.6 and the fact that $|c| = c$ to establish the result in this case.

(b) Let $\theta = \pi$. Then $\cos\theta = 1$ and $\mathbf{w} = c\mathbf{v}$ where $c > 0$. Write $|c|$ in this case and use Problem 2.6 to establish the result.

(a) Now suppose that $\mathbf{v}$ and $\mathbf{w}$ are not multiples of each other so that $0 < \theta < \pi$. Then you can form a triangle whose sides have lengths $\|\mathbf{v}\|$, $\|\mathbf{w}\|$ and $\|\mathbf{v} - \mathbf{w}\|$. Use the Law of Cosines and expand the side with $\|\mathbf{v} - \mathbf{w}\|^2$ and simplify to prove the result.

2.2 Basic Matrix Theory

One of the most fundamental tools to come from linear algebra is the *matrix*. It would be difficult to write down all the applications that involve matrices. For our purposes we use matrices to represent digital images and the transformations that we apply to digital images or signals in applications such as image enhancement, data compression, and denoising.

A matrix can be thought of as simply a table of numbers. More generally, we call these numbers elements since in some instances we want matrices comprised of functions or even other matrices. Unless stated otherwise, we will denote matrices by uppercase letters. Here are three examples of matrices:

$$A = \begin{bmatrix} 2 & 3 \\ 6 & -5 \end{bmatrix} \qquad B = \begin{bmatrix} b_{11} & b_{12} \\ b_{21} & b_{22} \\ b_{31} & b_{32} \end{bmatrix} \qquad C = \begin{bmatrix} 3 & 5 & 0 \\ 2 & \frac{1}{2} & 6 \end{bmatrix}.$$

We say the first matrix is a 2×2 matrix: two rows (horizontal lists) and two columns (vertical lists). The matrix A has four elements: The element in row 1, column 1 is 2; the element in row 1, column 2 is 3, and so on.

We give a generic 3×2 matrix for B. Here we use the common notation b_{ij}, $i = 1, 2, 3$, and $j = 1, 2$. The notation b_{ij} denotes the value in B located at row i, column j.

Matrix C is a 2×3 matrix. If we were to assign the variables c_{ij}, $i = 1, 2$ and $j = 1, 2, 3$ to the elements of C, we would have, for example, $c_{12} = 5$, $c_{21} = 2$, and $c_{23} = 6$.

We now give a formal definition of the dimension of a matrix.

Definition 2.5 (Dimension of a Matrix). *Let A be a matrix consisting of m rows and n columns. We say that the* dimension *of A is $m \times n$.* □

Note that the vectors introduced in Section 2.1 are actually $n \times 1$ matrices and $\mathbf{v}^T$ is a matrix with dimension $1 \times n$.

Matrix Arithmetic

Some of the most common operations we perform on matrices are addition, subtraction, and scalar multiplication. If A is a $m \times n$ matrix, then the only matrices that we can add to it or subtract from it must have the same dimension. If A and B are of the same dimension then addition and subtraction is straightforward. We simply add or subtract the corresponding elements.

Definition 2.6 (Addition and Subtraction of Matrices). *Let A and B be the $m \times n$ matrices*

$$A = \begin{bmatrix} a_{11} & a_{12} & \cdots & a_{1n} \\ a_{21} & a_{22} & \cdots & a_{2n} \\ \vdots & \vdots & \ddots & \vdots \\ a_{m1} & a_{m2} & \cdots & a_{mn} \end{bmatrix}, \quad B = \begin{bmatrix} b_{11} & b_{12} & \cdots & b_{1n} \\ b_{21} & b_{22} & \cdots & b_{2n} \\ \vdots & \vdots & \ddots & \vdots \\ b_{m1} & b_{m2} & \cdots & b_{mn} \end{bmatrix}.$$

The sum S of A and B is the $m \times n$ matrix given by

$$S = A + B$$

$$\begin{bmatrix} s_{11} & s_{12} & \cdots & s_{1n} \\ s_{21} & s_{22} & \cdots & s_{2n} \\ \vdots & \vdots & \ddots & \vdots \\ s_{m1} & s_{m2} & \cdots & s_{mn} \end{bmatrix} = \begin{bmatrix} a_{11}+b_{11} & a_{12}+b_{12} & \cdots & a_{1n}+b_{1n} \\ a_{21}+b_{21} & a_{22}+b_{22} & \cdots & a_{2n}+b_{2n} \\ \vdots & \vdots & \ddots & \vdots \\ a_{m1}+b_{m1} & a_{m2}+b_{m2} & \cdots & a_{mn}+b_{mn} \end{bmatrix}$$

and the difference D of A and B is the $m \times n$ matrix given by

$$D = A - B$$

$$\begin{bmatrix} d_{11} & d_{12} & \cdots & d_{1n} \\ d_{21} & d_{22} & \cdots & d_{2n} \\ \vdots & \vdots & \ddots & \vdots \\ d_{m1} & d_{m2} & \cdots & d_{mn} \end{bmatrix} = \begin{bmatrix} a_{11}-b_{11} & a_{12}-b_{12} & \cdots & a_{1n}-b_{1n} \\ a_{21}-b_{21} & a_{22}-b_{22} & \cdots & a_{2n}-b_{2n} \\ \vdots & \vdots & \ddots & \vdots \\ a_{m1}-b_{m1} & a_{m2}-b_{m2} & \cdots & a_{mn}-b_{mn} \end{bmatrix}.$$

□

Another basic operation we perform on matrices is scalar multiplication. This operation basically consists of multiplying every element in a matrix by the same number. Here is a formal definition.

Definition 2.7 (Matrices and Scalar Multiplication). *Let A be the $m \times n$ matrix*

$$A = \begin{bmatrix} a_{11} & a_{12} & \cdots & a_{1n} \\ a_{21} & a_{22} & \cdots & a_{2n} \\ \vdots & \vdots & \ddots & \vdots \\ a_{m1} & a_{m2} & \cdots & a_{mn} \end{bmatrix}$$

and let c be any real number. Then we define the scalar product cA *as follows:*

$$cA = c\begin{bmatrix} a_{11} & a_{12} & \cdots & a_{1n} \\ a_{21} & a_{22} & \cdots & a_{2n} \\ \vdots & \vdots & \ddots & \vdots \\ a_{m1} & a_{m2} & \cdots & a_{mn} \end{bmatrix} = \begin{bmatrix} ca_{11} & ca_{12} & \cdots & ca_{1n} \\ ca_{21} & ca_{22} & \cdots & ca_{2n} \\ \vdots & \vdots & \ddots & \vdots \\ ca_{m1} & ca_{m2} & \cdots & ca_{mn} \end{bmatrix}.$$

□

Matrices have all kinds of practical uses. The example that follows illustrates how matrix addition, subtraction, and multiplication by a scalar applies to images.

Example 2.3 (Matrix Arithmetic). *Let* $a = \frac{1}{5}$, $b = 2$, *and consider the following* 4×4 *matrices:*

$$A = \begin{bmatrix} 100 & 50 & 50 & 25 \\ 0 & 25 & 25 & 125 \\ 0 & 100 & 75 & 50 \\ 125 & 50 & 0 & 25 \end{bmatrix} \qquad B = \begin{bmatrix} 50 & 50 & 50 & 50 \\ 50 & 50 & 50 & 50 \\ 50 & 50 & 50 & 50 \\ 50 & 50 & 50 & 50 \end{bmatrix}$$

and

$$C = \begin{bmatrix} 255 & 255 & 255 & 255 \\ 255 & 255 & 255 & 255 \\ 255 & 255 & 255 & 255 \\ 255 & 255 & 255 & 255 \end{bmatrix}.$$

We can compute the sum and difference

$$A + B = \begin{bmatrix} 150 & 100 & 100 & 75 \\ 50 & 75 & 75 & 175 \\ 50 & 150 & 125 & 100 \\ 175 & 100 & 50 & 75 \end{bmatrix} \qquad C - A = \begin{bmatrix} 155 & 205 & 205 & 230 \\ 255 & 230 & 230 & 130 \\ 255 & 155 & 180 & 205 \\ 130 & 205 & 255 & 230 \end{bmatrix}$$

and the scalar products

$$aA = \frac{1}{5}A = \begin{bmatrix} 20 & 10 & 10 & 5 \\ 0 & 5 & 5 & 25 \\ 0 & 20 & 15 & 10 \\ 25 & 10 & 0 & 5 \end{bmatrix} \qquad bA = 2A = \begin{bmatrix} 200 & 100 & 100 & 50 \\ 0 & 50 & 50 & 250 \\ 0 & 200 & 150 & 100 \\ 250 & 100 & 0 & 50 \end{bmatrix}$$

easily enough, but when we view these results as images, we can see the effects of addition, subtraction, and scalar multiplication.

In Section 3.1 we will learn that pixels in grayscale (monochrome) images are often represented as integer values ranging from 0 *(black) to* 255 *(white). In Figure 2.5 we display the matrices* A, B, *and* C *and various arithmetic operations by shading their entries with the gray level that the value represents.*

In Figure 2.5 you can see that the sum $A + B$ *is about* 20% *lighter than* A. *The difference* $C - A$ *is more relevant. Subtracting each of the entries of* A *from* 255 *has*

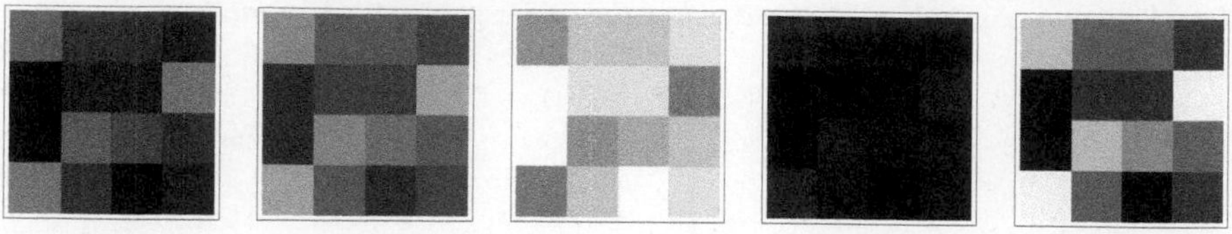

(a) A (b) $A+B$ (c) $C-A$ (d) $\frac{1}{5}A$ (e) $2A$

Figure 2.5 Matrices A, B, and C and various arithmetic operations with entries shaded to the appropriate gray level.

the effect of reversing the gray intensities. The scalar multiplication by $\frac{1}{5}$ has the effect of darkening each pixel of A by about 20%, and $2A$ has the effect of lightening each pixel in A by a factor of two.

(Live example: Visit `stthomas.edu/wavelets` *and click on Live Examples.)*

□

Matrix Multiplication

We next review matrix multiplication. It is not an elementwise operation as is the case for addition, subtraction, and scalar multiplication. Let's motivate the idea with an example. Let $\mathbf{v} = [150, 200, 250, 150]^T$ and assume that $\mathbf{v}$ represents the grayscale values in a column of some image. Suppose that we want to compute not only the average of the components of $\mathbf{v}$, but also some weighted averages.

If we want to average the components of $\mathbf{v}$, then we could dot $\mathbf{v}$ with the vector $\mathbf{a} = \left[\frac{1}{4}, \frac{1}{4}, \frac{1}{4}, \frac{1}{4}\right]^T$. Suppose that we want the middle two components weighted twice as much as the first and last. Then we could dot $\mathbf{v}$ with the vector $\mathbf{b} = \left[\frac{1}{6}, \frac{1}{3}, \frac{1}{3}, \frac{1}{6}\right]^T$. And finally, suppose that we want to average just the first and last components of $\mathbf{v}$. Then we would dot $\mathbf{v}$ with the vector $\mathbf{c} = \left[\frac{1}{2}, 0, 0, \frac{1}{2}\right]^T$.

So we compute three inner products: $\mathbf{a}^T\mathbf{v}$, $\mathbf{b}^T\mathbf{v}$, and $\mathbf{c}^T\mathbf{v}$. We store these row vectors in a matrix A and then consider the product $A\mathbf{v}$ as a three-vector containing our three inner products.

$$A\mathbf{v} = \begin{bmatrix} \mathbf{a}^T \\ \mathbf{b}^T \\ \mathbf{c}^T \end{bmatrix} \mathbf{v} = \begin{bmatrix} \mathbf{a}^T\mathbf{v} \\ \mathbf{b}^T\mathbf{v} \\ \mathbf{c}^T\mathbf{v} \end{bmatrix} = \begin{bmatrix} \mathbf{a}\cdot\mathbf{v} \\ \mathbf{b}\cdot\mathbf{v} \\ \mathbf{c}\cdot\mathbf{v} \end{bmatrix} = \begin{bmatrix} \frac{1}{4} & \frac{1}{4} & \frac{1}{4} & \frac{1}{4} \\ \frac{1}{6} & \frac{1}{3} & \frac{1}{3} & \frac{1}{6} \\ \frac{1}{2} & 0 & 0 & \frac{1}{2} \end{bmatrix} \begin{bmatrix} 100 \\ 200 \\ 250 \\ 150 \end{bmatrix}$$

$$= \begin{bmatrix} \frac{1}{4}\cdot 100 + \frac{1}{4}\cdot 200 + \frac{1}{4}\cdot 250 + \frac{1}{4}\cdot 150 \\ \frac{1}{6}\cdot 100 + \frac{1}{3}\cdot 200 + \frac{1}{3}\cdot 250 + \frac{1}{6}\cdot 150 \\ \frac{1}{2}\cdot 100 + 0\cdot 200 + 0\cdot 250 + \frac{1}{2}\cdot 150 \end{bmatrix} = \begin{bmatrix} 175 \\ 191\frac{2}{3} \\ 125 \end{bmatrix}.$$

Of course, we might want to apply our weighted averages to other columns in addition to $\mathbf{v}$. We would simply store these in a matrix V (with $\mathbf{v}$ as one of the columns) and dot each new column with the rows of A and store the results in the appropriate place. Suppose that we create the matrix

$$V = \begin{bmatrix} 100 & 150 & 150 \\ 200 & 100 & 200 \\ 250 & 50 & 250 \\ 150 & 200 & 50 \end{bmatrix}$$

and we wish to compute the product AV. Then we simply dot each row of A with each column of V. There are nine total dot products. We have three dot products for each of the three columns of V, so it makes sense to store the result in a 3×3 matrix Y with the element y_{ij} in row i, column j of Y obtained by dotting row i of A with column j of V. This corresponds to applying the ith weighted average to the jth vector in V. We have

$$\begin{aligned} AV &= \begin{bmatrix} \frac{1}{4} & \frac{1}{4} & \frac{1}{4} & \frac{1}{4} \\ \frac{1}{6} & \frac{1}{3} & \frac{1}{3} & \frac{1}{6} \\ \frac{1}{2} & 0 & 0 & \frac{1}{2} \end{bmatrix} \begin{bmatrix} 100 & 150 & 150 \\ 200 & 100 & 200 \\ 250 & 50 & 250 \\ 150 & 200 & 50 \end{bmatrix} \\ &= \begin{bmatrix} 175\frac{1}{2} & 125 & 162\frac{1}{2} \\ 191\frac{2}{3} & 108\frac{1}{3} & 183\frac{1}{3} \\ 125 & 175 & 100 \end{bmatrix}. \end{aligned}$$

We are now ready to define matrix multiplication. To compute AB, we dot the rows of A with the columns of B. Thus, for the inner products to make sense, the number of elements in each row of A must be the same as the number of elements in each column of B. Also note that while we can compute AB, the product BA might not make sense. For example, if A is 3×4 and B is 4×2, then AB will consist of six inner products stored in a 3×2 matrix. But the product BA makes no sense because the rows of B consist of two elements, whereas the columns of A consist of four elements and we cannot perform the necessary inner products.

Definition 2.8 (Matrix Multiplication). *Suppose that A is an $m \times p$ matrix and B is an $p \times n$ matrix. Then the product AB is the $m \times n$ matrix C whose entries c_{ij} are obtained by dotting row i of A with column j of B, where $i = 1, 2, \ldots, m$ and $j = 1, 2, \ldots, n$. That is,*

$$c_{ij} = \sum_{k=1}^{p} a_{ik} b_{kj}.$$

□

Example 2.4 (Matrix Multiplication). *Compute the matrix products (if they exist).*

(a) $\begin{bmatrix} 1 & 5 & 2 \\ -2 & 6 & 9 \end{bmatrix} \cdot \begin{bmatrix} 6 & 0 & 3 & 7 \\ 1 & 2 & 5 & 4 \\ 0 & -1 & 3 & -2 \end{bmatrix}$

(b) $\begin{bmatrix} 1 & 1 & 1 \\ 2 & 2 & 2 \\ 3 & 3 & 3 \end{bmatrix} \cdot \begin{bmatrix} 1 & 3 & -2 \\ 4 & 2 & -5 \\ -1 & 3 & 0 \end{bmatrix}$

(c) $\begin{bmatrix} 6 & 11 & 5 \\ 1 & 0 & -1 \end{bmatrix} \cdot \begin{bmatrix} 3 & -2 \\ 8 & -3 \end{bmatrix}$

(d) $\begin{bmatrix} 2 & 5 \\ 1 & 3 \end{bmatrix} \cdot \begin{bmatrix} 3 & -5 \\ -1 & 2 \end{bmatrix}$

Solution.

Since we are multiplying a 2×3 matrix by a 3×4 matrix, the product in part (a) will be a 2×4 matrix. We compute the eight inner products to obtain

$$\begin{bmatrix} 1 & 5 & 2 \\ -2 & 6 & 9 \end{bmatrix} \cdot \begin{bmatrix} 6 & 0 & 3 & 7 \\ 1 & 2 & 5 & 4 \\ 0 & -1 & 3 & -2 \end{bmatrix} = \begin{bmatrix} 11 & 8 & 34 & 23 \\ -6 & 3 & 51 & -8 \end{bmatrix}.$$

In part (b) we multiply two 3×3 matrices, so the result is a 3×3 matrix. We compute the nine inner products to obtain

$$\begin{bmatrix} 1 & 1 & 1 \\ 2 & 2 & 2 \\ 3 & 3 & 3 \end{bmatrix} \cdot \begin{bmatrix} 1 & 3 & -2 \\ 4 & 2 & -5 \\ -1 & 3 & 0 \end{bmatrix} = \begin{bmatrix} 4 & 8 & -7 \\ 8 & 16 & -14 \\ 12 & 24 & -21 \end{bmatrix}.$$

For the matrices in part (c), the product is undefined. We are unable to multiply a 2×3 matrix and a 2×2 matrix since the number of columns in the first matrix is not the same as the number of rows in the second matrix.

In part (d) we multiply two 2×2 matrices so that the result is a 2×2 matrix. We compute the four inner products to obtain

$$\begin{bmatrix} 2 & 5 \\ 1 & 3 \end{bmatrix} \cdot \begin{bmatrix} 3 & -5 \\ -1 & 2 \end{bmatrix} = \begin{bmatrix} 1 & 0 \\ 0 & 1 \end{bmatrix}.$$

(Live example: Visit `stthomas.edu/wavelets` *and click on Live Examples.)*

□

Part (b) of Example 2.4 gives us a chance to describe what a matrix multiplication actually does. First note that dotting each column of the second matrix by $\begin{bmatrix} 1 & 1 & 1 \end{bmatrix}$ (the top row of the first matrix) has the effect of summing the components of the column. Then it is easy to see that multiplying each column of the second matrix by the row $\begin{bmatrix} 2 & 2 & 2 \end{bmatrix}$ has the effect of summing the components of each column of the second matrix and doubling the result and multiplying each column of the second matrix by the row $\begin{bmatrix} 3 & 3 & 3 \end{bmatrix}$ has the effect of summing the components of each column of the second matrix and tripling the result.

Although the product in part (c) is undefined, it is true that we can compute

$$\begin{bmatrix} 3 & -2 \\ 8 & -3 \end{bmatrix} \cdot \begin{bmatrix} 6 & 11 & 5 \\ 1 & 0 & -1 \end{bmatrix} = \begin{bmatrix} 16 & 33 & 17 \\ 45 & 88 & 44 \end{bmatrix}.$$

We immediately see that matrix multiplication is not a commutative operation. It is a common error for students to perform the computation above and give it as the answer for the product requested in Example 2.4(c) – do not make this mistake. The product in Example 2.4(c) is undefined.

The Inverse of a Matrix

The computation in Example 2.4(d) produces a special matrix.

Definition 2.9 (Identity Matrix). *Let I_n be the $n \times n$ matrix whose kth column is given by the* standard basis vector

$$\mathbf{e}^k = \begin{bmatrix} 0, 0, \ldots, \underbrace{1}_{k\text{th}}, 0, \ldots, 0 \end{bmatrix}^T.$$

Then I_n is called the identity matrix of order n. □

We write three identity matrices:

$$I_2 = \begin{bmatrix} 1 & 0 \\ 0 & 1 \end{bmatrix} \qquad I_3 = \begin{bmatrix} 1 & 0 & 0 \\ 0 & 1 & 0 \\ 0 & 0 & 1 \end{bmatrix} \qquad I_4 = \begin{bmatrix} 1 & 0 & 0 & 0 \\ 0 & 1 & 0 & 0 \\ 0 & 0 & 1 & 0 \\ 0 & 0 & 0 & 1 \end{bmatrix}.$$

The special property about identity matrices is that for $n \times p$ matrix A and $m \times n$ matrix B, we have $I_n A = A$ and $BI_n = B$. That is, I_n is to matrix multiplication what one is to ordinary multiplication. Consider the product

$$I_m A = \begin{bmatrix} 1 & 0 & 0 & \cdots & 0 \\ 0 & 1 & 0 & \cdots & 0 \\ 0 & 0 & 1 & \cdots & 0 \\ \vdots & \vdots & \vdots & \ddots & \vdots \\ 0 & 0 & 0 & \cdots & 1 \end{bmatrix} \begin{bmatrix} a_{11} & a_{12} & \cdots & a_{1n} \\ a_{21} & a_{22} & \cdots & a_{2n} \\ \vdots & \vdots & \ddots & \vdots \\ a_{m1} & a_{m2} & \cdots & a_{mn} \end{bmatrix}.$$

The first row of I_n is $\begin{bmatrix} 1 & 0 & 0 & \cdots & 0 \end{bmatrix}$ and when we dot it with each column of A, the result is simply one times the first element in the column. So $\begin{bmatrix} 1 & 0 & 0 & \cdots & 0 \end{bmatrix} A$ gives the first row of A. In a similar manner we see that the kth row of I_n picks out the kth element of each column of A and thus returns the kth row of A.

We now turn our attention to the important class of nonsingular or invertible matrices. We start with a definition.

Definition 2.10 (Nonsingular Matrix). *Let A be an $n \times n$ matrix. We say that A is* nonsingular *or* invertible *if there exists an $n \times n$ matrix B such that*

$$AB = BA = I_n$$

In this case, B is called the inverse *of A and is denoted by $B = A^{-1}$. If no such matrix B exists, then A is said to be* singular. □

Let's take a closer look at $BA = I_n$ in the case where $B = A^{-1}$. We have

$$\begin{bmatrix} b_{11} & b_{12} & \cdots & b_{1n} \\ b_{21} & b_{22} & \cdots & b_{2n} \\ \vdots & \vdots & \ddots & \vdots \\ b_{n1} & b_{n2} & \cdots & b_{nn} \end{bmatrix} \begin{bmatrix} a_{11} & a_{12} & \cdots & a_{1n} \\ a_{21} & a_{22} & \cdots & a_{2n} \\ \vdots & \vdots & \ddots & \vdots \\ a_{n1} & a_{n2} & \cdots & a_{nn} \end{bmatrix} = \begin{bmatrix} 1 & 0 & 0 & \cdots & 0 \\ 0 & 1 & 0 & \cdots & 0 \\ 0 & 0 & 1 & \cdots & 0 \\ \vdots & \vdots & \vdots & \ddots & \vdots \\ 0 & 0 & 0 & \cdots & 1 \end{bmatrix}.$$

In particular, we see that the first row of B must dot with the first column of A and produce one and the first row of B dotted with any other column of A results in an answer of 0. We can generalize this remark and say that when we dot row j of B with column k of A, we get 1 when $j = k$ and 0 when $j \neq k$.

Some inverses are really easy to compute. Consider the following example.

Example 2.5 (Inverse of a Diagonal Matrix). *Let D be the $n \times n$ diagonal matrix*

$$D = \begin{bmatrix} d_1 & 0 & 0 & \cdots & 0 \\ 0 & d_2 & 0 & \cdots & 0 \\ \vdots & \vdots & \vdots & \ddots & \vdots \\ 0 & 0 & 0 & \cdots & d_n \end{bmatrix}$$

where $d_k \neq 0$, $k = 1, 2, \ldots, n$.

Let's think about computing D^{-1}. If we consider the product $DD^{-1} = I_n$, we see that the first row of D, $\begin{bmatrix} d_1 & 0 & 0 & \cdots & 0 \end{bmatrix}$ must dot with the first column of D^{-1} and produce one. The obvious selection for the first column of D^{-1} is to make the first element $\frac{1}{d_1}$ and set the rest of the elements in the column equal to 0. That is, the first column of the inverse is $\begin{bmatrix} \frac{1}{d_1} & 0 & \cdots & 0 & 0 \end{bmatrix}^T$.

Note also that if we take any other row of D and dot it with this column, we obtain 0. Thus, we have the first column of D^{-1}.

It should be clear how to choose the remaining columns of D^{-1}. The kth column of D^{-1} should have zeros in all positions except the kth position where we would want $\frac{1}{d_k}$. In this way we see that D is nonsingular with

$$D^{-1} = \begin{bmatrix} \frac{1}{d_1} & 0 & 0 & \cdots & 0 \\ 0 & \frac{1}{d_2} & 0 & \cdots & 0 \\ \vdots & \vdots & \vdots & \ddots & \vdots \\ 0 & 0 & 0 & \cdots & \frac{1}{d_n} \end{bmatrix}.$$

(Live example: Visit `stthomas.edu/wavelets` *and click on Live Examples.)*

□

Here are some basic properties obeyed by invertible matrices.

Proposition 2.1 (Properties of Nonsingular Matrices). *Suppose that A and B are $n \times n$ nonsingular matrices and c is any nonzero real number.*

(a) A^{-1} *is unique. That is, there is only one matrix* A^{-1} *such that*

$$A^{-1}A = AA^{-1} = I_n.$$

(b) A^{-1} *is nonsingular and* $(A^{-1})^{-1} = A$.

(c) AB *is nonsingular with* $(AB)^{-1} = B^{-1}A^{-1}$.

(d) cA *is nonsingular with* $(cA)^{-1} = \frac{1}{c}A^{-1}$.

□

Proof. The proof is left as Problem 2.19. □

Why are we interested in the class of nonsingular matrices? Suppose that an $m \times m$ matrix A represents some procedure that we want to apply to an $m \times n$ image matrix M. For example, suppose that AM is used to transform our image in order to facilitate compression. Then AM is encoded and transmitted over the internet. After decoding, the recipient has AM and since A is nonsingular, the recipient can compute

$$A^{-1}(AM) = (A^{-1}A)M = I_m M = M$$

and, in this process, recover the original image matrix M.

In effect, if A is nonsingular, we can solve the matrix equation $AX = B$ for unknown matrix X by computing $X = A^{-1}B$ or the vector equation $A\mathbf{x} = \mathbf{b}$ for unknown vector $\mathbf{x}$ by computing $\mathbf{x} = A^{-1}\mathbf{b}$. This is completely analogous to solving $ax = b$, where a, b are real numbers with $a \neq 0$. To solve this equation, we multiply both sides by a^{-1}, the multiplicative inverse of a, to obtain $x = a^{-1}b$.

Since A^{-1} is unique, there can only be one solution to $A\mathbf{x} = \mathbf{b}$ when A is nonsingular. What happens if A is singular? Let's look at an example.

Example 2.6 (Solving $\mathbf{Ax} = \mathbf{b}$). *Solve the equations*

$$\begin{bmatrix} 1 & 1 & 1 \\ 2 & 2 & 2 \\ 3 & 3 & 3 \end{bmatrix} \begin{bmatrix} x_1 \\ x_2 \\ x_3 \end{bmatrix} = \begin{bmatrix} 0 \\ 0 \\ 0 \end{bmatrix} \quad \textit{and} \quad \begin{bmatrix} 1 & 1 & 1 \\ 2 & 2 & 2 \\ 3 & 3 & 3 \end{bmatrix} \begin{bmatrix} x_1 \\ x_2 \\ x_3 \end{bmatrix} = \begin{bmatrix} 1 \\ 0 \\ 0 \end{bmatrix}.$$

Solution.
We show that no matrix B exists such that $AB = BA = I_3$. Thus A is singular. We proceed with a straightforward approach. For a general matrix B, it is easy to show that

$$AB = \begin{bmatrix} 1 & 1 & 1 \\ 2 & 2 & 2 \\ 3 & 3 & 3 \end{bmatrix} \cdot \begin{bmatrix} b_{11} & b_{12} & b_{13} \\ b_{21} & b_{22} & b_{23} \\ b_{31} & b_{32} & b_{33} \end{bmatrix} = \begin{bmatrix} s_1 & s_2 & s_3 \\ 2s_1 & 2s_2 & 2s_3 \\ 3s_1 & 3s_2 & 3s_3 \end{bmatrix}$$

where $s_1 = b_{11} + b_{21} + b_{31}$, $s_2 = b_{12} + b_{22} + b_{32}$ *and* $s_3 = b_{13} + b_{23} + b_{33}$. *Setting this product equal to* I_3 *gives*

$$\begin{bmatrix} s_1 & s_2 & s_3 \\ 2s_1 & 2s_2 & 2s_3 \\ 3s_1 & 3s_2 & 3s_3 \end{bmatrix} = \begin{bmatrix} 1 & 0 & 0 \\ 0 & 1 & 0 \\ 0 & 0 & 1 \end{bmatrix}$$

and we see that we can never solve for s_1, s_2, *or* s_3. *For example, equating corresponding elements in the first columns implies* $s_1 = 1$, $2s_1 = 0$, *and* $3s_1 = 0$. *Thus no inverse exists so that* A *is singular.*

To characterize the solution of the systems, we note that

$$\begin{bmatrix} 1 & 1 & 1 \\ 2 & 2 & 2 \\ 3 & 3 & 3 \end{bmatrix} = \begin{bmatrix} 1 & 0 & 0 \\ 0 & 2 & 0 \\ 0 & 0 & 3 \end{bmatrix} \begin{bmatrix} 1 & 1 & 1 \\ 1 & 1 & 1 \\ 1 & 1 & 1 \end{bmatrix}$$

and use the results of Example 2.5 to write

$$\begin{bmatrix} 1 & 0 & 0 \\ 0 & 2 & 0 \\ 0 & 0 & 3 \end{bmatrix}^{-1} = \begin{bmatrix} 1 & 0 & 0 \\ 0 & \frac{1}{2} & 0 \\ 0 & 0 & \frac{1}{3} \end{bmatrix}. \tag{2.3}$$

For the first system we have

$$\begin{bmatrix} 1 & 1 & 1 \\ 2 & 2 & 2 \\ 3 & 3 & 3 \end{bmatrix} \begin{bmatrix} x_1 \\ x_2 \\ x_3 \end{bmatrix} = \begin{bmatrix} 0 \\ 0 \\ 0 \end{bmatrix}$$

$$\begin{bmatrix} 1 & 0 & 0 \\ 0 & 2 & 0 \\ 0 & 0 & 3 \end{bmatrix} \begin{bmatrix} 1 & 1 & 1 \\ 1 & 1 & 1 \\ 1 & 1 & 1 \end{bmatrix} \begin{bmatrix} x_1 \\ x_2 \\ x_3 \end{bmatrix} = \begin{bmatrix} 0 \\ 0 \\ 0 \end{bmatrix}.$$

Multiplying both sides of this equation by the inverse found in (2.3) *gives*

$$\begin{bmatrix} 1 & 1 & 1 \\ 1 & 1 & 1 \\ 1 & 1 & 1 \end{bmatrix} \begin{bmatrix} x_1 \\ x_2 \\ x_3 \end{bmatrix} = \begin{bmatrix} 1 & 0 & 0 \\ 0 & \frac{1}{2} & 0 \\ 0 & 0 & \frac{1}{3} \end{bmatrix} \begin{bmatrix} 0 \\ 0 \\ 0 \end{bmatrix}$$

$$\begin{bmatrix} x_1 + x_2 + x_3 \\ x_1 + x_2 + x_3 \\ x_1 + x_2 + x_3 \end{bmatrix} = \begin{bmatrix} 0 \\ 0 \\ 0 \end{bmatrix}.$$

It is easy to see that we have an infinite number of solutions to this system – any three numbers that sum to zero work as a solution.

Using the same procedure for the second system gives

$$\begin{bmatrix} 1 & 1 & 1 \\ 1 & 1 & 1 \\ 1 & 1 & 1 \end{bmatrix} \begin{bmatrix} x_1 \\ x_2 \\ x_3 \end{bmatrix} = \begin{bmatrix} 1 & 0 & 0 \\ 0 & \frac{1}{2} & 0 \\ 0 & 0 & \frac{1}{3} \end{bmatrix} \begin{bmatrix} 1 \\ 0 \\ 0 \end{bmatrix}$$

$$\begin{bmatrix} x_1 + x_2 + x_3 \\ x_1 + x_2 + x_3 \\ x_1 + x_2 + x_3 \end{bmatrix} = \begin{bmatrix} 1 \\ 0 \\ 0 \end{bmatrix}.$$

and we see that this system cannot be solved – it is impossible to find three numbers that simultaneously sum to zero and one.

(Live example: Visit `stthomas.edu/wavelets` *and click on Live Examples.)*

□

Finding the Inverse of a Matrix

A formula for the inverse of a 2×2 nonsingular matrix is developed in Problem 2.20. One of the most common ways to find the inverse of a nonsingular $n \times n$ matrix is to use *Gaussian elimination*. This process is usually described in a sophomore linear algebra book such as Lay [65].

As motivated above, we need the inverse of nonsingular matrix W for applications such as image compression. Fortunately, the methods we use to build W incorporates the construction of W^{-1}. Thus we will not need Gaussian elimination or any other technique to compute W^{-1}. However, the construction of inverses we need does rely on the *transpose* of a matrix and *orthogonal matrices*. The transpose of a matrix is described next and orthogonal matrices are covered later in the section.

The Transpose of a Matrix

There is another matrix related to A that is important in our subsequent work. This matrix is called the *transpose* of A and denoted by A^T. Basically, we form A^T by taking the rows of A and making them the columns of A^T. So if A is an $m \times n$ matrix, A^T will be an $n \times m$ matrix. Here is the formal definition:

Definition 2.11 (Transpose of a Matrix). *Let A be an $m \times n$ matrix with elements a_{ij}, $i = 1, 2, \ldots, m$, $j = 1, 2, \ldots, n$. Then the* transpose *of A, denoted A^T, is the $n \times m$ matrix whose elements are given by a_{ji} for $j = 1, 2, \ldots, n$ and $i = 1, 2, \ldots, m$. That is, if*

$$A = \begin{bmatrix} a_{11} & a_{12} & \cdots & a_{1n} \\ a_{21} & a_{22} & \cdots & a_{2n} \\ \vdots & \vdots & \ddots & \vdots \\ a_{m1} & a_{m2} & \cdots & a_{mn} \end{bmatrix}, \quad \textit{then} \quad A^T = \begin{bmatrix} a_{11} & a_{21} & \cdots & a_{m1} \\ a_{12} & a_{22} & \cdots & a_{m2} \\ \vdots & \vdots & \ddots & \vdots \\ a_{1n} & a_{2n} & \cdots & a_{mn} \end{bmatrix}.$$

□

Next we present some elementary examples of transposes of matrices.

Example 2.7 (Matrix Transposes). *Find the transposes of*

$$A = \begin{bmatrix} 1 & 6 & -2 \\ 2 & 3 & 0 \end{bmatrix} \qquad B = \begin{bmatrix} 2 & 3 \\ -1 & 1 \end{bmatrix} \qquad C = \begin{bmatrix} 2 & 6 & 3 \\ 6 & 4 & -1 \\ 3 & -1 & 1 \end{bmatrix}.$$

Solution.

We have

$$A^T = \begin{bmatrix} 1 & 2 \\ 6 & 3 \\ -2 & 0 \end{bmatrix} \qquad B^T = \begin{bmatrix} 2 & -1 \\ 3 & 1 \end{bmatrix} \qquad C^T = \begin{bmatrix} 2 & 6 & 3 \\ 6 & 4 & -1 \\ 3 & -1 & 1 \end{bmatrix}.$$

(Live example: Visit `stthomas.edu/wavelets` *and click on Live Examples.)*

□

Note that in Example 2.7 C^T was the same as C. We call such matrices *symmetric*.

Definition 2.12 (Symmetric Matrix). *We say that A is* symmetric *if $A^T = A$.* □

Determinants

Determinants are an important concept in linear algebra. They can be used to find eigenvalues of matrices or indicate whether or not a matrix is invertible. In this book, we will have need for determinants of 2×2 matrices in Section 11.3. We state the definition of a 2×2 matrix determinant, provide some examples where determinants are computes and state a proposition that contains some properties of determinants.

Definition 2.13 (Determinant of a 2 × 2 Matrix). *Suppose $A = \begin{bmatrix} a & b \\ c & d \end{bmatrix}$. Then the* determinant *of A is defined as*

$$\det(A) = ad - bc.$$

□

As illustrated by the next example, computing determinants of 2×2 matrices is straightforward.

Example 2.8 (Computing Determinants). *Compute the determinants of the following matrices.*

(a) $A = \begin{bmatrix} 1 & 4 \\ 2 & 8 \end{bmatrix}$

(b) $I = \begin{bmatrix} 1 & 0 \\ 0 & 1 \end{bmatrix}$

(c) $B = \begin{bmatrix} -2 & 1 \\ 3 & 5 \end{bmatrix}$

(d) $C = \begin{bmatrix} -2 & 1 \\ 6 & 10 \end{bmatrix}$

(e) $D = BE$, *where* $E = \begin{bmatrix} 2 & 0 \\ 4 & 1 \end{bmatrix}$*; also compute* $\det(B) \cdot \det(E)$

Solution. *We have*

$$\begin{aligned} \det(A) &= 1 \cdot 8 - 4 \cdot 2 = 0 \\ \det(I) &= 1 - 0 = 1 \\ \det(B) &= -2 \cdot 5 - 1 \cdot 3 = -13 \\ \det(C) &= -2 \cdot 10 - 1 \cdot 6 = -26 = -2\det(B) \end{aligned}$$

For (e) we have $D = \begin{bmatrix} 0 & 1 \\ 26 & 5 \end{bmatrix}$ *so that* $\det(D) = 0 - 26 = -26$. *We have* $\det(B) = -13$ *and* $\det(E) = 2$ *so that* $\det(B)\det(E) = -26 = \det(BE)$.

(Live example: Visit `stthomas.edu/wavelets` *and click on Live Examples.)*

□

The following proposition lists a few properties obeyed by determinants that we will need later in the book.

Proposition 2.2 (Properties of Determinants). *Suppose* $A = \begin{bmatrix} a & b \\ c & d \end{bmatrix}$.

(a) $\det(AB) = \det(A)\det(B)$, *where* B *is any* 2×2 *matrix*

(b) $\det(A) = \det\left(A^T\right)$

(c) If $\det(A) = 0$ *where at least one of* a *or* b *is not zero, then* $c = ka$ *and* $d = kb$ *for some number* k

(d) *Suppose B is constructed from A by multiplying a row or column of A by α. Then* $\det(B) = \alpha \det(A)$.

□

Proof. We prove (a) and (c) and leave (b) and (d) as exercises.

For (a), we do a direct computation. Let $B = \begin{bmatrix} p & q \\ r & s \end{bmatrix}$. Then

$$\det(A)\det(B) = (ad - bc)(ps - qr) = apds - bcps - abqr + brcq.$$

Multiplying the matrices gives us

$$AB = \begin{bmatrix} a & b \\ c & d \end{bmatrix}\begin{bmatrix} p & q \\ r & s \end{bmatrix} = \begin{bmatrix} ap + br & aq + bs \\ cp + dr & cq + ds \end{bmatrix}$$

which results in

$$\begin{aligned}\det(AB) &= (ap + br)(cq + ds) - (aq + bs)(cp + dr) \\ &= apcq + brcq + apds + brds - aqcp - aqdr - bscp - bsdr \\ &= brcq + apds - aqdr - bscp \\ &= \det(A)\det(B).\end{aligned}$$

For (c), we note that $0 = \det A = ad - bc$ implies $ad = bc$. Now assume $b \neq 0$ (the case where $a \neq 0$ follows similarly). Then $c = \left(\frac{d}{b}\right) a$. Since we can write $d = \left(\frac{d}{b}\right) b$, we take $k = \frac{d}{b}$ and the proof is complete. □

Do you see how Example 2.8 illustrates the various properties in Proposition 2.2?

Orthogonal Matrices

We conclude this section by reviewing the class of matrices called *orthogonal matrices*. Consider the matrix

$$U = \begin{bmatrix} \frac{1}{2} & \frac{\sqrt{3}}{2} & 0 \\ 0 & 0 & 1 \\ -\frac{\sqrt{3}}{2} & \frac{1}{2} & 0 \end{bmatrix}.$$

Notice that if you dot any row (column) with any different row (column), you get 0. In addition, any row (column) dotted with itself returns one. So if we are interested in finding U^{-1}, it seems that the rows of U would work very well as the columns of U^{-1}. Indeed, if we compute UU^T, we have

$$UU^T = \begin{bmatrix} \frac{1}{2} & \frac{\sqrt{3}}{2} & 0 \\ 0 & 0 & 1 \\ -\frac{\sqrt{3}}{2} & \frac{1}{2} & 0 \end{bmatrix}\begin{bmatrix} \frac{1}{2} & 0 & -\frac{\sqrt{3}}{2} \\ \frac{\sqrt{3}}{2} & 0 & \frac{1}{2} \\ 0 & 1 & 0 \end{bmatrix} = \begin{bmatrix} 1 & 0 & 0 \\ 0 & 1 & 0 \\ 0 & 0 & 1 \end{bmatrix}.$$

The matrix U above is an example of an orthogonal matrix. There are a couple of ways to classify such matrices. A good thing to remember about orthogonal matrices is that their rows (or columns), taken as a set of vectors, forms an orthogonal set. We give the standard definition below.

Definition 2.14 (Orthogonal Matrix). *Suppose that U is an $n \times n$ matrix. We say U is an* orthogonal matrix *if*

$$U^{-1} = U^T. \tag{2.4}$$

□

Recall that the columns of U^T are the rows of U. So when we say that $UU^T = I$, we are saying that if row j of U is dotted with row k of U (or column k of U^T), then we must either get 1 if $j = k$ or 0 otherwise – this is the condition on the rows of U given in the text preceding Definition 2.14.

Orthogonal matrices also possess another important property related to vector norms:

Theorem 2.1 (Orthogonal Matrices Preserve Distance). *Suppose that U is an $n \times n$ orthogonal matrix and $\mathbf{x}$ is an n-vector. Then*

$$\|U\mathbf{x}\| = \|\mathbf{x}\|.$$

□

Proof. From Problem 2.4 in Section 2.1 we have

$$\|U\mathbf{x}\|^2 = (U\mathbf{x})^T U\mathbf{x}.$$

Starting with this identity, Problem 2.23 from Section 2.2, and (2.4), we have

$$\|U\mathbf{x}\|^2 = (U\mathbf{x})^T U\mathbf{x} = \mathbf{x}^T U^T U\mathbf{x} = \mathbf{x}^T U^{-1} U\mathbf{x} = \mathbf{x}^T I_n \mathbf{x} = \mathbf{x}^T \mathbf{x} = \|\mathbf{x}\|^2.$$

□

The following example further describes the geometry resulting from Theorem 2.1.

Example 2.9 (Rotations by an Orthogonal Matrix). *Consider the set of vectors $C = \left\{ \mathbf{x} \in \mathbb{R}^2 \,\middle|\, \|\mathbf{x}\| = 1 \right\}$. If we plot all these vectors in C, we would have a picture of a circle centered at $(0,0)$ with radius one. Suppose that U is the orthogonal matrix*

$$U = \begin{bmatrix} \frac{\sqrt{2}}{2} & \frac{\sqrt{2}}{2} \\ -\frac{\sqrt{2}}{2} & \frac{\sqrt{2}}{2} \end{bmatrix} = \frac{\sqrt{2}}{2} \begin{bmatrix} 1 & 1 \\ -1 & 1 \end{bmatrix}.$$

Then we know from Theorem 2.1 that $\|U\mathbf{x}\| = 1$ for any $\mathbf{x} \in C$. This means that applying U to some vector $\mathbf{x} \in C$ will result in some other vector in C.

In particular, if we apply U to $\mathbf{x} = [1, 0]^T$, $\mathbf{y} = \left[\frac{1}{2}, \frac{\sqrt{3}}{2}\right]^T$, and $\mathbf{z} = \left[-\frac{\sqrt{2}}{2}, \frac{\sqrt{2}}{2}\right]^T$ we obtain the vectors

$$U\mathbf{x} = \begin{bmatrix} \frac{\sqrt{2}}{2} \\ -\frac{\sqrt{2}}{2} \end{bmatrix} \qquad U\mathbf{y} = \begin{bmatrix} \frac{\sqrt{2}+\sqrt{6}}{4} \\ \frac{\sqrt{3}-1}{2\sqrt{2}} \end{bmatrix} \qquad U\mathbf{z} = \begin{bmatrix} 0 \\ 1 \end{bmatrix}.$$

The vectors $\mathbf{x}, \mathbf{y}, \mathbf{z}$, and $U\mathbf{x}, U\mathbf{y}, U\mathbf{z}$ are plotted in Figure 2.6. Do you see that U rotated each vector $\frac{\pi}{4}$ radians clockwise?

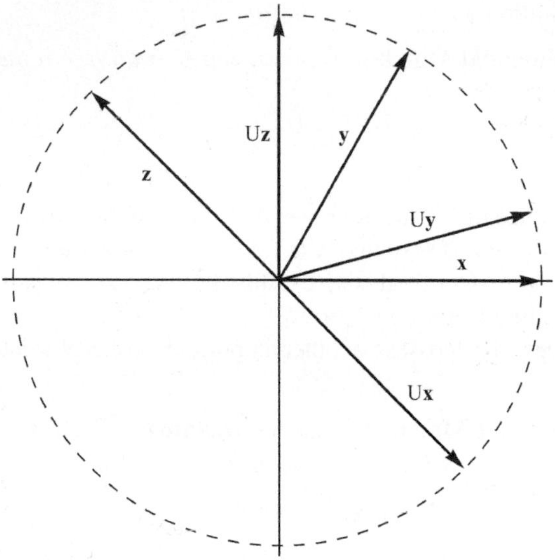

Figure 2.6 The vectors $\mathbf{x}, \mathbf{y}, \mathbf{z}$, and $U\mathbf{x}, U\mathbf{y}, U\mathbf{z}$ on the unit circle.

(Live example: Visit `stthomas.edu/wavelets` *and click on Live Examples.)*

□

PROBLEMS

2.16 In Example 2.4(c) we learned that matrix multiplication is not commutative. In this case the dimensions for the two matrices were incompatible. This is not the only reason matrix multiplication is not commutative. Can you find two 2×2 matrices A and B for which $AB \neq BA$?

2.17 Suppose that D is an $n \times n$ diagonal matrix with diagonal entries d_{ii}, $i = 1, 2, \ldots, n$. Let A be any $n \times n$ matrix.

(a) Describe in words the product DA.

(b) Describe in words the product AD.

2.18 Suppose that A is an $m \times n$ matrix. Explain how to write cA, where c is some real number, as a matrix product CA, where C is an $m \times m$ matrix.

2.19 Prove Proposition 2.1.

★**2.20** Let $A = \begin{bmatrix} a & b \\ c & d \end{bmatrix}$, where $ad - bc \neq 0$. Show that

$$A^{-1} = \frac{1}{ad - bc} \begin{bmatrix} d & -b \\ -c & a \end{bmatrix}.$$

★**2.21** Use Problem 2.20 to show that the inverse of $A = \begin{bmatrix} 0 & 1 \\ 1 & -a \end{bmatrix}$ is

$$A^{-1} = \begin{bmatrix} a & 1 \\ 1 & 0 \end{bmatrix}.$$

2.22 Let A be an $m \times n$ matrix. Show that AA^T is a symmetric matrix.

★**2.23** Show that $(A^T)^T = A$ and $(AB)^T = B^T A^T$.

★**2.24** Explain why $(A^{-1})^T = (A^T)^{-1}$.

2.25 Suppose that A and B are $m \times n$ matrices. Is it true that $A^T + B^T = (A + B)^T$? If so, prove the result. If not, find a counterexample.

2.26 Prove (b) and (d) of Proposition 2.2.

★**2.27** Suppose A is an invertible 2×2 matrix. Show that $\det\left(A^{-1}\right) = \frac{1}{\det(A)}$.

★**2.28** Suppose $A = \begin{bmatrix} a & b \\ c & d \end{bmatrix}$ and $B = \begin{bmatrix} a & b \\ e & f \end{bmatrix}$ with $\det(A) = \det(B)$. Show that $e = c + \alpha a$ and $f = d + \alpha b$ for some $\alpha \in \mathbb{R}$. The following steps will help you organize your work.

(a) Compute the determinants of both matrices, set them equal, and move all terms to one side obtaining $a(d - f) - b(c - e) = 0$.

(b) The identity in (a) is the zero determinant of some matrix. Find this matrix.

(c) Use Proposition 2.2(c) to complete the proof.

★**2.29** Suppose U is a 2×2 orthogonal matrix with first row $\begin{bmatrix} a & b \end{bmatrix}$. Since the second row must be orthogonal to the first row, it must be of the form $\begin{bmatrix} -b & a \end{bmatrix}$. In this case, U is of the form

$$U = \begin{bmatrix} a & b \\ -b & a \end{bmatrix}.$$

(a) Compute $UU^T = I_2$ to determine an equation satisfied by a and b.

(b) Use (a) and your knowledge of trigonometry to write a, b in terms of angle θ.

(c) Using (b), write down the general form of all 2×2 orthogonal matrices for a given angle θ.

(d) Construct a 2×2 orthogonal matrix U that rotates the vectors in C from Example 2.9 $\frac{\pi}{3}$ radians counterclockwise.

(e) Construct a 2×2 orthogonal matrix U that rotates the vectors in C from Example 2.9 $\frac{\pi}{2}$ radians clockwise.

(f) Describe in words what $U^{-1} = U^T$ does to vectors in C if U rotates the vectors in C by θ radians.

2.30 Let U be an orthogonal matrix. Use Proposition 2.2 to prove $\det(U) = \pm 1$.

2.31 Let U_1 and U_2 be orthogonal matrices of equal dimension with $P = U_1 U_2$. Show that P is an orthogonal matrix. Generalize this result to the case where $U_1, U_2, \ldots, U_n$ are orthogonal matrices of equal dimension and $P = U_1 U_2 \cdots U_n$.

2.32 Let $\mathbf{v} = [1, 2, 3, 4, 5, 6, 7, 8, 9, 10]^T$ and $\mathbf{w} = [1, 3, 5, 7, 9, 1, 1, 1, 1, 1]^T$.

(a) Find a 10×10 nonsingular matrix A such that $A\mathbf{v} = \mathbf{w}$. *Hint:* Do not use Gaussian elimination. What are "obvious" choices for the first five rows of A?

(b) Find A^{-1}.

★**2.33** Let n be an even positive integer, and let $\mathbf{v} = \begin{bmatrix} v_1 \\ v_2 \\ \vdots \\ v_n \end{bmatrix}$.

(a) Find a matrix H so that

$$H\mathbf{v} = \frac{1}{2} \begin{bmatrix} v_1 + v_2 \\ v_3 + v_4 \\ \vdots \\ v_{n-1} + v_n \end{bmatrix}.$$

(b) Note that the rows of H are orthogonal to each other. However, any inner product of a row with itself does not result in the value one. Can you find c such that the matrix $\widetilde{H} = cH$ has rows that are orthogonal to each other and of unit length?

2.3 Block Matrix Arithmetic

We discussed matrix arithmetic in Section 2.2 and we will learn how matrices can be utilized in image processing in Example 3.5. In many instances it is desirable to process only part of an image. In such a case, knowledge of *block matrices* is invaluable. Block matrices have applications throughout mathematics and sciences, and their manipulation is an important skill to master. Along with image processing applications, we will see that block matrices are very useful when developing the wavelet transformations that first appear in Chapter 4.

Partitioning Matrices

The idea of block matrices is quite simple. Given an $N \times M$ matrix A, we partition it into different blocks. There is no set rule for how partitions are created – it is driven by the application. These blocks are called *submatrices* because they are themselves matrices. If you have learned to compute the determinant of a matrix by the cofactor expansion method, then you have already worked with block matrices.

We denote our submatrices with capital letters, and often, we will add subscripts to indicate their position in the original matrix. Let's look at a simple example.

Example 2.10 (Block Matrices). *Consider the matrix*

$$A = \begin{bmatrix} -3 & 2 & 0 & 1 \\ 2 & 6 & 3 & 5 \\ 0 & -1 & -1 & 9 \end{bmatrix}.$$

We can partition A in a variety of ways. We present two possibilities.

(a) We partition A into two 3×2 matrices R and L. Thus,

$$A = \left[\begin{array}{cc|cc} -3 & 2 & 0 & 1 \\ 2 & 6 & 3 & 5 \\ 0 & -1 & -1 & 9 \end{array}\right] = [L \mid R]$$

where

$$L = \begin{bmatrix} -3 & 2 \\ 2 & 6 \\ 0 & -1 \end{bmatrix} \quad \text{and} \quad R = \begin{bmatrix} 0 & 1 \\ 3 & 5 \\ -1 & 9 \end{bmatrix}.$$

(b) We partition A into two 2×2 blocks and two 1×2 blocks. That is,

$$A = \left[\begin{array}{cc|cc} -3 & 2 & 0 & 1 \\ 2 & 6 & 3 & 5 \\ \hline 0 & -1 & -1 & 9 \end{array}\right] = \left[\begin{array}{c|c} A_{11} & A_{12} \\ \hline A_{21} & A_{22} \end{array}\right]$$

where

$$A_{11} = \begin{bmatrix} -3 & 2 \\ 2 & 6 \end{bmatrix} \qquad A_{12} = \begin{bmatrix} 0 & 1 \\ 3 & 5 \end{bmatrix}$$

and

$$A_{21} = \begin{bmatrix} 0 & -1 \end{bmatrix} \qquad A_{22} = \begin{bmatrix} -1 & 9 \end{bmatrix}.$$

The notation should be intuitive – we can view A as a 2×2 block matrix. In this case, A_{11} is the submatrix in the first "row," first "column," A_{12} is positioned at row 1, column 2; A_{21} occupies row 2, column one; and A_{22} is positioned at row 2, column 2.

□

Adding and Subtracting Block Matrices

If A and B are matrices whose dimensions are equal, then we can always compute $A + B$ and $A - B$. In many instances it is desirable to partition A and B and then write the sum or difference as sums or differences of the blocks. We can only add and subtract block by block if A and B are *partitioned in exactly the same way*.

Let's consider two 4×4 matrices:

$$A = \left[\begin{array}{ccc|c} a_{11} & a_{12} & a_{13} & a_{14} \\ a_{21} & a_{22} & a_{23} & a_{24} \\ \hline a_{31} & a_{32} & a_{33} & a_{34} \\ a_{41} & a_{42} & a_{43} & a_{44} \end{array}\right] = \left[\begin{array}{c|c} A_{11} & A_{12} \\ \hline A_{21} & A_{22} \end{array}\right]$$

and

$$B = \left[\begin{array}{ccc|c} b_{11} & b_{12} & b_{13} & b_{14} \\ b_{21} & b_{22} & b_{23} & b_{24} \\ \hline b_{31} & b_{32} & b_{33} & b_{34} \\ b_{41} & b_{42} & b_{43} & b_{44} \end{array}\right] = \left[\begin{array}{c|c} B_{11} & B_{12} \\ \hline B_{21} & B_{22} \end{array}\right].$$

Disregarding the partitioning, it is a straightforward task to compute

$$A + B = \begin{bmatrix} a_{11} + b_{11} & a_{12} + b_{12} & a_{13} + b_{13} & a_{14} + b_{14} \\ a_{21} + b_{21} & a_{22} + b_{22} & a_{23} + b_{23} & a_{24} + b_{24} \\ a_{31} + b_{31} & a_{32} + b_{32} & a_{33} + b_{33} & a_{34} + b_{34} \\ a_{41} + b_{41} & a_{42} + b_{42} & a_{43} + b_{43} & a_{44} + b_{44} \end{bmatrix} \tag{2.5}$$

and since A and B are partitioned exactly the same, we can simply add blocks

$$A + B = \left[\begin{array}{c|c} A_{11} + B_{11} & A_{12} + B_{12} \\ \hline A_{21} + B_{21} & A_{22} + B_{22} \end{array}\right],$$

noting that $A_{11} + B_{11}$, $A_{12} + B_{12}$, $A_{21} + B_{21}$, and $A_{22} + B_{22}$ are all valid matrix operations themselves. In particular,

$$A_{11} + B_{11} = \begin{bmatrix} a_{11} & a_{12} & a_{13} \\ a_{21} & a_{22} & a_{23} \end{bmatrix} + \begin{bmatrix} b_{11} & b_{12} & b_{13} \\ b_{21} & b_{22} & b_{23} \end{bmatrix}$$

$$= \begin{bmatrix} a_{11} + b_{11} & a_{12} + b_{12} & a_{13} + b_{13} \\ a_{21} + b_{21} & a_{22} + b_{22} & a_{23} + b_{23} \end{bmatrix}$$

is exactly the upper 2×3 corner of $A + B$ in (2.5). The other sums can be easily computed and verified as well.

Block Matrix Multiplication

We have learned that two matrices must be identically partitioned in order to add or subtract them. What about multiplication with blocks? Is it possible to partition matrices A and B into blocks so that the product AB can be computed by multiplying the blocks of A and B? Let's consider a couple of matrices as we formulate answers to these questions.

Example 2.11 (Block Matrix Multiplication). *Let*

$$A = \begin{bmatrix} 2 & 1 & -1 \\ 0 & 3 & 1 \end{bmatrix} \quad \textit{and} \quad B = \begin{bmatrix} 2 & 1 \\ -1 & 0 \\ 5 & 4 \end{bmatrix}.$$

Certainly we can compute

$$AB = \begin{bmatrix} -2 & -2 \\ 2 & 4 \end{bmatrix}. \tag{2.6}$$

What if we partition A as

$$A = \left[\begin{array}{cc|c} 2 & 1 & -1 \\ 0 & 3 & 1 \end{array}\right] = \left[\begin{array}{c|c} A_{11} & A_{12} \end{array}\right]?$$

The dimension of A, in terms of blocks, is 1×2. If we were to suppose for the moment that we could indeed partition B so that we could compute AB by multiplying blocks, then in block form, B must have two "rows" of blocks. Since A_{11} has two rows, the blocks in the first row of B must have two columns so we need to partition B as

$$B = \left[\begin{array}{c|c} 2 & 1 \\ -1 & 0 \\ \hline 5 & 4 \end{array}\right] = \left[\begin{array}{c|c} B_{11} & B_{12} \\ \hline B_{21} & B_{22} \end{array}\right] \quad \textit{or} \quad B = \left[\begin{array}{cc} 2 & 1 \\ -1 & 0 \\ \hline 5 & 4 \end{array}\right] = \left[\begin{array}{c} B_{11} \\ \hline B_{21} \end{array}\right]. \tag{2.7}$$

In the case of the first partitioning of B in (2.7), we have

$$AB = \left[\begin{array}{c|c} A_{11} & A_{12} \end{array}\right] \cdot \left[\begin{array}{c|c} B_{11} & B_{12} \\ \hline B_{21} & B_{22} \end{array}\right].$$

A is 1×2 blocks and B is 2×2 blocks, so the result should be 1×2 blocks:

$$AB = \left[\begin{array}{c|c} A_{11} & A_{12} \end{array}\right] \cdot \left[\begin{array}{c|c} B_{11} & B_{12} \\ \hline B_{21} & B_{22} \end{array}\right] = \left[\begin{array}{c|c} A_{11}B_{11} + A_{12}B_{21} & A_{11}B_{12} + A_{12}B_{22} \end{array}\right].$$

It is straightforward to verify that the four block matrix products $A_{11}B_{11}$, $A_{12}B_{21}$, $A_{11}B_{12}$, *and* $A_{12}B_{22}$ *are all well defined. Indeed,*

$$\begin{aligned} A_{11}B_{11} + A_{12}B_{21} &= \begin{bmatrix} 2 & 1 \\ 0 & 3 \end{bmatrix} \cdot \begin{bmatrix} 2 \\ -1 \end{bmatrix} + \begin{bmatrix} -1 \\ 1 \end{bmatrix} \cdot [5] \\ &= \begin{bmatrix} 3 \\ -3 \end{bmatrix} + \begin{bmatrix} -5 \\ 5 \end{bmatrix} = \begin{bmatrix} -2 \\ 2 \end{bmatrix}. \end{aligned}$$

This is indeed the first column of AB, *and if you think about it, the block procedure is simply another way to rewrite the multiplication* $\begin{bmatrix} 2 & 1 & -1 \\ 0 & 3 & 1 \end{bmatrix} \begin{bmatrix} 2 \\ -1 \\ 5 \end{bmatrix}$ *that produces the first column of* AB.

In a similar manner we see that

$$\begin{aligned} A_{11}B_{12} + A_{12}B_{22} &= \begin{bmatrix} 2 & 1 \\ 0 & 3 \end{bmatrix} \cdot \begin{bmatrix} 1 \\ 0 \end{bmatrix} + \begin{bmatrix} -1 \\ 1 \end{bmatrix} \cdot [4] \\ &= \begin{bmatrix} 2 \\ 0 \end{bmatrix} + \begin{bmatrix} -4 \\ 4 \end{bmatrix} = \begin{bmatrix} -2 \\ 4 \end{bmatrix} \end{aligned}$$

and this result is the second column of AB. *In Problem 2.34 you will verify that the second method for partitioning* B *in* (2.7) *also produces in a block matrix multiplication that results in* AB.

What would happen if we partitioned the columns of A *as before but partitioned the rows of* A *so that*

$$A = \left[\begin{array}{cc|c} 2 & 1 & -1 \\ \hline 0 & 3 & 1 \end{array}\right]?$$

You will verify in Problem 2.35 that either partition (2.7) *will result in well-defined block matrix multiplication products that give* AB *as the answer.*

What happens if we switch from $A = \left[\begin{array}{cc|c} 2 & 1 & -1 \\ 0 & 3 & 1 \end{array}\right]$ *to*

$$A = \left[\begin{array}{c|cc} 2 & 1 & -1 \\ 0 & 3 & 1 \end{array}\right] = \left[\, A_{11} \,\middle|\, A_{12} \,\right]? \tag{2.8}$$

In this case the partitions (2.7) *of* B *do not work. Since* A_{11} *is now a* 2×1 *matrix, we cannot multiply it against either* B_{11} *in* (2.7). *To make the block multiplication product work when* A *is partitioned as in* (2.8), *we must adjust the row partitions of* B. *That is, we must use either*

$$B = \left[\begin{array}{c|c} 2 & 1 \\ \hline -1 & 0 \\ 5 & 4 \end{array}\right] = \left[\begin{array}{c|c} B_{11} & B_{12} \\ \hline B_{21} & B_{22} \end{array}\right] \quad \text{or} \quad B = \left[\begin{array}{cc} 2 & 1 \\ \hline -1 & 0 \\ 5 & 4 \end{array}\right] = \left[\begin{array}{c} B_{11} \\ \hline B_{21} \end{array}\right]. \tag{2.9}$$

In Problem 2.36 you will verify that the partitions of B given in (2.9) along with the partition (2.8) of A will result in block matrix multiplication products that yield AB. (Live example: Visit `stthomas.edu/wavelets` *and click on Live Examples.)*

□

The results of Example 2.11 generalize to matrices A and B where the product AB is defined. The main point of Example 2.11 is to illustrate that to successfully partition A and B so that we can use block matrix multiplication to compute AB, we need the *column partitioning* of A to be the same as the *row partitioning* of B. That is, if A has been partitioned into $n_1, n_2, \ldots, n_k$ columns, then in order to make the block multiplications well-defined, B must be partitioned into $n_1, n_2, \ldots, n_k$ rows. The row partitioning of A and the column partitioning of B can be discretionary. Indeed, the row partitioning of A and the column partitioning of B are usually functions of the application at hand.

A Useful Way to Interpret Matrix Multiplication

In practice, we use some matrix M to process an input matrix A by way of MA to produce a desired effect. In particular, it is helpful to view the product MA as a way of *processing the columns* of A. Consider the following example.

Example 2.12 (Interpreting Matrix Multiplication). *Let M be the 2×6 matrix*

$$M = \begin{bmatrix} \frac{1}{3} & \frac{1}{3} & \frac{1}{3} & 0 & 0 & 0 \\ 0 & 0 & 0 & \frac{1}{3} & \frac{1}{3} & \frac{1}{3} \end{bmatrix}$$

and suppose that A is any matrix that consists of six rows. For the purposes of this example, suppose that

$$A = \begin{bmatrix} a_{11} & a_{12} \\ a_{21} & a_{22} \\ a_{31} & a_{32} \\ a_{41} & a_{42} \\ a_{51} & a_{52} \\ a_{61} & a_{62} \end{bmatrix} = \left[\begin{array}{c|c} a_{11} & a_{12} \\ a_{21} & a_{22} \\ a_{31} & a_{32} \\ a_{41} & a_{42} \\ a_{51} & a_{52} \\ a_{61} & a_{62} \end{array}\right] = \left[\mathbf{a}^1 \mid \mathbf{a}^2\right].$$

We can easily compute the 2×2 product MA and see that

$$MA = \begin{bmatrix} \frac{a_{11}+a_{21}+a_{31}}{3} & \frac{a_{12}+a_{22}+a_{32}}{3} \\ \frac{a_{41}+a_{51}+a_{61}}{3} & \frac{a_{42}+a_{52}+a_{62}}{3} \end{bmatrix}.$$

Let's look at the first column of this product. The first entry, $\frac{a_{11}+a_{21}+a_{31}}{3}$ is simply the average of the first three elements of $\mathbf{a}^1$, and the second entry is the average of the last three elements of $\mathbf{a}^1$. The same holds true for the second column of the product. The first entry of the second column is the average of the first three elements of $\mathbf{a}^2$, and the second entry is the average of the last three elements of $\mathbf{a}^2$.

So we can see that M processes the columns of A by computing averages of elements of each column and we can easily verify that

$$MA = \left[M\mathbf{a}^1 \mid M\mathbf{a}^2 \right].$$

If A consists of p columns, the interpretation is the same – the only difference is that the product would consist of p columns of averages rather than the two that resulted in this example.

(Live example: Visit `stthomas.edu/wavelets` *and click on Live Examples.)*

□

In general, suppose that M is an $m \times p$ matrix and A is a $p \times n$ matrix and we are interested in interpreting the product MA. We begin by partitioning A in terms of its columns. That is, if $\mathbf{a}^1, \mathbf{a}^2, \ldots, \mathbf{a}^n$ are the columns of A, then block matrix multiplication allows us to write

$$MA = M\left[\mathbf{a}^1 \mid \mathbf{a}^2 \mid \cdots \mid \mathbf{a}^n \right]$$

and view this product as a 1×1 block (M) times $1 \times n$ blocks. Thus,

$$MA = M\left[\mathbf{a}^1 \mid \mathbf{a}^2 \mid \cdots \mid \mathbf{a}^n \right] = \left[M\mathbf{a}^1 \mid M\mathbf{a}^2 \mid \cdots \mid M\mathbf{a}^n \right] \tag{2.10}$$

and we see that when we compute MA, we simply apply M to each column of A. Thus, when we multiply A on the left by M, we can view this product as processing the columns of A. Problem 2.39 allows you to further investigate this idea.[1]

Alternatively, suppose that we are given $p \times n$ matrix P and we wish to apply it to input matrix A with dimensions $m \times p$ by way of the product AP. Using block matrix multiplication, we can view this product as a way of *processing the rows* of A with matrix P. In Problem 2.40 you will produce the block matrix multiplication that supports this claim.

Transposes of Block Matrices

As we will see in subsequent chapters, it is important to understand how to transpose block matrices. As with block matrix multiplication, let's start with a generic matrix and its transpose

$$A = \begin{bmatrix} a_{11} & a_{12} & a_{13} & a_{14} \\ a_{21} & a_{22} & a_{23} & a_{24} \\ a_{31} & a_{32} & a_{33} & a_{34} \\ a_{41} & a_{42} & a_{43} & a_{44} \end{bmatrix} \quad \text{and} \quad A^T = \begin{bmatrix} a_{11} & a_{21} & a_{31} & a_{41} \\ a_{12} & a_{22} & a_{32} & a_{42} \\ a_{13} & a_{23} & a_{33} & a_{43} \\ a_{14} & a_{24} & a_{34} & a_{44} \end{bmatrix}.$$

[1]Many linear algebra texts define matrix multiplication by first defining the matrix/column product $M\mathbf{v}$ and then the matrix product MA as $\left[M\mathbf{a}^1 \mid M\mathbf{a}^2 \mid \cdots \mid M\mathbf{a}^n \right]$. See for example, Lay [65].

Suppose that we partition A as

$$A = \left[\begin{array}{cc|cc} a_{11} & a_{12} & a_{13} & a_{14} \\ a_{21} & a_{22} & a_{23} & a_{24} \\ \hline a_{31} & a_{32} & a_{33} & a_{34} \\ a_{41} & a_{42} & a_{43} & a_{44} \end{array}\right] = \left[\begin{array}{c|c} A_{11} & A_{12} \\ \hline A_{21} & A_{22} \end{array}\right].$$

If we partition A^T as

$$A^T = \left[\begin{array}{cc|cc} a_{11} & a_{21} & a_{31} & a_{41} \\ a_{12} & a_{22} & a_{32} & a_{42} \\ \hline a_{13} & a_{23} & a_{33} & a_{43} \\ a_{14} & a_{24} & a_{34} & a_{44} \end{array}\right]$$

then it is easy to see that in block form,

$$A^T = \left[\begin{array}{c|c} A_{11}^T & A_{21}^T \\ \hline A_{12}^T & A_{22}^T \end{array}\right].$$

We can try a different partition

$$A = \left[\begin{array}{ccc|c} a_{11} & a_{12} & a_{13} & a_{14} \\ a_{21} & a_{22} & a_{23} & a_{24} \\ a_{31} & a_{32} & a_{33} & a_{34} \\ \hline a_{41} & a_{42} & a_{43} & a_{44} \end{array}\right] = \left[\begin{array}{c|c} A_{11} & A_{12} \\ \hline A_{21} & A_{22} \end{array}\right]$$

and we can again easily verify that if we partition A^T as

$$A^T = \left[\begin{array}{ccc|c} a_{11} & a_{21} & a_{31} & a_{41} \\ a_{12} & a_{22} & a_{32} & a_{42} \\ a_{13} & a_{23} & a_{33} & a_{43} \\ \hline a_{14} & a_{24} & a_{34} & a_{44} \end{array}\right],$$

we have

$$A^T = \left[\begin{array}{c|c} A_{11}^T & A_{21}^T \\ \hline A_{12}^T & A_{22}^T \end{array}\right].$$

The results of this simple example hold in general. That is, if we partition A as

$$A = \left[\begin{array}{c|c|c|c} A_{11} & A_{12} & \cdots & A_{1q} \\ \hline A_{21} & A_{22} & & A_{2q} \\ \hline \vdots & & \ddots & \vdots \\ \hline A_{p1} & A_{p2} & \cdots & A_{pq} \end{array}\right], \quad \text{then} \quad A^T = \left[\begin{array}{c|c|c|c} A_{11}^T & A_{21}^T & \cdots & A_{p1}^T \\ \hline A_{12}^T & A_{22}^T & & A_{p2}^T \\ \hline \vdots & & \ddots & \vdots \\ \hline A_{1q}^T & A_{2q}^T & \cdots & A_{pq}^T \end{array}\right].$$

PROBLEMS

2.34 Confirm that the second partition of B in (2.7) results in a block matrix multiplication that gives AB in (2.6).

2.35 Verify that either partition given in (2.7) will result in well-defined block matrix multiplication products that give AB in (2.6) as the answer.

2.36 Verify that the partitions of B given in (2.9) along with the partition (2.8) of A will result in a block matrix multiplication product that yields AB in (2.6).

2.37 Let

$$A = \left[\begin{array}{cc|cc|c} 3 & -2 & 4 & 0 & 1 \\ 1 & 0 & 1 & -2 & 5 \\ -2 & 4 & 3 & 0 & 1 \\ \hline 1 & 2 & -2 & 3 & 3 \end{array}\right] \quad \text{and} \quad B = \begin{bmatrix} 2 & -2 & 1 \\ 1 & 0 & -1 \\ 0 & 0 & 5 \\ -3 & 1 & -2 \\ 4 & 1 & 0 \end{bmatrix}.$$

Partition B in an appropriate way so that AB can be computed in block matrix form.

2.38 Matrix multiplication as we defined it in Section 2.2 is nothing more than block matrix multiplication. Suppose that A is an $m \times p$ matrix and B is a $p \times n$ matrix. Let $\mathbf{a}^{k^T}$, $k = 1, \ldots, m$, denote the rows of A, and $\mathbf{b}^j$, $j = 1, \ldots, n$, denote the columns of B. Partition A by its rows and B by its columns and then write down the product AB in block matrix form. Do you see why this block matrix form of the product is exactly matrix multiplication as stated in Definition 2.8?

2.39 Let

$$M = \begin{bmatrix} 1 & 5 & 2 \\ -2 & 6 & 9 \end{bmatrix} \quad \text{and} \quad A = \begin{bmatrix} 6 & 0 & 3 & 7 \\ 1 & 2 & 5 & 4 \\ 0 & -1 & 3 & -2 \end{bmatrix}.$$

Verify (2.10) by showing that

$$MA = \left[M\mathbf{a}^1 \mid M\mathbf{a}^2 \mid M\mathbf{a}^3 \mid M\mathbf{a}^4 \right]$$

$$= \left[M\begin{bmatrix} 6 \\ 1 \\ 0 \end{bmatrix} \middle| M\begin{bmatrix} 0 \\ 2 \\ -1 \end{bmatrix} \middle| M\begin{bmatrix} 3 \\ 5 \\ 3 \end{bmatrix} \middle| M\begin{bmatrix} 7 \\ 4 \\ -2 \end{bmatrix} \right].$$

2.40 In this problem you will use block matrix multiplication to show that we can view the matrix R as processing the rows of matrix A when we compute the product AR.

(a) Suppose that R is a $p \times m$ matrix and A is an $n \times p$ matrix so that the product AR is defined. Suppose that $\mathbf{a}^{1^T}, \mathbf{a}^{2^T}, \ldots, \mathbf{a}^{n^T}$ are the rows of A (remember that each $\mathbf{a}^k$ is a vector consisting of p elements, so we need to transpose it to get a row). Partition A by rows and then use this block version of A to compute AR.

(b) Let

$$A = \begin{bmatrix} 3 & 1 \\ 0 & 5 \end{bmatrix} \quad \text{and} \quad R = \begin{bmatrix} -2 & 3 & 0 \\ 1 & 1 & -2 \end{bmatrix}.$$

Use (a) to verify that

$$AR = \begin{bmatrix} [3\ 1]\,R \\ \hline [0\ 5]\,R \end{bmatrix}.$$

★2.41 Suppose that H and G are $\frac{N}{2} \times N$ matrices and that $\mathbf{v}$ and $\mathbf{w}$ are vectors of length $\frac{N}{2}$. Compute

$$\begin{bmatrix} H \\ \hline G \end{bmatrix}^T \cdot \begin{bmatrix} \mathbf{v} \\ \hline \mathbf{w} \end{bmatrix}.$$

★2.42 Consider the 8×8 matrix

$$W = \begin{bmatrix} H \\ \hline G \end{bmatrix}$$

where

$$H = \begin{bmatrix} 1 & 1 & 0 & 0 & 0 & 0 & 0 & 0 \\ 0 & 0 & 1 & 1 & 0 & 0 & 0 & 0 \\ 0 & 0 & 0 & 0 & 1 & 1 & 0 & 0 \\ 0 & 0 & 0 & 0 & 0 & 0 & 1 & 1 \end{bmatrix}$$

and

$$G = \begin{bmatrix} 1 & -1 & 0 & 0 & 0 & 0 & 0 & 0 \\ 0 & 0 & 1 & -1 & 0 & 0 & 0 & 0 \\ 0 & 0 & 0 & 0 & 1 & -1 & 0 & 0 \\ 0 & 0 & 0 & 0 & 0 & 0 & 1 & -1 \end{bmatrix}.$$

Verify that $W^T = \left[\, H^T \mid G^T \,\right]$ and that

$$W^T W = H^T H + G^T G = 2I_8$$

where I_8 is the 8×8 identity matrix.

2.43 Let

$$A = \left[\begin{array}{c|c} A_{11} & A_{12} \\ \hline A_{21} & A_{22} \end{array}\right].$$

(a) Describe a block matrix L so that LA results in the product

$$LA = \left[\begin{array}{c|c} 3A_{11} & 3A_{12} \\ \hline 5A_{21} & 5A_{22} \end{array}\right].$$

(b) Describe a block matrix R so that AR results in the product

$$AR = \left[\begin{array}{c|c} 2A_{11} & 7A_{12} \\ \hline 2A_{21} & 7A_{22} \end{array}\right].$$

(c) What is LAR?

★**2.44** Suppose that A is an $N \times N$ matrix with N even and further suppose that A has the block structure

$$A = \left[\begin{array}{c|c} W & 0 \\ \hline 0 & I \end{array}\right]$$

where 0, I and W are $\frac{N}{2} \times \frac{N}{2}$ matrices. 0 is a zero matrix and I is an identity matrix.

Let $\mathbf{v} \in \mathbb{R}^N$ be partitioned as

$$\mathbf{v} = \left[\begin{array}{c} \mathbf{v}^{\ell} \\ \hline \mathbf{v}^{h} \end{array}\right]$$

where $\mathbf{v}^{\ell}, \mathbf{v}^{h} \in \mathbb{R}^{\frac{N}{2}}$. Write $A\mathbf{v}$ in block form.

2.45 Show that

$$\begin{bmatrix} A & B \\ B^T & C \end{bmatrix}$$

has the inverse

$$\begin{bmatrix} D & -DBC^{-1} \\ -C^{-1}B^TD & C^{-1}+C^{-1}B^TDBC^{-1} \end{bmatrix}$$

whenever C and $D = (A - BC^{-1}B^T)^{-1}$ are nonsingular.

★**2.46** Suppose that A is an $N \times N$ matrix written in block form as

$$A = \left[\begin{array}{c|c|c|c} A_1 & 0 & 0 & 0 \\ \hline 0 & A_2 & 0 & 0 \\ \hline \vdots & \vdots & \ddots & \vdots \\ \hline 0 & 0 & \cdots & A_k \end{array}\right]$$

where A_i, $i = 1, \ldots, k$ are square matrices with dimensions $n_i \times n_i$ and $n_1 + \cdots + n_k = N$. Assume further that each A_i is invertible. Show that A is invertible with

$$A^{-1} = \left[\begin{array}{c|c|c|c} A_1^{-1} & 0 & 0 & 0 \\ \hline 0 & A_2^{-1} & 0 & 0 \\ \hline \vdots & \vdots & \ddots & \vdots \\ \hline 0 & 0 & \cdots & A_k^{-1} \end{array}\right].$$

2.4 Convolution and Filters

In Section 2.1 we reviewed some basic ideas involving vectors. Vector addition and subtraction are examples of *binary operations* since they act on two vectors and produce a vector. Convolution is also a binary operator. For the time being we use bi-infinite sequences instead of vectors, but the idea is the same. *Convolution* takes two bi-infinite sequences $\mathbf{h}$ and $\mathbf{x}$ and produces a new bi-infinite sequence $\mathbf{y}$.

You should remember sequences from calculus; $\mathbf{a} = (a_1, a_2, a_3, \ldots)$ is the standard notation for a sequence. You can think of $\mathbf{a}$ as a function on the natural numbers $1, 2, 3, \ldots$ and instead of writing the result using normal function notation $a(1), a(2), \ldots$, we instead write $a_1, a_2, \ldots$. A bi-infinite sequence $\mathbf{b}$ is simply a sequence that is a function on the integers with components $\ldots, b_{-2}, b_{-1}, b_0, b_1, b_2, \ldots$, so we will write $\mathbf{b} = (\ldots, b_{-2}, b_{-1}, b_0, b_1, b_2, \ldots)$.

We use bi-infinite sequences to present a formal definition of convolution.

Definition 2.15 (Convolution). *Let* $\mathbf{h}$ *and* $\mathbf{x}$ *be two bi-infinite sequences. Then the convolution product* $\mathbf{y}$ *of* $\mathbf{h}$ *and* $\mathbf{x}$*, denoted by* $\mathbf{h} * \mathbf{x}$*, is the bi-infinite sequence* $\mathbf{y} = \mathbf{h} * \mathbf{x}$*, whose* nth *component is given by*

$$\boxed{y_n = \sum_{k=-\infty}^{\infty} h_k x_{n-k}.} \tag{2.11}$$

□

Certainly we can see from (2.11) that unless conditions are placed on the sequences $\mathbf{h}$ and $\mathbf{x}$, the series will diverge. In this book, either the components of $\mathbf{h}$ will be zero except for a finite number of terms or the terms in $\mathbf{h}$ and $\mathbf{x}$ will decay rapidly enough to ensure that the convolution product will converge. We assume throughout the book that for all integers n, the right-hand side of (2.11) is a convergent series.

Applications for the convolution operator abound throughout mathematics and engineering. You may not realize it, but you have been convolving sequences ever since you have known how to multiply – the multiplication algorithm you learned in elementary school is basically convolution. Using the convolution operator and the fast Fourier transformation (e.g. Kammler [62]), algorithms have been developed to quickly multiply numbers consisting of thousands of digits. Discrete density functions can be constructed using convolution, and many digital filters for processing audio are built using convolution.

We next present a series of examples of convolution products. We start with some basic computational examples and then proceed to some applications.

Example 2.13 (Convolution Products). *Compute each of the following convolution products.*

(a) $\mathbf{h} * \mathbf{x}$*, where* $\mathbf{x}$ *is any bi-infinite sequence and* $\mathbf{h}$ *is the sequence whose components satisfy*

$$h_k = \begin{cases} 1, & k = 0 \\ -1, & k = 1 \\ 0, & \textit{otherwise.} \end{cases}$$

(b) $\mathbf{h} * \mathbf{x}$*, where* $\mathbf{x}$ *is any bi-infinite sequence and* $\mathbf{h}$ *is the sequence whose components satisfy*

$$h_k = \begin{cases} \frac{1}{2}, & k = 0, k = 1 \\ 0, & \textit{otherwise.} \end{cases}$$

(c) Let $\mathbf{h}$ *be the sequence whose components are given by*

$$h_k = \begin{cases} 1, & k = 0, k = 1 \\ 0, & \textit{otherwise.} \end{cases}$$

Compute $\mathbf{h} * \mathbf{h}$ *and* $\mathbf{h} * \mathbf{h} * \mathbf{h}$.

Solution. *For part (a) we note that*

$$y_n = \sum_{k=-\infty}^{\infty} h_k x_{n-k} = \sum_{k=0}^{1} h_k x_{n-k} = h_0 x_n + h_1 x_{n-1} = x_n - x_{n-1}.$$

so that y_n can be obtained by subtracting x_{n-1} from x_n.

For part (b) we have

$$y_n = \sum_{k=-\infty}^{\infty} h_k x_{n-k} = \sum_{k=0}^{1} h_k x_{n-k} = \frac{1}{2} x_n + \frac{1}{2} x_{n-1}$$

so that y_n can be obtained by averaging x_{n-1} with x_n.

In part (c), we denote $\mathbf{y} = \mathbf{h} * \mathbf{h}$ and compute

$$y_n = \sum_{k=-\infty}^{\infty} h_k h_{n-k} = \sum_{k=0}^{1} h_{n-k} = h_n + h_{n-1}.$$

Now all the components of $\mathbf{y}$ will be zero except for $y_0 = h_0 + h_{-1} = 1$, $y_1 = h_1 + h_0 = 2$, and $y_2 = h_2 + h_1 = 1$. So

$$\mathbf{y} = (\ldots, 0, 0, 1, 2, 1, 0, 0, \ldots).$$

Now let's compute the components z_n of $\mathbf{z} = \mathbf{h} * \mathbf{y}$. We have

$$z_n = \sum_{k=-\infty}^{\infty} h_k y_{n-k} = \sum_{k=0}^{1} y_{n-k} = y_n + y_{n-1}.$$

Now all the components of $\mathbf{z}$ will be zero except for $z_0 = y_0 + y_{-1} = 1$, $z_1 = y_1 + y_0 = 3$, $z_2 = y_2 + y_1 = 3$, and $z_3 = y_3 + y_2 = 1$. So

$$\mathbf{z} = (\ldots, 0, 0, 1, 3, 3, 1, 0, 0, \ldots).$$

(Live example: Visit `stthomas.edu/wavelets` *and click on Live Examples.)*

□

Convolution as Shifting Inner Products

If the concept of convolution product is still difficult to grasp, perhaps this alternative look at convolution will help.

Recall that to compute the inner (dot) product of vectors $\mathbf{h} = [h_1, \ldots, h_N]^T$ and $\mathbf{x} = [x_1, \ldots, x_N]^T$, we simply compute $\sum_{k=1}^{N} h_k x_k$.

Now suppose that $\mathbf{h} = (\ldots, h_{-1}, h_0, h_1, \ldots)$ and $\mathbf{x} = (\ldots, x_{-1}, x_0, x_1, \ldots)$ are bi-infinite sequences and we wish to compute their inner product. Assuming the series converges, we have

$$\mathbf{h} \cdot \mathbf{x} = \sum_{k=-\infty}^{\infty} h_k x_k.$$

Now the inner product above looks quite similar to the convolution $\mathbf{h} * \mathbf{x}$ given in Definition 2.15. The only difference is that the subscript of the components of $\mathbf{x}$ in the convolution are $n-k$ rather than k. But we can at least agree that the convolution looks like an inner product.

How do we view the components x_{n-k} in the convolution product? If $n = 0$, then we have components x_{-k}. If we write these components in a bi-infinite sequence, we have

$$(\ldots, x_2, x_1, x_0, x_{-1}, x_{-2}, \ldots). \tag{2.12}$$

This sequence is simply a reflection of $\mathbf{x}$ about the x_0 term. Thus, to compute $y_0 = \sum_{k=-\infty}^{\infty} h_k x_{0-k}$, we simply reflect $\mathbf{x}$ and dot it with $\mathbf{h}$.

If $n = 1$, then we have components x_{1-k}. It is easy to see what is going on here – we have simply shifted the reflected sequence given in (2.12) one unit to the right. So to compute $y_1 = \sum_{k=-\infty}^{\infty} h_k x_{1-k}$, we simply reflect $\mathbf{x}$, shift all the elements one unit to the right, and dot the resulting sequence with $\mathbf{h}$.

We can certainly handle negative values of n as well. To compute y_{-13}, we reflect $\mathbf{x}$, shift the result 13 units to the *left*, and dot the resulting sequence with $\mathbf{h}$.

In general, to compute y_n, we reflect $\mathbf{x}$, shift it n units (left if n is negative, right if n is positive), and dot the resulting sequence with $\mathbf{h}$. Let's look at an example.

Example 2.14 (Convolution as Shifting Inner Products). *Let* $\mathbf{h}$ *be the bi-infinite sequence whose only nonzero terms are* $h_0 = 6$, $h_1 = 1$, $h_2 = 2$, *and* $h_3 = 3$. *Suppose that* $\mathbf{x}$ *is a bi-infinite sequence whose only nonzero components are* $x_{-2} = 3$, $x_{-1} = 2$, $x_0 = 1$, *and* $x_1 = 5$. *Use shifting inner products to compute the convolution product* $\mathbf{y} = \mathbf{h} * \mathbf{x}$.

Solution. *We will compute this product visually. Figure 2.7 shows* $\mathbf{h}$ *and a reflected version of* $\mathbf{x}$. *A sufficient number of zeros have been added to each sequence so that we can shift* $\mathbf{x}$ *far enough in either direction to see what is happening.*

	0	0	0	0	5	1	2	3	0	0	0	0	0	0
h:	0	0	0	0	0	6	1	2	3	0	0	0	0	0

Figure 2.7 The values of $\mathbf{h}$ on the bottom and reflected values of $\mathbf{x}$ on top. The box contains h_0 and x_0.

If we dot these two vectors, we obtain $y_0 = 6 \cdot 1 + 1 \cdot 2 + 2 \cdot 3 = 14$. *To compute* y_1, *we simply shift the reflected version of* $\mathbf{x}$ *one unit right and compute the inner product between* $\mathbf{h}$ *and the resulting sequence. To compute* y_2, *we shift the reflected version of* $\mathbf{x}$ *two units to the right and compute the inner product between* $\mathbf{h}$ *and the resulting sequence. The sequences needed to compute* y_1, *and* y_2 *are plotted in Figure 2.8. The inner products that result from Figure 2.8 are*

$$\begin{aligned} y_1 &= 6 \cdot 5 + 1 \cdot 1 + 2 \cdot 2 + 3 \cdot 3 = 44 \\ y_2 &= 1 \cdot 5 + 2 \cdot 1 + 3 \cdot 2 = 13. \end{aligned}$$

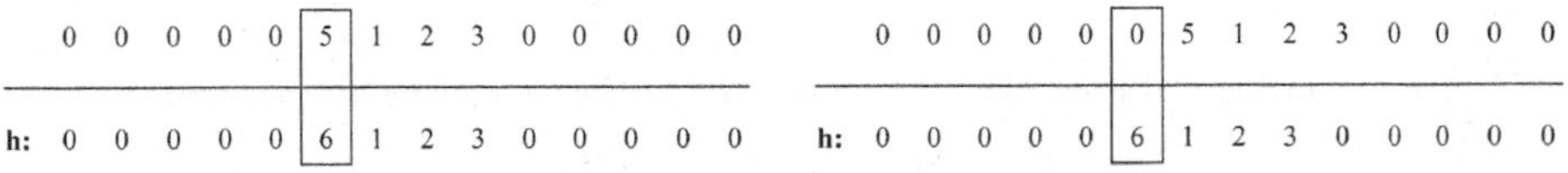

(a) Reflected $\mathbf{x}$ translated one unit right (b) Reflected $\mathbf{x}$ translated two units right

Figure 2.8 Plots of $\mathbf{h}$ (bottom of each graph) and translates of the reflection of $\mathbf{x}$.

If we look at Figure 2.8, we see that we can shift the reflected $\mathbf{x}$ *sequence two more times to the right – after that, the inner product values of subsequent positive shifts and* $\mathbf{h}$ *are* 0. *From Figure 2.7 we see that we can shift the reflected* $\mathbf{x}$ *sequence two units left before subsequent shifts lead to inner product values of* 0. *The remaining four shifts that result in nonzero inner products are plotted in Figure 2.9.*

The inner products resulting from dotting the sequences in Figure 2.9 are

$$\begin{aligned} y_3 &= 5 \cdot 2 + 1 \cdot 3 = 13 & y_4 &= 5 \cdot 3 = 15 \\ y_{-1} &= 6 \cdot 2 + 1 \cdot 3 = 15 & y_{-2} &= 6 \cdot 3 = 18. \end{aligned}$$

All other inner product values are 0. *Thus, the sequence* $\mathbf{y}$ ($y_0 = 14$ *is given in bold type) that results from convolving* $\mathbf{h}$ *and* $\mathbf{x}$ *is*

$$\mathbf{y} = (\ldots, 0, 0, 0, 18, 15, \mathbf{14}, 44, 13, 13, 15, 0, 0, 0, \ldots).$$

□

Filters

We now take convolution a bit further toward applications in signal and digital image processing. The idea of filtering data is in many ways analogous to what some of us do with our coffee makers in the morning. We run water through a paper filter containing ground coffee beans with the result that the paper filter allows the molecules

	0	0	0	0	0	0	0	5	1	2	3	0	0	0
h:	0	0	0	0	0	6	1	2	3	0	0	0	0	0

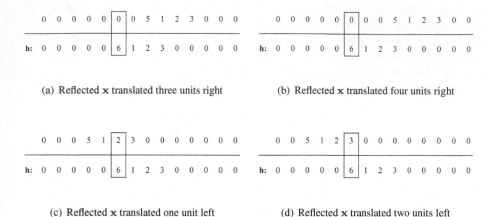

(a) Reflected $\mathbf{x}$ translated three units right

	0	0	0	0	0	0	0	0	5	1	2	3	0	0
h:	0	0	0	0	0	6	1	2	3	0	0	0	0	0

(b) Reflected $\mathbf{x}$ translated four units right

	0	0	0	5	1	2	3	0	0	0	0	0	0	0
h:	0	0	0	0	0	6	1	2	3	0	0	0	0	0

(c) Reflected $\mathbf{x}$ translated one unit left

	0	0	5	1	2	3	0	0	0	0	0	0	0	0
h:	0	0	0	0	0	6	1	2	3	0	0	0	0	0

(d) Reflected $\mathbf{x}$ translated two units left

Figure 2.9 Plots of $\mathbf{h}$ (bottom of each graph) and translates of the reflection of $\mathbf{x}$.

of caffeine to pass and holds back the grounds. The result is a pot of coffee. We basically want to do the same with data, but obviously we are not going to make coffee with it. Indeed, we are interested in filtering data so that we can easily identify characteristics such as locally constant segments, noise, large differences, or other artifacts. For example, think about data compression in a naive way. We would like to replace segments of similar data with a much smaller amount of information. Filtering will help us perform this task.

Note: Filters are sequences of numbers. In this book we consider only filters comprised of real numbers.

Let's start with an example of an elementary filter.

Example 2.15 (Averaging Filter). *Given a sequence of numbers*

$$\mathbf{x} = (\ldots, x_{-2}, x_{-1}, x_0, x_1, x_2, \ldots),$$

we form the sequence $\mathbf{y}$ *by defining* y_n *to be the average of the corresponding element* x_n *and its predecessor* x_{n-1} *of* $\mathbf{x}$. *That is,*

$$y_n = \frac{x_n + x_{n-1}}{2}.$$

This example should look familiar to you. In Example 2.13(b), the sequence $\mathbf{y}$ *was obtained via the convolution product* $\mathbf{y} = \mathbf{h} * \mathbf{x}$, *where* $h_0 = h_1 = \frac{1}{2}$ *and all other* $h_k = 0$. *We can think of* $\mathbf{h}$ *as filtering the elements of* $\mathbf{x}$ *to produce the elements of* $\mathbf{y}$.

□

In general, we perform filtering by choosing a filter sequence $\mathbf{h}$ and then producing the output $\mathbf{y}$ by convolving it with the input sequence $\mathbf{x}$. So filtering $\mathbf{x}$ by *filter* $\mathbf{h}$ means simply to compute

$$\mathbf{y} = \mathbf{h} * \mathbf{x}.$$

Obviously, how we choose $\mathbf{h}$ will entirely influence how the input $\mathbf{x}$ is processed. Think about the filter in Example 2.15. What happens when you average numbers? Generally speaking, you get one number that in some sense represents both original numbers. More specifically, when you average two identical values, the result is the same value. If you average a and $-a$, the result is zero. So we can say that this filter will do a suitably good job at reproducing consecutive values of $\mathbf{x}$ that are similar and returning values close to zero when consecutive values of $\mathbf{x}$ are close to being opposites of each other. This type of filter is very important and we formally define this class of filters later in the section.

Causal Filter and FIR Filters

We are primarily interested in filters that possess certain properties. We first consider the notion of a *causal filter.*

Definition 2.16 (Causal Filter). *Let* $\mathbf{h}$ *be a filter. We say that* $\mathbf{h}$ *is a* causal filter *if* $h_k = 0$ *for* $k < 0$. □

The $\mathbf{h}$ from Example 2.15 is an example of a causal filter. Definition 2.16 tells us how to identify a causal filter, but what does it really mean for a filter to be causal? Let's consider y_n from $\mathbf{y} = \mathbf{h} * \mathbf{x}$, where $\mathbf{h}$ is causal. We have

$$y_n = \sum_{k=-\infty}^{\infty} h_k x_{n-k} = \sum_{k=0}^{\infty} h_k x_{n-k} = h_0 x_n + h_1 x_{n-1} + h_2 x_{n-2} + h_3 x_{n-3} + \cdots$$

So if $\mathbf{h}$ is causal, then component y_n of $\mathbf{y}$ is formed from x_n and components of $\mathbf{x}$ that *precede* x_n. If you think about filtering in the context of sound, the causal requirement on $\mathbf{h}$ is quite natural. If we are receiving a digital sound and filtering it, it would make sense that we have only current and prior information about the signal.

For a large portion of this book, we will consider only causal filters. More specifically, we are primarily interested in causal filters $\mathbf{h}$, where $h_k = 0$ for $0 < k < L$. That is, our filters will typically have the form

$$\mathbf{h} = (\ldots, 0, 0, h_0, h_1, h_2, \ldots, h_L, 0, 0, \ldots).$$

Thus, the computation of y_n reduces to the finite sum

$$y_n = \sum_{k=0}^{L} h_k x_{n-k} = h_0 x_n + h_1 x_{n-1} + \cdots + h_L x_{n-L}.$$

Such causal filters with a finite number of nonzero components are called *finite impulse response* (FIR)filters.

Definition 2.17 (Finite Impulse Response Filter). *Let* $\mathbf{h}$ *be a causal filter and assume* L *is a positive integer. If* $h_k = 0$ *for* $k > L$*, and* $h_0, h_L \neq 0$*, then we say* $\mathbf{h}$ *is a* finite impulse response (FIR) filter *and write*

$$\mathbf{h} = (h_0, h_1, \ldots, h_L).$$

□

Note: Although we have motivated filters as bi-infinite sequences, we often identify only a finite number of filter elements inside ( ). Our convention is that the only possible nonzero elements of a filter will be enclosed in the parentheses.

It is often useful to think of FIR filters as *moving inner products* – we are simply forming y_n by computing an inner product of the FIR filter and the numbers $x_n, x_{n-1}, \ldots, x_{n-L}$.

Annihilating Filters

We are interested in filters that *annihilate* signals. More precisely, we have the following definition.

Definition 2.18 (Annihilating Filter). *Let* $\mathbf{g} = (g_\ell, \ldots, g_L)$ *be an FIR filter and* $\mathbf{v}$ *a bi-infinite sequence. We say* $\mathbf{g}$ annihilates $\mathbf{v}$ *if* $\mathbf{g} * \mathbf{v} = \mathbf{0}$. □

The following example illustrates an annihilating filter.

Example 2.16 (An Annihilating Filter). *Let* $\mathbf{g} = (g_{-1}, g_0, g_1, g_2) = (1, -3, 3, -1)$. *Show that* $\mathbf{g}$ *annihilates the sequences* $\mathbf{u}$ *where* $u_k = c$, $c \in \mathbb{R}$ *and* $\mathbf{v}$ *where* $v_k = k$, $k \in \mathbb{Z}$.

Solution. *We first consider* $\mathbf{y} = \mathbf{g} * \mathbf{u}$ *so that*

$$y_n = \sum_{k=-\infty}^{\infty} g_k u_{n-k} = \sum_{k=-1}^{2} g_k c = c\,(1 - 3 + 3 - 1) = 0.$$

For the second sequence $\mathbf{v}$*, we write* $\mathbf{z} = \mathbf{g} * \mathbf{v}$ *whose elements are*

$$\begin{aligned} z_k = \sum_{k=-\infty}^{\infty} g_{n-k} v_k &= \sum_{k=n-2}^{n+1} k g_{n-k} \\ &= (n-2)g_2 + (n-1)g_1 + n g_0 + (n+1)g_{-1} \\ &= -1(n-2) + 3(n-1) - 3n + (n+1) \\ &= 0. \end{aligned}$$

(Live example: Visit `stthomas.edu/wavelets` *and click on Live Examples.)*

□

In Chapters 5 and 7 we will consider filters that annihilate *polynomial data*. That is, for nonnegative integer m, $\Delta t \in \mathbb{R}$ with $\Delta t \neq 0$, we define the bi-infinite sequence $\mathbf{v}$ element–wise by

$$v_k = \begin{cases} 1, & m = 0 \\ (k\Delta t)^m, & m > 0. \end{cases}$$

When $m = 0, 1$, we will refer to $\mathbf{v}$ as constant data and linear data, respectively. The sequence $\mathbf{v}$ consists of uniform samples of the function $f(t) = t^m$. Figure 2.10 shows polynomial data for $m = 0, 1, 2, 3$.

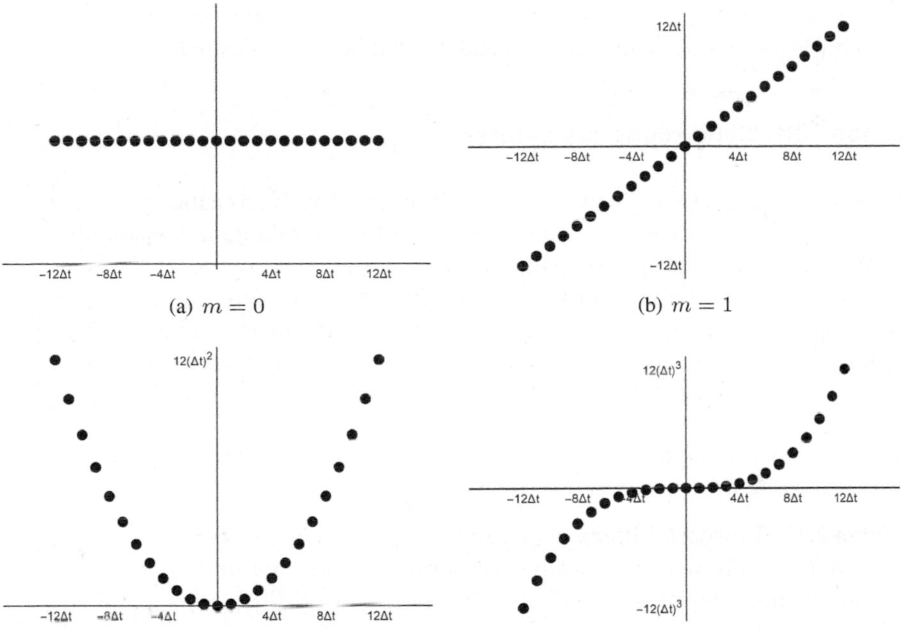

(a) $m = 0$ (b) $m = 1$ (c) $m = 2$ (d) $m = 3$

Figure 2.10 Plots of polynomial data for $m = 0, 1, 2, 3$.

The proposition below will help us with filter construction in Chapters 5 and 7.

Proposition 2.3 (Filters that Annihilate Polynomial Data). *Suppose* $\mathbf{g} = (g_\ell, \ldots, g_L)$ *is an FIR filter. Then* $\mathbf{g}$ *annihilates polynomial data of degree* m *if and only if for all* $n \in \mathbb{Z}$,

$$0 = \begin{cases} \sum_{k=\ell}^{L} g_k, & m = 0 \\ \sum_{k=\ell}^{L} g_k (n-k)^m, & m > 0. \end{cases} \tag{2.13}$$

□

Proof. Consider the case $m = 0$ so that $v_k = 1$, $k \in \mathbb{Z}$. Then the nth component of $\mathbf{y} = \mathbf{g} * \mathbf{v}$ is

$$y_n = \sum_k g_k v_{n-k} = \sum_{k=\ell}^{L} g_k.$$

Certainly $y_n = 0$ if and only if $\mathbf{g}$ annihilates constant data. In the case where $m > 0$ and $v_k = (k\Delta t)^m$, the nth component of $\mathbf{y} = \mathbf{g} * \mathbf{v}$ is

$$y_n = \sum_k g_k v_{n-k} = \sum_{k=\ell}^{L} g_k \left((n-k)\Delta t\right)^m = (\Delta t)^m \sum_{k=\ell}^{L} g_k (n-k)^m.$$

If $\mathbf{g}$ annihilates $\mathbf{v}$, then $y_n = 0$ for all $n \in \mathbb{Z}$ which implies (2.13) since $(\Delta t)^m \neq 0$. If (2.13) holds for all $n \in \mathbb{Z}$, then $y_n = 0$ and $\mathbf{g}$ annihilates $\mathbf{v}$ as desired. □

Lowpass Filters and Highpass Filters

In this book we are interested in two types of FIR filters. The first is called a *lowpass filter* and the second is called a *highpass filter*. Loosely speaking, a lowpass filter will process a vector that is essentially constant and return a vector of the same structure (i.e., a vector that is mostly constant). The same filter will process a vector whose components are largely oscillatory and attenuate the values. A highpass filter does just the opposite – an essentially constant vector will be attenuated and a vector whose components are highly oscillatory will result in output that is also highly oscillatory.

We define lowpass/highpass filters in terms of convolution but in Chapter 8, we will learn alternate ways to describe these filters.

Definition 2.19 (Lowpass Filter). *Suppose* $\mathbf{h} = (h_\ell, \ldots, h_L)$ *is an FIR filter. Let* $\mathbf{u}$ *and* $\mathbf{v}$ *be bi-infinite sequences whose components are* $u_k = 1$ *and* $v_k = (-1)^k$, $k \in \mathbb{Z}$. *We say* $\mathbf{h}$ *is a* lowpass filter *if*

$$\mathbf{h} * \mathbf{u} \neq \mathbf{0} \qquad \textit{and} \qquad \mathbf{h} * \mathbf{v} = \mathbf{0}.$$

□

We can further refine the characteristics of a lowpass filter by simplifying the convolution products in Definition 2.19. We have

$$0 \neq (\mathbf{h} * \mathbf{u})_n = \sum_{k=-\infty}^{\infty} h_k u_{n-k} = \sum_{k=\ell}^{L} h_k = \left(\sum_{k=\ell}^{L} h_k\right) \cdot 1$$

and

$$0 = (\mathbf{h} * \mathbf{v})_n = \sum_{k=-\infty}^{\infty} h_k v_{n-k} = \sum_{k=\ell}^{L} h_k (-1)^{n-k} = (-1)^n \sum_{k=\ell}^{L} (-1)^k h_k.$$

Thus we see that $\mathbf{h}$ is a lowpass filter if its elements do not sum to zero but the alternating sum of its elements is zero. Alternatively, $\mathbf{h} * \mathbf{u}$ returns a nonzero multiple of $\mathbf{u}$ and $\mathbf{h}$ annihilates the oscillatory data that comprises $\mathbf{v}$.

Note that the averaging filter from Example 2.15 is a lowpass filter. Let's look at some other examples.

Example 2.17 (Lowpass Filters). *Consider the following filters:*

(a) $\mathbf{h} = (h_0, h_1, h_2, h_3) = \left(1 + \sqrt{3}, 3 + \sqrt{3}, 3 - \sqrt{3}, 1 - \sqrt{3}\right)$

(b) $\mathbf{h} = (h_{-1}, h_0, h_1) = \left(\frac{1}{6}, \frac{1}{3}, \frac{1}{6}\right)$

Show that these filters are lowpass filters.

Solution. *Let* $\mathbf{u}$ *and* $\mathbf{v}$ *be given in Definition 2.19. Then*

$$(\mathbf{h} * \mathbf{u})_n = \sum_{k=-\infty}^{\infty} h_k \cdot 1 = 1 + \sqrt{3} + 3 + \sqrt{3} + 3 - \sqrt{3} + 1 - \sqrt{3} = 8 = 8u_n$$

and

$$(\mathbf{h} * \mathbf{v})_n = \sum_{k=-\infty}^{\infty} h_k(-1)^{n-k} = (-1)^n \sum_{k=0}^{4} (-1)^k h_k = h_0 - h_1 + h_2 - h_3 = 0.$$

For (b) we have

$$(\mathbf{h} * \mathbf{u})_n = \sum_{k=-\infty}^{\infty} h_k \cdot 1 = \frac{1}{6} + \frac{1}{3} + \frac{1}{6} = \frac{2}{3} = \frac{2}{3} u_n$$

and

$$(\mathbf{h} * \mathbf{v})_n = \sum_{k=-\infty}^{\infty} h_k(-1)^{n-k} = (-1)^n \sum_{k=-1}^{2} (-1)^k h_k = -\frac{1}{6} + \frac{1}{3} - \frac{1}{6} = 0.$$

(Live example: Visit `stthomas.edu/wavelets` *and click on Live Examples.)*

□

Highpass filters are the opposite of lowpass filters. We have the following definition.

Definition 2.20 (Highpass Filter). *Suppose* $\mathbf{g} = (g_\ell, \ldots, g_L)$ *is an FIR filter. Let* $\mathbf{u}$ *and* $\mathbf{v}$ *be bi-infinite sequences whose components are* $u_k = 1$ *and* $v_k = (-1)^k$, $k \in \mathbb{Z}$. *We say* $\mathbf{g}$ *is a* highpass filter *if*

$$\mathbf{g} * \mathbf{u} = \mathbf{0} \quad \text{and} \quad \mathbf{g} * \mathbf{v} \neq \mathbf{0}.$$

□

As was the case with lowpass filters, we can simplify the convolution products in Definition 2.20 to better characterize highpass filters. We have

$$0 = (\mathbf{g} * \mathbf{u})_n = \sum_{k=-\infty}^{\infty} g_k u_{n-k} = \sum_{k=\ell}^{L} g_k$$

and

$$0 \neq (\mathbf{g} * \mathbf{v})_n = \sum_{k=-\infty}^{\infty} g_k v_{n-k} = \sum_{k=\ell}^{L} g_k (-1)^{n-k} = (-1)^n \sum_{k=\ell}^{L} g_k (-1)^k = c(-1)^n.$$

We see that $\mathbf{g}$ is a highpass filter if its elements sum to zero and the alternating sum of its elements is not zero. In fact $\mathbf{g} * \mathbf{v}$ is a nonzero multiple of $\mathbf{v}$.

The filter in Example 2.13(a) is an example of a highpass filter. Here are some other examples of highpass filters.

Example 2.18 (Highpass Filters). *Consider the following filters:*

(a) $\mathbf{g} = (g_{-1}, g_0, g_1) = (-1, 2, -1)$

(b) $\mathbf{g} = (g_0, g_1, g_2, g_3) = (-1, 1, -1, 1)$

Show that these filters are highpass filters.

Solution. *Let* $\mathbf{u}$ *and* $\mathbf{v}$ *be given in Definition 2.20. Then*

$$(\mathbf{g} * \mathbf{u})_n = \sum_{k=-\infty}^{\infty} g_k u_{n-k} = \sum_{k=0}^{3} g_k \cdot 1 = -1 + 1 - 1 + 1 = 0$$

and

$$\begin{aligned}(\mathbf{g} * \mathbf{v})_n &= \sum_{k=-\infty}^{\infty} g_k v_{n-k} \\ &= \sum_{k=-1}^{1} g_k (-1)^{n-k} \\ &= -1(-1)^{n+1} + 2(-1)^n - 1(-1)^{n-1} \quad = (-1)^n (1 + 2 + 1) = 4v_n.\end{aligned}$$

For the filter in (b), we have

$$(\mathbf{g} * \mathbf{u})_n = \sum_{k=-\infty}^{\infty} g_k u_{n-k} = \sum_{k=-1}^{1} g_k \cdot 1 = -1 + 2 - 1 = 0$$

and

$$
\begin{aligned}
(\mathbf{g} * \mathbf{v})_n &= \sum_{k=-\infty}^{\infty} g_k v_{n-k} \\
&= \sum_{k=0}^{3} g_k (-1)^{n-k} \\
&= -1(-1)^n + 1(-1)^{n-1} - 1(-1)^{n-2} - 1(-1)^{n-3} \\
&= -4(-1)^n = -4v_n.
\end{aligned}
$$

(Live example: Visit `stthomas.edu/wavelets` *and click on Live Examples.)*

□

The following proposition lists properties of and relationships between lowpass and highpass filters.

Proposition 2.4 (Properties and Relationships of Lowpass/Highpass Filters). *Suppose* $\mathbf{a} = (a_\ell, \ldots, a_L)$ *is an FIR filter. Then*

(a) (Negation.) If $\mathbf{h}$ *is lowpass (highpass), then so is* $-\mathbf{h}$.

(b) (Translation.) Let $m \in \mathbb{Z}$ *and define filter* $\mathbf{b}$ *componentwise by* $b_k = a_{k-m}$. *If* $\mathbf{a}$ *is lowpass (highpass) then so is* $\mathbf{b}$.

(c) (Reflection.) Define filter $\mathbf{b}$ *componentwise by* $b_k = a_{-k}$. *If* $\mathbf{a}$ *is lowpass (highpass) then so is* $\mathbf{b}$.

(d) Define filter $\mathbf{b}$ *componentwise by* $b_k = (-1)^k a_k$. *If* $\mathbf{a}$ *is lowpass (highpass), then* $\mathbf{b}$ *is highpass (lowpass).*

□

Proof. The proof is left as Problem 2.65. □

Convolution as a Matrix Product

The wavelet transformations we develop in Chapters 4, 5, and 7 are based on convolution products. Since the wavelet transformations are represented at matrices, it is worthwhile to write the convolution $\mathbf{h} * \mathbf{x}$ as a matrix/vector product $H\mathbf{x}$. Since $\mathbf{x}$ is a bi-infinite sequence, H must be an infinite–dimensional matrix.

Note: In the development of convolution and filters, we have viewed the filter, input, and output as sequences and used ( ) to enclose entries of these sequences. In what follows, it is sometimes convenient to view these sequences as vectors, and in these cases we employ the vector notation [ ].

Let's start by writing $\mathbf{y} = \mathbf{h} * \mathbf{x}$ as a matrix H times $\mathbf{x}$. We will consider only FIR filters $\mathbf{h} = (h_0, h_1, \ldots, h_L)$. To see how this is done, let's recall the definition of component y_n of $\mathbf{y}$:

$$y_n = \sum_{k=0}^{L} h_k x_{n-k} = h_0 x_n + h_1 x_{n-1} + h_2 x_{n-2} + \cdots + h_L x_{n-L}. \tag{2.14}$$

Now think about writing $\mathbf{y}$ as a matrix H times $\mathbf{x}$. It is fair to ask if we can even write a convolution product as a matrix. In Problem 2.57 we learn that convolution is a linear operator and a well-known theorem from linear algebra states (see, e. g. Strang [86]) that such a linear operator on a vector space can always be represented by a matrix. Let's consider the components of $\mathbf{y}, \mathbf{x}$ in $\mathbf{y} = H\mathbf{x}$:

$$\begin{bmatrix} \vdots \\ y_{-2} \\ y_{-1} \\ y_0 \\ y_1 \\ \vdots \\ y_n \\ \vdots \end{bmatrix} = H \begin{bmatrix} \vdots \\ x_{-2} \\ x_{-1} \\ x_0 \\ x_1 \\ \vdots \\ x_n \\ \vdots \end{bmatrix}$$

We see then that y_n is the result of dotting row n of H with $\mathbf{x}$. But we already know what this inner product is supposed to look like – it is given in (2.14). Consider the first term in this expression: $h_0 x_n$. In terms of an inner product, h_0 must then be the nth element in the row vector, or in terms of a matrix, h_0 is the nth element of row n. This puts h_0 on the diagonal of row n. So every row n has h_0 on the diagonal. Similarly, since x_{n-1} is multiplied by h_1, then h_1 must be the $(n-1)$th element in row n. We say that h_1 is on the first subdiagonal of H. Continuing in this manner, we see that h_{n-L} must be the $(n-L)$th element in row n, so that h_{n-L} is on the Lth subdiagonal of row n. Thus, row n consists entirely of zeros except for column n back to column $n-L$. So to form H, we simply put h_0 on the main diagonal of each row and then left from the diagonal with $h_1, h_2, \ldots, h_L$, respectively. We have

$$H = \begin{bmatrix} \ddots & \ddots & \ddots & \ddots & \ddots & & & & & & & & & \\ & h_L & \ldots & h_2 & h_1 & \mathbf{h_0} & 0 & 0 & 0 & 0 & 0 & 0 & 0 & \\ & 0 & h_L & \ldots & h_2 & h_1 & \mathbf{h_0} & 0 & 0 & 0 & 0 & 0 & 0 & \\ \ldots & 0 & 0 & h_L & \ldots & h_2 & h_1 & \mathbf{h_0} & 0 & 0 & 0 & 0 & 0 & \ldots \\ & 0 & 0 & 0 & h_L & \ldots & h_2 & h_1 & \mathbf{h_0} & 0 & 0 & 0 & 0 & \\ & 0 & 0 & 0 & 0 & h_L & \ldots & h_2 & h_1 & \mathbf{h_0} & 0 & 0 & 0 & \\ & & & & & & \ddots & \ddots & \ddots & \ddots & \ddots & & & \ddots \end{bmatrix} \tag{2.15}$$

where we have indicated the main diagonal using bold face symbols.

There are many properties obeyed by H. Since $\mathbf{h}$ is causal, only zeros will appear in any row right of the diagonal. We say that H is *lower triangular* since the elements above the diagonal are zeros. Mathematically, we say that H is lower triangular since

$H_{ij} = 0$ for $j > i$, $i, j \in \mathbb{Z}$. H is also called a *banded matrix*. Along with the main diagonal, the first subdiagonal through the Lth subdiagonal have nonzero elements, but those are the only *bands* that are not zero.

Of course, it is computationally intractable to consider utilizing a matrix with infinite dimensions. Moreover, it is realistic to insist that all input sequences $\mathbf{x}$ are of finite length, so at some point we need to think about truncating H. We discuss truncation for wavelet transformation matrices in Chapters 5 and 7.

Example 2.19 (Convolution as a Matrix). *Let* $\mathbf{h} = (h_0, h_1, h_2) = \left(\frac{1}{3}, \frac{1}{2}, \frac{1}{6}\right)$. *Write down the matrix* H *for this filter.*

Solution. *We have*

$$H = \begin{bmatrix} \ddots & & \ddots & \ddots & \ddots & & & & & & & & & & \\ & 0 & 0 & \frac{1}{6} & \frac{1}{2} & \mathbf{\frac{1}{3}} & 0 & 0 & 0 & 0 & 0 & 0 & 0 & 0 & \\ & 0 & 0 & 0 & \frac{1}{6} & \frac{1}{2} & \mathbf{\frac{1}{3}} & 0 & 0 & 0 & 0 & 0 & 0 & 0 & \\ \cdots & 0 & 0 & 0 & 0 & \frac{1}{6} & \frac{1}{2} & \mathbf{\frac{1}{3}} & 0 & 0 & 0 & 0 & 0 & 0 & \cdots \\ & 0 & 0 & 0 & 0 & 0 & \frac{1}{6} & \frac{1}{2} & \mathbf{\frac{1}{3}} & 0 & 0 & 0 & 0 & 0 & \\ & 0 & 0 & 0 & 0 & 0 & 0 & \frac{1}{6} & \frac{1}{2} & \mathbf{\frac{1}{3}} & 0 & 0 & 0 & 0 & \\ & & & & & & & \ddots & \ddots & \ddots & & & & & \ddots \end{bmatrix}$$

We have marked the main diagonal element $h_0 = \frac{1}{3}$ *in bold.* □

PROBLEMS

★2.47 Suppose $\mathbf{s}, \mathbf{t}$ are bi-infinite sequences whose only nonzero components are $t_{-1} = t_0 = \frac{1}{2}$ and $s_0 = s_1 = \frac{1}{4}$. For bi-infinite sequences $\mathbf{o}, \mathbf{d}$, show that the components of $\mathbf{t} * \mathbf{o}$ are $(\mathbf{t} * \mathbf{o})_n = \frac{1}{2}(o_n + o_{n+1})$ and $(\mathbf{s} * \mathbf{d})_n = \frac{1}{4}(d_n + d_{n-1})$.

2.48 Let $\mathbf{h}^n = \underbrace{\mathbf{h} * \mathbf{h} * \cdots * \mathbf{h}}_{n+1 \text{ times}}$, where $\mathbf{h}_0 = \mathbf{h}$ and $\mathbf{h}$ is the bi-infinite sequence given in Example 2.13(c). In the example we computed $\mathbf{h}^1$ and $\mathbf{h}^2$. Compute $\mathbf{h}^3$ and $\mathbf{h}^4$. Do you see a pattern? Can you write down the components of $\mathbf{h}^n$ for general n? Have you seen this pattern in any of your other mathematics courses?

2.49 Let $\mathbf{d}$ be the bi-infinite sequence whose terms are given by

$$d_k = \begin{cases} \frac{1}{6}, & 1 \le k \le 6 \\ 0, & \text{otherwise.} \end{cases}$$

Compute $\mathbf{y} = \mathbf{d} * \mathbf{d}$. If we think of $\mathbf{d}$ as a sequence whose kth component gives the probability of rolling a k with a fair die, how could you interpret $\mathbf{y}$? What about $\mathbf{y} * \mathbf{d}$?

2.50 This problem is just for fun. Using paper and pencil, verify that $212 \cdot 231 = 48,972$. Note that no carrying took place. Now $212 = 2 \cdot 10^0 + 1 \cdot 10^1 + 2 \cdot 10^2$ and we could easily write 231 in an analogous manner. Let's load the coefficient of

10^k from 212 into the kth component of the bi-infinite sequence $\mathbf{x}$ ($x_0 = 2$ is given in bold type), so that

$$\mathbf{x} = (\ldots, 0, 0, \mathbf{2}, 1, 2, 0, 0, \ldots).$$

Let's do the same for 231 and load these coefficients into $\mathbf{y}$ ($y_0 = 1$ is given in bold type), so that

$$\mathbf{y} = (\ldots, 0, 0, \mathbf{1}, 3, 2, 0, 0, \ldots).$$

Now convolve $\mathbf{z} = \mathbf{x} * \mathbf{y}$. What can you claim about the components of $\mathbf{z}$?

2.51 Let $\mathbf{e}^0$ be the bi-infinite sequence defined by

$$e_k^0 = \begin{cases} 1, & k = 0 \\ 0, & \text{otherwise.} \end{cases}$$

Show that for any bi-infinite sequence $\mathbf{x}$, we have $\mathbf{e}^0 * \mathbf{x} = \mathbf{x}$.

2.52 Let $\mathbf{e}^m$ be the bi-infinite sequence defined by

$$e_k^m = \begin{cases} 1, & k = m \\ 0, & \text{otherwise.} \end{cases}$$

and let $\mathbf{x}$ be any bi-infinite sequence. Compute $\mathbf{y} = \mathbf{e}^m * \mathbf{x}$.

2.53 Let $\mathbf{h}$ and $\mathbf{x}$ be bi-infinite sequences. Show element–wise that

$$(\mathbf{h} * \mathbf{x})_n = \sum_{k=-\infty}^{\infty} h_k x_{n-k} = \sum_{k=-\infty}^{\infty} h_{n-k} x_k = (\mathbf{x} * \mathbf{h})_n.$$

In this way we see that the convolution product is commutative (i.e., $\mathbf{h} * \mathbf{x} = \mathbf{x} * \mathbf{h}$). *Hint:* Do the index substitution $j = n - k$ on the first sum above.

2.54 Use the method of shifting inner products to compute the following convolution products.

(a) $\mathbf{h} * \mathbf{x}$, where the only nonzero elements of $\mathbf{h}$ are $h_{-1} = -1, h_0 = 2$, and $h_1 = 3$ and the only nonzero elements of $\mathbf{x}$ are $x_0 = 5$ and $x_1 = 7$.

(b) $\mathbf{h} * \mathbf{x}$, where $x_k = 1$ for $k \in \mathbb{Z}$ and the only nonzero elements of $\mathbf{h}$ are $h_0 = 0.299$, $h_1 = 0.587$, and $h_2 = 0.114$.

(c) $\mathbf{h} * \mathbf{h}$, where the only nonzero elements of $\mathbf{h}$ are $h_{-1} = h_1 = \frac{1}{4}$ and $h_0 = \frac{1}{2}$.

(d) In Problem 2.53 you showed that convolution is commutative. Swap the roles of $\mathbf{h}$ and $\mathbf{x}$ from (a) in the shifting inner product method and verify the commutative property for this example.

2.55 Suppose that the only nonzero components of $\mathbf{h}$ are $h_0, h_1, \ldots, h_n$ and the only nonzero elements of $\mathbf{x}$ are $x_0, x_1, \ldots, x_m$, where $m > n > 0$. Suppose further that $h_0, \ldots, h_n, x_0, \ldots, x_m > 0$. Let $\mathbf{y} = \mathbf{h} * \mathbf{x}$. Use the method of shifting inner products to determine which components of $\mathbf{y}$ are nonzero.

2.56 Find a bi-infinite sequence $\mathbf{h}$ so that when we compute $\mathbf{y} = \mathbf{h} * \mathbf{x}$ (for any bi-infinite sequence $\mathbf{x}$), we have y_n is the average of x_{n-1}, x_{n-2}, and x_{n-3}.

2.57 Suppose that we fix a sequence $\mathbf{h}$ and then consider $\mathbf{y} = \mathbf{h} * \mathbf{x}$ as a function of the sequence $\mathbf{x}$. In this problem you will show that in this context, the convolution operator is linear. That is,

(a) for any $a \in \mathbb{R}$, we have

$$\mathbf{y} = \mathbf{h} * (a\mathbf{x}) = (a\mathbf{h}) * \mathbf{x}.$$

(b) for any sequences $\mathbf{x}^1$ and $\mathbf{x}^2$, we have

$$\mathbf{y} = \mathbf{h} * (\mathbf{x}^1 + \mathbf{x}^2) = \mathbf{h} * \mathbf{x}^1 + \mathbf{h} * \mathbf{x}^2.$$

2.58 Suppose that we create $\mathbf{v}^m$ from vector $\mathbf{v}$ by translating the components of $\mathbf{v}$ right m units if $m > 0$ and left m units if $m < 0$. In the case that $m = 0$, we set $\mathbf{v}^0 = \mathbf{v}$. Assume for a given $\mathbf{h}$ and $\mathbf{x}$, we compute $\mathbf{y} = \mathbf{h} * \mathbf{x}$. Now form $\mathbf{y}^m$ and $\mathbf{x}^m$. Show that $\mathbf{y}^m = \mathbf{h} * \mathbf{x}^m$. That is, show that translating $\mathbf{x}$ by m units and convolving it with $\mathbf{h}$ results in translating $\mathbf{y}$ by m units.

2.59 Suppose that $\mathbf{h} = (h_0, \ldots, h_9)$ is a lowpass filter whose elements sum to one.

(a) Show that

$$h_0 + h_2 + h_4 + h_6 + h_8 = \frac{1}{2} = h_1 + h_3 + h_5 + h_7 + h_9.$$

(b) In general, suppose that L is an odd integer $\mathbf{h} = (h_0, \ldots, h_L)$ is a lowpass filter whose elements sum to one. Show that

$$h_0 + h_2 + \cdots + h_{L-1} = \frac{1}{2} = h_1 + h_3 + \cdots + h_L.$$

2.60 Suppose that $\mathbf{h}$ is a causal filter with $h_k \neq 0$, $k = 0, \ldots, 5$ and $h_k = -h_{9-k}$ for all integers k. Is $\mathbf{h}$ an FIR filter? Can $\mathbf{h}$ ever be a lowpass filter? Explain.

2.61 Let L be an odd positive integer and suppose that $\mathbf{h}$ is a causal filter with $h_k \neq 0$, $k = 0, \ldots, \frac{L+1}{2}$ and $h_k = h_{L-k}$ for all integers k. Can $\mathbf{h}$ ever be highpass? Explain.

2.62 Find a highpass filter $\mathbf{h} = (h_0, h_1, h_2, h_3, h_4)$ that annihilates linear data. Here $h_k \neq 0$, $k = 0, \ldots, 4$.

★**2.63** Recall Problem 2.11 from Section 2.1. In terms of filters, we started with $\mathbf{h} = (h_0, h_1, \ldots, h_L)$ where L is odd and constructed $\mathbf{g} = (g_0, g_1, \ldots, g_L)$ by the rule $g_k = (-1)^k h_{L-k}$. We showed in the problem that $\mathbf{h} \cdot \mathbf{g} = 0$. Assume that $\mathbf{h}$ is a lowpass filter. Use Proposition 2.4 to show that $\mathbf{g}$ is a highpass filter. We use this idea in Chapters 5 and 7 to construct highpass filter $\mathbf{g}$ from lowpass filter $\mathbf{h}$ that is also orthogonal to $\mathbf{h}$.

2.64 Suppose that $\mathbf{h}^1$ and $\mathbf{h}^2$ are lowpass filters and $\mathbf{g}^1$ and $\mathbf{g}^2$ are highpass filters.

(a) Show that $\mathbf{h} = \mathbf{h}^1 * \mathbf{h}^2$ is a lowpass filter.

(b) Show that $\mathbf{g} = \mathbf{g}^1 * \mathbf{g}^2$ is a highpass filter.

(c) Is $\mathbf{h}^1 * \mathbf{g}^1$ lowpass, highpass, or neither?

2.65 Prove Proposition 2.4.

2.66 Consider the matrix

$$H = \begin{bmatrix} \ddots & & \ddots & \ddots & \ddots & & & & & & & \\ & 0 & 0 & -\frac{1}{4} & \frac{1}{2} & \mathbf{\frac{3}{4}} & 0 & 0 & 0 & 0 & 0 & \\ & 0 & 0 & 0 & -\frac{1}{4} & \frac{1}{2} & \mathbf{\frac{3}{4}} & 0 & 0 & 0 & 0 & \\ \cdots & 0 & 0 & 0 & 0 & -\frac{1}{4} & \frac{1}{2} & \mathbf{\frac{3}{4}} & 0 & 0 & 0 & \cdots \\ & 0 & 0 & 0 & 0 & 0 & -\frac{1}{4} & \frac{1}{2} & \mathbf{\frac{3}{4}} & 0 & 0 & \\ & 0 & 0 & 0 & 0 & 0 & 0 & -\frac{1}{4} & \frac{1}{2} & \mathbf{\frac{3}{4}} & 0 & \\ & & & & & & & & \ddots & \ddots & \ddots & \ddots \end{bmatrix}$$

where the main diagonal is denoted using boldface numbers.

(a) Write down the corresponding filter $\mathbf{h}$.

(b) Is $\mathbf{h}$ lowpass, highpass, or neither?

(c) For infinite-length vector $\mathbf{x}$, compute $\mathbf{h} * \mathbf{x}$. Verify that the answer is the same when we compute $H \cdot \mathbf{x}$.

2.67 Write down the convolution matrix for the following filters.

(a) $\mathbf{h} = (h_0, h_1, h_2) = (2, 0, 1)$

(b) $\mathbf{g} = (g_0, g_1) = \left(\frac{3}{4}, \frac{1}{4}\right)$

CHAPTER 3

AN INTRODUCTION TO DIGITAL IMAGES

One of the main application areas of wavelet transforms is image processing. Wavelet transforms can be used in processes designed to compress images, search for edges in images, or enhance image features. Since several sections of this book are dedicated to image compression and examples involving images are featured throughout the text, we include this chapter on the basics of digital images.

It is quite tempting to dedicate many pages to the various aspects of image processing. In this chapter we discuss the basic material necessary to successfully navigate through the remainder of the text. If you are interested in more thorough treatments on the numerous topics that comprise digital image processing, I might suggest you start by consulting Russ [79], Gonzalez and Woods [48], Gonzalez, Woods, and Eddins [49], or Wayner [101].

This chapter begins with a section that describes digital grayscale images. We learn how digital images can be interpreted by computer software and how matrices are important tools in image processing applications. We also discuss intensity transformations. These transformations are commonly used to improve the image display. Section 3.2 deals with digital color images as well as some popular color spaces that are useful for image processing. An important step in image compression is that of data encoding. Here the values of the (transformed) image are given new represen-

Discrete Wavelet Transformations: An Elementary Approach With Applications, Second Edition.
Patrick J. Van Fleet.

tations with the goal of requiring less space to store them. A simple method of data encoding was introduced by David Huffman [59] in 1952 and a more sophisticated version of this method is used to create JPEG images. In Section 3.3 we outline his encoding method. In Section 3.4, we learn how to calculate the cumulative energy of a signal or image and introduce the qualitative measures entropy and peak signal-to-noise ratio. Cumulative energy is useful in quantizing data for the purposes of compressing it and entropy and peak signal-to-noise ratio are measures that assess the effectiveness of an image compression algorithm.

3.1 The Basics of Grayscale Digital Images

In this section we learn how a *grayscale* (monochrome) image can be interpreted by computer software, how we can represent a grayscale image using a matrix, and how matrices can be used to manipulate images. The section concludes with an introduction to some basic intensity transformations.

Bits and Bytes

The fundamental unit on a computer is the *bit*. The value of a bit is either 0 or 1. A *byte* is the standard amount of information needed to store one keyboard character. There are eight bits in a byte. Considering that each bit can assume one of two values, we see that there are $2^8 = 256$ different bytes. Not coincidentally, there are 256 characters on a standard keyboard and each character is assigned a value from the range 0 to 255.[1] The American Standard Code for Information Interchange (ASCII) is the standard used to assign values to characters. For example, the ASCII standard for the characters *Y*, *y*, *9*, *?*, and *sp* (the space key) are 89, 121, 57, 63, and 32, respectively. The ASCII codes for the characters that appear on a common computer keyboard are given in Table 3.1. The table of all 256 ASCII codes can be found in numerous books or on many web sites. One possible reference is the book by Oualline [73].

Since computers work with bits, it is often convenient to write the ASCII codes in base 2 where now the intensities range from $0 = 0000000_2$ to $255 = 11111111_2$. When an intensity is expressed in base 2, it is said to be in *binary form*. The following example recalls how the conversion works.

Example 3.1 (Converting from Base 10 to Base 2). *The character Y has ASCII value* 89. *Convert this value to base* 2.

Solution.
The largest power of two that is less than or equal to 89 *is* $2^6 = 64$. *The the eighth bit, representing* $2^7 = 128$, *is* 0. *So we start with the seventh bit.* $89 - 64 = 25$ *and*

[1]If you take a quick glance at your keyboard, it is not clear that there are 256 characters. If you have access to a PC, open a new text file in Notepad or MS Word. You can display the characters by holding down the `Alt` key and then typing a number between 0 and 255 on the numeric keypad.

Table 3.1 ASCII character codes 32 through 126 in decimal and binary (base 2) format.

Char.	Dec.	Bin.	Char.	Dec.	Bin.	Char.	Dec.	Bin.
sp	32	00100000	@	64	01000000	`	96	01100000
!	33	00100001	A	65	01000001	a	97	01100001
“	34	00100010	B	66	01000010	b	98	01100010
#	35	00100011	C	67	01000011	c	99	01100011
$	36	00100100	D	68	01000100	d	100	01100100
%	37	00100101	E	69	01000101	e	101	01100101
&	38	00100110	F	70	01000110	f	102	01100110
’	39	00100111	G	71	01000111	g	103	01100111
(	40	00101000	H	72	01001000	h	104	01101000
)	41	00101001	I	73	01001001	i	105	01101001
*	42	00101010	J	74	01001010	j	106	01101010
+	43	00101011	K	75	01001011	k	107	01101011
,	44	00101100	L	76	01001100	l	108	01101100
-	45	00101101	M	77	01001101	m	109	01101101
.	46	00101110	N	78	01001110	n	110	01101110
/	47	00101111	O	79	01001111	o	111	01101111
0	48	00110000	P	80	01010000	p	112	01110000
1	49	00110001	Q	81	01010001	q	113	01110001
2	50	00110010	R	82	01010010	r	114	01110010
3	51	00110011	S	83	01010011	s	115	01110011
4	52	00110100	T	84	01010100	t	116	01110100
5	53	00110101	U	85	01010101	u	117	01110101
6	54	00110110	V	86	01010110	v	118	01110110
7	55	00110111	W	87	01010111	w	119	01110111
8	56	00111000	X	88	01011000	x	120	01111000
9	57	00111001	Y	89	01011001	y	121	01111001
:	58	00111010	Z	90	01011010	z	122	01111010
;	59	00111011	[	91	01011011	{	123	01111011
<	60	00111100	\	92	01011100	|	124	01111100
=	61	00111101	]	93	01011101	}	125	01111101
>	62	00111110	^	94	01011110	~	126	01111110
?	63	00111111	_	95	01011111			

the largest power of two that is less than or equal to 25 *is* $2^4 = 16$. $25 - 16 = 9$, *and we can write* 9 *as* $2^3 + 2^0$. *Thus, we see that*

$$\begin{aligned} 89 &= \mathbf{0} \cdot 2^7 + \mathbf{1} \cdot 2^6 + \mathbf{0} \cdot 2^5 + \mathbf{1} \cdot 2^4 + \mathbf{1} \cdot 2^3 + \mathbf{0} \cdot 2^2 + \mathbf{0} \cdot 2^1 + \mathbf{1} \cdot 2^0 \\ &= 01011001_2. \end{aligned}$$

(Live example: Visit `stthomas.edu/wavelets` *and click on Live Examples.)*

□

Storing Images on Computers

The basic unit of composition of a digital image is a *pixel*.[2] The *resolution* of a rectangular digital image is the number of rows times the number of columns that comprise the image. In order to work with a grayscale digital image, we must know the location of the pixel (i.e., the row and column of the pixel) and the *intensity* of the pixel. Since the fundamental unit on a computer is a bit, it is natural to measure these intensities using bits. Thus, the intensity scale of a grayscale image is typically 0 to $2^m - 1$, $m = 0, 1, 2, \ldots$. The value 0 typically corresponds to black, and $2^m - 1$ corresponds to white. Thus, the intensities between 0 and $2^m - 1$ represent the spectrum of gray values between black and white in increments of $1/2^m$. We say that an image is an *m-bit image* where the intensity scale ranges from 0 to $2^m - 1$.

A common value for use on computers is $m = 8$ for an intensity scale of 256 values that range from 0 (black) to 255 (white). This intensity scale is plotted in Figure 3.1.

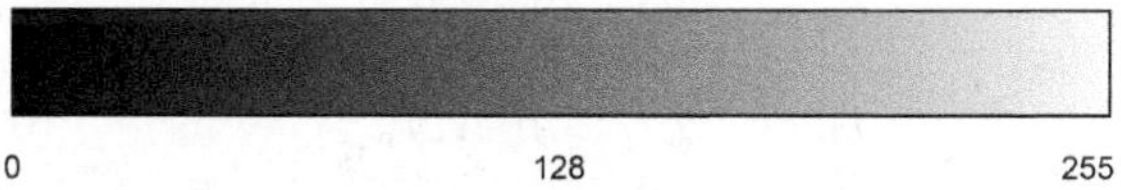

Figure 3.1 The eight-bit grayscale intensity scale.

Certainly, you can see why an eight-bit image is convenient for computer graphics. One byte is eight bits and we know that we can store the numbers 0 through 255 as single characters from the ASCII table (see Table 3.1 or [73]). For example, suppose that the first row of pixels in a digital image had the intensities $50, 100, 80$, and 124. Then we would only need the four characters *2*, *d*, *p*, and | to store these intensities. This is four bytes, compared to the minimum of 10 bytes needed for the number of digits that comprise the intensities in the row. This is exactly how file formats such as *portable gray maps* (PGM) store digital images.

When a finer intensity scale is desired, pixels can be stored using 16 bits. In this case, there are $2^{16} = 65{,}536$ possible intensity levels. Now $65{,}536 = 256^2$, so for storage purposes we only need to use two characters. Let's illustrate this statement with an example.

Example 3.2 (**16-Bit Intensities as ASCII Characters**). *Write* 31,785 *as two ASCII characters.*

[2]Pixel is short for *picture* (pix) *element* (el).

Solution.
We write 31,785 *in base* 2*:*

$$\begin{aligned}31{,}785 &= \mathbf{1}\cdot 2^{14} + \mathbf{1}\cdot 2^{13} + \mathbf{1}\cdot 2^{12} + \mathbf{1}\cdot 2^{11} + \mathbf{1}\cdot 2^{10} + \mathbf{1}\cdot 2^{5} + \mathbf{1}\cdot 2^{3} + \mathbf{1}\cdot 2^{0}\\ &= 0111110000101001_2\\ &= \underline{01111100}_{\,2}\ \underline{00101001}_{\,2}\end{aligned}$$

Now the leftmost eight bits can be viewed as a number between 0 *and* 255 *and indeed,* $01111100_2 = 124$. *Using Table 3.1, we see that ASCII code for* 124 *is* |. *The second eight bits are* 00101001_2*, and in base* 10*, the value is* 41. *The ASCII code for* 41 *is*). *So we could store the five-digit number* 31,785 *as* |)*.*

Converting to base 10 *is quite simple. We can think of the base* 2 *representation as the sum*

$$0111110000101001_2 = \mathbf{01111100}00000000_2 + 00000000\mathbf{00101001}_2$$

Now the second term is simply 41*, and to get the first term we simply shift the bits* $01111100_2 = 124$ *eight bits to the left. This is nothing more than the multiplication* $124 \cdot 2^8 = 31{,}744$.

So to obtain the original base 10 *intensity level, we find the ASCII codes* 124 *and* 41 *for the characters* | *and*)*, respectively, and then compute* $124 \cdot 2^8 + 41$.

(Live example: Visit `stthomas.edu/wavelets` *and click on Live Examples.)*

□

Note: All images in this book are 8-bit images.

Grayscale Images as Matrices

A natural tool with which to describe the locations and intensities of pixels in a digital image is a matrix. The dimensions of a matrix correspond directly to the resolution of an image and the location of a pixel in an image is nothing more than the (i, j) element of a matrix. If we denote the matrix by A, then a_{ij} is the intensity value. Thus, we can store a digital grayscale image in a matrix A whose integer entries range from 0 (black) to 255 (white).

Figure 3.2 illustrates this concept. In the figure, we have plotted a 512×768 image. Although we do not display a 512×768 matrix, we choose a particular 10×10 region, zoom in on it, and display those values in a matrix.

Matrices and Basic Image Processing

Basic operations involving matrices are important in image processing. Let's look at a few simple examples.

$$\begin{bmatrix} 170 & 150 & 163 & 171 & 165 & 146 & 148 & 193 & 133 & 121 \\ 202 & 226 & 210 & 201 & 187 & 166 & 144 & 196 & 146 & 174 \\ 213 & 220 & 228 & 230 & 220 & 181 & 167 & 132 & 190 & 86 \\ 189 & 196 & 202 & 199 & 157 & 117 & 69 & 31 & 29 & 46 \\ 212 & 213 & 211 & 201 & 184 & 146 & 89 & 48 & 41 & 37 \\ 240 & 231 & 211 & 147 & 95 & 90 & 68 & 49 & 36 & 48 \\ 236 & 233 & 223 & 191 & 110 & 62 & 41 & 32 & 56 & 44 \\ 203 & 204 & 181 & 138 & 85 & 49 & 53 & 61 & 72 & 54 \\ 217 & 201 & 192 & 158 & 100 & 52 & 52 & 53 & 53 & 53 \\ 222 & 224 & 176 & 145 & 99 & 49 & 54 & 48 & 49 & 58 \end{bmatrix}$$

Figure 3.2 Storing a digital image as a matrix.

Example 3.3 (Image Negation). *One common image processing tool involves a simple matrix operation. Image negation is the process of inverting values on the intensity scale. For example,* 0 *maps to* 255, 255 *maps to* 0, 100 *maps to* 155, *and* 20 *maps to* 235. *In general, pixel* p *is mapped to* $255 - p$.

We can easily perform image negation using matrix subtraction. Consider the 512×768 *image plotted in Figure 3.3(a). Suppose that we represent this image in matrix* A.

(a) A 512×768 grayscale image

(b) Negation of the image

Figure 3.3 Image negation.

Define S *to be the* 512×768 *matrix where each element of* S *is set to* 255. *If we were to plot* S *as an image, it would be a white square. To produce the negative image, we simply compute* $S - A$. *The result is plotted on the right in Figure 3.3.*
(Live example: Visit `stthomas.edu/wavelets` *and click on Live Examples.)*

□

Example 3.4 (Inner Products and Grayscale Images). *Suppose that we have an image that is* 10 *pixels wide and that the pixel values in the top row are* 100, 102, 90, 94, 91, 95, 250, 252, 220, *and* 210. *We can load these pixel values into a vector* $\mathbf{v}$*:*

$$\mathbf{v} = [50, 62, 90, 94, 26, 18, 250, 252, 220, 210]^T$$

Practically speaking the row starts out a medium dark gray and then the last four pixels are quite close to white. Pictorially, the vector is shown in Figure 3.4.

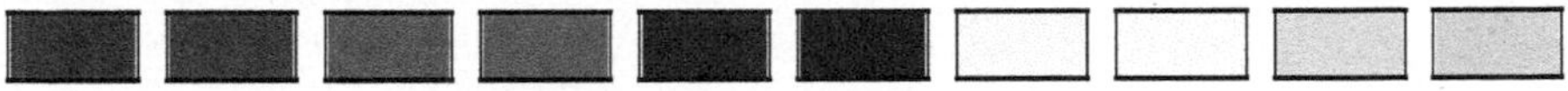

Figure 3.4 A pictorial representation of $\mathbf{v}$.

Suppose that our task is to produce a five-vector $\mathbf{u}$ *that consists of the averages of two consecutive elements of* $\mathbf{v}$*. This is a common task in image compression – the average values of some pixels are used to represent a string of values. So we want as output the vector*

$$\mathbf{u} = [56, 92, 22, 251, 215]^T$$

It is easy to compute these averages, but how do we express the calculations mathematically? In particular, can we want to write $\mathbf{u}$ *as the product of a matrix* A *and* $\mathbf{v}$*? Given the dimensions of* $\mathbf{v}$ *and* $\mathbf{u}$*, we know the dimensions of* A *must be* 5×10*. Each row of* A *should dot with* $\mathbf{v}$ *to produce the desired average. If we take*

$$\mathbf{a}^1 = \left[\frac{1}{2}, \frac{1}{2}, 0, 0, 0, 0, 0, 0, 0, 0\right]^T$$

then the inner product $\mathbf{a}^T\mathbf{v}$ *gives* $u_1 = 56$*. Similarly, the vector*

$$\mathbf{a}^2 = \left[0, 0, \frac{1}{2}, \frac{1}{2}, 0, 0, 0, 0, 0, 0\right]^T$$

can be dotted with $\mathbf{v}$ *to obtain the value* $u_2 = 92$*. We can continue this process and additionally create the vectors*

$$\mathbf{a}^3 = \left[0, 0, 0, 0, \frac{1}{2}, \frac{1}{2}, 0, 0, 0, 0\right]^T \qquad \mathbf{a}^4 = \left[0, 0, 0, 0, 0, 0, \frac{1}{2}, \frac{1}{2}, 0, 0\right]^T$$

and

$$\mathbf{a}^5 = \left[0, 0, 0, 0, 0, 0, 0, 0, \frac{1}{2}, \frac{1}{2}\right]^T$$

and dot them each in turn with $\mathbf{v}$ *to obtain* $u_3 = 22$*,* $u_4 = 251$*, and* $u_5 = 215$*. A pictorial representation of* $\mathbf{u}$ *appears in Figure 3.5.*

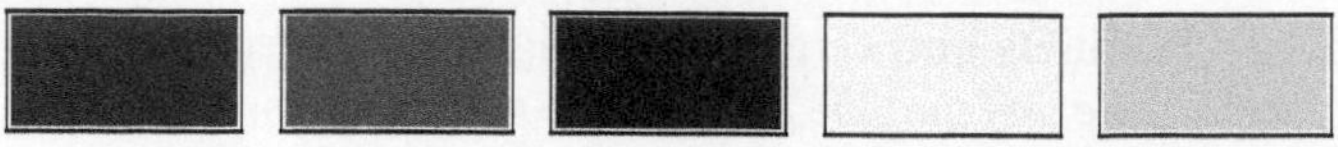

Figure 3.5 A pictorial representation of **u**.

We create A from the vectors $\mathbf{a}^1, \ldots, \mathbf{a}^5$ *and then compute the matrix product*

$$A\mathbf{v} = \frac{1}{2}\begin{bmatrix} 1 & 1 & 0 & 0 & 0 & 0 & 0 & 0 & 0 & 0 \\ 0 & 0 & 1 & 1 & 0 & 0 & 0 & 0 & 0 & 0 \\ 0 & 0 & 0 & 0 & 1 & 1 & 0 & 0 & 0 & 0 \\ 0 & 0 & 0 & 0 & 0 & 0 & 1 & 1 & 0 & 0 \\ 0 & 0 & 0 & 0 & 0 & 0 & 0 & 0 & 1 & 1 \end{bmatrix} \cdot \begin{bmatrix} 50 \\ 62 \\ 90 \\ 94 \\ 26 \\ 18 \\ 250 \\ 252 \\ 220 \\ 210 \end{bmatrix} = \begin{bmatrix} 56 \\ 92 \\ 22 \\ 251 \\ 215 \end{bmatrix} = \mathbf{u}.$$

(Live example: Visit `stthomas.edu/wavelets` *and click on Live Examples.)*

□

We next look at a practical example where we illustrate the use of matrices to perform naive edge detection.

Example 3.5 (Matrices and Naive Edge Detection). *Let M be the* 512×512 *matrix that represents the image plotted in Figure 3.6. Then M has integer entries ranging from* 0 *to* 255 *and is of the form*

$$M = \begin{bmatrix} m_{11} & m_{12} & \cdots & m_{1,512} \\ m_{21} & m_{22} & \cdots & m_{2,512} \\ \vdots & \vdots & \ddots & \vdots \\ m_{512,1} & m_{512,2} & \cdots & m_{512,512} \end{bmatrix}$$

Now consider the 512×512 *matrix*

$$H = \begin{bmatrix} 1 & -1 & 0 & 0 & 0 & \cdots & 0 & 0 & 0 \\ 0 & 1 & -1 & 0 & 0 & \cdots & 0 & 0 & 0 \\ 0 & 0 & 1 & -1 & 0 & \cdots & 0 & 0 & 0 \\ \vdots & \vdots & \vdots & \vdots & \vdots & \ddots & \vdots & \vdots & \vdots \\ 0 & 0 & 0 & 0 & 0 & \cdots & 1 & -1 & 0 \\ 0 & 0 & 0 & 0 & 0 & \cdots & 0 & 1 & -1 \\ -1 & 0 & 0 & 0 & 0 & \cdots & 0 & 0 & 1 \end{bmatrix}.$$

Figure 3.6 The image represented by matrix M.

What happens when we compute HM? It is certainly easy enough to write down the answer, HM:

$$\begin{bmatrix} 1 & -1 & 0 & \cdots & 0 & 0 \\ 0 & 1 & -1 & \cdots & 0 & 0 \\ 0 & 0 & 1 & \cdots & 0 & 0 \\ \vdots & \vdots & \vdots & \ddots & \vdots & \vdots \\ 0 & 0 & 0 & \cdots & -1 & 0 \\ 0 & 0 & 0 & \cdots & 1 & -1 \\ -1 & 0 & 0 & \cdots & 0 & 1 \end{bmatrix} \cdot \begin{bmatrix} m_{11} & m_{12} & \cdots & m_{1,512} \\ m_{21} & m_{22} & \cdots & m_{2,512} \\ \vdots & \vdots & \ddots & \vdots \\ m_{512,1} & m_{512,2} & \cdots & m_{512,512} \end{bmatrix}$$

$$= \begin{bmatrix} m_{11}-m_{21} & m_{12}-m_{22} & \cdots & m_{1,512}-m_{2,512} \\ m_{21}-m_{31} & m_{22}-m_{32} & \cdots & m_{2,512}-m_{3,512} \\ \vdots & \vdots & \ddots & \vdots \\ m_{512,1}-m_{11} & m_{512,2}-m_{12} & \cdots & m_{512,512}-m_{1,512} \end{bmatrix}.$$

Can you describe the output that results from computing HM? The first row of HM is obtained by subtracting row two of M from row one of M. In general, for $j = 1, \ldots, 511$, *the* jth *row of HM is obtained by subtracting row* $j+1$ *from row* j *of* M. *Row* 512 *of HM is obtained by subtracting the first row of M from the last* *row of M. The value of an element in HM will depend on the size of the difference in the associated values in M. If these associated values are close in size, then the resulting value in HM will small, producing a gray intensity near* 0 *(black). On the other hand, a large difference will produce an intensity that, in magnitude (absolute value), is large (nearer to white). Thus the role of H is to detect large changes along rows in M. The magnitudes of the values of HM are plotted in Figure 3.7(a).*

Recall that the product HM can be viewed as H processing the columns of M. That is, $HM = \left[H\mathbf{m}^1, H\mathbf{m}^2, \ldots, H\mathbf{m}^{512}\right]$ where $\mathbf{m}^k$ is the kth column of M. In this case, we process a column of M by subtracting consecutive values in the column (with the exception of the last value). In this way we detect large differences, or edges, along the rows of M.

We can use a similar idea to process the columns of M. Taking the transpose of M places rows into columns and in this way the product HM^T processes the rows of M. We just need an additional transpose to reset the orientation of the image. That is, the computation we need is $\left(HM^T\right)^T$, which by Problem 2.23 is $\left(M^T\right)^T H^T = MH^T$. Element-wise we have

$$MH^T = \begin{bmatrix} m_{11} & m_{12} & \cdots & m_{1,512} \\ m_{21} & m_{22} & \cdots & m_{2,512} \\ \vdots & \vdots & \ddots & \vdots \\ m_{512,1} & m_{512,2} & \cdots & m_{512,512} \end{bmatrix} \cdot \begin{bmatrix} 1 & 0 & \cdots & 0 & -1 \\ -1 & 1 & \cdots & 0 & 0 \\ 0 & -1 & \cdots & 0 & 0 \\ \vdots & \vdots & \ddots & \vdots & \vdots \\ 0 & 0 & \cdots & 0 & 0 \\ 0 & 0 & \cdots & 1 & 0 \\ 0 & 0 & \cdots & -1 & 1 \end{bmatrix}$$

$$= \begin{bmatrix} m_{11}-m_{12} & m_{12}-m_{13} & \cdots & m_{1,512}-m_{11} \\ m_{21}-m_{22} & m_{22}-m_{23} & \cdots & m_{2,512}-m_{21} \\ \vdots & \vdots & \ddots & \vdots \\ m_{512,1}-m_{512,2} & m_{512,2}-m_{512,3} & \cdots & m_{512,512}-m_{512,1} \end{bmatrix}.$$

The magnitudes of the values of MH^T are plotted in Figure 3.7(b).

It is worthwhile for you to compare the matrix product HM with the convolution product detailed in Example 2.13(a).

(Live example: Visit `stthomas.edu/wavelets` *and click on Live Examples.)*

□

Block matrix multiplication is also quite useful in image processing.

Example 3.6 (Block Matrices and Image Masking). *Consider the* 450×450 *grayscale image plotted in Figure 3.8. Denote the matrix of grayscale values by A.*

Suppose that we are interested in displaying the center of the image. That is, we wish to display the region enclosed by the square with corners $(151, 151)$, $(300, 151)$, $(300, 300)$, *and* $(151, 300)$ *(see Figure 3.8) and convert the remaining pixels to black. We denote the matrix enclosed by this square as* A_{22}.

(a) HM

(b) MH^T

Figure 3.7 Plots of HM and MH^T.

(a) The original image A

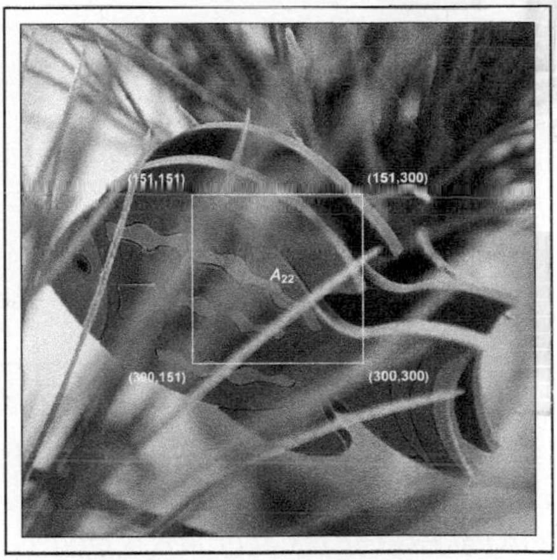

(b) The isolated portion A_{22}

Figure 3.8 A digital image and marking it for masking.

We can mask this portion of the image using block matrices. Consider the 450×450 *matrix* M *that has the block structure*

$$M = \left[\begin{array}{c|c|c} Z_{150} & Z_{150} & Z_{150} \\ \hline Z_{150} & I_{150} & Z_{150} \\ \hline Z_{150} & Z_{150} & Z_{150} \end{array}\right],$$

where Z_{150} *is a* 150×150 *matrix whose entries are* 0 *and* I_{150} *is the* 150×150 *identity matrix.*

We partition A *into nine blocks each of size* 150×150 *and label these blocks as* A_{jk}, $j, k = 1, 2, 3$. *The partitioning is illustrated in Figure 3.9.*

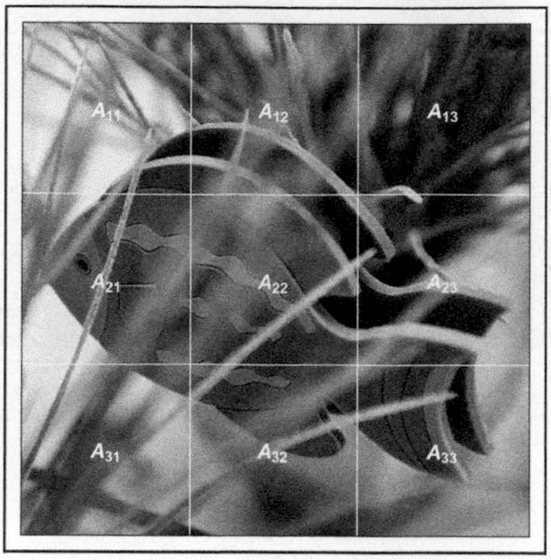

Figure 3.9 Partitioning of A into nine blocks A_{jk} each of size 150×150.

If we compute MA *using block matrix multiplication, we obtain*

$$MA = \left[\begin{array}{c|c|c} Z_{150} & Z_{150} & Z_{150} \\ \hline Z_{150} & I_{150} & Z_{150} \\ \hline Z_{150} & Z_{150} & Z_{150} \end{array}\right] \cdot \left[\begin{array}{c|c|c} A_{11} & A_{12} & A_{13} \\ \hline A_{21} & A_{22} & A_{23} \\ \hline A_{31} & A_{32} & A_{33} \end{array}\right] = \left[\begin{array}{c|c|c} Z_{150} & Z_{150} & Z_{150} \\ \hline A_{21} & A_{22} & A_{23} \\ \hline Z_{150} & Z_{150} & Z_{150} \end{array}\right].$$

Thus, multiplication on the left by M *leaves only the middle row of blocks of* A *as nonzero blocks. The result of* MA *is plotted in Figure 3.10(a).*

We now multiply MA *on the right by* $M^T = M$ *to obtain*

$$MAM^T = \left[\begin{array}{c|c|c} Z_{150} & Z_{150} & Z_{150} \\ \hline A_{21} & A_{22} & A_{23} \\ \hline Z_{150} & Z_{150} & Z_{150} \end{array}\right] \cdot \left[\begin{array}{c|c|c} Z_{150} & Z_{150} & Z_{150} \\ \hline Z_{150} & I_{150} & Z_{150} \\ \hline Z_{150} & Z_{150} & Z_{150} \end{array}\right] = \left[\begin{array}{c|c|c} Z_{150} & Z_{150} & Z_{150} \\ \hline Z_{150} & A_{22} & Z_{150} \\ \hline Z_{150} & Z_{150} & Z_{150} \end{array}\right].$$

The product MAM^T *accomplishes our goal of leaving the middle block* A_{22} *as the only nonzero block. It is plotted in Figure 3.10(b).*

(Live example: Visit `stthomas.edu/wavelets` *and click on Live Examples.)*

□

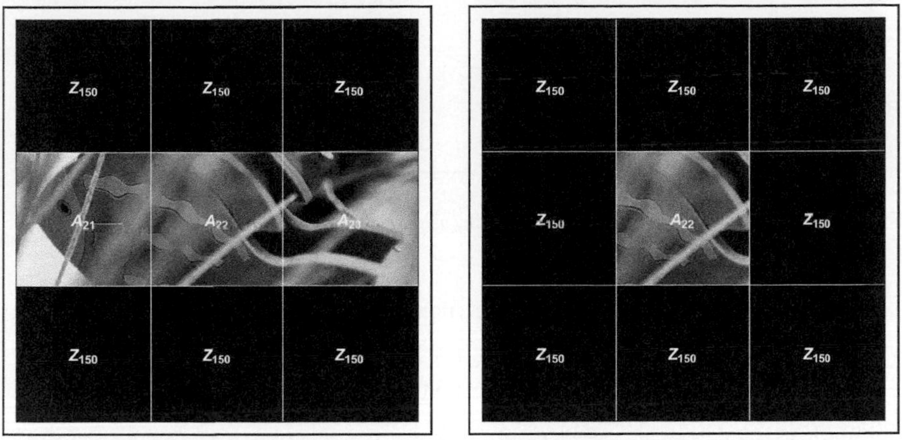

(a) The product MA (b) The product MAM^T

Figure 3.10 Masking an image.

Intensity Transformations

Intensity transformations are useful in several areas. We can alter the intensities in an image to present a more balanced display. If we are given an image whose intensities are disproportionately dark or light, an intensity transformation might give us a better idea of the true image. In this book, we learn how to transform images for the purposes of compressing the information necessary to store them or to remove noise. In some cases it is desirable to view the transformation and intensity transformations can prove useful in this regard.

We discuss two different intensity transformations. Each transformation is quite easy to implement on a computer and Computer Lab 3.1 guides you through the process of writing and testing code for implementing the transformations.

Gamma Correction

Perhaps the simplest intensity transformation is *gamma correction*. The idea is to pick an exponent $\gamma > 0$ and then create a new pixel p_{new} by raising the old pixel to the power of γ. That is

$$p_{new} = p_{orig}^{\gamma}.$$

To ensure that the data stay within the $[0, 255]$ range, we first map the pixels in the image to the interval $[0, 1]$. We can perform this map by simply dividing all of our pixels by 255. Figure 3.11 shows gamma correction for various values of γ.

Note that $\gamma = 1$ is a linear map of the intensities. Exponents $\gamma > 1$ create concave-up functions on $(0, 1]$ that tend to compress the lower intensities toward 0,

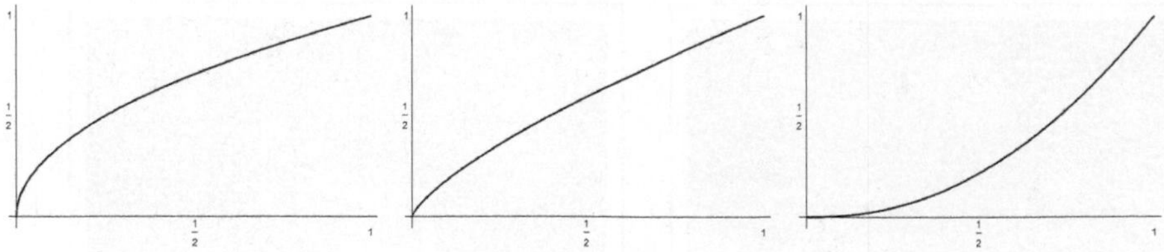

(a) $\gamma = 0.5$ (b) $\gamma = 0.75$ (c) $\gamma = 2.2$

Figure 3.11 Gamma correction transformations.

while $\gamma < 1$ result in concave-down functions on $(0, 1]$ that tend to compress higher intensities toward 1.

The transformation is quite simple. The intensities in an image are first divided by 255 so that they reside in the interval $[0, 1]$. Then each pixel is raised to the γ power. We next multiply by 255 and round each product to the nearest integer. Mathematically, we have

$$p_{new} = \text{Round}\left(255 \cdot \left(\frac{p_{orig}}{255}\right)^{\gamma}\right). \tag{3.1}$$

Figure 3.12 shows the results of gamma corrections for the values of gamma given in Figure 3.11.

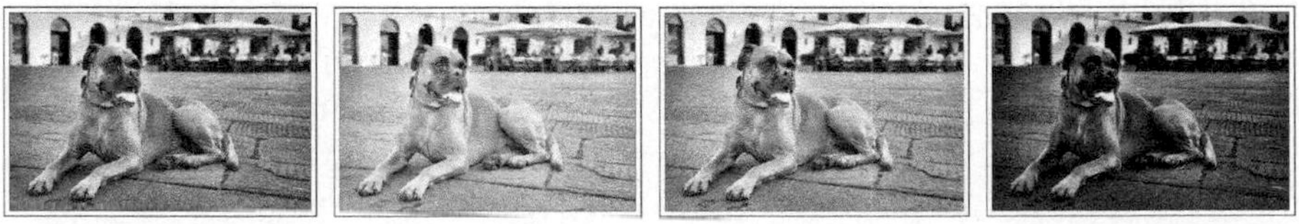

(a) The original image (b) $\gamma = 0.5$ (c) $\gamma = 0.75$ (d) $\gamma = 2.2$

Figure 3.12 Gamma correction transformations applied to a digital image.

Histogram Equalization

We can learn much about the contrast of intensities if we plot a histogram of the frequencies at which the intensities occur. In Figures 3.13(a) – 3.13(c) we have plotted three 512×768 images. Most of the intensities in Figure 3.13(a) are located near the bottom of the $[0, 255]$ intensity scale. In Figure 3.13(b), the intensities are bunched near the top of the intensity scale. The majority of its intensities in Figure 3.13(c) are located away from the ends of the intensity scale. In Figures 3.13(d) – 3.13(f) we have plotted histograms for each image. These histograms show how often intensities occur in the image.

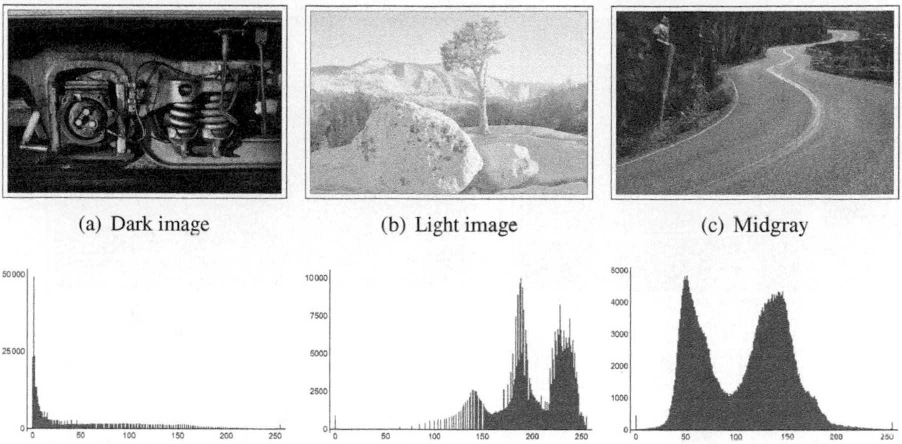

(a) Dark image (b) Light image (c) Midgray

(d) Histogram for (a) (e) Histogram for (b) (f) Histogram for (c)

Figure 3.13 Images and histograms of their gray-level intensity distributions. The image in Figure 3.13(a) has a large concentration of dark pixels, the image in Figure 3.13(b) has a large concentration of light pixels and the image in Figure 3.13(c) has a large concentration of midgray pixels.

We work with digital images throughout the remainder of the book. You are encouraged to download or create your own digital images and apply the techniques discussed in this book to them. Some of the images you might want to use may well look like those plotted in Figure 3.13. *Histogram equalization* is a process that, when applied to an image, produces an adjusted image that has a more balanced distribution of the grayscale intensities. Figure 3.14 illustrates the desired effect after histogram equalization is applied to the image in Figure 3.13(c).

How do we get to the image in Figure 3.14(b) from the image on the upper left in Figure 3.14(a)? We define a function $T(k)$ that sends intensity k, $k = 0, \ldots, 255$, to a value $T(k) \in [0, 1]$. The function will depend on the frequency of intensity k.

The value $T(k)$ measures the fraction of intensity k that we wish to retain. The final step in the process is to multiply $T(k)$ by 255. This stretches the values from $[0, 1]$ to $[0, 255]$. Finally, we use the *floor function* $\lfloor \cdot \rfloor$ or *greatest integer function* to convert $255 \cdot T(k)$ to an integer. Mathematically we have for $r \in \mathbb{R}$

$$\lfloor r \rfloor = m, \quad m \in \mathbb{Z}, m \leq r < m + 1. \tag{3.2}$$

Once we have established the map $T(k)$, we use the mapping

$$eq(k) = \lfloor 255 \cdot T(k) \rfloor \tag{3.3}$$

to send intensity k to its new intensity value.

Let's think about the design of $T(k)$. If we want to spread the frequencies across the $[0, 255]$ spectrum, then $T(k)$ should be nondecreasing. We also want $T(k)$ to

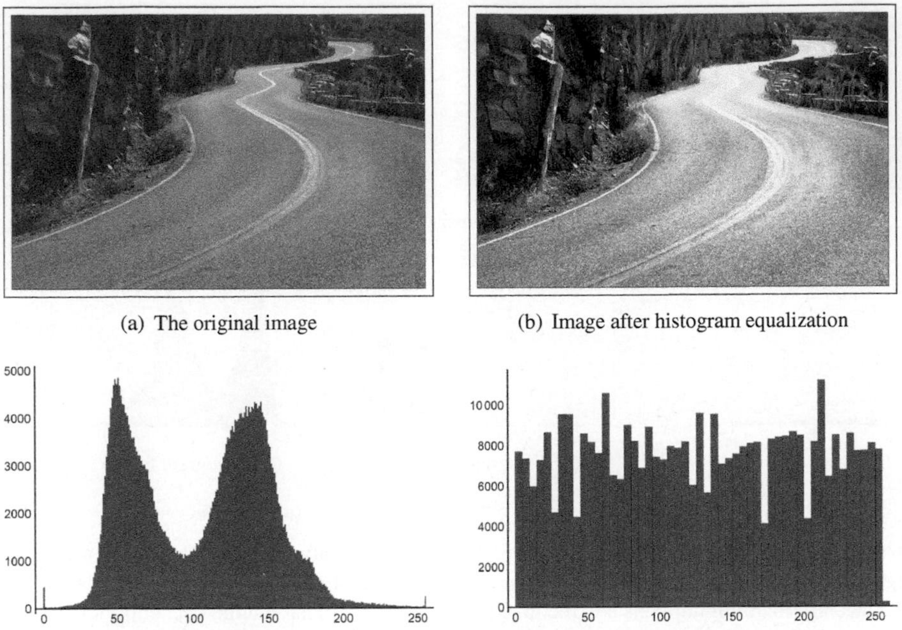

(a) The original image

(b) Image after histogram equalization

(c) Histogram for (a)

(d) Histogram for (b)

Figure 3.14 Histogram equalization applied to a digital image.

depend on the image histogram. For an $N \times M$ image, let $h(k)$ be the frequency of intensity k and for $k = 0, 1, \ldots, 255$, and define the function

$$T(k) = \frac{1}{NM} \cdot \sum_{j=0}^{k} h(j) \tag{3.4}$$

We can easily write down the first few values of T.[3] We have $T(0) = h(0)/NM$, $T(1) = \frac{1}{NM}\big(h(0) + h(1)\big)$, $T(2) = \frac{1}{NM}\big(h(0) + h(1) + h(2)\big)$. Thus we see that the summation in (3.4) creates a *cumulative total* of the frequencies. Since the sum must add to the total number of pixels, we divide by NM so that $T(k)$ maps $[0, 255]$ to $[0, 1]$. In Problem 3.10 you will show that $T(k) \leq T(k+1)$. In Figure 3.15 we have plotted T for the image in Figure 3.13(c).

A schematic of the histogram equalization process appears in Figure 3.16. The horizontal axis represents all the intensities possible in the original image. For each value k, $255 \cdot T(k)$ is computed and the result floored to produce the new intensity k'.

[3]The reader with some background in statistics will recognize T as a cumulative distribution function (CDF) for the probability density function that describes the intensity levels of the pixels in an image. Gonzalez and Woods [48] argue that output intensity levels using T follow a uniform distribution. This is a more rigorous argument for why histogram equalization works than the one we present here.

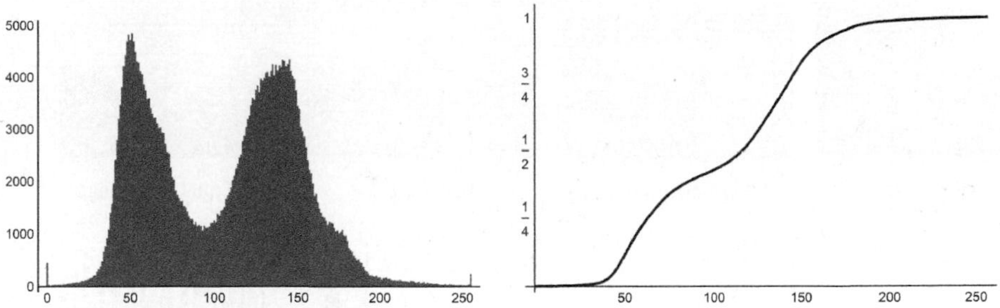

(a) Histogram for image in Figure 3.13(a)

(b) $T(k)$ for the histogram

Figure 3.15 Histogram and the cumulative total function.

Thus in the original image, each pixel with intensity k is changed to have intensity k'. For example, the histogram equalization of the image in Figure 3.13(c) sends the intensities 50, 100 and 175 to 34, 112 and 244, respectively.

The histogram-equalized versions of the images in Figure 3.13(a) and Figure 3.13(b) appear in Figure 3.17.

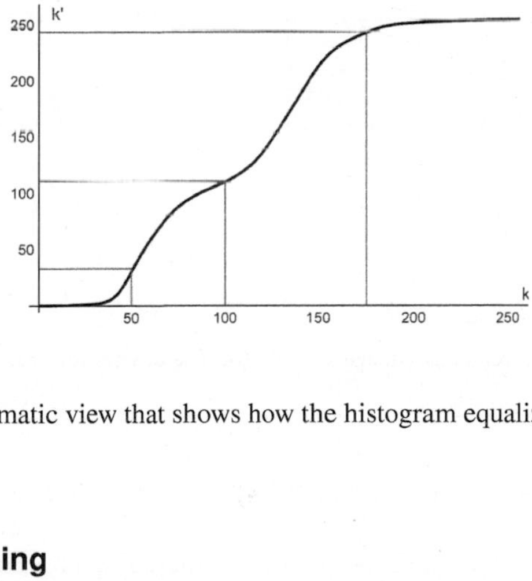

Figure 3.16 A schematic view that shows how the histogram equalization process works.

Histogram Matching

Histogram matching is another process that can be used to adjust the intensities in a grayscale image. It is in some sense a generalization of histogram equalization in that it uses cumulative distribution functions from two images to determine new intensity values. In Section 7.4, we employ histogram matching as part of the algorithm we use to perform image pansharpening.

Histogram matching utilizes a *reference image* A and a *target image* B. We begin by using (3.4) to create cumulative density functions $a(k)$ and $b(k)$ for each of A and B, respectively. For each $k = 0, \ldots, 255$, we map the value $b(k)$ of the target

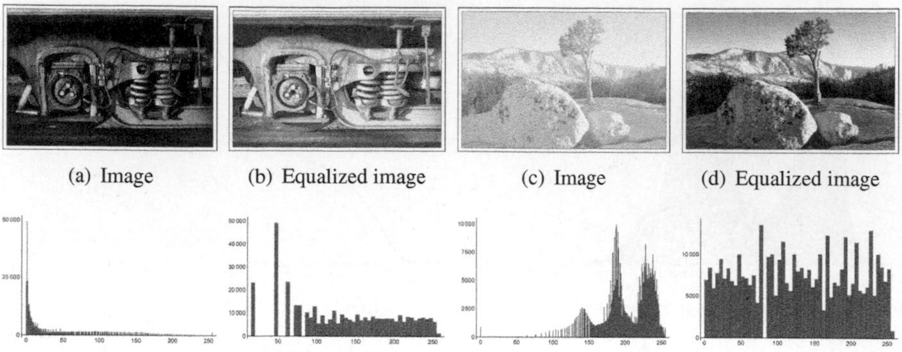

(a) Image (b) Equalized image (c) Image (d) Equalized image

Figure 3.17 Histogram-equalized versions of the images in Figure 3.13(a) and Figure 3.13(b). The histograms for each image are shown below the image.

cumulative density function horizontally to the nearest corresponding value $a(j)$ of the reference cumulative density function. The process is completed by mapping intensity values k in the target image to j. Figure 3.18 illustrates the process.

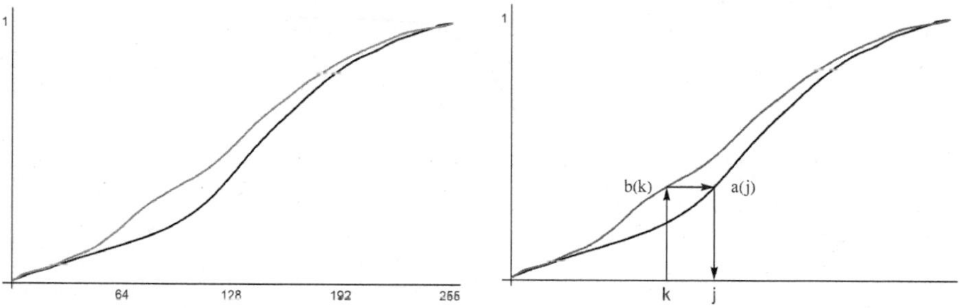

(a) The cdf functions for the reference image (black) and target image (gray)

(b) The process for matching intensities

Figure 3.18 Cumulative distribution functions and the histogram matching process.

The following example illustrates the histogram matching process.

Example 3.7 (Histogram Matching). *Consider the reference image and target image given in Figure 3.19. Use histogram matching to bring the intensity values of the target image more in line with those of the reference image.*

Solution.
We begin by computing cumulative density functions for the reference and target images. They are displayed in Figure 3.20(a). The matching process is illustrated in Figure 3.20(b) for a couple of intensity values. The intensities 170 and 220 in the target image are mapped to 49 and 139, respectively.

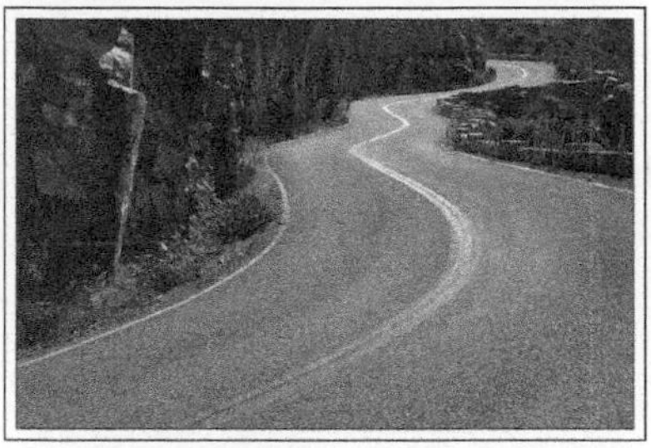

(a) Reference image A

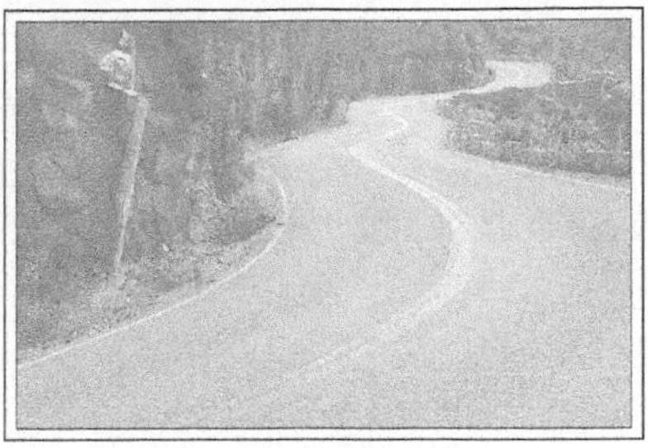

(b) Target image B

Figure 3.19 The reference and target images for Example 3.7.

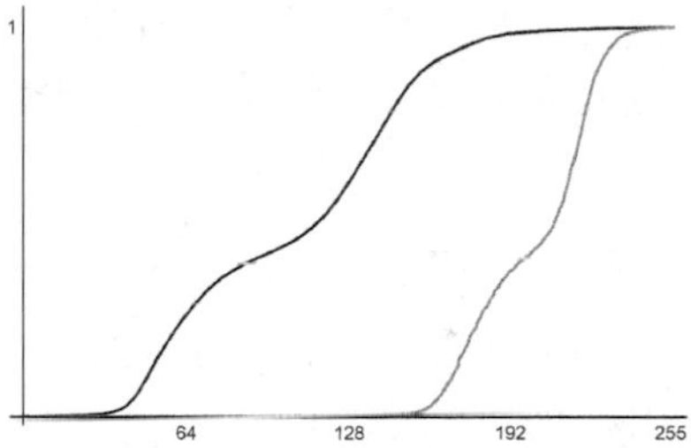

(a) Cumulative distribution functions

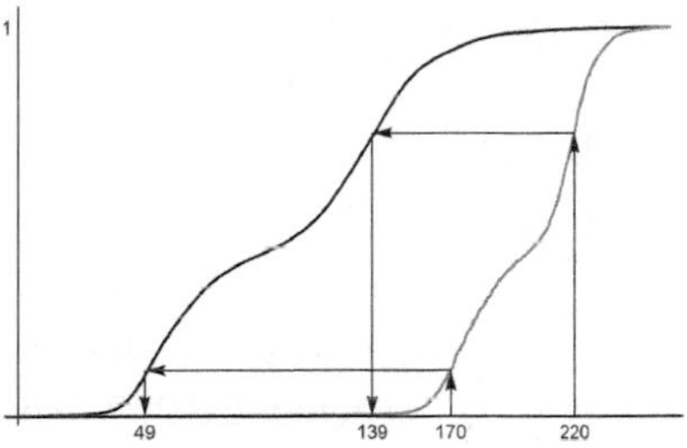

(b) The matching process

Figure 3.20 The image at left shows the reference cdf (black) and the target cdf (gray). At right, the process for matching intensities is illustrated using the intensity values 170 and 220.

The resulting image formed by matching all intensity values using histogram matching is displayed in Figure 3.21.

Figure 3.21 The histogram-matched image from Example 3.7.

(Live example: Visit `stthomas.edu/wavelets` *and click on Live Examples.)*

PROBLEMS

3.1 Convert each of 46, 184 and 219 to base 2.

3.2 Using the ideas from Example 3.2, write each of the following intensities as two ASCII characters.

(a) $24,173$

(b) 9288

(c) 1888

3.3 Find the ASCII characters (see Table 3.1) for each of the following base 2 numbers.

(a) 00110101_2

(b) 01100001_2

(c) 01100100_2

3.4 In Example 3.3 we learned how to produce the negative of an image. For each pixel intensity p, we computed $255 - p$.

(a) The negative of $p = 79$ is 176. Write the base 2 representations for each of these numbers.

(b) Repeat (a) with $p = 112$.

(c) What is the relationship between the bits of a pixel and its negative?

(d) Prove your assertion from (c) by writing 255 in base 2 and then subtracting an intensity of the form $(b_7b_6b_5b_4b_3b_2b_1b_0)_2$, where $b_k \in \{0, 1\}$ for $k = 0, \ldots, 7$.

3.5 In some cases, 24-bit images are desired.

(a) What is the intensity range in this case?

(b) Write $2,535,208$ as three ASCII characters.

3.6 Describe the matrix and corresponding image that results from the product AM^T in Example 3.6.

3.7 Suppose that matrix A represents an 256×256 image. What matrix operation(s) would you use to produce a plot of A with

(a) the last 128 rows converted to black?

(b) the first 128 columns colored gray (ASCII code 128)?

(c) its elements reflected across the main diagonal?

(d) its elements reflected about the $j = 128$ row?

(e) its elements reflected about the $k = 128$ column?

(f) the image is rotated 90° counterclockwise?

You may need to create special matrices to use in conjunction with A.

3.8 Let $\gamma > 0$ and define $f(t) = t^\gamma$ for $t \in (0, 1]$. Show that $f(t)$ is concave down on $(0, 1]$ whenever $0 < \gamma < 1$ and $f(t)$ is concave up on $(0, 1]$ for $\gamma > 1$.

3.9 Consider the 6×6 image plotted below (with intensity values displayed rather than gray shading) and assume that gray levels range from 0 (black) to 5 (white).

0	2	2	2	2	4
0	2	1	1	2	3
0	2	1	1	2	3
1	1	1	1	3	3
0	1	1	2	3	4
1	1	2	2	4	5

.

(a) Fill in the following chart and sketch the associated histogram.

Intensity	Frequency
0	
1	
2	
3	
4	
5	

(b) Using (3.4), fill in the following chart and then plot $T(k)$, $k = 0, \ldots, 5$.

k	$T(k)$
0	
1	
2	
3	
4	
5	

(c) Using (b) and (3.3), perform histogram equalization and write the new intensity values in the following chart.

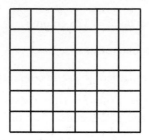

You may want to use a **CAS** to plot the "images" from this problem.

(d) Using the new "image" you constructed in (c), fill in the following chart and sketch the associated histogram.

Intensity	**Frequency**
0	
1	
2	
3	
4	
5	

3.10 Show that $T(k)$ defined in (3.4) satisfies $0 \leq T(k) \leq T(k+1)$ for $k = 0, 1, \ldots, 254$.

3.11 What must be true about the histogram of an image to ensure that $T(k) < T(k+1)$ for all $k = 0, \ldots, 254$?

3.12 *True or False:* If image matrix B is the histogram-equalized version of image A, then at least one element of B has the value 255. If the result is true, explain why. If the result is false, provide a counterexample to support your claim.

3.13 If $eq(\ell) = 0$ for some ℓ with $0 \leq \ell \leq 255$, then what can you claim about the intensities $0, \ldots, \ell$ in the original image?

3.14 Suppose A and B are 10×10 matrices whose elements are integers in the range $[0, 9]$. The cumulative distribution functions are computed for each matrix and the results are displayed in Figure 3.22.

k	$a(k)$	$b(k)$
0	0.0077	0.0547
1	0.1204	0.3715
2	0.2837	0.5886
3	0.3848	0.7522
4	0.4713	0.8139
5	0.5751	0.8937
6	0.7418	0.9488
7	0.8948	0.9769
8	0.9925	0.9900
9	1.0000	1.0000

(a)

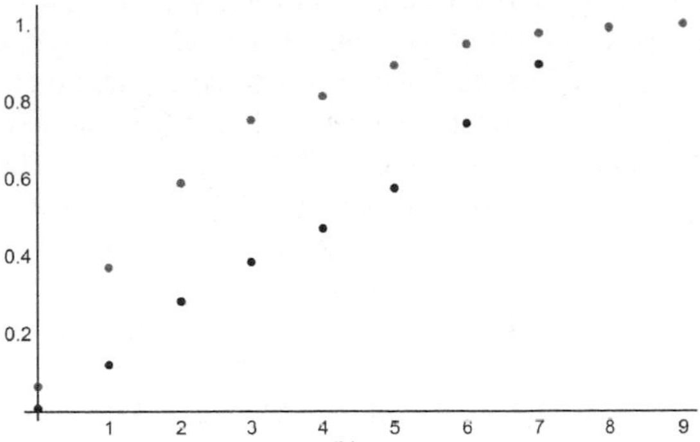

(b)

Figure 3.22 The table at left shows the cumulative distribution values for A and B. The values for $a(k)$ (black) and $b(k)$ (gray) are plotted on the right.

Using A as the reference matrix and B as the target matrix, perform histogram matching and describe how the elements of B will change.

Computer Lab

3.1 Exploring Intensity Transformations. From `stthomas.edu/wavelets`, access the file `IntensityTransforms`. In this lab you will develop and test software modules for gamma correction and histogram equalization.

3.2 Color Images and Color Spaces

In this section we discuss color images. We begin with a brief description of how these images might be stored on a computer and continue by illustrating how to write color images in terms of matrices. In subsequent chapters we apply wavelet transformations to color images. In these applications it is desirable to first transform the image to a different *color space.* Such color spaces allow the wavelet transformation to process the data in a setting that better describes how humans might interpret changes in color images. We conclude the section with a discussion of the YCbCr and HSI color spaces.

The Basics of a Color Image

Pixels in grayscale images assume an integer value p where $0 \leq p \leq 255$. In Section 3.1 we learned how to convert this value to and from an ASCII code and how to write an array of pixels as a matrix. To form the color of a pixel we use the concept of *additive mixing*. Here, suppose we have three flashlights each of a different *primary color* whose intensity levels we can control. By shining the flashlights to a common area in a dark room and varying each light's intensity, we can produce the different colors of the spectrum. For *RGB color images*, we use the primary colors red, green and blue each with intensity scales set exactly like grayscale images. That is, intensities are integer-valued with 0 indicating no presence of the primary color and 255 indicating maximum presence of the color. Thus the *RGB triples* $(255, 0, 0)$, $(0, 255, 0)$, and $(0, 0, 255)$ represent bright red, green, and blue, respectively. Black is represented by $(0, 0, 0)$ and white is represented by $(255, 255, 255)$. A *secondary color* is one that is formed by subtracting a primary color from white. Thus the secondary colors are magenta, yellow, and cyan their RGB triples are $(255, 0, 255)$, $(255, 255, 0)$ and $(0, 255, 255)$, respectively. Figure 3.23 shows how additive mixing uses the primary colors to generate white and the secondary colors.

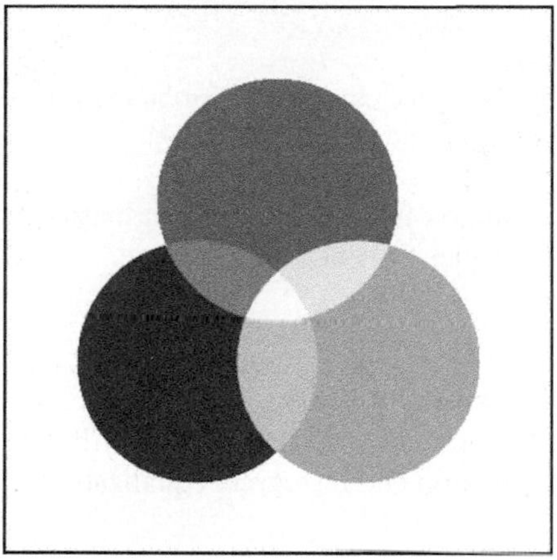

Figure 3.23 Using additive mixing to generate white and the secondary colors magenta, yellow, and cyan from the primary colors red, green, and blue. See the color plates for a color version of this image.

As the primary color values range from 0 to 255, different colors in the spectrum are created. For example, $(0, 50, 50)$ represents a dark cyan, brown is $(128, 42, 42)$, and maroon is $(176, 48, 96)$. If the three primary color intensity values are all the same, we obtain a grayscale value. Figure 3.24(a) illustrates the RGB color space, and Figure 3.24(b) gives a schematic of the cube from Figure 3.24(a).

We need three times as much information to store a color image as we do to store a grayscale image. We can still use ASCII codes to represent a color pixel. We just need three codes instead of one. For example, the color olive can be represented with the RGB triple $(59, 94, 43)$. If we consult Table 3.1, we see that this triple can be represented by the ASCII code triple ;^+. Raw (uncompressed) data file formats such as the *portable pixel map* (PPM) format use these ASCII code triples

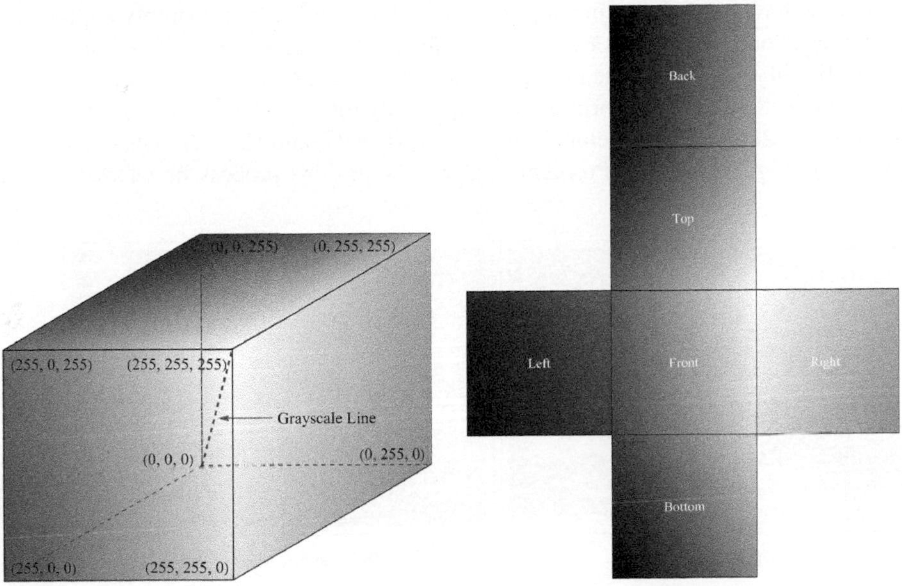

(a) RGB color cube (b) Color cube schematic

Figure 3.24 The image at left is the RGB color cube. The image on the right is a schematic of the color cube. See the color plates for color versions of these images.

to write a file to disk. After some header information that lists the bit depth and image dimensions, the pixels are stored as a long string of ASCII characters. Three characters are used for each pixel and they are written to file one row after the other.

We can easily represent color images using matrices, but now we need three matrices or *channels*. The intensity levels for each primary color are stored in a matrix. Figure 3.25 shows a color image and the three matrices used to represent the image. We have plotted the three matrices using shades of the primary colors, but you should understand that each matrix is simply an integer-valued matrix where the values range from 0 to 255.

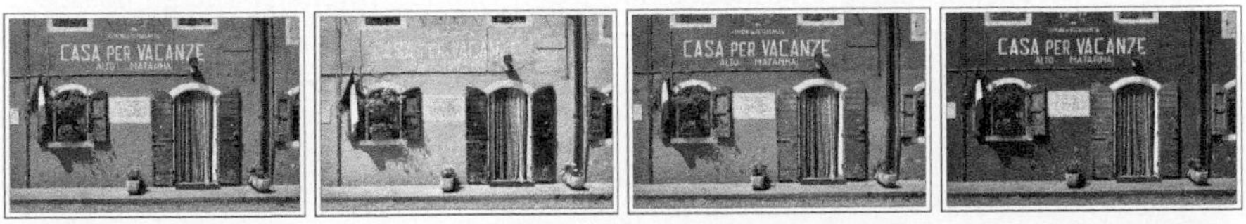

(a) A color image (b) The red channel (c) The green channel (d) The blue channel

Figure 3.25 A color image and the three primary color channels. See the color plates for color versions of these images.

We can perform some basic image processing on color images by simply applying transformations to each color channel. For example, suppose that R, G, and B represent the red, green, and blue intensity matrices, respectively, for the image plotted in Figure 3.25. We can perform image negation by forming the matrix W whose entries are each 255, and then computing $W - R$, $W - G$, and $W - B$. The results are plotted in Figure 3.26. In Problem 3.23 we discuss this process in a different context.

(a) A digital color image (b) The negative image

Figure 3.26 Color image negation. See the color plates for color versions of these images.

Luminance and Chrominance

Although Figure 3.26 illustrates that image negation operates on the individual red, green, and blue color channels, it is more natural for humans to view images in terms of *luminance* and *chrominance*. In fact, some color image-processing tools do not work well when they process the image in *RGB space* (see Problem 3.26). Thus, in many applications, it is desirable to first convert an image from RGB space to a space that stores information about pixels in terms of luminance and chrominance.

You can think of a luminance channel as one that carries information regarding the brightness and contrast of the pixel. *Chrominance* is the difference between a color and a reference color at the same brightness (see Poynton [75]). Grayscale images contain no chrominance channels since we only need intensity (brightness) values to identify them. The conversions we consider in this book usually quantize chrominance either as color differences or in terms of hue and saturation.

The YCbCr Color Space

Color spaces that describe chrominance via *color difference* channels are typically linear transformations and thus easy to use. One such space is the *YCbCr color space*. There are three channels in the YCbCr color space. Y is the luminance or intensity channel, and the Cb and Cr channels hold chrominance information.

As pointed out by Varma and Bell [97], images in RGB space typically have a more even distribution of energy than those stored in YCbCr space. Moreover in YCbCr space, most of the energy resides in the Y channel. Since the human eye is less sensitive to changes in chrominance, the chrominance channels are often *subsampled* so that less information can be used to adequately display the image. For this reason, the YCbCr color space is typically used in image compression techniques (see Chapter 12).

A straightforward way to describe the luminance would be to average the red, green, and blue channels. However, as you will investigate in Problem 3.15, it is better to use a weighted average when computing the luminance. Suppose that (r, g, b) are the intensity values of red, green, and blue, respectively, and further assume that these values have been scaled by $\frac{1}{255}$ so that they reside in the interval $[0, 1]$. We then define the intensity value y as

$$y = 0.299r + 0.587g + 0.114b \tag{3.5}$$

Note that the minimum value of y is 0 and it occurs when $r = g = b = 0$ and the maximum value of y is 1 and it occurs when $r = g = b = 1$.

Cb and Cr are the color difference channels. Cb is a multiple of the difference between blue and the intensity y given in (3.5), and Cr is a multiple of the difference between red and the intensity y. The exact values can be computed using

$$Cb = \frac{(b - y)}{1.772} \quad \text{and} \quad Cr = \frac{(r - y)}{1.402}. \tag{3.6}$$

Equation (3.5) is used to convert color to grayscale in television (see Problem 3.15). The interested reader is encouraged to consult Poynton [75] for some historical information on formulas (3.5) and (3.6).

In Problem 3.17 you will show that the range of the values of Cb and Cr is $\left[-\frac{1}{2}, \frac{1}{2}\right]$. In this problem you will also find the matrix C that maps the vector $[r, g, b]^T$ to $[y, Cb, Cr]^T$. The fact that this matrix is invertible gives us a way to convert from YCbCr space to RGB space.

For display purposes, values y, Cb and Cr are typically scaled and shifted so that $y \in [16, 235]$, $Cb, Cr \in [16, 240]$. The condensed ranges are used to protect against overflow due to roundoff error (see Poynton [75]). Thus, the transformation from (y, Cb, Cr) to (Y, C_b, C_r) we use for display is

$$\begin{aligned} y' &= 219y + 16 \\ C_b &= 224Cb + 128 \\ C_r &= 224Cr + 128 \end{aligned} \tag{3.7}$$

The algorithm for converting from RGB space to YCbCr space is as follows:

1. Divide the red, green, and blue intensity values by 255 to obtain $r, g, b \in [0, 1]$.
2. Compute the intensity y using (3.5).
3. Compute the chrominance values Cb and Cr using (3.6).

4. If we wish to display the results, employ (3.7) and round the results.

Figure 3.27 shows y', C_b, and C_r for the image plotted in Figure 3.26.

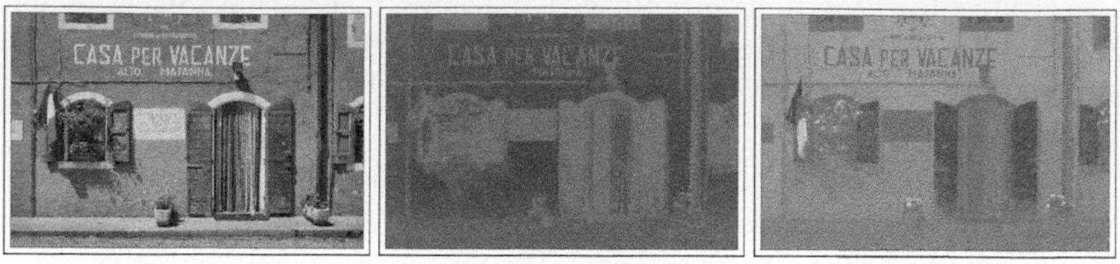

(a) Channel y' (b) Channel C_b (c) Channel C_r

Figure 3.27 Intensity and chrominance channels for the image in Figure 3.26(a).

The HSI Color Space

Many image processing applications require color images to be stored in the *HSI color space*. The three channels in the HSI color space are hue, saturation and intensity.

Hue designates the pure color present in the pixel and is measured as an angle with the primary colors red, green and blue located 0, $2\pi/3$ and $4\pi/3$ radians, respectively. The secondary colors yellow, cyan, and magenta are located at $\pi/3$, π, and $5\pi/3$ radians, respectively.

The saturation of a hue can be viewed as the measure of a color's purity. The larger the value, the more pure or vibrant the color. Low saturation values indicate the color is closer to gray. For example, pink has the same hue as red but is considered less saturated than red.

The intensity channel indicates the brightness of the color – it is similar to the Y channel in YCbCr space with values ranging from 0 (black) to 255 (white).

Summarizing, the HSI space can be envisioned as concentric circles centered in three-space from $(0, 0, 0)$ to $(0, 0, 1)$ whose radii grow linearly from 0 at $(0, 0, 0)$ to 1 at $\left(0, 0, \frac{1}{2}\right)$ and then decrease linearly back to 0 at $(0, 0, 1)$. The HSI space is illustrated graphically in Figure 3.28.

Converting points in RGB space to HSI space is equivalent to the mathematical problem of mapping points in the cube in Figure 3.24(a) to the conic solid in Figure 3.28. In the two sections below, we follow the derivation of conversion formulas between RGB space and HSI space given by Gonzalez and Woods [47].

Converting from RGB Space to HSI Space

Let's assume that a given (R, G, B) triple has been normalized so that all values are in the interval $[0, 1]$. This can be achieved by simply dividing the elements of a given

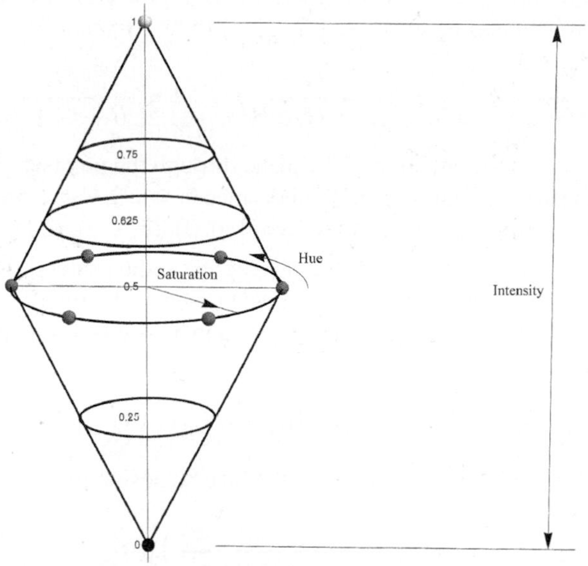

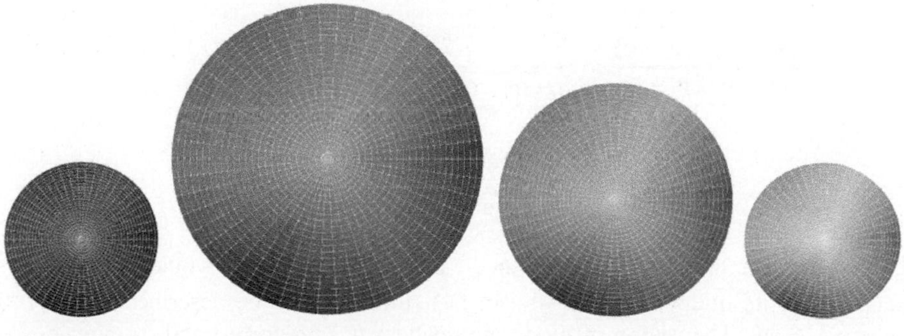

(a) $I = 0.25$ (b) $I = 0.5$ (c) $I = 0.625$ (d) $I = 0.75$.

Figure 3.28 The HSI solid with cross sections at various levels of intensities. See the color plates for color versions of these images.

triple from the color cube in Figure 3.24(a) by 255. We can easily obtain the intensity value I by the formula

$$I = \frac{1}{3}(R + G + B). \tag{3.8}$$

If $R = G = B$, we return the triple $(0, 0, I)$.[4] If this is not the case, we facilitate the ease of computation by mapping the triple (R, G, B) to its *barycentric coordinates*

$$r = \frac{R}{R+G+B}, \qquad g = \frac{G}{R+G+B}, \qquad b = \frac{B}{R+G+B}. \tag{3.9}$$

Note that $r + g + b = 1$ is the equation of the plane through the points $(1, 0, 0)$ (red), $(0, 1, 0)$ (green) and $(0, 0, 1)$ (blue). In particular, since $0 \le r, g, b \le 1$, we know that the triple (r, g, b) lies on the triangle with vertices $(1, 0, 0)$, $(0, 1, 0)$, and $(0, 0, 1)$. Let $\mathbf{c} = \left[\frac{1}{3}, \frac{1}{3}, \frac{1}{3}\right]^T$, $\mathbf{r} = [1, 0, 0]^T$, and $\mathbf{v} = [r, g, b]^T$ represent the center of the triangle, the unit vector in the direction of red and the given color triple, respectively. Then the hue is the angle θ_H between the vectors $\mathbf{r} - \mathbf{c}$ and $\mathbf{v} - \mathbf{c}$ (see Figure 3.29(b)). Using Problem 2.15, we see that

$$(\mathbf{v} - \mathbf{c}) \cdot (\mathbf{r} - \mathbf{c}) = \|\mathbf{v} - \mathbf{c}\| \|\mathbf{r} - \mathbf{c}\| \cos \theta_H.$$

Thus for $0 \le \theta_H \le \pi$, which is the domain of the inverse cosine function, we have

$$\theta_H = \cos^{-1}\left(\frac{(\mathbf{v} - \mathbf{c}) \cdot (\mathbf{r} - \mathbf{c})}{\|\mathbf{v} - \mathbf{c}\| \|\mathbf{r} - \mathbf{c}\|}\right). \tag{3.10}$$

In Problem 3.19 you will show that

$$\frac{(\mathbf{v} - \mathbf{c}) \cdot (\mathbf{r} - \mathbf{c})}{\|\mathbf{r} - \mathbf{c}\| \|\mathbf{v} - \mathbf{c}\|} = \frac{\frac{1}{2}(2R - G - B)}{\sqrt{(R-G)^2 + (R-B)(G-B)}}.$$

so that

$$\theta_H = \cos^{-1}\left(\frac{\frac{1}{2}(2R - G - B)}{\sqrt{(R-G)^2 + (R-B)(G-B)}}\right).$$

If we draw the line through R and $\left(\frac{1}{3}, \frac{1}{3}, \frac{1}{3}\right)$ in Figure 3.29(b), it would intersect segment $\overline{BG}$ at its midpoint $\left(0, \frac{1}{2}, \frac{1}{2}\right)$. In fact, this line can be described as the line where $b = g$. The region of the triangle where $b \le g$ corresponds to angles $0 \le \theta_H \le \pi$. When $b > g$, we take as our hue angle $2\pi - \theta_H$.

To find the saturation value s, consider the plot in Figure 3.30. The saturation is the ratio $\frac{|CV|}{|CV'|}$. The notation $|XY|$ denotes the length of segment $\overline{XY}$. Using the fact that $\triangle CVP$ is similar to $\triangle CV'M$, we see that

$$s = \frac{|CV|}{|CV'|} = \frac{|CP|}{|CM|} = \frac{|CM| - |PM|}{|CM|} = 1 - \frac{|PM|}{|CM|}. \tag{3.11}$$

We have drawn the figure so that V is in the *red/green sector*, i.e., in $\triangle CRB$. Thus the coordinates of P are $\left(\frac{1}{3}, \frac{1}{3}, b\right)$ so that $|PM| = b$. Since the coordinates of M are $\left(\frac{1}{3}, \frac{1}{3}, 0\right)$, we see that $|CM| = \frac{1}{3}$. Plugging these values into (3.11), we see that

[4]The value for hue makes no sense if saturation is 0. We set hue to be 0 for purposes of inverting the transformation.

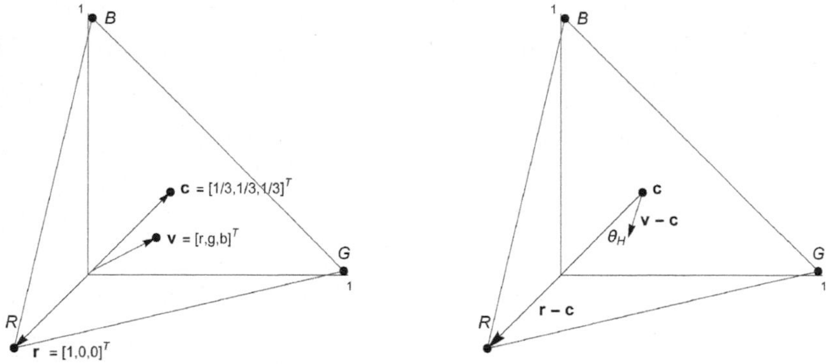

(a) The hue triangle (b) The hue angle

Figure 3.29 Map of the triple (r, g, b) to the unit triangle in $\mathbb{R}^3$ and the angle θ_H that represents hue.

$s = 1 - 3b$. Similar arguments show that if V would have been in the red/blue sector or the blue/green sector, the value for saturation would have been $s = 1 - 3g$ or $s = 1 - 3r$, respectively. In fact, the minimum value of (r, g, b) indicates the sector holding the point. This observation allows us to write

$$\begin{aligned} s &= 1 - 3\min(r, g, b) \\ &= 1 - 3\min\left(\frac{R}{R+G+B}, \frac{G}{R+G+B}, \frac{B}{R+G+B}\right) \\ &= 1 - \frac{3}{R+G+B}\min(R, G, B). \end{aligned} \tag{3.12}$$

Summarizing we see that the HSI representation of the given triple (R, G, B) with $0 \le R, G, B \le 1$ is

$$\boxed{\begin{aligned} \theta_H &= \begin{cases} \cos^{-1}\left(\frac{\frac{1}{2}(2R-G-B)}{\sqrt{(R-G)^2+(R-B)(G-B)}}\right), & B \le G \\ 2\pi - \cos^{-1}\left(\frac{\frac{1}{2}(2R-G-B)}{\sqrt{(R-G)^2+(R-B)(G-B)}}\right), & B > G \end{cases} \\ s &= 1 - \frac{3}{R+G+B}\min(R, G, B) \\ I &= \frac{1}{3}(R+G+B). \end{aligned}}$$

Figure 3.31 shows the θ_H, s, and I matrices for the image plotted in Figure 3.26.

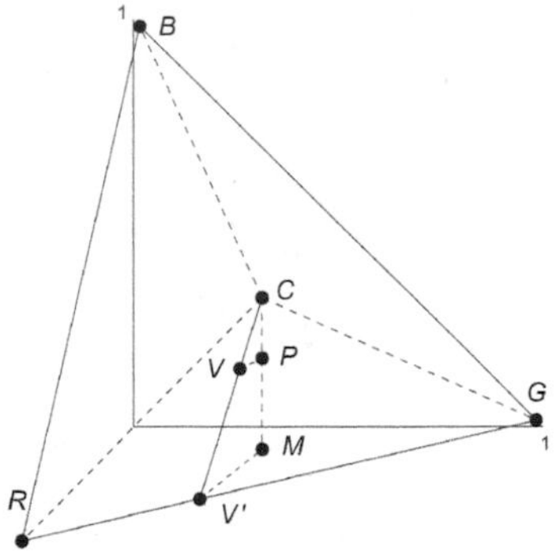

Figure 3.30 The triangles $\triangle CVP$ and $\triangle CV'M$ and the sectors red/green ($\triangle RCG$), red/blue ($\triangle RCB$) and blue/green ($\triangle BCG$) used to compute the saturation value.

(a) Hue channel (b) Saturation channel (c) Intensity channel

Figure 3.31 The hue, saturation and intensity channels for the image in Figure 3.26(a).

Converting from HSI Space to RGB Space

In order to convert an HSI triple to RGB, we note that saturation depends on the sector that holds the original triple (r, g, b). Thus we must consider three cases. We develop the conversion for the case where (r, g, b) is in the red/green sector (so that $0 \le \theta_H \le \frac{2\pi}{3}$ and $b \le g$) and note that the remaining two cases follow similarly.

In the red/green sector, we know that $\min(r, g, b) = b$ so (3.12) gives $s = 1 - 3b$ or $b = \frac{1}{3}(1 - s)$. Once we find r, we use the fact that $r + g + b = 1$ to obtain $g = 1 - r - b$. Equation (3.8) implies $R + G + B = 3I$ and inserting this value in the identities in (3.9) and simplifying gives the values $R = 3Ir$, $G = 3Ig$, and $B = 3Ib$. Thus our last remaining task is to find a formula for r.

Consider Figure 3.32(a) where we have now added several points to help us compute r. In particular, the point Q is the intersection of the line through R and C and the line through B and G. Figure 3.32(b) shows $\triangle ROQ$ with the dashed line indicating where the line that projects V onto $\overline{OR}$ intersects $\triangle ROQ$. The intersection points on $\overline{RQ}$ and $\overline{OR}$ are denoted by S and T, respectively. Note that $|OT| = r$. Recall that $\theta_H = \text{m}\angle SCV$ and from Figure 3.32(a), using the right triangle $\triangle CSV$ ($\text{m}\angle CSV = \frac{\pi}{2}$), we see that $\cos(\theta_H) = \frac{|SC|}{|CV|}$ or $|SC| = \cos(\theta_H)\,|CV|$.

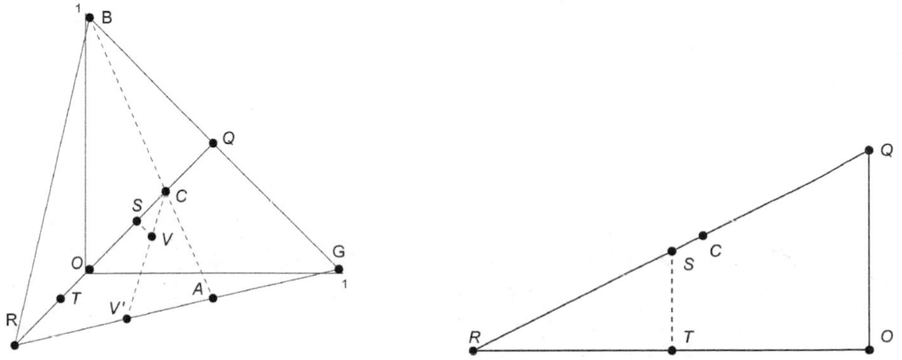

(a) The HSI region (b) $\triangle ROQ$

Figure 3.32 Plots used to find the value r.

Since $\triangle ROQ$ and $\triangle RST$ (Figure 3.32(b)) are similar and noting that $|OR| = 1$ and $|RT| = 1 - r$, we see that

$$\frac{|QR|}{|OR|} = |QR| = \frac{|RS|}{|RT|} = \frac{|RS|}{1-r} \quad \text{or} \quad |QR| = \frac{|RS|}{1-r}.$$

Solving for r above gives

$$r = \frac{|QR| - |RS|}{|QR|}$$

But $|RS| = |QR| - |SC| - |CQ| = |QR| - |CV|\cos(\theta_H) - |CQ|$ and inserting this identity into the above identity shows that

$$r = \frac{|QR| - |QR| + |CV|\cos(\theta_H) + |CQ|}{|QR|} = \frac{|CV|\cos(\theta_H)}{|QR|} + \frac{|CQ|}{|QR|}. \tag{3.13}$$

Problem 2.14 can be used to show that $\frac{|CQ|}{|QR|} = \frac{1}{3}$ and note from the saturation computation (3.11), we have $|CV| = s|CV'|$. Inserting these values into (3.13) gives

$$r = \frac{s|CV'|\cos(\theta_H)}{|QR|} + \frac{1}{3}. \tag{3.14}$$

$|QR|$ can be computed (we choose not to do so at this time) so our last step is to find an expression for $|CV'|$.

Consider $\triangle RGB$ from Figure 3.28(a). Since it is an equilateral triangle, we can infer that $m\angle RCA = \frac{\pi}{3}$ and as a result $m\angle V'CA = \frac{\pi}{3} - \theta_H$.

Using Figure 3.33, we see that $\cos\left(\frac{\pi}{3} - \theta_H\right) = \frac{|CA|}{|CV'|}$ or $|CV'| = |CA|\cos\left(\frac{\pi}{3} - \theta_H\right)$. Again using the fact that $\triangle RGB$ is equilateral gives $|CA| = |CQ|$. Thus we have

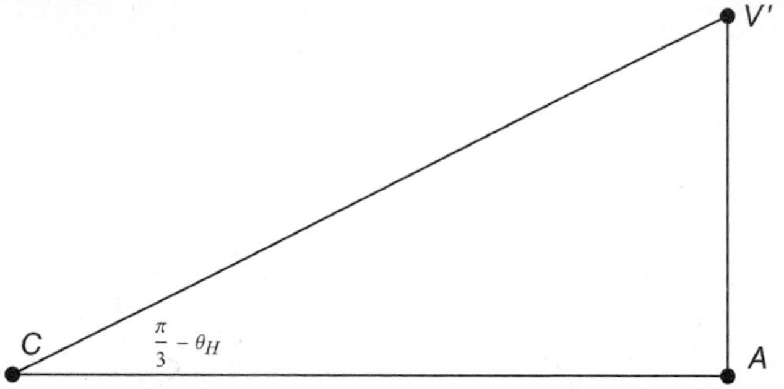

Figure 3.33 $\triangle CAV'$ for computing r.

$$|CV'| = \frac{|CQ|}{\cos\left(\frac{\pi}{3} - \theta_H\right)}.$$

Again using Problem 2.14 we see that (3.14) becomes

$$\begin{aligned} r &= \frac{s|CQ|\cos(\theta_H)}{|QR|\cos\left(\frac{\pi}{3} - \theta_H\right)} + \frac{1}{3} \\ &= \frac{1}{3}\frac{s\cos(\theta_H)}{\cos\left(\frac{\pi}{3} - \theta_H\right)} + \frac{1}{3} \\ &= \frac{1}{3}\left(1 + \frac{s\cos(\theta_H)}{\cos\left(\frac{\pi}{3} - \theta_H\right)}\right). \end{aligned}$$

Summarizing, the conversion formulas for θ_H in the red/green sector are

$$\begin{aligned} b &= \frac{1}{3}(1 - s) \\ r &= \frac{1}{3}\left(1 + \frac{s\cos(\theta_H)}{\cos\left(\frac{\pi}{3} - \theta_H\right)}\right) \\ g &= 1 - r - b. \end{aligned} \tag{3.15}$$

To recover R, G, and B we compute $R = 3Ir$, $G = 3Ig$, and $B = 3Ib$. To put these values in the RGB color cube shown in Figure 3.24(a), we multiply each of R, G and B by 255.

For θ_H in the green/blue sector, we note that $\frac{2\pi}{3} < \theta_H \leq \frac{4\pi}{3}$ and now $r = \frac{1}{3}(1 - s)$. The computation for g follows that of r above except for the adjustment

$\theta_H \to \theta_H - \frac{2\pi}{3}$. We obtain

$$g = \frac{1}{3}\left(1 + \frac{s\cos(\theta_H)}{\cos\left(\frac{\pi}{3} - \theta_H\right)}\right)$$

which leads to $b = 1 - r - g$.

For θ_H in the red/blue sector, we note that $g = \frac{1}{3}(1-s)$ and adjust θ_H by $\theta_H - \frac{4\pi}{3}$ to arrive at

$$b = \frac{1}{3}\left(1 + \frac{s\cos(\theta_H)}{\cos\left(\frac{\pi}{3} - \theta_H\right)}\right)$$

which leads to $r = 1 - g - b$.

PROBLEMS

3.15 In this problem you will learn how to use some basic matrix arithmetic to convert color images into grayscale images. We can view a color image as three matrices. Suppose that matrices R, G, and B contain the red, green, and blue intensities, respectively, of some color image.

(a) Possibly the most natural way to convert a color image to grayscale is to average corresponding elements of R, G, and B. Let M be the resulting grayscale matrix and write M as a linear combination of R, G, and B. (Recall that a linear combination of matrices R, G, and B is simply any expression of the form $aR + bG + cB$ where a, b, c are real numbers.)

(b) The National Television System Committee (NTSC) uses a different formula to convert color to grayscale. As Poynton points out [75], if red, green, and blue have the same luminescence in the color spectrum, then green will appear brightest, red is less bright, and blue will appear darkest. For this reason, the NTSC uses (3.5) to convert color images to grayscale. If M is the resulting grayscale matrix, write M as a linear combination of R, G, and B.

3.16 Sometimes it is useful to convert a grayscale image to a color image. In such cases we create a *pseudocolor map* to do the job. Suppose that you are given a grayscale image and wish to convert it to a color image where we use the "front" of the RGB color cube (see Figure 3.24(b)). In particular, we want magenta to represent the grayscale value 0 and yellow to represent the grayscale value 255. Let A be the matrix that represents the grayscale image, and let W be a matrix of the same dimensions as A, whose elements are each 255. Let R, G, and B be the red, green, and blue intensity matrices, respectively, that will house the pseudocolor-mapped result. Write R, G, and B in terms of W and A.

★3.17 In this problem we further investigate the conversion between RGB color space and YCbCr color space. Toward this end, suppose that r, g, and b are the intensities values of red, green, and blue, respectively, for some pixel. Assume further that these intensities have been scaled by $\frac{1}{255}$ so that they reside in the interval $[0, 1]$. Let y, Cb, Cr be the values defined in (3.5) and (3.6).

(a) Use (3.6) to show that if $r, g, b \in [0, 1]$, then $Cb, Cr \in \left[-\frac{1}{2}, \frac{1}{2}\right]$.

(b) Using (3.5) and (3.6), find the matrix S so that

$$\begin{bmatrix} y \\ Cb \\ Cr \end{bmatrix} = S \cdot \begin{bmatrix} r \\ g \\ b \end{bmatrix}$$

(c) Using a **CAS**, find S^{-1}.

(d) You can also find S^{-1} by first using (3.6) to express b and r in terms of Cb, y and Cr, y, respectively, and then using these results in conjunction with (3.5) to solve for g in terms of y, Cb, Cr.

(e) Use (3.7) and (b) to write a matrix equation that expresses y', C_b, C_r in terms of r, g, b.

(f) Use (c) and (e) to write a matrix equation that expresses r, g, b in terms of y', C_b, C_r.

3.18 What is the y', C_b, C_r triple that represents black? Which triple represents white?

3.19 In this problem, you will compute the value of $\cos\theta_H$ where θ_H is the hue angle given in (3.10).

(a) Let $\mathbf{r} = [1, 0, 0]^T$, $\mathbf{c} = \left[\frac{1}{3}, \frac{1}{3}, \frac{1}{3}\right]^T$ and $\mathbf{v} = [r, g, b]^T$. Show that $\|\mathbf{r} - \mathbf{c}\| = \sqrt{\frac{2}{3}}$, $\|\mathbf{v} - \mathbf{c}\| = \sqrt{r^2 + g^2 + b^2 - \frac{1}{3}}$ and $(\mathbf{v} - \mathbf{c}) \cdot (\mathbf{r} - \mathbf{c}) = \frac{2r-g-b}{3}$. Thus

$$\frac{(\mathbf{v} - \mathbf{c}) \cdot (\mathbf{r} - \mathbf{c})}{\|\mathbf{r} - \mathbf{c}\|\|\mathbf{v} - \mathbf{c}\|} = \frac{\sqrt{3}}{\sqrt{2}} \frac{2r - g - b}{3} \frac{1}{\sqrt{r^2 + g^2 + b^2 - \frac{1}{3}}}.$$

(b) Recall that $r = \frac{R}{R+G+B}$, $g = \frac{G}{R+G+B}$ and $b = \frac{B}{R+G+B}$. Use (a) to show that

$$\frac{(\mathbf{v} - \mathbf{c}) \cdot (\mathbf{r} - \mathbf{c})}{\|\mathbf{r} - \mathbf{c}\|\|\mathbf{v} - \mathbf{c}\|} = \frac{\frac{1}{2}(2R - G - B)}{\sqrt{(R - G)^2 + (R - B)(G - B)}}.$$

3.20 Use a calculator or a **CAS** to convert each of the RGB triples $(100, 50, 200)$, $(5, 200, 20)$, $(0, 50, 50)$ to HSI triples.

3.21 Suppose a color image has an RGB space representation whose entries in R are arbitrary and G and B are matrices whose entries are all 0. The image is converted to HSI space. Describe the hue, saturation and intensity matrices H, S and I.

3.22 Repeat the previous problem but now assume that R is matrix comprised of 0 entries and $G = B$.

3.23 Color printers typically use the *CYM color space*. CYM stands for the secondary colors cyan, yellow, and magenta. Suppose that R, G, and B represent the matrices of intensities for red, green, and blue, respectively, and let W be defined as in Problem 3.16. If C, Y, and M represent the matrices of intensities for cyan, yellow, and magenta, respectively, write C, Y, and M in terms of R, G, B, and W. Compare your answer to the process[5] used to create Figure 3.26.

3.24 The *YUV color space* is related to the YCbCr color space. Y is the luminance channel, and the chrominance channels U and V are color difference channels. If (y, u, v) is a triple in YUV space, then y is given by (3.5), $u = 0.492(b - y)$, and $v = 0.877(r - y)$, where $r, g, b \in [0, 1]$ represent the normalized intensity values of a pixel.

(a) Find matrix T so that

$$\begin{bmatrix} y \\ u \\ v \end{bmatrix} = T \cdot \begin{bmatrix} r \\ g \\ b \end{bmatrix}$$

(b) Use a **CAS** to find T^{-1}.

(c) Use the process described in (d) of Problem 3.17 with the formulas for u and v to find T^{-1}.

(d) To see how YUV relates to YCbCr, find diagonal matrix D such that

$$\begin{bmatrix} y \\ u \\ v \end{bmatrix} = D \cdot \begin{bmatrix} y \\ Cb \\ Cr \end{bmatrix} = D \cdot S \begin{bmatrix} r \\ g \\ b \end{bmatrix}$$

where S is given in (b) of Problem 3.17.

3.25 The NTSC also defines the YIQ color space for television sets. This space again uses the *luminance* (Y) channel. The other two channels are *intermodulation* (I) and *quadrature* (Q) as the additional channels.[6] If a pixel in RGB space is given by normalized values $r, g, b \in [0, 1]$, then the corresponding triple (y, i, q) has y given by (3.5), $i = 0.735514(r - y) - 0.267962(b - y)$, and $q = 0.477648(r - y) + 0.412626(b - y)$.

[5] In practice, a black channel is added to the model to better display darker colors. This color space is referred to as the *CYMK* color space.

[6] A black-and-white television set only uses the luminance channel.

(a) Find matrix V so that

$$\begin{bmatrix} y \\ i \\ q \end{bmatrix} = V \cdot \begin{bmatrix} r \\ g \\ b \end{bmatrix}$$

(b) Use a **CAS** to find V^{-1}.

(c) The I and Q channels of YIQ can be obtained from the U and V channels of YUV by rotating the U and V channels 33° counterclockwise and then swapping the results. Use this information in addition to Problem 2.29 from Section 2.2 to find a 2×2 orthogonal matrix R such that

$$\begin{bmatrix} y \\ i \\ q \end{bmatrix} = \left[\begin{array}{c|cc} 1 & 0 & 0 \\ \hline 0 & & \\ 0 & \multicolumn{2}{c}{R} \end{array}\right] \cdot \begin{bmatrix} y \\ u \\ v \end{bmatrix}$$

(d) What is the matrix (in block form) that converts y, u, v to y, i, q?

3.26 Color histogram equalization is a bit more complicated than histogram equalization of grayscale images.

(a) Explain what might go wrong if we perform color histogram equalization by simply performing histogram equalization on each of the red, green, and blue intensity matrices.

(b) In some cases, histogram equalization is applied to the matrix $M = (R + G + B)/3$, where R, G, and B are the red, green, and blue intensity matrices, respectively. The elements of R, G, B are modified using h, where h is the histogram returned by performing histogram equalization on M. Explain the advantage that this process has over the process described in (a).

Computer Lab

3.2 Color Space Transformations. From `stthomas.edu/wavelets`, access the file `ColorTransforms`. In this lab, you will write functions to convert images between RGB space and YCbCr and HSI spaces.

3.3 Huffman Coding

In this section, we discuss a simple method for reducing the number of bits needed to represent a signal or digital image.

Huffman coding, introduced by David Huffman [59], is an example of *lossless compression*.[7] The routine can be applied to integer-valued data and the basic idea

[7] *Lossless compression* is compression whose output can be uncompressed so that the original data are completely recovered.

is quite simple. The method exploits the fact that signals and digital images often contain some elements that occur with a much higher frequency than other elements (consider, for example, the digital image and its corresponding histogram in Figure 3.14). Recall that intensity values from grayscale images are integers that range from 0 to 255, and each integer can be represented with an 8-bit ASCII code (see Table 3.1 or Oualline [73]). Thus, if the dimensions of an image are $N \times M$, then we need $8 \cdot NM$ bits to store it. Instead of insisting that each intensity be represented with eight bits, Huffman suggested that we could use a variable number of bits for each intensity. That is, intensities that occur the most are represented with a low number of bits and intensities occurring infrequently are represented with a larger number of bits.

Generating Huffman Codes

Let's illustrate how to create Huffman codes via an example.

Example 3.8 (Creating Huffman Codes). *Consider the* 5×5 *digital grayscale image given in Figure 3.34. The bit stream for this image is created by writing each character in binary form and then listing them consecutively. Here is the bit stream:*

$$\begin{array}{ccccc} 00100001_2 & 01010000_2 & 01010000_2 & 01010000_2 & 00100001_2 \\ 01010000_2 & 00111000_2 & 01101000_2 & 00111000_2 & 01010000_2 \\ 00111000_2 & 01010000_2 & 01111110_2 & 01010000_2 & 00111000_2 \\ 01010000_2 & 00111000_2 & 01101000_2 & 00111000_2 & 01010000_2 \\ 00100001_2 & 01010000_2 & 01010000_2 & 01010000_2 & 00100001_2 \end{array}$$

Table 3.2 lists the binary representations, ASCII codes, frequencies, and relative frequencies for each intensity. To assign new codes to the intensities, we use a tree

Table 3.2 The ASCII codes, binary representation, frequencies, and relative frequencies for the intensities that appear in the image in Figure 3.34.

Intensity	ASCII Code	Binary Rep.	Frequency	Rel. Frequency
33	!	00100001_2	4	0.16
56	8	00111000_2	6	0.24
80	P	01010000_2	12	0.48
104	h	01101000_2	2	0.08
126	~	01111110_2	1	0.04

diagram. As is evident in Figure 3.35, the first tree is simply one line of nodes that lists the intensities and their relative frequencies in nondecreasing order. Note that for this example, all the relative frequencies were distinct. In the case that some characters have the same relative frequency, the order in which you list these characters does not matter – you need only ensure that all relative frequencies are listed in nondecreasing order.

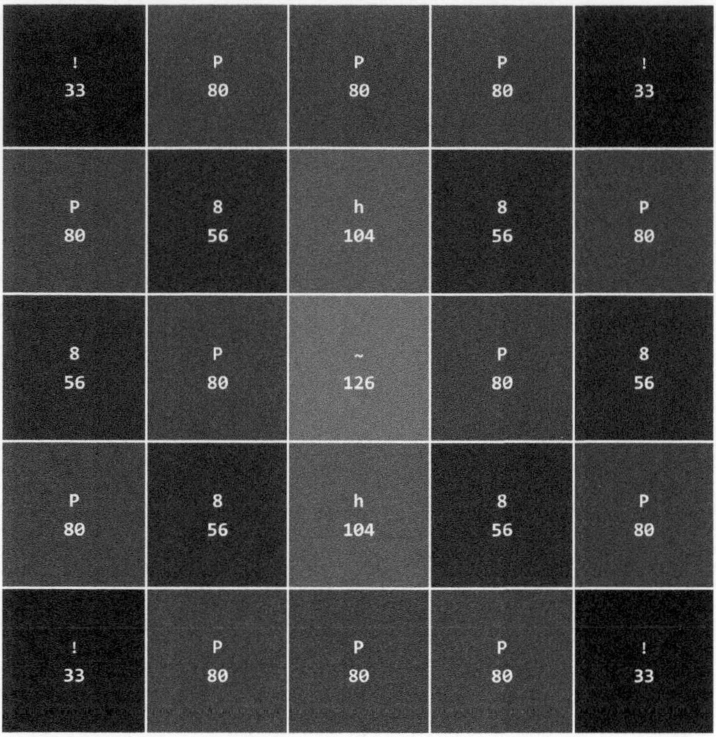

Figure 3.34 A 5×5 grayscale image with its intensity levels superimposed.

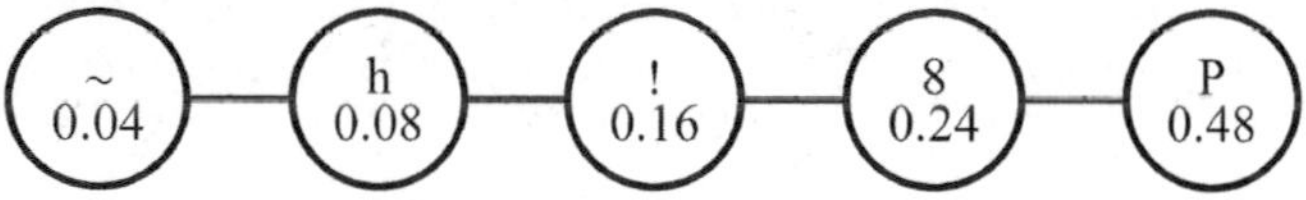

Figure 3.35 The first tree in the Huffman coding scheme. The characters and their associated relative frequencies are arranged so that the relative frequencies are in nondecreasing order.

The next step in the process is to create a new node from the two leftmost nodes in the tree. The probability of this new node is the sum of the probabilities of the two leftmost nodes. In our case, the new node is assigned the relative frequency $0.12 = 0.04 + 0.08$*, which is the sum of the relative frequencies of the two leftmost nodes. Thus, the first level of our tree now has four nodes (again arranged in nondecreasing order), and the leftmost node spawns two new nodes on a second level as illustrated in Figure 3.36.*

We repeat the process and again take the two leftmost nodes and replace them with a single node. The relative frequency of this new node is the sum of the relative

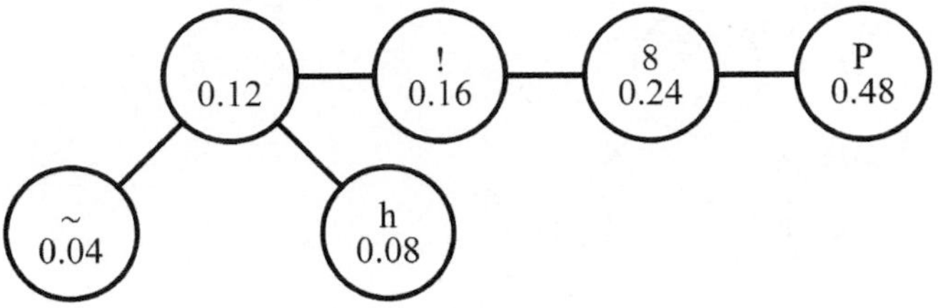

Figure 3.36 The second tree in the Huffman coding scheme.

frequencies of these two leftmost nodes. In our example, the new node is assigned the relative frequency $0.28 = 0.12 + 0.16$*. We again arrange the first level so that the relative frequencies are nondecreasing. Note that this new node is larger than the node with frequency* 0.24*. Our tree now consists of three levels, as illustrated in Figure 3.37(a). The fourth iteration is plotted in Figure 3.37(b).*

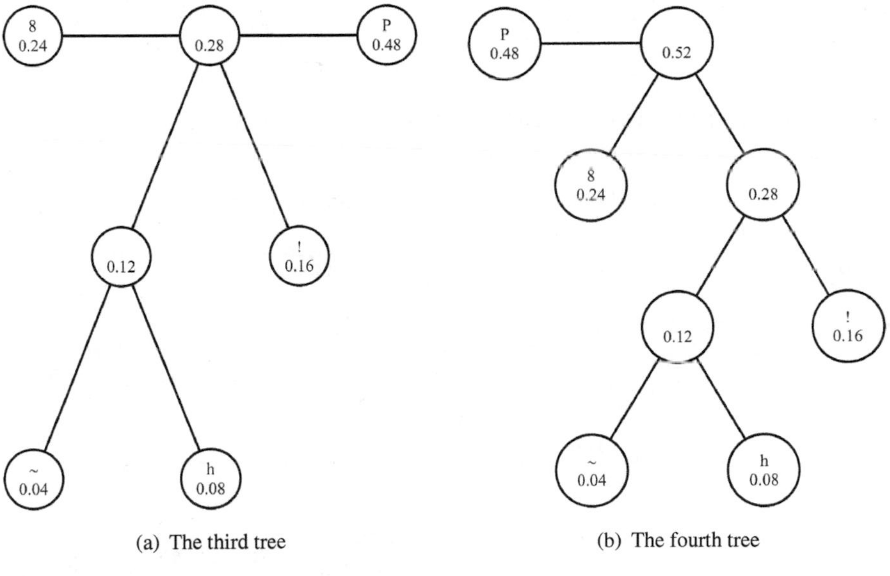

Figure 3.37 The third and fourth steps in the Huffman coding algorithm.

The final Huffman tree is plotted in Figure 3.38. Notice that we label branches at each level using 0 *for left branches and* 1 *for right branches. From the top of the tree, we can use these numbers to describe a path to a particular character. This description is nothing more than a binary number, and thus the Huffman code for the particular character. Table 3.3 shows the Huffman codes for each character.*

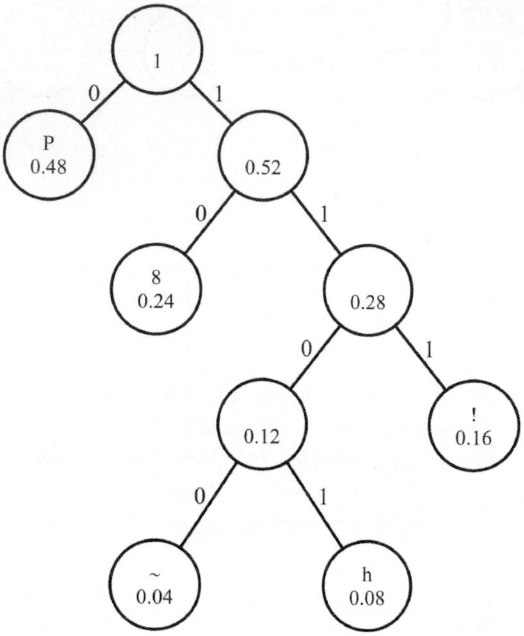

Figure 3.38 The final tree in the Huffman coding algorithm.

Table 3.3 The Huffman code, frequency, and total bits required are given for each character. The total bits needed to encode the image is given as well.

Character	Huffman Code	Frequency	Total Bits
!	111_2	4	12
8	10_2	6	12
P	0_2	12	12
h	1101_2	2	8
~	1100_2	1	4
	Bits needed to encode the image:		**48**

Here is the new bit stream using the Huffman codes:

$$\begin{array}{rrrrr} 111_2 & 0_2 & 0_2 & 0_2 & 111_2 \\ 0_2 & 10_2 & 1101_2 & 10_2 & 0_2 \\ 10_2 & 0_2 & 1100_2 & 0_2 & 10_2 \\ 0_2 & 10_2 & 1101_2 & 10_2 & 0_2 \\ 111_2 & 0_2 & 0_2 & 0_2 & 111_2 \end{array}$$

Note that the characters that appear the most have the least number of bits assigned to their new code. Moreover, the savings is substantial. Using variable-length codes for the characters, the Huffman scheme needs 48 *bits plus the code dictionary to represent the image. With no coding, each character requires eight bits. Since our image consists of* 25 *pixels, without Huffman coding, we would need* 200 *bits to represent it.*

(Live example: Visit `stthomas.edu/wavelets` *and click on Live Examples.)*

□

Algorithm for Generating Huffman Codes

The general algorithm for creating Huffman codes for the elements in a set appears below.

Algorithm 3.1 (Huffman Code Generation). *Given set A, this algorithm generates the Huffman codes for each distinct element in A. Here are the basic steps:*

1. *Compute the relative frequency of each distinct element in A.*
2. *Create a list of nodes where each node contains a distinct element and its relative frequency. Designate these nodes as* parentless.
3. *Sort the parentless nodes, lowest to highest, based on relative frequencies.*
4. *Select the two left-most nodes and create a new parentless node.*
5. *This new node is considered to be a* parent *and is connected to its children via* branches.
6. *Set the relative frequency of the new node as the sum of the children's relative frequencies.*
7. *Label the left branch with* 0 *and the right branch with* 1.
8. *Repeat steps three through six until only one parentless node remains. The relative frequency of this last parentless node is* 1.
9. *For each distinct element $a \in A$, assign a binary string based on the branches that connect the remaining parentless node to the node containing a.*

□

Not only is Huffman coding easy to implement, but as Gonzalez and Woods point out in [48], if we consider all the variable-bit-length coding schemes we could apply to a signal of fixed length, where the scheme is applied to one element of the signal at a time, then the Huffman scheme is optimal. There are several variations of Huffman coding as well as other types of coding. Please see Gonzalez and Woods [48] or Sayood [80] (or references therein) if you are interested in learning more about coding.

Decoding

Decoding Huffman codes is quite simple. You need only the code tree and the bit stream. Consider the following example.

Example 3.9 (Decoding Huffman Codes). *Consider the bit stream*

$$1001111001101011110011 00_2$$

that is built using the coding tree plotted in Figure 3.39. The stream represents some text. Determine the text.

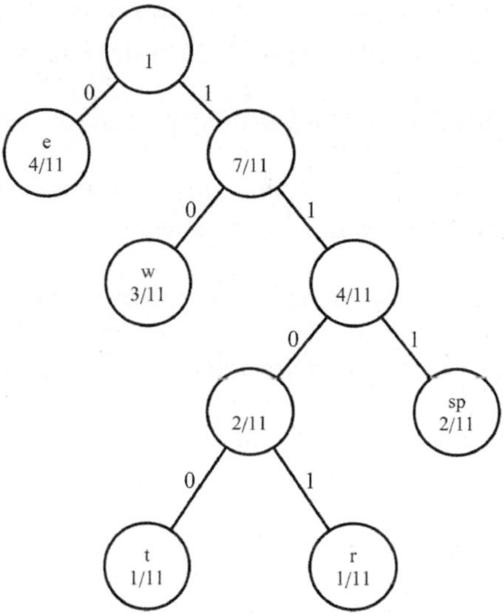

Figure 3.39 The Huffman code tree for Example 3.9. Here *"sp"* denotes a space.

Solution.
From the tree, we see that e $= 0_2$*, w* $= 10_2$*, sp* $=$ " " $= 111_2$*, r* $= 1101_2$*, and t* $= 1100_2$*. We begin the decoding process by examining the first few characters until we find a code for one of the letters above. Note that no character has* (1_2) *for a code, so we look at the first two characters* (10_2)*. This is the code for* w*, so we now have the first letter. We continue in this manner taking only as many bits as needed to represent a character in the tree. The next bit is* 0_2*, which is the code for e. The next code needs three bits –* 111_2 *is the space key. The next few characters are* w (10_2)*, e* (0_2)*, r* (1101_2)*, and e* (0_2)*. The last four characters are the space key* (111_2)*, then* w*, e, and t, so that the coded string is* "we were wet."

(Live example: Visit `stthomas.edu/wavelets` *and click on Live Examples.)*

□

Bits per Pixel

In this book, we are primarily interested in using Huffman codes to assess the effectiveness of wavelet transformations in image compression. We usually measure this effectiveness by computing the *bits per pixel* (bpp) needed to store the image.

Definition 3.1 (Bits per Pixel). *We define the* bits per pixel *of an image, denoted as* bpp, *as the number of bits used to represent the image divided by the number of pixels in the image.*

□

For example, the 5×5 image in Figure 3.34 is originally coded using 8 bpp. Since Huffman coding resulted in a bit stream of length 48, we say that we can code the image via the Huffman scheme using an average of $48/25 = 1.92$ bpp. As you will soon see, wavelets can drastically reduce this average.

PROBLEMS

3.27 Consider the string *mississippi*.

(a) Generate the Huffman code tree for the string.

(b) Look up the ASCII value for each character in *mississippi*, convert them to base 2, and then write the bit stream for *mississippi*.

(c) Write the bit stream for the string using the Huffman codes.

3.28 Repeat Problem 3.27 for the string *terminal*.

3.29 Consider the 4×4 image whose intensity matrix is

$$\begin{matrix} 100 & 100 & 120 & 100 \\ 100 & 50 & 50 & 40 \\ 100 & 40 & 40 & 50 \\ 120 & 120 & 100 & 100 \end{matrix}$$

(a) Generate the Huffman code tree for the image.

(b) Write the bit stream for the image using Huffman codes.

(c) Compute the bpp for this bit stream.

3.30 Suppose that you wish to draw the Huffman tree for a set of letters S. Let $s \in S$. Explain why the node for s cannot spawn new branches of the tree. Why is this fact important?

3.31 Suppose that a 40×50 image consists of the four intensities 50, 100, 150, and 200 with relative frequencies $\frac{1}{8}$, $\frac{1}{8}$, $\frac{1}{4}$, and $\frac{1}{2}$, respectively.

(a) Create the Huffman tree for the image.

(b) What is the bpp for the Huffman bit stream representation of the image?

3.32 Consider the Huffman codes $e = 0_2$, $n = 10_2$, $s = 111_2$, and $T = 110_2$. The frequencies for e, n, s and T are $4, 2, 2$ and 1, respectively.

(a) Draw the Huffman code tree for these codes.

(b) Are the given frequencies the only frequencies that would result in the tree you plotted in (a)? If not, find other frequencies for the letters that would result in a tree with the same structure.

(c) Can you find a proper noun that uses all these letters with the given frequencies?

3.33 Given the Huffman codes $g = 10_2$, $n = 01_2$, $o = 00_2$, *space key* $= 110_2$, $e = 1110_2$, and $i = 1111_2$, draw the Huffman code tree and use it to decode the bit stream

$$100011110110110100011110110110100001111 0_2.$$

3.4 Qualitative and Quantitative Measures

Image compression is a popular application of wavelet transformations. In lossless compression, the goal is simply to reduce the storage size of the compressed version of the original image. In lossy compression, we sacrifice resolution in hopes of further reducing the space necessary to store the compressed image. Some natural questions arise for both compression methods. For example, if we compress an image, how do we measure the efficacy of the compression scheme? Alternatively, is it possible to determine if our compression scheme is (near) optimal? Finally, can we find a way to quantize our data for use in a lossy compression method?

We study three such measures in this section. *Entropy* produces the optimal bits per pixel value for lossless compression of given data. The *peak signal-to-noise ratio* measures how well a compressed image approximates the original image. The *cumulative energy vector* of a matrix or vector describes how the intensities of the input elements are distributed and can be used in lossy compression methods.

Entropy

We start by defining the entropy of a set of numbers. Entropy is used to assess the effectiveness of compression algorithms, and its definition plays an important role in information theory. Claude Shannon pioneered much of the work in this area and in his landmark 1948 paper [81] he showed that the best that we can hope for when compressing data via a lossless compression method S is to encode the output of S where the average number of bits per character equals the entropy of S. Let's define entropy and then look at some examples.

Definition 3.2 (Entropy). *Let S be a set of n numbers. Suppose that there are k distinct values in S. Denote these distinct values by $a_1, \ldots, a_k$, then let A_i, $i = 1, \ldots, k$, be the set of elements in S that equal a_i. For each $i = 1, \ldots, k$, let $p(a_i)$ represent the relative frequency of a_i in $\mathbf{v}$. That is, $p(a_i)$ is the number of elements in A_i divided by n. We define the entropy of S by*

$$\boxed{\text{Ent}(S) = \sum_{i=1}^{k} p(a_i) \log_2 \left(1/p(a_i)\right).} \tag{3.16}$$

□

Note that (3.16) is equivalent to

$$\text{Ent}(S) = -\sum_{i=1}^{k} p(a_i) \log_2 \left(p(a_i)\right).$$

Before investigating entropy via examples, let's understand what it is trying to measure. The factor $\log_2(1/p(a_i))$ is an exponent – it measures the power of two we need to represent $1/p(a_i)$. Base 2 is used so that this factor approximates the number of bits we need to represent a_i. For example, if $p(a_i) = \frac{1}{16}$, then $\log_2(1/p(a_i)) = \log_2 16 = 4$. This makes sense if we consider the number of bits (four) needed to represent a_i if $p(a_i) = \frac{1}{16}$. Multiplying this value by $p(a_i)$ gives us the percentage of bits necessary to represent a_i throughout the entire set S. Summing over all distinct values gives us an optimal number of bits necessary to compress the values in S.

If all the elements of $S = \{s_1, \ldots, s_n\}$ are distinct, then $k = n$, $a_i = s_i$, and $p(a_i) = \frac{1}{n}$, $i = 1, \ldots, n$. In this case

$$\begin{aligned} \text{Ent}(S) &= -\sum_{i=1}^{n} \frac{1}{n} \log_2(1/n) \\ &= -n \cdot \frac{1}{n} \cdot \log_2(1/n) \\ &= \log_2(n). \end{aligned} \tag{3.17}$$

If all the elements of S have the same value c, then $k = 1$ with $a_1 = c$, $p(a_1) = 1$, and $\log_2(p(a_1)) = \log_2(1) = 0$ so that

$$\text{Ent}(\mathbf{v}) = 0 \tag{3.18}$$

In Problem 3.39 you will show that for any set S, $\text{Ent}(S) \geq 0$ so that sets with consisting of one distinct element minimizes entropy. In Problems 3.41 – 3.45 you will show that (3.17) gives the largest possible value for entropy. Thus sets whose elements are all distinct maximize entropy. This should make sense to you – vectors consisting of a single value should be easiest to encode, and vectors with distinct components should require the most information to encode. Let's look at some examples.

Example 3.10 (Entropy). *Find the entropy of each of the following vectors and matrix.*

(a) $\mathbf{u} = [1, 1, 1, 1, 2, 2, 3, 4]^T$

(b) $\mathbf{v} = [1, 0, 0, \ldots, 0]^T \in \mathbb{R}^{65}$

(c) $\mathbf{w} = [1, 2, 3, 4, 0, 0, 0, 0]^T$

(d) The matrix from Example 3.8.

Solution.
For $\mathbf{u}$, *we have* $k = 4$ *with* $p(a_1) = \frac{1}{2}$, $p(a_2) = \frac{1}{4}$, *and* $p(a_3) = p(a_4) = \frac{1}{8}$. *We can compute the entropy as*

$$\begin{aligned} Ent(\mathbf{u}) &= \frac{1}{2}\log_2(2) + \frac{1}{4}\log_2(4) + \frac{1}{8}\log_2(8) + \frac{1}{8}\log_2 8 \\ &= \frac{1}{2} + \frac{1}{2} + \frac{3}{8} + \frac{3}{8} \\ &= 1.75. \end{aligned}$$

For $\mathbf{v}$, *we have* $k = 2$ *with* $p(a_1) = \frac{1}{65}$ *and* $p(a_2) = \frac{64}{65}$. *The entropy is*

$$\begin{aligned} Ent(\mathbf{v}) &= \frac{1}{65}\log_2(65) + \frac{64}{65}\log_2\left(\frac{65}{64}\right) \\ &= \frac{1}{65}\log_2(65) + \left(\frac{64}{65}\right)(\log_2(65) - \log_2(64)) \\ &= \log_2(65) - \left(\frac{64}{65}\right)\log_2(64) \\ &= \log_2 65 - 6\cdot\frac{64}{65} \\ &\approx 0.1147. \end{aligned}$$

The vector $\mathbf{w}$ *has* $k = 5$ *with* $p(a_1) = \cdots = p(a_4) = \frac{1}{8}$ *and* $p(a_5) = \frac{1}{2}$. *We compute the entropy as*

$$Ent(\mathbf{w}) = 4\cdot\frac{1}{8}\log_2(8) + \frac{1}{2}\log_2(2) = \frac{3}{2} + \frac{1}{2} = 2$$

Finally, denote the matrix in Example 3.8 as A *and note that consists of* $k = 5$ *distinct elements with relative frequencies are given in Table 3.2. Thus the entropy is*

$$\begin{aligned} Ent(A) = &-(0.16\log_2(0.16) + 0.24\log_2(0.24) + 0.48\log_2(0.48) \\ &+0.08\log_2(0.08) + 0.04\log_2(0.04)) \\ \approx &\ 1.90268. \end{aligned}$$

Using Table 3.3, we can compute the bits per pixel of A to be $48/25 = 1.92$ *so that we see that the Huffman-coded version of A is very close to optimal.*
(Live example: Visit `stthomas.edu/wavelets` *and click on Live Examples.)*

□

Cumulative Energy

We now move to the idea of cumulative energy. Roughly speaking, the cumulative energy of vector $\mathbf{v}$ is a new vector that indicates how the size of the components of $\mathbf{v}$ are distributed.

Definition 3.3 (Cumulative Energy). *Let* $\mathbf{v} \in \mathbb{R}^n$, $\mathbf{v} \neq \mathbf{0}$. *Form a new vector* $\mathbf{y} \in \mathbb{R}^n$ *by taking the absolute value of each component of* $\mathbf{v}$ *and then sorting them largest to smallest. Thus,* $y_1 \geq y_2 \geq \cdots \geq y_n \geq 0$.

Define the cumulative energy *of* $\mathbf{v}$ *as the vector* $\mathbf{C}_e(\mathbf{v}) \in \mathbb{R}^n$ *whose components are given by*

$$\boxed{\mathbf{C}_e(\mathbf{v})_k = \sum_{i=1}^{k} \frac{y_i^2}{\|\mathbf{y}\|^2}, \qquad k = 1, \ldots, n.}$$

□

Some remarks about Definition 3.3 are in order. First, we observe that since the components of $\mathbf{y}$ are formed by taking the absolute value of the components of $\mathbf{v}$ and then ranking them, the norms are the same. That is $\|\mathbf{v}\| = \|\mathbf{y}\|$. Also, if we recall that $\|\mathbf{y}\|^2 = y_1^2 + y_2^2 + \cdots + y_n^2$, we see that

$$\mathbf{C}_e(\mathbf{v})_k = \frac{y_1^2 + y_2^2 + \cdots + y_k^2}{y_1^2 + y_2^2 + \cdots + y_k^2 + \ldots + y_n^2}$$

so that $\mathbf{C}_e(\mathbf{v})_k$ is basically the percentage of the largest k components of $\mathbf{v}$ (in absolute value) in terms of $\|\mathbf{v}\|^2$.

We also note that $0 \leq \mathbf{C}_e(\mathbf{v})_k \leq 1$ for every $k = 1, 2, \ldots, n$ with $\mathbf{C}_e(\mathbf{v})_n = 1$. Let's compute the cumulative energy of two vectors.

Example 3.11 (Cumulative Energy). *Find the cumulative energy of*

$$\mathbf{v} = [1, 2, 3, \ldots, 10]^T \qquad \textit{and} \qquad \mathbf{w} = [1, 3, 5, 7, 9, 1, 1, 1, 1, 1]^T$$

Solution.
Let's take a closer look at $\mathbf{w}$. *As you learned in Problem 2.32, there exists a nonsingular matrix A such that* $A\mathbf{v} = \mathbf{w}$. *Certainly* $\mathbf{w}$ *would have a shorter Huffman-encoded bit stream that* $\mathbf{v}$. *If we wished to store* $\mathbf{v}$*, it would be more efficient to map* $\mathbf{v}$ *to* $\mathbf{w}$ *and then Huffman-encode* $\mathbf{w}$. *The vector* $\mathbf{v}$ *could be recovered by reversing these steps. Let's see now how the cumulative energies of the two vectors compare.*

We have

$$\|\mathbf{v}\|^2 = 385 \qquad \textit{and} \qquad \|\mathbf{w}\|^2 = 170.$$

We compute the cumulative energies of $\mathbf{v}$ *and* $\mathbf{w}$ *to be*

$$\begin{aligned}\mathbf{C}_e(\mathbf{v}) &= \frac{1}{385}\left[100, 181, 245, 294, 330, 355, 371, 380, 384, 385\right]^T \\ &= \left[0.2597, 0.4701, 0.6364, 0.7636, 0.8571, 0.9221, 0.9636, 0.987, 0.9974, 1\right]^T\end{aligned}$$

and

$$\begin{aligned}\mathbf{C}_e(\mathbf{w}) &= \frac{1}{170}\left[81, 130, 155, 164, 165, 166, 167, 168, 169, 170\right]^T \\ &= \left[.4765, .7647, .9118, .9647, .9706, .9765, .9824, .9882, .9941, 1\right]^T.\end{aligned}$$

Plotting the cumulative energy vectors allows us to better see the distributions of the magnitudes of the components of $\mathbf{v}$ *and* $\mathbf{w}$*. These graphs appear in Figure 3.40.*

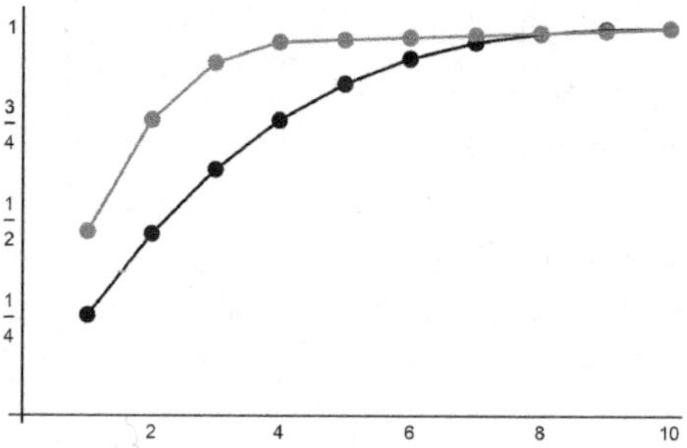

Figure 3.40 A plot of $\mathbf{C}_e(\mathbf{v})$ (black) and $\mathbf{C}_e(\mathbf{w})$ (gray).

We see that over 90% (91.1765%) of the energy of $\mathbf{w}$ *is contained in the largest three components of* $\mathbf{w}$*, whereas it takes the six largest components of* $\mathbf{v}$ *to produce over 90% (92.2078%) of the cumulative energy of* $\mathbf{v}$*.*

The energy is much more spread out in $\mathbf{v}$ *than is the case in* $\mathbf{w}$*.*

(Live example: Visit `stthomas.edu/wavelets` *and click on Live Examples.)*

□

A vector that is sparse (many elements with values at or near 0) results in a cumulative energy vector whose elements rapidly approach 1. We can exploit this fact and convert to 0 those elements that do not contribute to the energy of the vector. The application of Huffman coding at this point will be quite efficient. The trade off of course, is that we cannot recover our original signal.

We have defined cumulative energy for a vector but we can certainly compute the cumulative energy vector of a $M \times N$ matrix A by simply concatenating the rows of A to produce a vector of length MN.

Peak Signal-to-Noise Ratio

The last measure we discuss is used primarily in image processing and is called the peak signal-to-noise ratio (PSNR). In practice we will use the PSNR to compare an original image with a compressed version of itself.

Suppose that we reconstruct an image using some compression technique and we wish to measure the efficiency of the approximation. The PSNR is often used for this purpose.

Definition 3.4 (Peak Signal-to-Noise Ratio). *Let A and B be $M \times N$ matrices with entries a_{ij} and b_{ij}, respectively. We first define the* error function $\mathrm{Err}(A, B)$ *as*

$$\mathrm{Err}(A, B) = \frac{1}{MN}\sum_{i=1}^{M}\sum_{j=1}^{N}(a_{ij} - b_{ij})^2 \tag{3.19}$$

and next define the peak signal-to-noise ratio *(PSNR)as*

$$\mathrm{PSNR}(A, B) = 10\log_{10}\left(\frac{255^2}{\mathrm{Err}(A, B)}\right) \tag{3.20}$$

Here, PSNR *is measured in decibels (dB).* □

We use (3.20) when considering matrices A and B whose elements are integers that represent intensity values. These values are integers in the range $0, \ldots, 255$. The worst possible value for $\mathrm{Err}(A, B)$ occurs when *each* difference $a_{ij} - b_{ij} = \pm 255$ so that

$$\mathrm{Err}(A, B) = \frac{1}{MN}\sum_{i=1}^{M}\sum_{j=1}^{N}(a_{ij} - b_{ij})^2 = \frac{1}{MN}\sum_{i=1}^{M}\sum_{j=1}^{N}255^2 = 255^2.$$

The value 255 is the *peak* error and that is why 255^2 appears in the numerator of the argument of the log function in (3.20). For the worst possible error, the ratio of the peak value squared and the error function is 1. In this case the log is 0 as is the PSNR. As the error decreases in size, the ratio as well as the PSNR increases. So for image compression applications, we find larger values of PSNR desirable.

Example 3.12 (Peak Signal-To-Noise Ratio). *The image A in Figure 3.41(a) is the original 512×768 grayscale image. We have used two different compression methods to produce image B in Figure 3.41(b) and image C in Figure 3.41(c). In particular, to obtain B (C), we rounded all pixel values to the nearest multiple of 10 (20).*

We utilize (3.19) to compute $\mathrm{Err}(A, B) = 8.495$ and $\mathrm{Err}(A, C) = 32.603$. We then use (3.20) and find that $\mathrm{PSNR}(A, B) = 38.839$ and $\mathrm{PSNR}(A, C) = 32.998$. As you can see, the higher PSNR is associated with the middle image B and image B is a better representation of the original image than image C.

(Live example: Visit `stthomas.edu/wavelets` and click on Live Examples.)

□

(a) The digital image A (b) The approximate B (c) The approximate C

Figure 3.41 Peak signal-to-noise ratio.

PROBLEMS

3.34 Find the entropy of each of the following vectors. A calculator, **CAS**, or the software package that accompanies this book will be useful.

(a) $\mathbf{a} = [1, 1, 1, 1, 0, 0, 0, 0]^T$

(b) $\mathbf{b} = [1, 1, 0, 0, 1, 0, 1, 0]^T$

(c) $\mathbf{c} = [1, 1, 1, 2, 2, 3]^T$

(d) $\mathbf{d} = [8, 8, 8, 8, 0, 0, 0, 0]^T$

(e) $\mathbf{e} = [1, 2, 3, \ldots, 15, 16]^T$

(f) $\mathbf{f} = [0, 0, 43, 0, 0, 0]^T$

(g) $\mathbf{g} = [0, 0, 0, 51.265, 0, 0]^T$

Can you draw any conclusions about the entropies you have computed?

3.35 Consider the image described in Problem 3.31. You will need to have completed (b) of this exercise to continue.

(a) What is the entropy of the image?

(b) Recall in the discussion preceding Definition 3.2 that Shannon [81] showed that the best bpp we could hope for when compressing a signal is the signal's entropy. Does your answer to (a) agree with Shannon's postulate?

3.36 Let $\mathbf{v} \in \mathbb{R}^n$ and let $c \neq 0$. Show that $\text{Ent}(c\mathbf{v}) = \text{Ent}(\mathbf{v})$.

3.37 Suppose that $0 < v_1 < v_2 < \cdots < v_n$, where $n \geq 4$ is an even integer. From (3.17), we know that $\text{Ent}(\mathbf{v}) = \log_2(n)$. Suppose we form the first $\frac{n}{2}$ elements in $\mathbf{w} \in \mathbb{R}^n$ by the formula $w_k = \frac{v_{2k}+v_{2k-1}}{2}$, $k = 1, \ldots, \frac{n}{2}$ and complete $\mathbf{w}$ by making the remaining $\frac{n}{2}$ elements 0. Show that $\text{Ent}(\mathbf{w}) = \frac{1}{2}\text{Ent}(\mathbf{v}) + \frac{1}{2}$. Thus, the bpp needed to store $\mathbf{w}$ should be roughly half that needed to store $\mathbf{v}$.

★**3.38** Let $\mathbf{v} = [1, 2, 3, \ldots, n]^T$ where $n \geq 4$ is an even integer. Form the first $\frac{n}{2}$ components of $\mathbf{w}$ by the formula $w_k = \frac{v_{2k}+v_{2k-1}}{2}$, $k = 1, \ldots, \frac{n}{2}$ and the last $\frac{n}{2}$ components of $\mathbf{w}$ by the formula $w_{k+n/2} = \frac{v_{2k}-v_{2k-1}}{2}$, $k = 1, \ldots, \frac{n}{2}$. Show that $\text{Ent}(\mathbf{w}) = \frac{1}{2}\text{Ent}(\mathbf{v}) + \frac{1}{2}$. Would the result change if we multiplied each element of $\mathbf{v}$ by $m \neq 0$?

3.39 In this problem you will show that for all vectors $\mathbf{v} \in \mathbb{R}^n$, $\text{Ent}(\mathbf{v}) \geq 0$.

(a) For $i = 1, \ldots, k$, what can you say about the range of values of $p\,(a_i)$? The range of $\frac{1}{p(a_i)}$? The range of $\log_2\left(1/p\,(a_i)\right)$?

(b) Use (a) to show that $\text{Ent}(v) \geq 0$.

(c) We learned (see the discussion that leads to (3.18)) that vectors $\mathbf{v} \in \mathbf{R}^n$ consisting of a single value satisfy $\text{Ent}(\mathbf{v}) = 0$. Can any nonconstant vector $\mathbf{w}$ satisfy $\text{Ent}(\mathbf{w}) = 0$? Why or why not?

Note: Problems 3.40 – 3.45 were developed from a report by Conrad [27]. In the report, Conrad gives a nice argument that shows that the maximum entropy value given by (3.17) *is attained if and only if the elements of* $\mathbf{v}$ *are distinct.*[8]

3.40 Let $t > 0$ and consider the function $f(t) = t - 1 - \ln t$. Note that $f(1) = 0$. Use calculus to show that $f(t) > 0$ whenever $t \neq 1$.

3.41 Let $x, y > 0$. Show that the inequality

$$y - y\ln y \leq x - y\ln x \tag{3.21}$$

is equivalent to

$$\frac{x}{y} - 1 - \ln\left(\frac{x}{y}\right) \geq 0 \tag{3.22}$$

and then use Problem 3.40 to show that (3.22) (and hence (3.21)) is always true for $x, y > 0$ with equality if and only if $x = y$.

3.42 Suppose that $p_k > 0$ and $q_k > 0$, $k = 1, \ldots m$, and that the numbers satisfy

$$\sum_{k=1}^{m} p_k = \sum_{k=1}^{m} q_k = 1. \tag{3.23}$$

Show that

$$-\sum_{k=1}^{m} p_k \ln\left(p_k\right) \leq -\sum_{k=1}^{m} p_k \ln\left(q_k\right). \tag{3.24}$$

Hint: Apply (3.21) to each pair p_k, q_k and then sum over the indices $k = 1, \ldots, m$.

[8] I am indebted to Brad Dallas for pointing out some errors in the first edition's formulation of these problems and suggestions for improving the presentation in this edition.

3.43 In Problem 3.42 show that equality holds in (3.24) if and only if $p_k = q_k$. *Hint:* It is easy to observe that if $p_k = q_k$ then (3.24) becomes an equality. Now assume

$$-\sum_{k=1}^{m} p_k \ln(p_k) = -\sum_{k=1}^{m} p_k \ln(q_k). \tag{3.25}$$

(i) Use (3.23) and (3.25) to show that

$$\sum_{k=1}^{m} q_k - p_k \ln(q_k) - (p_k - p_k \ln(p_k)) = 0.$$

(ii) Explain how (3.21) allows us to infer that $p_k - p_k \ln(p_k) = q_k - p_k \ln(q_k)$.

(iii) Use Problem 3.41 to explain why $p_k = q_k$.

3.44 Let $q_k = \frac{1}{m}$, $p_k > 0$, $k = 1, \ldots, m$ with $\sum\limits_{k=1}^{m} p_k = 1$. Use Problems 3.42 and 3.43 to show that

$$-\sum_{k=1}^{m} p_k \log_2(p_k) \leq \log_2(m) \tag{3.26}$$

with equality if and only if $p_k = \frac{1}{m}$.

3.45 Let S be a set of n elements. We can view (3.26) as $\text{Ent}(S)$ where $m \leq n$ is the number of distinct elements in S. Use Problem 3.44 and the fact that $f(t) = \log_2 t$ is an increasing function to show that $\text{Ent}(S) \leq \log_2 n$ with equality if and only if the elements of S are distinct.

3.46 Can you create an image so that the bpp for the Huffman bit stream is equal to the entropy of the image?

3.47 Find $\mathbf{C}_e(\mathbf{v})$ for each of the following. A calculator, **CAS**, or the software package that accompanies this text will be useful.

(a) $\mathbf{v} = [\,2, -1, 3, 0, 6, -4\,]^T$

(b) $\mathbf{v} = [\,1, 1, 1, 1, 1, 1, 1, 1, 1, 1\,]^T$

(c) $\mathbf{v} = \left[\; 1, 2, 3, \ldots, 10, \underbrace{0, \ldots, 0}_{10 \text{ zeros}} \;\right]^T$

3.48 Find $\mathbf{C}_e(\mathbf{v})$ when $\mathbf{v} = [\,v_1, \ldots, v_n\,]^T$, where

(a) $v_k = c$, $k = 1, \ldots, n$, where $c \neq 0$ is any real number

(b) $v_k = k$, $k = 1, \ldots, n$

(c) $v_k = \sqrt{k}$, $k = 1, \ldots, n$

(*Hint:* Problem 2.3 from Section 2.1 will be useful.)

3.49 Let $\mathbf{v} \in \mathbb{R}^n$, $c \neq 0$ any real number, and define $\mathbf{w} = c\mathbf{v}$. Show that $\mathbf{C}_e(\mathbf{v}) = \mathbf{C}_e(\mathbf{w})$.

3.50 Let

$$A = \begin{bmatrix} 2 & -1 & 0 & 4 \\ 1 & 0 & -3 & 2 \\ -3 & 1 & 2 & 1 \\ 0 & 1 & 1 & 0 \end{bmatrix} \quad \text{and} \quad B = \begin{bmatrix} 3 & -1 & 1 & 6 \\ 0 & 0 & -4 & 1 \\ -4 & 3 & 0 & 0 \\ 0 & -1 & -1 & 1 \end{bmatrix}$$

Find $\text{Err}(A, B)$ and $\text{PSNR}(A, B)$.

3.51 Show that $\text{PSNR}(A, B) = 20\log_{10}(255) - 10\log_{10}(\text{Err}(A, B))$.

3.52 Suppose that A, C_1, and C_2 are $n \times n$ matrices. The elements of A are original grayscale values for a digital image, while C_1 and C_2 are approximations of A that have been created by two different compression schemes. Suppose that $\text{Err}\,(A, C_1) = \epsilon$ and $\text{Err}\,(A, C_2) = \epsilon/10$ for $\epsilon > 0$. Show that $\text{PSNR}\,(A, C_2) = \text{PSNR}\,(A, C_1) + 10$.

3.53 A good PSNR value for compressed images that are used in web applications is 35. Use a calculator or a **CAS** to approximate $\text{Err}(A, B)$ in this case.

Computer Labs

3.3 Entropy and PSNR. From `stthomas.edu/wavelets`, access the file `EntropyPSNR`. In this lab, you will write functions to compute the entropy of a vector or matrix and the PSNR of two matrices.

3.4 Cumulative Energy. From `stthomas.edu/wavelets`, access the file `CumulativeEnergy`. In this lab, you will write functions to compute the cumulative energy of a vector or matrix.

CHAPTER 4

THE HAAR WAVELET TRANSFORMATION

Consider the 512×512 image plotted in Figure 4.1(a). In its raw format, it is stored at 8 bpp for a total of $512^2 \cdot 8 = 2{,}097{,}152$ bits. Based on what we learned in Section 3.3, we should be able to reduce the space needed to store the image if we create Huffman codes for the intensities in the image. We compute the Huffman codes for the intensities in the image and find that the new bits per pixel for the encoded data is ... 7.585 bpp or a savings of only 5.18%.

What went wrong? The main reason for performing Huffman coding is to reduce the size of storage space. However, Huffman coding works best if relatively few intensities comprise the majority of the image. A look at Figure 4.1(b) shows that the distribution of the intensities of A is quite spread out. Thus Huffman coding will not provide a substantial savings in terms of storage space.

Suppose we construct an invertible transformation T such that $T(A)$ has the distribution of intensities show in Figure 4.2. Such a transformation exists and in this case, the Huffman-encoded version of $T(A)$ is 2.77 bpp. We could save the Huffman-encoded version of $T(A)$ and if we want to recover A, we simple decode the Huffman codes for $T(A)$ and then apply T^{-1}.

We have stipulated that our transformation be invertible and concentrate the distribution of intensities over a small interval. Since image files are often large, it is

Discrete Wavelet Transformations: An Elementary Approach With Applications, Second Edition.
Patrick J. Van Fleet.

(a) A digital image A.

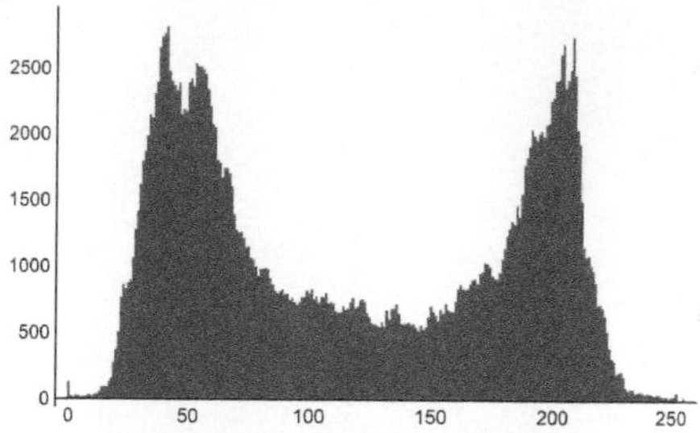

(b) The distribution of intensities in A.

Figure 4.1 An image and the distribution of its intensities.

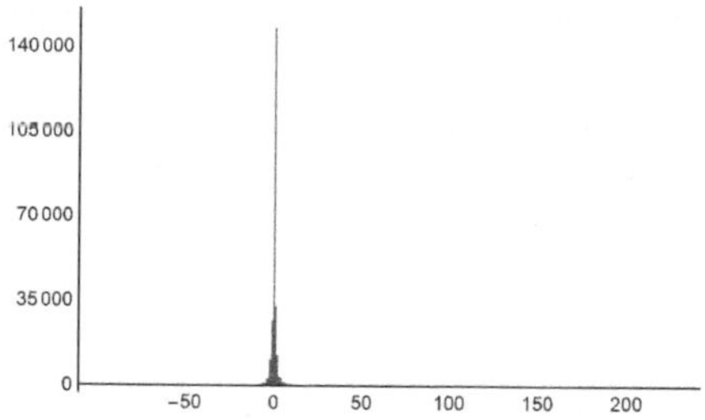

Figure 4.2 The distribution of intensities of the transformed data $T(A)$.

important that the transformation is computationally efficient. It turns out a type of transformation that satisfies all these criteria is the *discrete wavelet transform*. In fact, the *discrete Haar wavelet transform* resulted in the distribution shown in Figure 4.2.

In this chapter, we construct the discrete Haar wavelet transformation for vectors and matrices and illustrate its use in applications. The basic construction of the Haar wavelet transformation follows in the subsequent section. In Section 4.2 we fine-tune the Haar wavelet transformation so that it is even more useful in applications. Section 4.3 shows how the results of the previous two sections can be used to create a Haar wavelet transformation that can be applied to matrices. The chapter concludes with applications of the Haar wavelet transformation to the problems of image compression and image edge detection.

4.1 Constructing the Haar Wavelet Transformation

We start by constructing the discrete Haar wavelet transformation for vectors. To facilitate the construction, consider $\mathbf{v} = [v_1, v_2, \ldots, v_8]^T \in \mathbb{R}^8$. We wish to construct an approximation to $\mathbf{v}$ that consists of only four terms. A natural way to do this is to create $\mathbf{a} \in \mathbb{R}^4$ so that its components are simply the pairwise averages of the components of $\mathbf{v}$:

$$a_k = \frac{1}{2}v_{2k-1} + \frac{1}{2}v_{2k}, \qquad k = 1, 2, 3, 4.$$

Clearly the transformation $\mathbf{v} \mapsto \mathbf{a}$ is not invertible; it is impossible to determine the original eight values from the four pairwise averages without additional information. Figure 4.3 gives us an idea how to obtain the additional information needed to create an invertible transformation.

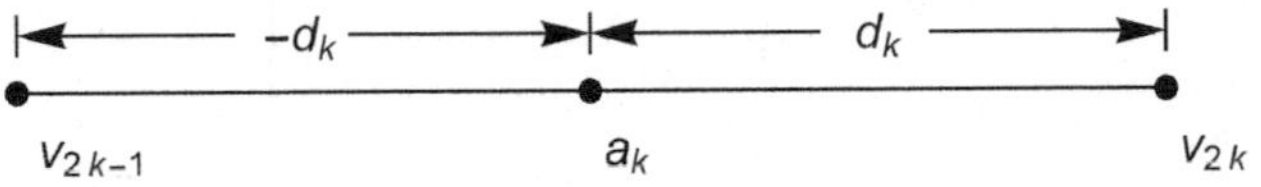

Figure 4.3 A geometric interpretation of the elements of the Haar wavelet transform.

The value a_k is the midpoint of the segment formed by v_{2k-1} and v_{2k} and it is clear, at least for the values in Figure 4.3, that $a_k + d_k = v_{2k}$ and $a_k - d_k = v_{2k-1}$ where

$$d_k = \frac{v_{2k} - v_{2k-1}}{2} = -\frac{1}{2}v_{2k-1} + \frac{1}{2}v_{2k}, \qquad k = 1, 2, 3, 4.$$

Thus the transformation

$$\mathbf{v} = \begin{bmatrix} v_1 \\ v_2 \\ v_3 \\ v_4 \\ v_5 \\ v_6 \\ v_7 \\ v_8 \end{bmatrix} \mapsto \begin{bmatrix} a_1 \\ a_2 \\ a_3 \\ a_4 \\ \hline d_1 \\ d_2 \\ d_3 \\ d_4 \end{bmatrix} = \begin{bmatrix} \mathbf{a} \\ \hline \mathbf{d} \end{bmatrix}$$

where

$$\begin{aligned} a_k &= \frac{1}{2}v_{2k-1} + \frac{1}{2}v_{2k} \\ d_k &= -\frac{1}{2}v_{2k-1} + \frac{1}{2}v_{2k}, \qquad k = 1, 2, 3, 4 \end{aligned} \tag{4.1}$$

is invertible since

$$\begin{aligned} v_{2k} &= a_k + d_k \\ v_{2k-1} &= a_k - d_k, \qquad k = 1, 2, 3, 4. \end{aligned}$$

Here is an example of how the transformation works.

Example 4.1 (Application of the Haar Transform). *Consider the vector*

$$\mathbf{v} = [4, 6, 7, 7, 3, 1, 0, 4]^T.$$

Apply the transform described by (4.1) *to* $\mathbf{v}$.

Solution.
The pairwise averages of the elements in $\mathbf{v}$ *are*

$$\frac{4+6}{2} = 5, \qquad \frac{7+7}{2} = 7, \qquad \frac{3+1}{2} = 2, \qquad \frac{0+4}{2} = 2,$$

and the pairwise differences are

$$\frac{6-4}{2} = 1, \qquad \frac{7-7}{2} = 0, \qquad \frac{1-3}{2} = -1, \qquad \frac{4-0}{2} = 2.$$

Thus the transformation is

$$\mathbf{v} = \begin{bmatrix} 4 \\ 6 \\ 7 \\ 7 \\ 3 \\ 1 \\ 0 \\ 4 \end{bmatrix} \mapsto \left[\begin{array}{c} 5 \\ 7 \\ 2 \\ 2 \\ \hline 1 \\ 0 \\ -1 \\ 2 \end{array}\right].$$

(Live example: Visit `stthomas.edu/wavelets` *and click on Live Examples.)*

□

Haar Wavelet Transform Defined

We can generalize our transform to operate on vectors of arbitrary even length. The transformation defined below is known as the *discrete Haar wavelet transformation.*

Definition 4.1 (Haar Wavelet Transformation). *Let* $\mathbf{v} \in \mathbb{R}^n$*, where* n *is an even positive integer. Then the* Haar wavelet transformation *is defined as the mapping*

$$\mathbf{v} \mapsto \left[\begin{array}{c} \mathbf{a} \\ \hline \mathbf{d} \end{array}\right]$$

where for $k = 1, \ldots, \frac{n}{2}$,

$$\boxed{a_k = \frac{1}{2}v_{2k} + \frac{1}{2}v_{2k-1} \quad \text{and} \quad d_k = \frac{1}{2}v_{2k} - \frac{1}{2}v_{2k-1}.} \tag{4.2}$$

We refer to **a** *as the* averages portion *and* **d** *as the* differences portion *of the Haar wavelet transform.* □

Let's look at an example that shows how the Haar wavelet transformation can concentrate the energy in a signal.

Example 4.2 (The Haar Wavelet Transform and Concentration of Energy). *Let* $f(t) = \cos(2\pi t)$. *We create the vector* $\mathbf{v} \in \mathbb{R}^{200}$ *by uniformly sampling* f *on the interval* $[0, 1)$ *with step size* $h = 0.005$. *That is,* $v_k = f(kh)$, $k = 0, \ldots, 199$. *The elements of* $\mathbf{v}$ *are plotted in Figure 4.4.*

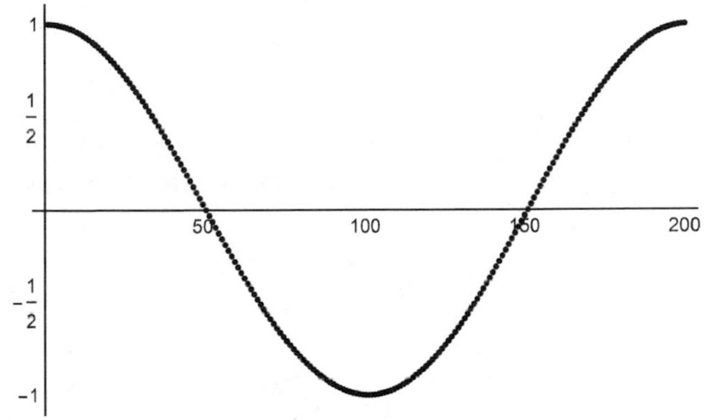

Figure 4.4 The vector $\mathbf{v} \in \mathbb{R}^{200}$ created by uniformly sampling $f(t) = \cos(2\pi t)$.

We apply the Haar wavelet transformation (4.2) *to* $\mathbf{v}$ *and plot the results in Figure 4.5(a). The values* $d_1, \ldots, d_{100}$ *are graphed again in Figure 4.5(b).*

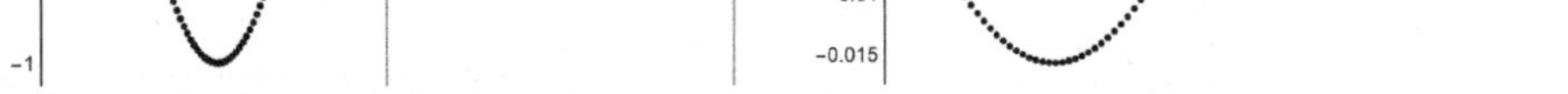

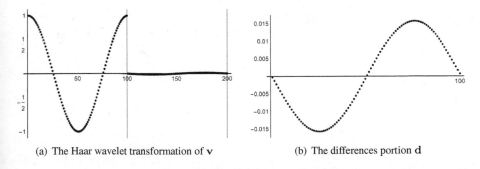

(a) The Haar wavelet transformation of $\mathbf{v}$

(b) The differences portion $\mathbf{d}$

Figure 4.5 The Haar wavelet transformation of $\mathbf{v}$ and the differences portion of the transform.

In Figure 4.6, we have plotted the cumulative energy vectors for $\mathbf{v}$ *(black) and its Haar wavelet transformation (gray). Note that the Haar wavelet transformation has more energy concentrated in fewer terms. This results from the differences portion* $\mathbf{d}$ *providing a relatively small contribution to the energy vector.*

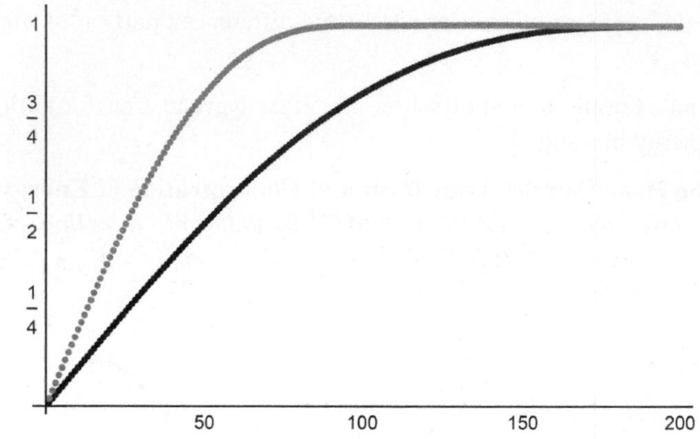

Figure 4.6 The cumulative energy vectors for **v** (black) and its Haar wavelet transformation (gray).

(Live example: Visit `stthomas.edu/wavelets` *and click on Live Examples.)*

□

Motivation for the Term *Wavelet*

We digress for a moment to give some insight into the word *wavelet*. In classical wavelet theory, we are interested in constructing a ladder of spaces V_k, $k \in \mathbb{Z}$ that can be used to approximate functions. One of the simplest spaces is the space V_0 that is comprised of all square-integrable[1] functions that can be formed from the function $\phi(t)$ in Figure 4.7(a) and its integer translates $\phi(t-k)$, $k \in \mathbb{Z}$. That is, V_0 is the space of square-integrable piecewise constant functions with jumps at the integers. Note that the fundamental generator is

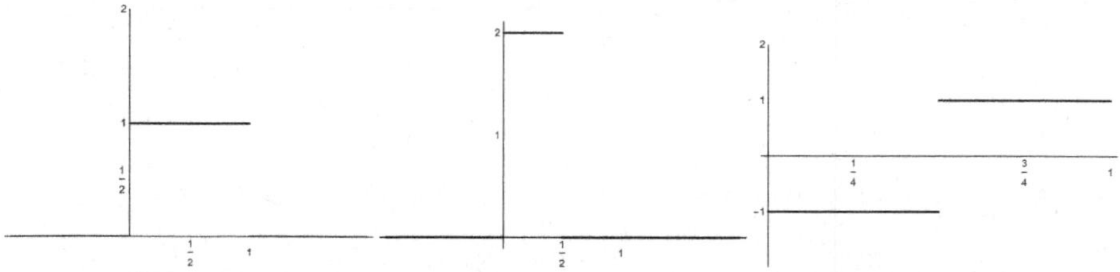

(a) The Haar function $\phi(t)$ (b) The function $2\phi(2t)$ (c) The wavelet function $\psi(t)$

Figure 4.7 The Haar functions $\phi(t)$ and $2\phi(2t)$ and the wavelet function $\psi(t)$.

[1]A function f is square-integrable if $\int_{\mathbb{R}} f^2(t)\, dt < \infty$.

$$\phi(t) = \begin{cases} 1, & t \in (0,1) \\ 0, & t \notin (0,1) \end{cases}$$

and that $\int\limits_{\mathbb{R}} \phi(t)\,\mathrm{d}t = 1$.

The function $\phi(t)$ is sometimes referred to as a *Haar function* since Alfred Haar studied them as part of his PhD dissertation in 1909. The results are published in [51].

We construct V_1 to be the space of square-integrable piecewise constant functions with jumps at the half-integers and take as the fundamental generator of this space $\phi_1(t) = 2\phi(2t)$ (see Figure 4.7(b)). The coefficient of 2 is included so that $\int\limits_{\mathbb{R}} \phi_1(t)\,\mathrm{d}t = 1$. Certainly $V_0 \subset V_1$.

We can create "finer" spaces V_k, $k > 0$, by taking as the generator $2^k\phi\left(2^k t\right)$. In this case V_k is the space of square-integrable piecewise constant functions with jumps at the integers divided by 2^k. We can create "coarser" spaces V_k, $k < 0$, where now the spaces are those of square-integrable piecewise constant functions with jumps at the integers times 2^k. Thus we have the ladder $\cdots \subset V_{-1} \subset V_0 \subset V_1 \subset \cdots$.

It is easy to see that

$$\phi(t) = \frac{1}{2} \cdot 2\phi(2t) + \frac{1}{2} \cdot 2\phi(2t-1), \tag{4.3}$$

where $2\phi(2t-1)$ is a half unit shift to the right of $\phi_1(t)$. The identity (4.3) is often called a *refinement* or *dilation equation*. Note that the coefficients in (4.3) are exactly the weights we use to compute a_k in (4.2).

Finally, we seek the "detail" space W_0 such that $V_1 = V_0 \oplus W_0$. That is W_0 is the space of all functions that are orthogonal[2] to those in V_0, and any function in V_1 can be written as the sum of a function from V_0 and a function for W_0. We can find the generator $\psi(t)$ of W_0 by computing the difference

$$\begin{aligned} \psi(t) &= \phi(t) - 2\phi(2t) \\ &= \begin{cases} -1, & t \in (0, \frac{1}{2}) \\ 1, & t \in (\frac{1}{2}, 1) \\ 0, & t \notin (0,1) \end{cases} \\ &= -\frac{1}{2} \cdot 2\phi(2t) + \frac{1}{2} \cdot 2\phi(2t-1). \end{aligned} \tag{4.4}$$

The function $\psi(t)$ is plotted in Figure 4.7(c). The coefficients of $\phi(2t), \phi(2t-1)$ in (4.4) are exactly the weights we use to compute d_k in (4.2).

It is easy to check that $\phi(t)$ and $\psi(t)$ are orthogonal and it can be verified that any function in V_1 can be written as the sum of a function in V_0 and a function in W_0. We can construct other detail spaces W_k so that $V_{k+1} = V_k \oplus W_k$.

[2]We say two functions are orthogonal in this setting if $\int_{\mathbb{R}} f(t)g(t)\,\mathrm{d}t = 0$.

The researchers who developed this theory were interested in creating small waves, or *wavelets*, that could be used to provide a good local approximation of a function. The function $\psi(t)$ is called a *wavelet function* and in particular, the *Haar wavelet function*, since it was constructed from the Haar function.

Ours is called a discrete Haar wavelet transformation since we are using the coefficients in the relations (4.3) and (4.4) to process discrete data. We will develop more sophisticated discrete wavelet transformations in later chapters. It should be noted that a more classical derivation of wavelet functions also leads to discrete wavelet transformations and the interested reader is encouraged to consult [7], [77], or [100].

The Inverse Haar Wavelet Transform

Inverting the Haar wavelet transformation is straightforward. Figure 4.3 provides the necessary inversion formulas

$$\boxed{v_{2k} = a_k + d_k \qquad \text{and} \qquad v_{2k-1} = a_k - d_k.} \tag{4.5}$$

Example 4.3 (Inverting the Haar Wavelet Transform). *Use* (4.5) *on the transformed data in Example 4.1 to recover* $\mathbf{v}$.

Solution.
Using (4.5)*, we see that* $v_1 = a_1 - d_1 = 5 - 1 = 4$ *and* $v_2 = a_1 + d_1 = 6$. *Similarly*

$$\begin{aligned} v_3 &= a_2 - d_2 = 7, & v_4 &= a_2 + d_2 = 7, \\ v_5 &= a_3 - d_3 = 3, & v_6 &= a_3 + d_3 = 1, \\ v_7 &= a_4 - d_4 = 0, & v_8 &= a_4 + d_4 = 4. \end{aligned}$$

(Live example: Visit `stthomas.edu/wavelets` *and click on Live Examples.)*

□

The next example illustrates how the Haar wavelet transformation and its inverse can be used to perform naive image compression.

Example 4.4 (Naive Data Compression). *We return to the samples of* $\cos(2\pi t)$ *from Example 4.2. Since the magnitudes of the elements of the differences portion* $\mathbf{d}$ *are bounded by approximately* 0.0157*, we convert all elements of* $\mathbf{d}$ *to zero. Thus, the modified wavelet transformation is*

$$\mathbf{w} = \left[\begin{array}{c} \mathbf{a} \\ \hline \mathbf{0} \end{array}\right].$$

We compute $Ent(\mathbf{v}) = 4.6121$ *and* $Ent\,(\mathbf{w}) = 2.9957$ *and see that for compression purposes,* $\mathbf{w}$ *should be used. Of course, if we compress and store* $\mathbf{w}$*, then we have no chance of recovering* $\mathbf{v}$*, but if* $\tilde{\mathbf{v}}$ *is the inverse Haar wavelet transformation of* $\mathbf{w}$ *and using* (4.5)*, we see that*

$$\tilde{v}_{2k} = a_k + 0 = a_k - 0 = \tilde{v}_{2k-1}.$$

Note that

$$|v_{2k} - \tilde{v}_{2k}| = |a_k + d_k - a_k - 0| = |d_k| \leq \max|d_k| \approx 0.0157,$$

and a similar argument holds for $|v_{2k-1} - \tilde{v}_{2k-1}|$. *Figure 4.8 shows* $\mathbf{v}$, *the approximate* $\tilde{\mathbf{v}}$ *and the error in approximation* $\mathbf{e}$. *The components of* $\mathbf{e}$ *are* $e_k = |v_k - \tilde{v}_k|$.

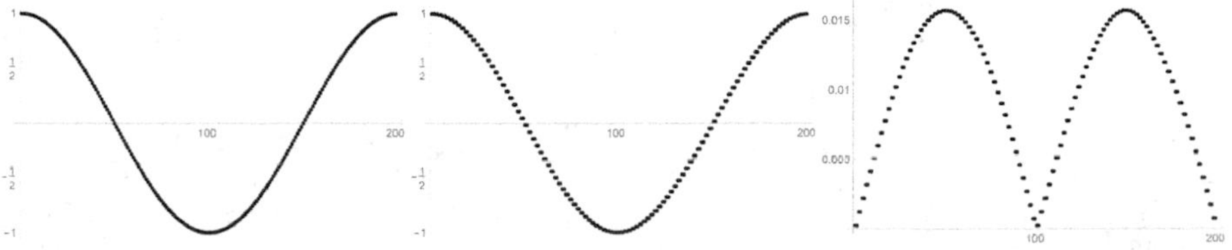

(a) $\mathbf{v}$ (b) $\tilde{\mathbf{v}}$ (c) $\mathbf{e}$

Figure 4.8 The vector $\mathbf{v}$, its approximate $\tilde{\mathbf{v}}$ and the error vector $\mathbf{e}$.

(Live example: Visit `stthomas.edu/wavelets` *and click on Live Examples.)*

□

Matrix Representation of the Haar Wavelet Transform

It is often useful to have a matrix formulation for the Haar wavelet transform. Let's consider the problem for $\mathbf{v} \in \mathbb{R}^8$. In particular, we seek an 8×8 matrix W_8 that satisfies

$$W_8\mathbf{v} = \begin{bmatrix} w_{11} & w_{12} & w_{13} & w_{14} & w_{15} & w_{16} & w_{17} & w_{18} \\ w_{21} & w_{22} & w_{23} & w_{24} & w_{25} & w_{26} & w_{27} & w_{28} \\ w_{31} & w_{32} & w_{33} & w_{34} & w_{35} & w_{36} & w_{37} & w_{38} \\ w_{41} & w_{42} & w_{43} & w_{44} & w_{45} & w_{46} & w_{47} & w_{48} \\ w_{51} & w_{52} & w_{53} & w_{54} & w_{55} & w_{56} & w_{57} & w_{58} \\ w_{61} & w_{62} & w_{63} & w_{64} & w_{65} & w_{66} & w_{67} & w_{68} \\ w_{71} & w_{72} & w_{73} & w_{74} & w_{75} & w_{76} & w_{77} & w_{78} \\ w_{81} & w_{82} & w_{83} & w_{84} & w_{85} & w_{86} & w_{87} & w_{88} \end{bmatrix} \cdot \begin{bmatrix} v_1 \\ v_2 \\ v_3 \\ v_4 \\ v_5 \\ v_6 \\ v_7 \\ v_8 \end{bmatrix} = \frac{1}{2} \left[\begin{array}{c} v_1 + v_2 \\ v_3 + v_4 \\ v_5 + v_6 \\ v_7 + v_8 \\ \hline -v_1 + v_2 \\ -v_3 + v_4 \\ -v_5 + v_6 \\ -v_7 + v_8 \end{array}\right].$$

It is clear that $w_{11} = w_{12} = \frac{1}{2}$ and the remaining elements of the first row of W_8 are zero. Rows 2 – 4 follow a similar pattern – all elements in these rows are zero except for $w_{23} = w_{24} = w_{35} = w_{36} = w_{47} = w_{48} = \frac{1}{2}$. Rows 5 – 8 are similar to the first four rows with $w_{51} = w_{63} = w_{75} = w_{87} = -\frac{1}{2}$, $w_{52} = w_{64} = w_{76} = w_{88} = \frac{1}{2}$ and the remaining elements zero. We have

$$W_8 = \left[\frac{H_4}{G_4}\right] = \left[\begin{array}{cccccccc} \frac{1}{2} & \frac{1}{2} & 0 & 0 & 0 & 0 & 0 & 0 \\ 0 & 0 & \frac{1}{2} & \frac{1}{2} & 0 & 0 & 0 & 0 \\ 0 & 0 & 0 & 0 & \frac{1}{2} & \frac{1}{2} & 0 & 0 \\ 0 & 0 & 0 & 0 & 0 & 0 & \frac{1}{2} & \frac{1}{2} \\ \hline -\frac{1}{2} & \frac{1}{2} & 0 & 0 & 0 & 0 & 0 & 0 \\ 0 & 0 & -\frac{1}{2} & \frac{1}{2} & 0 & 0 & 0 & 0 \\ 0 & 0 & 0 & 0 & -\frac{1}{2} & \frac{1}{2} & 0 & 0 \\ 0 & 0 & 0 & 0 & 0 & 0 & -\frac{1}{2} & \frac{1}{2} \end{array}\right]. \tag{4.6}$$

We can generalize the above matrix formulation for even positive integer N and write

$$W_N = \left[\frac{H_{N/2}}{G_{N/2}}\right] = \left[\begin{array}{ccccccccc} \frac{1}{2} & \frac{1}{2} & 0 & 0 & \cdots & 0 & 0 & 0 & 0 \\ 0 & 0 & \frac{1}{2} & \frac{1}{2} & \cdots & 0 & 0 & 0 & 0 \\ & & & & \ddots & & & & \\ 0 & 0 & 0 & 0 & \cdots & 0 & 0 & \frac{1}{2} & \frac{1}{2} \\ \hline -\frac{1}{2} & \frac{1}{2} & 0 & 0 & \cdots & 0 & 0 & 0 & 0 \\ 0 & 0 & -\frac{1}{2} & \frac{1}{2} & \cdots & 0 & 0 & 0 & 0 \\ & & & & \ddots & & & & \\ 0 & 0 & 0 & 0 & \cdots & 0 & 0 & -\frac{1}{2} & \frac{1}{2} \end{array}\right]. \tag{4.7}$$

Note that for $\mathbf{v} \in \mathbb{R}^N$, we have

$$W_N\mathbf{v} = \left[\frac{H_{N/2}}{G_{N/2}}\right] \cdot \mathbf{v} = \left[\frac{H_{N/2}\mathbf{v}}{G_{N/2}\mathbf{v}}\right] = \left[\frac{\mathbf{a}}{\mathbf{d}}\right].$$

Thus $H_{N/2}\mathbf{v}$ produces the averages portion and $G_{N/2}\mathbf{v}$ produces the differences portion of the Haar wavelet transform.

Connection to Filters and Convolution as a Matrix Product

Suppose we define $\mathbf{h} = (h_0, h_1) = \left(\frac{1}{2}, \frac{1}{2}\right)$ as a filter. Then the first row of $H_{N/2}$ in (4.7) can be viewed as

$$\begin{bmatrix} h_1 & h_0 & 0 & 0 & \cdots 0 & 0 \end{bmatrix},$$

and subsequent rows of $H_{N/2}$ are simply cyclic 2-shifts of this row. We have seen $\mathbf{h}$ in Examples 2.13(b) and 2.15 and from these examples we can easily infer that

$\mathbf{h}$ satisfies Definition 2.19 and is thus a lowpass filter. In a similar way, we see that $G_{N/2}$ is built from filter $\mathbf{g} = (g_0, g_1) = \left(\frac{1}{2}, -\frac{1}{2}\right)$. The first row of $G_{N/2}$ is

$$\begin{bmatrix} g_1 & g_0 & 0 & 0 & \cdots 0 & 0 \end{bmatrix},$$

and subsequent rows of $G_{N/2}$ are simply cyclic 2-shifts of this row. We have seen $\mathbf{g}$ before in Example 2.13(a) and from this example we can infer that $\mathbf{g}$ satisfies Definition 2.20 and is thus a highpass filter.

Now the matrix product representation of $\mathbf{h} * \mathbf{x}$ can be expressed using (2.15) as

$$H\mathbf{x} = \begin{bmatrix} \ddots & & & & & & & & & \\ & \frac{1}{2} & \frac{1}{2} & 0 & 0 & \cdots & 0 & 0 & 0 & \\ & 0 & \frac{1}{2} & \frac{1}{2} & 0 & \cdots & 0 & 0 & 0 & \\ & & & & & \ddots & & & & \\ & 0 & 0 & 0 & 0 & \cdots & \frac{1}{2} & \frac{1}{2} & 0 & \\ & 0 & 0 & 0 & 0 & \cdots & 0 & \frac{1}{2} & \frac{1}{2} & \\ & & & & & & & & & \ddots \end{bmatrix} \cdot \begin{bmatrix} \vdots \\ x_{-1} \\ x_0 \\ x_1 \\ \vdots \end{bmatrix}.$$

Notice that to get $H_{N/2}$ from H above, we simply discard every other row (this is called *downsampling*) and then truncate in the appropriate places to form an $N/2 \times N$ matrix.

$G_{N/2}$ is constructed in much the same way from the matrix G in the convolution product $G\mathbf{x} = \mathbf{g} * \mathbf{x}$. We call $\mathbf{h}$ and $\mathbf{g}$ defined by

$$\boxed{\mathbf{h} = (h_0, h_1) = \left(\frac{1}{2}, \frac{1}{2}\right) \quad \text{and} \quad \mathbf{g} = (g_0, g_1) = \left(\frac{1}{2}, -\frac{1}{2}\right)} \tag{4.8}$$

the *Haar lowpass filter* and the *Haar highpass filter*, respectively.

Matrix Representation of the Inverse Haar Wavelet Transformation

We could use Gaussian elimination to formulate the matrix that represents the inverse Haar wavelet transformation but we instead use (4.5) for the 8×8 case. We have

$$W_8^{-1} \left[\frac{\mathbf{a}}{\mathbf{d}}\right] = \mathbf{v}$$

$$\left[\begin{array}{cccc|cccc} 1 & 0 & 0 & 0 & -1 & 0 & 0 & 0 \\ 1 & 0 & 0 & 0 & 1 & 0 & 0 & 0 \\ 0 & 1 & 0 & 0 & 0 & -1 & 0 & 0 \\ 0 & 1 & 0 & 0 & 0 & 1 & 0 & 0 \\ 0 & 0 & 1 & 0 & 0 & 0 & -1 & 0 \\ 0 & 0 & 1 & 0 & 0 & 0 & 1 & 0 \\ 0 & 0 & 0 & 1 & 0 & 0 & 0 & -1 \\ 0 & 0 & 0 & 1 & 0 & 0 & 0 & 1 \end{array}\right] \cdot \frac{1}{2} \left[\begin{array}{c} v_1 + v_2 \\ v_3 + v_4 \\ v_5 + v_6 \\ v_7 + v_8 \\ \hline -v_1 + v_2 \\ -v_3 + v_4 \\ -v_5 + v_6 \\ -v_7 + v_8 \end{array}\right] = \begin{bmatrix} v_1 \\ v_2 \\ v_3 \\ v_4 \\ v_5 \\ v_6 \\ v_7 \\ v_8 \end{bmatrix}.$$

In general, for even positive integer N the matrix representation for the inverse Haar wavelet transform is

$$\boxed{W_N^{-1} = \left[\begin{array}{cccc|cccc} 1 & 0 & \cdots & 0 & -1 & 0 & \cdots & 0 \\ 1 & 0 & \cdots & 0 & 1 & 0 & \cdots & 0 \\ 0 & 1 & \cdots & 0 & 0 & -1 & \cdots & 0 \\ 0 & 1 & \cdots & 0 & 0 & 1 & \cdots & 0 \\ & & \ddots & & & & \ddots & \\ 0 & 0 & \cdots & 1 & 0 & 0 & \cdots & -1 \\ 0 & 0 & \cdots & 1 & 0 & 0 & \cdots & 1 \end{array}\right].} \tag{4.9}$$

Orthogonalizing the Haar Wavelet Matrix

Note that the left side of W_N^{-1} is $2H_{N/2}^T$ and the right side is $2G_{N/2}^T$ so that

$$W_N^{-1} = \left[2H_{N/2}^T \mid 2G_{N/2}^T\right] = 2\left[H_{N/2}^T \mid G_{N/2}^T\right] = 2W_N^T. \tag{4.10}$$

Recalling Definition 2.14, we see that W_N^{-1} is not an orthogonal matrix. However, if we scale W_N by $\sqrt{2}$, we see from Proposition 2.1(d) and (4.10) that

$$\left(\sqrt{2}W_N\right)^{-1} = \frac{1}{\sqrt{2}}W_N^{-1} = \frac{1}{\sqrt{2}} \cdot 2W_N^T = \left(\sqrt{2}W_N\right)^T.$$

We have the following definition.

Definition 4.2 (Orthogonal Haar Wavelet Transformation). *Let N be a positive even integer. We define the* orthogonal Haar wavelet transformation *as*

$$W_N^o = \sqrt{2}W_N \tag{4.11}$$

where W_N is given by (4.7). *The filter*

$$\boxed{\mathbf{h} = (h_0, h_1) = \left(\frac{\sqrt{2}}{2}, \frac{\sqrt{2}}{2}\right)} \tag{4.12}$$

is called the orthogonal Haar lowpass filter. □

The orthogonal version of the Haar wavelet transformation is useful in applications where we wish to preserve the magnitude of the transformed vector. We will also find orthogonality to be a useful tool when creating more sophisticated wavelet transformations in Chapter 5.

PROBLEMS

4.1 Show that the Haar wavelet transformation of the vector

$$\mathbf{v} = [\,200, 200, 200, 210, 40, 80, 100, 102\,]^T$$

is

$$\mathbf{y} = [200, 205, 60, 101 \mid 0, 5, 20, 1]^T .$$

4.2 In this problem, you will use the vectors $\mathbf{v}$ and $\mathbf{y}$ from Problem 4.1. Suppose we create $\widetilde{\mathbf{y}}$ by setting the differences portion $\mathbf{d}$ of $\mathbf{y}$ to $\mathbf{0}$. That is,

$$\widetilde{\mathbf{y}} = [200, 205, 60, 101 \mid 0, 0, 0, 0]^T .$$

(a) How does the entropy of $\widetilde{\mathbf{y}}$ compare to that of $\mathbf{v}$?

(b) Use a **CAS** to compute and plot the cumulative energy of $\mathbf{v}$ and $\widetilde{\mathbf{y}}$. Which energy vector is more concentrated?

(c) Compute the inverse Haar wavelet transformation of $\widetilde{\mathbf{y}}$ and compare the result to $\mathbf{v}$? In which components are the differences in the two vectors biggest?

4.3 Let $\mathbf{v} = [1, 1, 1, 1, 1, 1, 1, 1]^T$ and $\mathbf{w} = [-1, 1, -1, 1, -1, 1, -1, 1]^T$.

(a) Compute $W_8\mathbf{v}$ and $W_8\mathbf{w}$ and describe the output of both computations.

(b) Compute $W_N\mathbf{v}$ and $W_N\mathbf{w}$ in the case where N is even and $\mathbf{v}$ has elements $v_k = 1$ and $\mathbf{w}$ has elements $w_k = (-1)^k$, $k = 1, \ldots, N$. Describe the output of both computations.

4.4 Let $\mathbf{v} = [2, 6, -4, 2, 400, 402, -8, -6]^T$ and compute the Haar wavelet transformation of $\mathbf{v}$. There is a large jump in the values of $\mathbf{v}$ from v_4 to v_5 and from v_6 to v_7. Is this reflected in the differences portion of the transformed data? Explain why or why not.

4.5 Suppose N is an even positive integer, $\mathbf{v} \in \mathbb{R}^N$ and $\mathbf{y}$ the Haar wavelet transformation of $\mathbf{v}$. Show that

(a) if $\mathbf{v}$ is a constant vector (i.e., $v_k = c$ where c is any real number), then the components of the differences portion of $\mathbf{y}$ are zero.

(b) if $\mathbf{v}$ is a linear vector (i.e., $v_k = mk + b$ for real number $m \neq 0$ and b), then the components of the differences portion of $\mathbf{y}$ are constant. Find this constant value.

4.6 Consider the error vector in Example 4.4. Explain why the elements of this vector are (near) zero for elements e_0, e_{100} and e_{200}.

4.7 This problem is intended for those with a background in computer science. Let $\mathbf{v} \in \mathbb{R}^N$ where N is a positive even integer and suppose that $\mathbf{y}$ is the Haar wavelet transformation of $\mathbf{v}$. Write a `For` (`Do`) loop that computes y_k and $y_{k+N/2}$ for each $k = 1, \ldots, N/2$.

4.8 Suppose that $\mathbf{v} \in \mathbb{R}^8$, W_8 is the matrix given in (4.6), and $\mathbf{h}$, $\mathbf{g}$ are the vectors given by

$$\mathbf{h} = \begin{bmatrix} \frac{1}{2} \\ \frac{1}{2} \end{bmatrix} \quad \text{and} \quad \mathbf{g} = \begin{bmatrix} -\frac{1}{2} \\ \frac{1}{2} \end{bmatrix}.$$

(a) Find a 4×2 matrix X so that

$$W_8\mathbf{v} = \left[\begin{array}{c} X\mathbf{h} \\ \hline X\mathbf{g} \end{array}\right]. \tag{4.13}$$

For a **CAS** such as *Mathematica*, utilizing (4.13) will produce a faster algorithm than the matrix multiplication $W_8\mathbf{v}$ or the code from Problem 4.7.

(b) Generalize (a) for the case where N is a positive even integer.

(c) Use (a) to compute the Haar wavelet transformation $\mathbf{y}$ of the vector $\mathbf{v}$ given in Problem 4.4.

4.9 This problem is intended for those with a background in computer science. Let $\mathbf{y}$ be the Haar wavelet transformation of $\mathbf{v} \in \mathbb{R}^N$ where N is a positive even integer. Write a `For` (`Do`) loop that computes v_{2k-1} and v_{2k} for each $k = 1, \ldots, N/2$.

4.10 Suppose $\mathbf{y} \in \mathbb{R}^8$ is the Haar wavelet transformation of $\mathbf{v}$ and that $\mathbf{h}, \mathbf{g}$ are given by

$$\mathbf{h} = \begin{bmatrix} 1 \\ 1 \end{bmatrix} \quad \text{and} \quad \mathbf{g} = \begin{bmatrix} 1 \\ -1 \end{bmatrix}.$$

We know that

$$\mathbf{v} = \begin{bmatrix} v_1 \\ v_2 \\ v_3 \\ v_4 \\ v_5 \\ v_6 \\ v_7 \\ v_8 \end{bmatrix} = W_8^{-1}\mathbf{y} = \begin{bmatrix} -y_1 + y_5 \\ y_1 + y_5 \\ -y_2 + y_6 \\ y_2 + y_6 \\ -y_3 + y_7 \\ y_3 + y_7 \\ -y_4 + y_8 \\ y_4 + y_8 \end{bmatrix}.$$

(a) Find a 4×2 matrix X such that $X\mathbf{g}$ holds the odd components of $\mathbf{v}$ and $X\mathbf{h}$ holds the even components of $\mathbf{v}$.

(b) Find an 8×8 matrix P such that

$$\mathbf{v} = P\left[\begin{array}{c} X\mathbf{h} \\ \hline X\mathbf{g} \end{array}\right].$$

(c) Generalize (a) and (b) when $\mathbf{y} \in \mathbb{R}^N$ (N positive and even) and W_N^{-1} is given by (4.9).

(d) In (c) of Problem 4.8, you computed the Haar wavelet transformation $\mathbf{y}$ of the vector from Problem 4.4. Use (a) and (b) of this problem to find the inverse Haar wavelet transformation of $\mathbf{y}$.

4.11 Let $\mathbf{v} = [1, 2, 3, 4, 5, 6, 7, 8]^T$.

(a) Compute the discrete Haar wavelet transformation $\mathbf{y} = W_8\mathbf{v}$.

(b) What is the entropy of $\mathbf{v}$, $\mathbf{y}$?

(c) Suppose in general that the components of $\mathbf{v} \in \mathbb{R}^N$, ($N$ even and positive) are given by $v_k = k$, $k = 1, \ldots, N$. Compute the Haar wavelet transformation $\mathbf{y}$ and entropies for both $\mathbf{v}$ and $\mathbf{y}$. How do these results compare to those of Problems 3.45 and 3.38?

4.12 Sometimes the result of the Haar wavelet transformation can have higher entropy than the original vector. Suppose that $\mathbf{v} \in \mathbb{R}^N$ with N even and $v_k = c$ for $k = 1, \ldots, N$. Then by Problem 3.39, $\mathrm{Ent}(\mathbf{v}) = 0$.

(a) Compute $\mathbf{y} = W_N\mathbf{v}$.

(b) What is $\mathrm{Ent}(\mathbf{y})$?

4.13 Use (4.11) to write down W_2^o. Explain geometrically the application of the 2×2 orthogonal Haar wavelet transformation to vectors in $\mathbb{R}^2$. *Hint:* See Example 2.9.

4.14 In this problem, we develop the orthogonal Haar wavelet transformation in a different manner. To facilitate the construction, we consider the 4×4 matrix

$$W_4^o = \left[\begin{array}{c} H \\ \hline G \end{array}\right] = \left[\begin{array}{cccc} a & b & 0 & 0 \\ 0 & 0 & a & b \\ \hline c & d & 0 & 0 \\ 0 & 0 & c & d \end{array}\right]$$

where $a, b, c, d \in \mathbb{R}$ are nonzero numbers. We will find values for a, b, c, d that ensure W_4^o is orthogonal and that the rows of G, when applied to a constant vector, return a zero vector.

(a) Verify that in order for W_4^o to be orthogonal, we must have

$$a^2 + b^2 = 1 \tag{4.14}$$

$$c^2 + d^2 = 1 \tag{4.15}$$

$$ac + bd = 0. \tag{4.16}$$

(b) We can view the left side of (4.16) as the orthogonal inner product of vectors $\begin{bmatrix} a \\ b \end{bmatrix}$ and $\begin{bmatrix} c \\ d \end{bmatrix}$. Show that (4.16) is satisfied when $\begin{bmatrix} c \\ d \end{bmatrix} = k \begin{bmatrix} -b \\ a \end{bmatrix}$ for $k \neq 0$. If we then assume that (4.14) holds, use (4.15) to show that $k = \pm 1$. Going forward, we take $k = 1$. Write W_4^o in terms of a and b.

(c) It is desirable if the differences portion G of W_4^o annihilate constant data (the Haar wavelet matrix (4.7) satisfies this condition). That is, if $v = e\,[1, 1, 1, 1]^T$, $e \neq 0$, we insist that $G\mathbf{v} = \mathbf{0}$, where $\mathbf{0} \in \mathbb{R}^2$. What additional condition does this impose on a and b? Write W_4^o in terms of a.

(d) Use (c) to find a positive value for a. Verify this solution results in the 4×4 orthogonal Haar matrix we can form using (4.11).

Certainly, the steps above work for W_N^o where N is any even positive integer. The construction outlined here will serve as a template for constructing orthogonal wavelet transformations in Chapter 5.

Computer Lab

4.1 The Haar Wavelet Transform. From `stthomas.edu/wavelets`, access the file `HWT`. In this lab you will write code that computes the (inverse) Haar wavelet transformation of a vector.

4.2 Iterating the Process

The Haar wavelet transformation takes vector $\mathbf{v} \in \mathbb{R}^N$ and returns $\mathbf{y} = W_N\mathbf{v}$, where the top half of $\mathbf{y}$ consists of the averages portion $\mathbf{a}$ of the elements in $\mathbf{v}$ and the bottom half is comprised of differences portion $\mathbf{d}$ of the elements of $\mathbf{v}$. We have seen in Problem 4.11 that application of W_N can result in a vector $\mathbf{y}$ that has lower entropy and more concentrated cumulative energy than the original input $\mathbf{v}$. These measures are both indicators that we might obtain a higher compression ratio using $\mathbf{y}$ rather than $\mathbf{v}$.

It is natural to ask if it is possible to improve these measures even further. The answer is yes, and all that we need to do is make further use of the Haar wavelet transform. Suppose that N is divisible by four and let $\mathbf{v} \in \mathbb{R}^N$. We apply W_N to $\mathbf{v}$ and obtain the vector

$$\mathbf{y}^1 = \left[\begin{array}{c} \mathbf{a}^1 \\ \hline \mathbf{d}^1 \end{array}\right]. \tag{4.17}$$

Superscripts have been added to the vectors in (4.17) to facilitate the ease of notation in the ensuing discussion. Since $\mathbf{a}^1 \in \mathbb{R}^{\frac{N}{2}}$ and N is divisible by four, we know that $\frac{N}{2}$ is divisible by two. Since $\mathbf{a}^1$ is an approximation of $\mathbf{v}$ and is itself of even length, it makes sense to compute its Haar wavelet transform. We have

$$\mathbf{y}^2 = \left[\begin{array}{c} W_{N/2}\mathbf{a}^1 \\ \hline \mathbf{d}^1 \end{array}\right] = \left[\begin{array}{c} \mathbf{a}^2 \\ \hline \mathbf{d}^2 \\ \hline \mathbf{d}^1 \end{array}\right]. \tag{4.18}$$

We have thus performed a second iteration of the Haar wavelet transform. Let's look at an example to see how the process works.

Example 4.5 (Two Iterations of the Haar Wavelet Transform). *Suppose*

$$\mathbf{v} = [80, 140, 160, 180, 120, 120, 40, 120]^T \in \mathbb{R}^8.$$

Use (4.18) *to compute two iterations of the Haar wavelet transform.*

Solution.
We begin by computing the Haar wavelet transformation of $\mathbf{v}$ *and then the Haar wavelet transformation of the vector* $\mathbf{a}^1$*:*

$$\mathbf{y}^1 = \left[\begin{array}{c} \mathbf{a}^1 \\ \hline \mathbf{d}^1 \end{array}\right] = \left[\begin{array}{c} 110 \\ 170 \\ 120 \\ 80 \\ \hline 30 \\ 10 \\ 0 \\ 40 \end{array}\right], \qquad W_4\mathbf{a}^1 = \left[\begin{array}{c} \mathbf{a}^2 \\ \hline \mathbf{d}^2 \end{array}\right] = \left[\begin{array}{c} 140 \\ 100 \\ \hline 30 \\ -20 \end{array}\right].$$

Combining these two vectors gives

$$\mathbf{y}^2 = \left[\begin{array}{c} \mathbf{a}^2 \\ \hline \mathbf{d}^2 \\ \hline \mathbf{d}^1 \end{array}\right] = \left[\begin{array}{c} 140 \\ 100 \\ \hline 30 \\ -20 \\ \hline 30 \\ 10 \\ 0 \\ 40 \end{array}\right].$$

(Live example: Visit `stthomas.edu/wavelets` *and click on Live Examples.)*

□

To understand the effects the iterative Haar wavelet transformation has in applications involving data processing consider the following example.

Example 4.6 (The Iterated Haar Wavelet Transform and Cumulative Energy). *Let* $\mathbf{v} \in \mathbb{R}^{512}$ *be the first column of gray intensities of the image in Figure 4.1. We compute the Haar wavelet transformation and then the second iteration of the Haar wavelet transformation of* $\mathbf{v}$*. The results are plotted in Figure 4.9.*

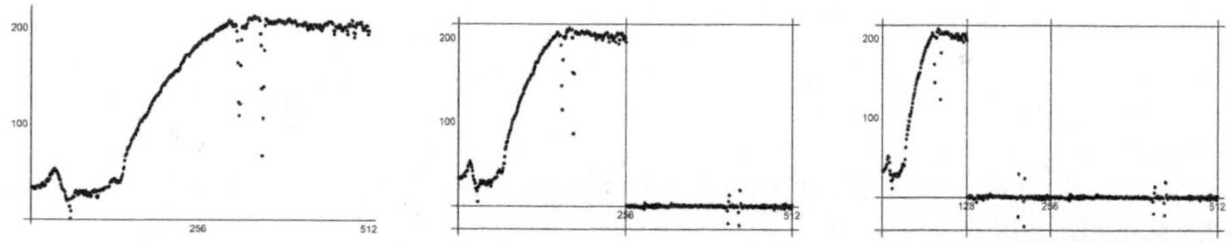

(a) $\mathbf{v}$

(b) The Haar wavelet transformation $\mathbf{y}^1$

(c) The iterated Haar wavelet transformation $\mathbf{y}^2$

Figure 4.9 A vector $\mathbf{v}$, its Haar wavelet transformation and the iterated Haar wavelet transform.

We next compute the cumulative energy of each vector plotted in Figure 4.9. The results are plotted in Figure 4.10.

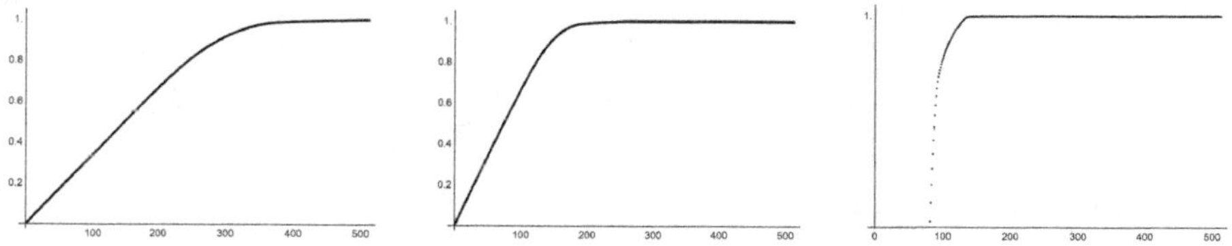

(a) $\mathbf{v}$

(b) The Haar wavelet transformation $\mathbf{y}^1$

(c) The iterated Haar wavelet transformation $\mathbf{y}^2$

Figure 4.10 Cumulative energy vectors for $\mathbf{v}$, its Haar wavelet transformation and the iterated Haar wavelet transform.

The iterated Haar wavelet transformation $\mathbf{y}^2$ *produces a much more concentrated energy vector than either* $\mathbf{v}$ *or* $\mathbf{y}^1$*. For example, the largest (in absolute value) 100 components of each of* $\mathbf{v}$, $\mathbf{y}^1$, *and* $\mathbf{y}^2$ *hold* 34.1%, 66.6% *and* 98.9% *of the energy of each vector, respectively.*
(Live example: Visit `stthomas.edu/wavelets` *and click on Live Examples.)*

□

We can certainly iterate the Haar wavelet transformation of $\mathbf{v} \in \mathbb{R}^N$ as many times as the length of the data allow. In particular, if N is divisible by 2^j, then we can compute j iterations of the Haar wavelet transform. We have

$$\mathbf{y}^j = \left[\begin{array}{c} \mathbf{a}^j \\ \hline \mathbf{d}^j \\ \hline \vdots \\ \hline \mathbf{d}^1 \end{array}\right] \tag{4.19}$$

where $\mathbf{a}^j \in \mathbb{R}^{N/2^j}$ and $\mathbf{d}^k \in \mathbb{R}^{N/2^k}$, $k = 1, \ldots, j$.

The Iterated Haar Wavelet Transform as a Product of Matrices

It is sometimes convenient to express the iterated Haar wavelet transformation (4.19) as a product of matrices. Consider for example when N is divisible by four:

$$\mathbf{y}^2 = \left[\begin{array}{c} \mathbf{a}^2 \\ \mathbf{d}^2 \\ \mathbf{d}^1 \end{array}\right] = \left[\begin{array}{c} W_{N/2}\mathbf{a}^1 \\ \hline \mathbf{d}^1 \end{array}\right] = \left[\begin{array}{c} W_{N/2}\mathbf{a}^1 + 0_{N/2}\mathbf{d}^1 \\ \hline 0_{N/2}\mathbf{a}^1 + I_{N/2}\mathbf{d}^1 \end{array}\right] \tag{4.20}$$

where $O_{N/2}$ and $I_{N/2}$ are the $\frac{N}{2} \times \frac{N}{2}$ zero and identity matrices, respectively. The rightmost term in (4.20) can be written as the block matrices product

$$\left[\begin{array}{c} W_{N/2}\mathbf{a}^1 + 0_{N/2}\mathbf{d}^1 \\ \hline 0_{N/2}\mathbf{a}^1 + I_{N/2}\mathbf{d}^1 \end{array}\right] = \left[\begin{array}{c|c} W_{N/2} & 0_{N/2} \\ \hline 0_{N/2} & I_{N/2} \end{array}\right] \cdot \left[\begin{array}{c} \mathbf{a}^1 \\ \hline \mathbf{d}^1 \end{array}\right]. \tag{4.21}$$

Combining (4.17), (4.20), and (4.21), we see that

$$\begin{aligned} \mathbf{y}^2 &= \left[\begin{array}{c|c} W_{N/2} & 0_{N/2} \\ \hline 0_{N/2} & I_{N/2} \end{array}\right] \cdot \left[\begin{array}{c} \mathbf{a}^1 \\ \hline \mathbf{d}^1 \end{array}\right] \\ &= \left[\begin{array}{c|c} W_{N/2} & 0_{N/2} \\ \hline 0_{N/2} & I_{N/2} \end{array}\right] \cdot \mathbf{y}^1 \\ &= \left[\begin{array}{c|c} W_{N/2} & 0_{N/2} \\ \hline 0_{N/2} & I_{N/2} \end{array}\right] W_N \mathbf{v}. \end{aligned} \tag{4.22}$$

In the case where N is divisible by eight and we perform three iterations of the Haar wavelet transform, we have

$$\mathbf{y}^3 = \left[\begin{array}{c|c} \begin{array}{c|c} W_{N/4} & 0_{N/4} \\ \hline 0_{N/4} & I_{N/4} \end{array} & 0_{N/2} \\ \hline 0_{N/2} & I_{N/2} \end{array}\right] \left[\begin{array}{c|c} W_{N/2} & 0_{N/2} \\ \hline 0_{N/2} & I_{N/2} \end{array}\right] W_N \mathbf{v}. \tag{4.23}$$

Notation for Iterated Wavelet Transforms

As we can see, writing down the matrix products gets quite cumbersome. We introduce some new notation so that we can easily write down iterated wavelet transformations in terms of matrices.

We add an index to W_N to identify the iteration. The transform matrix will be indexed both by the length N of the input vector $\mathbf{v}$ and the iteration. Suppose that N is divisible by 2^j for some positive integer j. We set

$$W_{N,\,1} = W_N \qquad \text{and} \qquad W_{N,\,2} = \left[\begin{array}{c|c} W_{N/2} & 0_{N/2} \\ \hline 0_{N/2} & I_{N/2} \end{array}\right].$$

In a similar fashion, we have

$$W_{N,\,3} = \left[\begin{array}{c|c} \begin{array}{c|c} W_{N/4} & 0_{N/4} \\ \hline 0_{N/4} & I_{N/4} \end{array} & 0_{N/2} \\ \hline 0_{N/2} & I_{N/2} \end{array}\right].$$

Using our new notation, (4.22) becomes

$$\mathbf{y}^2 = W_{N,\,2} W_N \mathbf{v}$$

and (4.23) can be expressed as

$$\mathbf{y}^3 = W_{N,\,3} W_{N,\,2} W_N \mathbf{v}.$$

If we were to continue in this fashion, it would be quite difficult to display $W_{N,\,k}$, where $1 \le k \le j$, in matrix form. Note that $W_{N,\,k}$ is a block matrix where the only nonzero blocks occur on the main diagonal. In particular, the first block is $W_{N/2^k}$ and the remaining diagonal blocks are $I_{N/2^k}, I_{N/2^{k-1}}, \ldots, I_{N/4}, I_{N/2}$.

Notation. For the kth iteration, $1 \le k \le j$, of the Haar wavelet transform, we can write

$$\boxed{W_{N,\,k} = \begin{cases} W_N, & k = 1 \\ \operatorname{diag}\left[W_{N/2^{k-1}}, I_{N/2^{k-1}}, I_{N/2^{k-2}}, \ldots, I_{N/2}\right], & 2 \le k \le j. \end{cases}} \tag{4.24}$$

where $\operatorname{diag}[C_1, C_2, \ldots, C_n]$ is a matrix that has the blocks $C_1, \ldots, C_n$ on its diagonal and zeros elsewhere.

If we wish to compute j iterations of the wavelet transformation applied to $\mathbf{v} \in \mathbb{R}^N$, where N is divisible by 2^j, we calculate

$$\mathbf{y}^j = W_{N,\,j} W_{N,\,j-1} \cdots W_{N,\,2} W_N \mathbf{v}. \tag{4.25}$$

Iterating the Inverse Haar Wavelet Transformation

The matrix product formulation of the iterated Haar wavelet transformation allows us to easily write down the iterated inverse Haar wavelet transform. In Problem 4.22 you will show that

$$\frac{1}{2^{k-1}} + \frac{1}{2^{k-1}} + \frac{1}{2^{k-2}} + \cdots + \frac{1}{4} + \frac{1}{2} = 1$$

so that the sum of the dimensions of each block in (4.24) is

$$N = \frac{N}{2^{k-1}} + \frac{N}{2^{k-1}} + \frac{N}{2^{k-2}} + \cdots + \frac{N}{4} + \frac{N}{2}.$$

This fact allows us to use Problem 2.46 from Section 2.3 to write

$$\begin{aligned} W_{N,\,k}^{-1} &= \operatorname{diag}\left[W_{N/2^{k-1}}^{-1}, I_{N/2^{k-1}}^{-1}, I_{N/2^{k-2}}^{-1}, \ldots, I_{N/4}^{-1}, I_{N/2}^{-1}\right] \\ &= \operatorname{diag}\left[2W_{N/2^{k-1}}^{T}, I_{N/2^{k-1}}, I_{N/2^{k-2}}, \ldots, I_{N/4}, I_{N/2}\right]. \end{aligned} \tag{4.26}$$

Here we have used (4.10) from the previous section. We can use (4.26), (4.25), and repeated use of Proposition 2.1(c) to recover $\mathbf{v}$:

$$\begin{aligned} \mathbf{v} &= \left(W_{N,\,j} W_{N,\,j-1} \cdots W_{N,\,2} W_N\right)^{-1} \mathbf{y}^j \\ &= W_N^{-1} W_{N,\,2}^{-1} \cdots W_{N,\,j-1}^{-1} W_{N,\,j}^{-1} \mathbf{y}^j. \end{aligned} \tag{4.27}$$

The notation for the iterated Haar wavelet transformation and its inverse can be a bit overwhelming. Let's look at an example.

Example 4.7 (Iterating the Inverse Haar Wavelet Transform). *Suppose* $\mathbf{v} \in \mathbb{R}^{40}$ *and that we have computed* $\mathbf{y}^3 = W_{40,\,3} W_{40,\,2} W_{40} \mathbf{v}$. *We want to recover* $\mathbf{v}$ *from* $\mathbf{y}^3$. *Recall that*

$$\mathbf{y}^3 = \begin{bmatrix} \mathbf{a}^3 \\ \hline \mathbf{d}^3 \\ \hline \mathbf{d}^2 \\ \hline \mathbf{d}^1 \end{bmatrix}.$$

The first thing we need to do is recover $\mathbf{a}^2 \in \mathbb{R}^{10}$ *from* $\mathbf{a}^3, \mathbf{d}^3 \in \mathbb{R}^5$. *We compute*

$$\mathbf{a}^2 = W_{10}^{-1} \begin{bmatrix} \mathbf{a}^3 \\ \hline \mathbf{d}^3 \end{bmatrix}.$$

Thus, we have

$$\mathbf{y}^2 = \begin{bmatrix} \mathbf{a}^2 \\ \hline \mathbf{d}^2 \\ \hline \mathbf{d}^1 \end{bmatrix} = \begin{bmatrix} W_{10}^{-1} \begin{bmatrix} \mathbf{a}^3 \\ \hline \mathbf{d}^3 \end{bmatrix} \\ \hline \mathbf{d}^2 \\ \hline \mathbf{d}^1 \end{bmatrix} = \begin{bmatrix} \left[\begin{array}{c|c} W_{10}^{-1} & 0_{10} \\ \hline 0_{10} & I_{10} \end{array}\right] \cdot \begin{bmatrix} \mathbf{a}^3 \\ \hline \mathbf{d}^3 \\ \hline \mathbf{d}^2 \end{bmatrix} \\ \hline \mathbf{d}^1 \end{bmatrix}$$

$$= \left[\begin{array}{c|c} \begin{array}{c|c} W_{10}^{-1} & 0_{10} \\ \hline 0_{10} & I_{10} \end{array} & 0_{20} \\ \hline 0_{20} & I_{20} \end{array}\right] \cdot \begin{bmatrix} \mathbf{a}^3 \\ \hline \mathbf{d}^3 \\ \hline \mathbf{d}^2 \\ \hline \mathbf{d}^1 \end{bmatrix} = W_{40,3}^{-1}\mathbf{y}^3.$$

In a similar fashion,

$$\mathbf{y}^1 = W_{40,2}^{-1}\mathbf{y}^2 = W_{40,2}^{-1}W_{40,3}^{-1}\mathbf{y}^3$$

so that we can recover $\mathbf{v}$ *by computing*

$$\mathbf{v} = W_{40}^{-1}\mathbf{y}^1 = W_{40}^{-1}W_{40,2}^{-1}\mathbf{y}^2 = W_{40}^{-1}W_{40,2}^{-1}W_{40,3}^{-1}\mathbf{y}^3.$$

□

PROBLEMS

4.15 Let $\mathbf{v} = [1, 2, 3, 4, 5, 6, 7, 8]^T$. Compute three iterations of the Haar wavelet transform. You should work this problem by hand.

4.16 Compute three iterations of the inverse Haar wavelet transformation on the vector $\mathbf{y}$ from Problem 4.15. You should work this problem by hand.

4.17 Suppose that $\mathbf{v} \in \mathbf{R}^N$ where $N = 2^p$ for some $p = 1, 2, \ldots$ Further suppose that $v_k = c$ for $k = 1, \ldots, N$ where c is any real number. Suppose that we apply p iterations of the Haar wavelet transformation to $\mathbf{v}$. What is the resulting vector?

4.18 Let $\mathbf{v} \in \mathbb{R}^N$ where N is divisible by four and suppose that $v_k = k$ for $k = 1, \ldots, N$. Then by Problem 3.45 in Section 3.4, we know that $\text{Ent}(\mathbf{v}) = \log_2(N)$.

(a) Compute two iterations of the Haar wavelet transform.

(b) If we call the result in (a) $\mathbf{y}$, what is $\text{Ent}(\mathbf{y})$? Compare this result with Problem 4.11 of Section 4.1.

4.19 Let $\mathbf{v} \in \mathbb{R}^N$ where $N = 2^p$ for some $p = 1, 2, \ldots$, and suppose that $v_k = k$, $k = 1, \ldots, N$. Let $\mathbf{y}$ be the vector that results from computing p iterations of the Haar wavelet transform. What is $\text{Ent}(\mathbf{y})$?

4.20 Suppose that $\mathbf{v} \in \mathbb{R}^N$, where N is divisible by four and the elements in $\mathbf{v}$ are quadratic (i.e., $v_k = k^2$, $k = 1, \ldots, N$). Let $\mathbf{y}^{(2)}$ be the result of applying two iterations of the Haar wavelet transformation to $\mathbf{v}$. Show that the elements of the first differences portion $\mathbf{d}^{(1)} \in \mathbb{R}^{N/2}$ are linear (i.e., $d_k^{(1)} = m_1 k + b_1$ for some real numbers m_1 and b_1). Find the values of m_1 and b_1. Show that the elements of the second differences portion $\mathbf{d}^{(2)}$ are linear as well and find m_2, b_2 so that $d_k^{(2)} = m_2 k + b_2$.

4.21 Create a vector $\mathbf{v}$ of length eight and compute three iterations of the Haar wavelet transformation to obtain $\mathbf{y}^3$. Then $\mathbf{a}^3$ is a vector with one element, say a. Can you describe a in terms of the elements of $\mathbf{v}$? Repeat the problem with a different vector. Does the same description of a hold in this case? In general, in the case where $\mathbf{v}$ is of length 2^p and p iterations of the Haar wavelet transformation are computed, does the description of a hold?

4.22 Show that

$$\frac{1}{2^{k-1}} + \frac{1}{2^{k-1}} + \frac{1}{2^{k-2}} + \cdots + \frac{1}{4} + \frac{1}{2} = 1.$$

Hint: Get a common denominator for the left-hand side and then use the closed formula for geometric sums found in a calculus book (see for example [84]) to simplify the numerator.

4.23 Suppose we used the orthogonal Haar wavelet transformation (4.11) to form the matrix factors (4.24). How would the inverse matrix (4.26) change? What is an alternate formulation of (4.27)?

Computer Lab

4.2 Iterating the Haar Wavelet Transform. From `stthomas.edu/wavelets`, access the file `HWTIterated`. In this lab you will write code that computes the iterated (inverse) Haar wavelet transformation of a vector.

4.3 The Two-Dimensional Haar Wavelet Transformation

In Section 4.2, we learned how to apply the Haar wavelet transformation iteratively to a vector. We can use this transform on a variety of applications involving one-dimensional data. In this section we learn how to extend the transform so that we can apply it to two-dimensional data. Such a transform can be used in applications involving images. In Chapter 3 we learned that a grayscale digital image can be expressed as a matrix. We first learn how to perform one iteration of the two-dimensional Haar wavelet transformation and the iterative process will follow in a very natural way.

Processing the Columns of a Matrix with the Haar Wavelet Transform

Let's suppose that we have an $N \times N$ matrix A, where N is even (later, we consider rectangular matrices). We denote the columns of A by $\mathbf{a}^1, \mathbf{a}^2, \ldots, \mathbf{a}^N$. If we compute $W_N\mathbf{a}^1$, we are simply transforming the first column of A into weighted averages and differences. We continue by applying W_N to the other columns of A to obtain $W_N\mathbf{a}^1, W_N\mathbf{a}^2, \ldots, W_N\mathbf{a}^N$. But these computations are precisely what we would get if we computed $W_N A$. So applying W_N to A processes all the columns of A.

Example 4.8 (Processing Matrix Columns with the Haar Wavelet Transform). *Apply W_4 to the matrix*

$$A = \begin{bmatrix} 16 & 20 & 14 & 8 \\ 8 & 10 & 0 & 12 \\ 11 & 12 & 9 & 20 \\ 9 & 10 & 15 & 10 \end{bmatrix}$$

Solution.

We have

$$W_4 A = \left[\begin{array}{cccc} \frac{1}{2} & \frac{1}{2} & 0 & 0 \\ 0 & 0 & \frac{1}{2} & \frac{1}{2} \\ \hline -\frac{1}{2} & \frac{1}{2} & 0 & 0 \\ 0 & 0 & -\frac{1}{2} & \frac{1}{2} \end{array}\right] \begin{bmatrix} 16 & 20 & 14 & 8 \\ 8 & 10 & 0 & 12 \\ 11 & 12 & 9 & 20 \\ 9 & 10 & 15 & 10 \end{bmatrix} = \left[\begin{array}{cccc} 12 & 15 & 7 & 10 \\ 10 & 11 & 12 & 15 \\ \hline -4 & -5 & -7 & 2 \\ -1 & -1 & 3 & -5 \end{array}\right].$$

Note that each column of A has indeed been transformed to a corresponding column of weighted averages and differences.

(Live example: Visit `stthomas.edu/wavelets` *and click on Live Examples.)*

□

Now that we have seen how W_N works on a small example, let's try it on a image.

Example 4.9 (Processing Image Columns with the Haar Wavelet Transform). *A 512×512 grayscale image A is plotted in Figure 4.11(a). The plot in Figure 4.11(b) is the product $W_{512}A$. Notice that the bottom half of $W_{512}A$ represents the differences portion of each column of A. If there is not much difference in consecutive pixel values, the value is close to zero and plotted as black. The bigger the value, the lighter the pixel intensity.*

(Live example: Visit `stthomas.edu/wavelets` *and click on Live Examples.)*

□

Processing Matrix Rows with the Haar Wavelet Transform

We now know how to process the columns of a image. Since our images are two-dimensional, it is natural to expect our transformation to process the rows as well.

(a) The image A

(b) $W_{512}A$

Figure 4.11 The Haar wavelet transformation applied to the columns of an image.

We learned in Chapter 2 that we can process the rows of $N \times N$ matrix A if we right-multiply A by some matrix. Since we want averages and differences, a natural choice for this product is W_N.

Let's consider a simple case. Suppose A is a 4×4 matrix. If we compute AW_4, we have

$$AW_4 = \begin{bmatrix} a_{11} & a_{12} & a_{13} & a_{14} \\ a_{21} & a_{22} & a_{23} & a_{24} \\ a_{31} & a_{32} & a_{33} & a_{34} \\ a_{41} & a_{42} & a_{43} & a_{44} \end{bmatrix} \left[\begin{array}{cccc} \frac{1}{2} & \frac{1}{2} & 0 & 0 \\ 0 & 0 & \frac{1}{2} & \frac{1}{2} \\ \hline -\frac{1}{2} & \frac{1}{2} & 0 & 0 \\ 0 & 0 & -\frac{1}{2} & \frac{1}{2} \end{array}\right]$$

$$= \frac{1}{2} \begin{bmatrix} a_{11} - a_{13} & a_{11} + a_{13} & a_{12} - a_{14} & a_{12} + a_{14} \\ a_{21} - a_{23} & a_{21} + a_{23} & a_{22} - a_{24} & a_{22} + a_{24} \\ a_{31} - a_{33} & a_{31} + a_{33} & a_{32} - a_{34} & a_{32} + a_{34} \\ a_{41} - a_{43} & a_{41} + a_{43} & a_{42} - a_{44} & a_{42} + a_{44} \end{bmatrix}$$

and this result is not what we want. If we are really going to process rows, our output should be

$$\frac{1}{2} \left[\begin{array}{cc|cc} a_{11} + a_{12} & a_{13} + a_{14} & -a_{11} + a_{12} & -a_{13} + a_{14} \\ a_{21} + a_{22} & a_{23} + a_{24} & -a_{21} + a_{22} & -a_{23} + a_{24} \\ a_{31} + a_{32} & a_{33} + a_{34} & -a_{31} + a_{32} & -a_{33} + a_{34} \\ a_{41} + a_{42} & a_{43} + a_{44} & -a_{41} + a_{42} & -a_{43} + a_{44} \end{array}\right].$$

To obtain this result, we must multiply A on the right by W_4^T instead of W_4. In the case that A is an $N \times N$ matrix, the product AW_N^T processes the rows of A into weighted averages and differences. Let's look at an example.

Example 4.10 (Processing Image Rows with the Haar Wavelet Transform). *A* 512×512 *gray scale image is plotted in Figure 4.12(a). The plot in Figure 4.12(b) is the product* AW_{512}^T*. Notice that the right half of* AW_{512}^T *represents the differences portion of each row of A. If there is not much difference in consecutive pixel values, the value is close to zero and plotted as black. The bigger the value, the lighter the pixel intensity.*

(a) The image A (b) AW_{512}^T

Figure 4.12 The Haar wavelet transformation applied to the rows of an image.

(Live example: Visit `stthomas.edu/wavelets` *and click on Live Examples.)*

□

The Two-Dimensional Transform as a Product of Matrices

How do we put all this together? We want to process the rows and the columns of A so we need to multiply on the left by W_N and on the right by W_N^T. Thus our two-dimensional transformation seems to take the form $W_N A W_N^T$. Does this make sense? Let's try it out on an arbitrary 4×4 matrix and analyze the result.

$$W_4 A W_4^T = \left[\begin{array}{cccc} \frac{1}{2} & \frac{1}{2} & 0 & 0 \\ 0 & 0 & \frac{1}{2} & \frac{1}{2} \\ \hline -\frac{1}{2} & \frac{1}{2} & 0 & 0 \\ 0 & 0 & -\frac{1}{2} & \frac{1}{2} \end{array}\right] \begin{bmatrix} a_{11} & a_{12} & a_{13} & a_{14} \\ a_{21} & a_{22} & a_{23} & a_{24} \\ a_{31} & a_{32} & a_{33} & a_{34} \\ a_{41} & a_{42} & a_{43} & a_{44} \end{bmatrix} \left[\begin{array}{cc|cc} \frac{1}{2} & 0 & -\frac{1}{2} & 0 \\ \frac{1}{2} & 0 & \frac{1}{2} & 0 \\ 0 & \frac{1}{2} & 0 & -\frac{1}{2} \\ 0 & \frac{1}{2} & 0 & \frac{1}{2} \end{array}\right].$$

We multiply $W_4 A$ first to obtain

$$W_4 A W_4^T = \begin{bmatrix} \frac{a_{11}+a_{21}}{2} & \frac{a_{12}+a_{22}}{2} & \frac{a_{13}+a_{23}}{2} & \frac{a_{14}+a_{24}}{2} \\ \frac{a_{31}+a_{41}}{2} & \frac{a_{32}+a_{42}}{2} & \frac{a_{33}+a_{43}}{2} & \frac{a_{34}+a_{44}}{2} \\ \frac{-a_{11}+a_{21}}{2} & \frac{-a_{12}+a_{22}}{2} & \frac{-a_{13}+a_{23}}{2} & \frac{-a_{14}+a_{24}}{2} \\ \frac{-a_{31}+a_{41}}{2} & \frac{-a_{32}+a_{42}}{2} & \frac{-a_{33}+a_{43}}{2} & \frac{-a_{34}+a_{44}}{2} \end{bmatrix} \left[\begin{array}{cc|cc} \frac{1}{2} & 0 & -\frac{1}{2} & 0 \\ \frac{1}{2} & 0 & \frac{1}{2} & 0 \\ 0 & \frac{1}{2} & 0 & -\frac{1}{2} \\ 0 & \frac{1}{2} & 0 & \frac{1}{2} \end{array}\right].$$

Computing the final product is better understood if we group the elements into four 2×2 blocks. Since these blocks are fundamental in understanding the geometric nature of the wavelet transformation both here and in subsequent chapters, we use special notation for them. We have

$$W_4 A W_4^T = \left[\begin{array}{c|c} \mathcal{B} & \mathcal{V} \\ \hline \mathcal{H} & \mathcal{D} \end{array}\right]$$

where

$$\begin{aligned} \mathcal{B} &= \frac{1}{4}\begin{bmatrix} a_{11}+a_{12}+a_{21}+a_{22} & a_{13}+a_{14}+a_{23}+a_{24} \\ a_{31}+a_{32}+a_{41}+a_{42} & a_{33}+a_{34}+a_{43}+a_{44} \end{bmatrix} \\ \mathcal{V} &= \frac{1}{4}\begin{bmatrix} (a_{12}+a_{22})-(a_{11}+a_{21}) & (a_{14}+a_{24})-(a_{23}+a_{13}) \\ (a_{32}+a_{42})-(a_{31}+a_{41}) & (a_{34}+a_{44})-(a_{33}+a_{43}) \end{bmatrix} \\ \mathcal{H} &= \frac{1}{4}\begin{bmatrix} (a_{21}+a_{22})-(a_{11}+a_{12}) & (a_{23}+a_{24})-(a_{13}+a_{14}) \\ (a_{41}+a_{42})-(a_{31}+a_{32}) & (a_{43}+a_{44})-(a_{33}+a_{34}) \end{bmatrix} \\ \mathcal{D} &= \frac{1}{4}\begin{bmatrix} (a_{11}+a_{22})-(a_{12}+a_{21}) & (a_{13}+a_{24})-(a_{14}+a_{23}) \\ (a_{31}+a_{42})-(a_{32}+a_{41}) & (a_{33}+a_{44})-(a_{34}+a_{43}) \end{bmatrix}. \end{aligned} \tag{4.28}$$

To analyze these blocks, we organize A into blocks and name each block:

$$A = \left[\begin{array}{cc|cc} a_{11} & a_{12} & a_{13} & a_{14} \\ a_{21} & a_{22} & a_{23} & a_{24} \\ \hline a_{31} & a_{32} & a_{33} & a_{34} \\ a_{41} & a_{42} & a_{43} & a_{44} \end{array}\right] = \left[\begin{array}{c|c} A_{11} & A_{12} \\ \hline A_{21} & A_{22} \end{array}\right].$$

Look at the $(1,1)$ element in each of $\mathcal{B}$, $\mathcal{V}$, $\mathcal{H}$, and $\mathcal{D}$. They are all constructed using the elements in A_{11}. In general, the i,j elements of $\mathcal{B}$, $\mathcal{V}$, $\mathcal{H}$, and $\mathcal{D}$ are constructed from elements of A_{ij}, $i,j = 1,2$.

Let's start with block $\mathcal{B}$. Note that each element b_{ij} of $\mathcal{B}$ is constructed by averaging the values of A_{ij}, $i, j = 1, 2$. For example,

$$A_{21} = \begin{bmatrix} a_{31} & a_{32} \\ a_{41} & a_{42} \end{bmatrix} \longrightarrow \frac{a_{31} + a_{32} + a_{41} + a_{42}}{4} = b_{21}.$$

We can thus think of $\mathcal{B}$ as an approximation or *blur* of A.

What about the other blocks? Let's look at the upper right block $\mathcal{V}$. Not only is each element v_{ij} constructed from the elements of A_{ij}, $i, j = 1, 2$, but each v_{ij} is constructed in exactly the same way. We first compute the sum of each of the two columns of A_{ij}. We subtract these sums and divide by four. For example,

$$A_{12} = \left[\boxed{\begin{matrix} a_{13} \\ a_{23} \end{matrix}} \quad \boxed{\begin{matrix} a_{14} \\ a_{24} \end{matrix}} \right] \longrightarrow \frac{(a_{14} + a_{24}) - (a_{23} + a_{13})}{4} = v_{12}.$$

We can interpret this computation geometrically as a weighted difference of the column sums in the block. With regards to a image, we can think of $\mathcal{V}$ as describing the *vertical changes* between the original image A and the blur $\mathcal{B}$.

In a similar way, we can view $\mathcal{H}$ as describing the *horizontal changes* between the original image A and the blur $\mathcal{B}$. Finally, the lower left block $\mathcal{D}$ measures *diagonal differences* between the original image A and the blur $\mathcal{B}$.

To provide a more formal explanation for the geometric significance of these blocks, suppose that A is an $N \times N$ matrix (N even). We write W_N in block format

$$W_N = \left[\begin{array}{c} H_{N/2} \\ \hline G_{N/2} \end{array} \right]$$

where $H_{N/2}$ and $G_{N/2}$ are $\frac{N}{2} \times N$ matrices. Matrices $H_{N/2}$ and $G_{N/2}$ are the averaging and differencing blocks, respectively, of W_N given in (4.7). We have

$$W_N A W_N^T = \left[\begin{array}{c} H \\ \hline G \end{array} \right] \cdot A \cdot \left[\begin{array}{c} H \\ \hline G \end{array} \right]^T = \left[\begin{array}{c} HA \\ \hline GA \end{array} \right] \left[H^T \,\middle|\, G^T \right]$$

$$= \left[\begin{array}{c|c} HAH^T & HAG^T \\ \hline GAH^T & GAG^T \end{array} \right] = \left[\begin{array}{c|c} \mathcal{B} & \mathcal{V} \\ \hline \mathcal{H} & \mathcal{D} \end{array} \right]. \tag{4.29}$$

Recall that H averages along the columns of A and H^T averages along the rows of A. So, if we multiply left to right, HAH^T first averages the columns of A and then computes the averages along the rows of HA. This creates the approximation (or blur) $\mathcal{B}$ of A.

The matrix G^T differences along the rows of A. So if we again multiply left to right, we see that the product HAG^T first computes the averages along columns of A and then differences along rows of HA. This is exactly the vertical differences described above.

The matrix GAH^T first differences along columns of A and then averages along rows of GA. This product produces the horizontal differences stored in $\mathcal{H}$. Finally the matrix GAG^T first differences along the columns of A and next differences along the rows of GA, forming the diagonal differences found in $\mathcal{D}$. The schematic in Figure 4.13 summarizes our discussion to this point.

$$A \longmapsto W_N A W_N^T = \left[\begin{array}{c|c} \mathcal{B} & \mathcal{V} \\ \hline \mathcal{H} & \mathcal{D} \end{array}\right] = \left[\begin{array}{c|c} \text{blur} & \begin{array}{c}\text{vertical}\\ \text{differences}\end{array} \\ \hline \begin{array}{c}\text{horizontal}\\ \text{differences}\end{array} & \begin{array}{c}\text{diagonal}\\ \text{differences}\end{array} \end{array}\right]$$

Figure 4.13 The two-dimensional Haar wavelet transformation takes matrix A and maps it to a new matrix consisting of four blocks. Block $\mathcal{B}$ is a blur or approximation of A, block $\mathcal{V}$ represents the vertical differences between the original and the blur, and blocks $\mathcal{H}$ and $\mathcal{D}$ represent the horizontal and diagonal differences between A and the blur, respectively.

The following example further illustrates the block structure of the two-dimensional Haar wavelet transform.

Example 4.11 (Applying the Haar Wavelet Transform to an Image). *We return to the image used in Examples 4.9 and 4.10. We plot it and its two-dimensional Haar wavelet transformation in Figure 4.14. The upper left block is a blur of the original. The upper right block represents vertical differences between A and the blur, the lower left block gives horizontal differences between A and the blur, and the bottom right block show diagonal differences between A and the blur.*

(a) The image A (b) $W_{512}AW_{512}^T$

Figure 4.14 The Haar wavelet transformation applied to an image.

(Live example: Visit `stthomas.edu/wavelets` *and click on Live Examples.)*

□

Rectangular Matrices and the Haar Wavelet Transform

Rectangular matrices present no problems for the two-dimensional Haar wavelet transformation as long as the dimensions are even. Suppose that A is an $N \times M$ matrix with N and M even. Since each column of A has N elements, we multiply on the left by W_N. The result, $W_N A$, is an $N \times M$ matrix. Each row in $W_N A$ has M elements, so we need to process the rows by multiplying $W_N A$ on the right by W_M^T. The result will be an $N \times M$ matrix.

Here is an example.

Example 4.12 (Applying the Haar Wavelet Transform to a Rectangular Image). *The image A shown in Figure 4.15(a) is a 512×768 matrix. It is plotted along with its two-dimensional Haar wavelet transformation $W_{512} A W_{768}^T$ in Figure 4.15(b).*

(a) The image A (b) $W_{512} A W_{768}^T$

Figure 4.15 The Haar wavelet transformation applied to a rectangular image.

(Live example: Visit `stthomas.edu/wavelets` *and click on Live Examples.)*

□

Inverting the Two-Dimensional Transform

Inverting the two-dimensional Haar wavelet transform is a simple process. Suppose that A is an $N \times M$ matrix and we have computed the Haar wavelet transform

$$C = W_N A W_M^T. \tag{4.30}$$

We can exploit the fact that W_N and W_M are near orthogonal matrices with inverses $2W_N^T$ and $2W_M^T$, respectively, to recover A. We multiply both sides of (4.30) on the left by $2W_N^T$ and $2W_M$ on the right to obtain

$$\begin{aligned}(2W_N^T)\, C\, (2W_M) &= (2W_N^T)\, (W_N A W_M^T)\, (2W_M) \\ &= (2W_N^T W_N) A (2W_M^T W_M) \\ &= I_N A I_M \\ &= A. \end{aligned} \tag{4.31}$$

Here I_N and I_M are $N \times N$ and $M \times M$ identity matrices, respectively.

Now that we know how to perform a two-dimensional Haar wavelet transformation and its inverse, let's write algorithms for both operations. We start with the two-dimensional Haar wavelet transform.

Iterating the Process

In Section 4.2 we learned how to iterate the one-dimensional Haar wavelet transform. We applied the Haar wavelet transformation successively to the averages portion of the output of the preceding step. We can easily extend this iterative process to two-dimensional data.

Consider the $N \times M$ (N,M divisible by four) matrix A and its Haar wavelet transformation $C_1 = W_N A W_M^T$. In block form,

$$W_N A W_M^T = \left[\begin{array}{c|c} \mathcal{B}_1 & \mathcal{V}_1 \\ \hline \mathcal{H}_1 & \mathcal{D}_1 \end{array}\right]$$

where $\mathcal{B}_1$, $\mathcal{V}_1$, $\mathcal{H}_1$, and $\mathcal{D}_1$ are summarized in Figure 4.13. In particular, $\mathcal{B}_1$ is the blur (or approximation) of A. We also know that $\mathcal{B}_1$ is obtained by applying the averages portions of W_N and W_M^T to A. We perform a second iteration by applying the Haar wavelet transformation to $\mathcal{B}_1$. That is, we compute

$$W_{N/2} \mathcal{B}_1 W_{M/2}^T \tag{4.32}$$

and overwrite $\mathcal{B}_1$ with this product. This process is entirely invertible – to recover $\mathcal{B}_1$, we pre- and post-multiply (4.32) by $2W_{N/2}^T$ and $2W_{M/2}$, respectively. In block matrix form, the two iterations of the Haar wavelet transformation look like

$$A \mapsto \left[\begin{array}{c|c} \mathcal{B}_1 & \mathcal{V}_1 \\ \hline \mathcal{H}_1 & \mathcal{D}_1 \end{array}\right] \mapsto \left[\begin{array}{c|c} \begin{array}{c|c} \mathcal{B}_2 & \mathcal{V}_2 \\ \hline \mathcal{H}_2 & \mathcal{D}_2 \end{array} & \mathcal{V}_1 \\ \hline \mathcal{H}_1 & \mathcal{D}_1 \end{array}\right].$$

Let's look at an example.

(a) The image A (b) The first iteration $W_{512}AW_{512}^T$

Figure 4.16 One iteration of the Haar wavelet transformation applied to an image.

Example 4.13 (Iterating the Two-Dimensional Haar Wavelet Transform). *We return to the* 512×512 *image introduced in Example 4.9. We first plot the original image (denoted by A) and its Haar wavelet transformation in Figure 4.16.*

We now extract the upper left corner $\mathcal{B}_1$ *which is a blur of A. We compute the Haar wavelet transformation of* $\mathcal{B}_1$, $W_{256}\mathcal{B}_1W_{256}^T$. *An enlarged version of the blur* $\mathcal{B}_1$ *and the resulting transform are plotted in Figure 4.17.*

(a) The blur $\mathcal{B}_1$ (b) The product $W_{256}\mathcal{B}_1W_{256}^T$

Figure 4.17 A second application of the Haar wavelet transformation to $\mathcal{B}_1$. Note that $\mathcal{B}_1$ has been enlarged for purposes of illustration.

Finally, we replace the upper left corner $\mathcal{B}_1$ *with* $W_{256}\mathcal{B}_1W_{256}^T$ *to produce a composite image of two iterations of the Haar wavelet transform. The result is plotted in Figure 4.18(a). The cumulative energies for A, and the two transformations are plotted in Figure 4.18(b).*

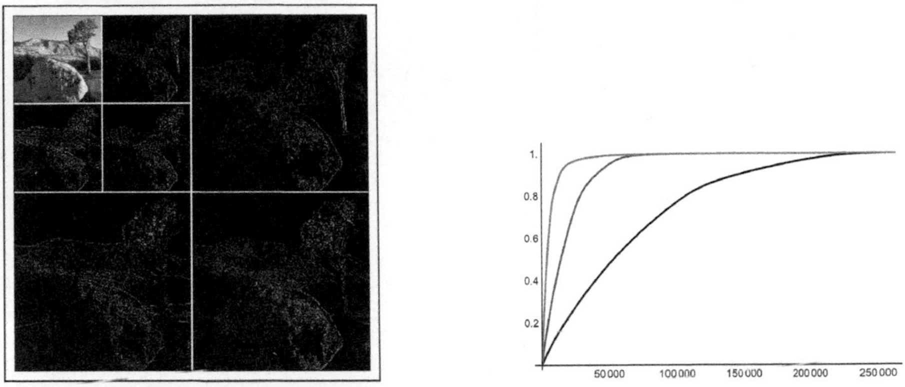

(a) Two iterations of the Haar wavelet transform

(b) Cumulative energies

Figure 4.18 Plot of two iterations of the Haar wavelet transformation of matrix A and the cumulative energies for A (black), the first iteration of the Haar wavelet transformation (gray), and the second iteration of the Haar wavelet transformation (light gray).

Clearly, the second iteration of the Haar wavelet transformation needs fewer elements to store more of the energy than does the first iteration or the original image. We would probably do better if we were to iterate a third time.

(Live example: Visit `stthomas.edu/wavelets` *and click on Live Examples.)*

□

Let A be an $N \times M$ matrix with N, M divisible by 2^i. Then we can continue the iterative process as many as i times. For $i = 3$, we have the iterated transform

$$A \mapsto \left[\begin{array}{c|c} \mathcal{B}_1 & \mathcal{V}_1 \\ \hline \mathcal{H}_1 & \mathcal{D}_1 \end{array}\right] \mapsto \left[\begin{array}{c|c} \begin{array}{c|c} \mathcal{B}_2 & \mathcal{V}_2 \\ \hline \mathcal{H}_2 & \mathcal{D}_2 \end{array} & \mathcal{V}_1 \\ \hline \mathcal{H}_1 & \mathcal{D}_1 \end{array}\right] \mapsto \left[\begin{array}{c|c} \begin{array}{c|c} \begin{array}{c|c} \mathcal{B}_3 & \mathcal{V}_3 \\ \hline \mathcal{H}_3 & \mathcal{D}_3 \end{array} & \mathcal{V}_2 \\ \hline \mathcal{H}_2 & \mathcal{D}_2 \end{array} & \mathcal{V}_1 \\ \hline \mathcal{H}_1 & \mathcal{D}_1 \end{array}\right]$$

and the inverse

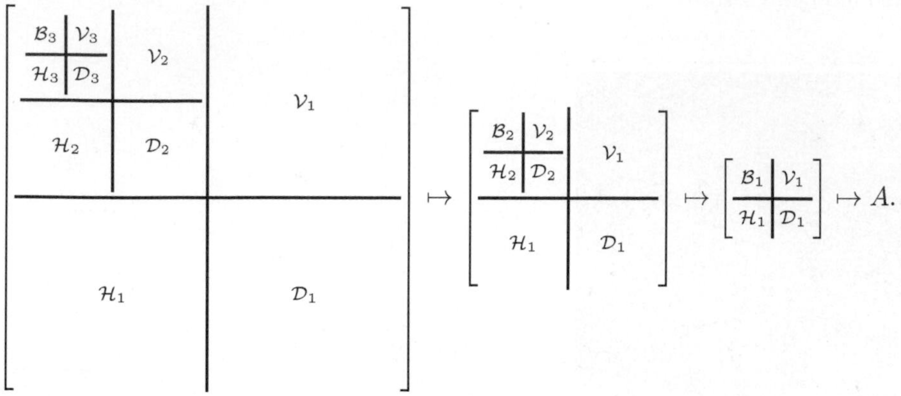

$\mathcal{B}_3$ $\mathcal{V}_3$ $\mathcal{H}_3$ $\mathcal{D}_3$ $\mathcal{V}_2$ $\mathcal{H}_2$ $\mathcal{D}_2$ $\mathcal{V}_1$ $\mathcal{H}_1$ $\mathcal{D}_1$
$\mapsto$
$\mathcal{B}_2$ $\mathcal{V}_2$ $\mathcal{H}_2$ $\mathcal{D}_2$ $\mathcal{V}_1$ $\mathcal{H}_1$ $\mathcal{D}_1$
$\mapsto$
$\mathcal{B}_1$ $\mathcal{V}_1$ $\mathcal{H}_1$ $\mathcal{D}_1$
$\mapsto A.$

Transforming Color Images

We conclude this section with a discussion of how to apply the Haar wavelet transformation to color images. Recall from Chapter 3 that a color image is constructed using three matrices. The elements of matrix R correspond to the red intensities of the image (with range $0-255$), elements of matrix G hold the green intensities, and matrix B houses the blue intensities.

There are some applications where i iterations of the Haar wavelet transformation are applied directly to the R, G, B channels of a color image, but a more common method used in applications such as image compression is to first convert the R, G, and B components to YCbCr space and then apply the iterated Haar wavelet transformation to each of Y, Cb, and Cr.

Example 4.14 (The Haar Wavelet Transform of a Color Image). *Consider the* 256×384 *image and red, green, and blue color channels plotted in Figure 3.25. We apply two iterations of the Haar wavelet transformation to each of R, G, and B. The resulting transformations are plotted in Figure 4.19(a) – 4.19(c).*

For many applications, we first convert the image to YCbCr space. The conversion is plotted in Figure 3.27. In Figure 4.19(d) – 4.19(f) we have plotted the Haar wavelet transformation of each of the Y, Cb, *and* Cr *channels.*

(Live example: Visit `stthomas.edu/wavelets` *and click on Live Examples.)*

□

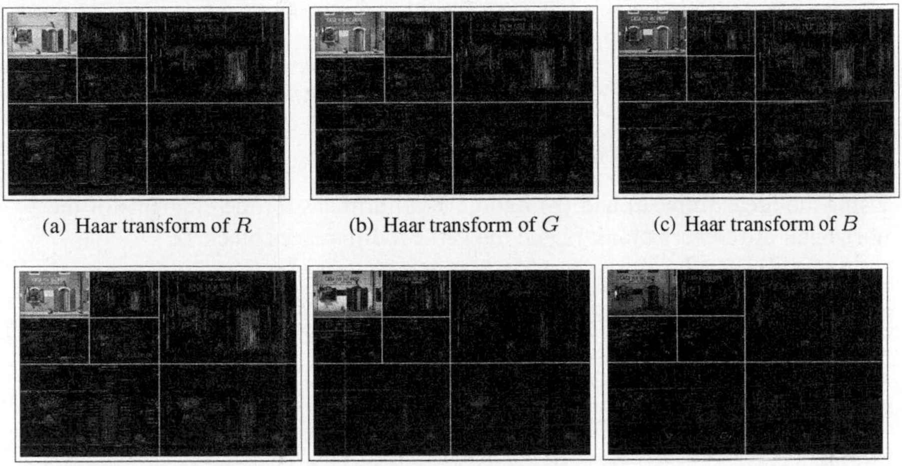

(a) Haar transform of R (b) Haar transform of G (c) Haar transform of B

(d) Haar transform of Y (e) Haar transform of Cb (f) Haar transform of Cr

Figure 4.19 The Haar wavelet transformation of each of the color channels R, G, and B in Figure 3.25(b) – 3.25(d) and the Y, Cb, and Cr channels. See the color plates for color versions of these images.

PROBLEMS

★**4.24** Consider the matrix $A = \begin{bmatrix} a & b \\ c & d \end{bmatrix}$. Find the Haar wavelet transformation $C = W_2 A W_2^T$ of A.

4.25 Consider the matrix

$$A = \begin{bmatrix} 100 & 200 & 100 & 120 \\ 100 & 160 & 120 & 120 \\ 0 & 80 & 100 & 80 \\ 0 & 40 & 80 & 0 \end{bmatrix}$$

Compute two iterations of the two-dimensional Haar wavelet transform. You should do this computation by hand.

4.26 Consider the $M \times N$ matrix A with entries a_{jk}, $j = 1, \ldots, M$, $k = 1, \ldots, N$ where N and M are even. Let C denote one iteration of the Haar wavelet transformation applied to A. That is, $C = W_M A W_N^T$.

(a) Write down a general formula that describes all the c_{jk} in the blur or approximation block $\mathcal{B}$. The indices for c_{jk} in this block are $j = 1, \ldots, M/2$ and $k = 1, \ldots, N/2$. *Hint:* Each value c_{jk} is computed using the elements the 2×2 block matrix A_{jk}. Here

$$A_{jk} = \begin{bmatrix} a_{2j-1,2k-1} & a_{2j-1,2k} \\ a_{2j,2k-1} & a_{2j,2k} \end{bmatrix}. \tag{4.33}$$

The identity involving $\mathcal{B}$ in (4.28) can serve as a guide in the case where $M = N = 4$ or you might find Problem 4.24 helpful.

(b) Repeat (a) for the block $\mathcal{V}$ of vertical differences. *Hint:* The elements in $\mathcal{V}$ are $c_{j,k+N/2}$ where $j = 1, \ldots, M/2$ and $k = 1, \ldots, N/2$. As in (a), these values are computed using the elements of A_{jk}.

(c) Using the ideas from (a) and (b) find general formulas for the elements of the horizontal differences block $\mathcal{H}$ and the vertical differences block $\mathcal{D}$.

4.27 Suppose one iteration of the Haar wavelet transformation is applied to 4×8 matrix A to obtain

$$B = \left[\begin{array}{cccc|cccc} 23 & 13 & 14 & 11 & -5 & 3 & 2 & 3 \\ 12 & 15 & 19 & 17 & 0 & -5 & -5 & 1 \\ \hline -1 & -13 & -2 & -7 & -5 & -3 & -6 & 1 \\ 6 & -7 & -7 & 7 & -2 & -3 & -7 & 3 \end{array}\right].$$

(a) Use B to compute the second iteration of the Haar wavelet transform.

(b) Use B and the inverse Haar wavelet transformation to recover A.

You should do these computations by hand.

4.28 Let $C = \begin{bmatrix} w & x \\ y & z \end{bmatrix}$. You should think of C as the Haar transform of matrix A. That is $C = W_2^T A W_2$. We can recover A by using the formula (4.31) to write

$$A = \left(2W_2^T\right) C \left(2W_2\right) = 4W_2^T C W_2.$$

Write down the elements of A using this formula.

4.29 Suppose that C is an $M \times N$ matrix with elements c_{jk}, $j = 1, \ldots, M$ and $k = 1, \ldots, N$ where M and N are even. Let A be the matrix that results from the application of one iteration of the inverse Haar wavelet transformation to C. Then $A = \left(2W_M^T\right) C \left(2W_N\right) = 4W_M^T C W_N$.

(a) Write the entries of C in terms of the entries of A in the case where $M = 4$ and $N = 6$.

(b) You should notice in (a) that each 2×2 block A_{jk} (see (4.33)) is constructed using the same four elements of C. Use this fact to write formulas for $a_{2j-1,2k-1}$, $a_{2j-1,2k}$, $a_{2j,2k-1}$ and $a_{2j,2k}$ in terms of the correct four elements of C where $j = 1, \ldots, M/2$ and $k = 1, \ldots, N/2$.

★**4.30** Suppose that A is an $N \times N$ matrix (N even) that represents the intensity levels of some image and we apply one iteration of the Haar wavelet transformation to obtain C. In order to crudely compress the image, all elements of vertical, horizontal, and diagonal blocks of C are set to zero. That is, $\widetilde{C}$ has the block form

$$\widetilde{C} = \left[\begin{array}{c|c} \mathcal{B} & 0 \\ \hline 0 & 0 \end{array}\right]$$

We can approximate A by applying one iteration of the inverse Haar wavelet transformation to $\widetilde{C}$. Call the result $\widetilde{A}$.

(a) For $N = 4$, write down the elements $\widetilde{a}_{jk}$, $j, k = 1, \ldots, 4$ of $\widetilde{A}$ in terms of the elements $\widetilde{c}_{jk}$ of $\widetilde{C}$.

(b) Generalize your result for arbitrary even N. *Hint:* What is the value of $\widetilde{a}_{2j,2k}$ where $j, k = 1, \ldots, N/2$ in terms an element or elements of $\widetilde{C}$? Once you have that, find values for $\widetilde{a}_{2j-1,2k}$, $\widetilde{a}_{2j-1,2k-1}$ and $\widetilde{a}_{2j,2k-1}$?

4.31 Repeat Problem 4.30, but this time assume that N is divisible by four and apply two iterations of the Haar wavelet transformation to A. Again, we set the elements of the vertical, horizontal, and diagonal elements (of both iterations) to zero. Thus, the elements of $\widetilde{C}$ are zero save for the $\frac{N}{4} \times \frac{N}{4}$ block in its upper right-hand corner. Write down the elements of $\widetilde{A}$ in terms of the elements of $\widetilde{C}$. How does this compare to the result from Problem 4.30? What would happen if N were divisible by eight and we applied three iterations of the Haar wavelet transform?

4.32 In Section 4.2, we used block matrices given by (4.24) to compute two iterations of the Haar wavelet transformation of vector $\mathbf{v} \in \mathbb{R}^N$, where N is divisible by four. In particular, $\mathbf{y}^2 = W_{N,2} W_N \mathbf{v}$. It is tempting to use these block matrices to construct a matrix product that represents the iterated Haar wavelet transformation applied to an $N \times M$ matrix A where N, M are both divisible by four. Indeed, for one iteration, the product product

$$W_N A W_M^T = \left[\begin{array}{c|c} \mathcal{B}_1 & \mathcal{V}_1 \\ \hline \mathcal{H}_1 & \mathcal{D}_1 \end{array}\right]$$

produces the correct result. Why does the product $W_{N,2} W_N A W_M^T W_{M,2}^T$ fail to produce the second iteration of the Haar wavelet transform?

Computer Lab

4.3 The Two-Dimensional Haar Transform From `stthomas.edu/wavelets`, access the file `HWT2d`. In this lab you will write code that computes the iterated (inverse) Haar wavelet transformation of a matrix.

4.4 Applications: Image Compression and Edge Detection

We conclude Chapter 4 with two applications of the Haar wavelet transformation. The first application is image compression and the second is the detection of edges in digital images. It is not our intent to produce state-of-the-art procedures in either case. Indeed, we will see that much better wavelet transformations exist for image compression. The Haar wavelet transformation is easy to implement and understand, and thus it is a good tool for providing a basic understanding of how the wavelet transformation applies to these problems.

Image Compression

We conduct a detailed study of image compression when we learn about the FBI Wavelet Scalar Quantization specification in Section 10.3 and the JPEG2000 compression format in Chapter 12. For now, we study the basic algorithm for compressing an image.

The idea is as follows. Suppose that we want to send a digital image file to a friend via the internet. To expedite the process, we wish to *compress* the contents of the file. There are two forms of compression. The first form is *lossless compression.* Data compressed by lossless schemes can be recovered exactly with no loss of information. *Lossy compression* schemes alter the data in some way. Generally speaking, smaller bits per pixel ratios can be obtained using lossy schemes, but the savings come at the expense of information lost from the original data. We demonstrate a naive lossy compression scheme that utilizes the Haar wavelet transform. We will describe the process for two-dimensional data, but the method works just as well for univariate data.

Suppose that the intensities of a grayscale image of dimension $N \times M$ are stored in matrix A. Further assume that for some positive integer p, both N and M are divisible by 2^p. The first step in our process is to "center" the values in A. Recall that the elements of A are integers in the range $0, \ldots, 255$. We subtract 128 from each element and store the result in $\widetilde{A}$. Thus the values in $\widetilde{A}$ are integers ranging from -128 to 127. The purpose of this step is explained at the end of this subsection.

Next we compute i iterations, $1 \leq i \leq p$, of the Haar wavelet transformation of $\widetilde{A}$. The idea is that most of the information in the transform will be stored in the blur or approximation block while the vertical, horizontal, and diagonal components are sparse by comparison. The reason that this process works is that images are typically comprised of large homogeneous regions (think of a background in a still or the sky in an outdoor photograph). If there is little change in some part of an image, we expect the difference portions for that part to be comprised primarily of small values.

The third step is to *quantize* the components of the Haar wavelet transformation of $\widetilde{A}$. Any values of the transform that are "small" in absolute value are converted to zero. Those values not converted to zero are floored to the nearest integer. The final step is to Huffman code the integer values in the quantized transform. The resulting bit stream and the Huffman dictionary (i.e., the new codes for each gray intensity) constitute the compressed version of our original image.

There are a three questionable points about this algorithm. The first is the choice of wavelet transform. For now, we use the Haar wavelet transformation since that is what we have developed to this point. In Chapters 5 and 7 and in Section 9.5, we learn about filters that do a better job transforming the data so that coding is more efficient.

The second point is the quantization process. In particular, how do we determine if a value is "small enough" to convert to zero? In Chapters 6 and 12 we learn about some quantization schemes, but for this application we use cumulative energy as our quantization tool. That is, we first choose (somewhat arbitrarily) a percentage of energy that we wish to preserve in the transform and then retain only the m largest elements (in absolute value) that constitute this level of energy. The remaining elements in the transformed data are converted to zero.

The final item we need to address is the efficacy of Huffman coding with respect to the range of values in the quantized wavelet transform. A grayscale image consists of at most $256 = 2^8$ intensities and each of these intensities can be encoded using 8 bits for each. The idea of Huffman coding is to use variable length bit representations for each intensity in the quantized wavelet transform. If we do not implement the centering step described above, then the elements of quantized transform are integers in $[-128, 255]$ (see Problem 4.33(a)). In this case, it is possible that there are more distinct values (384) in the quantized transform than in the original image. In Problem 4.33(c), you will show that if the initial range of values is $[-128, 127]$ (the result of the centering step) then the range of values in the resulting Haar wavelet transformation is $[-128, 127.5]$. Flooring these values to the nearest integer results in the same range for the input values and the transformed values. Undoubtedly, the flooring step during quantization alters the transformed values (one iteration of the Haar transform returns integers, half- or quarter-integers), but this is to be expected in a lossy compression method.

Image Compression Example

Let's try the procedure on a digital image file.

Example 4.15 (Image Compression with the Haar Wavelet Transform). *Consider the 256×384 image with intensity matrix A plotted in Figure 4.20(a). Our goal is to compress A. The number of bits needed to store A is $256 \times 384 \times 8 = 786{,}432$. If we use a **CAS** to apply Huffman coding to the original image, we find that the new bit stream has length 759,868 or a coding rate of 7.72978 bpp. The entropy for the original image is 7.71119. Remember, the entropy serves as a guide for the best bpp we can expect if we encode the given data. We will perform lossy compression in hopes of reducing the bit stream length of A while maintaining an acceptable degree of image resolution.*

We center A by subtracting 128 from each entry of A to form $\widetilde{A}$. Let C represent two iterations of the Haar wavelet transformation of $\widetilde{A}$. This matrix is plotted in Figure 4.20(b).

In Figure 4.21, the cumulative energy vectors are plotted for both the original image and its wavelet transform. Note that the vectors each have $256 \cdot 384 = 98{,}304$ elements.

*Suppose we choose $r = 99.7\%$ for our energy threshold. We then use a **CAS** to find that the 39,419 largest elements (in absolute value) in the wavelet transformation C produce 99.7% of the energy. We retain these largest elements and floor them to the nearest integer. The remaining 58,885 elements of C are set to zero. Let's call this quantized wavelet transformation C_1. For comparison purposes, we repeat the process with energy levels $r = 99.9\%$ and $r = 99.99\%$ and produce quantized transformations C_2 and C_3. The results are summarized in Table 4.1.*

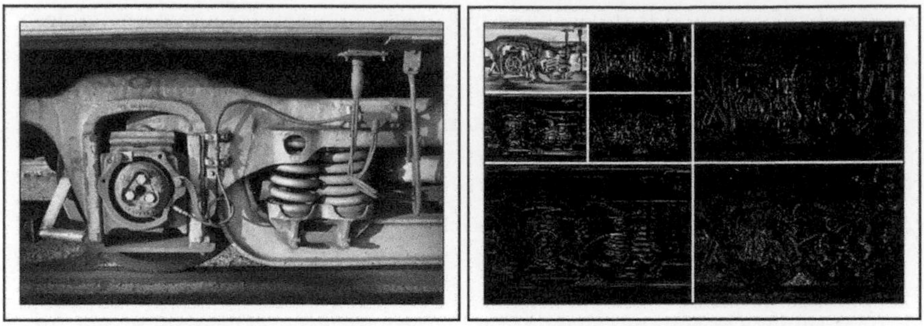

(a) Image A (b) Haar wavelet transform

Figure 4.20 An image and two iterations of the Haar wavelet transform.

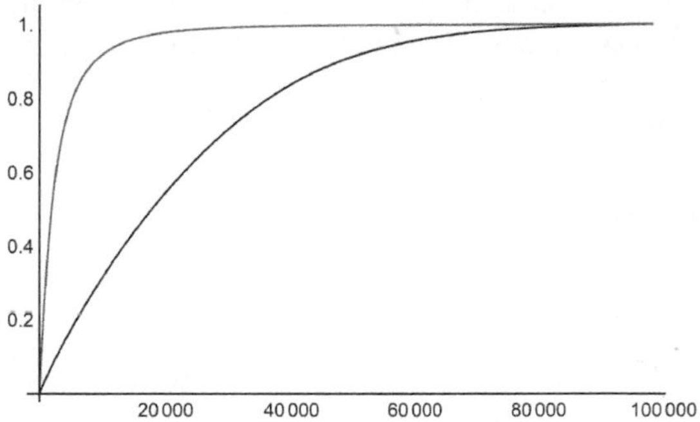

Figure 4.21 The cumulative energy vectors for the original image (black) and the wavelet transformation (gray).

The quantized wavelet transformations are plotted in Figure 4.22. Note that the amount of detail increases as we move from C_1 to C_3.

In Figure 4.23 we plot the distribution of pixel intensities for the original image and the quantized wavelet transformation C_3. The histograms for C_1 and C_2 are similar to that of C_3.

Before we transmit the image over the internet, we first convert the elements of the quantized wavelet transformation to Huffman codes and then create the bit stream for the file. This bit stream along with some header information (e.g., the new code for each intensity value) make up the compressed file. The length of the bit stream for each transform is given in Table 4.2. With regard to the original bit stream length of $8 \times 98{,}304 = 786{,}432$, the new bit stream lengths suggest that the quantized transformations will create storage sizes of roughly 43.6%, 52.6%, and 59.9%, respectively, of the original. The entropy gives an idea of the best bpp we could hope for when compressing each quantized transform. The bpp is given for each quantized transformation.

Table 4.1 Data for the quantized wavelet transformations C_1, C_2, and C_3.[a]

Transform	Energy Level %	Nonzero Terms	% Zero
C_1	99.7	34,419	59.9
C_2	99.9	53,197	43.9
C_3	99.99	69,998	28.8

[a]The second column indicates the amount of energy conserved by the transform, the third column lists the number of nonzero terms still present in the quantized transformations and the last column gives the percentage of elements that are converted to zero in each quantized transform. Note that the numbers in the nonzero terms column might include multiple copies of the smallest values (in absolute value) that comprise the desired energy level. Since it is impossible to know which of these values to keep or discard, the routine keeps all of them.

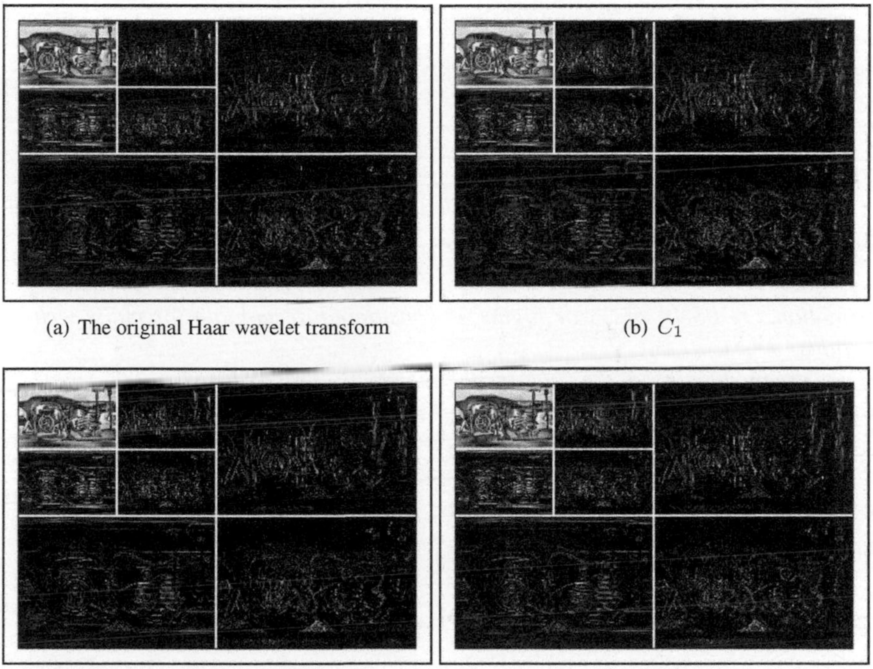

(a) The original Haar wavelet transform

(b) C_1

(c) C_2

(d) C_3

Figure 4.22 The Haar wavelet transformation and different quantized versions of it.

After inspecting the results in Table 4.2, we might be inclined to use either C_1 or C_2 to compress our image. Of course, it is very important to weigh this selection against the quality of the compressed image.

We now invert each of the quantized transformations to obtain the compressed images A_1, A_2, and A_3. These images, along with their PSNR when compared to A, are plotted in Figure 4.24. Recall that the higher the PSNR, the better the

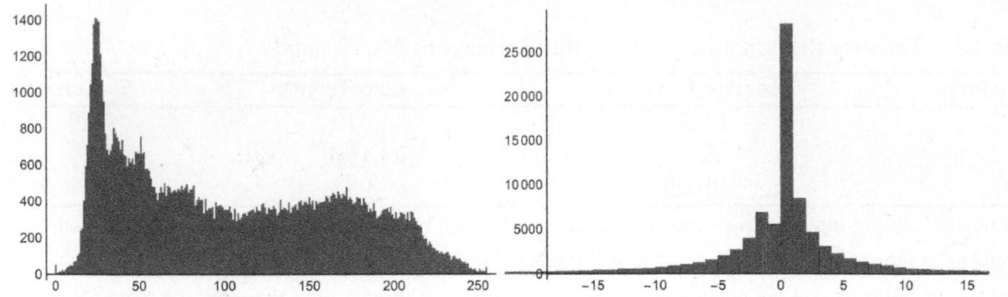

(a) Original image (b) C_3

Figure 4.23 Histograms of the original image and the quantized Haar wavelet transformation C_3.

Table 4.2 Huffman bit stream length, bpp, and entropy for each quantized transform C_1, C_2, and C_3.

Transform	Bit Stream Length	bpp	Entropy
C_1	342,669	3.48581	3.44703
C_2	413,846	4.20986	4.18646
C_3	471,167	4.79296	4.76188

approximation. A PSNR of 35 *or higher is considered acceptable for many web applications.*

As you can see by the plots of the compressed images, C_3 *produced a result that most resembles the original image. One of the most difficult problems in image compression is deciding how and when to sacrifice good compression numbers for resolution. Often, the application drives the decision.*

(Live example: Visit `stthomas.edu/wavelets` *and click on Live Examples.)*

□

Compression of Color Digital Images

In our next example we consider image compression for color images. What follows is a naive version of color image compression. In Chapter 12 we study the JPEG2000 image compression standard. This standard employs wavelet transformations that are better suited to the task of image compression.

Example 4.16 (Color Image Compression). *Consider the* 256×384 *image plotted in Figure 3.25. Rather than perform compression on each of the red, green, and blue channels (matrices), we first convert the image to YCbCr space using* (3.7). *By* (3.5) *and Problem 3.17 we know the elements of* Y *are integers in* $[16, 235]$ *and the elements of* Cb, Cr *are integers that reside in* $[16, 240]$. *We denote the resulting channels by* $\widetilde{Y}$, $\widetilde{Cb}$ *and* $\widetilde{Cr}$ *and "center" these channels by subtracting* 128 *from each value in each channel. Now each channel is centered with values in the interval*

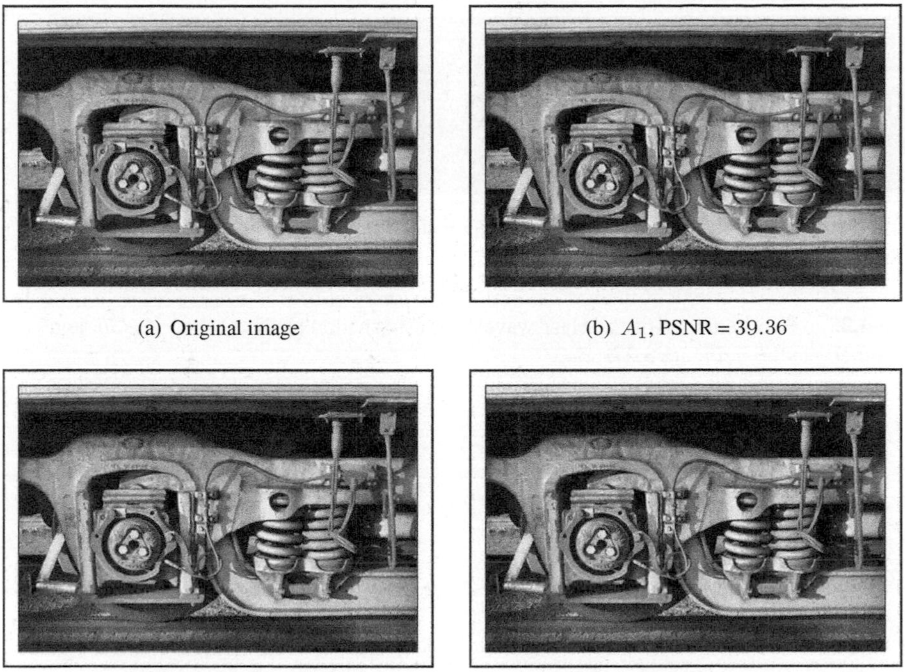

(a) Original image

(b) A_1, PSNR = 39.36

(c) A_2, PSNR = 43.38

(d) A_3, PSNR = 45.92

Figure 4.24 The original images and three compressed images.

$[-128, 127]$ and by Problem 4.33, we know the floored values of all Haar wavelet transformations are integers in the $[-128, 127]$ as well. As discussed in Section 3.2, most of the energy is stored in the Y channel and the human eye is less sensitive to changes in the Cb and Cr channels. The $\widetilde{Y}$, $\widetilde{Cb}$, and $\widetilde{Cr}$ channels for the image are plotted in Figure 3.27.

We begin by performing two iterations of the Haar wavelet transformation on each of the $\widetilde{Y}$, $\widetilde{Cb}$, and $\widetilde{Cr}$ channels. The transformations are plotted in Figure 4.25.

We next compute the cumulative energy vectors for each of the wavelet transformations. We keep the largest intensities (in absolute value), that constitute 99.2% of the energy contained in each channel. These values are floored to the nearest integer. The remaining intensities are set to zero. Figure 4.26 shows the cumulative energies for each transformed channel, and Table 4.3 summarizes the number of values we retain for compression. The quantized wavelet transformations are plotted in Figure 4.27.

Each of the R, G, and B channels of the original image consists of $256 \times 384 =$ 98,304 pixels so without any compression strategy, we must store $8 \cdot 98,304 =$ 786,432 bits or 8 bpp for each channel. The entropy values of the R, G, and B channels are 7.5819, 7.5831, and 7.3431, respectively.

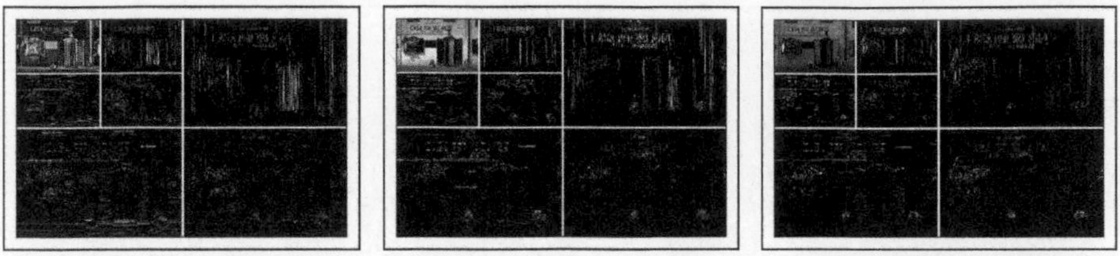

(a) Haar transform of $\widetilde{Y}$ (b) Haar transform of $\widetilde{Cb}$ (c) Haar transform of $\widetilde{Cr}$

Figure 4.25 Two iterations of the Haar wavelet transformation applied to the $\widetilde{Y}$, $\widetilde{Cb}$, and $\widetilde{Cr}$ channels plotted in Figure 3.27.

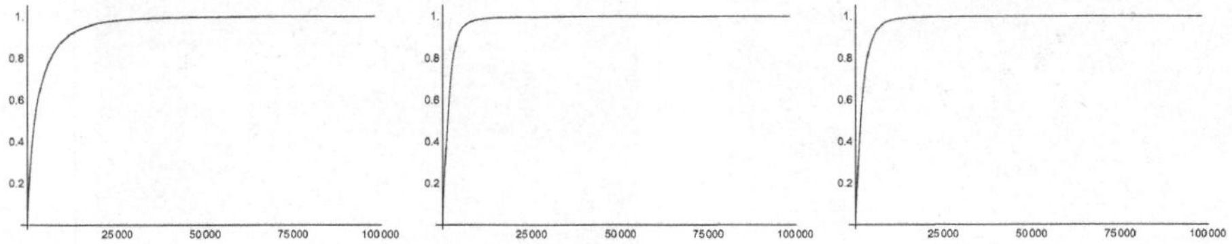

(a) $\widetilde{Y}$ (b) $\widetilde{Cb}$ (c) $\widetilde{Cr}$

Figure 4.26 The cumulative energy vectors for the Haar wavelet transformation of each of $\widetilde{Y}, \widetilde{Cb}$, and $\widetilde{Cr}$.

Table 4.3 Values retained for compression.[a]

Transform of	Nonzero Terms	Zero Terms	% Zero
$\widetilde{Y}$	34,880	63,424	64.42
$\widetilde{Cb}$	13,986	84,318	85.77
$\widetilde{Cr}$	14,934	83,370	84.81

[a]The second column shows the number of largest elements (in absolute value) that we must retain to preserve 99.99% of the energy of each Haar wavelet transform, the third column lists the number of terms we convert to zero in each transform, and the final column gives the percentage of elements in each transform that are zero. Note that the numbers in the nonzero terms column might include multiple copies of the smallest values (in absolute value) that comprise the 99.99% energy level. Since it is impossible to know which of these values to keep or discard, the selection routine keeps all of them.

To create the compressed image, we convert the elements of each quantized transform plotted in Figure 4.27 to Huffman codes. The bit stream length, bpp, and entropy for each of the quantized wavelet transformations are given in Table 4.4.

The total bit stream length for the uncompressed image is $3 \times 8 \times 256 \times 384 = 2{,}359{,}296$. *Compression via the Haar wavelet results in a bit stream length of* $657{,}943$. *This length is about* 28% *of the original bit stream length or* 2.231 *bpp.*

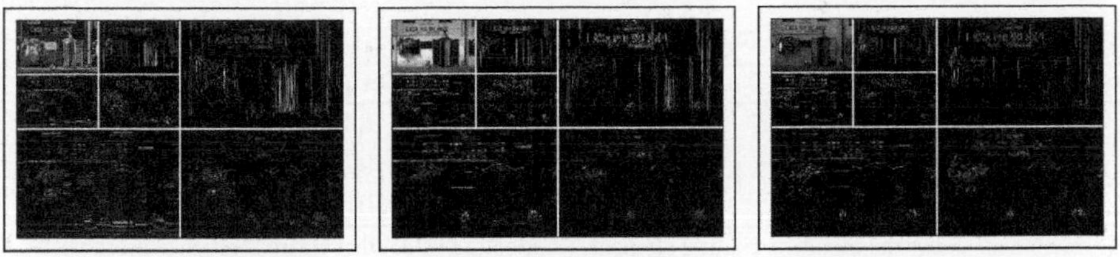

(a) $\widetilde{Y}$ (b) $\widetilde{Cb}$ (c) $\widetilde{Cr}$

Figure 4.27 The quantized Haar wavelet transformations for each of $\widetilde{Y}$, $\widetilde{Cb}$, and $\widetilde{Cr}$.

Table 4.4 Huffman bit stream length, bpp, and entropy for each quantized transform of $\widetilde{Y}$, $\widetilde{Cb}$, and $\widetilde{Cr}$.

Transform	Bit Stream Length	bpp	Entropy
$\widetilde{Y}$	305,568	3.10840	3.03793
$\widetilde{Cb}$	170,534	1.73476	1.32018
$\widetilde{Cr}$	181,841	1.84987	1.46085

To recover the approximate image we convert the Huffman codes in each channel to intensities, apply two iterations of the inverse Haar wavelet transform, add 128 *to all elements (to undo the centering) and finally convert back to RGB space. The approximate image is plotted in Figure 4.29(b) and the PSNR between it and the original image is* 35.68. *Considering that the bit stream length is about* 28% *of the original, the reconstructed image does a good job of approximating the original.*

If we look at the Haar wavelet transformations of the $\widetilde{Cb}$ *and* $\widetilde{Cr}$ *channels, we can see why it is desirable to convert RGB to YCbCr space. The detail portions of these channels' transformations are quite sparse, so we might be able to obtain even better compression rates if we lower the percent of energy we wish to preserve in these transformations. We can make the transformations even more sparse if we perform more iterations.*

We set the number of iterations to three and use energy levels 99.2%, 95%, *and* 95% *for the transformations of* $\widetilde{Y}$, $\widetilde{Cb}$, *and* $\widetilde{Cr}$, *respectively. Table 4.5 lists the number of values we will retain for compression, and Figure 4.28 shows the quantized transformations.*

Table 4.6 gives bit stream information for compressing the quantized wavelet transformations into Huffman codes.

Recall the total bit stream length is 2, 359, 296. *Compression via the Haar wavelet transformation and the varied energy levels results in a bit stream length of* 575, 326. *This length is just over* 24% *of the original bit stream length or* 1.951 *bpp.*

The approximate image using the energy levels given in Table 4.5 is plotted in Figure 4.29(c). While the reconstructed image is not as good an approximate as that

Table 4.5 Values retained for compression.[a]

Transform	Energy %	Nonzero Terms	Zero Terms	% Zero
$\widetilde{Y}$	99.2	34,880	63,424	64.42
$\widetilde{Cb}$	95	5,966	92,338	93.93
$\widetilde{Cr}$	95	6,922	91,382	92.96

[a]The third column shows the number of largest elements (in absolute value) that we must retain to preserve the energy (listed in the second column) of each Haar wavelet transform. The fourth column shows the number of terms we convert to zero in each transform and the last column gives the percentage of elements in each transform that are zero. Note that the numbers in the nonzero terms column might include multiple copies of the smallest values (in absolute value) that comprise the desired energy level. Since it is impossible to know which of these values to keep or discard, the selection routine keeps all of them.

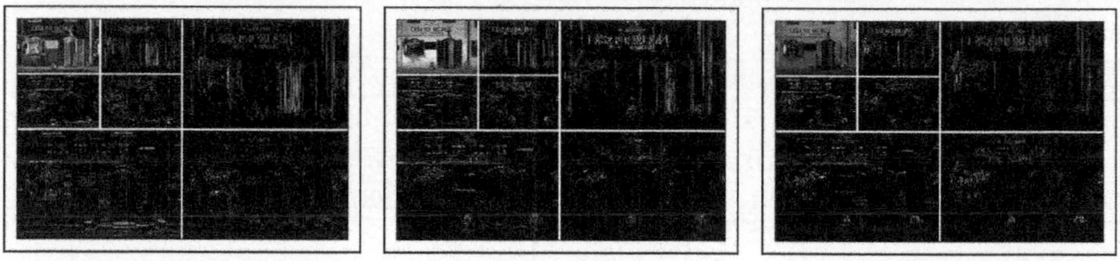

(a) $\widetilde{Y}$ (b) $\widetilde{Cb}$ (c) $\widetilde{Cr}$

Figure 4.28 The quantized Haar wavelet transformations for each of $\widetilde{Y}$, $\widetilde{Cb}$, and $\widetilde{Cr}$ for the energy levels given in Table 4.5.

Table 4.6 Huffman bit stream length, bpp, and entropy for each quantized transform of $\widetilde{Y}$, $\widetilde{Cb}$, and $\widetilde{Cr}$ for the energy levels given in Table 4.5.

Transform	Bit Stream Length	bpp	Entropy
$\widetilde{Y}$	307,095	3.1239	3.05138
$\widetilde{Cb}$	130,193	1.3244	0.65284
$\widetilde{Cr}$	138,038	1.4042	0.76962

shown in Figure 4.29(b), it does fairly well considering that the bit stream length is about one-fourth of the original bit stream length.
(Live example: Visit `stthomas.edu/wavelets` and click on Live Examples.)

□

Edge Detection

The final application we discuss is the detection of edges in digital images. There are many different ways to perform edge detection. Some of the best known methods

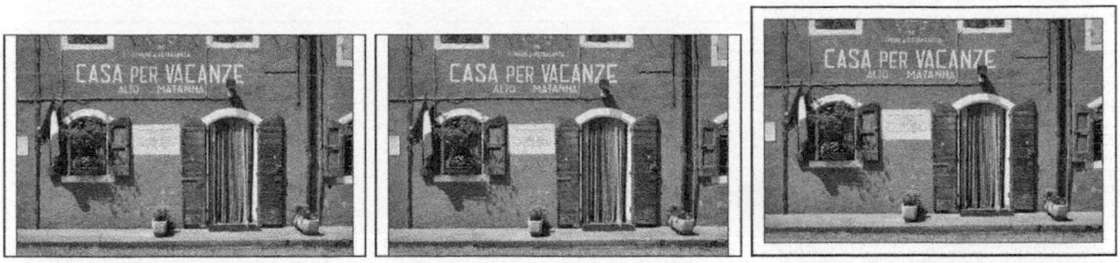

(a) Original image (b) 99.2% energy (c) Variable energy rates

Figure 4.29 The original and reconstructed images. The reconstructed image that uses 99.2% energy across each channel has PSNR = 35.68, while the reconstructed image that uses the energy rates given in Table 4.5 has PSNR = 31.04. See the color plates for color versions of these images.

are those of Marr and Hildreth [70], Canny [18], and Sobel (see the description in Gonzalez and Woods [48]). Other methods for edge detection are outlined in Gonzalez, Woods, and Eddins [49].

A Brief Review of Popular Edge Detection Methods

Roughly speaking, edges occur when there is an abrupt change in pixel intensity values (see Figure 4.30). As you know from elementary calculus, the derivative is used to measure rate of change. Thus for two-dimensional images, edge detectors attempt to estimate the gradient at each pixel and then identify large gradient values as edges. Recall from multivariate calculus that the gradient of $f(x, y)$ is defined to be

$$\nabla f(x,y) = f_x(x,y)\begin{bmatrix}1\\0\end{bmatrix} + f_y(x,y)\begin{bmatrix}0\\1\end{bmatrix}$$

where f_x and f_y are the partial derivatives of $f(x, y)$. See Stewart [84] for more details.

The Sobel method is straightforward. The method uses two 3×3 *mask* matrices that are convolved[3] with the image. One mask is designed to approximate the partial derivative in the horizontal (x) direction, while the other approximates the partial derivative in the vertical (y) direction. The results of the convolutions are used to estimate the magnitude of the gradient value at each pixel. Large gradient values correspond to edges.

Many digital images contain noise, and one of the problems with the Sobel detector is that it often identifies noisy pixels as edge pixels. The Canny method [18]

[3] We do not discuss two-dimensional convolution in this book. It is quite straightforward to extend the ideas from Section 2.4 to higher dimensions. See Gonzalez and Woods [48] and Gonzalez, Woods, and Eddins [49] for more details.

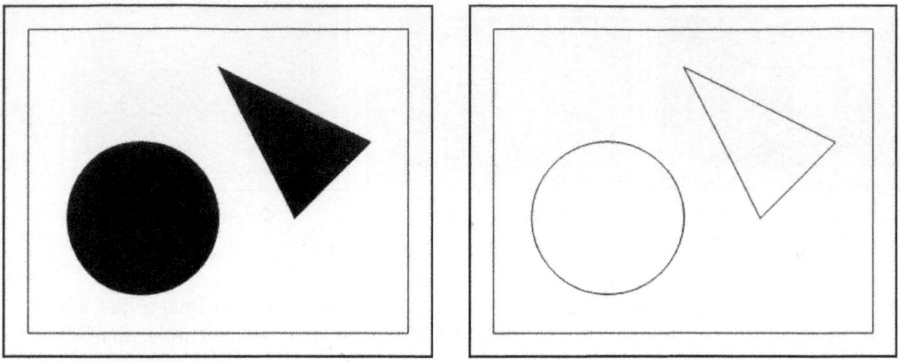

(a) An image (b) Image edges

Figure 4.30 An image and its edges.

addresses this problem by first convolving the image with a *smoothing mask*. The method then estimates the magnitude of the gradient at each pixel. The Canny detector then applies an algorithm to refine the edges. Finally, two threshold values are created and used to determine *strong edges* and *weak edges*. Weak edges that are not connected to strong edges are discarded. Since the Canny detector addresses problems such as noise and refinement of edges, it is typically considered one of the most efficient edge detectors.

Rather than search for large magnitudes of the gradient, the Marr-Hildreth detector looks for zero values (called zero crossings) in the Laplacian. Recall the Laplacian (see Stewart [84]) is defined to be

$$\nabla^2 f(x,y) = \frac{\partial^2 f}{\partial x^2} + \frac{\partial^2 f}{\partial y^2}.$$

A threshold is applied to the set of zero crossings to determine those pixels that represent edges.

Using Wavelets to Detect Edges

If you worked through the "Things to Try" section of the software module accompanying Example 4.2, then you know that the differences portion of the Haar wavelet transformation approximates (up to scale) the derivative of a sampled function (for further details, see Problem 4.38). The idea extends to two-dimensional wavelet transformations. Information about the edges in an image is contained in $\mathcal{V}$, $\mathcal{H}$ and $\mathcal{D}$. Mallat and Hwang [69] have shown that a maximum of the modulus of the wavelet transformation can be used to detect "irregular structures" – in particular, the modulus of the differences portions of the wavelet transformation can be used to detect edges in images.

In the remainder of this section we give a brief overview of how wavelets can be used in edge detection. In the discussion above, it is noted that digital images are typically denoised or smoothed before application of an edge detector. We do not perform denoising in this application (see Chapter 6 for a detailed discussion of wavelet-based denoising). We also consider only one iteration of the wavelet transformation when detecting edges. Wavelet-based edge detection can be improved if more iterations of the transform are performed (see Mallat [69] or Li [67]).

Naive Edge Detection

Possibly the simplest thing we could do to detect edges in a digital image is to compute the images's Haar wavelet transform, convert the values in the blur block $\mathcal{B}$ to zero, and then apply the inverse Haar wavelet transform. Let's try it on an example.

Example 4.17 (Naive Edge Detection). *Consider the* 512×512 *image* A *plotted in Figure 4.31(a). One iteration of the Haar wavelet transformation of* A *is also plotted in Figure 4.31(b).*

(a) An image (b) One iteration of the Haar transform

Figure 4.31 An image and its Haar wavelet transform.

We now convert all the values in block $\mathcal{B}$ *to zero. Let's call the modified wavelet transformation* $\widetilde{C}$*. We then apply the inverse Haar wavelet transformation to* $\widetilde{C}$ *to obtain a matrix that is comprised only of the vertical, horizontal, and diagonal differences between* A *and* $\mathcal{B}$*. To determine the significance of the edges, we also compute the absolute value of each element of the inverse transform. Let* $E = W_{512}^T \widetilde{C} W_{512}$ *denote the inverse transform of* $\widetilde{C}$*. Both* $\widetilde{C}$ *and* E *are plotted in Figure 4.32.*

(Live example: Visit `stthomas.edu/wavelets` *and click on Live Examples.)*

□

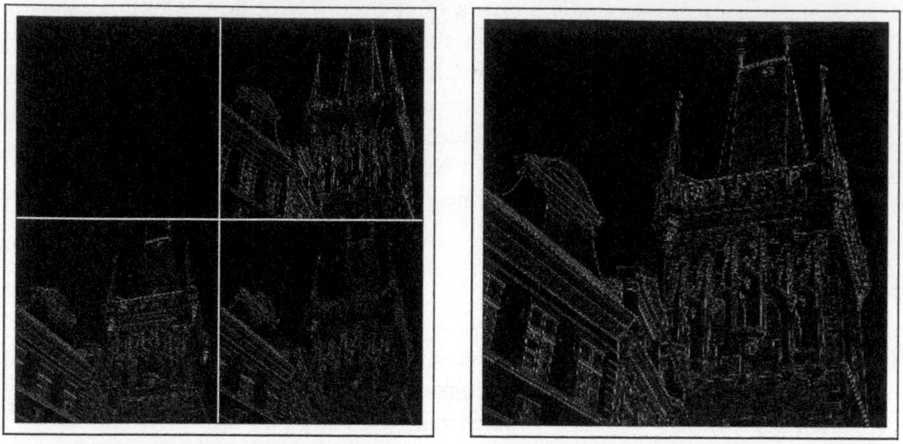

(a) Modified Haar wavelet transform (b) The resulting edges

Figure 4.32 The modified Haar transform and the resulting edges.

The naive process does a fair job at detecting edges in the image. Notice that it has problems distinguishing edges in areas where light and dark pixels are interspersed. It is natural to assume that not every element in the detail matrices $\mathcal{V}, \mathcal{H}$, and $\mathcal{D}$ constitutes an edge in the original image. We can improve our edge detector by keeping only those detail elements that do indeed correspond to edge pixels. The mathematical procedure for selecting only particular elements from a set of numbers is called *thresholding*.

Thresholding

To apply a threshold to the elements of our transform matrix C, we simply choose a value $\tau > 0$ and create a new matrix $\widetilde{C}$ whose entries are

$$\widetilde{c}_{ij} = \begin{cases} c_{ij}, & |\, c_{ij} \,| > \tau \\ 0, & |\, c_{ij} \,| \leq \tau \end{cases}$$

Unfortunately, there is no proven way to choose τ. We could use trial and error and try different values of τ until we obtain a satisfactory edge matrix. Alternatively, we could employ some elementary statistics to determine τ.

Automating the Threshold Choice

To automate the process, we use an iterative method for determining τ that is described in Gonzalez and Woods [48]. The process creates a series of thresholds

$\tau_1, \tau_2, \ldots$ and stops only when the difference $|\tau_{n+1} - \tau_n|$ is smaller than some prescribed $\alpha > 0$.

The method in Gonzalez and Woods [48] for choosing a threshold for a set of numbers S is as follows.

Algorithm 4.1 (Automated Threshold Algorithm). *Given set S consisting of nonnegative real numbers and a stopping criteria $\alpha > 0$ this algorithm uses a "divide and conquer" approach to determine a threshold τ for dividing the data. The algorithm begins by computing τ_1 as the mean of the smallest and largest values in S. The values of S are divided into two sets: S_1 holds values smaller than τ_1 and S_2 holds values greater than or equal to τ_1. The means of these sets are computed and their means give us τ_2. If $|\tau_1 - \tau_2| < \alpha$, the algorithm terminates and returns τ_2. If not, the algorithm proceeds constructing $\tau_3, \ldots, \tau_n$ until $|\tau_{n-1} - \tau_n| < \alpha$.*

1. *Pick a tolerance $\alpha > 0$ that will determine the stopping criteria for the process.*
2. *Set τ_1 to be the average of the largest and smallest values in S.*
3. *Divide the values of S into two sets S_1 and S_2. The values in S_1 are all smaller than τ_1 and the values in S_2 are greater than or equal to τ_1.*
4. *Compute the means $\bar{s}_1$ for S_1 and $\bar{s}_2$ for S_2 and let $\tau_2 = \frac{1}{2}(\bar{s}_1 + \bar{s}_2)$.*
5. *Repeat steps 3 and 4 replacing τ_1 with the new threshold obtained from step 4. Continue until $|\tau_{n+1} - \tau_n| < \alpha$.*

□

Let's look at the thresholding process in the following example.

Example 4.18 (Edge Detection with Automated Thresholding). *Consider again the image from Example 4.17. We choose the stopping criteria value to be $\alpha = 1$ pixel intensity value. We first compute $C = W_{512}AW_{512}^T$. Next, we create a set S that is comprised of the absolute values of all elements of $\mathcal{V}$, $\mathcal{H}$, and $\mathcal{D}$. We set τ_1 to be the average of the largest and smallest values in S. For the image plotted in Figure 4.31, we have $\max(S) = 127.5$ and $\min(S) = 0$, so that $\tau_1 = 63.75$.*

We are now ready to employ the method given in Algorithm 4.1. We begin by creating set S_1 that is comprised of all values from S whose values are smaller than $\tau_1 = 63.75$. The set S_2 is the complement of S_1 in S. The mean of S_1 is $\overline{s}_1 = 4.82454$ and the mean of S_2 is $\overline{s}_2 = 79.9845$, so that $\tau_2 = (\overline{s}_1 + \overline{s}_2)/2 = 42.4045$. Since $|\tau_2 - \tau_1| = 21.3455 \geq 1 = \alpha$, we repeat the process. As we can see from Table 4.7, we need seven iterations to reach a threshold of $\tau = \tau_7 = 22.3063$.

We now set to zero each element in $\mathcal{V}$, $\mathcal{H}$, and $\mathcal{D}$ whose absolute value is smaller than $\tau = \tau_7 = 22.3063$ and convert the elements in block $\mathcal{B}$ to zero. If we denote the modified Haar wavelet transformation by $\widetilde{C}$, we obtain the edges by using the inverse Haar wavelet transformation to compute the edge matrix E.

In Figure 4.33 we plot the edge matrix E and the edge matrix from Figure 4.32 side by side so that we can compare the results. The automated method does a little

Table 4.7 Iterations to obtain the threshold value $\tau = 22.3063$.[a]

n	$\overline{s}_1$	$\overline{s}_2$	τ_n	$\vert\tau_{n+1} - \tau_n\vert$
1	—	—	63.7500	21.3455
2	4.82454	79.9845	42.4045	9.87007
3	4.15590	60.9130	32.5345	5.10283
4	3.65093	51.2123	27.4316	2.71026
5	3.33842	46.1043	24.7214	1.53116
6	3.14817	43.2322	23.1902	0.88390
7	3.03367	41.5789	22.3063	—

[a]The first column is the iteration number, the next two numbers are the means of S_1 and S_2, respectively, the third value is the average τ_n of $\overline{s}_1$ and $\overline{s}_2$; and the last column gives the stopping criterion $|\tau_{n+1} - \tau_n|$. The method stops when the value in the last column is smaller than $\alpha = 1$.

(a) Naive method (b) Automated with $\tau = 22.3063$

Figure 4.33 Different edge matrices. The image on the left is from Example 4.17 and the image on the right was constructed in Example 4.18.

better job at clearing up the edges. In Problem 4.40 you are asked to think about the consequences of increasing the size of α.

(Live example: Visit `stthomas.edu/wavelets` and click on Live Examples.)

□

Although the methods we present for compression and edge detection are quite naive, they do illustrate the possibilities that exist when we decompose data into average and differences portions. In later chapters, we learn how to construct wavelet transformations that outperform the Haar wavelet transformation in the applications of this section.

PROBLEMS

4.33 In this problem, you consider the range of the output of the two-dimensional Haar wavelet transform. Suppose A is a matrix of even dimensions $N \times M$ and let $C = W_N A W_M^T$ be the Haar wavelet transformation of A.

(a) Assume that the elements of A are in $[0, 255]$. Show that the elements of C are in $[-127.5, 255]$. *Hint:* Problem 4.24 may be helpful for this exercise.

(b) More generally, assume the elements of A are in $[\ell, L]$ and show that the elements of C are in $[m, M]$, where

$$m = \min\left(\ell, \frac{\ell - L}{2}\right) \quad \text{and} \quad M = \max\left(L, \frac{L-\ell}{2}\right).$$

(c) Use part (b) to show that if the elements in A are in $[-128, 127]$, then the elements in C are in the interval $[-127.5, 127.5]$.

4.34 Suppose we compute a second iteration of the Haar wavelet transformation on the matrix in Problem 4.33. Do the values of the second iteration also reside in $[-128, 127]$? Explain why or why not.

4.35 In the second half of Example 4.16, a lower energy level was used for the Cb channel. Can you explain why the lower level makes sense for this particular example?

★**4.36** The Haar wavelet transformation is constructed by taking two numbers, a and b and mapping them to the average $\frac{1}{2}a + \frac{1}{2}b$ and the difference value $\frac{1}{2}b - \frac{1}{2}a$. We could multiply these values by two and easily modify the transform so that integer-valued input is mapped to integers. The modified transform is

$$\begin{bmatrix} a \\ b \end{bmatrix} \to \begin{bmatrix} a+b \\ b-a \end{bmatrix}.$$

The output of this transform is certainly integer-valued if a and b are integers. To invert the process, we simply divide the inverse transform by two.

The problem with this transform is that it increases the range of the averages and differences portion of the data. That is, for a, b integers from $\{0, \ldots, 255\}$, the average values could range from 0 to 510 and the difference values could range from -255 to 255. It is desirable to keep the range of average values the same as that of the original input. Thus Sweldens [92] suggests the modified Haar wavelet transform

$$\begin{bmatrix} a \\ b \end{bmatrix} \to \begin{bmatrix} \frac{a+b}{2} \\ b-a \end{bmatrix}. \tag{4.34}$$

Certainly, the average of a and b is contained in $[a, b]$, but the problem is that transform doesn't necessarily map integers to integers.

In this problem, we see how to compute the transform (4.34) via a process known as *lifting* and then convert the results to integers in such a way that the original input can be recovered. (We will study the concept of lifting in detail in Chapter 11.) Throughout this problem, we assume that $\mathbf{v} = [v_0, \ldots, v_{N-1}]^T$ where N is even with $v_k \in \{0, \ldots, 255\}, k = 1, \ldots, N$.

(a) The transform (4.34) sends

$$\mathbf{v} \to \left[\frac{v_0 + v_1}{2}, \ldots, \frac{v_{N-2} + v_{N-1}}{2} \,\middle|\, v_1 - v_0, \ldots, v_{N-1} - v_{N-2} \right]^T.$$

If we let $\mathbf{a}$ and $\mathbf{d}$ denote the averages and differences portions of the transform, respectively, then for $k = 0, \ldots, \frac{N}{2} - 1$,

$$a_k = \frac{v_{2k} + v_{2k+1}}{2} \qquad \text{and} \qquad d_k = v_{2k+1} - v_{2k}.$$

Let $\mathbf{e} = \left[e_0, \ldots, e_{\frac{N}{2}-1} \right]^T = [v_0, v_2, \ldots, v_{N-2}]^T$ hold the even-indexed elements of $\mathbf{v}$, and let $\mathbf{o} = \left[o_1, \ldots, o_{\frac{N}{2}-1} \right]^T = [v_1, v_3, \ldots, v_{N-1}]^T$ hold the odd-indexed elements of $\mathbf{v}$. Show that we can obtain $\mathbf{d}$ and $\mathbf{a}$ by first computing

$$d_k = o_k - e_k \tag{4.35}$$

for $k = 0, \ldots, \frac{N}{2} - 1$, and then computing

$$a_k = e_k + \frac{1}{2} d_k \tag{4.36}$$

for $k = 0, \ldots, \frac{N}{2} - 1$.

(b) What is the process for inverting (4.35) and (4.36)?

(c) The transformation (4.35) and (4.36) in (a) can be modified so that it maps integers to integers. We compute d_k using (4.35), but we apply the floor function (3.2) to the second term in (4.36) to write

$$a_k^* = e_k + \left\lfloor \frac{1}{2} d_k \right\rfloor. \tag{4.37}$$

for $k = 0, \ldots, \frac{N}{2} - 1$. What is the inverse process for this transformation?

(d) Suppose that we apply (4.35) followed by (4.37) to $\mathbf{v}$. Explain why the inverse process you found in (c) returns the original $\mathbf{e}$ and $\mathbf{o}$.

4.37 Let $\mathbf{v} = [\,10, 20, 25, 105, 0, 100, 100, 55, 60, 75\,]^T$.

(a) Use (4.35) and (4.37) to compute an integer-valued Haar wavelet transform.

(b) Use the inverse process you found in (d) of Problem 4.36 on the transformed data from (a) to recover $\mathbf{v}$.

4.38 Recall from calculus that one way to define the derivative is

$$f'(t) = \lim_{h \to 0} \frac{f(t) - f(t-h)}{h}$$

So for "small" values of h, the quantity $\frac{f(t)-f(t-h)}{h}$ serves as an approximation of $f'(t)$. Suppose that we form a vector $\mathbf{v}$ of length N (where N is even) by sampling some function $f(t)$. That is,

$$v_k = f(a + (k-1)\,\Delta x)$$

where $k = 1, \ldots, N$, a is the starting value, and Δx is the sampling width.

(a) Let $\mathbf{y}$ denote one iteration of the Haar wavelet transform. Using the fact that the elements of the differences portion of $\mathbf{y}$ are given by

$$y_{N/2+k} = \frac{1}{2}(v_{2k} - v_{2k-1})$$

for $k = 1, \ldots, \frac{N}{2}$, show that

$$y_{N/2+k} \approx \frac{1}{2}\Delta x\, f'(a + (2k-1)\,\Delta x).$$

(b) Consider the function

$$f(t) = \begin{cases} 1, & t < 0 \\ t, & t \geq 0. \end{cases}$$

Create the 10-vector $\mathbf{v}$ by sampling $f(t)$ with starting point $a = -0.5$ and sampling width $\Delta x = 0.1$. Then compute the differences portion of the Haar wavelet transformation of $\mathbf{v}$ and verify the result you obtained in (a).

(c) How could you use the differences portion of the Haar wavelet transformation to detect jumps in the derivative of $f(t)$?

(d) The function

$$f(t) = \begin{cases} t^2, & t < 0 \\ t^3, & t \geq 0 \end{cases}$$

is continuously differentiable at $a = 0$ but has a jump at zero in the second derivative. Can you use the Haar wavelet transformation to detect this jump? Why or why not? *Hint:* Problem 4.20 might offer some insight.

4.39 Use Algorithm 4.1 with $\alpha = 1$ to find the threshold value τ for

$$S = [0, 0, 0, 0, 0, 0, 0, 0, 0, 0, 51, 77, 170, 204, 220, 268, 292, 316, 380, 410]^T$$

4.40 What do you suppose would happen if we increased $\alpha = 1$ to $\alpha = 25$ in Example 4.18? From $\alpha = 1$ to $\alpha = 100$?

4.41 In forming the Haar wavelet transformation matrix W_N (4.7), we performed what is known as *downsampling*. That is, in forming $W_N = \left[\frac{H}{G}\right]$, each row of H and G is formed by cyclically shifting its predecessor by two. This makes sense in part because the resulting W_N is nonsingular. Recall in the edge detection application, we set the elements of the approximation portion $\mathcal{B}$ of the transform to zero, performed thresholding on the detail matrices, and then inverted the result to obtain an edge matrix.

In this problem we consider a method for edge detection that neither utilizes downsampling nor requires matrix invertibility in the usual sense. A **CAS** will be helpful to complete this problem.

Suppose that A is an $N \times N$ matrix, $N \geq 2$, that represents some grayscale image.

(a) Form the *undecimated* Haar wavelet matrix U_N as follows. The first row of U_N has N elements and is of the form $\begin{bmatrix} \frac{1}{2} & \frac{1}{2} & 0 & 0 & \cdots & 0 & 0 \end{bmatrix}$. Rows two through N are formed by cyclically shifting the previous row one unit to the right. Row N will have $\frac{1}{2}$ in the first and last positions. Rows $N+1$ to $2N$ are formed in exactly the same manner except the starting row is $\begin{bmatrix} -\frac{1}{2} & \frac{1}{2} & 0 & 0 & \cdots & 0 & 0 \end{bmatrix}$. Write down U_4 and U_5.

(b) The top (bottom) half of U_N is the averages (differences) portion of the transform. Denote these blocks by H and G, respectively. To transform A, we compute

$$B = U_N A U_N^T = \left[\frac{H}{G}\right] A \left[H^T \middle| G^T \right].$$

Further expansion gives a block form similar to (4.29) with blocks $\mathcal{B}, \mathcal{V}, \mathcal{H}$, and $\mathcal{D}$. What are the dimensions of each of these blocks?

(c) Can you state any advantages that this transformation might have over the conventional Haar wavelet transformation in terms of edge detection?

(d) Certainly U_N is not invertible since it is $2N \times N$. Suppose there exists $N \times 2N$ matrix V that satisfies $VU_N = I_N$. Explain how we could use V to recover A from $B = U_N A U_N^T$.

(e) Find V for U_4 so that $VU_4 = I_4$. *Hint:* It turns out there are infinitely many possibilities for V but one is not difficult to obtain. Let the columns of U_4 be denoted by $\mathbf{u}_k$, $k = 1, \ldots, 4$. Show that $\mathbf{u}_k \cdot \mathbf{u}_k = 1$. What is true about $\mathbf{u}_k \cdot \mathbf{u}_j$ when $k \neq j$?

(f) Generalize your result in (e) for arbitrary N.

Note: V is called a *pseudoinverse* of U_N. For a nice introduction to pseudoinverses, please see Strang [87]. Since we know how to "invert" the undecimated Haar wavelet transform, we can use it in the edge detection methods described in this section.

Computer Labs

4.4 Image Edge Detection. From `stthomas.edu/wavelets`, access the file `EdgeDetect`. In this lab, you will investigate the use of the undecimated Haar wavelet transformation described in Problem 4.41 in the application of edge detection in images.

4.5 Lifting. From `stthomas.edu/wavelets`, access the file `HaarLifting`. In this lab you will write a module to implement the transform described in Problem 4.36.

CHAPTER 5

DAUBECHIES WAVELET TRANSFORMATIONS

The discrete Haar wavelet transformation introduced in Chapter 4 serves as an excellent introduction to the theory and construction of *orthogonal wavelet transformations*. We were able to see in Section 4.4 how the wavelet transformation can be utilized in applications such as image compression and edge detection. There are advantages and limitations to the Haar wavelet transformation. Conceptually, the Haar wavelet transformation is easy to understand. The big advantage of the Haar wavelet transformation is computational speed. Each row of the Haar wavelet transformation matrix consists of only two nonzero terms so it straightforward to write fast algorithms to apply the transformation to a vector or matrix. This same advantage is also a disadvantage. Consider the vector $\mathbf{v} \in \mathbb{R}^{40}$ where

$$v_k = \begin{cases} 2, & k = 1, \ldots, 20 \\ 12, & k = 21, \ldots, 40. \end{cases} \tag{5.1}$$

The vector $\mathbf{v}$ and its Haar wavelet transformation are plotted in Figure 5.1.

Recall in Section 4.4 we searched for edges in images by replacing the blur of the transformation with a zero matrix and then inverting the modified transformation. We could certainly do the same for vectors but in this case, if we set the values of the averages portion to 0, then the modified transformation vector is the zero

Patrick J. Van Fleet.

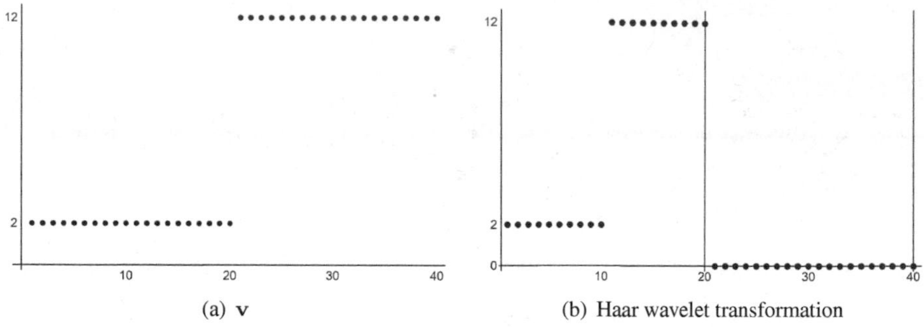

Figure 5.1 A vector and its Haar wavelet transformation.

vector and inverting it returns the zero vector. Thus the differences portion of the Haar transformation failed to identified the jump that exists between v_{20} and v_{21}. Do you see why the Haar wavelet transformation failed to pick up this jump? The fact that there are only two nonzero terms located at even/odd positions in each row and the cyclic two-shift of the rows in the averages and differences portions of the transformation results in a *decoupling* of the output data.

We could fix this problem by either cyclically shifting by one in the averages and differences portions of the transformation or by retaining the two-shift structure and creating a transformation matrix with more than two nonzero elements that generate lowpass/highpass rows. We considered cyclic shifts by one in Problem 4.41 and learned that the resulting matrix is rectangular and thus not invertible. In the sections that follow we consider the notion of adding more nonzero terms to each row of the wavelet transformation.

We named the filters $\mathbf{h} = (h_0, h_1) = \left(\frac{1}{2}, \frac{1}{2}\right)$ and $\mathbf{g} = (g_0, g_1) = \left(\frac{1}{2}, -\frac{1}{2}\right)$ the Haar lowpass and highpass filters, respectively, in (4.8) and in this chapter we wish to construct filters of length greater than two. We have to determine what we want from our longer filters. Certainly, we do not want them to be so long as to markedly affect computational speed. We require our transformation matrix W to be orthogonal and we retain the averages/differences (lowpass/highpass) structure of W. Suppose that we were to make $\mathbf{h}$ and $\mathbf{g}$ longer. We are in some sense adding more flexibility to the transformation. How can we best take advantage of this flexibility?

In her landmark 1988 paper [29], Ingrid Daubechies describes a family of lowpass filters whose members each give rise to an orthogonal transformation matrix. It turns out that the first member of this family is the *orthogonal Haar lowpass filter*

$$\mathbf{h} = (h_0, h_1) = \left(\frac{\sqrt{2}}{2}, \frac{\sqrt{2}}{2}\right)$$

given by (4.12) in Section 4.1. This filter gives rise to the orthogonal transformation matrix (4.11). Daubechies shows how to construct other family members (all lowpass filters) of arbitrary even length and the accompanying highpass filters. In Section 5.1

we derive Daubechies' filters of length 4 and in Section 5.2 we derive the length six filter and revisit the image compression application of Section 4.4. We conclude the chapter by extending the ideas of Sections 5.1 and 5.2 to arbitrary even-length filters in Section 5.3.

5.1 Daubechies Filter of Length 4

In this section we construct Ingrid Daubechies' orthogonal filter [30] of length 4. The filter is orthogonal in the sense that the wavelet transformation matrix W_N we construct satisfies $W_N^{-1} = W_N^T$. The construction in this section and the next will serve as motivation for building orthogonal filters of arbitrary even length.

The construction here is different from the one that Daubechies gives in [29] and [32]. We start with a matrix that has the same structure as that associated with the orthogonal Haar wavelet transformation matrix (4.11), but constructed with filters $\mathbf{h} = (h_0, h_1, h_2, h_3)$ and $\mathbf{g} = (g_0, g_1, g_2, g_3)$. We then construct a system of linear and quadratic equations that the elements of $\mathbf{h}$ must satisfy. These equations arise from conditions required for the transformation matrix W to be orthogonal and from some additional conditions we impose on the differences portion of the transformation. Surprisingly, the accompanying highpass filter $\mathbf{g}$ can be easily constructed during this process.

The Daubechies Four-Term Orthogonal Filter

In the introduction to this chapter, we stated that Daubechies filters are of arbitrary even length. The orthogonal Haar filter $\mathbf{h} = (h_0, h_1) = \left(\frac{\sqrt{2}}{2}, \frac{\sqrt{2}}{2}\right)$ is of length 2. The next even number is four, so let's construct the lowpass filter $\mathbf{h} = (h_0, h_1, h_2, h_3)$ and the highpass filter $\mathbf{g} = (g_0, g_1, g_2, g_3)$. In Problem 5.1 you investigate why we do not consider odd-length filters for this construction.

Note: We insist that all filter elements are nonzero.

To motivate the construction, let's create the wavelet transformation using filters $\mathbf{h} = (h_0, h_1, h_2, h_3)$ and $\mathbf{g} = (g_0, g_1, g_2, g_3)$ that we can apply to vectors of length 8. We could choose any even length larger than 4 and obtain the same results that follow, but find that $N = 8$ is large enough to understand the process but small enough to make the computations manageable. We construct the matrix

$$W_8 = \left[\frac{H}{G}\right] = \left[\begin{array}{cccccccc} h_3 & h_2 & h_1 & h_0 & 0 & 0 & 0 & 0 \\ 0 & 0 & h_3 & h_2 & h_1 & h_0 & 0 & 0 \\ 0 & 0 & 0 & 0 & h_3 & h_2 & h_1 & h_0 \\ h_1 & h_0 & 0 & 0 & 0 & 0 & h_3 & h_2 \\ \hline g_3 & g_2 & g_1 & g_0 & 0 & 0 & 0 & 0 \\ 0 & 0 & g_3 & g_2 & g_1 & g_0 & 0 & 0 \\ 0 & 0 & 0 & 0 & g_3 & g_2 & g_1 & g_0 \\ g_1 & g_0 & 0 & 0 & 0 & 0 & g_3 & g_2 \end{array}\right]. \tag{5.2}$$

We have built W_8 in much the same way that we constructed the matrix for the Haar wavelet transformation. Note that the filter elements loaded in each row are in reverse order. This is due to the fact that (5.2) can be viewed as a truncated convolution matrix (2.15) with every other row deleted. Unlike the Haar wavelet transformation, the matrix above has two "wrapping rows." The coefficients in the fourth and eighth rows have "wrapped" from the end of the row to the beginning of the row. We did not have to worry about this with the Haar filter – since there were only two coefficients per row, there was no need to wrap. What effect does the wrapping have? Consider

$$W_8\mathbf{v} = \left[\begin{array}{cccccccc} h_3 & h_2 & h_1 & h_0 & 0 & 0 & 0 & 0 \\ 0 & 0 & h_3 & h_2 & h_1 & h_0 & 0 & 0 \\ 0 & 0 & 0 & 0 & h_3 & h_2 & h_1 & h_0 \\ h_1 & h_0 & 0 & 0 & 0 & 0 & h_3 & h_2 \\ \hline g_3 & g_2 & g_1 & g_0 & 0 & 0 & 0 & 0 \\ 0 & 0 & g_3 & g_2 & g_1 & g_0 & 0 & 0 \\ 0 & 0 & 0 & 0 & g_3 & g_2 & g_1 & g_0 \\ g_1 & g_0 & 0 & 0 & 0 & 0 & g_3 & g_2 \end{array}\right] \left[\begin{array}{c} v_1 \\ v_2 \\ v_3 \\ v_4 \\ v_5 \\ v_6 \\ v_7 \\ v_8 \end{array}\right] = \left[\begin{array}{c} h_3v_1+h_2v_2+h_1v_3+h_0v_4 \\ h_3v_3+h_2v_4+h_1v_5+h_0v_6 \\ h_3v_5+h_2v_6+h_1v_7+h_0v_8 \\ h_3v_7+h_2v_8+h_1v_1+h_0v_2 \\ \hline g_3v_1+g_2v_2+g_1v_3+g_0v_4 \\ g_3v_3+g_2v_4+g_1v_5+g_0v_6 \\ g_3v_5+g_2v_6+g_1v_7+g_0v_8 \\ g_3v_7+g_2v_8+g_1v_1+g_0v_2 \end{array}\right].$$

The fourth and eighth components of $W_8\mathbf{v}$ are built using v_1, v_2, v_7 and v_8. If the elements of $\mathbf{v}$ were sampled from a periodic function, our matrix W_8 would be ideal. Typically, this is not the case, so we might have some problems. One idea is to discard h_0 and h_1 from row 4 and g_0 and g_1 from row eight, but then we would not be using the entire filter to process the data. It is possible to build wavelet matrices such as W_8 in different ways (some intertwine the averages and differences rows!), but we will stick with the form in (5.2) for our presentation.

In the averages (differences) portion of W_8, elements of each row overlap with two elements of the rows above and below it (see Figure 5.2). This will help us avoid the problem with the Haar transformation that was discussed in the chapter introduction. As we will see, the longer we make $\mathbf{h}$ and $\mathbf{g}$, the more rows in the averages (differences) portions that contain overlapping elements.

Characterizing the Orthogonality Conditions

We will create the elements of W_8 in a few stages. Let's first write W_8 in block format and then multiply these blocks to understand what effect the orthogonality of

$$\begin{array}{cccc|cc|cccc} \cdots & h_3 & h_2 & \mathbf{h_1} & \mathbf{h_0} & 0 & 0 & \cdots \\ & 0 & 0 & \mathbf{h_3} & \mathbf{h_2} & h_1 & h_0 & \end{array}$$

Figure 5.2 Two overlapping rows in the averages portion of W_8.

W_8 has on the elements of $\mathbf{h}$. We write

$$W_8 = \left[\begin{array}{cccccccc} h_3 & h_2 & h_1 & h_0 & 0 & 0 & 0 & 0 \\ 0 & 0 & h_3 & h_2 & h_1 & h_0 & 0 & 0 \\ 0 & 0 & 0 & 0 & h_3 & h_2 & h_1 & h_0 \\ h_1 & h_0 & 0 & 0 & 0 & 0 & h_3 & h_2 \\ \hline g_3 & g_2 & g_1 & g_0 & 0 & 0 & 0 & 0 \\ 0 & 0 & g_3 & g_2 & g_1 & g_0 & 0 & 0 \\ 0 & 0 & 0 & 0 & g_3 & g_2 & g_1 & g_0 \\ g_1 & g_0 & 0 & 0 & 0 & 0 & g_3 & g_2 \end{array}\right] = \left[\begin{array}{c} H \\ \hline G \end{array}\right]. \tag{5.3}$$

Using (5.3) to compute $W_8 W_8^T$ and insisting that W_8 is orthogonal gives

$$W_8 W_8^T = \left[\begin{array}{c} H \\ \hline G \end{array}\right] \left[\begin{array}{c|c} H^T & G^T \end{array}\right] = \left[\begin{array}{c|c} HH^T & HG^T \\ \hline GH^T & GG^T \end{array}\right] = \left[\begin{array}{c|c} I_4 & 0_4 \\ \hline 0_4 & I_4 \end{array}\right] \tag{5.4}$$

where I_4 is the 4×4 identity matrix and 0_4 is the 4×4 zero matrix.

Let's analyze HH^T from (5.4). Direct computation gives

$$I_4 = HH^T = \begin{bmatrix} h_3 & h_2 & h_1 & h_0 & 0 & 0 & 0 & 0 \\ 0 & 0 & h_3 & h_2 & h_1 & h_0 & 0 & 0 \\ 0 & 0 & 0 & 0 & h_3 & h_2 & h_1 & h_0 \\ h_1 & h_0 & 0 & 0 & 0 & 0 & h_3 & h_2 \end{bmatrix} \begin{bmatrix} h_3 & 0 & 0 & h_1 \\ h_2 & 0 & 0 & h_0 \\ h_1 & h_3 & 0 & 0 \\ h_0 & h_2 & 0 & 0 \\ 0 & h_1 & h_3 & 0 \\ 0 & h_0 & h_2 & 0 \\ 0 & 0 & h_1 & h_3 \\ 0 & 0 & h_0 & h_2 \end{bmatrix} = \begin{bmatrix} a & b & 0 & b \\ b & a & b & 0 \\ 0 & b & a & b \\ b & 0 & b & a \end{bmatrix} = \begin{bmatrix} 1 & 0 & 0 & 0 \\ 0 & 1 & 0 & 0 \\ 0 & 0 & 1 & 0 \\ 0 & 0 & 0 & 1 \end{bmatrix}$$

where

$$a = h_0^2 + h_1^2 + h_2^2 + h_3^2 \quad \text{and} \quad b = h_0 h_2 + h_1 h_3. \tag{5.5}$$

We see that $a = 1$ and $b = 0$, so we obtain *orthogonality constraints* on $\mathbf{h}$:

$$\begin{aligned} h_0^2 + h_1^2 + h_2^2 + h_3^2 &= 1 \\ h_0 h_2 + h_1 h_3 &= 0. \end{aligned} \tag{5.6}$$

Conditions for the three other blocks of (5.4) must be met and these remaining blocks all depend on the highpass filter $\mathbf{g}$. It turns out that the construction of $\mathbf{g}$ is quite straightforward.

Creating the Highpass Filter $\mathbf{g}$

Assume that $\mathbf{h}$ is known and satisfies (5.6). To obtain $\mathbf{g}$, we first recall Problem 2.11 from Section 2.1. In this problem we constructed a vector $\mathbf{v}$ orthogonal to a given even-length vector $\mathbf{u}$ by reversing the elements of $\mathbf{u}$ and alternating the signs. Thus, we have a candidate for $\mathbf{g}$. We consider

$$\mathbf{g} = (g_0, g_1, g_2, g_3) = (h_3, -h_2, h_1, -h_0).$$

Let's insert these values into G in (5.3) and then compute the product HG^T. You should verify the easy computation that follows.

$$HG^T = \begin{bmatrix} h_3 & h_2 & h_1 & h_0 & 0 & 0 & 0 & 0 \\ 0 & 0 & h_3 & h_2 & h_1 & h_0 & 0 & 0 \\ 0 & 0 & 0 & 0 & h_3 & h_2 & h_1 & h_0 \\ h_1 & h_0 & 0 & 0 & 0 & 0 & h_3 & h_2 \end{bmatrix} \cdot \begin{bmatrix} -h_0 & 0 & 0 & -h_2 \\ h_1 & 0 & 0 & h_3 \\ -h_2 & -h_0 & 0 & 0 \\ h_3 & h_1 & 0 & 0 \\ 0 & -h_2 & -h_0 & 0 \\ 0 & h_3 & h_1 & 0 \\ 0 & 0 & -h_2 & -h_0 \\ 0 & 0 & h_3 & h_1 \end{bmatrix} = \begin{bmatrix} 0 & 0 & 0 & 0 \\ 0 & 0 & 0 & 0 \\ 0 & 0 & 0 & 0 \\ 0 & 0 & 0 & 0 \end{bmatrix}.$$

So $HG^T = 0_4$ as desired and we can easily verify $GH^T = 0_4$ as well. Using Problem 2.23, we have

$$GH^T = (HG^T)^T = 0_4^T = 0_4.$$

All that remains is to verify $GG^T = I_4$, and we have at least picked $\mathbf{h}$ and $\mathbf{g}$ so that W_8 is orthogonal. We compute

$$GG^T = \begin{bmatrix} -h_0 & h_1 & -h_2 & h_3 & 0 & 0 & 0 & 0 \\ 0 & 0 & -h_0 & h_1 & -h_2 & h_3 & 0 & 0 \\ 0 & 0 & 0 & 0 & -h_0 & h_1 & -h_2 & h_3 \\ -h_2 & h_3 & 0 & 0 & 0 & 0 & -h_0 & h_1 \end{bmatrix} \cdot \begin{bmatrix} -h_0 & 0 & 0 & -h_2 \\ h_1 & 0 & 0 & h_3 \\ -h_2 & -h_0 & 0 & 0 \\ h_3 & h_1 & 0 & 0 \\ 0 & -h_2 & -h_0 & 0 \\ 0 & h_3 & h_1 & 0 \\ 0 & 0 & -h_2 & -h_0 \\ 0 & 0 & h_3 & h_1 \end{bmatrix} = \begin{bmatrix} a & b & 0 & b \\ b & a & b & 0 \\ 0 & b & a & b \\ b & 0 & b & a \end{bmatrix} = I_4$$

where a and b are as given in (5.5).

So we can construct an orthogonal matrix by finding $\mathbf{h} = (h_0, h_1, h_2, h_3)$ that satisfies (5.6) and then choosing

$$\boxed{\mathbf{g} = (g_0, g_1, g_2, g_3) = (h_3, -h_2, h_1, -h_0).} \tag{5.7}$$

Note that

$$g_k = (-1)^k h_{3-k}, \quad k = 0, 1, 2, 3. \tag{5.8}$$

Solving the System Involving Two Orthogonality Conditions

Let's attempt to solve (5.6). The second equation in (5.6) implies the two-dimensional vectors $[h_0, h_1]^T$ and $[h_2, h_3]^T$ are orthogonal. You will show in Problem 5.4 that in this case we must have

$$[h_2, h_3]^T = c[-h_1, h_0]^T \tag{5.9}$$

for some real number $c \neq 0$. If we insert this identity into the first equation in (5.6) and simplify, we obtain

$$h_0^2 + h_1^2 = \frac{1}{1 + c^2}. \tag{5.10}$$

Geometrically, it is easy to see what is happening with this system. Equation (5.10) says that the values h_0 and h_1 must lie on a circle of radius $\frac{1}{\sqrt{1+c^2}}$ (see Figure 5.3). Since $c \neq 0$ is arbitrary, there are an infinite number of solutions to this system.

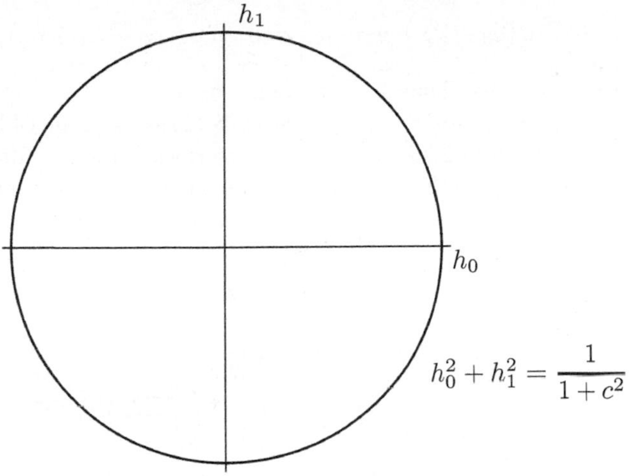

Figure 5.3 Solutions of the system (5.6).

The orthogonality conditions ensure that the filters **h** and **g** produce an orthogonal transformation matrix. Since there are infinitely many solutions to the system, we have the flexibility to add more constraints to our system. Note that we have imposed no conditions that ensure the averages (or differences) portions of the transformation produces an approximation (or differences) of input data. We continue by imposing conditions on the highpass filter **g**.

A First Condition on the Highpass Filter

Since we require that **g** given in (5.7) is a highpass filter, it must annihilate constant data. Using Proposition 2.3 with $m = 0$ we see that

$$0 = g_0 + g_1 + g_2 + g_3.$$

Using (5.8) we see that the above equation is really a condition on the lowpass filter **h**:

$$h_3 - h_2 + h_1 - h_0 = 0. \tag{5.11}$$

To this point, we have made no requirements of the constant c except that it is nonzero. In Problem 5.5, you will show that $c = \pm 1$ leads us to filters that contain zero elements. Going forward, we assume that $c \neq 0, \pm 1$.

In Problem 5.2, you will show that (5.6) and (5.11) imply

$$h_0 + h_1 + h_2 + h_3 = \pm\sqrt{2}. \tag{5.12}$$

Note that (5.11) and (5.12) imply that $\mathbf{h}$ is a lowpass filter. In order for $\mathbf{g}$ to be a highpass filter, it must satisfy $\mathbf{g} * \mathbf{v} = c\mathbf{v}$, where the entries of $\mathbf{v}$ are $v_k = (-1)^k$, in addition to annihilating constant data. The nth element in the convolution product is

$$\pm (g_0 - g_1 + g_2 - g_3) = \pm (h_3 + h_2 + h_1 + h_0)$$

which by (5.12) is $\pm\frac{\sqrt{2}}{2}$. Thus $\mathbf{g}$ is a highpass filter.

We can think of the sum on the left side of (5.12) as the result of $\mathbf{h} * \mathbf{u} = c\mathbf{u}$ (with $c = \pm\sqrt{2}$) in Definition 2.19. Thus (5.12) describes a lowpass filter with the scale $\pm\sqrt{2}$ necessary to ensure that the transformation matrix is orthogonal.

If we insert (5.9) into (5.11) and simplify we obtain

$$\begin{aligned} h_0 - h_1 + h_2 - h_3 &= 0 \\ h_0 - h_1 - ch_1 - ch_0 &= 0 \\ h_0(1-c) - h_1(1+c) &= 0 \\ h_1 &= \left(\frac{1-c}{1+c}\right) h_0. \end{aligned} \tag{5.13}$$

It is easy to update our system. Equations (5.10) and (5.13) are a circle with radius $\dfrac{1}{\sqrt{1+c^2}}$ and a line through the origin with slope $\dfrac{1-c}{1+c}$, respectively. They are plotted in Figure 5.4.

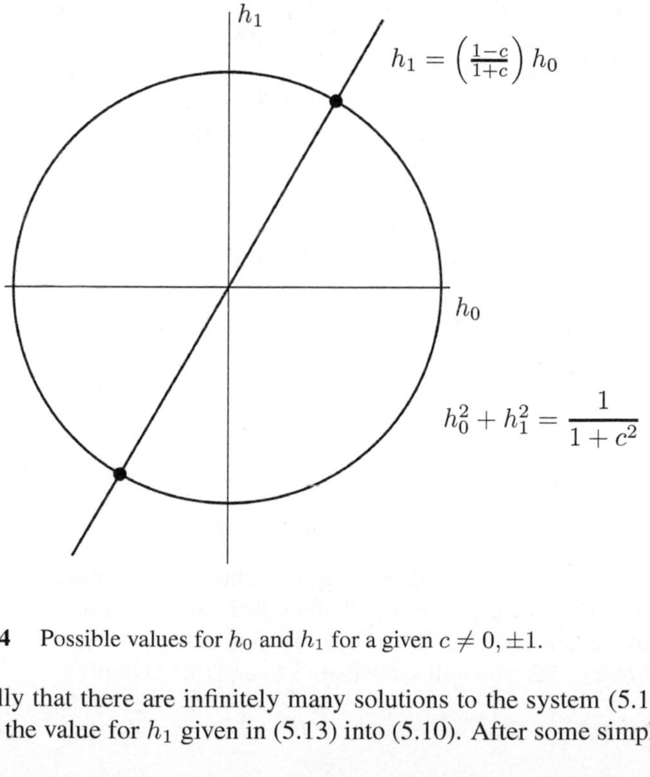

Figure 5.4 Possible values for h_0 and h_1 for a given $c \neq 0, \pm 1$.

To see algebraically that there are infinitely many solutions to the system (5.10) and (5.13), we insert the value for h_1 given in (5.13) into (5.10). After some simplification we obtain a formula for h_0 in terms of c:

$$\begin{aligned} h_0^2 + h_1^2 &= \frac{1}{1+c^2} \\ h_0^2 + \left(\frac{1-c}{1+c}\right)^2 h_0^2 &= \frac{1}{1+c^2} \\ h_0^2\left(1 + \frac{(1-c)^2}{(1+c)^2}\right) &= \frac{1}{1+c^2} \\ 2h_0^2\left(\frac{1+c^2}{(1+c)^2}\right) &= \frac{1}{1+c^2} \\ h_0^2 &= \frac{(1+c)^2}{2(1+c^2)^2}. \end{aligned}$$

Taking square roots gives

$$h_0 = \pm\frac{1+c}{\sqrt{2}\,(1+c^2)}. \tag{5.14}$$

Let's choose the positive root. Then

$$h_0 = \frac{1+c}{\sqrt{2}\,(1+c^2)} \tag{5.15}$$

and we can find values for h_1, h_2, and h_3. From (5.13) we have

$$h_1 = \frac{1-c}{\sqrt{2}\,(1+c^2)} \tag{5.16}$$

and using (5.9), we find that

$$h_2 = -\frac{c(1-c)}{\sqrt{2}(1+c^2)} \quad \text{and} \quad h_3 = \frac{c(1+c)}{\sqrt{2}(1+c^2)}. \tag{5.17}$$

Since c can be any real number other than $0, \pm 1$, equations (5.15) – (5.17) tell us that there are an infinite number of solutions to the system (5.6) and (5.11). Of course, you can also deduce this fact from Figure 5.4; the line whose slope is $(1-c)/(1+c)$ that passes through the origin must intersect the circle twice, so for every $c \neq 0, \pm 1$, there are two solutions.

An Additional Condition on the Highpass Filter

Since our system still has an infinite number of solutions, we add a final condition on the highpass filter **g**. We know that **g** annihilates constant data. What if we insist that **g** also annihilate linear data?

Using Proposition 2.3 with $m = 1$ and (5.8), we have for all $n \in \mathbb{Z}$,

$$0 = \sum_{k=0}^{3} g_k(n-k) = \sum_{k=0}^{3}(-1)^k h_{3-k}(n-k) = h_3 n - h_2(n-1) + h_1(n-2) - h_0(n-3).$$

We note that the above equation holds in particular for $n = 3$. Making this substitution gives

$$0 = h_1 - 2h_2 + 3h_3.$$

System for Finding the Daubechies Four-Term Orthogonal Filter

Thus, to find $\mathbf{h}$, we seek a solution to the nonlinear system of equations

$$\boxed{\begin{aligned} h_0^2 + h_1^2 + h_2^2 + h_3^2 &= 1 \\ h_0 h_2 + h_1 h_3 &= 0 \\ h_0 - h_1 + h_2 - h_3 &= 0 \\ h_1 - 2h_2 + 3h_3 &= 0. \end{aligned}} \tag{5.18}$$

Solving the Daubechies Four-Term System

Let's solve this system. In Problem 5.10 you will employ a solution method that is algebraic in nature. We continue with a geometric solution method. Recall that we used (5.9) in conjunction with the first and third equations of (5.18) to arrive at the two equations

$$h_0^2 + h_1^2 = \frac{1}{1 + c^2} \qquad \text{and} \qquad h_1 = \left(\frac{1 - c}{1 + c}\right) h_0.$$

These equations are plotted in Figure 5.4. Now if we use (5.9) and insert $h_2 = -ch_1$ and $h_3 = ch_0$ into the last equation of (5.18), we obtain

$$0 = h_1 - 2h_2 + 3h_3 = h_1 + 2ch_1 + 3ch_0 = (1 + 2c)h_1 + 3ch_0.$$

If we solve this last equation for h_1, we obtain

$$h_1 = -\left(\frac{3c}{1 + 2c}\right) h_0 \tag{5.19}$$

for $c \neq -\frac{1}{2}$.

Now (5.19) is a line through the origin. If we are to solve the system, then the slopes of the lines given in (5.13) and (5.19) must be the same. Thus we seek c such that

$$\frac{1 - c}{1 + c} = -\frac{3c}{1 + 2c}$$

provided that $c \neq 0, \pm 1, -\frac{1}{2}$. If we cross-multiply and expand, we obtain

$$1 + c - 2c^2 = -3c - 3c^2.$$

This leads to the quadratic equation

$$c^2 + 4c + 1 = 0$$

and using the quadratic formula, we find that

$$c = -2 \pm \sqrt{3}.$$

If we take $c = -2 + \sqrt{3}$ and insert it into (5.13), we obtain

$$h_1 = \left(\frac{1-c}{1+c}\right) h_0 = \left(\frac{3-\sqrt{3}}{-1+\sqrt{3}}\right) h_0.$$

Multiplying top and bottom of the right side of this last equation by $-1-\sqrt{3}$ and simplifying gives

$$h_1 = \sqrt{3}\, h_0. \tag{5.20}$$

We insert this identity into the left side of (5.10) to obtain

$$h_0^2 + h_1^2 = h_0^2 + (\sqrt{3}\, h_0)^2 = 4h_0^2. \tag{5.21}$$

For $c = -2 + \sqrt{3}$, the right side of (5.10) becomes

$$\frac{1}{1+c^2} = \frac{1}{1+(-2+\sqrt{3})^2} = \frac{1}{4(2-\sqrt{3})} = \frac{2+\sqrt{3}}{4}. \tag{5.22}$$

Combining (5.21) and (5.22) gives

$$h_0^2 = \frac{2+\sqrt{3}}{16}. \tag{5.23}$$

Now we need only to take square roots of both sides to find values for h_0. If we take square roots of both sides of (5.23), the right side will be quite complicated. Before taking roots, we note that

$$(1+\sqrt{3})^2 = 4 + 2\sqrt{3} = 2(2+\sqrt{3})$$

so that

$$2 + \sqrt{3} = \frac{(1+\sqrt{3})^2}{2}.$$

Inserting the previous identity into (5.23) gives

$$h_0^2 = \frac{(1+\sqrt{3})^2}{32}.$$

Taking square roots of both sides of the previous equation, we obtain

$$h_0 = \pm\frac{1+\sqrt{3}}{4\sqrt{2}}. \tag{5.24}$$

If we take the positive value in (5.24) and use (5.20), we have

$$h_1 = \frac{3+\sqrt{3}}{4\sqrt{2}}. \tag{5.25}$$

Finally, we use (5.24), (5.25), and $c = -2 + \sqrt{3}$ in conjunction with (5.9) to write

$$h_2 = \frac{3-\sqrt{3}}{4\sqrt{2}} \quad \text{and} \quad h_3 = \frac{1-\sqrt{3}}{4\sqrt{2}}. \tag{5.26}$$

We take the positive value in (5.24) and the values given by (5.25) – (5.26) as our Daubechies length 4 orthogonal filter.

Definition 5.1 (Daubechies Four-Term Orthogonal Filter). *The* Daubechies filter of length 4 *(denoted by* D4*) is* $\mathbf{h} = (h_0, h_1, h_2, h_3)$*, where* h_0*,* h_1*,* h_2*, and* h_3 *are given by*

$$\boxed{\begin{array}{ll} h_0 = \frac{1}{4\sqrt{2}}\left(1+\sqrt{3}\right) & h_1 = \frac{1}{4\sqrt{2}}\left(3+\sqrt{3}\right) \\ h_2 = \frac{1}{4\sqrt{2}}\left(3-\sqrt{3}\right) & h_3 = \frac{1}{4\sqrt{2}}\left(1-\sqrt{3}\right). \end{array}} \tag{5.27}$$

□

In Problem 5.7 you will find the other solutions by finding h_1, h_2, and h_3 using the negative value from (5.24) and also the solutions associated with $c = -2 - \sqrt{3}$.

The highpass filter $\mathbf{g}$ for D4 is given by the rule

$$\boxed{g_k = (-1)^k h_{3-k}, \qquad k = 0, 1, 2, 3,}$$

so that

$$\boxed{\begin{array}{ll} g_0 = h_3 = \frac{1}{4\sqrt{2}}\left(1-\sqrt{3}\right) & g_1 = -h_2 = -\frac{1}{4\sqrt{2}}\left(3-\sqrt{3}\right) \\ g_2 = h_1 = \frac{1}{4\sqrt{2}}\left(3+\sqrt{3}\right) & g_3 = -h_0 = -\frac{1}{4\sqrt{2}}\left(1+\sqrt{3}\right). \end{array}} \tag{5.28}$$

Constructing D4 Wavelet Transformation Matrices

The derivation of the D4 filter was done using an 8×8 matrix. Do you see how to form matrices of larger dimension? Of course, N must be even for the construction of W_N, but the basic form is

$$W_N = \left[\frac{H}{G}\right] = \left[\begin{array}{cccccccccccc} h_3 & h_2 & h_1 & h_0 & 0 & 0 & \cdots & 0 & 0 & 0 & 0 \\ 0 & 0 & h_3 & h_2 & h_1 & h_0 & & 0 & 0 & 0 & 0 \\ \vdots & & & & & & \ddots & & & & \vdots \\ 0 & 0 & 0 & 0 & & & & h_3 & h_2 & h_1 & h_0 \\ h_1 & h_0 & 0 & 0 & & & \cdots & 0 & 0 & h_3 & h_2 \\ \hline -h_0 & h_1 & -h_2 & h_3 & 0 & 0 & \cdots & 0 & 0 & 0 & 0 \\ 0 & 0 & -h_0 & h_1 & -h_2 & h_3 & & 0 & 0 & 0 & 0 \\ \vdots & & & & & & \ddots & & & & \vdots \\ 0 & 0 & 0 & 0 & & & & -h_0 & h_1 & -h_2 & h_3 \\ -h_2 & h_3 & 0 & 0 & & & \cdots & 0 & 0 & -h_0 & h_1 \end{array}\right]. \tag{5.29}$$

Now that we know how to construct W_N, we can compute one- and two-dimensional iterated transformations just as we did with the (orthogonal) Haar wavelet transformation in Chapter 4. We conclude this section with an example that illustrates features of the Daubechies four-term orthogonal wavelet transformation.

Example 5.1 (Computing the D4 Wavelet Transformation of a Vector). *Let* $\mathbf{v}$ *be the vector given in* (5.1) *and* $\mathbf{w} \in \mathbb{R}^{40}$ *be defined component-wise by* $w_k = k$ *where* $k = 1, \ldots, 40$. *We compute* $W_{40}\mathbf{v}$ *and* $W_{40}\mathbf{w}$ *where* W_{40} *is given by* (5.29). *The vectors and transformations are plotted in Figure 5.5.*

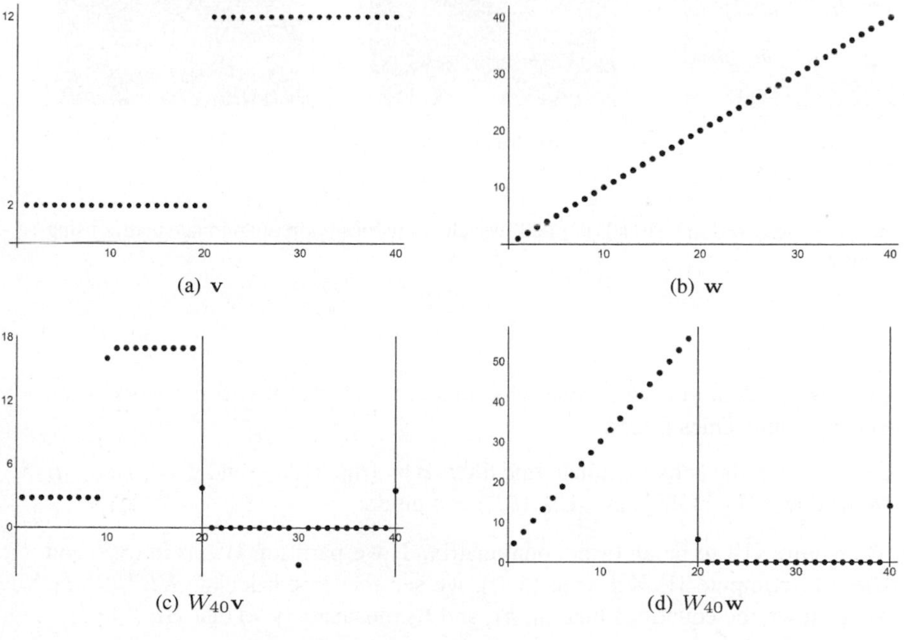

Figure 5.5 Two vectors and their D4 wavelet transformations.

Note that the jump between v_{20} and v_{21} is reflected in the differences portion of the wavelet transformation in Figure 5.5(c) with jumps between w_{29}, w_{30} and w_{30}, sw_{31}. Note that the differences portion of the wavelet transformation, plotted in Figure 5.5(d) of $\mathbf{w}$ *is largely zero thus illustrating that* $\mathbf{g}$ *does annihilate linear data. The averages portions of both transformations are scaled by $\sqrt{2}$. This is due to* (5.12) *with the positive root. Finally, we can see the effects of the wrapping rows (rows 20 and 40 of W_{40}) as the 20th and 40th components of each transformation are computed using the first two and last two elements of the input vector.*
(Live example: Visit `stthomas.edu/wavelets` *and click on Live Examples.)*

□

Two-Dimensional Wavelet Transformations

We can use the D4 filter to compute two-dimensional wavelet transformations. The process is exactly the same as described in Section 4.3 but uses W_N given in (5.29). Given an $M \times N$ matrix A, we compute the two-dimensional wavelet transformation $B = W_M A W_N^T$. We can iterate the process in exactly the same manner as was done in Section 4.3. Figure 5.6 illustrates the results of computing a two-dimensional (iterated) wavelet transformation using the D4 filter.

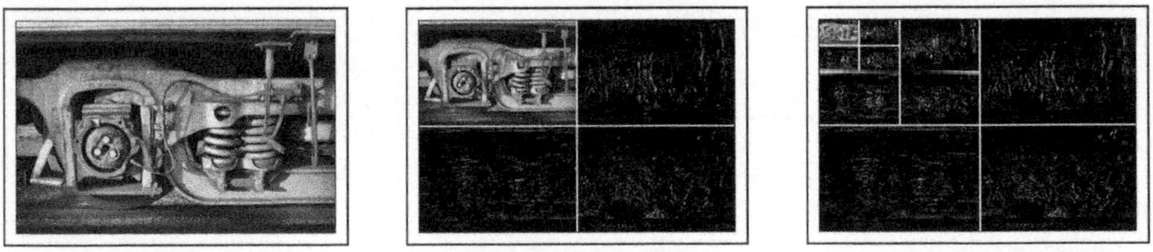

(a) A (b) One iteration (c) Three iterations

Figure 5.6 The two-dimensional (iterated) wavelet transformation of an image matrix using the D4 filter.

PROBLEMS

5.1 In this problem you will show why filters of odd length will not work when constructing Daubechies filters.

(a) Let's attempt the construction to find filters $\mathbf{h} = (h_0, h_1, h_2)$ and $\mathbf{g} = (g_0, g_1, g_2)$. Write down W_8 in this case. Use (5.2) as a guide.

(b) We require W_8 to be an orthogonal matrix. If we partition W_8 as in (5.3) and use it to compute $W_8 W_8^T$ (see (5.4)), we see that in particular, $HH^T = I_4$. Write down the equations that h_0, h_1, and h_2 must satisfy so that $HH^T = I_4$.

(c) What do these equations say about either h_0 or h_2? In either case, what are the values of the other filter coefficients? See Problem 4.14.

(d) Generalize the argument from (a) – (c) to any arbitrary odd length.

5.2 In this problem, you will show that if h_0, h_1, h_2, and h_3 satisfy the orthogonality conditions (5.6) and condition (5.11), then

$$h_0 + h_1 + h_2 + h_3 = \pm\sqrt{2}$$

(a) Use (5.6) to show that

$$(h_0 + h_1 + h_2 + h_3)^2 = 1 + 2\left(h_0h_1 + h_0h_3 + h_1h_2 + h_2h_3\right)$$

(b) Square both sides of (5.11) and then use (5.6) to show that

$$0 = (h_0 - h_1 + h_2 - h_3)^2 = 1 - 2\left(h_0h_1 + h_0h_3 + h_1h_2 + h_2h_3\right).$$

(c) Add the equations in (a) and (b) to obtain the desired result.

5.3 In this problem you will use Theorem 2.1 from Section 2.2 to show that

$$h_0 + h_1 + h_2 + h_3 = \pm\sqrt{2}$$

when W_8 given in (5.2) is orthogonal. Theorem 2.1 says that for all nonzero $\mathbf{v} \in \mathbb{R}^8$, we have $\|W_8\mathbf{v}\| = \|\mathbf{v}\|$. Suppose that

$$h_0 + h_1 + h_2 + h_3 = c \tag{5.30}$$

for some nonzero $c \in \mathbb{R}$.

(a) For $\mathbf{v} = [\,1, 1, 1, 1, 1, 1, 1, 1\,]^T$, verify that $\|\mathbf{v}\| = 2\sqrt{2}$.

(b) For $\mathbf{v}$ in (a), use (5.30) to compute $\mathbf{y} = W_8\mathbf{v}$.

(c) Find $\|\mathbf{y}\|$.

(d) Use (a) and (c) along with Theorem 2.1 to find all possible values for c.

5.4 Suppose that the two-dimensional vectors $[h_0, h_1]^T$ and $[h_2, h_3]^T$ are orthogonal. Draw a picture in the plane that illustrates this fact. Show that $[h_2, h_3]^T = c[-h_1, h_0]^T$ for any $c \in \mathbb{R}$, $c \neq 0$.

5.5 In deriving (5.9) and (5.10), the only condition imposed on c is that it is nonzero. We need to further restrict the possible values of c once we add (5.11). In this problem, you will show that if $c = \pm 1$, then we end up with filters with zero elements.

(a) Let $c = 1$. What are (5.9) and (5.10) in this case?

(b) Plug your results from (a) into (5.11) substituting for h_2, h_3. Show that $h_1 = h_2 = 0$ and $h_0 = h_3 = \pm\frac{\sqrt{2}}{2}$.

(c) Repeat (a) and (b) when $c = -1$. Show that $h_0 = h_3 = 0$ and $h_1 = h_2 = \pm\frac{\sqrt{2}}{2}$.

5.6 Take the negative root in (5.14) and show that we obtain an infinite number of solutions to the system (5.10) and (5.13).

5.7 In this problem, we find the other solutions to the system given in (5.18).

(a) Take the negative root in (5.24) and insert it into (5.10), (5.13), and (5.19) to find another solution to (5.18).

(b) Take $c = -2 - \sqrt{3}$ and use it to find two values for h_0. For each value of h_0, find a solution to the system.

(c) If $\mathbf{h} = (h_0, h_1, h_2, h_3)$ is the filter given in Definition 5.1, how can you characterize the other solutions of (5.18) in terms of $\mathbf{h}$?

5.8 In this problem you will show that the conditions

$$\begin{aligned} h_0 + h_1 + h_2 + h_3 &= \sqrt{2} \\ h_0 - h_1 + h_2 - h_3 &= 0 \\ h_0 h_2 + h_1 h_3 &= 0 \end{aligned} \tag{5.31}$$

imply

$$h_0^2 + h_1^2 + h_2^2 + h_3^2 = 1.$$

To simplify computations, let

$$a = h_0^2 + h_1^2 + h_2^2 + h_3^2.$$

(a) Square each of the first two equations in (5.31) and simplify to obtain

$$\begin{aligned} a + 2\sum_{j \neq k} h_j h_k &= 2 \\ a + 2\sum_{j \neq k} (-1)^{j+k} h_j h_k &= 0. \end{aligned}$$

(b) Use the third equation in (5.31) to simplify both equations from (a). That is, show that

$$\begin{aligned} a + 2\left(h_0 h_1 + h_0 h_3 + h_1 h_2 + h_2 h_3\right) &= 2 \\ a - 2\left(h_0 h_1 + h_0 h_3 + h_1 h_2 + h_2 h_3\right) &= 0 \end{aligned}$$

(c) Combine the two equations in (b) to show that $a = 1$.

(d) Verify that we can replace $\sqrt{2}$ by $-\sqrt{2}$ in (5.31) and the result still holds.

5.9 In this problem you will show how to use one solution of (5.18) to obtain other solutions.

(a) Suppose that $\mathbf{h} = (h_0, h_1, h_2, h_3) = (a, b, c, d)$ solves (5.18). Show that $-\mathbf{h} = (-a, -b, -c, -d)$ also solves (5.18).

(b) Suppose that $\mathbf{h} = (h_0, h_1, h_2, h_3) = (a, b, c, d)$ solves (5.18). Show that $\mathbf{h}^r = (h_0^r, h_1^r, h_2^r, h_3^r) = (d, c, b, a)$ also solves (5.18). *Hint:* The first three equations are straightforward to verify. To obtain the fourth, you need to show $3a - 2b + c = 0$. Remember Proposition 2.3 holds for all $n \in \mathbb{Z}$. Can you find a particular n that gives you the desired result?

5.10 In this problem you will use an algebraic approach to solve for the D4 filter. In Problem 5.8 we showed that the orthogonality condition

$$h_0^2 + h_1^2 + h_2^2 + h_3^2 = 1.$$

can be replaced by

$$h_0 + h_1 + h_2 + h_3 = \pm\sqrt{2}.$$

in (5.18). Thus, we can find the D4 filter coefficients by solving the system

$$\begin{aligned} h_0 + h_1 + h_2 + h_3 &= \sqrt{2} \\ h_0 - h_1 + h_2 - h_3 &= 0 \\ h_1 - 2h_2 + 3h_3 &= 0 \\ h_0 h_2 + h_1 h_3 &= 0. \end{aligned} \tag{5.32}$$

(a) Show that the first two equations in (5.32) can be replaced by the linear equations

$$h_0 + h_2 = \frac{\sqrt{2}}{2} \qquad \text{and} \qquad h_1 + h_3 = \frac{\sqrt{2}}{2}.$$

(b) Use (a) and show that the linear equations in (5.32) can be replaced by the equivalent system

$$\begin{aligned} h_0 &= \frac{\sqrt{2}}{4} - h_3 \\ h_1 &= \frac{\sqrt{2}}{2} - h_3 \\ h_2 &= \frac{\sqrt{2}}{4} + h_3. \end{aligned}$$

(c) Insert the expressions for h_0, h_1, and h_2 from (b) into the last equation of (5.32) to show that

$$h_3^2 - \frac{\sqrt{2}}{4}h_3 - \frac{1}{16} = 0.$$

(d) Use the quadratic formula to find two values for h_3. For each value of h_3, find values for h_0, h_1, and h_2.

(e) Replace $\sqrt{2}$ with $-\sqrt{2}$ in (5.32) and repeat the problem to find two other solutions to the system.

(f) Compare the solutions you obtained in this problem with the results of Problem 5.9.

5.11 Although the condition

$$h_0 + h_1 + h_2 + h_3 = \sqrt{2}$$

can be derived using other equations from the system (5.18), what would be the effect of adding the above equation to the system (5.18)? *Hint:* Problem 5.10 will be helpful here.

5.12 Suppose $\mathbf{v} \in \mathbb{R}^N$ with N even. We wish compute $W_N\mathbf{v} = \left[\begin{array}{c}\mathbf{a}\\ \hline \mathbf{d}\end{array}\right]$ where W_N is given in (5.29). Computing the product $W_N\mathbf{v}$ involves a large number of multiplications by zero due to the sparse nature of W_N. We seek a more expedient way to compute $W_N\mathbf{v}$. Can you identify a $N/2 \times 4$ matrix X such that

$$X\begin{bmatrix} h_3 \\ h_2 \\ h_1 \\ h_0 \end{bmatrix} = \mathbf{a} \quad \text{and} \quad X\begin{bmatrix} g_3 \\ g_2 \\ g_1 \\ g_0 \end{bmatrix} = \mathbf{d}?$$

Hint: Problem 4.8 will be helpful.

5.13 Suppose we have computed

$$W_N\mathbf{v} = \left[\begin{array}{c}\mathbf{a}\\ \hline \mathbf{d}\end{array}\right]$$

where W_N is given in (5.29). Here N is even and we let $L = N/2$. To recover $\mathbf{v}$, we write W_N in block form and compute

$$\mathbf{v} = W_N^T\left[\begin{array}{c}\mathbf{a}\\ \hline \mathbf{d}\end{array}\right] = \left[\begin{array}{c|c} H_N^T & G_N^T \end{array}\right]\left[\begin{array}{c}\mathbf{a}\\ \hline \mathbf{d}\end{array}\right] = H_N^T\mathbf{a} + G_N^T\mathbf{b}.$$

We can find efficient representations of $H_N^T\mathbf{a}$ and $G_N^T\mathbf{d}$ and sum up the results in order to compute the inverse transformation. We consider $H_N^T\mathbf{a}$ and note that the process for computing $G_N^T\mathbf{d}$ is similar.

(a) Let $\mathbf{a} = \begin{bmatrix} a_1 \\ a_2 \\ \vdots \\ a_L \end{bmatrix}$. Find an $L \times 2$ matrix A, constructed from the elements of $\mathbf{a}$, such that

$$A\begin{bmatrix} h_1 & h_0 \\ h_3 & h_2 \end{bmatrix} = [\mathbf{o} \quad \mathbf{e}]$$

where $\mathbf{o}$, $\mathbf{e}$, hold the odd and even components, respectively, of $H_N^T \mathbf{a}$. *Hint:* Write A in the $L = 4$ case. It should be straightforward to generalize the construction.

(b) A *permutation matrix* is an $N \times N$ matrix formed by permuting the rows of the identity matrix I_N. Find permutation matrix P such that $P \left[\frac{\mathbf{o}}{\mathbf{e}}\right] = H_N^T \mathbf{a}$.

5.14 This problem is adapted from results in [17]. For illustrative purposes, consider the 8×8 wavelet transformation W_8 given in (5.29) where the values h_0, h_1, h_2, and h_3 are given in Definition 5.1 and $g_k = (-1)^{k+1} h_{3-k}$ for $k = 0, 1, 2, 3$. In many applications, the averages and differences rows of W_8 are often intertwined:

$$\widehat{W}_8 = \begin{bmatrix} h_3 & h_2 & h_1 & h_0 & 0 & 0 & 0 & 0 \\ g_3 & g_2 & g_1 & g_0 & 0 & 0 & 0 & 0 \\ 0 & 0 & h_3 & h_2 & h_1 & h_0 & 0 & 0 \\ 0 & 0 & g_3 & g_2 & g_1 & g_0 & 0 & 0 \\ 0 & 0 & 0 & 0 & h_3 & h_2 & h_1 & h_0 \\ 0 & 0 & 0 & 0 & g_3 & g_2 & g_1 & g_0 \\ h_1 & h_0 & 0 & 0 & 0 & 0 & h_3 & h_2 \\ g_1 & g_0 & 0 & 0 & 0 & 0 & g_3 & g_2 \end{bmatrix}.$$

(a) Find the 8×8 permutation matrix P so that

$$W_8 = P\widehat{W}_8.$$

The definition of a permutation matrix is given in Problem 5.13(b).

(b) Find P^{-1} and describe the actions of P, P^{-1}.

(c) $\widehat{W}_8$ has a nice block structure. Let

$$H_1 = \begin{bmatrix} h_3 & h_2 \\ g_3 & g_2 \end{bmatrix} \qquad H_0 = \begin{bmatrix} h_1 & h_0 \\ g_1 & g_0 \end{bmatrix} \qquad 0_4 = \begin{bmatrix} 0 & 0 \\ 0 & 0 \end{bmatrix}$$

Using these matrices, write $\widehat{W}_8$ in block form.

(d) Verify that

$$H_1 = U \begin{bmatrix} 1 & 0 \\ 0 & 0 \end{bmatrix} V \qquad H_0 = U \begin{bmatrix} 0 & 0 \\ 0 & 1 \end{bmatrix} V \tag{5.33}$$

where

$$U = \frac{1}{2\sqrt{2}} \begin{bmatrix} -1+\sqrt{3} & 1+\sqrt{3} \\ -1-\sqrt{3} & -1+\sqrt{3} \end{bmatrix} \qquad V = \frac{1}{2} \begin{bmatrix} -1 & \sqrt{3} \\ \sqrt{3} & 1 \end{bmatrix}$$

(e) Show that U and V from (d) are orthogonal matrices. The factorizations in (5.33) are called *singular value decompositions* of H_1 and H_0 (see Strang [86]).

(f) What rotation angle is associated with U? With V? See Problem 2.29 from Section 2.2.

(g) Replace H_0 and H_1 in the block formulation of $\widehat{W}_8$ you found in (c) by the factorizations given in (5.33) and then write $\widehat{W}_8$ as $\mathcal{UPV}$, where $\mathcal{U}$ and $\mathcal{V}$ are block diagonal matrices and P is a permutation matrix.

(h) Show that $\mathcal{U}$ and $\mathcal{V}$ from (e) are orthogonal and describe in words the action of permutation matrix P.

The factorization in (e) is used in Calderbank et al. [17] to show how to modify the D4 wavelet transformation so that it maps integers to integers. Recall that an orthogonal matrix rotates the vector to which it is applied and a permutation matrix simply reorders the elements of a vector. Thus, $\mathcal{UPV}$ rotates $\mathbf{v}$, permutes $\mathcal{V}\mathbf{v}$ and then rotates this product!

5.15 One of the problems with the D4 wavelet transformation is the presence of wrapping rows. If our data are periodic, then the wrapping rows make sense, but we typically do not think of digital audio or images as periodic data.

In this problem you will learn how to construct a wavelet transformation for the D4 filter that has no wrapping rows. The technique is known as *matrix completion.* The resulting transformation is orthogonal, and some of the highpass structure is retained. The process is described in Strang and Nguyen [88] and Keinert [63].[1] A **CAS** will be helpful for calculations in the latter parts of this problem. For illustrative purposes, we consider completion of the 8×8 matrix initially constructed from the D4 filter.

Strang and Nguyen [88] suggest replacing the first and last rows of each of the lowpass and highpass portions of the wavelet transformation matrix given in (5.2). We denote our modified wavelet transformation by

$$\widehat{W}_8 = \left[\begin{array}{cccccccc} a_2 & a_1 & a_0 & 0 & 0 & 0 & 0 & 0 \\ 0 & h_3 & h_2 & h_1 & h_0 & 0 & 0 & 0 \\ 0 & 0 & 0 & h_3 & h_2 & h_1 & h_0 & 0 \\ 0 & 0 & 0 & 0 & 0 & b_2 & b_1 & b_0 \\ \hline c_2 & c_1 & c_0 & 0 & 0 & 0 & 0 & 0 \\ 0 & g_3 & g_2 & g_1 & g_0 & 0 & 0 & 0 \\ 0 & 0 & 0 & g_3 & g_2 & g_1 & g_0 & 0 \\ 0 & 0 & 0 & 0 & 0 & d_2 & d_1 & d_0 \end{array}\right] = \left[\begin{array}{cccccccc} a_2 & a_1 & a_0 & 0 & 0 & 0 & 0 & 0 \\ 0 & h_3 & h_2 & h_1 & h_0 & 0 & 0 & 0 \\ 0 & 0 & 0 & h_3 & h_2 & h_1 & h_0 & 0 \\ 0 & 0 & 0 & 0 & 0 & b_2 & b_1 & b_0 \\ \hline c_2 & c_1 & c_0 & 0 & 0 & 0 & 0 & 0 \\ 0 & -h_0 & h_1 & -h_2 & h_3 & 0 & 0 & 0 \\ 0 & 0 & 0 & -h_0 & h_1 & -h_2 & h_3 & 0 \\ 0 & 0 & 0 & 0 & 0 & d_2 & d_1 & d_0 \end{array}\right] \tag{5.34}$$

where $\mathbf{h} = (h_0, h_1, h_2, h_3)$ is the D4 filter given in (5.27).

(a) Why do you suppose that only three terms are used in the first and last rows? Also, why do you think the second rows in each of the averages and differences

[1]Keinert also gives other methods for creating discrete wavelet transformations with no wrapping rows.

portion are shifted one unit right instead of two (as is the case for the Haar and D4 transformations)?

(b) What condition should be imposed on the elements of $\mathbf{c} = (c_0, c_1, c_2)$ if we want $\mathbf{c}$ to be a highpass filter?

(c) We first find $\mathbf{a} = (a_0, a_1, a_2)$ and $\mathbf{c}$. Since $\widehat{W}_8^{-1} = \widehat{W}_8^T$, we know that

$$a_0^2 + a_1^2 + a_2^2 = c_0^2 + c_1^2 + c_2^2 = 1. \tag{5.35}$$

Using (5.34), write down the other three nontrivial orthogonality conditions that the elements of $\mathbf{c}$ must satisfy.

(d) Two of the equations you found in (c) involve only c_0 and c_1. Without solving the system, can you think of two values that work? *Hint:* Think about the orthogonality conditions satisfied by the "conventional rows" in $\widehat{W}_8$.

(e) Use (d) and (b) to find c_2.

(f) We know from (c) that $a_0^2 + a_1^2 + a_2^2 = 1$. Write down the other three orthogonality equations that the elements of $\mathbf{a}$ must satisfy. Two of these equations involve only a_0 and a_1. Use the ideas from (d) to find a_0 and a_1.

(g) The remaining orthogonality condition is $\mathbf{a} \cdot \mathbf{c} = 0$. You should be able to use this equation and your previous work to show that $a_2 = \frac{\sqrt{2}}{2}$.

(h) The values in $\mathbf{a}$ and $\mathbf{c}$ do not satisfy (5.35). To finish this part of the completion problem, normalize $\mathbf{a}$ and $\mathbf{c}$ (see Problem 2.12 in Section 2.1).

(i) Repeat (b) – (h) to find values for $\mathbf{b} = (b_0, b_1, b_2)$ and $\mathbf{d} = (d_0, d_1, d_2)$.

(j) Is this matrix completion unique? That is, can you find other values for $\mathbf{a}, \mathbf{b}, \mathbf{c}$, and $\mathbf{d}$ that preserve orthogonality and the highpass conditions on $\mathbf{c}$ and $\mathbf{d}$?

Computer Lab

5.1 Constructing the D4 Filter. From `stthomas.edu/wavelets`, access the file `D4Filter`. In this lab, you will write code that allows you to solve for the D4 filter coefficients. In addition, you will write code that uses the D4 filter to compute the (inverse) wavelet transformations of a vector or a matrix.

5.2 Daubechies Filter of Length 6

It is straightforward to construct the Daubechies filter of length 6 now that we have constructed one of length 4. Since the filter we seek is longer in length, let's use a

10×10 matrix this time:

$$W_{10} = \left[\begin{array}{cccccccccc} h_5 & h_4 & h_3 & h_2 & h_1 & h_0 & 0 & 0 & 0 & 0 \\ 0 & 0 & h_5 & h_4 & h_3 & h_2 & h_1 & h_0 & 0 & 0 \\ 0 & 0 & 0 & 0 & h_5 & h_4 & h_3 & h_2 & h_1 & h_0 \\ h_1 & h_0 & 0 & 0 & 0 & 0 & h_5 & h_4 & h_3 & h_2 \\ h_3 & h_2 & h_1 & h_0 & 0 & 0 & 0 & 0 & h_5 & h_4 \\ \hline g_5 & g_4 & g_3 & g_2 & g_1 & g_0 & 0 & 0 & 0 & 0 \\ 0 & 0 & g_5 & g_4 & g_3 & g_2 & g_1 & g_0 & 0 & 0 \\ 0 & 0 & 0 & 0 & g_5 & g_4 & g_3 & g_2 & g_1 & g_0 \\ g_1 & g_0 & 0 & 0 & 0 & 0 & g_5 & g_4 & g_3 & g_2 \\ g_3 & g_2 & g_1 & g_0 & 0 & 0 & 0 & 0 & g_5 & g_4 \end{array}\right]. \tag{5.36}$$

Orthogonality Conditions

Notice that overlapping rows now come in groups of two. We can write W_{10} in block form as

$$W_{10} = \left[\begin{array}{c} H_5 \\ \hline G_5 \end{array}\right]$$

and then compute $W_{10}W_{10}^T$. As was the case in (5.4), we see that W_{10} is orthogonal if

$$\begin{aligned} H_5 H_5^T &= G_5 G_5^T = I_5 \\ H_5 G_5^T &= G_5 H_5^T = 0_5. \end{aligned} \tag{5.37}$$

You can verify by direct multiplication of H_5 and H_5^T in $W_{10}W_{10}^T$ that the three orthogonality conditions are

$$\begin{aligned} h_0^2 + h_1^2 + h_2^2 + h_3^2 + h_4^2 + h_5^2 &= 1 \\ h_0 h_2 + h_1 h_3 + h_2 h_4 + h_3 h_5 &= 0 \\ h_0 h_4 + h_1 h_5 &= 0. \end{aligned} \tag{5.38}$$

We again use the ideas from Problem 2.11 in Section 2.1 to define the highpass filter. It can be verified that if

$$\mathbf{g} = (g_0, g_1, g_2, g_3, g_4, g_5) = (h_5, -h_4, h_3, -h_2, h_1, -h_0), \tag{5.39}$$

then the remaining identities in (5.37) are satisfied thus making W_{10} orthogonal. Note as in the case of the D4 filter, we have

$$g_k = (-1)^k h_{5-k}, \quad k = 0, \ldots, 5. \tag{5.40}$$

Conditions on the Highpass Filter

We mimic the construction of the D4 filter and require that our six-term filter annihilate both constant and linear data. Using Proposition 2.3 with $m = 0$ and (5.40), we have

$$\begin{aligned} 0 &= \sum_{k=0}^{5} g_k \\ &= \sum_{k=0}^{5} (-1)^k h_{5-k} \\ &= h_5 - h_4 + h_3 - h_2 + h_1 - h_0 \end{aligned}$$

or

$$h_0 - h_1 + h_2 - h_3 + h_4 - h_5 = 0. \tag{5.41}$$

In Problem 5.16, you will show that $h_0 + h_1 + \cdots + h_5 = \pm\sqrt{2}$ follows from (5.38). This result along with (5.41) guarantee that $\mathbf{h}$ is a lowpass filter. Since $\mathbf{h}$ is a lowpass filter and $\mathbf{g}$ is constructed using (5.39), Proposition 2.4 implies $\mathbf{g}$ is a highpass filter.

Using Proposition 2.3 with $m = 1$ and (5.40), we have for all $n \in \mathbb{Z}$,

$$\begin{aligned} 0 &= \sum_{k=0}^{5} g_k (n-k) \\ &= \sum_{k=0}^{5} (-1)^k h_{5-k} (n-k) \\ &= h_5 n - h_4 (n-1) + h_3 (n-2) - h_2 (n-3) + h_1 (n-4) - h_0 (n-5). \end{aligned}$$

Since we can choose any $n \in \mathbb{Z}$ in the above equation, we take $n = 5$ to obtain

$$h_1 - 2h_2 + 3h_3 - 4h_4 + 5h_5 = 0. \tag{5.42}$$

If we use a **CAS** such as *Mathematica* to solve (5.38), (5.41) and (5.42), we find there are infinitely many solutions. Thus we are motivated to add another condition on the highpass filter. Continuing the pattern we have established with the Haar and D4 filters, we will insist that the six-term filter annihilate quadratic data. Again using Proposition 2.3 with $m = 2$ and (5.40), we have

$$\begin{aligned} 0 &= \sum_{k=0}^{5} g_k (n-k)^2 \\ &= \sum_{k=0}^{5} (-1)^k h_{5-k} (n-k)^2 \\ &= h_5 n^2 - h_4 (n-1)^2 + h_3 (n-2)^2 - h_2 (n-3)^2 + h_1 (n-4)^2 - h_0 (n-5). \end{aligned}$$

Choosing $n = 5$ in the above equation gives

$$h_1 - 4h_2 + 9h_3 - 16h_4 + 25h_5 = 0. \tag{5.43}$$

The System for Finding Daubechies Six-Term Orthogonal Filters

We seek a solution to the three orthogonality equations (5.38) and the three highpass filter conditions (5.41) – (5.43):

$$\boxed{\begin{aligned} h_0^2 + h_1^2 + h_2^2 + h_3^2 + h_4^2 + h_5^2 &= 1 \\ h_0 h_2 + h_1 h_3 + h_2 h_4 + h_3 h_5 &= 0 \\ h_0 h_4 + h_1 h_5 &= 0 \\ h_0 - h_1 + h_2 - h_3 + h_4 - h_5 &= 0 \\ h_1 - 2h_2 + 3h_3 - 4h_4 + 5h_5 &= 0 \\ h_1 - 4h_2 + 9h_3 - 16h_4 + 25h_5 &= 0. \end{aligned}} \tag{5.44}$$

In Problem 5.19 you will produce two real solutions to the system (5.44). We define one of these solutions to be the Daubechies filter of length 6.

Definition 5.2 (Daubechies Six-Term Orthogonal Filter). *The* Daubechies filter of length 6 *(denoted by* D6*) is* $\mathbf{h} = (h_0, h_1, h_2, h_3, h_4, h_5)$ *where*

$$\boxed{\begin{aligned} h_0 &= \tfrac{\sqrt{2}}{32}\left(1 + \sqrt{10} + \sqrt{5 + 2\sqrt{10}}\right) \approx 0.332671 \\ h_1 &= \tfrac{\sqrt{2}}{32}\left(5 + \sqrt{10} + 3\sqrt{5 + 2\sqrt{10}}\right) \approx 0.806892 \\ h_2 &= \tfrac{\sqrt{2}}{32}\left(10 - 2\sqrt{10} + 2\sqrt{5 + 2\sqrt{10}}\right) \approx 0.459878 \\ h_3 &= \tfrac{\sqrt{2}}{32}\left(10 - 2\sqrt{10} - 2\sqrt{5 + 2\sqrt{10}}\right) \approx -0.135011 \\ h_4 &= \tfrac{\sqrt{2}}{32}\left(5 + \sqrt{10} - 3\sqrt{5 + 2\sqrt{10}}\right) \approx -0.085441 \\ h_5 &= \tfrac{\sqrt{2}}{32}\left(1 + \sqrt{10} - \sqrt{5 + 2\sqrt{10}}\right) \approx 0.035226. \end{aligned}} \tag{5.45}$$

□

The highpass filter $\mathbf{g}$ for D6 is given by the rule

$$\boxed{g_k = (-1)^k h_{5-k}, \qquad k = 0, \ldots, 5.} \tag{5.46}$$

so that

$$
\boxed{
\begin{aligned}
g_0 &= h_5 = \tfrac{\sqrt{2}}{32}\left(1 + \sqrt{10} - \sqrt{5+2\sqrt{10}}\right) \approx 0.035226\\
g_1 &= -h_4 = -\tfrac{\sqrt{2}}{32}\left(5 + \sqrt{10} - 3\sqrt{5+2\sqrt{10}}\right) \approx 0.085441\\
g_2 &= h_3 = \tfrac{\sqrt{2}}{32}\left(10 - 2\sqrt{10} - 2\sqrt{5+2\sqrt{10}}\right) \approx -0.135011\\
g_3 &= -h_2 = -\tfrac{\sqrt{2}}{32}\left(10 - 2\sqrt{10} + 2\sqrt{5+2\sqrt{10}}\right) \approx -0.459878\\
g_4 &= h_1 = \tfrac{\sqrt{2}}{32}\left(5 + \sqrt{10} + 3\sqrt{5+2\sqrt{10}}\right) \approx 0.806892\\
g_5 &= -h_0 = -\tfrac{\sqrt{2}}{32}\left(1 + \sqrt{10} + \sqrt{5+2\sqrt{10}}\right) \approx -0.332671.
\end{aligned}
}
$$

Constructing D6 Wavelet Transformation Matrices

The construction of matrices needed to compute the D6 wavelet transformation is similar to that used for D4 filters. Given $\mathbf{v} \in \mathbb{R}^N$, where N is even, we use (5.45) and (5.46) to naturally extend (5.36) to create

$$
W_N = \left[\frac{H}{G}\right]
$$
$$
= \left[\begin{array}{ccccccccccccccc}
h_5 & h_4 & h_3 & h_2 & h_1 & h_0 & 0 & 0 & \cdots & 0 & 0 & 0 & 0 & 0 & 0\\
0 & 0 & h_5 & h_4 & h_3 & h_2 & h_1 & h_0 & & 0 & 0 & 0 & 0 & 0 & 0\\
\vdots & & & & & & & \ddots & & & & & & & \vdots\\
0 & 0 & 0 & 0 & 0 & 0 & & & & h_5 & h_4 & h_3 & h_2 & h_1 & h_0\\
h_1 & h_0 & 0 & 0 & 0 & 0 & & & & 0 & 0 & h_5 & h_4 & h_3 & h_2\\
h_3 & h_2 & h_1 & h_0 & 0 & 0 & & & \cdots & 0 & 0 & 0 & 0 & h_5 & h_4\\
\hline
-h_0 & h_1 & -h_2 & h_3 & -h_4 & h_5 & 0 & 0 & \cdots & 0 & 0 & 0 & 0 & 0 & 0\\
0 & 0 & -h_0 & h_1 & -h_2 & h_3 & -h_4 & h_5 & & 0 & 0 & 0 & 0 & 0 & 0\\
\vdots & & & & & & & \ddots & & & & & & & \vdots\\
0 & 0 & 0 & 0 & 0 & 0 & & & & -h_0 & h_1 & -h_2 & h_3 & -h_4 & h_5\\
-h_4 & h_5 & 0 & 0 & 0 & 0 & & & & 0 & 0 & -h_0 & h_1 & -h_2 & h_3\\
-h_2 & h_3 & -h_4 & h_5 & 0 & 0 & & & \cdots & 0 & 0 & 0 & 0 & -h_0 & h_1
\end{array}\right]. \tag{5.47}
$$

The D6 wavelet transformation $\mathbf{y}$ of $\mathbf{v}$ is then computed via the product $\mathbf{y} = W_N\mathbf{v}$. The inverse transformation is simply $\mathbf{v} = W_N^T\mathbf{y}$. Assuming that N is divisible by 2^i, we can iterate the transformation at most i times in exactly the same manner we did for the Haar and D4 wavelet transformations.

For two-dimensional transformations of an $M \times N$ matrix A, we use (5.47) to construct W_M and W_N and compute the transformation $B = W_M A W_N^T$. The inverse transformation is computed by the product $A = W_M^T B W_N$. We can iterate this transformation in exactly the same way we did for the Haar and D4 wavelet transformations.

Example: Data Compression

Let's look at some examples and compare the filters that we have constructed to date.

Example 5.2 (Signal Compression Using Haar, D4 and D6). *We consider the digital audio file plotted in Figure 5.7. The file consists of* $23,616$ *samples. The duration of the signal is* 2.142 *seconds.*

Figure 5.7 Plot of the audio signal for Example 5.2.

In this example we first compute three wavelet transformations (six iterations each) using the Haar filter $\mathbf{h} = \left(\frac{\sqrt{2}}{2}, \frac{\sqrt{2}}{2}\right)$*, the D4 filter* (5.27)*, and the D6 filter* (5.45)*, respectively.*

We plot the cumulative energy of the original audio file as well as the energy for each of the wavelet transformations in Figure 5.8. As you can see, the D6 slightly outperforms the D4, and both are more efficient at storing the energy of the signal than the Haar transformation. The conservation of energy of all transformations is better than that of the original data.

(Live example: Visit `stthomas.edu/wavelets` *and click on Live Examples.)*

□

Example – Image Compression

Our final example revisits the image compression application considered in Example 4.15.

Example 5.3 (Image Compression). *We return to Example 4.15. In that example we used the Haar wavelet to perform image compression on the* 256×384 *image plotted in Figure 4.20(a). In this example we repeat the process used in Example 4.15, but this time we will use the orthogonal Haar, D4, and D6 filters.*

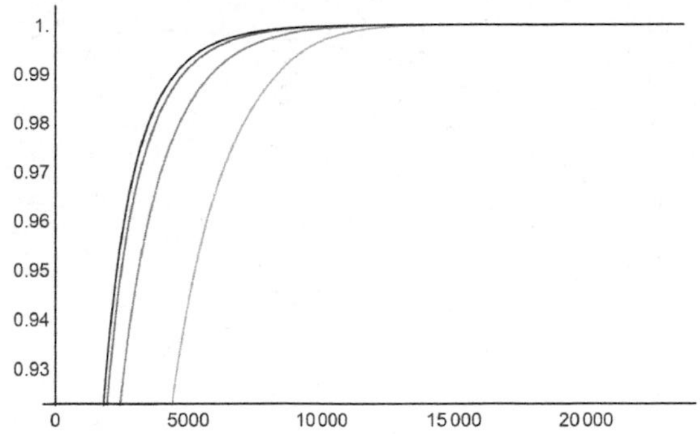

Figure 5.8 Plot of the cumulative energies of the original audio file (light gray), the Haar transformation (gray), the D4 transformation (dark gray) and the D6 transformation (black).

Our first step is to compute the wavelet transformations using each of the three filters. We center A by subtracting 128 from each value in the matrix and then perform two iterations of each transformation on the resulting matrix. We then construct W_{256} and W_{384} for the respective filters and initially compute $B = W_{256}AW_{384}^T$. We next extract the 128×192 upper left corner $\mathcal{B}$ of B and compute $W_{128}\mathcal{B}W_{192}^T$. The results are plotted in Figure 5.9.

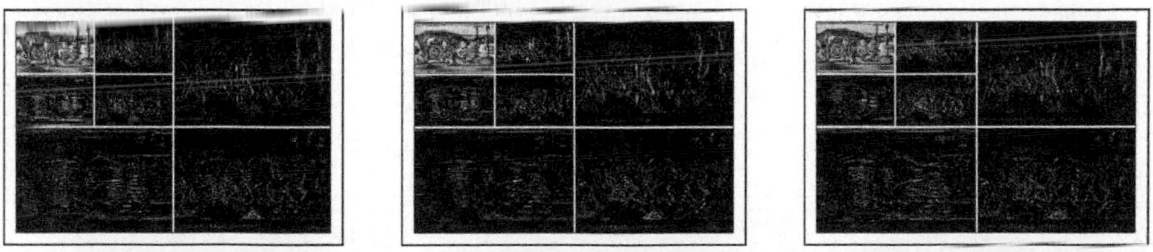

(a) Haar (b) D4 (c) D6

Figure 5.9 Two iterations of the wavelet transformation using the orthogonal Haar, D4, and D6 filters.

We next compute the cumulative energy for each wavelet transformation. These cumulative energies are plotted in Figure 5.10. Note that the cumulative energies for the D4 and D6 filters are virtually identical and slightly outperform the orthogonal Haar filter.

Instead of performing quantization for three different energy levels as we did in Example 4.15, we use just one energy level. We use a **CAS** *to find the largest elements (in absolute value) in each wavelet transformation that constitutes $r = 99.7\%$ of the energy. These largest values are floored to the nearest integer. All remaining elements in the transformation matrix are set to zero. Table 5.1 summarizes the*

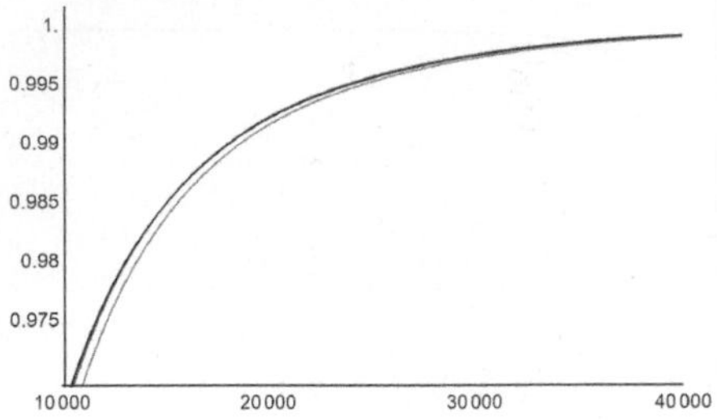

Figure 5.10 The cumulative energies for the Haar transformation (gray), the D4 filter (dark gray), and the D6 filter (black).

compression information for each wavelet transformation. The quantized transformations are plotted in Figure 5.11.

Table 5.1 Data for the quantized wavelet transformations using the Haar, D4, and D6 filters.[a]

Filter	Nonzero Terms	% Zero
Haar	29,604	69.89
D4	28,670	70.84
D6	28,969	70.53

[a] The second column indicates the number of nonzero terms still present in the quantized transformations, and the third column lists the percentage of elements that are zero in each quantized transformation. Note that the numbers in the nonzero terms column might include multiple copies of the smallest values (in absolute value) that comprise the 99.7% energy level. Since it is impossible to know which of these values to keep or discard, the selection routine keeps all of them.

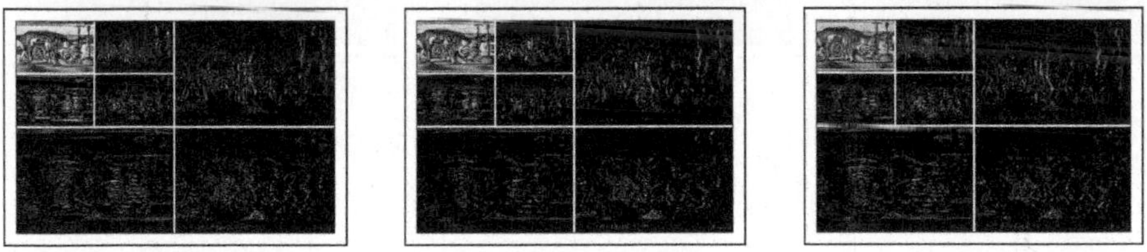

(a) Haar (b) D4 (c) D6

Figure 5.11 The quantized wavelet transformation using the orthogonal Haar, D4, and D6 filters.

The original image from Example 4.15 consists of 98,304 *pixels, so without any compression strategy, we must store pixels at* 8 *bpp for a total of 786,432 bits. The entropy for the original image is* 7.711.

Recall that we modified the Haar wavelet transformation so that it mapped integers to integers. The process for modifying the orthogonal Haar, D4, and D6 filters so that they map integers to integers is a difficult problem (see [17] and Problem 5.14), so we simply round the elements in each transformation to produce integer-valued matrices that can be Huffman-encoded. We convert the elements of each quantized and rounded transformation to Huffman codes and then create a bit stream for each transformation. The length of the bit stream for each transformation is given in Table 5.2.

Table 5.2 Huffman bit stream length, bpp, and entropy for each quantized transformation.

Filter	Bit Stream Length	bpp	Entropy
Haar	337,284	3.432	3.304
D4	329,928	3.356	3.219
D6	331,047	3.368	3.233

For each quantized transformation, we compute the inverse transformation and add 128 to each element in the resulting matrix. The compressed images are plotted in Figure 5.12.

From Table 5.2 we note that the D4 and D6 filters slightly outperformed the orthogonal Haar filter in terms of bits per pixel. The captions in Figure 5.12 contain PSNR information and show that the D4 and D6 filters barely outperform the orthogonal Haar filter.

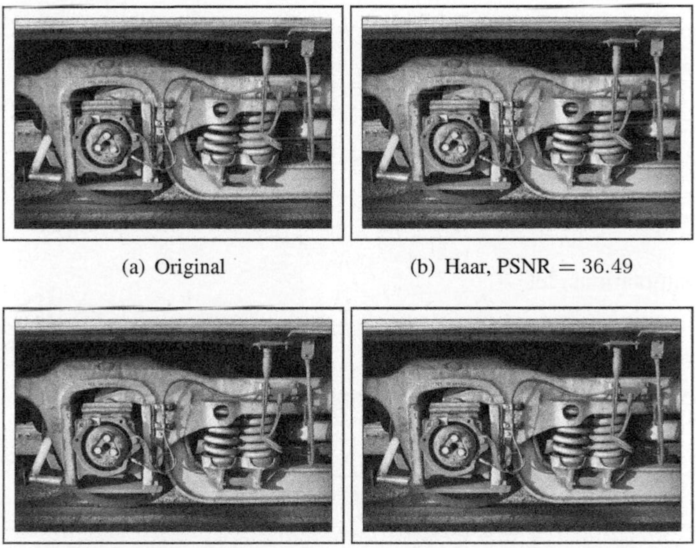

(a) Original

(b) Haar, PSNR $= 36.49$

(c) D4, PSNR $= 36.53$

(d) D6, PSNR $= 36.53$

Figure 5.12 The compressed images using the orthogonal Haar, D4, and D6 filters.

(Live example: Visit `stthomas.edu/wavelets` *and click on Live Examples.)*

□

PROBLEMS

5.16 In this problem you will show that if $h_0, \ldots, h_5$ satisfy the orthogonality conditions (5.38) in addition to (5.41), then

$$h_0 + h_1 + h_2 + h_3 + h_4 + h_5 = \pm\sqrt{2}.$$

(a) Use (5.38) to show that

$$(h_0 + h_1 + h_2 + h_3 + h_4 + h_5)^2 = 1 + 2b$$

where

$$b = h_0h_1 + h_0h_3 + h_0h_5 + h_1h_2 + h_1h_4 + h_2h_3 + h_2h_5 + h_3h_4 + h_4h_5.$$

(b) Square both sides of (5.41) and then use (5.38) to show that

$$0 = (h_0 - h_1 + h_2 - h_3 + h_4 - h_5)^2 = 1 - 2b.$$

(c) Add the equations in (a) and (b) to obtain the desired result.

5.17 Suppose $h_0, \ldots, h_5$ satisfy

$$\begin{aligned} h_0 + h_1 + h_2 + h_3 + h_4 + h_5 &= \sqrt{2} \\ h_0 - h_1 + h_2 - h_3 + h_4 - h_5 &= 0 \\ h_0h_2 + h_1h_3 + h_2h_4 + h_3h_5 &= 0 \\ h_0h_4 + h_1h_5 &= 0. \end{aligned} \tag{5.48}$$

Show that

$$h_0^2 + h_1^2 + h_2^2 + h_3^2 + h_4^2 + h_5^2 = 1.$$

To simplify computations, let

$$a = h_0^2 + h_1^2 + h_2^2 + h_3^2 + h_4^2 + h_5^2.$$

(a) Square both sides of the first two equations of (5.48) and simplify to show that

$$a + 2\sum_{j \neq k} h_jh_k = 2$$

$$a + 2\sum_{j \neq k} (-1)^{j+k} h_jh_k = 0$$

(b) Use the orthogonality conditions given by the last two equations of (5.48) to reduce the equations in (a) to

$$\begin{aligned} a + 2b &= 2 \\ a - 2b &= 0 \end{aligned}$$

where

$$b = h_0h_1 + h_0h_3 + h_0h_5 + h_1h_2 + h_1h_4 + h_2h_3 + h_2h_5 + h_3h_4 + h_4h_5$$

(c) Combine the equations in (b) to show that $a = 1$.

(d) Verify that the result still holds if we replace $\sqrt{2}$ by $-\sqrt{2}$ in (5.48).

5.18 Suppose we insist **g** annihilate cubic data in addition to polynomial data of degree $0, 1, 2$. Find the resulting equation. Add this equation to the system (5.44) and use a **CAS** to solve the system. What happens when you do this?

5.19 In this problem you will find two real solutions to the system given by (5.44). The process outlined in this problem is algebra intensive. A **CAS** may come in handy with some of the computations. We use Problem 5.17 to replace the first equation of (5.38) with the condition

$$h_0 + h_1 + h_2 + h_3 + h_4 + h_5 = \sqrt{2}.$$

in (5.44).

(a) Show that the system of linear equations in (5.44) is equivalent to

$$\begin{aligned} h_0 &= \frac{\sqrt{2}}{8} - h_4 + 2h_5 \\ h_1 &= \frac{3\sqrt{2}}{8} - 2h_4 + 3h_5 \\ h_2 &= \frac{3\sqrt{2}}{8} - 2h_5 \\ h_3 &= \frac{\sqrt{2}}{8} + 2h_4 - 4h_5. \end{aligned}$$

(b) Use Problem 5.4 and the last equation of (5.38) to write

$$h_0 = -ch_5 \qquad h_1 = ch_4 \tag{5.49}$$

where $c \neq 0$.

(c) Plug the results from (b) in for h_0 and h_1 in (a) and show that

$$h_4 = \frac{3\sqrt{2}(1+c)}{8(1+4c+c^2)} \qquad h_5 = \frac{\sqrt{2}(1-c)}{8(1+4c+c^2)}. \tag{5.50}$$

(d) Plug the results from (c) into the last two equations from (a) to show that

$$h_2 = \frac{\sqrt{2}}{8} \cdot \frac{1+14c+3c^2}{1+4c+c^2} \qquad h_3 = \frac{\sqrt{2}}{8} \cdot \frac{3+14c+c^2}{1+4c+c^2}. \tag{5.51}$$

Note that $c \neq -2 \pm \sqrt{3}$ since then the denominators in (5.51) would be zero. Also, $c = \pm 1$ leads to filters consisting of only four nonzero terms. These filters have been considered in Section 5.1. Going forward, assume that $c \neq 0, \pm 1, -2 \pm \sqrt{3}$.

(e) We have now written all five variables (5.49) – (5.51) in terms of a parameter c. From (b) we know that $c \neq 0$. What other nonzero condition must c satisfy? Insert the parametrized versions of $h_0, \ldots, h_5$ from (5.49) – (5.51) into the second equation of (5.38) to show that

$$3 + 32c + 38c^2 + 32c^3 + 3c^4 = 0.$$

(f) Verify that the quartic polynomial

$$p(c) = 3 + 32c + 38c^2 + 32c^3 + 3c^4$$

can be factored as

$$p(c) = 3\left(c^2 + \frac{4}{3}(4 + \sqrt{10})c + 1\right)\left(c^2 + \frac{4}{3}(4 - \sqrt{10})c + 1\right).$$

(g) Use (f) to find the two real roots of $p(c)$.

(h) For each real value of c from (g), use (5.49) – (5.51) to compute the values $h_0, \ldots, h_5$.

(i) For each real value of c from (g), verify that the values for $h_0, \ldots, h_5$ from (h) satisfy the first equation of (5.38).

5.20 Suppose that $\mathbf{h} = (a, b, c, d, e, f)$ satisfies (5.44). Use the ideas from Problem 5.9 to show that $-\mathbf{h} = (-a, -b, -c, -d, -e, -f)$ and $\mathbf{h}^r = (f, e, d, c, b, a)$ both satisfy (5.44).

5.21 What is the effect of adding the redundant equation

$$h_0 + h_1 + h_2 + h_3 + h_4 + h_5 = \sqrt{2}$$

to (5.44)? See Problem 5.20.

5.22 Suppose $\mathbf{v} \in \mathbb{R}^N$ with N even. We wish to compute $W_N\mathbf{v} = \left[\frac{\mathbf{a}}{\mathbf{d}}\right]$ where W_N is given by (5.47). Computing the product $W_N\mathbf{v}$ involves a large number of multiplications by zero due to the sparse nature of W_N. We seek a more expedient way to compute $W_N\mathbf{v}$. Can you identify a $N/2 \times 6$ matrix X such that

$$X\begin{bmatrix} h_5 \\ h_4 \\ h_3 \\ h_2 \\ h_1 \\ h_0 \end{bmatrix} = \mathbf{a} \quad \text{and} \quad X\begin{bmatrix} g_5 \\ g_4 \\ g_3 \\ g_2 \\ g_1 \\ g_0 \end{bmatrix} = \mathbf{d}?$$

Hint: Problem 5.12 will be helpful.

5.23 Suppose we have computed

$$W_N \mathbf{v} = \left[\frac{\mathbf{a}}{\mathbf{d}}\right]$$

where W_N is given by (5.47). Here N is even and we let $L = N/2$. To recover $\mathbf{v}$, we write W_N in block form and compute

$$\mathbf{v} = W_N^T \left[\frac{\mathbf{a}}{\mathbf{d}}\right] = \left[H_N^T \middle| G_N^T \right] \left[\frac{\mathbf{a}}{\mathbf{d}}\right] = H_N^T \mathbf{a} + G_N^T \mathbf{d}.$$

We can find efficient representations of $H_N^T \mathbf{a}$ and $G_N^T \mathbf{d}$ and sum up the results in order to compute the inverse transformation. We consider $H_N^T \mathbf{a}$ and note that the process for computing $G_N^T \mathbf{d}$ is similar.

(a) Let $\mathbf{a} = \begin{bmatrix} a_1 \\ a_2 \\ \vdots \\ a_L \end{bmatrix}$. Find an $L \times 3$ matrix A, constructed from the elements of $\mathbf{a}$, such that

$$A \begin{bmatrix} h_1 & h_0 \\ h_3 & h_2 \\ h_5 & h_4 \end{bmatrix} = \begin{bmatrix} \mathbf{o} & \mathbf{e} \end{bmatrix}$$

where $\mathbf{o}$, $\mathbf{e}$, hold the odd and even components, respectively, of $H_N^T \mathbf{a}$.

(b) Find permutation matrix P such that $P \left[\frac{\mathbf{o}}{\mathbf{e}}\right] = H_N^T \mathbf{a}$.

Computer Lab

5.2 Constructing the D6 Filter. From `stthomas.edu/wavelets`, access the file `D6Filter`. In this lab, you will write code that allows you to solve for the D6 filter coefficients. In addition, you will write code that uses the D6 filter to compute the (inverse) wavelet transformations of a vector or a matrix.

5.3 Daubechies Filters of Even Length

We now derive and solve the system of equations needed to find the Daubechies filter $\mathbf{h} = (h_0, h_1, \ldots, h_L)$, where L is an odd positive integer. We denote the filter length by

$$\boxed{M = L + 1.}$$

As we have seen with both the D4 and D6 filters, the equations should come from the orthogonality of the transformation matrix and conditions on the highpass filter $\mathbf{g}$.

Orthogonality Conditions

We found the orthogonality conditions for the Haar, D4, and D6 filters by writing W_N for sufficiently large N and then analyzing $W_N W_N^T = I_N$. We could just as easily look only at the matrix W_N itself. As you can see, we have several wrapped rows in the general case.

$$W_N = \left[\begin{array}{c} H_{N/2} \\ \hline G_{N/2} \end{array}\right] = \left[\begin{array}{cccccccccccc} h_L & h_{L-1} & h_{L-2} & h_{L-3} & \cdots & h_1 & h_0 & 0 & 0 & \cdots & 0 & 0 \\ 0 & 0 & h_L & h_{L-1} & \cdots & h_3 & h_2 & h_1 & h_0 & \cdots & 0 & 0 \\ 0 & 0 & 0 & 0 & \cdots & h_5 & h_4 & h_3 & h_2 & \cdots & 0 & 0 \\ & & & & \ddots & & & & & & & \\ 0 & 0 & 0 & 0 & \cdots & h_L & h_{L-1} & h_{L-2} & h_{L-3} & \cdots & 0 & 0 \\ & & & & \ddots & & & & & & & \\ h_{L-2} & h_{L-3} & h_{L-4} & h_{L-5} & \cdots & & & & & \cdots & h_L & h_{L-1} \\ \hline g_L & g_{L-1} & g_{L-2} & g_{L-3} & \cdots & g_1 & g_0 & 0 & 0 & \cdots & 0 & 0 \\ 0 & 0 & g_L & g_{L-1} & \cdots & g_3 & g_2 & g_1 & g_0 & \cdots & 0 & 0 \\ 0 & 0 & 0 & 0 & \cdots & g_5 & g_4 & g_3 & g_2 & \cdots & 0 & 0 \\ & & & & \ddots & & & & & & & \\ 0 & 0 & 0 & 0 & \cdots & g_L & g_{L-1} & g_{L-2} & g_{L-3} & \cdots & 0 & 0 \\ & & & & \ddots & & & & & & & \\ g_{L-2} & g_{L-3} & g_{L-4} & g_{L-5} & \cdots & & & & & \cdots & g_L & g_{L-1} \end{array}\right]. \tag{5.52}$$

To meet some of the orthogonality conditions, the inner product of each row of W_N with itself must be one and with any other row must be zero.

We can further simplify our analysis by recalling that we need only find the filter $\mathbf{h} = (h_0, h_1, \ldots, h_L)$ since the highpass vector $\mathbf{g}$ needed to complete W_N can be found by the rule $g_k = (-1)^k h_{L-k}$, $k = 0, \ldots, L$ (see Problem 2.11).

Consider row $\mathbf{w}_i$ from $H_{N/2}$. The only nonzero elements in $\mathbf{w}_i$ are $h_0, h_1, \ldots, h_L$, so that when we compute $\mathbf{w}_i \cdot \mathbf{w}_i$, we obtain our first orthogonality condition

$$h_0^2 + h_1^2 + \cdots + h_L^2 = \sum_{k=0}^{L} h_k^2 = 1. \tag{5.53}$$

We now need to establish the *zero orthogonality conditions* that result when we dot row $\mathbf{w}_i$ and $\mathbf{w}_j$, $i \neq j$ from $H_{N/2}$.

If you recall, there were no zero orthogonality conditions when deriving the orthogonal Haar filter (the $L = 1$ case). The inner product of the distinct rows of $H_{N/2}$ (4.7) was zero since no filter coefficients overlapped.

Now think about the zero orthogonality condition for D4. There was only one zero orthogonality condition (see (5.6)) that resulted from one row that overlapped with the first row.

Finally, for D6, there were two zero orthogonality conditions in the system (5.38), resulting from two rows that overlapped with the first row.

There seems to be a pattern linking the numbers of zero orthogonality conditions to the filter length. We summarize these observations in Table 5.3. Certainly, this pattern is linear and it is easy to guess that if the filter length is M, then the number

Table 5.3 The number of zero orthogonality conditions for the orthogonal Haar, D4, and D6 wavelet transformations.

Filter	Length	Zero Orthogonality Conditions
Haar	2	0
D4	4	1
D6	6	2

of zero orthogonality conditions is

$$\frac{M}{2} - 1 = \frac{L+1}{2} - 1 = \frac{L-1}{2}.$$

Now we know that we need $\frac{L-1}{2}$ zero orthogonality conditions. We can use the first $\frac{M}{2}$ rows of the top half of W_N

$$\begin{array}{lcccccccccc}
\text{Row 1:} & h_L & h_{L-1} & h_{L-2} & h_{L-3} & h_{L-4} & \cdots & h_1 & h_0 & 0 & \cdots \\
\text{Row 2:} & & & h_L & h_{L-1} & h_{L-2} & \cdots & h_3 & h_2 & h_1 & \cdots \\
\text{Row 3:} & & & & & h_L & \cdots & h_5 & h_4 & h_3 & \cdots \\
\vdots & & & & & & & & & & \vdots \\
\text{Row } M/2\text{:} & & & & & & \cdots & h_L & h_{L-1} & h_{L-2} & \cdots
\end{array} \tag{5.54}$$

to describe them.

We need to write down the dot products of row one with rows $2, 3, \ldots, \frac{M}{2}$ from (5.54). As in the cases of the derivation of the D4 and D6 filters, we will obtain all orthogonality conditions for $h_0, \ldots, h_L$ by considering only inner products involving the first row. We begin by dotting row one with row two. This inner product

$$h_0 h_2 + h_1 h_3 + h_2 h_4 + \cdots + h_{L-3} h_{L-1} + h_{L-2} h_L \tag{5.55}$$

has $M - 2 = L - 1$ terms, while the inner product of row three and row one

$$h_0 h_4 + h_1 h_5 + h_2 h_6 + \cdots + h_{L-5} h_{L-1} + h_{L-4} h_L \tag{5.56}$$

has $M - 4 = L - 3$ terms. Each subsequent inner product requires two fewer terms than the one before it. The final inner product is row one with row $\frac{M}{2}$ and this inner product is

$$h_0 h_{L-1} + h_1 h_L. \tag{5.57}$$

and has only two terms. Any remaining rows in the top half of W_N either do not have filter terms that overlap with the first row or the overlaps are the same as the ones we have previously recorded.

We have written each inner product (5.55) – (5.57) in "reverse order" such that the first factor in each term is from row one. Writing the inner products in this way makes it easier to derive a general formula for the $\frac{M}{2} - 1$ zero orthogonality conditions.

For each row $m = 2, \ldots, \frac{M}{2}$, the inner product formula can be written in summation notation. Let's use k as the summing index variable. We need to determine the starting and stopping points for k as well as the subscripts for the two factors involving elements of $\mathbf{h}$ that appear in each term of the summation. Symbolically, we must find values for each of the boxes in

$$\sum_{k=\square}^{\square} h_{\square} h_{\square}.$$

The number of terms in each inner product decreases by two for each subsequent row. We need to incorporate this relationship into our summation formula. There are a couple of ways to proceed. We can either build our summing variable k to reflect the fact that the first factor in the first term of the inner product is always h_0 or we can observe that the second factor in the last term of each inner product is h_L. In the former case we start our sum at $k = 0$ and vary its stopping point, and in the latter, we can vary the starting point and always stop our sum at $k = L$. We adopt the latter scheme and leave the former for Problem 5.24.

Our stopping point will be at $k = L$ and the starting point of the sum depends on the index of the second factor of the first term for each row. Row two starts with h_2, row three starts with h_4, row four starts with h_6 and so on. Clearly, there is a pattern relating the row number to this index. You should verify that the second factor of the first term of row m is $h_{2(m-1)}$ for $m = 2, 3, \ldots, \frac{M}{2}$.

We have determined the sum limits of each inner product as well as the second factor of the first term of each inner product. Now we need to determine how to write the first factor of each term. This is easy since we are translating the filter $\mathbf{h}$ in row two by two units, row three by four units, and so on. This is exactly the same relation that we had for the index of the second factor. In general, for rows $m = 2, 3, \ldots, \frac{M}{2}$, we are translating $\mathbf{h}$ by $2(m-1)$ units. Since we always want our first factor to start with h_0 and we know that k starts at $2(m-1)$, we see that the general form of our first factor is $h_{k-2(m-1)}$.

We thus have the following $\frac{M}{2} - 1$ orthogonality conditions.

$$\sum_{k=2(m-1)}^{L} h_{k-2(m-1)} h_k = 0, \qquad m = 2, 3, \ldots, \frac{M}{2}.$$

It is a bit less cumbersome if we replace $m-1$ by m above. In this case we note that $\frac{M}{2} - 1 = \frac{L-1}{2}$ and write

$$\sum_{k=2m}^{L} h_k h_{k-2m} = 0, \qquad m = 1, 2, \ldots, \frac{L-1}{2}. \tag{5.58}$$

Constructing the Filter $\mathbf{g}$

The filter $\mathbf{g}$ is constructed so that W_N in (5.52) is orthogonal. We mimic the approach in the previous sections and take

$$g_k = (-1)^k h_{L-k}, \quad k = 0, \ldots, L \tag{5.59}$$

so that

$$\mathbf{g} = (g_0, g_1, \ldots, g_{L-1}, g_L) = (h_L, -h_{L-1}, \ldots, h_1, -h_0).$$

We first consider orthogonality conditions on $\mathbf{g}$. In order for W_N to be orthogonal, it is enough to show that

$$\sum_k g_k g_{k-2m} = \begin{cases} 1, & m = 0 \\ 0, & m \neq 0 \end{cases} \tag{5.60}$$

and

$$\sum_k g_k h_{k-2m} = 0. \tag{5.61}$$

The inner products in (5.60) are necessary for $G_{N/2}G_{N/2}^T = I_{N/2}$ and the inner products in (5.61) ensure that $G_{N/2}H_{N/2}^T = 0_{N/2}$. It follows then that

$$H_{N/2}G_{N/2}^T = \left(G_{N/2}H_{N/2}^T\right)^T = 0_{N/2}.$$

We show (5.60) holds and leave (5.61) as Problem 5.25. Since $\mathbf{g}$ is an FIR filter, we can establish limits on the sum in (5.60). We have using (5.59)

$$\begin{aligned}
\sum_k g_k g_{k-2m} &= \sum_{k=0}^{L} g_k g_{k-2m} = \sum_{k=2m}^{L} g_k g_{k-2m} \\
&= \sum_{k=2m}^{L} (-1)^k h_{L-k} (-1)^{k-2m} h_{L-(k-2m)} = \sum_{k=2m}^{L} h_{L-k} h_{L-k+2m}.
\end{aligned}$$

Next make the index substitution $j = L - k + 2m$ so that $k = L - j + 2m$. When $k = 2m$, $j = L$ and when $k = L$, $j = 2m$. We have

$$\begin{aligned}
\sum_k g_k g_{k-2m} &= \sum_{k=2m}^{L} h_{L-k} h_{L-k+2m} \\
&= \sum_{j=L}^{2m} h_{L-(L-j+2m)} h_{L-(L-j+2m)+2m} = \sum_{j=L}^{2m} h_{j-2m} h_j.
\end{aligned}$$

The last term is 1 if $m = 0$ and zero otherwise by (5.53) and (5.58) so that (5.60) holds.

Conditions on the Highpass Filter

When constructing the orthogonal Haar, D4, and D6 filters, we required the companion highpass filter to annihilate polynomial data up to a certain degree. Table 5.4 summarizes these requirements.

Table 5.4 The number of zero orthogonality conditions for the orthogonal Haar, D4, and D6 wavelet transformations.

Filter	Length	Degree of Polynomial Data Annihilated
Haar	2	0
D4	4	1
D6	6	2

Like the zero orthogonality conditions in Table 5.3, we surmise that the highpass filter $\mathbf{g} = (g_0, \ldots, g_L)$ should annihilate polynomial data up to and including degree $\frac{L-1}{2}$. For $m = 0$, Proposition 2.3 and (5.59) imply

$$0 = \sum_{k=0}^{L} g_k = \sum_{k=0}^{L} (-1)^k h_{L-k} = h_L - h_{L-1} + \cdots + h_1 - h_0. \tag{5.62}$$

For $m = 1, \ldots, \frac{L-1}{2}$, we use Proposition 2.3 with $n = L$ and (5.59) to write

$$0 = \sum_{k=0}^{L} g_k (L-k)^m = \sum_{k=0}^{L} (-1)^k h_{L-k} (L-k)^m.$$

Replace $L - k$ with k in the above sum and use the fact that L is odd to write

$$0 = \sum_{k=0}^{L} (-1)^k h_{L-k} (L-k)^m = -\sum_{k=L}^{0} (-1)^k h_k k^m.$$

Thus for $m = 1, \ldots, \dfrac{L-1}{2}$, since $k^m = 0$ when $k = 0$, we have the additional conditions on $\mathbf{g}$:

$$\sum_{k=1}^{L} (-1)^k h_k k^m = 0.$$

In Problem 5.26 you will show that if $\mathbf{h}$ satisfies the orthogonality conditions (5.53) in addition to (5.58), then

$$h_0 + \cdots + h_L = \pm\sqrt{2}. \tag{5.63}$$

This result along with (5.62) guarantee that $\mathbf{h}$ is a lowpass filter. Since $\mathbf{h}$ is lowpass and $\mathbf{g}$ is constructed using (5.59), Proposition 2.4 ensures that $\mathbf{g}$ is a highpass filter.

System for Finding Daubechies Filters

Our system thus far consists of $\frac{M}{2}$ orthogonality conditions and $\frac{M}{2}$ lowpass conditions:

$$
\begin{aligned}
\sum_{k=0}^{L} h_k^2 &= 1 && \text{(orthogonality)} \\
\sum_{k=2m}^{L} h_k h_{k-2m} &= 0, \quad m = 1, 2, \ldots, \frac{L-1}{2} && \text{(orthogonality)} \\
\sum_{k=0}^{L} (-1)^k h_k &= 0 && \text{(constant data)} \\
\sum_{k=1}^{L} (-1)^k k^m h_k &= 0, \quad m = 1, 2, \ldots, \frac{L-1}{2} && \text{(polynomial data)}
\end{aligned}
\tag{5.64}
$$

It should be clear that if $\mathbf{h} = (h_0, \ldots, h_L)$ is a solution to (5.64), then $-\mathbf{h}$ is a solution as well. Now only one of $\mathbf{h}$ or $-\mathbf{h}$ can satisfy (5.63) when we choose $\sqrt{2}$ for the right side. If we add the extra condition (5.63) to our system, then we can reduce the number of solutions to our system by a factor of 2.

Thus the $M + 1$ equations we must solve to obtain Daubechies lowpass filter $\mathbf{h} = (h_0, h_1, \ldots, h_L)$, $M = L + 1$, are given by (5.64) and (5.63). We state the complete system below.

$$
\boxed{
\begin{aligned}
\sum_{k=0}^{L} h_k^2 &= 1 && \text{(orthogonality)} \\
\sum_{k=2m}^{L} h_k h_{k-2m} &= 0, \quad m = 1, 2, \ldots, \tfrac{L-1}{2} && \text{(orthogonality)} \\
\sum_{k=0}^{L} h_k &= \sqrt{2} && \text{(redundancy condition)} \\
\sum_{k=0}^{L} (-1)^k h_k &= 0 && \text{(constant data)} \\
\sum_{k=1}^{L} (-1)^k k^m h_k &= 0, \quad m = 1, 2, \ldots, \tfrac{L-1}{2} && \text{(polynomial data)}
\end{aligned}
}
\tag{5.65}
$$

Characterizing Solutions to the Daubechies System

We conclude this section with a discussion about solving the system (5.65). We have already solved the system for $L = 1, 3, 5$. Can we solve it for larger L? If so, how many solutions exist for a given L?

Daubechies [29] gives a complete characterization of the solutions to the system (5.65). Her derivation is different from the one we present here. We state a couple of results and use them to discuss a method that Daubechies developed to choose a particular solution to the system (5.65).

Proofs of Theorems 5.1 and 5.2 are beyond the scope of this book. The reader with some background in advanced analysis is referred to the proofs given by Daubechies in [29, 30].

It turns out that the system (5.65) always has real solutions. Daubechies [30] proved the following result.

Theorem 5.1 (Number of Solutions: Daubechies [30]). *The system* (5.65) *has* $2^{\lfloor \frac{L+2}{4} \rfloor}$ *real solutions. Here,* $\lfloor \cdot \rfloor$ *is the floor function defined by* (3.2) *in Section 3.1.* □

Table 5.5 shows how the number of real solutions $2^{\lfloor \frac{L+2}{4} \rfloor}$ for the system (5.65) grows as L gets larger.

Table 5.5 The number of real solutions to the system (5.65) for a given L.

L	1	3	5	7	9	11	13
No. of Solutions	1	2	2	4	4	8	8

The solutions to the system (5.65) also obey a basic symmetry property. We have the following result.

Proposition 5.1 (Reflections are Solutions). *Suppose that L is an odd positive integer and assume that* $\mathbf{h} = (h_0, h_1, \ldots, h_L)$ *solves the system* (5.65). *Then the reflection* $\mathbf{h}^r$, *whose components are given by* $h_k^r = h_{L-k}$, $k = 0, \ldots, L$, *solves* (5.65) *as well.* □

Proof. The proof is left as Problem 5.28. □

Proposition 5.1 tells us that we effectively have half as many, or $2^{\lfloor \frac{L+2}{4} \rfloor - 1}$ solutions to find. The remaining solutions are simply reflections. Daubechies was interested in choosing a particular solution for the system (5.65). We give an outline of this method now. Before we begin, we need a definition.

Definition 5.3 (Multiplicity of Roots). *Suppose that* $p(t)$ *is a polynomial of degree* n, *and let* ℓ *be a nonnegative integer with* $\ell \leq n$. *We say that* $t = a$ *is a root of multiplicity* ℓ *if* $p(t) = (t - a)^\ell q(t)$, *where* $q(t)$ *is a polynomial of degree* $n - \ell$ *and* $q(a) \neq 0$. □

Example 5.4 (Finding Multiplicity of Roots). *Consider the degree five polynomial*

$$p(t) = t^5 - t^4 - 24t^3 + 88t^2 - 112t + 48.$$

Since $p(t) = (t-2)^3(t^2+5t-6)$ *and* $q(t) = t^2+5t-6 = (t-1)(t+6)$ *satisfies* $q(2) \neq 0$, *we see that* $a = 2$ *is a root of multiplicity* 3 *of* $p(t)$. *We can also note that* $a = 1$, $a = -6$ *are roots of multiplicity* 1 *for* $p(t)$. □

In Problem 5.31 you will show that $t = a$ is a root of multiplicity ℓ if and only if $p^{(m)}(a) = 0$ for $m = 0, 1, \ldots, \ell - 1$.

Returning to Daubechies' selection of a particular solution to (5.65), suppose that $\mathbf{h} = (h_0, h_1, \ldots, h_L)$, $L \geq 3$, is a solution to (5.65). We consider the polynomial

$$P(z) = \sum_{k=0}^{L} h_k z^k. \tag{5.66}$$

The third equation of (5.64) implies $P(-1) = 0$ and it is straightforward to check, using the last equation of (5.64) with $m = 1$, that $P'(-1) = 0$. As a matter of fact, using Problem 5.29 we see that $P^{(m)}(-1) = 0$ for $m = 0, 1, \ldots, \frac{L-1}{2}$. Thus, $z = -1$ is a root of *at least* multiplicity $\frac{L+1}{2}$ for $P(z)$. This means that

$$P(z) = (z+1)^{\frac{L+1}{2}} Q(z) \tag{5.67}$$

where $Q(z)$ is a polynomial of degree $\frac{L-1}{2}$. Daubechies [29] showed $Q(-1) \neq 0$ so $z = -1$ is indeed a root of multiplicity $\frac{L+1}{2}$ for $P(z)$. In fact, when she derived the filters, she showed the following result.

Theorem 5.2 (Characterizing Solutions: Daubechies [29]). *For each* $L = 1, 3, 5, \ldots$ *there is exactly one solution for* (5.65) *where the roots* z_k, $k = 1, 2, \ldots, \frac{L-1}{2}$, *of* $Q(z)$ *given in* (5.67) *satisfy* $|z_k| \geq 1$. □

You have previously encountered polynomials that have complex roots $r = a + bi$. We will review the complex plane in Chapter 8, but for now, plotting $r = a + bi$ in the complex plane can be viewed as plotting the ordered pair (a, b) in the xy-plane. Then the distance formula $|r| = \sqrt{a^2 + b^2}$ is a natural extension of the absolute value of a complex number.

Theorem 5.2 gives us a method for choosing a particular solution of (5.65). We know that $P(z)$ is a polynomial of degree L, and we also know that $z = -1$ is a root of multiplicity $\frac{L+1}{2}$, so we find the roots z_k, $k = 1, \ldots, L$ of $P(z)$ and compute $|z_k|$. If $|z_k| \geq 1$ for $k = 1, \ldots, L$, then the solution we used to form $P(z)$ is the Daubechies filter. We demonstrate this result for the D4 filter.

Example 5.5 ($\mathbf{P(z)}$ for the D4 Filter). *We first construct* $P(z)$ *using the coefficients given in* (5.27)*:*

$$P(z) = \frac{1}{4\sqrt{2}} \left(1 + \sqrt{3} + (3+\sqrt{3})z + (3-\sqrt{3})z^2 + (1-\sqrt{3})z^3\right).$$

Since $P(-1) = 0$ and $P'(-1) = 0$, we know that -1 is a root of multiplicity two of $P(z)$, so that

$$P(z) = (z+1)^2 Q(z).$$

Thus $Q(z)$ must be a linear polynomial. Long division gives

$$Q(z) = \frac{1-\sqrt{3}}{4\sqrt{2}}\left(z - (2+\sqrt{3})\right).$$

The root of $Q(z)$ is $2+\sqrt{3}$ and the absolute value of this root is larger than one.

Using Theorem 5.1, we know that there are two real solutions to the system (5.65) *for $L = 3$, and by Proposition 5.1 we know that the other solution is merely a reflection of the D4 filter. If we form $P(z)$ for this solution we have*

$$P(z) = \frac{1}{4\sqrt{2}}\left(1-\sqrt{3}+(3-\sqrt{3})z+(3+\sqrt{3})z^2+(1+\sqrt{3})z^3\right).$$

We know that $Q(z)$ is a linear polynomial and if we divide $P(z)$ by $(z+1)^2$, we find that

$$Q(z) = \frac{1+\sqrt{3}}{4\sqrt{2}}\left(z - (2-\sqrt{3})\right).$$

The root in this case is $2-\sqrt{3}$, and the absolute value of this number is certainly smaller than one. □

Let's look at an example using a longer filter.

Example 5.6 (Finding the D10 Filter). *Let's find the Daubechies filter of length* 10. *By Theorem 5.1 we know that there are four real solutions. We set up the system* (5.65) *for $L = 9$ and solve it using a* ***CAS***. *The solutions (rounded to eight decimal places) are given in Table 5.6.*

Table 5.6 Solutions to (5.65) for $L = 9$.

	Solution 1	Solution 2	Solution 3	Solution 4
h_0	0.00333573	0.02733307	0.01953888	0.16010240
h_1	–0.01258075	0.02951949	–0.02110183	0.60382927
h_2	–0.00624149	–0.03913425	–0.17532809	0.72430853
h_3	0.07757149	0.19939753	0.01660211	0.13842815
h_4	–0.03224487	0.72340769	0.63397896	–0.24229489
h_5	–0.24229489	0.63397896	0.72340769	–0.03224487
h_6	0.13842815	0.01660211	0.19939753	0.07757149
h_7	0.72430853	–0.17532809	–0.03913425	–0.00624149
h_8	0.60382927	–0.02110183	0.02951949	–0.01258075
h_9	0.16010240	0.01953888	0.02733307	0.00333573

Now for solution j, $j = 1, \ldots, 4$, we create the degree nine polynomial $P_j(z) = \sum_{k=0}^{9} h_k z^k$ and then find its roots. We know that -1 is a root of multiplicity $\frac{9+1}{2} = 5$

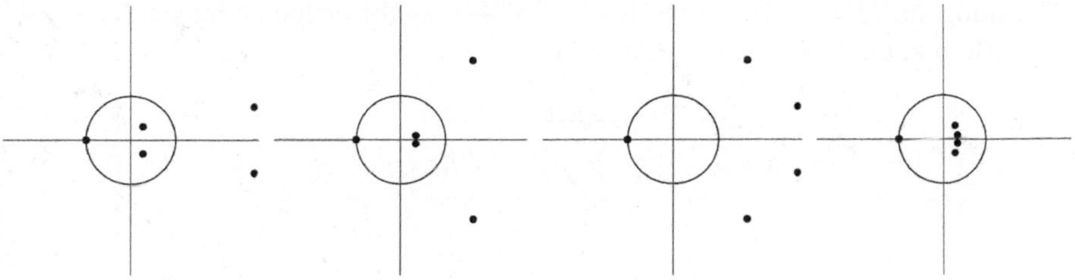

(a) Solution $j = 1$ (b) Solution $j = 2$ (c) Solution $j = 3$ (d) Solution $j = 4$

Figure 5.13 The roots of the degree nine polynomials. Note that -1 is a root of multiplicity five for each polynomial.

for each of these polynomials. The roots for each polynomial are plotted in Figure 5.13.

Note that the roots z_k for solution 4 satisfy $|z_k| \geq 1$, so this is the solution we use. Note that solution 1 is a reflection of solution 4, and solutions 2 and 3 are reflections of each other as well.

(Live example: Visit `stthomas.edu/wavelets` *and click on Live Examples.)*

□

PROBLEMS

5.24 Show that the orthogonality conditions given by (5.58) can be written as

$$\sum_{k=0}^{L-2m} h_k h_{k+2m} = 0$$

for $m = 1, \ldots, \frac{L-1}{2}$.

5.25 In this problem, you establish the inner product given in (5.61). The following steps will help you organize your work.

(a) Use the fact that $\mathbf{h}$, $\mathbf{g}$ are FIR filters and (5.59) to show that (5.61) can be written as

$$\sum_k g_k h_{k-2m} = \sum_{k=2m}^{L} (-1)^k h_{L-k} h_{k-2m}.$$

(b) The midpoint of the summation limits is $\frac{2m+L}{2}$. Is this value an integer? Use this information to rewrite the sum in (a) as

$$\sum_{k=2m}^{L} (-1)^k h_{L-k} h_{k-2m} = \sum_{k=2m}^{\frac{L+2m-1}{2}} (-1)^k h_{L-k} h_{k-2m} + \sum_{k=\frac{L+2m+1}{2}}^{L} (-1)^k h_{L-k} h_{k-2m}.$$

(c) Write down the first (last) two terms of the first (last) sums above. What did you observe?

(d) The argument concludes by showing the second sum in (b) is the opposite of the first sum. Establish this result by making the index substitution $j = L - k + 2m$ in the second sum in (b).

5.26 In this problem you will show that if $\mathbf{h} = (h_0, h_1, \ldots, h_L)$ satisfies (5.53), (5.58), and the third equation of (5.64), then

$$h_0 + h_1 + h_2 + \cdots + h_L = \pm\sqrt{2}.$$

(a) Use (5.53) to show that

$$(h_0 + \cdots + h_L)^2 = 1 + 2\sum_{j \neq k} h_j h_k.$$

(b) Use (5.53) and the third equation of (5.64) to show that

$$0 = (h_0 - h_1 + h_2 - h_3 + \cdots - h_L)^2 = 1 + 2\sum_{j \neq k} (-1)^{j+k} h_j h_k.$$

(c) Split the summations in (a) and (b) into two summations. One summation goes over the $j + k$ odd terms and one over the $j + k$ even terms. What do the orthogonality conditions (5.58) say about one of these two sums?

(d) Use (c) to simplify the expressions in (a) and (b).

(e) Finally, combine the simplified expressions that you obtained in (d) to obtain the desired result.

5.27 Suppose that $\mathbf{h} = (h_0, \ldots, h_L)$ satisfies the orthogonality conditions (5.58), the third equation of (5.64) and

$$h_0 + h_1 + \cdots + h_L = \sqrt{2} \tag{5.68}$$

Show that $\mathbf{h}$ satisfies (5.53). The following steps will help you organize your work. To simplify notation, let

$$a = h_0^2 + h_1^2 + \cdots + h_L^2.$$

(a) Square both sides of (5.68) and show that

$$2 = a + 2\sum_{j \neq k} h_j h_k.$$

(b) Square both sides of the third equation of (5.64) and show that

$$0 = a + 2\sum_{j \neq k} (-1)^{j+k} h_j h_k.$$

(c) Now use the ideas from (c) of Problem 5.26 to simplify the expressions you verified in (a) and (b).

(d) Combine the simplified expressions from (c) to obtain the desired result.

(e) Replace the $\sqrt{2}$ by $-\sqrt{2}$ on the right side of (5.68) and repeat the problem.

5.28 Prove Proposition 5.1. (Problem 5.9 will be helpful.)

5.29 Let L be an odd integer with $L \geq 3$ and suppose $\mathbf{h} = (h_0, \ldots, h_L)$ satisfies the last two equations of (5.64). Show that the polynomial $P(z)$ given in (5.66) satisfies $P^{(r)}(-1) = 0$ where $0 \leq r \leq \frac{L-1}{2}$. The following steps will help you organize your work.

(a) Use the third equation of (5.64) to observe that $P(-1) = 0$ and the last equation of (5.64) with $m = 1$ to show that $P'(-1) = 0$.

(b) Suppose r is an integer with $1 \leq r \leq \frac{L-1}{2}$. Show that

$$P^{(r)}(z) = \sum_{k=r}^{L} h_k k(k-1)\cdots(k-r+1)z^{k-r}$$

so that

$$P^{(r)}(-1) = (-1)^r \sum_{k=r}^{L} h_k (-1)^k k(k-1)\cdots(k-r+1).$$

(c) Rewrite $P^{(r)}(-1)$ as

$$P^{(r)}(-1) = (-1)^r \sum_{k=1}^{L} h_k (-1)^k k(k-1)\cdots(k-r+1)$$
$$- (-1)^r \sum_{k=1}^{r-1} h_k (-1)^k k(k-1)\cdots(k-r+1)$$

and explain why the second sum on the right side above is zero. Thus we have

$$P^{(r)}(-1) = (-1)^r \sum_{k=1}^{L} h_k (-1)^k k(k-1)\cdots(k-r+1). \tag{5.69}$$

(d) Verify that if we expand $k(k-1)\cdots(k-r+1)$ we obtain a polynomial of degree r in k:

$$k(k-1)\cdots(k-r+1) = \sum_{j=1}^{r} c_j k^j$$

for real numbers $c_1, \ldots, c_r$. Insert this polynomial into (5.69), interchange sums and use the last equation of (5.64) to explain why $P^{(r)}(-1)$ given in (5.69) is zero.

★5.30 In this problem you will find a general formula[2] for $(fg)^{(k)}(t)$, where $k = 0, 1, 2, \ldots$. Use mathematical induction to show that

$$(fg)^{(k)}(t) = \sum_{m=0}^{k} \binom{k}{m} f^{(m)}(t) g^{(k-m)}(t).$$

where the *binomial coefficient* is defined as

$$\binom{n}{m} = \frac{n!}{m!\,(n-m)!} \tag{5.70}$$

with $n! = 1 \cdot 2 \cdots n$ and $0! = 1$. *Hint:* Pascal's identity $\binom{n}{k} = \binom{n-1}{k-1} + \binom{n-1}{k}$ will be helpful.

5.31 Suppose that $p(t)$ is a polynomial of degree n and let ℓ be a nonnegative integer satisfying $\ell \leq n$. Show that $t = a$ is a root of multiplicity ℓ if and only if $p^{(m)}(a) = 0$ for $m = 0, 1, \ldots, \ell - 1$. (*Hint:* For the necessity assertion, you will find it useful to write out a Taylor's series for $p(t)$. For the sufficiency assertion, Problem 5.30 will be helpful).

5.32 In this problem you will find the Daubechies lowpass filter of length 8.

(a) Set up the system of equations that we must solve in order to find a Daubechies lowpass filter for length 8.

(b) Use a **CAS** to solve your system.

(c) Verify that -1 is a root of multiplicity 4. Use the method of Example 5.6 to find the D8 filter.

[2]This rule is often referred to as the *Leibniz generalized product rule*.

Computer Lab

5.3 Constructing Daubechies Filters. From `stthomas.edu/wavelets`, access the file `DLFilters`. In this lab, you will write code that allows you to solve for even-length Daubechies filter coefficients. In addition, you will write code that uses these filters to compute the (inverse) wavelet transformations of a vector or a matrix.

CHAPTER 6

WAVELET SHRINKAGE: AN APPLICATION TO DENOISING

We now consider the problem of denoising a digital signal. We present a method for denoising called *wavelet shrinkage*. This method was developed largely by Stanford statistician David Donoho and his collaborators. In a straightforward argument, Donoho [35] explains why wavelet shrinkage works well for denoising problems, and the advantages and disadvantages of wavelet shrinkage have been discussed by Taswell [93]. Vidakovic [98] has authored a nice book that discusses wavelet shrinkage in detail and also covers several other applications of wavelets in the area of statistics.

Note: The material developed in this section makes heavy use of ideas from statistics. If you are not familiar with concepts such as random variables, distributions, and expected values, you might first want to read Appendix A. You might also wish to work through the problems at the end of each section of the appendix. Two good sources for the material we use in this section are DeGroot and Schervish [33] and Wackerly et al. [99].

In the next section we present a basic overview of wavelet shrinkage and its application to signal denoising. In the final two sections of this chapter, we discuss

Discrete Wavelet Transformations: An Elementary Approach With Applications, Second Edition.
Patrick J. Van Fleet.

two wavelet-based methods used to denoise signals. The *VisuShrink* method [36] is described in Section 6.2 and the *SureShrink* method [37] is developed in Section 6.3. Section 6.2 contains an application of wavelets and denoising to the problem of image segmentation.

6.1 An Overview of Wavelet Shrinkage

The simple process of taking a picture can create noise. Scanning a picture can produce a digital image where noise is present. Recording a voice or instrument to an audio file often produces noise. Noise can also be incurred when we transmit a digital file.

Let's suppose that the true signal is stored in an N-vector $\mathbf{v}$. In practice, we never know $\mathbf{v}$. Through acquisition or transmission, we create the noisy N-vector

$$\mathbf{y} = \mathbf{v} + \mathbf{e}$$

Gaussian White Noise

The N-vector $\mathbf{e} = [e_1, e_2, \ldots, e_N]^T$ is the noise vector. We assume that the entries e_k are *Gaussian white noise*. That is, you can think of the e_k as *independent samples* that are *normally distributed* with mean zero and variance σ^2. See Definition A.10 in Section A.3 for a definition of an independent random variable and Definition A.14 in Section A.5 for a definition of the normal distribution.

The mean is zero since we would expect the noisy components y_k, $k = 1, \ldots, N$, of $\mathbf{y}$ to both underestimate and overestimate the true values. The variance σ^2 estimates how much the data are spread out. For this reason we also refer to σ as the *noise level*. Of course, σ is usually unknown. In some variations of wavelet shrinkage, we will need to estimate σ.

Examples of Noisy Signals and Images

For an example of a noisy signal, consider the N-vector $\mathbf{v}$ formed by uniformly sampling the *heavisine function* (see Donoho and Johnstone [36])

$$f(t) = 4\sin(4\pi t) - \text{sgn}(t - .3) - \text{sgn}(.72 - t) \tag{6.1}$$

on the interval $[0, 1]$. The *sign function*, denoted $\text{sgn}(t)$, is defined in (6.2). It is simply a function that indicates the sign of the given input.

$$\text{sgn}(t) = \begin{cases} 1, & t > 0 \\ 0, & t = 0 \\ -1, & t < 0. \end{cases} \tag{6.2}$$

Since the sgn function has a jump discontinuity at $t = 0$, it is often used to introduce jump discontinuities in a function. The function $f(t)$ has jumps at .3 and at .72.

A signal $\mathbf{v}$ and a noisy version $\mathbf{y}$ of it are plotted in Figure 6.1. Our task is to produce an N-vector $\widehat{\mathbf{v}}$ that does a good job estimating $\mathbf{v}$.

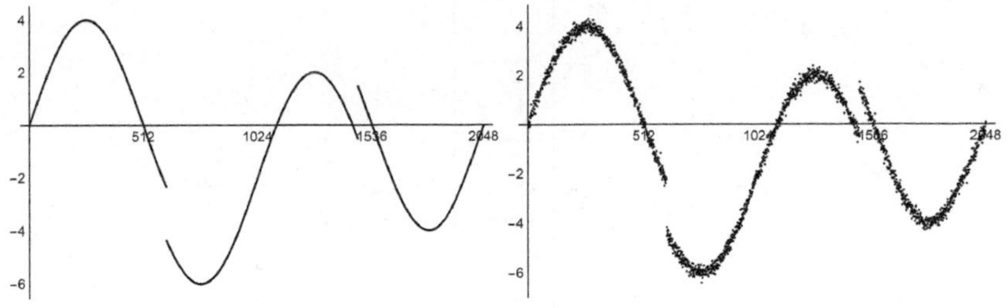

(a) $\mathbf{v}$ (b) $\mathbf{y}$

Figure 6.1 The heavisine function is uniformly sampled 2048 times and stored in $\mathbf{v}$. The vector $\mathbf{y}$ is formed by adding noise to $\mathbf{v}$ with $\sigma = 0.15$.

The Wavelet Shrinkage Algorithm

The wavelet shrinkage method for denoising can be outlined as follows:

Algorithm 6.1 (Wavelet Shrinkage Algorithm). *This algorithm takes a noisy N-vector $\mathbf{y} = \mathbf{v} + \mathbf{e}$, where $\mathbf{v}$ is the true (unknown) signal and $\mathbf{e}$ is the noise vector, and returns the (denoised) estimate $\widehat{\mathbf{v}}$ of $\mathbf{v}$.*

1. *Compute i iterations of the wavelet transformation on $\mathbf{y}$ to obtain the transformed vector*

$$\mathbf{z} = \begin{bmatrix} \mathbf{a}^i \\ \hline \mathbf{d}^i \\ \hline \mathbf{d}^{i-1} \\ \hline \vdots \\ \hline \mathbf{d}^2 \\ \hline \mathbf{d}^1 \end{bmatrix}.$$

2. *Apply a threshold rule (described below) to the differences portions $\mathbf{d}^k$, $k = 1, \ldots, i$ of $\mathbf{z}$. The rule will either "shrink" or set to zero values in $\mathbf{d}^k$. Denote the modified differences portions by $\widehat{\mathbf{d}}^k$.*

3. *Rejoin these modified differences portions with the original averages portion* $\mathbf{a}^i$ *to form a modified transform vector*

$$\widehat{\mathbf{z}} = \begin{bmatrix} \mathbf{a}^i \\ \hline \widehat{\mathbf{d}}^i \\ \hline \widehat{\mathbf{d}}^{i-1} \\ \hline \vdots \\ \hline \widehat{\mathbf{d}}^2 \\ \hline \widehat{\mathbf{d}}^1 \end{bmatrix}.$$

4. *Compute* i *iterations of the inverse wavelet transformation of* $\widehat{\mathbf{z}}$ *to obtain* $\widehat{\mathbf{v}}$.

□

Measuring the Effectiveness of the Denoising Method

How do we decide if the shrinkage method does a good job denoising the signal? The shrinkage method, in particular the threshold rule, is constructed so that the estimator $\widehat{\mathbf{v}}$ of $\mathbf{v}$ is optimal in the sense of the *mean squared error*.

Suppose that $\mathbf{v} = [v_1, \ldots, v_N]^T$ and $\widehat{\mathbf{v}} = [\widehat{v}_1, \ldots, \widehat{v}_N]^T$. Statisticians define the mean squared error as the *expected value* of the quantity

$$\sum_{k=1}^{N} (v_k - \hat{v}_k)^2.$$

The above quantity is nothing more than the square of the norm of $\mathbf{v} - \widehat{\mathbf{v}}$, so we can write the mean squared error as

$$\text{mean squared error} = E(\|\mathbf{v} - \widehat{\mathbf{v}}\|^2).$$

Here, $E(\cdot)$ is the notation we use for the expected value. See Definition A.11 in Appendix A for a definition of the expected value of a random variable.

The closer $\widehat{\mathbf{v}}$ is to $\mathbf{v}$, the lower the expected value. The expected value can then be interpreted as the mean value of the summands $(v_k - \hat{v}_k)^2$. The threshold rule we describe below attempts to minimize this average value.

Threshold Rules

There are two types of threshold rules. The simplest rule is a *hard threshold rule*. To apply a hard threshold rule to vector $\mathbf{z}$, we first choose a tolerance $\lambda > 0$. We retain z_k if $|z_k| > \lambda$ and we set $z_k = 0$ if $|z_k| \leq \lambda$. Note that we used the hard threshold rule in Example 5.3 in Section 5.2. When we performed image compression, we applied the hard threshold rule to the our transformed data after using the cumulative energy vector to determine a hard threshold value. Although denoising is certainly an application diffcrent than compression, we see in some applications that the compressed images either exhibit ringing or are somewhat blocky. To avoid this problem we use a *soft threshold rule*.

The Soft Threshold Function

To apply the soft threshold rule to vector $\mathbf{z}$, we first choose a tolerance $\lambda > 0$. Like the hard threshold rule, we set to zero any element z_k where $|z_k| \leq \lambda$. The soft threshold rule differs from the hard threshold rule in the case where $|z_k| > \lambda$. Here we *shrink* the value z_k so that its value is λ units closer to zero.

Recall that $|z_k| > \lambda$ means that either $z_k > \lambda$ or $z_k < -\lambda$. If $z_k > \lambda$, we replace z_k with $z_k - \lambda$. If $z_k < -\lambda$, then we replace z_k with $z_k + \lambda$ to bring it λ units closer to zero.

We can easily write down the function $s_\lambda(t)$ that describes the soft threshold rule:

$$s_\lambda(t) = \begin{cases} t - \lambda, & t > \lambda \\ 0, & -\lambda \leq t \leq \lambda \\ t + \lambda, & t < -\lambda \end{cases} \tag{6.3}$$

The soft threshold function $s_\lambda(t)$ is plotted in Figure 6.2.

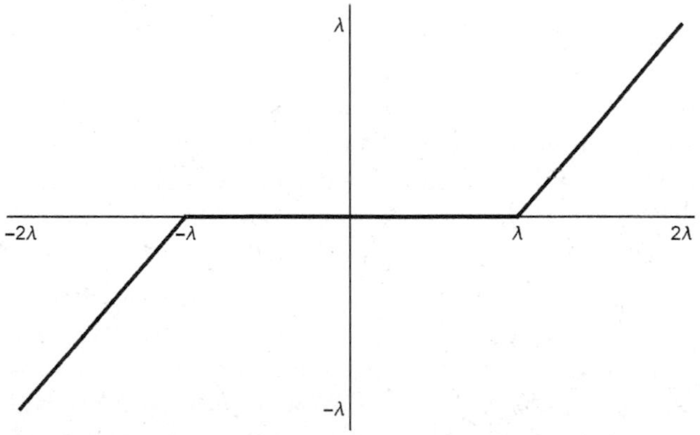

Figure 6.2 The soft threshold rule function $s_\lambda(t)$.

Note that $s_\lambda(t)$ is a piecewise linear function. The piece between $-\lambda$ and λ is the zero function. The other two pieces are lines with slope 1 (so that the values $s_\lambda(t)$ are neither expanded or contracted) that have been vertically shifted so that they are λ units closer to the t-axis.

Example 6.1 (Soft Threshold $s_\lambda(t)$). *Suppose that* $\lambda = 2$ *and we apply the soft threshold rule* (6.3) *to* $\mathbf{z} = [1, 3.5, -1.25, 6, -3.6, 1.8, 2.4, -3.5]$. *The resulting vector is* $\mathbf{w} = [0, 1.5, 0, 4, -1.6, 0, .4, -1.5]$ *and both* $\mathbf{z}$ *and* $\mathbf{w}$ *are plotted in Figure 6.3. Note that all the black dots in* $[-2, 2]$ *are set to zero, while those black dots outside* $[-2, 2]$ *are shrunk so that they are two units closer to the t-axis.*

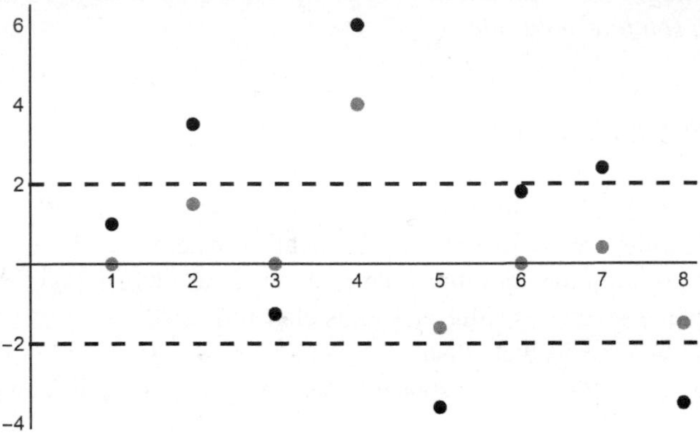

Figure 6.3 The vector $\mathbf{z}$ (black). Application of the soft threshold rule (6.3) to $\mathbf{z}$ gives $\mathbf{w}$ (gray).

(Live example: Visit `stthomas.edu/wavelets` *and click on Live Examples.)*

□

Why Does Wavelet Shrinkage Work?

Before we describe two methods for choosing the tolerance λ, let's try to better understand why wavelet shrinkage works. Suppose that we perform one iteration of a wavelet transformation to a noisy vector $\mathbf{y} = \mathbf{v} + \mathbf{e}$. We obtain

$$\mathbf{z} = W_N \mathbf{y} = W_N(\mathbf{v} + \mathbf{e}) = W_N \mathbf{v} + W_N \mathbf{e} \tag{6.4}$$

Recall the basic structure of $W_N \mathbf{v}$. This vector contains the averages (differences) portion of the transformation in the first (second) half of $\mathbf{z}$. As we have seen throughout the book, most of the energy in $\mathbf{z}$ is stored in the averages portion $\mathbf{a}$, while $\mathbf{d}$ is a sparse contributor to energy by comparison. What about $W_N \mathbf{e}$? You will show in Problem 6.1 that application of an orthogonal matrix W_N to a Gaussian white noise vector $\mathbf{e}$ returns a Gaussian white noise vector. In particular, the variance of the resultant is unchanged, so the noise level for $W_N \mathbf{e}$ is still σ. As Donoho points out in [35], since the differences portion $\mathbf{d}$ is sparse, it is comprised primarily of noise. So application of the soft threshold rule (6.3) to the elements in $\mathbf{d}$ should serve to denoise the signal.

In the next two sections, we discuss methods for choosing a tolerance λ for the soft threshold rule.

PROBLEMS

6.1 Let $\mathbf{e} = [e_1, \ldots, e_N]^T$ be a Gaussian white noise vector. That is, for $k = 1, \ldots, N$, the entries e_k are independent and normally distributed with mean $E(e_k) = 0$ and variance $E(e_k^2) = \sigma^2$. In this problem, you will show that if W is an $N \times N$ orthogonal matrix, then the entries of y_j, $j = 1, \ldots, N$, of $\mathbf{y} = W\mathbf{e}$ will each have mean zero and variance σ^2 as well.

The following steps will help you organize your proof.

(a) Let w_{jk} denote the j,kth element of W, $j, k = 1, \ldots, N$. For a fixed $j = 1, \ldots, N$, write y_j in terms of w_{jk} and e_k, $k = 1, \ldots, N$. Now take the expectation of each side and utilize (d) of Proposition A.3 from Appendix A.

(b) From (a) we know that $\text{Var}(y_j) = E(y_j^2) = E\left(\left(\sum\limits_{k=1}^{N} w_{jk}e_k\right)^2\right)$. Use the fact that for numbers $a_1, \ldots, a_N$ and $b_1, \ldots, b_N$, we have

$$\left(\sum_{k=1}^{N} a_k\right) \cdot \left(\sum_{m=1}^{N} b_m\right) = \sum_{k=1}^{N}\sum_{m=1}^{N} a_k b_m$$

to write this expectation as a double sum and then use (d) of Proposition A.3 to pass the expectation through the double sum.

(c) Use the fact that e_k and e_m are independent for $k \neq m$ to reduce the double sum of (b) to a single sum.

(d) Use what you know about orthogonal matrices and the fact that $E(e_k^2) = \sigma^2$ to show that $E(y_j^2) = \sigma^2$.

6.2 Consider the vector $\mathbf{v} = [-3, 2.6, 0.35, -5, -0.3, 5, 3, 1.6, 6.2, 0.5, -0.9, 0]^T$.

(a) Apply the soft threshold function (6.3) to $\mathbf{v}$ using $\lambda = 1.5$.

(b) Repeat (a) but use $\lambda = 3.5$.

6.3 Compute $\lim\limits_{\lambda \to 0} s_\lambda(t)$ and $\lim\limits_{\lambda \to \infty} s_\lambda(t)$ and interpret the results in terms of the wavelet shrinkage process.

6.4 In this problem we develop an alternative formula for the soft threshold rule (6.3). Define the function

$$t_+ = \max\{0, t\}$$

(a) Graph t_+, $(t - \lambda)_+$, and $(|t| - \lambda)_+$ for some $\lambda > 0$.

(b) Observe that $(|t| - \lambda)_+ = s_\lambda(t)$ whenever $t \geq -\lambda$. What's wrong with the formula when $t < -\lambda$?

(c) Show that $s_\lambda(t) = \text{sgn}(t)(|t| - \lambda)_+$.

The formula in (c) is useful for coding $s_\lambda(t)$ with a **CAS**.

6.5 Hard thresholding was mentioned in this section. To apply hard thresholding with tolerance $\lambda > 0$ to elements of vector $\mathbf{z}$, we simply replace z_k with zero whenever $|z_k| \le \lambda$ and leave z_k unchanged otherwise. There are instances (see Yoon and Vaidyanathan [104]) when hard thresholding yields desirable results.

(a) Using (6.3) as a model, write a piecewise-defined function $f_\lambda(t)$ that performs hard thresholding on input t.

(b) Write $f_\lambda(t)$ in terms of the shrinkage function $s_\lambda(t)$ and $\text{sgn}(t)$.

(c) Sketch a graph of $f_\lambda(t)$ for $\lambda = 1, 1.5, 2$.

6.6 Repeat Problem 6.2 but use the hard thresholding function developed in Problem 6.5.

Computer Lab

6.1 An Overview of Wavelet Shrinkage. From `stthomas.edu/wavelets`, access the file `ShrinkageIntro`. In this lab you will learn how to generate Gaussian white noise and add it to signals and images. You will then use the wavelet shrinkage module to experiment with denoising various data sets and audio signals.

6.2 VisuShrink

VisuShrink utilizes the wavelet shrinkage algorithm (Algorithm 6.1) with a *universal threshold* λ^{univ}. To motivate the theorem of Donoho and Johnstone [36] that defines λ^{univ}, suppose that we have observed

$$\mathbf{y} = \mathbf{v} + \mathbf{e}.$$

Suppose further that we decide to estimate $\mathbf{v}$ with $\widehat{\mathbf{w}} = (\widehat{w}_1, \ldots, \widehat{w}_N)$ by keeping certain values of y_k and setting other values $y_k = 0$. Let A be the set of indices for which we retain the values y_k. We have

$$\widehat{w}_k = \begin{cases} y_k, & k \in A \\ 0, & k \notin A. \end{cases}$$

We can compute the expected value $E(\|\widehat{\mathbf{w}} - \mathbf{v}\|^2)$. We know (see Proposition A.3 in Appendix A) that

$$E\left(\|\widehat{\mathbf{w}} - \mathbf{v}\|^2\right) = E\left(\sum_{k=1}^{N} (\widehat{w}_k - v_k)^2\right) = \sum_{k=1}^{N} E\left((\widehat{w}_k - v_k)^2\right).$$

Now if $k \in A$, we have

$$E\left((\widehat{w}_k - v_k)^2\right) = E\left((y_k - v_k))^2\right) = E\left((v_k + e_k - v_k)^2\right) = E\left(e_k^2\right).$$

Recall that the elements e_k are normally distributed with mean zero and variance σ^2. From Theorem A.1 we know that if we define random variable $Z = e_k/\sigma$, then Z^2 has a χ^2 distribution with one degree of freedom. Proposition A.7 then tells us (with $n = 1$) that

$$1 = E\left(Z^2\right) = E\left(\frac{e_k^2}{\sigma^2}\right) = \frac{1}{\sigma^2}E\left(e_k^2\right)$$

or

$$E\left(e_k^2\right) = \sigma^2. \tag{6.5}$$

Now if $k \notin A$, we have $\widehat{w}_k = 0$, so that

$$E\left(\left(\widehat{w}_k - v_k\right)^2\right) = E\left(v_k^2\right) = v_k^2. \tag{6.6}$$

We can use (6.5) and (6.6) to define the function

$$f(k) = \begin{cases} \sigma^2, & k \in A \\ v_k^2, & k \notin A \end{cases}$$

and write the expected value

$$E(\|\widehat{\mathbf{w}} - \mathbf{v}\|^2) = \sum_{k=1}^{N} f(k).$$

Now ideally, we would form A by choosing all the indices k so that $v_k^2 > \sigma^2$. The reason we would do this is that if $v_k^2 \le \sigma^2$, then the true value v_k is below (in absolute value) the noise level and we aren't going to be able to recover it. Of course such an ideal rule is impossible to implement since we don't know the values v_k, but it serves as a good theoretical measure for our threshold rule. If we do form A in this manner, then our *ideal mean squared error* is

$$E\left(\|\widehat{\mathbf{w}} - \mathbf{v}\|^2\right) = \sum_{k=1}^{N} \min(v_k^2, \sigma^2). \tag{6.7}$$

To better understand the ideal mean squared error, let's think about how we will use it. We apply our threshold rule to the highpass portion of one iteration of the wavelet transformation of $\mathbf{y}$. We have

$$\mathbf{z} = W_N\mathbf{y} = W_N(\mathbf{v} + \mathbf{e}) = W_N\mathbf{v} + W_N\mathbf{e}$$

Let $\mathbf{d}$ denote the differences (highpass) portion of $W_N\mathbf{v}$. Remember that $W_N\mathbf{e}$ is again Gaussian white noise, so let's denote the components of $\mathbf{e}$ corresponding to $\mathbf{d}$ as $\mathbf{e}_d$. Then we will apply our threshold rule to the vector $\mathbf{d} + \mathbf{e}_d$.

We expect that most of the components of $\mathbf{d}$ are zero so that most of the components of $\mathbf{d} + \mathbf{e}_d$ will be at or below the noise level σ. In such a case, a large number of the summands of the ideal mean squared error (6.7) will be σ^2.

An Accuracy Result for Wavelet Shrinkage

The entire point of the discussion thus far is to provide motivation for the result we now state.

Theorem 6.1 (Measuring the Error: Donoho and Johnstone [36]). *Suppose that* $\mathbf{u}$ *is any vector in* $\mathbb{R}^N$ *and that we are given the data* $\mathbf{y} = \mathbf{u} + \mathbf{e}$ *where* $\mathbf{e}$ *is Gaussian white noise with noise level* σ. *Let* $s_\lambda(t)$ *be the soft threshold rule* (6.3) *with* $\lambda = \sigma\sqrt{2\ln(N)}$. *Let* $\widehat{\mathbf{u}}$ *be the vector formed by applying* $s_\lambda(t)$ *to* $\mathbf{u}$. *Then*

$$E(\|\widehat{\mathbf{u}} - \mathbf{u}\|^2) \leq (2\ln(N) + 1)\left(\sigma^2 + \sum_{k=1}^{N} \min(u_k^2, \sigma^2)\right).$$

□

Proof. The proof requires some advanced ideas from analysis and statistics and is thus omitted. The interested reader is encouraged to see the proof given by Vidakovic ([98], Theorem 6.3.1). □

Definition 6.1 (Universal Threshold). *The tolerance*

$$\boxed{\lambda = \sigma\sqrt{2\ln(N)}} \tag{6.8}$$

given in Theorem 6.1 is called the universal threshold *and hereafter denoted by* λ^{univ}.

□

It is important to understand what Theorem 6.1 is telling us. The choice of $\lambda^{\text{univ}} = \sigma\sqrt{2\ln(N)}$ gives us something of a *minimax* solution to the problem of minimizing the ideal mean squared error. For this λ^{univ} and *any* choice of $\mathbf{u}$, the soft threshold rule (6.3) produces a mean squared error that is always smaller than a constant times the noise level squared plus the ideal mean squared error. It is worth noting that the constant $2\ln(N) + 1$ is quite small relative to N (see Table 6.1).

Table 6.1 The values of $2\ln(N) + 1$ for various values of N.

N	$2\ln(N) + 1$
100	10.21
1,000	14.82
10,000	19.42
100,000	24.03

Estimating the Noise Level

Note that the universal threshold λ^{univ} given in Definition 6.1 depends on the size of the vector N and the noise level σ. In practice, we typically do not know σ so we have to estimate it.

Hampel [52] has shown that the *median absolute deviation* (MAD) converges to 0.6745σ as the sample size goes to infinity. MAD is defined in Definition A.4 in Appendix A.

Donoho and Johnstone [36] suggest that we should estimate σ using the elements of $\mathbf{d}^1$, the first iteration of the differences portion, of the wavelet transformation, since this portion of the transformation is predominantly noise. We have the following estimator for σ:

$$\boxed{\widehat{\sigma} = MAD(\mathbf{d}^1)/0.6745.} \tag{6.9}$$

VisuShrink Examples

Next we look at two examples that implement VisuShrink. The first example illustrates the method on the test signal plotted in Figure 6.1. This example is somewhat artificial since we set the noise level when we form the noisy vector. We do, however, use (6.9) to estimate σ. The second example introduces the idea of *image segmentation*. In this application, pixels are separated into different regions based on similarity. Denoising the image before segmenctation is applied can sometimes increase the effectiveness of the algorithm.

Example 6.2 (VisuShrink). *We consider the one-dimensional signal and the noisy version plotted in Figure 6.1. To obtain the original signal, we set* $N = 2048$ *and uniformly sample the function* $f(t) = 4\sin(4\pi t) - \text{sgn}(t-0.3) - \text{sgn}(0.72-t)$ *on the interval* $[0,1)$. *Thus, the components of our signal* $\mathbf{v}$ *are* $v_k = f(k/2048)$, $k = 0,\ldots,2047$. *We set the noise level to* $\sigma = 0.15$ *and then create the Gaussian white noise vector* $\mathbf{e}$ *by generating* 2048 *random normal samples. We add this noise vector to the original to get the noisy signal* $\mathbf{y}$.

We use the D4 and D6 wavelet filters and perform five iterations of each wavelet transformation on the noisy signal $\mathbf{y}$. *After computing five iterations of the wavelet transformation, we construct the estimate* $\widehat{\sigma}$ *using* (6.9). *Recall that we use* $\mathbf{d}^1$ *from each transformation to compute the median absolute deviation. We then divide this number by* 0.6745 *to obtain the estimates of the noise. These values are summarized in Table 6.2.*

Now the length of $\mathbf{a}^5$, *the averages portion of the transformation, is* $2048/2^5 = 64$, *so the length of all the differences portions* $\mathbf{d}^1,\ldots,\mathbf{d}^5$ *together is* $2048 - 64 = 1984$. *Using the formula* (6.8) *for* λ^{univ} *from Definition 6.1, we find our VisuShrink tolerances and list them in Table 6.2.*

We have plotted the combined differences portions for each wavelet transformation in Figure 6.4. The horizontal lines in the figure are at heights $\pm\lambda^{\text{univ}}$. *Note that*

Table 6.2 Noise estimates and threshold values using the D4 and D6 filters.

Filter	$\widehat{\sigma}$	λ^{univ}
D4	0.151225	0.589306
D6	0.154137	0.600655

most of the elements fall between these two lines, so we are shrinking only a few of these values.

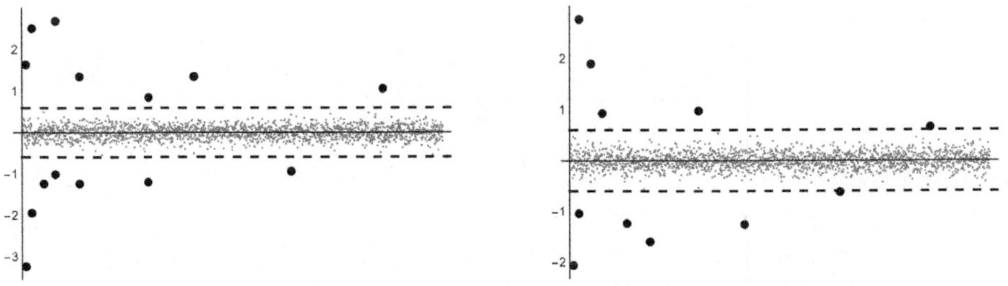

(a) D4 (b) D6

Figure 6.4 The combined difference portions for each wavelet transformation. The gray values are those that will be set to zero by the soft threshold rule. The black values will be shrunk so that they are λ units closer to zero.

We apply the soft threshold rule (6.3) *to the difference portions of each transformation and then join the results with their respective averages portion. Finally, we compute five iterations of the inverse wavelet transformation of each modified transformation to arrive at our denoised signals. The results are plotted in Figure 6.5 along with the mean squared error between the original signal* $\mathbf{v}$ *and the denoised signal* $\widehat{\mathbf{v}}$.

(Live example: Visit `stthomas.edu/wavelets` *and click on Live Examples.)*

□

The signal on the right (D6) appears a bit smoother than the one on the left (D4). It is no coincidence that the longer filter produces a smoother approximation. In the classical derivation of wavelet theory (see, e.g. Walnut [100]), these filters are derived from special functions called *scaling functions*. The smoothness of these functions is directly proportional to the filter length associated with them.

In Computer Lab 6.2, you will continue to investigate VisuShrink using different filters and different test signals.

Our last example demonstrates an application of VisuShrink to the problem of image segmentation.

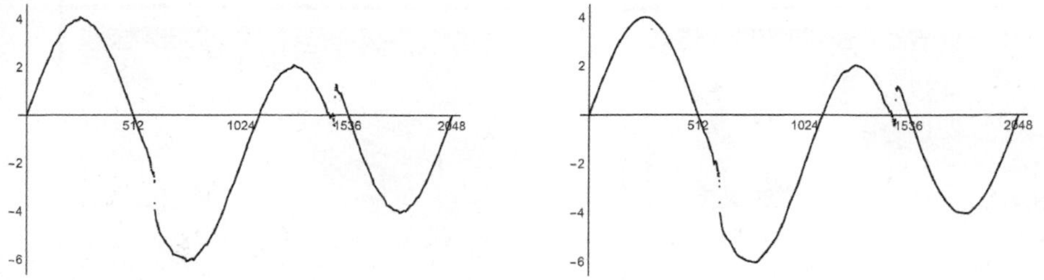

(a) $D4$, $\text{MSE}(\mathbf{v}, \widehat{\mathbf{v}}) = 0.0038$

(b) $D6$, $\text{MSE}(\mathbf{v}, \widehat{\mathbf{v}}) = 0.0031$

Figure 6.5 The denoised signals using D4 and D6 filters.

Example 6.3 (Image Segmentation). *It is sometimes desirable to group pixels in an image into like groups or segments. This process is called image segmentation. Consider the* 128×192 *image, denoted by A, plotted in Figure 6.6(a). Our goal is to isolate the goat in the image and color those pixels white and color the remaining pixels black.*

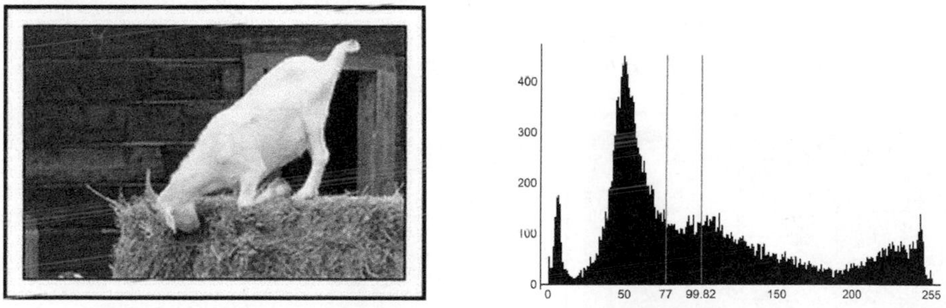

(a) Image A

(b) Distribution of intensities

Figure 6.6 A digital image and a distribution of its intensity values. The image median $a = 77$ and mean $\overline{a} = 99.82$ are identified as well.

Figure 6.6(b) shows the distribution of the gray intensities for the image A. The median intensity value is $a = 77$. *We could naively divide using this value mapping all intensities in* A *less than* a *to zero and the remaining intensities to* 255. *Figure 6.7(a) shows that this is not a sound strategy as many mid-gray regions are converted to white. We seek a better threshold that will better identify those intensities at or near* 255. *To find this value in an automated way, we compute the mean value* $\overline{a} = 99.82$ *of all intensities in* A *and use that as a starting point in Algorithm 4.1 in Section 4.4. This algorithm should give more weight to the larger intensity values and indeed it returns* $\tau = 128.08$ *as a threshold. The segmented image using this* τ *is plotted in Figure 6.7(b).*

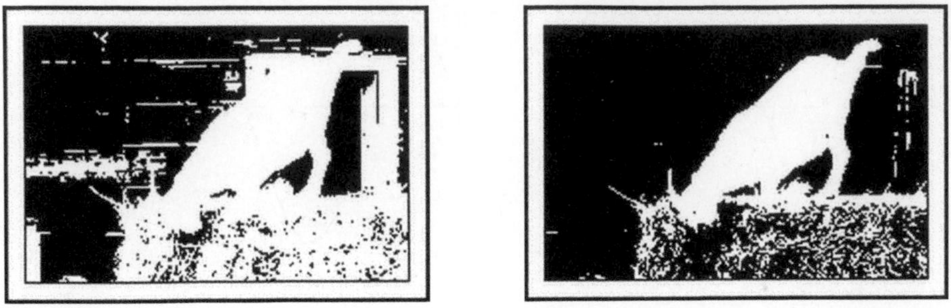

(a) $\tau = 77$ (b) $\tau = 128.08$

Figure 6.7 Two attempts to perform image segmentation on the image from Figure 6.6(a). The image on the left results when we determine regions by simply converting all pixels smaller than the median $\tau = 77$ to 9 and the remaining values to 255. The image on the right uses the threshold $\tau = 128.08$ generated by Algorithm 4.1.

The goat is identified but unfortunately so is much of the hay below the goat. For our purposes, we think of this region as "noise" and before computing τ, we denoise A using VisuShrink. For this application, we compute four iterations of the orthogonal wavelet transformation using the D10 filter. The denoised image B and its histogram of intensities is plotted in Figure 6.8.

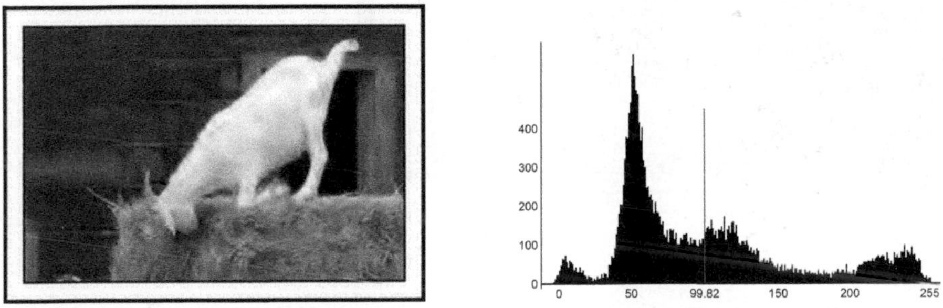

(a) Denoised image (b) Distribution of intensities

Figure 6.8 The denoised image and a distribution of its intensity values. The image mean $\overline{a} = 99.82$ is identified as well.

We apply Algorithm 4.1 to B and find that $\tau = 134.43$. We use this threshold to perform segmentation on the denoised image and plot the result in Figure 6.9.

The denoised image is better suited for segmentation. More sophisticated segmentation routines exist (see e.g. [49]) and indeed some employ wavelets (see e.g. [44]).

(Live example: Visit `stthomas.edu/wavelets` *and click on Live Examples.)*

□

Figure 6.9 Segmentation of the denoised image in Figure 6.8(a) with all intensities less than $\tau = 134.76$ converted to zero and the remaining pixels converted to 255.

PROBLEMS

6.7 Suppose that we observe $\mathbf{y} = \mathbf{v} + \mathbf{e}$, where the length of $\mathbf{y}$ is N, $\mathbf{v}$ is the true signal, and $\mathbf{e}$ is Gaussian white noise with noise level $\sigma > 0$. Further, suppose that $\|\mathbf{v}\| = 1$. Let n_σ denote the number of elements in $\mathbf{v}$ such that $|v_k| > \sigma$, $k = 1, \ldots, N$. Show that the ideal mean squared error given in (6.7) satisfies

$$\sum_{k=1}^{N} \min(v_k^2, \sigma^2) = \sigma^2 n_\sigma + (1 - C_e(\mathbf{v})_{n_\sigma})$$

where $C_e(\mathbf{v})$ is the cumulative energy vector for $\mathbf{v}$ (see Definition 3.3). In this way we see, as Donoho points out [35], that the ideal mean squared error "is explicitly a measure of the extent to which the energy is compressed into a few big coefficients" of $\mathbf{v}$.

6.8 Suppose we add Gaussian white noise with $\sigma = 1.5$ to $\mathbf{v} \in \mathbb{R}^{20}$ and then compute one iteration of the wavelet transformation using the D6 filter. The resulting differences portion of the transformation is

$$\begin{aligned}\mathbf{d} = [&0.243756, 1.34037, 2.34499, 0.474407, -0.475891\\ &-0.697551, -2.31985, 1.27441, -2.29948, -0.781795]^T.\end{aligned}$$

(a) Using (6.9) and a **CAS** or a calculator, compute the estimated noise level $\widehat{\sigma}$.

(b) Use $\widehat{\sigma}$ from (a) to compute λ^{univ} for this data set.

6.9 The universal threshold λ^{univ} is data dependent and it is possible that for large enough vector length N, the choice of λ^{univ} will result in all highpass values being shrunk to zero.

Suppose that we are given a noisy vector $\mathbf{y} = [y_1, \ldots, y_N]^T$ with $|y_k| \leq M$ for some $M > 0$ and each $k = 1, \ldots, N$. Suppose that $\mathbf{z}$ is the result of applying one iteration of the wavelet tranform using the Haar filter $\mathbf{h} = (h_0, h_1) = \left(\frac{\sqrt{2}}{2}, \frac{\sqrt{2}}{2}\right)$. Let $\mathbf{d} = \left[d_1, \ldots, d_{N/2}\right]^T$ denote the difference portion of $\mathbf{z}$.

(a) The triangle inequality states that for all $a, b \in \mathbb{R}$, $|a + b| \leq |a| + |b|$. Use the triangle inequality to show that $|d_k| \leq \sqrt{2}M$, $k = 1, \ldots, N/2$.

(b) Suppose that we have computed the noise level estimate $\widehat{\sigma}$. How large must N be so that $\lambda^{\text{univ}} \geq \sqrt{2}M$? In this case, we literally discard the entire highpass portion of the transformation.

(c) Find N in the case where $M = 4$ and $\widehat{\sigma} = 1.8$.

Computer Lab

6.2 Denoising with VisuShrink. From `stthomas.edu/wavelets`, access the file `VisuShrink`. In this lab you will use the wavelet shrinkage module with λ^{univ} to apply the VisuShrink method to various test signals, audio signals, and digital images using different orthogonal filters. You will also write a module to compute the NormalShrink tolerance and compare the results with those you obtain using the VisuShrink tolerance λ^{univ}.

6.3 SureShrink

The SureShrink method utilizes Stein's unbiased risk estimator (SURE) [83]. Donoho and Johnstone [37] suggest minimizing this estimator to find what they call λ^{sure}. Unlike λ^{univ}, the tolerance λ^{sure} does not depend on the size of the input vector. The SureShrink method also calls for a tolerance to be computed for *each* highpass portion of the wavelet transformation. A test is performed on each highpass portion to determine its sparseness. If a highpass portion is deemed sparse, we compute λ^{univ} for that portion and use it in conjunction with the soft threshold rule (6.3). If the highpass portion is determined not to be sparse, we compute λ^{sure} for the portion and use it with (6.3).

Before we develop the SureShrink method, we need to discuss some ideas about vector-valued functions that we need to develop λ^{sure}. If you have had a course in multivariable calculus, then you are familiar with functions that map vectors to vectors. We are only concerned with functions that map $\mathbb{R}^N$ to $\mathbb{R}^N$. Let $\mathbf{x} \in \mathbb{R}^N$. We denote such functions by

$$\mathbf{f}(\mathbf{x}) = (f_1(\mathbf{x}), \ldots, f_N(\mathbf{x}))$$

where the functions $f_1, \ldots, f_N$ that map $\mathbb{R}^N$ to $\mathbb{R}$ are called *coordinate functions.*

We have actually been using vector-valued functions throughout the book without actually calling them by name. In Problem 6.10 you will show that one iteration of the Haar wavelet transformation applied to $\mathbf{x} \in \mathbb{R}^N$ can be expressed as a vector-valued function.

The soft threshold rule (6.3) we utilize in this section can be expressed as a vector-valued function. Suppose that $\mathbf{x} \in \mathbb{R}^N$. Define

$$\mathbf{s}_\lambda(\mathbf{x}) = (s_1(\mathbf{x}), \ldots, s_N(\mathbf{x})) \tag{6.10}$$

where all the coordinate functions are the same,

$$s_k(\mathbf{x}) = s_\lambda(x_k) \tag{6.11}$$

and $s_\lambda(t)$ is the soft threshold rule given in (6.3).

Review of Partial Differentiation

We will have occasion to differentiate coordinate functions. To perform this differentiation, we need the concept of a *partial derivative.*

Definition 6.2. *Let* $g(\mathbf{x}) = g(x_1, x_2, \ldots, x_N)$*. We define the* partial derivative of g with respect to x_k *as*

$$\frac{\partial}{\partial x_k} g(x_1, \ldots, x_N) = \lim_{h \to 0} \frac{g(x_1, \ldots, x_k + h, \ldots, x_N) - g(x_1, \ldots, x_N)}{h}.$$

□

Thus to compute a partial derivative $\frac{\partial g}{\partial x_k}$, we simply hold all other variables constant and differentiate with respect to x_k. We can often employ differentiation rules from elementary calculus to compute partial derivatives. For example, if

$$g(x_1, x_2, x_3) = 2x_1 x_2^2 - \sin(x_1) + e^{x_3}$$

then

$$\begin{aligned}
\frac{\partial}{\partial x_1} g(x_1, x_2, x_3) &= 2x_2^2 - \cos(x_1) \\
\frac{\partial}{\partial x_2} g(x_1, x_2, x_3) &= 4x_1 x_2 \\
\frac{\partial}{\partial x_3} g(x_1, x_2, x_3) &= e^{x_3}.
\end{aligned}$$

Consult a calculus book (e.g. Stewart [84]) for more information regarding partial derivatives.

Multivariable Soft Threshold Function

Our primary purpose for introducing the partial derivative is so that we can compute the partial derivatives of the coordinate functions $s_k(\mathbf{x})$, $k = 1, \ldots, N$ given by

(6.11). We have

$$s_k(\mathbf{x}) = s_\lambda(x_k) = \begin{cases} x_k - \lambda, & x_k > \lambda \\ 0, & -\lambda \le x_k \le \lambda \\ x_k + \lambda, & x_k < -\lambda. \end{cases}$$

so that we need to simply differentiate s_λ with respect to x_k:

$$\frac{\partial}{\partial x_k} s_k(\mathbf{x}) = s'_\lambda(x_k) = \begin{cases} 1, & x_k > \lambda \\ 0, & -\lambda < x_k < \lambda \\ 1, & x_k < -\lambda \end{cases} = \begin{cases} 1, & |x_k| > \lambda \\ 0, & |x_k| < \lambda. \end{cases} \tag{6.12}$$

Note that the partial derivative in (6.12) is undefined at $\pm\lambda$.

Stein's Unbiased Risk Estimator

We are now ready to begin our development of the SureShrink method. For completeness, we first state Stein's theorem without proof.

Theorem 6.2 (SURE: Stein's Unbiased Risk Estimator [83]). *Suppose that the N-vector $\mathbf{w}$ is formed by adding N-vectors $\mathbf{z}$ and $\boldsymbol{\epsilon}$. That is, $\mathbf{w} = \mathbf{z} + \boldsymbol{\epsilon}$, where $\boldsymbol{\epsilon} = (\epsilon_1, \ldots, \epsilon_N)$ and each ϵ_k is normally distributed with variance 1. Let $\widehat{\mathbf{z}}$ be the estimator formed by*

$$\widehat{\mathbf{z}} = \mathbf{w} + \mathbf{g}(\mathbf{w}). \tag{6.13}$$

where the coordinate functions $g_k : \mathbb{R}^N \to \mathbb{R}$ of the vector-valued function $\mathbf{g} : \mathbb{R}^N \to \mathbb{R}^N$ are differentiable except at a finite number of points.[1] *Then*

$$E\left(\|\widehat{\mathbf{z}} - \mathbf{z}\|^2\right) = E\left(N + \|\mathbf{g}(\mathbf{w})\|^2 + 2\sum_{k=1}^{N} \frac{\partial}{\partial w_k} g_k(\mathbf{w})\right). \tag{6.14}$$

□

Application of SURE to Wavelet Shrinkage

We want to apply Theorem 6.2 to the wavelet shrinkage method. We can view $\mathbf{w}$ as the highpass portion of the wavelet transformation of our observed signal $\mathbf{y} = \mathbf{v} + \mathbf{e}$ so that $\mathbf{z}$ is the highpass portion of the wavelet transformation of the true signal $\mathbf{v}$. The vector $\boldsymbol{\epsilon}$ is the highpass portion of the wavelet transformation of the noise $\mathbf{e}$ (see the discussion following (6.4)). Then $\widehat{\mathbf{z}}$ is simply the threshold function (6.10). We just need to define $\mathbf{g}$ in (6.13) so that $\widehat{\mathbf{z}} = \mathbf{s}_\lambda(\mathbf{w})$. This is straightforward. We take

[1] The theorem that appears in Stein [83] calls for g_k to be only *weakly differentiable*. The concept of weakly differentiable functions is beyond the scope of this book and we do not require a statement of the theorem in its weakest form.

$$\mathbf{g}(\mathbf{w}) = \mathbf{s}_\lambda(\mathbf{w}) - \mathbf{w} \tag{6.15}$$

so that $\widehat{\mathbf{z}} = \mathbf{w} + \mathbf{g}(\mathbf{w}) = \mathbf{w} + \mathbf{s}_\lambda(\mathbf{w}) - \mathbf{w} = \mathbf{s}_\lambda(\mathbf{w})$.

What Theorem 6.2 gives us is an alternative way to write the mean squared error $E\left(\|\widehat{\mathbf{z}} - \mathbf{z}\|^2\right)$. We can certainly simplify the right-hand side of (6.14) to obtain an equation with only λ unknown. If we want to minimize the mean squared error, then we must first simplify

$$f(\lambda) = N + \|\mathbf{g}(\mathbf{w})\|^2 + 2\sum_{k=1}^{N} \frac{\partial}{\partial w_k} g_k(\mathbf{w}) \tag{6.16}$$

and then find a value of λ that minimizes it. This minimum value will be our λ^{sure}. Let's simplify $f(\lambda)$. We have

$$\|\mathbf{g}(\mathbf{w})\|^2 = \sum_{k=1}^{N} g_k(\mathbf{w})^2.$$

Now

$$g_k(\mathbf{w}) = s_\lambda(w_k) - w_k = \begin{Bmatrix} w_k - \lambda, & w_k > \lambda \\ 0, & |w_k| \le \lambda \\ w_k + \lambda, & w_k < -\lambda \end{Bmatrix} - w_k = \begin{cases} -\lambda, & w_k > \lambda \\ -w_k, & |w_k| \le \lambda \\ \lambda, & w_k < -\lambda \end{cases}$$

so that

$$g_k(\mathbf{w})^2 = \begin{cases} \lambda^2, & |w_k| > \lambda \\ w_k^2, & |w_k| \le \lambda \end{cases} = \min\left(\lambda^2, w_k^2\right).$$

Thus

$$\|\mathbf{g}(\mathbf{w})\|^2 = \sum_{k=1}^{N} \min\left(\lambda^2, w_k^2\right). \tag{6.17}$$

From (6.15) we know that $g_k(\mathbf{w}) = s_\lambda(w_k) - w_k$. Using (6.12), we have

$$\begin{aligned} \frac{\partial}{\partial w_k} g_k(\mathbf{w}) &= s_\lambda'(w_k) - 1 \\ &= \begin{Bmatrix} 1, & |w_k| > \lambda \\ 0, & |w_k| < \lambda \end{Bmatrix} - 1 \\ &= \begin{cases} 0, & |w_k| > \lambda \\ -1, & |w_k| < \lambda. \end{cases} \end{aligned} \tag{6.18}$$

Now we need to sum up all the partial derivatives given by (6.18). We have

$$2\sum_{k=1}^{N} \frac{\partial}{\partial w_k} g_k(\mathbf{w}) = -2\sum_{k=1}^{N} \begin{cases} 0, & |w_k| > \lambda \\ 1, & |w_k| < \lambda. \end{cases} \tag{6.19}$$

The last sum is simply producing a count of the number of w_k's that are smaller (in absolute value) than λ. We will write this count as

$$\#\{k\colon |w_k| < \lambda\}. \tag{6.20}$$

Inserting (6.20) into (6.19) gives

$$2\sum_{k=1}^{N} \frac{\partial}{\partial w_k} g_k(\mathbf{w}) = -2 \cdot \#\{k\colon |w_k| < \lambda\}. \tag{6.21}$$

Combining (6.17) and (6.21), we see that (6.16) can be written as

$$\boxed{f(\lambda) = N - 2 \cdot \#\{k\colon |w_k| < \lambda\} + \sum_{k=1}^{N} \min\left(\lambda^2, w_k^2\right).} \tag{6.22}$$

Simplifying the Estimator Function $f(\lambda)$

We want to find, if possible, a λ that minimizes $f(\lambda)$. While the formula (6.22) might look complicated, it is a straightforward process to minimize it.

The value of the function $f(\lambda)$ in (6.22) depends on the elements in $\mathbf{w}$. Note that neither the factor $\#\{k\colon |w_k| < \lambda\}$ nor the term $\sum_{k=1}^{N} \min\left(\lambda^2, w_k^2\right)$ depend on the *ordering* of the elements of $\mathbf{w}$. Thus, we assume in our subsequent analysis of $f(\lambda)$ that the elements of $\mathbf{w}$ are ordered so that

$$|w_1| \le |w_2| \le \cdots \le |w_N|.$$

Let's first consider $0 \le \lambda \le |w_1|$. In this case,

$$\#\{k\colon |w_k| < \lambda\} = 0 \qquad \text{and} \qquad \sum_{k=1}^{N} \min\left(\lambda^2, w_k^2\right) = \sum_{k=1}^{N} \lambda^2 = N\lambda^2$$

so that

$$f(\lambda) = N + N\lambda^2. \tag{6.23}$$

Now if $\lambda > |w_N|$, we have

$$\#\{k\colon |w_k| < \lambda\} = N \qquad \text{and} \qquad \sum_{k=1}^{N} \min\left(\lambda^2, w_k^2\right) = \sum_{k=1}^{N} w_k^2$$

so that

$$f(\lambda) = N - 2N + \sum_{k=1}^{N} w_k^2 = -N + \sum_{k=1}^{N} w_k^2. \tag{6.24}$$

What about $|w_1| < \lambda \le |w_N|$? Note that the term $N - 2 \cdot \#\{k\colon |w_k| < \lambda\}$ is piecewise constant and will change only whenever λ moves past a $|w_k|$ that is bigger than the preceding $|w_{k-1}|$. We can write the second term as

$$\sum_{k=1}^{N} \min\left(\lambda^2, w_k^2\right) = \sum_{|w_k|<\lambda} w_k^2 + \#\{k\colon |w_k| \geq \lambda\}\,\lambda^2 \tag{6.25}$$

and it is easy to see that this term is piecewise quadratic and it will also change whenever λ moves past a $|w_k|$ that is bigger than the preceding $|w_{k-1}|$. So $f(\lambda)$ is a piecewise quadratic function with breaks occurring at *distinct* values of $|w_k|$. For $\lambda > |w_N|$, $f(\lambda)$ is the constant value given by (6.24). We can use (6.23) and (6.25) to note that whenever λ is between distinct values of $|w_k|$, $f'(\lambda) > 0$, so that $f(\lambda)$ is increasing between these values.

Examples of the Estimator Function $\mathrm{f}(\lambda)$

Let's look at a simple example that summarizes our analysis of $f(\lambda)$ thus far.

Example 6.4 (Plotting $\mathbf{f}(\lambda)$). *Let* $\mathbf{w} = [1, 1.1, 2, 2, 2.4]^T$. *Note that the elements of* $\mathbf{w}$ *are nonnegative and ordered smallest to largest. For* $0 < \lambda \leq 1 = w_1$, *we can use* (6.23) *and write* $f(\lambda) = 5 + 5\lambda^2$. *We can also use* (6.24) *and write* $f(\lambda) = -5 + (1^2 + 1.1^2 + 2^2 + 2^2 + 2.4^2) = 10.97$ *for* $\lambda > 2.4 = w_5$.

For $1 < \lambda \leq 1.1$, $\#\{k\colon |w_k| < \lambda\} = 1$ *and we can use* (6.22) *and write*

$$f(\lambda) = 5 - 2\cdot 1 + 1^2 + 4\lambda^2 = 4 + 4\lambda^2.$$

Note that the endpoint $|w_2| = 1.1$ *is included in this interval – the function doesn't change definitions again until* λ *moves past* $|w_2| = 1.1$. *We can use* (6.22) *for the remaining two intervals and describe* $f(\lambda)$:

$$f(\lambda) = \begin{cases} 5 + 5\lambda^2, & 0 < \lambda \leq 1 \\ 4 + 4\lambda^2, & 1 < \lambda \leq 1.1 \\ 3.21 + 3\lambda^2, & 1.1 < \lambda \leq 2 \\ 7.21 + \lambda^2, & 2 < \lambda \leq 2.4 \\ 10.97, & 2.4 < \lambda. \end{cases}$$

Thus we see that $f(\lambda)$ *is a left continuous function. The function is plotted in Figure 6.10.*

(Live example: Visit `stthomas.edu/wavelets` *and click on Live Examples.)*

□

In Figure 6.11 we have plotted $f(\lambda)$ for a different vector $\mathbf{w}$. From the graphs in Figures 6.10 and 6.11, it would seem that the desired minimum value of $f(\lambda)$ would either be at the left endpoint of some interval or the constant value $\sum_{k=1}^{N} w_k^2 - N$, but $f(\lambda)$ is left continuous at each breakpoint $|w_\ell|$. Theoretically, we could compute

$$m_k = \lim_{\lambda \to |w_k|^+} f(\lambda)$$

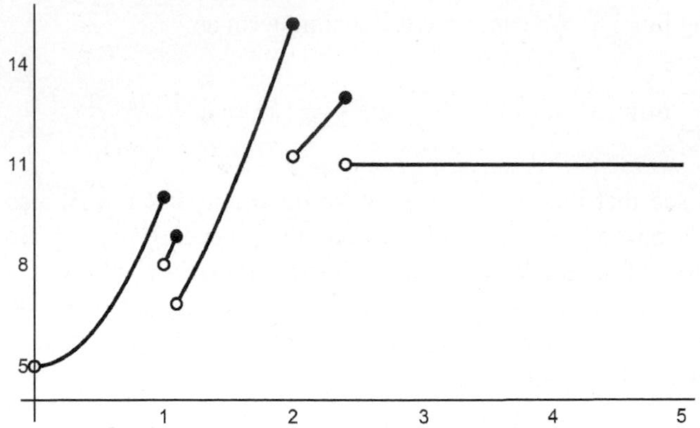

Figure 6.10 A plot of $f(\lambda)$. The function is left continuous at the breakpoints (noted by the black points) but is not right continuous at the breakpoints (noted by the open circle points).

for each of $|w_1|, \ldots, |w_N|$ and then pick the $|w_k|$ that corresponds to the minimum value of $\{m_1, m_2, \ldots, m_N, \sum_{k=1}^{N} w_k^2 - N\}$, but this is hardly tractable on a computer.

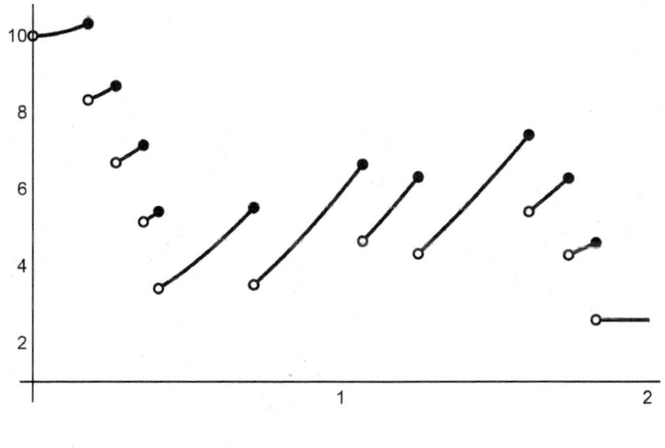

Figure 6.11 $f(\lambda)$ for $\mathbf{w} = [0.18, 0.27, 0.36, 0.41, 0.72, 1.07, 1.25, 1.61, 1.74, 1.83]^T$.

Minimizing $f(\lambda)$ on a Computer

A more feasible solution would be simply to change our function $f(\lambda)$ so that the quadratic pieces are defined on intervals $|w_\ell| \leq \lambda < |w_{\ell+1}|$, where $|w_\ell| < |w_{\ell+1}|$. The easiest way to do this is simply to change our counting function to include $|w_\ell|$ in the count but exclude $|w_{\ell+1}|$ from the count. That is, we consider minimizing the function[2]

$$\boxed{f(\lambda) = N - 2 \cdot \#\{k \colon |w_k| \leq \lambda\} + \sum_{k=1}^{N} \min\left(\lambda^2, w_k^2\right)} \tag{6.26}$$

Note: We use (6.26) in all subsequent examples, problem sets, and computer labs.

In Figure 6.12 we have used (6.26) and plotted $f(\lambda)$ using the vectors from Example 6.4 and Figure 6.11. Note that the new formulation sets the constant value $\sum_{k=1}^{N} w_k^2 - N$ as $f\left(|w_N|\right)$.

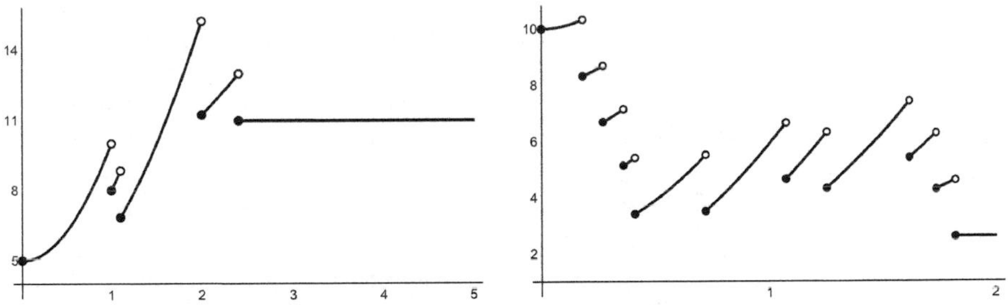

(a) $\mathbf{w}$ from Example 6.4

(b) $\mathbf{w}$ from Figure 6.11

Figure 6.12 Two plots of $f(\lambda)$ using (6.26). The graph in part (a) uses vector $\mathbf{w} = [1, 1.1, 2, 2, 2.4]^T$ from Example 6.4, while the graph in part (b) uses vector $\mathbf{w} = [0.18, 0.27, 0.36, 0.41, 0.72, 1.07, 1.25, 1.61, 1.74, 1.83]^T$ from Figure 6.11. The functions are now right continuous at the breakpoints.

Choosing the SUREShrink Tolerance λ^{sure}

Now our minimum can simply be chosen to be $\lambda^{\text{sure}} = |w_\ell|$ where $f(|w_\ell|) \leq f\left(|w_k|\right)$, $k = 1, \ldots, N$ and $\lambda^{\text{sure}} \leq \lambda^{\text{univ}}$. Computationally, λ^{sure} is quite straightforward. In Problem 6.14 you will derive a recursive formula for quickly computing $f\left(|w_k|\right)$, and in Computer Lab 6.3 you will develop the routine for computing λ^{sure} as given in the software package WaveLab [38].

The reason we insist that λ^{sure} is no larger than λ^{univ} is that the SureShrink method has some problems when the input vector $\mathbf{y}$ is sparse. If $\mathbf{y} = \mathbf{v} + \mathbf{e}$, where $\mathbf{e}$ is Gaussian white noise, then $\mathbf{y}$ is considered to be sparse if only a small number of elements of $\mathbf{v}$ are nonzero. As Donoho and Johnstone point out [37], when the input

[2]The reader with some background in analysis may realize that if the integrand of the expected value in (6.14) is viewed as an $L^2(\mathbb{R})$ function, we could have written (6.26) immediately instead of (6.16).

vector is sparse, the coefficients that are essentially noise dominate the relatively small number of remaining terms.

We need some way to determine if an input vector $\mathbf{y}$ is sparse. Donoho and Johnstone [37] suggest computing the values

$$s = \frac{1}{N}\sum_{k=1}^{N}(y_k^2 - 1) \qquad \text{and} \qquad u = \frac{1}{\sqrt{N}}\left(\log_2(N)\right)^{3/2} \tag{6.27}$$

If $s \le u$, then $\mathbf{y}$ is sparse. In this case, we use λ^{univ} with the soft threshold rule (6.3). Otherwise, we use λ^{sure} with the soft threshold function.

Examples of SUREShrink

The following example illustrates an application of the SUREShrink method.

Example 6.5 (SUREShrink). *We return to the heavisine function* (6.1) *to create a vector* $\mathbf{v}$ *of* $n = 2048$ *samples. This vector and its noisy counterpart* $\mathbf{y}$ *(with* $\sigma = 0.15$*) are plotted in Figure 6.1.*

We compute five iterations of the D6 wavelet transformation $\mathbf{z}$ *of* $\mathbf{y}$ *and use the difference portion* $\mathbf{d}^1$ *and* (6.9) *to estimate* σ *by* $\widehat{\sigma} = 0.14517$.

Recall that in order to find λ^{sure}*, Theorem 6.2 maintains that the noise vector have variance* $\sigma^2 = 1$. *We thus use* $\mathbf{d}^1/widehat\sigma, \ldots, \mathbf{d}^5/\widehat{\sigma}$ *to find our thresholds. We use* (6.27) *to determine the sparseness of each of* $\mathbf{d}^1\widehat{\sigma}, \ldots, \mathbf{d}^5\widehat{\sigma}$. *If a highpass portion is found to be sparse, we compute* λ^{univ} *for it; otherwise we compute* λ^{sure}. *Table 6.3 summarizes our results and gives the appropriate thresholds.*

Table 6.3 Checking the sparseness of the highpass components and computing the appropriate threshold.

Iteration	s	u	Sparse	λ^{univ}	λ^{sure}
1	0.0480	0.9882	Yes	0.5629	
2	0.2239	1.1932	Yes	0.5340	
3	0.5623	1.4142	Yes	0.5035	
4	1.9357	1.6370	No		2.0260
5	8.9140	1.8371	No		1.3922

Figure 6.13 shows each of $\mathbf{d}^k$ *along with the horizontal lines at* $\pm\lambda_k$, $k = 1, \ldots, 5$. *Points between the lines are colored gray and will be converted to zero by the soft threshold rule. Values outside the region, plotted in black, will be shrunk by* λ_k.

We apply the shrinkage function (6.3) *to each* $\mathbf{d}^k$ *using* λ_k, $k = 1, \ldots, 5$. *The modified highpass portions* $\widehat{\mathbf{d}}^k$ *are then joined with the averages portion* $\mathbf{a}^5$ *to form the modified wavelet transformation* $\widehat{\mathbf{z}}$. *To obtain the denoised signal, we compute five iterations of the inverse D6 wavelet transformation. The result is plotted in*

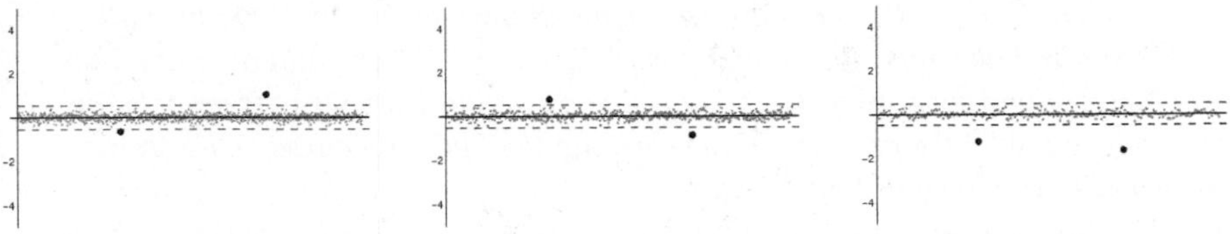

(a) $\mathbf{d}^1$, $\lambda_1 = 0.5629$ (b) $\mathbf{d}^2$, $\lambda_2 = 0.5340$ (c) $\mathbf{d}^3$, $\lambda_3 = 0.5035$

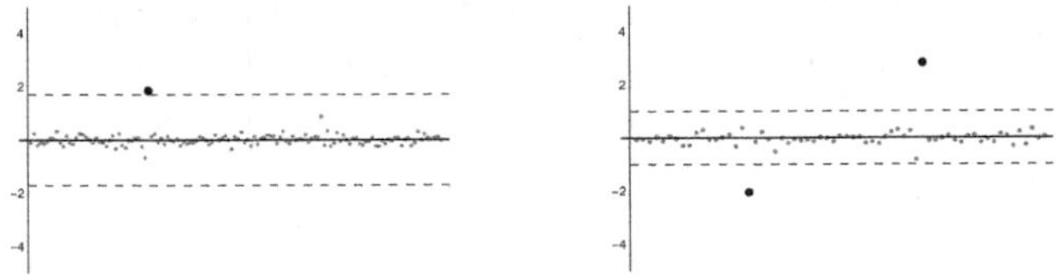

(d) $\mathbf{d}^4$, $\lambda_4 = 2.0260$ (e) $\mathbf{d}^5$, $\lambda_5 = 1.3922$

Figure 6.13 The highpass portions of the wavelet transformation along with the threshold regions.

Figure 6.14(a). For comparative purposes, we include the plot (Figure 6.5(b)) from Example 6.2 as Figure 6.14(b). The mean square errors for both denoising methods are listed as well.

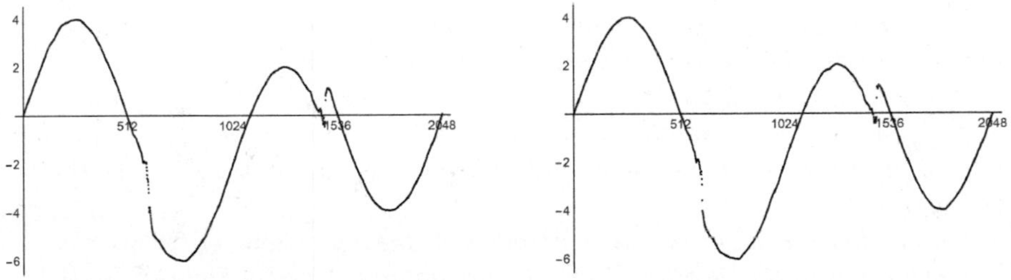

(a) SUREShrink, $\text{MSE}(\mathbf{v}, \widehat{\mathbf{v}}) = 0.0044$ (b) VisuShrink, $\text{MSE}(\mathbf{v}, \widehat{\mathbf{v}}) = 0.0031$

Figure 6.14 The denoised signals using SUREShrink and Visushrink.

(Live example: Visit `stthomas.edu/wavelets` *and click on Live Examples.)*

□

Our next example utilizes a "real" data set.

Example 6.6 (Denoising the Speed Readings of a Sailboat). *The real data in this example come from research done in the School of Engineering at the University of St. Thomas. Faculty members Christopher Greene, Michael Hennessey, and Jeff Jalkio and student Colin Sullivan were interested in determining the optimal path of a sailboat given factors such as changing wind speed, boat type, sail configura-*

tion, and current. They collected data from June 28 through July 1, 2004 by sailing preset courses on Lake Superior near Bayfield, Wisconsin. They gathered data from multiple sensors and then used these data in conjunction with ideas from calculus of variations to model the problem of determining the optimal course. Their results may be found in Hennessey et. al. [55].

The data, stored in vector $\mathbf{v}$ *and plotted in Figure 6.15, are boat speeds measured during one of their course runs. The speed was measured in knots* (1 *knot* ≈ 1.15 *mph) once per second. The total number of observations is* 4304*, so the experiment was conducted for* 71 *minutes and* 44 *seconds. The maximum speed measured during the experiment was* 7.7 *knots and the minimum speed was* 1.1 *knots.*

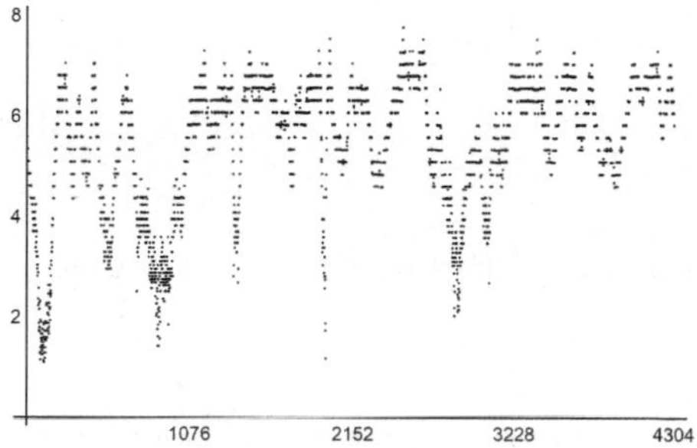

Figure 6.15 The boat speeds (recorded in knots at one second intervals).

Several factors can contribute to noise when attempting to record the boat speed. Factors such as sail trim, wind speed, steering, sail configuration, waves, and current can all affect the sensor's ability to record the boat speed accurately. It is important to point out that the noise we observe in this experiment cannot be classified as Gaussian white noise. Thus we cannot quantify how well wavelet shrinkage works on this example.

We compute four iterations (as many iterations as the data allow) of the wavelet transformation $\mathbf{z}$ *using the D4 filter.. This transformation is plotted in Figure 6.16(a). We use* $\mathbf{d}^1$*, the first iteration highpass portion, in conjunction with* (6.9) *to obtain the noise estimate* $\widehat{\sigma} = 0.179$*. We check the sparseness of each of* $\mathbf{d}^1, \ldots, \mathbf{d}^4$ *in order to determine how to compute the threshold tolerances. We find that* $\mathbf{d}^4$ *is not sparse so we compute* $\lambda_4^{sure} = 0.6882$ *for the threshold. The other portions are not sparse and use* (6.8) *to compute* $\lambda_1^{univ} = 0.7013, \lambda_2^{univ} = 0.6689$*, and* $\lambda_3^{univ} = 0.6882$*. We use these tolerances with the shrinkage function* (6.3) *to perform soft thresholding on each of the highpass portions to obtain* $\widehat{\mathbf{d}}^1, \ldots \widehat{\mathbf{d}}^4$*. The modified wavelet transformation* $\widehat{\mathbf{z}}$ *is plotted in Figure 6.16(b).*

Finally, we join the lowpass portion $\mathbf{a}^4$ *of the wavelet transformation* $\mathbf{z}$ *with the modified highpass portions* $\widehat{\mathbf{d}}^1, \ldots, \widehat{\mathbf{d}}^4$ *to create a "new" wavelet transformation* $\widehat{\mathbf{z}}$*. Four iterations of the inverse wavelet transformation of* $\widehat{\mathbf{z}}$ *produces a vector* $\widehat{\mathbf{v}}$ *containing the denoised boatspeeds. This vector is plotted in Figure 6.17.*

(Live example: Visit `stthomas.edu/wavelets` *and click on Live Examples.)*

□

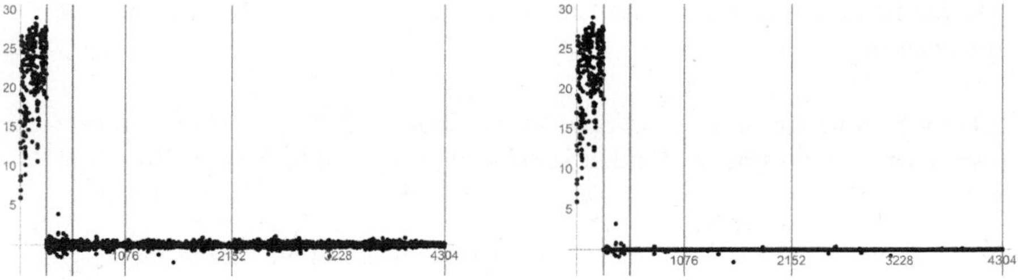

Figure 6.16 The wavelet transformation $\mathbf{z}$ of the boat speeds and the modified wavelet transformation $\widehat{\mathbf{z}}$.

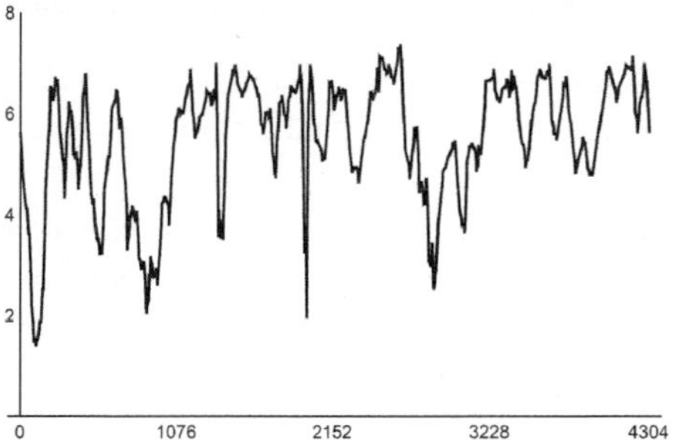

Figure 6.17 The denoised boat speeds.

PROBLEMS

6.10 Let $\mathbf{x} \in \mathbb{R}^N$. Write one iteration of the Haar wavelet transformation applied to $\mathbf{x}$ as a vector-valued function $\mathbf{f}(\mathbf{x})$. Identify the coordinate functions $f_1, \ldots, f_N$.

6.11 Let $\mathbf{w} = [0.2, -0.5, 0.3, -0.1, 0.2, -0.1]^T$. Find an expression for $f(\lambda)$ (see Example 6.4) and plot the result.

6.12 In this problem you will investigate the location of λ^{sure} for vectors of different lengths whose *distinct* elements are the same.

(a) Let $\mathbf{v} = [\,\frac{1}{2}, \frac{1}{2}\,]^T$. Find an expression for $f(\lambda)$ (see Example 6.4) and plot the result. Where does the minimum of $f(\lambda)$ occur?

(b) Repeat (a) with $\mathbf{w} = [\,\frac{1}{2}, 1, 1, 2\,]^T$.

(c) Can you create a vector $\mathbf{x}$ whose elements are selected from $1/2, 1, 2$ so that the minimum of $f(\lambda)$ occurs at $\lambda = 0$? (*Hint:* A tolerance of $\lambda = 0$ means that we are keeping all coefficients – that is, there is no noise to remove. What must a vector look like in this case?

(d) Can you create a vector $\mathbf{x}$ whose elements are selected from $1/2, 1, 2$ so that the minimum of $f(\lambda)$ occurs at $\lambda = 2$?

6.13 Let $\mathbf{v}$ be a vector of length N and let us denote (6.26) by $f_{\mathbf{v}}(\lambda)$ to indicate that it was constructed using $\mathbf{v}$. Further suppose that $|v_m|$ minimizes (6.26) where $1 \leq m \leq N$.

(a) Construct an N-vector $\mathbf{w}$ by permuting the elements of $\mathbf{v}$. Show that $|v_m|$ minimizes $f_{\mathbf{w}}(\lambda)$.

(b) Suppose that $\mathbf{x}$ is an N-vector whose entries are $\pm v_k$. Show that $|v_m|$ minimizes $f_{\mathbf{x}}(\lambda)$.

(c) *True or False:* $|v_m|$ minimizes $f_{\mathbf{y}}(\lambda)$ where $y_k = cv_k$ for $c \neq 0$. If the statement is true, prove it. If the statement is false, find a counterexample.

(d) *True or False:* $|v_m|$ minimizes $f_{\mathbf{z}}$ where $z_k = v_k + c$ for $c \neq 0$. If the statement is true, prove it. If the statement is false, find a counterexample.

6.14 In this problem you will derive a recursive formula that will allow you quickly to compute the values $f\left(|w_\ell|\right)$, $\ell = 2, \ldots, N$. For this problem, assume that $|w_1| \leq |w_2| \leq \cdots \leq |w_N|$.

(a) Using (6.26), show that

$$f\left(|w_\ell|\right) = N - 2\ell + \sum_{k=1}^{\ell} w_k^2 + (N - \ell)w_\ell^2. \tag{6.28}$$

(b) Replace ℓ by $\ell + 1$ in (a) to find an analogous expression for $f\left(|w_{\ell+1}|\right)$.

(c) Using (a) and (b), show that

$$f\left(|w_{\ell+1}|\right) = f\left(|w_\ell|\right) - 2 + (N - \ell)(w_{\ell+1}^2 - w_\ell^2).$$

To employ this recursion formula, we must first use (6.26) and compute $f\left(|w_1|\right) = N\left(1 + |w_1|^2\right)$. Then we can compute all the values $f(|w_\ell|)$, $\ell = 2, \ldots, N$ using the formula in (c). To find λ^{sure}, we simply pick the minimum of these values. The algorithm for computing λ^{sure} in the WaveLab software [38] is based on the formula in (a).

6.15 We create the vector $\mathbf{w}$ by adding Gaussian white noise with $\sigma = 1$ to the vector $\mathbf{v} = [0, 0.5, 1, 1.5, 2, 2.5, 3, 3.5]^T$. We have

$$\mathbf{w} = [-0.159354, -1.62378, 1.22804, 1.57117, 3.15511, 2.67804, 2.38617, 3.98616]^T.$$

Use (6.27) and a **CAS** or a calculator to determine if $\mathbf{w}$ is sparse.

6.16 Repeat Problem 6.15 using

$$\mathbf{w} = [\,0.0885254, 0.689566, -0.707012, 0.33514, 1.22295, \\ 2.68978, 0.831615, 1.16285\,]^T.$$

In this case, $\mathbf{w}$ was created by adding Gaussian white noise with $\sigma = 1$ to the vector $\mathbf{v} = [\,0, 0.25, 0.5, 0.75, 1, 1.25, 1.5, 1.75\,]^T$.

6.17 Yoon and Vaidyanathan [104] note that hard thresholding can "yield abrupt artifacts in the denoised signal." This is caused by the jump discontinuity in the threshold function $f_\lambda(t)$ (see Problem 6.5 in Section 6.1). Moreover, Bruce and Gao [14] argue that "soft shrink tends to have a bigger bias" because the large elements in the highpass portion are scaled toward zero. Some researchers have developed customized threshold functions for the purpose of alleviating the problems with hard and soft thresholding. In this problem you will study such a function.

(a) Bruce and Gao [13], introduced a *firm threshold function*. Let λ_1 and λ_2 be two positive numbers with $\lambda_1 < \lambda_2$. Define the firm threshold function by

$$f_{\lambda_1,\lambda_2}(t) = \begin{cases} 0, & |\,t\,| \le \lambda_1 \\ \operatorname{sgn}(t)\frac{\lambda_2(|\,t\,|-\lambda_1)}{(\lambda_2-\lambda_1)}, & \lambda_1 < |\,t\,| \le \lambda_2 \\ t, & |\,t\,| > \lambda_2 \end{cases}$$

Sketch $f_{\lambda_1,\lambda_2}(t)$ for the following threshold pairs: $\lambda_1 = 1/2, \lambda_2 = 1$, $\lambda_1 = 3/4, \lambda_2 = 1$, $\lambda_1 = .9, \lambda_2 = 1$, and $\lambda_1 = 1, \lambda_2 = 100$.

(b) Write $f_{\lambda_1,\lambda_2}(t)$ as a piecewise function that does not utilize $|\cdot|$ or sgn. How many pieces are there? What type of function is each piece?

(c) For which values of λ_1, λ_2 is $f_{\lambda_1,\lambda_2}(t)$ continuous? Continuously differentiable?

(d) What happens to $f_{\lambda_1,\lambda_2}(t)$ as λ_2 approaches infinity? (*Hint:* To see the result geometrically, use (b) to draw $f_{\lambda_1,\lambda_2}(t)$ with a fixed λ_1 and progressively larger values of λ_2. To find the result analytically, view $f_{\lambda_1,\lambda_2}(t)$ as a function of λ_2 and use L'Hospital's rule.)

(e) What happens to $f_{\lambda_1,\lambda_2}(t)$ as λ_1 approaches λ_2? (*Hint:* Use the hint from (c) and also consider what happens to $|t|$ as $\lambda_1 \to \lambda_2$.

6.18 The *customized soft threshold function* developed in Yoon and Vaidyanathan [104] is related to the firm threshold function defined in Problem 6.17. Let λ, γ, and α be real numbers such that $0 < \gamma < \lambda$, and $0 \le \alpha \le 1$. The customized shrinkage function is then defined by

$$f_{\lambda,\gamma,\alpha}(t) = \begin{cases} t - \operatorname{sgn}(t)(1-\alpha)\lambda, & |t| \ge \lambda \\ 0, & |t| \le \gamma \\ \operatorname{sgn}(t)\,\alpha\left(\frac{|t|-\gamma}{\lambda-\gamma}\right)^2\left((\alpha-3)\left(\frac{|t|-\gamma}{\lambda-\gamma}\right)+4-\alpha\right), & \text{otherwise.} \end{cases} \tag{6.29}$$

(a) Plot $f_{1,\frac{1}{2},\alpha}(t)$ for $\alpha = 0, \frac{1}{4}, \frac{1}{2}, \frac{3}{4}, 1$.

(b) Write $f_{\lambda,\gamma,\alpha}(t)$ as a piecewise-defined function that does not utilize $|\cdot|$ and sgn. How many pieces are there? What type of function is each piece?

(c) Show that $\lim\limits_{\alpha\to 0} f_{\lambda,\gamma,\alpha}(t)$ is the soft threshold function (6.3).

(d) What is $\lim\limits_{\gamma\to\lambda} f_{\lambda,\gamma,1}(t)$? The hints in Problem 6.17 will be useful for computing this limit.

(e) For which values of λ, γ, and α is $f_{\lambda,\gamma,\alpha}(t)$ continuous? Continuously differentiable? Compare your results to (c) of Problem 6.17.

6.19 Consider the soft threshold function $s_\lambda(t)$ defined in (6.3) and the hard threshold function $f_\lambda(t)$ given in Problem 6.5 from Section 6.1.

(a) Plot the following linear combinations

(i) $\frac{1}{3}f_\lambda(t) + \frac{2}{3}s_\lambda(t)$

(ii) $\frac{1}{4}f_\lambda(t) + \frac{3}{4}s_\lambda(t)$

(iii) $\frac{3}{4}f_\lambda(t) + \frac{1}{4}s_\lambda(t)$

(b) Find a piecewise representation of the function

$$\alpha f_\lambda(t) + (1-\alpha)s_\lambda(t).$$

(c) Use (b) to show that for $|t| > \lambda$ and $|t| \le \gamma$, $f_{\lambda,\gamma,\alpha}(t) = \alpha f_\lambda(t) + (1-\alpha)s_\lambda(t)$. In this way we see that the customized threshold function is a linear combination[3] of hard and soft threshold functions except on the interval from $(\gamma,\lambda]$. On this interval, the quadratic piece given in (6.29) serves to connect the other two pieces smoothly.

Computer Labs

6.3 Constructing λ^{sure}. From `stthomas.edu/wavelets`, access the development package `lambdaSURE`. In this development lab you will write a module that implements Donoho's scheme (see (6.28) in Problem 6.14) for computing λ^{sure}.

6.4 Denoising with SUREShrink. From `stthomas.edu/wavelets`, access the file `SUREShrink`. In this lab you will use the wavelet shrinkage module with λ^{sure} to apply the SureShrink method to various audio signals and digital images. You will also experiment with the customized threshold functions described in Problems 6.17 and 6.18.

[3]The linear combination here is special; it is also known as a *convex combination* of the functions $f_\lambda(t)$ and $s_\lambda(t)$ since the multipliers α and $1-\alpha$ are nonnegative and sum to 1. You may have seen convex combinations in analytic geometry – given two points $\mathbf{a}$ and $\mathbf{b}$, the line segment connecting the points is given by $t\mathbf{a} + (1-t)\mathbf{b}$ for $0 \le t \le 1$.

CHAPTER 7

BIORTHOGONAL WAVELET TRANSFORMATIONS

The Daubechies orthogonal filters constructed in Chapter 5 have proven useful in applications such as image compression, signal denoising and image segmentation. The transformation matrix constructed from a filter is orthogonal and thus easily invertible. The orthogonality and sparse structure of the matrix facilitates the creation of fast and efficient algorithms for applying the transformation and its inverse. The matrix has nice structure. The top half of the matrix is constructed from a short even-length filter $\mathbf{h}$ that behaves like a lowpass filter and returns an approximation of the input data. The bottom half is also constructed from a filter $\mathbf{g}$ that is built from $\mathbf{h}$ and behaves like a highpass filter annihilating constant and, as the construction allows, polynomial data.

There are some issues with the orthogonal filters. For image compression, it is desirable that integer-valued intensities are mapped to integers in order to efficiently perform Huffman coding. The Haar filter $\mathbf{h} = \left(\frac{1}{2}, \frac{1}{2}\right)$ at least mapped integers to half-integers, but the orthogonal filters we constructed (the orthogonal Haar, D4, and D6) were comprised of irrational numbers. It turns out Daubechies filters of longer length are built entirely from irrational numbers as well. Another disadvantage was the wrapping rows that arose in our construction of orthogonal transformation matrices built from filters of length 4 or greater.

Discrete Wavelet Transformations: An Elementary Approach With Applications, Second Edition. Patrick J. Van Fleet.

Researchers have long known that *symmetric filters* work best in applications such as image compression. As we learn in Section 7.3, symmetric filters can negate the effect of wrapping rows in the transformation matrix. The (orthogonal) Haar filter is symmetric since the first half of the filter (albeit a single value) is a mirror reflection of the last half of the filter. However, Daubechies [30] proved that no other member of her orthogonal family is a symmetric filter.

Unfortunately we cannot obtain all the desired qualities (lowpass/highpass filters, efficient computation, orthogonality, and symmetry) with one filter. Of these properties, orthogonality is the one that might be considered the most "artificial." Constructing an orthogonal matrix provides an immediate representation of the inverse and given that the original transformation matrix is sparse, the inverse will be as well. But once we compute an inverse, we never need to derive it again. Suppose we could construct *non-orthogonal* W_N from a symmetric filter that retained the sparse lowpass/highpass structure, the ability in the highpass portion of the transformation to annihilate constant or polynomial data and possessed an inverse that is the transpose of matrix $\widetilde{W}_N$ that obeys the same properties? The only difference would be that $\widetilde{W}_N$ was constructed from a *different* symmetric filter.

Our objective in this chapter is to construct a *biorthogonal filter pair* $\left(\mathbf{h}, \widetilde{\mathbf{h}}\right)$ that can be used to generate wavelet transformation matrices W_N and $\widetilde{W}_N$ with the property that $W_N^{-1} = \widetilde{W}_N^T$. Our construction will be ad hoc in nature and intractable with regards to constructing biorthogonal filter pairs whose lengths are relatively long. On the other hand, the construction does allow us to see the advantages of biorthogonal filter pairs in applications. This chapter will thus be a transitional one. In the next two sections, we construct *short* biorthogonal filter pairs. In Section 7.3, we show how to exploit the symmetry of the biorthogonal filters to offset the issues caused by wrapping rows in the wavelet transformation matrix. The last section contains applications of the biorthogonal wavelet transformation to image compression and image pansharpening.

We are not yet able to solve the problem of mapping integers to integers but in Chapters 8 – 9 we will construct (bi-)orthogonal filters in a more systematic (and mathematically elegant) way. The mathematics we learn in these chapters leads us to Chapter 11 which illustrates a method for mapping the output of wavelet transformations of integer-valued data to integers.

7.1 The $(5, 3)$ Biorthogonal Spline Filter Pair

In this section, we construct a *biorthogonal filter pair* of lengths 5 and 3. The construction is quite ad hoc and draws on some ideas from Section 5.1 and the discussion at the beginning of this chapter.

Symmetric Filters Defined

Since all filters we consider are of finite length, we need to clarify what we mean by a symmetric filter. In order to state this definition, we need to extend the range of indices on our filter elements. In the case when the filter length is odd, we require our filter to be symmetric about index 0 and when the filter length is even, we ask that our filters be symmetric about (nonexistent) index $\frac{1}{2}$. We have the following definition.

Definition 7.1 (Symmetric Filters). *Suppose $\ell, L \in \mathbb{Z}$ with $\ell \leq 0$ and $L > 0$. Let $\mathbf{h} = (h_\ell, \ldots, h_L)$ be a finite-length filter with length $N = L - \ell + 1$. We say that $\mathbf{h}$ is* symmetric *if*

(a) $h_k = h_{-k}$ for all $k \in \mathbb{Z}$ whenever N is odd,

(b) $h_k = h_{1-k}$ for all $k \in \mathbb{Z}$ whenever N is even.

□

Note that the orthogonal Haar filter $\mathbf{h} = (h_0, h_1) = \left(\frac{\sqrt{2}}{2}, \frac{\sqrt{2}}{2}\right)$ is a symmetric filter satisfying $h_k = h_{1-k}$ for $k = 0, 1$. An example of a symmetric filter of odd length is $\mathbf{h} = (h_{-1}, h_0, h_1) = (1, 2, 1)$.

It is clear that symmetry imposes some restrictions on ℓ and L in the case where $\mathbf{h} = (h_\ell, \ldots, h_L)$. We have the following proposition.

Proposition 7.1 (Starting and Stopping Indices for Symmetric FIR Filters). *Suppose that $\mathbf{a} = (a_\ell, \ldots, a_L)$ is a symmetric finite-length filter. Let $N = L - \ell + 1$ denote the length of $\mathbf{a}$. If N is odd, then $\ell = -L$. If N is even, then $\ell = -L + 1$.* □

Proof. The proof is left as Problem 7.1. □

Convention. We often refer to the *length* of symmetric filter $\mathbf{a}$. If $\mathbf{a}$ has even length $2L$, then using Proposition 7.1, we mean that $a_L\ (= a_{-L+1}) \neq 0$ and if $\mathbf{a}$ has odd length $2L + 1$, then $a_L\ (= a_{-L}) \neq 0$.

Constructing a Symmetric Lowpass Filter of Length 3.

It is clear that a two-term symmetric filter must be a multiple of the Haar filter. In Problem 7.12, you will construct biorthogonal filter pairs $\left(\widetilde{\mathbf{h}}, \mathbf{h}\right)$ where $\widetilde{\mathbf{h}} = \left(\widetilde{h}_0, \widetilde{h}_1\right) = \left(\frac{\sqrt{2}}{2}, \frac{\sqrt{2}}{2}\right)$.

Our construction starts with $\widetilde{\mathbf{h}} = \left(\widetilde{h}_{-1}, \widetilde{h}_0, \widetilde{h}_1\right)$ and since $\widetilde{\mathbf{h}}$ is symmetric, we know that $\widetilde{h}_{-1} = \widetilde{h}_1$. Since we want $\widetilde{\mathbf{h}}$ to process data as a lowpass filter, we use

Definition 2.19 to write

$$\sum_{k=-1}^{1} \widetilde{h}_k \cdot 1 = c \qquad \text{or} \qquad \widetilde{h}_{-1} + \widetilde{h}_0 + \widetilde{h}_1 = \widetilde{h}_0 + 2\widetilde{h}_1 = c \tag{7.1}$$

and

$$\sum_{k=-1}^{1} \widetilde{h}_k (-1)^{n-k} = (-1)^n \left(-\widetilde{h}_{-1} + \widetilde{h}_0 - \widetilde{h}_1 \right) = 0 \quad \text{or} \quad \widetilde{h}_0 - 2\widetilde{h}_1 = 0. \tag{7.2}$$

Recall that the elements of the orthogonal filters in Chapter 5 summed to $\sqrt{2}$. For comparative purposes, we choose $c = \sqrt{2}$ in (7.1). In this case, adding (7.1) and (7.2) gives $2\widetilde{h}_0 = \sqrt{2}$ or $\widetilde{h}_0 = \frac{\sqrt{2}}{2}$. Plugging this value into (7.2) gives $\widetilde{h}_{-1} = \widetilde{h}_1 = \frac{\sqrt{2}}{4}$. So our three-term filter is

$$\boxed{\widetilde{\mathbf{h}} = \left(\widetilde{h}_{-1}, \widetilde{h}_0, \widetilde{h}_1 \right) = \left(\widetilde{h}_1, \widetilde{h}_0, \widetilde{h}_1 \right) = \left(\frac{\sqrt{2}}{4}, \frac{\sqrt{2}}{2}, \frac{\sqrt{2}}{4} \right).}$$

Constructing $\widetilde{\mathbf{W}}_N$

Now that we have $\widetilde{\mathbf{h}}$, we can start the construction of $\widetilde{W}_N$. We want $\widetilde{W}_N$ to have the averages/differences blocks so we start with

$$\widetilde{W}_N = \left[\begin{array}{c} \widetilde{H}_{N/2} \\ \hline \widetilde{G}_{N/2} \end{array} \right].$$

We don't know $\widetilde{\mathbf{g}}$ so we can't construct $\widetilde{G}_{N/2}$ but we can build $\widetilde{H}_{N/2}$. For the Daubechies filters $(h_0, \ldots, h_L)$, we reversed the order of the filter elements and created the first row of the transformation matrix by padding enough zeros onto $h_L, \ldots, h_0$ to form a row of length N. Subsequent rows were created by cyclically two-shifting the previous row.

Since $\widetilde{\mathbf{h}}$ is symmetric, there is no need to reverse the order of its elements. We also now have a negative index. These observations lead to some changes with regards to defining the wavelet transformation matrix.

Note. We adopt the convention that $\widetilde{h}_0$ will be placed in the $(1, 1)$ position of the matrix $\widetilde{H}_{N/2}$. Elements with positive subscripts will follow and those with negative subscripts will be wrapped at the end of the row.

We illustrate throughout the remainder of the section with $N = 8$.

$$\widetilde{H}_4 = \begin{bmatrix} \widetilde{h}_0 & \widetilde{h}_1 & 0 & 0 & 0 & 0 & 0 & \widetilde{h}_{-1} \\ 0 & \widetilde{h}_{-1} & \widetilde{h}_0 & \widetilde{h}_1 & 0 & 0 & 0 & 0 \\ 0 & 0 & 0 & \widetilde{h}_{-1} & \widetilde{h}_0 & \widetilde{h}_1 & 0 & 0 \\ 0 & 0 & 0 & 0 & 0 & \widetilde{h}_{-1} & \widetilde{h}_0 & \widetilde{h}_1 \end{bmatrix}$$

$$= \begin{bmatrix} \widetilde{h}_0 & \widetilde{h}_1 & 0 & 0 & 0 & 0 & 0 & \widetilde{h}_1 \\ 0 & \widetilde{h}_1 & \widetilde{h}_0 & \widetilde{h}_1 & 0 & 0 & 0 & 0 \\ 0 & 0 & 0 & \widetilde{h}_1 & \widetilde{h}_0 & \widetilde{h}_1 & 0 & 0 \\ 0 & 0 & 0 & 0 & 0 & \widetilde{h}_1 & \widetilde{h}_0 & \widetilde{h}_1 \end{bmatrix}$$

$$= \begin{bmatrix} \frac{\sqrt{2}}{2} & \frac{\sqrt{2}}{4} & 0 & 0 & 0 & 0 & 0 & \frac{\sqrt{2}}{4} \\ 0 & \frac{\sqrt{2}}{4} & \frac{\sqrt{2}}{2} & \frac{\sqrt{2}}{4} & 0 & 0 & 0 & 0 \\ 0 & 0 & 0 & \frac{\sqrt{2}}{4} & \frac{\sqrt{2}}{2} & \frac{\sqrt{2}}{4} & 0 & 0 \\ 0 & 0 & 0 & 0 & 0 & \frac{\sqrt{2}}{4} & \frac{\sqrt{2}}{2} & \frac{\sqrt{2}}{4} \end{bmatrix}. \tag{7.3}$$

Finding the Filter $\mathbf{h}$

Now that we have $\widetilde{\mathbf{h}}$, we construct the companion lowpass filter $\mathbf{h}$. We start by insisting that the elements of $\mathbf{h}$ satisfy conditions similar to (7.1) and (7.2). That is, the elements of $\mathbf{h}$ should satisfy

$$\sum_k h_k = \sqrt{2} \qquad \text{and} \qquad \sum_k (-1)^k h_k = 0. \tag{7.4}$$

To determine feasible filters $\mathbf{h}$, we consider the orthogonality conditions that exist between $\widetilde{W}_8$ and W_8. We require $\widetilde{W}_8^{-1} = W_8^T$ so that
$\widetilde{W}_8 W_8^T = I_8$ and in block form we have

$$\widetilde{W}_8 W_8^T = \left[\begin{array}{c} \widetilde{H}_4 \\ \hline \widetilde{G}_4 \end{array}\right] \cdot \left[\begin{array}{c|c} H_4^T & G_4^T \end{array}\right] = \left[\begin{array}{c|c} \widetilde{H}_4 H_4^T & \widetilde{H}_4 G_4^T \\ \hline \widetilde{G}_4 H_4^T & \widetilde{G}_4 G_4^T \end{array}\right] = \left[\begin{array}{c|c} I_4 & 0_4 \\ \hline 0_4 & I_4 \end{array}\right] \tag{7.5}$$

so in particular, $\widetilde{H}_4 H_4^T = I_4$. We seek to create the shortest filter $\mathbf{h}$ possible. If we take $\mathbf{h} = (h_0, h_1) = (a, a)$, then

$$H_4 = \begin{bmatrix} a & a & 0 & 0 & 0 & 0 & 0 & 0 \\ 0 & 0 & a & a & 0 & 0 & 0 & 0 \\ 0 & 0 & 0 & 0 & a & a & 0 & 0 \\ 0 & 0 & 0 & 0 & 0 & 0 & a & a \end{bmatrix}$$

and it is easily verified that

$$I_4 = \begin{bmatrix} \frac{\sqrt{2}}{2} & \frac{\sqrt{2}}{4} & 0 & 0 & 0 & 0 & 0 & \frac{\sqrt{2}}{4} \\ 0 & \frac{\sqrt{2}}{4} & \frac{\sqrt{2}}{2} & \frac{\sqrt{2}}{4} & 0 & 0 & 0 & 0 \\ 0 & 0 & 0 & \frac{\sqrt{2}}{4} & \frac{\sqrt{2}}{2} & \frac{\sqrt{2}}{4} & 0 & 0 \\ 0 & 0 & 0 & 0 & 0 & \frac{\sqrt{2}}{4} & \frac{\sqrt{2}}{2} & \frac{\sqrt{2}}{4} \end{bmatrix} \cdot \begin{bmatrix} a & 0 & 0 & 0 \\ a & 0 & 0 & 0 \\ 0 & a & 0 & 0 \\ 0 & a & 0 & 0 \\ 0 & 0 & a & 0 \\ 0 & 0 & a & 0 \\ 0 & 0 & 0 & a \\ 0 & 0 & 0 & a \end{bmatrix}$$

$$= \frac{\sqrt{2}a}{4} \begin{bmatrix} 3 & 0 & 0 & 1 \\ 1 & 3 & 0 & 0 \\ 0 & 1 & 3 & 0 \\ 0 & 0 & 1 & 3 \end{bmatrix}.$$

The matrix above can never be I_4 so the length two symmetric filter does not work. In Problem 7.2 you will show in order for $\mathbf{h} = (h_{-1}, h_0, h_1)$ to generate matrix H_4 that satisfies $\widetilde{H}_4 H_4^T = I_4$, we must have $h_{-1} = h_1 = 0$ and $h_0 = \sqrt{2}$. This filter satisfies the first equation in (7.4) but not the second one. In Problem 7.3, you will show that symmetric $\mathbf{h} = (h_{-1}, h_0, h_1, h_2) = (a, b, b, a)$ cannot produce a matrix H_4 such that $\widetilde{H}_4 H_4^T = I_4$. Thus $\mathbf{h}$ cannot be of length 4.

We consider then the length 5 symmetric filter

$$\mathbf{h} = (h_{-2}, h_{-1}, h_0, h_1, h_2) = (h_2, h_1, h_0, h_1, h_2)$$

and create

$$H_4 = \begin{bmatrix} h_0 & h_1 & h_2 & 0 & 0 & 0 & h_2 & h_1 \\ h_2 & h_1 & h_0 & h_1 & h_2 & 0 & 0 & 0 \\ 0 & 0 & h_2 & h_1 & h_0 & h_1 & h_2 & 0 \\ h_2 & 0 & 0 & 0 & h_2 & h_1 & h_0 & h_1 \end{bmatrix}.$$

We next compute

$$I_4 = \widetilde{H}_4 H_4^T$$

$$= \begin{bmatrix} \frac{\sqrt{2}}{2} & \frac{\sqrt{2}}{4} & 0 & 0 & 0 & 0 & 0 & \frac{\sqrt{2}}{4} \\ 0 & \frac{\sqrt{2}}{4} & \frac{\sqrt{2}}{2} & \frac{\sqrt{2}}{4} & 0 & 0 & 0 & 0 \\ 0 & 0 & 0 & \frac{\sqrt{2}}{4} & \frac{\sqrt{2}}{2} & \frac{\sqrt{2}}{4} & 0 & 0 \\ 0 & 0 & 0 & 0 & 0 & \frac{\sqrt{2}}{4} & \frac{\sqrt{2}}{2} & \frac{\sqrt{2}}{4} \end{bmatrix} \cdot \begin{bmatrix} h_0 & h_2 & 0 & h_2 \\ h_1 & h_1 & 0 & 0 \\ h_2 & h_0 & h_2 & 0 \\ 0 & h_1 & h_1 & 0 \\ 0 & h_2 & h_0 & h_2 \\ 0 & 0 & h_1 & h_1 \\ h_2 & 0 & h_2 & h_0 \\ h_1 & 0 & 0 & h_1 \end{bmatrix}$$

$$= \frac{\sqrt{2}}{2} \begin{bmatrix} h_0 + h_1 & \frac{h_1+2h_2}{2} & 0 & \frac{h_1+2h_2}{2} \\ \frac{h_1+2h_2}{2} & h_0 + h_1 & \frac{h_1+2h_2}{2} & 0 \\ 0 & \frac{h_1+2h_2}{2} & h_0 + h_1 & \frac{h_1+2h_2}{2} \\ \frac{h_1+2h_2}{2} & 0 & \frac{h_1+2h_2}{2} & h_0 + h_1 \end{bmatrix}.$$

The last matrix above leads to the two equations

$$\begin{aligned} h_0 + h_1 \qquad &= \sqrt{2} \\ h_1 + 2h_2 &= 0. \end{aligned} \tag{7.6}$$

We need one more equation to produce a square system. If we add the equations in (7.6) together, we obtain the first equation in (7.4), so we plug our filter into the second equation of (7.4) to obtain

$$h_0 - 2h_1 + 2h_2 = 0. \tag{7.7}$$

Solving (7.6) together with (7.7) (see Problem 7.4) gives (see Problem 7.4) gives

$$\boxed{\mathbf{h} = (h_{-2}, h_{-1}, h_0, h_1, h_2) = \left(-\frac{\sqrt{2}}{8}, \frac{\sqrt{2}}{4}, \frac{3\sqrt{2}}{4}, \frac{\sqrt{2}}{4}, -\frac{\sqrt{2}}{8}\right).}$$

Thus

$$H_4 = \begin{bmatrix} \frac{3\sqrt{2}}{4} & \frac{\sqrt{2}}{4} & -\frac{\sqrt{2}}{8} & 0 & 0 & 0 & -\frac{\sqrt{2}}{8} & \frac{\sqrt{2}}{4} \\ -\frac{\sqrt{2}}{8} & \frac{\sqrt{2}}{4} & \frac{3\sqrt{2}}{4} & \frac{\sqrt{2}}{4} & -\frac{\sqrt{2}}{8} & 0 & 0 & 0 \\ 0 & 0 & -\frac{\sqrt{2}}{8} & \frac{\sqrt{2}}{4} & \frac{3\sqrt{2}}{4} & \frac{\sqrt{2}}{4} & -\frac{\sqrt{2}}{8} & 0 \\ -\frac{\sqrt{2}}{8} & 0 & 0 & 0 & -\frac{\sqrt{2}}{8} & \frac{\sqrt{2}}{4} & \frac{3\sqrt{2}}{4} & \frac{\sqrt{2}}{4} \end{bmatrix}. \tag{7.8}$$

Finding the Filters $\widetilde{\mathrm{g}}$ and g

To find the highpass filters $\widetilde{\mathbf{g}}$ and $\mathbf{g}$, we return to the block representations of $\widetilde{W}_8$ and W_8 given in (7.5). We need filters $\widetilde{\mathbf{g}}$ and $\mathbf{g}$ that generate matrices $\widetilde{G}_4$ and G_4, respectively, that satisfy $\widetilde{H}_4 G_4^T = \widetilde{G}_4 H_4^T = 0_4$ and $\widetilde{G}_4 G_4^T = I_4$.

Let's start with the upper right hand block $\widetilde{H}_4 G_4^T = 0_4$. This identity says that each row of $\widetilde{H}_4$ must be orthogonal to each row of G_4. The nonzero elements in a typical row of $\widetilde{H}_4$ are the values $\frac{\sqrt{2}}{4}, \frac{\sqrt{2}}{2}, \frac{\sqrt{2}}{4}$. The process that worked in Chapter 5 for constructing the highpass filter was to reverse the order of elements in the lowpass filter and then multiply each by alternating ± 1. The multiplication of elements by ± 1 creates a highpass filter and since the length of the lowpass filter is even, Problem 2.11 in Section 2.1 ensures that the filters are orthogonal. Mimicking this process here does not quite work. That is, if we set $\mathbf{g} = (g_{-1}, g_0, g_1) = \left(\frac{\sqrt{2}}{4}, -\frac{\sqrt{2}}{2}, \frac{\sqrt{2}}{4}\right)$ and compute $\widetilde{\mathbf{h}} \cdot \mathbf{g}$, we find the value of the inner product is $-\frac{1}{4} \neq 0$.

Notice, however if we were to slide the entries $\mathbf{g}$ left (or right) by one, then the filters are orthogonal. That is, we take $\mathbf{g} = (g_0, g_1, g_2) = \left(\frac{\sqrt{2}}{4}, -\frac{\sqrt{2}}{2}, \frac{\sqrt{2}}{4}\right)$ and compute $\widetilde{\mathbf{h}} \cdot \mathbf{g}$ we obtain

$$\widetilde{\mathbf{h}} \cdot \mathbf{g} = \widetilde{h}_0 g_0 + \widetilde{h}_1 g_1 = \frac{\sqrt{2}}{2} \cdot \frac{\sqrt{2}}{4} + \frac{\sqrt{2}}{4}\left(-\frac{\sqrt{2}}{2}\right) = 0.$$

Thus we construct G_4 as

$$G_4 = \begin{bmatrix} g_0 & g_1 & g_2 & 0 & 0 & 0 & 0 & 0 \\ 0 & 0 & g_0 & g_1 & g_2 & 0 & 0 & 0 \\ 0 & 0 & 0 & 0 & g_0 & g_1 & g_2 & 0 \\ g_2 & 0 & 0 & 0 & 0 & 0 & g_0 & g_1 \end{bmatrix}$$

$$= \begin{bmatrix} \frac{\sqrt{2}}{4} & -\frac{\sqrt{2}}{2} & \frac{\sqrt{2}}{4} & 0 & 0 & 0 & 0 & 0 \\ 0 & 0 & \frac{\sqrt{2}}{4} & -\frac{\sqrt{2}}{2} & \frac{\sqrt{2}}{4} & 0 & 0 & 0 \\ 0 & 0 & 0 & 0 & \frac{\sqrt{2}}{4} & -\frac{\sqrt{2}}{2} & \frac{\sqrt{2}}{4} & 0 \\ \frac{\sqrt{2}}{4} & 0 & 0 & 0 & 0 & 0 & \frac{\sqrt{2}}{4} & -\frac{\sqrt{2}}{2} \end{bmatrix} \tag{7.9}$$

and easily verify that $\widetilde{H}_4 G_4^T = 0_4$ (see Problem 7.5). To summarize, we created the elements of $\mathbf{g}$ by reversing the elements of $\widetilde{\mathbf{h}}$, translating them, and alternating their signs. Mathematically, we have

$$g_k = (-1)^k \widetilde{h}_{1-k}. \tag{7.10}$$

Given this definition of the filter $\mathbf{g}$ and the fact that $\widetilde{\mathbf{h}}$ is a lowpass filter, Proposition 2.4 tells us that $\mathbf{g}$ is a highpass filter.

We try the same "trick" to construct $\widetilde{\mathbf{g}}$. That is, we take $\widetilde{g}_k = (-1)^k h_{1-k}$, for $k = -1, \ldots, 3$. These indices are the only ones for which $h_{1-k} \neq 0$. We have

$$\begin{aligned} \widetilde{\mathbf{g}} &= (\widetilde{g}_{-1}, \widetilde{g}_0, \widetilde{g}_1, \widetilde{g}_2, \widetilde{g}_3) \\ &= (-h_2, h_1, -h_0, h_1, -h_2) \\ &= \left(\frac{\sqrt{2}}{8}, \frac{\sqrt{2}}{4}, -\frac{3\sqrt{2}}{4}, \frac{\sqrt{2}}{4}, \frac{\sqrt{2}}{8} \right). \end{aligned}$$

We use the fact that $\mathbf{h}$ is lowpass in conjunction with Proposition 2.4 to infer that $\widetilde{\mathbf{g}}$ is a highpass filter. We compute, considering only terms where both h_k, $\widetilde{g}$ are nonzero,

$$\mathbf{h} \cdot \widetilde{\mathbf{g}} = \sum_{k=-1}^{2} h_k \widetilde{g}_k = \sum_{k=-1}^{2} h_k (-1)^k h_{1-k} = -h_{-1}h_2 + h_0 h_1 - h_0 h_1 + h_2 h_{-1} = 0.$$

Next we construct

$$\widetilde{G}_4 = \begin{bmatrix} \widetilde{g}_0 & \widetilde{g}_1 & \widetilde{g}_2 & \widetilde{g}_3 & 0 & 0 & 0 & \widetilde{g}_{-1} \\ 0 & \widetilde{g}_{-1} & \widetilde{g}_0 & \widetilde{g}_1 & \widetilde{g}_2 & \widetilde{g}_3 & 0 & 0 \\ 0 & 0 & 0 & \widetilde{g}_{-1} & \widetilde{g}_0 & \widetilde{g}_1 & \widetilde{g}_2 & \widetilde{g}_3 \\ \widetilde{g}_2 & \widetilde{g}_3 & 0 & 0 & 0 & \widetilde{g}_{-1} & \widetilde{g}_0 & \widetilde{g}_1 \end{bmatrix}$$
$$= \begin{bmatrix} h_1 & -h_0 & h_1 & -h_2 & 0 & 0 & 0 & -h_2 \\ 0 & -h_2 & h_1 & -h_0 & h_1 & -h_2 & 0 & 0 \\ 0 & 0 & 0 & -h_2 & h_1 & -h_0 & h_1 & -h_2 \\ h_1 & -h_2 & 0 & 0 & 0 & -h_2 & h_1 & -h_0 \end{bmatrix}$$
$$= \begin{bmatrix} \frac{\sqrt{2}}{4} & -\frac{3\sqrt{2}}{4} & \frac{\sqrt{2}}{4} & \frac{\sqrt{2}}{8} & 0 & 0 & 0 & \frac{\sqrt{2}}{8} \\ 0 & \frac{\sqrt{2}}{8} & \frac{\sqrt{2}}{4} & -\frac{3\sqrt{2}}{4} & \frac{\sqrt{2}}{4} & \frac{\sqrt{2}}{8} & 0 & 0 \\ 0 & 0 & 0 & \frac{\sqrt{2}}{8} & \frac{\sqrt{2}}{4} & -\frac{3\sqrt{2}}{4} & \frac{\sqrt{2}}{4} & \frac{\sqrt{2}}{8} \\ \frac{\sqrt{2}}{4} & \frac{\sqrt{2}}{8} & 0 & 0 & 0 & \frac{\sqrt{2}}{8} & \frac{\sqrt{2}}{4} & -\frac{3\sqrt{2}}{4} \end{bmatrix}. \tag{7.11}$$

In Problem 7.6, you will show that $\widetilde{G}_4 H_4^T = 0_4$ and by direct computation, we have

$$\widetilde{G}_4 G_4^T = \begin{bmatrix} \widetilde{g}_0 & \widetilde{g}_1 & \widetilde{g}_2 & \widetilde{g}_3 & 0 & 0 & 0 & \widetilde{g}_{-1} \\ 0 & \widetilde{g}_{-1} & \widetilde{g}_0 & \widetilde{g}_1 & \widetilde{g}_2 & \widetilde{g}_3 & 0 & 0 \\ 0 & 0 & 0 & \widetilde{g}_{-1} & \widetilde{g}_0 & \widetilde{g}_1 & \widetilde{g}_2 & \widetilde{g}_3 \\ \widetilde{g}_2 & \widetilde{g}_3 & 0 & 0 & 0 & \widetilde{g}_{-1} & \widetilde{g}_0 & \widetilde{g}_1 \end{bmatrix} \cdot \begin{bmatrix} g_0 & 0 & 0 & g_2 \\ g_1 & 0 & 0 & 0 \\ g_2 & g_0 & 0 & 0 \\ 0 & g_1 & 0 & 0 \\ 0 & g_2 & g_0 & 0 \\ 0 & 0 & g_1 & 0 \\ 0 & 0 & g_2 & g_0 \\ 0 & 0 & 0 & g_1 \end{bmatrix}$$

$$= \begin{bmatrix} a & b & 0 & c \\ c & a & b & 0 \\ 0 & c & a & b \\ b & 0 & c & a \end{bmatrix}$$

where

$$\begin{aligned} a &= \widetilde{g}_0 g_0 + \widetilde{g}_1 g_1 + \widetilde{g}_2 g_2 \\ &= h_1 \widetilde{h}_1 + (-h_0) \cdot \left(-\widetilde{h}_0\right) + h_1 \widetilde{h}_1 \\ &= \widetilde{h}_0 h_0 + 2\widetilde{h}_1 h_1 \\ &= \frac{3\sqrt{2}}{4} \cdot \frac{\sqrt{2}}{2} + 2\frac{\sqrt{2}}{4} \cdot \frac{\sqrt{2}}{4} = 1, \end{aligned}$$

$$b = \widetilde{g}_2 g_0 + \widetilde{g}_3 g_1 = h_1 \widetilde{h}_1 + h_2 \widetilde{h}_0 = \frac{\sqrt{2}}{4} \cdot \frac{\sqrt{2}}{4} - \frac{\sqrt{2}}{8} \cdot \frac{\sqrt{2}}{2} = 0$$

and

$$c = \widetilde{g}_0 g_2 + \widetilde{g}_{-1} g_1 = h_1 \widetilde{h}_1 + h_2 \widetilde{h}_0 = b = 0.$$

Thus $\widetilde{G}_4 \cdot G_4^T = I_4$ so that $\widetilde{W}_8 W_8^T = I_8$ as desired. In Problem 7.9 you will show that $\widetilde{W}_N W_N^T = I_N$ for all even $N \geq 6$.

The $(5,3)$ Biorthogonal Spline Filter Pair

Let's summarize our work to this point. We have constructed a biorthogonal filter pair $\left(\mathbf{h},\widetilde{\mathbf{h}}\right)$ of lengths five and three, respectively. In Section 9.4, we will learn that this pair belongs to a family of biorthogonal filter pairs Daubechies constructed in [30]. These filter pairs are called *biorthogonal spline filter pairs* and we define the $(5,3)$ pair at this time.

Definition 7.2 (($5,3$) Biorthogonal Spline Filter Pair). *We define the* $(5,3)$ biorthogonal spline filter pair $\left(\mathbf{h},\widetilde{\mathbf{h}}\right)$ *as*

$$\boxed{\mathbf{h}=(h_{-2},h_{-1},h_0,h_1,h_2)=\left(-\frac{\sqrt{2}}{8},\frac{\sqrt{2}}{4},\frac{3\sqrt{2}}{4},\frac{\sqrt{2}}{4},-\frac{\sqrt{2}}{8}\right)} \tag{7.12}$$

and

$$\boxed{\widetilde{\mathbf{h}}=\left(\widetilde{h}_{-1},\widetilde{h}_0,\widetilde{h}_1\right)=\left(\frac{\sqrt{2}}{4},\frac{\sqrt{2}}{2},\frac{\sqrt{2}}{4}\right).} \tag{7.13}$$

□

The elements of the accompanying highpass filters $\widetilde{\mathbf{g}}$ and $\mathbf{g}$ are given by the formulas

$$\boxed{\widetilde{g}_k=(-1)^k h_{1-k},\quad k=-1,\ldots,3} \tag{7.14}$$

and

$$\boxed{g_k=(-1)^k \widetilde{h}_{1-k},\quad k=0,1,2.} \tag{7.15}$$

The wavelet transformation matrices for the $(5, 3)$ biorthogonal spline filter pair are

$$
\widetilde{W}_N = \left[\begin{array}{c} \widetilde{H}_{N/2} \\ \hline \widetilde{G}_{N/2} \end{array}\right]
= \left[\begin{array}{cccccccccccc}
\widetilde{h}_0 & \widetilde{h}_1 & 0 & 0 & 0 & 0 & \cdots & 0 & 0 & 0 & 0 & \widetilde{h}_{-1} \\
0 & \widetilde{h}_{-1} & \widetilde{h}_0 & \widetilde{h}_1 & 0 & 0 & \cdots & 0 & 0 & 0 & 0 & 0 \\
0 & 0 & 0 & \widetilde{h}_{-1} & \widetilde{h}_0 & \widetilde{h}_1 & \cdots & 0 & 0 & 0 & 0 & 0 \\
\vdots & & & & & & \ddots & & & & & \vdots \\
0 & 0 & 0 & 0 & 0 & 0 & \cdots & 0 & 0 & \widetilde{h}_{-1} & \widetilde{h}_0 & \widetilde{h}_1 \\
\hline
\widetilde{g}_0 & \widetilde{g}_1 & \widetilde{g}_2 & \widetilde{g}_3 & 0 & 0 & \cdots & 0 & 0 & 0 & 0 & \widetilde{g}_{-1} \\
0 & \widetilde{g}_{-1} & \widetilde{g}_0 & \widetilde{g}_1 & \widetilde{g}_2 & \widetilde{g}_3 & \cdots & 0 & 0 & 0 & 0 & 0 \\
\vdots & & & & & & \ddots & & & & & \vdots \\
0 & 0 & 0 & 0 & 0 & 0 & \cdots & \widetilde{g}_{-1} & \widetilde{g}_0 & \widetilde{g}_1 & \widetilde{g}_2 & \widetilde{g}_3 \\
\widetilde{g}_2 & \widetilde{g}_3 & 0 & 0 & 0 & 0 & \cdots & 0 & 0 & \widetilde{g}_{-1} & \widetilde{g}_0 & \widetilde{g}_1
\end{array}\right] \tag{7.16}
$$

and

$$
W_N = \left[\begin{array}{c} H_{N/2} \\ \hline G_{N/2} \end{array}\right]
= \left[\begin{array}{cccccccccccc}
h_0 & h_1 & h_2 & 0 & 0 & \cdots & 0 & 0 & 0 & 0 & h_{-2} & h_{-1} \\
h_{-2} & h_{-1} & h_0 & h_1 & h_2 & \cdots & 0 & 0 & 0 & 0 & 0 & 0 \\
\vdots & & & & & \ddots & & & & & & \vdots \\
0 & 0 & 0 & 0 & 0 & \cdots & h_{-2} & h_{-1} & h_0 & h_1 & h_2 & 0 \\
h_2 & 0 & 0 & 0 & 0 & \cdots & 0 & 0 & h_{-2} & h_{-1} & h_0 & h_1 \\
\hline
g_0 & g_1 & g_2 & 0 & 0 & \cdots & 0 & 0 & 0 & 0 & 0 & 0 \\
0 & 0 & g_0 & g_1 & g_2 & \cdots & 0 & 0 & 0 & 0 & 0 & 0 \\
\vdots & & & & & \ddots & & & & & & \vdots \\
0 & 0 & 0 & 0 & 0 & \cdots & 0 & 0 & g_0 & g_1 & g_2 & 0 \\
g_2 & 0 & 0 & 0 & 0 & \cdots & 0 & 0 & 0 & 0 & g_0 & g_1
\end{array}\right]. \tag{7.17}
$$

Examples

We conclude this section with two examples that utilize the $(5, 3)$ biorthogonal spline filter pair.

Example 7.1 (Application of the Biorthogonal Wavelet Transformation to a Vector). *We take $f(t)$ to be the heavisine function given by* (6.1) *and create the vector* $\mathbf{v}$ *by uniformly sampling f with $n = 128$ on the interval $[0, 1)$. That is $v_k = f(k/128)$, $k = 0, \ldots, 127$. The vector is plotted in Figure 7.1.*

We construct the biorthogonal wavelet transformation $\widetilde{W}_{128}$ using (7.16) *and apply it to* $\mathbf{v}$*. The result is plotted in Figure 7.1(a). In practice, we exploit the sparse nature of $\widetilde{W}_{128}$ and utilize an efficient algorithm rather than the matrix/vector product to do the computation. One such set of algorithms is developed in Computer Lab 7.1.*

The highpass portion of the biorthogonal wavelet transformation is plotted in Figure 7.1(c). Once we have the transformation matrix $\widetilde{W}_{128}$, we can certainly iterate the transformation as was done in Chapters 4 and 5.

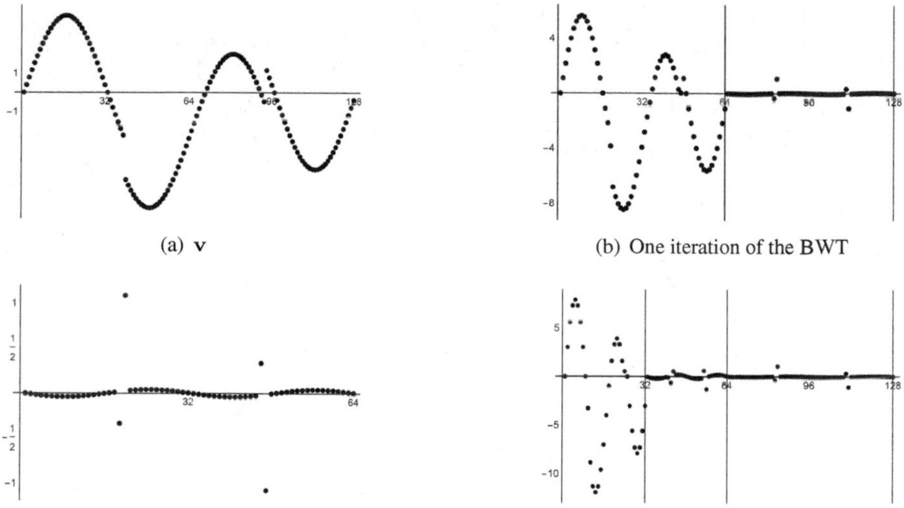

(a) $\mathbf{v}$

(b) One iteration of the BWT

(c) Highpass portion of the BWT

(d) Two iterations of the BWT

Figure 7.1 A vector $\mathbf{v}$, its biorthogonal wavelet transformation (BWT) using the $(5, 3)$ biorthogonal spline filter and the iterated BWT.

To recover $\mathbf{v}$ from $\mathbf{y} = \widetilde{W}_{128}\mathbf{v}$, we multiply both sides of the equation by $W_{128}^T = \widetilde{W}_{128}^{-1}$ to obtain $\mathbf{v} = W_{128}^T\mathbf{y}$. There is no rule that designates $\widetilde{W}_{128}$ as the "transformation" matrix and W_{128} as the inverse. Indeed we could have used W_N to compute the transformation and $\widetilde{W}_{128}^T$ to compute the inverse transformation.

(Live example: Visit `stthomas.edu/wavelets` *and click on Live Examples.)*

□

Our final example considers the computation of the biorthogonal wavelet transformation of a matrix.

Example 7.2 (Application of the Biorthogonal Wavelet Transformation to a Matrix). *Consider the image from Example 4.15 that is plotted in Figure 4.20(a). The intensities are stored in* 256×384 *matrix A. To compute the biorthogonal wavelet transformation of A using the* $(5,3)$ *biorthogonal spline filter pair, we use*

$$B = \widetilde{W}_{256} A W_{384}^T.$$

This product is similar to the products used by two-dimensional wavelet transformations in Chapters 4 and 5 where we pre- and post-multiplied A by W_{256} *and the inverse* W_{384}^T *of* $\widetilde{W}_{384}$*. Note that we could have also computed*

$$B = W_{256} A W_{384}^{-1} = W_{256} A \widetilde{W}_{384}^T.$$

We can also iterate the transformation like we did in Chapters 4 and 5. The biorthogonal wavelet transformation and an iterated biorthogonal transformation of A are plotted in Figure 7.2.

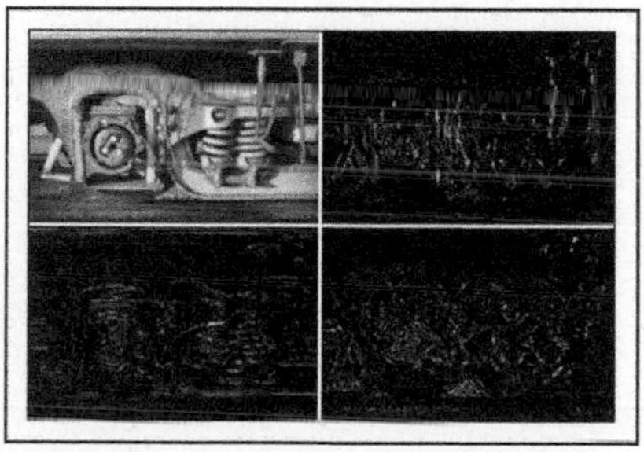

(a) *BWT*

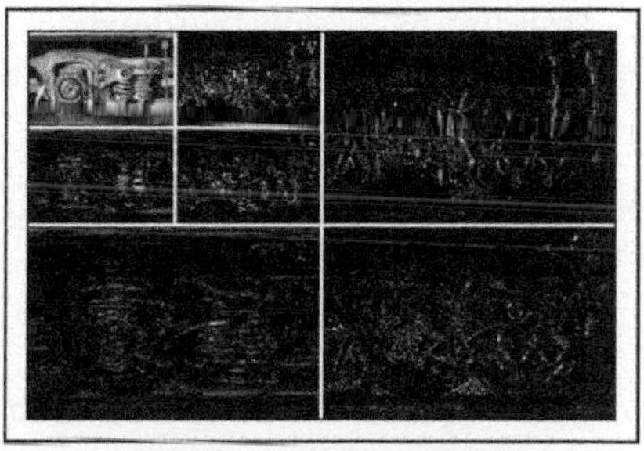

(b) Two iterations of the BWT

Figure 7.2 The biorthogonal wavelet transformation (BWT) of A using the $(5,3)$ biorthogonal spline filter pair and two iterations of the transformation.

(Live example: Visit `stthomas.edu/wavelets` *and click on Live Examples.)*

□

PROBLEMS

7.1 Prove Proposition 7.1.

7.2 Let $\widetilde{\mathbf{h}}$ be given by (7.13) and assume $\mathbf{h} = (h_{-1}, h_0, h_1)$ is a symmetric filter (so $h_{-1} = h_1$). Write down H_4 and use the identity $\widetilde{H}_4 H_4^T = I_4$, where $\widetilde{H}_4$ is given by (7.3), to show that $h_{-1} = h_1 = 0$ and $h_0 = \sqrt{2}$.

7.3 Let $\widetilde{H}$ be given by (7.13) and assume $\mathbf{h} = (h_{-1}, h_0, h_1, h_2) = (a, b, b, a)$ is a symmetric filter for $a, b \in \mathbb{R}$. Write down H_4 and use the identity $\widetilde{H}_4 H_4^T = I_4$, where $\widetilde{H}_4$ is given by (7.3), to show that values for a, b do not exist.

7.4 Solve the system given by (7.6) and (7.7).

7.5 Let $\widetilde{H}_4$ and G_4 be given by (7.3) and (7.9), respectively. Show by direct computation that $\widetilde{H}_4 G_4^T = 0_4$.

7.6 Let $\widetilde{G}_4$ and H_4 be given by (7.11) and (7.8), respectively. Show by direct computation that $\widetilde{G}_4 H_4^T = 0_4$.

7.7 Compute by hand the biorthogonal wavelet transformations of the following vectors.

(a) $\mathbf{v} \in \mathbb{R}^8$ where $v_k = k$, $k = 1, \ldots, 8$

(b) $\mathbf{v} \in \mathbb{R}^8$ where $v_k = \frac{\sqrt{2}}{2}$, $k = 1, \ldots, 8$

(c) $\mathbf{v} \in \mathbb{R}^8$, where $v_k = (-1)^{k+1}\frac{\sqrt{2}}{2}$, $k = 1, \ldots, 8$

7.8 As pointed out at the end of Example 7.1, it does not matter which of $\widetilde{W}_N$ or W_N you use as the left biorthogonal transformation matrix. Can you think of instances where one might be more advantageous than the other?

7.9 In this problem, you will explain why $\widetilde{W}_N$ and W_N given by (7.16) and (7.17), respectively, satisfy $\widetilde{W}_N W_N^T = I_N$. The following steps will help you organize your work.

(a) There are four products to consider: $\widetilde{H}_{N/2} H_{N/2}^T$, $\widetilde{H}_{N/2} G_{N/2}^T$, $\widetilde{G}_{N/2} H_{N/2}^T$ and $\widetilde{G}_{N/2} G_{N/2}^T$. Start with the first product and in particular, the "non-wrapping" rows in each of $\widetilde{H}_{N/2}$ and $H_{N/2}$. We need to show that the dot product of row j of $\widetilde{H}_{N/2}$ with row k of $H_{N/2}$ (column k of $H_{N/2}^T$) is 1 if $j = k$ and 0 otherwise. Show that for $2 \le j \le \frac{N}{2}$, the nonzero elements in row j of $\widetilde{H}_{N/2}$ occur at positions $2j - 2$, $2j - 1$, and $2j$ and the values of these elements are $\widetilde{h}_1$, $\widetilde{h}_0$, and $\widetilde{h}_1$, respectively.

(b) Show that for $2 \le k \le \frac{N}{2} - 1$, the nonzero elements in row k of $H_{N/2}$ occur at positions $2k - 3$, $2k - 2$, $2k - 1$, $2k$, and $2k + 1$ and the values of these elements are h_2, h_1, h_0, h_1, and h_2, respectively.

(c) Show that for $2 \le j \le \frac{N}{2}$ and $2 \le k \le \frac{N}{2} - 1$, the only rows in $\widetilde{H}_{N/2}$ and $H_{N/2}$ that have nonzero elements that overlap are when $j = k - 1$, $j = k$, $j = k + 1$. Show that the dot product is 1 when $j = k$ and 0 in the other two cases. Observe that for all other cases, dotting row j of $\widetilde{H}_{N/2}$ with row k of $H_{N/2}$ is 0.

(d) Dot row one of $\widetilde{H}_{N/2}$ with rows one, two, and $N/2$ of $H_{N/2}$ and show that the products are $1, 0$ and 0, respectively. Observe that row one of $\widetilde{H}_{N/2}$ and the remaining rows of $H_{N/2}$ have no overlap of nonzero elements so these dot products are 0.

(e) Repeat steps similar to (a) – (d) for the products $\widetilde{H}_{N/2}G^T_{N/2}$, $\widetilde{G}_{N/2}H^T_{N/2}$.

(f) For the final product $\widetilde{G}_{N/2}G^T_{N/2}$, replace the $\widetilde{g}_k$ and g_k in (7.16) and (7.17) with the appropriate values from $\mathbf{h}$ and $\widetilde{\mathbf{h}}$. How do the rows of $\widetilde{G}_{N/2}$ and $G_{N/2}$ relate to the rows of $H_{N/2}$ and $\widetilde{H}_{N/2}$, respectively? Use this relationship to show that $\widetilde{G}_{N/2}G^T_{N/2} = I_{N/2}$.

★**7.10** In some applications (see Example 7.9 and Section 12.2), it is desirable to map integers to rational numbers. Due to the presence of $\sqrt{2}$, it is not possible to do so if we use the $(5, 3)$ biorthogonal spline filter pair to construct $\widetilde{W}_N$ and W_N. In this exercise, we modify the transformation matrices given in (7.16) and (7.17) that can be used to map integers to rational numbers.

(a) In the previous exercise, we learned that $\widetilde{W}_N$ is nonsingular with $\widetilde{W}_N^{-1} = W^T_N$. Use this fact to show that W_N is nonsingular with $W_N^{-1} = \widetilde{W}^T_N$.

(b) Let $r > 0$ and define the (invertible) $N \times$N diagonal matrix D_N with diagonal elements $d_{ii} = r$, $i = 1, \ldots, \dfrac{N}{2}$ and $d_{ii} = \dfrac{1}{r}$, $i = \frac{N}{2} + 1, \ldots, N$. Define new matrices $\widetilde{U}_N = D_N\widetilde{W}_N$ and $U_N = D_N^{-1}W_N$. Show that $\widetilde{U}$ is invertible with $\widetilde{U}_N^{-1} = U^T_N$.

(c) Show that U_N is invertible with $U^T_N = \widetilde{U}_N^{-1}$.

(d) Let A be an $N \times M$ matrix and partition it into $N/2 \times M/2$ blocks A_{ij}, $i, j = 1, 2$. Show that

$$D_N^{-1}AD_M = \left[\begin{array}{c|c} A_{11} & \frac{1}{r^2}A_{12} \\ \hline r^2 A_{21} & A_{22} \end{array}\right] \quad \text{and} \quad D_N A D_M^{-1} = \left[\begin{array}{c|c} A_{11} & r^2 A_{12} \\ \hline \frac{1}{r^2}A_{21} & A_{22} \end{array}\right].$$

(e) How do can the $(5, 3)$ biorthogonal spline filter pair $\widetilde{\mathbf{h}}, \mathbf{h}$ be modified in order to product $\widetilde{U}_N, U_N$?

7.11 The filters $\widetilde{\mathbf{h}}$ and $\mathbf{h}$ are symmetric filters but the highpass filters $\widetilde{\mathbf{g}}$ and $\mathbf{g}$ are not. That said, the highpass filters do exhibit some symmetry.

(a) Write down the elements of $\widetilde{\mathbf{g}}$ in terms of the elements of $\mathbf{h}$ and determine if any symmetry exists among the elements of $\widetilde{\mathbf{g}}$.

(b) Repeat (a) using $\mathbf{g}$ and $\widetilde{\mathbf{h}}$. Does your claim still hold?

(c) Suppose that $\widetilde{\mathbf{h}}$ is any odd-length symmetric filter and the elements of $\mathbf{g}$ are defined by $g_k = (-1)^k \widetilde{h}_{1-k}$. Prove your conjectures from (a) and (b) hold in this general case.

7.12 Let $\widetilde{\mathbf{h}} = \left(\widetilde{h}_0, \widetilde{h}_1\right) = \left(\frac{\sqrt{2}}{2}, \frac{\sqrt{2}}{2}\right)$. In this problem you will investigate some possible candidates for the companion filter $\mathbf{h}$.

(a) Suppose $\mathbf{h}$ is a symmetric 4-term filter. Then $\mathbf{h} = (h_{-1}, h_0, h_1, h_2) = (a, b, b, a)$ for $a, b \in \mathbb{R}$. Let $N = 8$ and write down $\widetilde{H}_4$ and H_4. Recall $\widetilde{H}_4 H_4^T = I_4$. Use this fact to show that $\mathbf{h} = \widetilde{\mathbf{h}}$.

(b) Suppose $\mathbf{h}$ is a symmetric 6-term filter. Then $\mathbf{h} = (h_{-2}, h_{-1}, h_0, h_1, h_2, h_3) = (a, b, c, c, b, a)$ for $a, b, c \in \mathbb{R}$. Form $\widetilde{H}_4$ and H_4 as in (a) and use these matrices to show that $c = \frac{\sqrt{2}}{2}$ and $b = -a$.

(c) Does the filter $\mathbf{h} = \left(a, -a, \frac{\sqrt{2}}{2}, \frac{\sqrt{2}}{2}, -a, a\right)$ satisfy (7.4)?

(d) One way to attempt to find a value for a is to insist the filter $\widetilde{\mathbf{g}}$ whose elements are $\widetilde{g}_k = (-1)^k h_{1-k}$, $k = -2, \ldots, 3$, annihilate linear data. To see if this is the case, we compute $\widetilde{\mathbf{g}} \cdot \mathbf{x} = 0$ where $\mathbf{x} = [0, 1, 2, 3, 4, 5]^T$ (or any other list of linear data) and hopefully solve for a. Does this computation computation allow us to solve for a? Can $\widetilde{\mathbf{g}}$ annihilate linear data?

7.13 Suppose we wish to compute

$$W_N \mathbf{v} = W_N \begin{bmatrix} v_1 \\ v_2 \\ \vdots \\ v_N \end{bmatrix} = \begin{bmatrix} \mathbf{a} \\ \hline \mathbf{d} \end{bmatrix}$$

where W_N is given by (7.17) and N is an even number. Since W_N is sparse matrix, we reformulate the product to create a more efficient means of computing the transformation. Towards that end, let $L = N/2$. Using the elements of $\mathbf{v}$, construct an $L \times 5$ matrix A and an $L \times 3$ matrix B such that $A\mathbf{h} = \mathbf{a}$ and $B\mathbf{g} = \mathbf{d}$, where $\mathbf{h}$ and $\mathbf{g}$ are given by (7.12) and (7.10), respectively.

7.14 Suppose we have computed the biorthogonal wavelet transformation $W_N \mathbf{v} = \begin{bmatrix} \mathbf{a} \\ \hline \mathbf{d} \end{bmatrix}$, N even, and we wish to recover $\mathbf{v}$. We know $W_N^{-1} = \widetilde{W}_N^T$, where $\widetilde{W}_N$ is given by (7.16). We compute

$$\mathbf{v} = \widetilde{W}_N^T \begin{bmatrix} \mathbf{a} \\ \hline \mathbf{d} \end{bmatrix} = \left[\widetilde{H}_N^T \middle| \widetilde{G}_N^T \right] \begin{bmatrix} \mathbf{a} \\ \hline \mathbf{d} \end{bmatrix} = \widetilde{H}_N^T \mathbf{a} + \widetilde{G}_N^T \mathbf{d}.$$

Both $\widetilde{H}_N^T$ and $\widetilde{G}_N^T$ are sparse matrices so we reformulate the computation to improve efficiency. Towards this end, let $L = N/2$ and consider the products

$$\widetilde{H}_N^T\mathbf{a} = \begin{bmatrix} \widetilde{h}_0 & 0 & \cdots & 0 & 0 \\ \widetilde{h}_1 & \widetilde{h}_1 & & 0 & 0 \\ 0 & \widetilde{h}_0 & & 0 & 0 \\ 0 & \widetilde{h}_1 & & 0 & 0 \\ \vdots & \vdots & \ddots & \vdots & \vdots \\ 0 & 0 & & \widetilde{h}_1 & \widetilde{h}_1 \\ 0 & 0 & & 0 & \widetilde{h}_0 \\ \widetilde{h}_1 & 0 & \cdots & 0 & \widetilde{h}_1 \end{bmatrix} \begin{bmatrix} a_1 \\ a_2 \\ \vdots \\ a_L \end{bmatrix}$$

and

$$\widetilde{G}_N^T\mathbf{d} = \begin{bmatrix} \widetilde{g}_0 & 0 & 0 & \cdots & 0 & 0 & \widetilde{g}_2 \\ \widetilde{g}_1 & \widetilde{g}_{-1} & 0 & & 0 & 0 & \widetilde{g}_3 \\ \widetilde{g}_2 & \widetilde{g}_0 & 0 & & 0 & 0 & 0 \\ \widetilde{g}_3 & \widetilde{g}_1 & \widetilde{g}_{-1} & & 0 & 0 & 0 \\ \vdots & & & \ddots & & & \vdots \\ 0 & 0 & 0 & & \widetilde{g}_3 & \widetilde{g}_1 & \widetilde{g}_{-1} \\ 0 & 0 & 0 & & 0 & \widetilde{g}_2 & \widetilde{g}_0 \\ \widetilde{g}_{-1} & 0 & 0 & \cdots & 0 & \widetilde{g}_3 & \widetilde{g}_1 \end{bmatrix} \begin{bmatrix} d_1 \\ d_2 \\ \vdots \\ d_L \end{bmatrix}.$$

(a) Observe that $\mathbf{o}^1 = \widetilde{h}_0\mathbf{a}$ contains the odd-indexed elements of $\widetilde{H}_N^T\mathbf{a}$. Using the elements of $\mathbf{a}$, find an $L \times 2$ matrix A such that $A \begin{bmatrix} \widetilde{h}_1 \\ \widetilde{h}_1 \end{bmatrix} = \mathbf{e}^1$, where the elements of $\mathbf{e}^1$ are the even-indexed elements of $\widetilde{H}_N^T\mathbf{a}$.

(b) Using the elements of $\mathbf{d}$, find an $L \times 2$ matrix X such that $X \begin{bmatrix} \widetilde{g}_2 \\ \widetilde{g}_0 \end{bmatrix} = \mathbf{o}^2$, where the elements of $\mathbf{o}^2$ are the odd-indexed elements of $\widetilde{G}_N^T\mathbf{d}$.

(c) Using the elements of $\mathbf{d}$, find an $L \times 3$ matrix Y such that $Y \begin{bmatrix} \widetilde{g}_3 \\ \widetilde{g}_1 \\ \widetilde{g}_{-1} \end{bmatrix} = \mathbf{e}^2$, where the elements of $\mathbf{e}^2$ are the odd-indexed elements of $\widetilde{G}_N^T\mathbf{d}$.

(d) Find permutation matrix P so that

$$P\left(\begin{bmatrix} \mathbf{o}^1 \\ \mathbf{e}^1 \end{bmatrix} + \begin{bmatrix} \mathbf{o}^2 \\ \mathbf{e}^2 \end{bmatrix}\right) = \widetilde{H}_N^T\mathbf{a} + \widetilde{G}_N^T\mathbf{d}.$$

Computer Lab

7.1 The Biorthogonal Wavelet Transformation Using the $(5,3)$ Filter Pair. From `stthomas.edu/wavelets`, access the file `Biorth53`. In this lab, you will develop and test efficient algorithms for computing the one- and two-dimensional biorthogonal wavelet transformations using the $(5,3)$ biorthogonal spline filter pair.

7.2 The $(8,4)$ Biorthogonal Spline Filter Pair

In this section, we construct a biorthogonal spline filter pair $\left(\mathbf{h}, \widetilde{\mathbf{h}}\right)$ whose lengths are 8 and 4. The construction will mimic much of what was done in Section 7.1. Ultimately we will build the necessary filters (including $\widetilde{\mathbf{g}}$ and $\mathbf{g}$) to create a biorthogonal wavelet transformation, but the process leaves many questions unanswered. We will quickly build the 4-term filter $\widetilde{\mathbf{h}}$ but then have to determine which filter lengths (do not) work for $\mathbf{h}$. The $\mathbf{h}$ we construct will be of length 8, but in the problem set, you will see that we could have constructed $\mathbf{h}$ to be a length 4 filter. In the process, we also learn that lengths 2, 3, 5, and 6 do not work. This is our last "ad hoc" construction. In Chapters 8 and 9 we will learn about mathematics that will allow us to systematically construct orthogonal filters and biorthogonal filter pairs.

Constructing $\widetilde{\mathbf{h}}$

Let's start by taking a look back at our symmetric lowpass filters of lengths 2 and 3. We have

$$\begin{aligned} \left(\widetilde{h}_0, \widetilde{h}_1\right) &= \left(\tfrac{\sqrt{2}}{2}, \tfrac{\sqrt{2}}{2}\right) &= \tfrac{\sqrt{2}}{2}(1,1) \\ \left(\widetilde{h}_{-1}, \widetilde{h}_0, \widetilde{h}_1\right) &= \left(\tfrac{\sqrt{2}}{4}, \tfrac{\sqrt{2}}{2}, \tfrac{\sqrt{2}}{4}\right) &= \tfrac{\sqrt{2}}{4}(1,2,1). \end{aligned} \tag{7.18}$$

We see the emergence of Pascal's triangle on the right column of (7.18). The rows of this triangle are symmetric so a natural choice for $\widetilde{\mathbf{h}}$ is $(1,3,3,1)$ which would be the next row in the triangle. We still require that $\widetilde{\mathbf{h}}$ satisfy (7.4) and while the second equation is satisfied, the sum of the components is not $\sqrt{2}$. We can easily satisfy the first equation in (7.4) by dividing each term of $(1,3,3,1)$ by 8 (the total of the components) and then multiplying each element by $\sqrt{2}$. We have

$$\boxed{\widetilde{\mathbf{h}} = \left(\widetilde{h}_{-1}, \widetilde{h}_0, \widetilde{h}_1, \widetilde{h}_2\right) = \left(\frac{\sqrt{2}}{8}, \frac{3\sqrt{2}}{8}, \frac{3\sqrt{2}}{8}, \frac{\sqrt{2}}{8}\right).} \tag{7.19}$$

It can be quickly verified that $\widetilde{\mathbf{h}}$ satisfies Definition 2.19 and is thus a lowpass filter. In Problems 7.15 and 7.17, you will construct lowpass filters $\widetilde{\mathbf{h}}$ from Pascal's triangle of arbitrary length.

We display $\widetilde{H}_6$ below. We use $N = 12$ as wavelet matrix dimensions throughout the section to construct $\mathbf{h}$, $\widetilde{\mathbf{g}}$, and $\mathbf{g}$.

$$\widetilde{H}_6 = \begin{bmatrix} \widetilde{h}_0 & \widetilde{h}_1 & \widetilde{h}_2 & 0 & 0 & 0 & 0 & 0 & 0 & 0 & 0 & \widetilde{h}_{-1} \\ 0 & \widetilde{h}_{-1} & \widetilde{h}_0 & \widetilde{h}_1 & \widetilde{h}_2 & 0 & 0 & 0 & 0 & 0 & 0 & 0 \\ 0 & 0 & 0 & \widetilde{h}_{-1} & \widetilde{h}_0 & \widetilde{h}_1 & \widetilde{h}_2 & 0 & 0 & 0 & 0 & 0 \\ 0 & 0 & 0 & 0 & 0 & \widetilde{h}_{-1} & \widetilde{h}_0 & \widetilde{h}_1 & \widetilde{h}_2 & 0 & 0 & 0 \\ 0 & 0 & 0 & 0 & 0 & 0 & 0 & \widetilde{h}_{-1} & \widetilde{h}_0 & \widetilde{h}_1 & \widetilde{h}_2 & 0 \\ \widetilde{h}_2 & 0 & 0 & 0 & 0 & 0 & 0 & 0 & 0 & \widetilde{h}_{-1} & \widetilde{h}_0 & \widetilde{h}_1 \end{bmatrix}$$

$$= \frac{\sqrt{2}}{8} \begin{bmatrix} 3 & 3 & 1 & 0 & 0 & 0 & 0 & 0 & 0 & 0 & 0 & 1 \\ 0 & 1 & 3 & 3 & 1 & 0 & 0 & 0 & 0 & 0 & 0 & 0 \\ 0 & 0 & 0 & 1 & 3 & 3 & 1 & 0 & 0 & 0 & 0 & 0 \\ 0 & 0 & 0 & 0 & 0 & 1 & 3 & 3 & 1 & 0 & 0 & 0 \\ 0 & 0 & 0 & 0 & 0 & 0 & 0 & 1 & 3 & 3 & 1 & 0 \\ 1 & 0 & 0 & 0 & 0 & 0 & 0 & 0 & 0 & 1 & 3 & 3 \end{bmatrix} \tag{7.20}$$

Constructing $\mathbf{h}$

The first thing we must determine in constructing $\mathbf{h}$ is its length. In Problem 7.18, you will show that the length of $\mathbf{h}$ cannot be odd. In Problems 7.20 and 7.24, you show that symmetric $\mathbf{h}$ cannot be of lengths 2 and 6, respectively. In Problem 7.21, you will find a symmetric length 4 filter such that $\widetilde{H}_6 H_6^T = I_6$. (The result holds if 6 is replaced by $N/2$.) Thus we attempt to construct a symmetric filter $\mathbf{h}$ of length 8. We have

$$\begin{aligned} \mathbf{h} &= (h_{-3}, h_{-2}, h_{-1}, h_0, h_1, h_2, h_3, h_4) \\ &= (h_4, h_3, h_2, h_1, h_1, h_2, h_3, h_4) . \end{aligned}$$

The choice of length 8 is not entirely arbitrary – it is desirable that the lengths of the filters $\widetilde{\mathbf{h}}, \mathbf{h}$ are comparable but at the same time, longer filters provide more flexibility for tasks such as annihilating polynomial data.

Since we want $\mathbf{h}$ to be a lowpass filter, we insist that

$$\sum_{k=-3}^{4} h_k = \sqrt{2} \qquad \text{and} \qquad \sum_{k=-3}^{4} (-1)^k h_k = 0. \tag{7.21}$$

Using the fact that $\mathbf{h}$ is symmetric, the first equation in (7.21) can be rewritten as

$$h_1 + h_2 + h_3 + h_4 = \frac{\sqrt{2}}{2} \tag{7.22}$$

and the second equation in (7.21) is automatically satisfied due to symmetry and the fact that the length of $\mathbf{h}$ is even.

We use $\mathbf{h}$ to form

$$H_6 = \begin{bmatrix} h_0 & h_1 & h_2 & h_3 & h_4 & 0 & 0 & 0 & 0 & h_{-3} & h_{-2} & h_{-1} \\ h_{-2} & h_{-1} & h_0 & h_1 & h_2 & h_3 & h_4 & 0 & 0 & 0 & 0 & h_{-3} \\ 0 & h_{-3} & h_{-2} & h_{-1} & h_0 & h_1 & h_2 & h_3 & h_4 & 0 & 0 & 0 \\ 0 & 0 & 0 & h_{-3} & h_{-2} & h_{-1} & h_0 & h_1 & h_2 & h_3 & h_4 & 0 \\ h_4 & 0 & 0 & 0 & 0 & h_{-3} & h_{-2} & h_{-1} & h_0 & h_1 & h_2 & h_3 \\ h_2 & h_3 & h_4 & 0 & 0 & 0 & 0 & h_{-3} & h_{-2} & h_{-1} & h_0 & h_1 \end{bmatrix}$$

$$= \begin{bmatrix} h_1 & h_1 & h_2 & h_3 & h_4 & 0 & 0 & 0 & 0 & h_4 & h_3 & h_2 \\ h_3 & h_2 & h_1 & h_1 & h_2 & h_3 & h_4 & 0 & 0 & 0 & 0 & h_4 \\ 0 & h_4 & h_3 & h_2 & h_1 & h_1 & h_2 & h_3 & h_4 & 0 & 0 & 0 \\ 0 & 0 & 0 & h_4 & h_3 & h_2 & h_1 & h_1 & h_2 & h_3 & h_4 & 0 \\ h_4 & 0 & 0 & 0 & 0 & h_4 & h_3 & h_2 & h_1 & h_1 & h_2 & h_3 \\ h_2 & h_3 & h_4 & 0 & 0 & 0 & 0 & h_4 & h_3 & h_2 & h_1 & h_1 \end{bmatrix}$$

and then directly compute $I_6 = \widetilde{H}_6 H_6^T$ to obtain

$$\begin{bmatrix} 1 & 0 & 0 & 0 & 0 & 0 \\ 0 & 1 & 0 & 0 & 0 & 0 \\ 0 & 0 & 1 & 0 & 0 & 0 \\ 0 & 0 & 0 & 1 & 0 & 0 \\ 0 & 0 & 0 & 0 & 1 & 0 \\ 0 & 0 & 0 & 0 & 0 & 1 \end{bmatrix} = \begin{bmatrix} a & b & c & 0 & c & b \\ b & a & b & c & 0 & c \\ c & b & a & b & c & 0 \\ 0 & c & b & a & b & c \\ c & 0 & c & b & a & b \\ b & c & 0 & c & b & a \end{bmatrix}$$

where

$$\begin{aligned} a = 1 &= \frac{3h_1 + h_2}{2\sqrt{2}} \\ b = 0 &= \frac{h_1 + 3h_3 + 3h_3 + h_4}{4\sqrt{2}} \\ c = 0 &= \frac{h_3 + 3h_4}{4\sqrt{2}}. \end{aligned}$$

or

$$\begin{aligned} 2\sqrt{2} &= 3h_1 + h_2 \\ 0 &= h_1 + 3h_2 + 3h_3 + h_4 \\ 0 &= h_3 + 3h_4. \end{aligned} \tag{7.23}$$

It can be shown that the system (7.23) has infinitely many solutions (see Problem 7.19). Adding one more linear equation could result in a system with a unique solution. We can't add (7.22) since it can be derived by summing the three equations in (7.23). As was the case in Section 7.1, we will construct the highpass filter $\widetilde{\mathbf{g}}$ from $\mathbf{h}$ using the formula

$$\widetilde{g}_k = (-1)^k h_{1-k}, \quad k = -3, \ldots, 4$$

and we will insist this filter annihilate linear data. Using Proposition 2.3 with $m = 1$ and $n = 1$ and the fact that $\mathbf{h}$ is symmetric, we have

$$\begin{aligned} 0 &= \sum_{k=-3}^{4} g_k(1-k) \\ &= \sum_{k=-3}^{4} (-1)^k h_{1-k}(1-k) \\ &= -4h_{-3} + 3h_{-2} - 2h_{-1} + h_0 - h_2 + 2h_3 - 3h_4 \\ &= h_1 - 3h_2 + 5h_3 - 7h_4. \end{aligned} \tag{7.24}$$

In Problem 7.26 you are asked to show that the solution of the system (7.23) – (7.24) is $h_1 = \frac{45\sqrt{2}}{64}$, $h_2 = -\frac{14\sqrt{2}}{64}$, $h_3 = -\frac{9\sqrt{2}}{64}$, and $h_4 = \frac{3\sqrt{2}}{64}$. Thus our eight-term symmetric lowpass filter is

$$\boxed{\begin{aligned} \mathbf{h} &= (h_{-3}, h_{-2}, h_{-1}, h_0, h_1, h_2, h_3, h_4) \\ &= \left(\frac{3\sqrt{2}}{64}, -\frac{9\sqrt{2}}{64}, -\frac{7\sqrt{2}}{64}, \frac{45\sqrt{2}}{64}, \frac{45\sqrt{2}}{64}, -\frac{7\sqrt{2}}{64}, -\frac{9\sqrt{2}}{64}, \frac{3\sqrt{2}}{64}\right). \end{aligned}} \tag{7.25}$$

Finding $\widetilde{\mathbf{g}}$ and $\mathbf{g}$

We use the same approach for finding the highpass filters as we did in Section 7.1. That is, we define $\widetilde{\mathbf{g}}$ and $\mathbf{g}$ element-wise by the formulas

$$g_k = (-1)^k \widetilde{h}_{1-k},\ k = -1, 0, 1, 2 \qquad \text{and} \qquad \widetilde{g}_k = (-1)^k h_{1-k},\ k = -3, \ldots, 4.$$

The indices k are those for which $\widetilde{h}_{1-k}$ and h_{1-k} are nonzero and note that the ranges are the same. That is, $\mathbf{g}$ and $\widetilde{\mathbf{h}}$ both range from -1 to 2 and $\widetilde{\mathbf{g}}$ and $\mathbf{h}$ both range from -3 to 4. We have

$$\boxed{\begin{aligned} \mathbf{g} &= (g_{-1}, g_0, g_1, g_2) \\ &= \left(-\widetilde{h}_2, \widetilde{h}_1, -\widetilde{h}_0, \widetilde{h}_{-1}\right) \\ &= \left(-\widetilde{h}_2, \widetilde{h}_1, -\widetilde{h}_1, \widetilde{h}_2\right) \\ &= \left(-\frac{\sqrt{2}}{8}, \frac{3\sqrt{2}}{8}, -\frac{3\sqrt{2}}{8}, \frac{\sqrt{2}}{8}\right) \end{aligned}} \tag{7.26}$$

and

$$\boxed{\begin{aligned} \widetilde{\mathbf{g}} &= (\widetilde{g}_{-3}, \widetilde{g}_{-2}, \widetilde{g}_{-1}, \widetilde{g}_0, \widetilde{g}_1, \widetilde{g}_2, \widetilde{g}_3, \widetilde{g}_4) \\ &= (-h_4, h_3, -h_2, h_1, -h_0, h_{-1}, -h_{-2}, h_{-3}) \\ &= (-h_4, h_3, -h_2, h_1, -h_1, h_2, -h_3, h_4) \\ &= \left(-\tfrac{3\sqrt{2}}{64}, -\tfrac{9\sqrt{2}}{64}, \tfrac{7\sqrt{2}}{64}, \tfrac{45\sqrt{2}}{64}, -\tfrac{45\sqrt{2}}{64}, -\tfrac{7\sqrt{2}}{64}, \tfrac{9\sqrt{2}}{64}, \tfrac{3\sqrt{2}}{64}\right). \end{aligned}} \tag{7.27}$$

Since $\mathbf{h}$ and $\widetilde{\mathbf{h}}$ are lowpass filters, Proposition 2.4 implies that $\widetilde{\mathbf{g}}$ and $\mathbf{g}$ are highpass filters. Thus both annihilate constant data and $\widetilde{\mathbf{g}}$ was designed so that it annihilates linear data. In Problem 7.27, you will show it also annihilates quadratic data. The highpass filter $\mathbf{g}$ annihilates both linear and quadratic data (see Problem 7.28).

The $(8, 4)$ Biorthogonal Spline Filter Pair

The filters $\mathbf{h}$ and $\widetilde{\mathbf{h}}$ constructed in this section are also members of Daubechies biorthogonal spline filter family. We have the following definition.

Definition 7.3 (**$(8, 4)$ Biorthogonal Spline Filter Pair**). *We define the* $(8, 4)$ biorthogonal spline filter pair $\left(\mathbf{h}, \widetilde{\mathbf{h}}\right)$ *as those given in* (7.25) *and* (7.19), *respectively.* □

The elements of the accompanying highpass filters $\widetilde{\mathbf{g}}$ and $\mathbf{g}$ are given by (7.27) and (7.26), respectively. As was the case for Daubechies orthogonal filters, longer highpass filters are able to annihilate higher degree polynomial data.

The wavelet transformation matrices for the $(8, 4)$ biorthogonal spline filter pair are

$$\widetilde{W}_N = \left[\begin{array}{c} \widetilde{H}_{N/2} \\ \hline \widetilde{G}_{N/2} \end{array}\right] \tag{7.28}$$

$$= \left[\begin{array}{ccccccccccccccc}
\widetilde{h}_0 & \widetilde{h}_1 & \widetilde{h}_2 & 0 & 0 & 0 & 0 & \cdots & 0 & 0 & 0 & 0 & 0 & 0 & \widetilde{h}_{-1} \\
0 & \widetilde{h}_{-1} & \widetilde{h}_0 & \widetilde{h}_1 & \widetilde{h}_2 & 0 & 0 & \cdots & 0 & 0 & 0 & 0 & 0 & 0 & 0 \\
\vdots & & & & & & & & & \ddots & & & & & \vdots \\
0 & 0 & 0 & 0 & 0 & 0 & 0 & \cdots & 0 & 0 & \widetilde{h}_{-1} & \widetilde{h}_0 & \widetilde{h}_1 & \widetilde{h}_2 & 0 \\
\widetilde{h}_2 & 0 & 0 & 0 & 0 & 0 & 0 & \cdots & 0 & 0 & 0 & 0 & \widetilde{h}_{-1} & \widetilde{h}_0 & \widetilde{h}_1 \\
\hline
\widetilde{g}_0 & \widetilde{g}_1 & \widetilde{g}_2 & \widetilde{g}_3 & \widetilde{g}_4 & 0 & 0 & \cdots & 0 & 0 & 0 & 0 & \widetilde{g}_{-3} & \widetilde{g}_{-2} & \widetilde{g}_{-1} \\
\widetilde{g}_{-2} & \widetilde{g}_{-1} & \widetilde{g}_0 & \widetilde{g}_1 & \widetilde{g}_2 & \widetilde{g}_3 & \widetilde{g}_4 & \cdots & 0 & 0 & 0 & 0 & 0 & 0 & \widetilde{g}_{-3} \\
\vdots & & & & & & & \ddots & & & & & & & \vdots \\
\widetilde{g}_4 & 0 & 0 & 0 & 0 & 0 & 0 & \cdots & \widetilde{g}_{-3} & \widetilde{g}_{-2} & \widetilde{g}_{-1} & \widetilde{g}_0 & \widetilde{g}_1 & \widetilde{g}_2 & \widetilde{g}_3 \\
\widetilde{g}_2 & \widetilde{g}_3 & \widetilde{g}_4 & 0 & 0 & 0 & 0 & \cdots & 0 & 0 & \widetilde{g}_{-3} & \widetilde{g}_{-2} & \widetilde{g}_{-1} & \widetilde{g}_0 & \widetilde{g}_1
\end{array}\right]$$

and

$$
W_N = \left[\begin{array}{c} H_{N/2} \\ \hline G_{N/2} \end{array}\right] \tag{7.29}
$$

$$
= \left[\begin{array}{cccccccccccccccc}
h_0 & h_1 & h_2 & h_3 & h_4 & 0 & 0 & 0 & \cdots & 0 & 0 & 0 & 0 & h_{-3} & h_{-2} & h_{-1} \\
h_{-2} & h_{-1} & h_0 & h_1 & h_2 & h_3 & h_4 & 0 & \cdots & 0 & 0 & 0 & 0 & 0 & 0 & h_{-3} \\
\vdots & & & & & & & & \ddots & & & & & & & \vdots \\
h_4 & 0 & 0 & 0 & 0 & 0 & 0 & 0 & \cdots & h_{-3} & h_{-2} & h_{-1} & h_0 & h_1 & h_2 & h_3 \\
h_2 & h_3 & h_4 & 0 & 0 & 0 & 0 & 0 & \cdots & 0 & 0 & h_{-3} & h_{-2} & h_{-1} & h_0 & h_1 \\
\hline
g_0 & g_1 & g_2 & 0 & 0 & 0 & 0 & 0 & \cdots & 0 & 0 & 0 & 0 & 0 & 0 & g_{-1} \\
0 & g_{-1} & g_0 & g_1 & g_2 & 0 & 0 & 0 & \cdots & 0 & 0 & 0 & 0 & 0 & 0 & 0 \\
\vdots & & & & & & & & \ddots & & & & & & & \vdots \\
0 & 0 & 0 & 0 & 0 & 0 & 0 & 0 & \cdots & 0 & 0 & g_{-1} & g_0 & g_1 & g_2 & 0 \\
g_2 & 0 & 0 & 0 & 0 & 0 & 0 & 0 & \cdots & 0 & 0 & 0 & 0 & g_{-1} & g_0 & g_1
\end{array}\right].
$$

In Problem 7.31, you will show that $\widetilde{H}_6 G_6^T = G_6 \widetilde{H}_6^T = 0_6$ and $\widetilde{G}_6 G_6^T = I_6$.

PROBLEMS

7.15 In this problem you will use the ideas for creating $\widetilde{\mathbf{H}}$ in (7.19) to construct lowpass filters $\widetilde{\mathbf{h}}$ of lengths 5 and 6.

(a) From (7.18), we know the first three rows of Pascal's triangle are $(1,1)$, $(1,2,1)$ and $(1,3,3,1)$. Write down the next two rows of the triangle.

(b) Create filters $\widetilde{\mathbf{h}}$ (of lengths 5 and 6) for each of the rows in (a) and normalize them so that the elements of each filter sums to $\sqrt{2}$. Show that these filters satisfy $\sum\limits_k (-1)^k h_k = 0$ and are thus lowpass filters.

★7.16 The Nth row of Pascal's triangle is generated by expanding $(a+b)^N$ and identifying the coefficients in the expansion. The *binomial theorem* gives us a closed formula for this expansion. We have

---

Binomial Theorem: For all real numbers a, b and nonnegative integer N, we have

$$
(a+b)^N = \sum_{k=0}^{N} \binom{N}{k} a^k b^{N-k} \tag{7.30}
$$

where the binomial coefficient $\binom{N}{k}$ is defined in Problem 5.30.

(a) The binomial coefficients given in (5.70) make up the Nth row of Pascal's triangle. Show that the binomial coefficients are symmetric about $k = N/2$. That is, show that $\binom{N}{k} = \binom{N}{N-k}$, $k = 0, \ldots, N$.

(b) The well-known Pascal's identity is

$$\binom{N+1}{k} = \binom{N}{k-1} + \binom{N}{k} \tag{7.31}$$

where N and k are nonnegative integers with $N \geq k$. Use the definition of the binomial coefficient (5.70) to prove Pascal's identity. Note that Pascal's identity is what we used to generate rows three through five in constructing the filters in (7.19) and Problem 7.15.

(c) Use Pascal's identity (7.31) and mathematical induction to prove the binomial theorem.

7.17 In this problem, you will use row N of Pascal's triangle to construct a symmetric lowpass filter.

(a) Make appropriate choices for a and b in the binomial theorem to find $\sum\limits_{k=0}^{N} \binom{N}{k}$.

(b) Make appropriate choices for a and b in the binomial theorem to show that $0 = \sum\limits_{k=0}^{N} \binom{N}{k}(-1)^k$.

(c) Under the assumption that the sum of the filter elements is $\sqrt{2}$, use the Nth row of Pascal's triangle and (a) to write down the element $\widetilde{h}_k$ of length $N+1$ symmetric filter $\widetilde{\mathbf{h}}$. *Hint:* You will have different formulas for $\widetilde{h}_k$ depending on whether N is even (so that k runs from $-N/2$ to $N/2$) or N is odd. What are the index limits on k when N is odd?

7.18 In this problem, you will show that no odd-length filter $\mathbf{h}$ can be used to construct $H_{N/2}$ so that $\widetilde{H}_{N/2}H_{N/2}^T = I_{N/2}$, where $\widetilde{H}_{N/2}$ constructed with $\widetilde{\mathbf{h}}$ given in (7.19). We prove this result by induction. Here is a general symmetric odd-length filter $\mathbf{h}$:

$$\begin{aligned}\mathbf{h} &= (h_{-L}, h_{-L+1}, \ldots, h_{-1}, h_0, h_1, \ldots, h_{L-1}, h_L) \\ &= (h_L, h_{L-1}, \ldots, h_1, h_0, h_1, \ldots, h_{L-1}, h_L).\end{aligned}$$

(a) Let $L = 1$. When we dot row j of $\widetilde{H}_{N/2}$ with row j of $H_{N/2}$ (column k of $H_{N/2}^T$), we compute

$$\begin{array}{cccccccc} \cdots & 0 & \widetilde{h}_{-1} & \widetilde{h}_0 & \widetilde{h}_1 & \widetilde{h}_2 & 0 & \cdots \\ \cdots & 0 & h_{-1} & h_0 & h_1 & 0 & 0 & \cdots \end{array}$$

and this product must be 1. What must the dot product be if we shift the second row above by two left? What does this say about $h_{-1} = h_1$?

(b) Now assume the negative result when symmetric $\mathbf{h}$ has length $2L + 1$. Show that the negative result holds for symmetric filters of length $2L + 3$. *Hint:* Use the idea of (a) – depending on L, you may have to shift a sufficient amount of units either left or right to obtain the contradiction that $h_L = 0$.

7.19 Show that the system (7.23) has infinitely many solutions.

7.20 Suppose symmetric $\mathbf{h} = (h_0, h_1) = (a, a)$ for $a \in \mathbb{R}$, $a \neq 0$. Write down H_6 for this filter. Show that we cannot obtain the desired result $\widetilde{H}_6 H_6^T = I_6$, where $\widetilde{H}_6$ is given in (7.20).

7.21 Suppose $\mathbf{h} = (h_{-1}, h_0, h_1, h_2) = (a, b, b, a)$ for $a, b \in \mathbb{R}$. Write down H_6 and compute $\widetilde{H}_6 H_6^T$, where $\widetilde{H}_6$ is given in (7.20), and show that the product is I_6 when $a = -\frac{\sqrt{2}}{4}$ and $b = \frac{3\sqrt{2}}{4}$. The pair $\left(\mathbf{h}, \widetilde{\mathbf{h}}\right)$ where $\widetilde{\mathbf{h}}$ is given in (7.19) is called the *(4,4) biorthogonal spline filter pair.* We will learn more about these pairs in Section 9.4.

7.22 Construct matrices $\widetilde{W}_N$ and W_N for the biorthogonal filter pair from Problem 7.21 and then by hand, compute the biorthogonal wavelet transformations of the following vectors.

(a) $\mathbf{v} \in \mathbb{R}^{12}$ where $v_k = k$, $k = 1, \ldots, 12$

(b) $\mathbf{v} \in \mathbb{R}^{12}$ where $v_k = \frac{\sqrt{2}}{2}$, $k = 1, \ldots, 12$

(c) $\mathbf{v} \in \mathbb{R}^{12}$, where $v_k = (-1)^{k+1}\frac{\sqrt{2}}{2}$, $k = 1, \ldots, 12$

7.23 Using the formulas $g_k = (-1)^k \widetilde{h}_{1-k}$ and $\widetilde{g}_k = (-1)^k h_{1-k}$, $k = -1, \ldots, 3$, write down $\widetilde{\mathbf{g}}$ and $\mathbf{g}$ for the filter pair found in Problem 7.21.

(a) Do $\widetilde{\mathbf{g}}$ and $\mathbf{g}$ annihilate constant data?

(b) Do $\widetilde{\mathbf{g}}$ and $\mathbf{g}$ annihilate linear data? (Dot each filter with $[k, k+1, k+2, k+3]^T$ to check.)

(c) Write down wavelet matrices $\widetilde{W}_8$ and W_8 and verify that $\widetilde{W}_8 W_8^T = I_8$.

7.24 Suppose symmetric $\mathbf{h} = (h_{-2}, h_{-1}, h_0, h_1, h_2, h_3) = (a, b, c, c, b, a)$ for $a, b, c \in \mathbb{R}$. Write down H_6 for this filter. Show that insisting $\widetilde{H}_6 H_6^T = I_6$, where $\widetilde{H}_6$ is given in (7.20), implies that $a = 0$ and b, c are the values obtained for the 4-term filter in Problem 7.21. Thus $\mathbf{h}$ cannot be a length six symmetric filter.

7.25 Suppose that $\mathbf{h}$ is a symmetric even-length filter. That is

$$\mathbf{h} = (h_{-L+1}, h_{-L+2}, \ldots, h_0, h_1, \ldots, h_{L-1}, h_L).$$

Show that

$$\sum_{k=-L+1}^{L} (-1)^k h_k = 0.$$

Hint: Split the sum into two sums, the first running from $-L+1$ to 0 and the second running from 1 to L. Then use the fact that $h_k = h_{1-k}$.

7.26 Use either Gaussian elimination or a **CAS** to solve the system (7.23) – (7.24) and verify the answer given in (7.25).

7.27 Show that the filter $\widetilde{\mathbf{g}}$ given in (7.27) annihilates quadratic data.

7.28 Show that the filter $\mathbf{g}$ given in (7.26) annihilates both linear and quadratic data.

7.29 Repeat Problem 7.22 for the $(8, 4)$ biorthogonal spline filter pair.

7.30 Suppose that $\widetilde{\mathbf{h}}$ is an even-length symmetric filter and the elements of $\mathbf{g}$ are defined by $g_k = (-1)^k \widetilde{h}_{1-k}$. Show that $g_k = -g_{1-k}$.

7.31 Show by direct computation that $\widetilde{H}_6 G_6^T = \widetilde{G}_6 H_6^T = 0_6$ and $\widetilde{G}_6 G_6^T = I_6$.

7.32 Suppose we wish to compute

$$W_N \mathbf{v} = W_N \begin{bmatrix} v_1 \\ v_2 \\ \vdots \\ v_N \end{bmatrix} = \left[\frac{\mathbf{a}}{\mathbf{d}}\right]$$

where W_N is given by (7.29) and N is an even number. Since W_N is sparse matrix, we reformulate the product to create a more efficient means of computing the transformation. Toward that end, let $L = N/2$. Using the elements of $\mathbf{v}$, construct an $L \times 8$ matrix A and an $L \times 4$ matrix B such that $A\mathbf{h} = \mathbf{a}$ and $B\mathbf{g} = \mathbf{d}$, where $\mathbf{h}$ and $\mathbf{g}$ are given by (7.25) and (7.26), respectively.

7.33 Suppose we have computed the biorthogonal wavelet transformation $W_N \mathbf{v} = \left[\frac{\mathbf{a}}{\mathbf{d}}\right]$, N even, and we wish to recover $\mathbf{v}$. We know $W_N^{-1} = \widetilde{W}_N^T$, where $\widetilde{W}_N$ is given by (7.28). We compute

$$\mathbf{v} = \widetilde{W}_N^T \left[\frac{\mathbf{a}}{\mathbf{d}}\right] = \left[\widetilde{H}_N^T \middle| \widetilde{G}_N^T \right] \left[\frac{\mathbf{a}}{\mathbf{d}}\right] = \widetilde{H}_N^T \mathbf{a} + \widetilde{G}_N^T \mathbf{d}.$$

Both $\widetilde{H}_N^T$ and $\widetilde{G}_N^T$ are sparse matrices so we reformulate the computation to improve efficiency. Toward this end, let $L = N/2$ and consider the products

$$\widetilde{H}_N^T\mathbf{a} = \begin{bmatrix} \widetilde{h}_0 & 0 & 0 & \cdots & 0 & \widetilde{h}_2 \\ \widetilde{h}_1 & \widetilde{h}_{-1} & 0 & & 0 & 0 \\ \widetilde{h}_2 & \widetilde{h}_0 & 0 & & 0 & 0 \\ 0 & \widetilde{h}_1 & \widetilde{h}_{-1} & & 0 & 0 \\ \vdots & & & \ddots & & \vdots \\ 0 & 0 & 0 & & \widetilde{h}_1 & \widetilde{h}_{-1} \\ 0 & 0 & 0 & & \widetilde{h}_2 & \widetilde{h}_0 \\ \widetilde{h}_{-1} & 0 & 0 & \cdots & 0 & \widetilde{h}_1 \end{bmatrix} \begin{bmatrix} a_1 \\ a_2 \\ \vdots \\ a_L \end{bmatrix}$$

and

$$\widetilde{G}_N^T\mathbf{d} = \begin{bmatrix} \widetilde{g}_0 & \widetilde{g}_{-2} & 0 & 0 & \cdots & 0 & 0 & \widetilde{g}_4 & \widetilde{g}_2 \\ \widetilde{g}_1 & \widetilde{g}_{-1} & \widetilde{g}_{-3} & 0 & & 0 & 0 & 0 & \widetilde{g}_3 \\ \widetilde{g}_2 & \widetilde{g}_0 & \widetilde{g}_{-2} & 0 & & 0 & 0 & 0 & \widetilde{g}_4 \\ \widetilde{g}_3 & \widetilde{g}_1 & \widetilde{g}_{-1} & \widetilde{g}_{-3} & & 0 & 0 & 0 & 0 \\ \widetilde{g}_4 & \widetilde{g}_2 & \widetilde{g}_0 & \widetilde{g}_{-2} & & 0 & 0 & 0 & 0 \\ \vdots & & & & \vdots & & & & \vdots \\ 0 & 0 & 0 & 0 & & \widetilde{g}_4 & \widetilde{g}_2 & \widetilde{g}_0 & \widetilde{g}_{-2} \\ \widetilde{g}_{-3} & 0 & 0 & 0 & & 0 & \widetilde{g}_3 & \widetilde{g}_1 & \widetilde{g}_{-1} \\ \widetilde{g}_{-2} & 0 & 0 & 0 & & 0 & \widetilde{g}_4 & \widetilde{g}_2 & \widetilde{g}_0 \\ \widetilde{g}_{-1} & \widetilde{g}_{-3} & 0 & 0 & \cdots & 0 & 0 & \widetilde{g}_3 & \widetilde{g}_1 \end{bmatrix} \begin{bmatrix} d_1 \\ d_2 \\ \vdots \\ d_L \end{bmatrix}.$$

(a) Using the elements of $\mathbf{a}$, find an $L \times 2$ matrix A such that $A \begin{bmatrix} \widetilde{h}_2 \\ \widetilde{h}_0 \end{bmatrix} = \mathbf{o}^1$ contains the odd-indexed elements of $\widetilde{H}_N^T\mathbf{a}$.

(b) Using the elements of $\mathbf{a}$, find an $L \times 2$ matrix B such that $B \begin{bmatrix} \widetilde{h}_1 \\ \widetilde{h}_{-1} \end{bmatrix} = \mathbf{e}^1$, where the elements of $\mathbf{e}^1$ are the even-indexed elements of $\widetilde{H}_N^T\mathbf{a}$.

(c) Using the elements of $\mathbf{d}$, find an $L \times 4$ matrix X such that $X \begin{bmatrix} \widetilde{g}_4 \\ \widetilde{g}_2 \\ \widetilde{g}_0 \\ \widetilde{g}_{-2} \end{bmatrix} = \mathbf{o}^2$, where the elements of $\mathbf{o}^2$ are the odd-indexed elements of $\widetilde{G}_N^T\mathbf{d}$.

(d) Using the elements of $\mathbf{d}$, find an $L \times 4$ matrix Y such that $Y \begin{bmatrix} \widetilde{g}_3 \\ \widetilde{g}_1 \\ \widetilde{g}_{-1} \\ \widetilde{g}_{-3} \end{bmatrix} = \mathbf{e}^2$, where the elements of $\mathbf{e}^2$ are the odd-indexed elements of $\widetilde{G}_N^T\mathbf{d}$.

(e) Find permutation matrix P so that

$$P\left(\begin{bmatrix}\mathbf{o}^1\\ \mathbf{e}^1\end{bmatrix}+\begin{bmatrix}\mathbf{o}^2\\ \mathbf{e}^2\end{bmatrix}\right)=\widetilde{H}_N^T\mathbf{a}+\widetilde{G}_N^T\mathbf{d}.$$

Computer Lab

7.2 The Biorthogonal Wavelet Transformation Using the $(8,4)$ Filter Pair. From `stthomas.edu/wavelets`, access the file `Biorth84`. In this lab, you will develop and test efficient algorithms for computing the one- and two-dimensional biorthogonal wavelet transformations using the $(8,4)$ biorthogonal spline filter pair.

7.3 Symmetry and Boundary Effects

In our construction of orthogonal W_N in Chapter 5, we wrapped rows near the bottom parts of the averages and differences portions of the transformation matrix. The averages portion H_5 of W_{10} using the D6 filter is

$$H_5=\begin{bmatrix} h_5 & h_4 & h_3 & h_2 & h_1 & h_0 & 0 & 0 & 0 & 0\\ 0 & 0 & h_5 & h_4 & h_3 & h_2 & h_1 & h_0 & 0 & 0\\ 0 & 0 & 0 & 0 & h_5 & h_4 & h_3 & h_2 & h_1 & h_0\\ h_1 & h_0 & 0 & 0 & 0 & 0 & h_5 & h_4 & h_3 & h_2\\ h_3 & h_2 & h_1 & h_0 & 0 & 0 & 0 & 0 & h_5 & h_4 \end{bmatrix} \tag{7.32}$$

where $\mathbf{h}=(h_0,h_1,h_2,h_3,h_4,h_5)$ is given by Definition 5.2.

As we move our filter two to the right in each subsequent row, it finally reaches the last column in row three. In rows four and five, we wrap the filter around the first two and four columns, respectively, of each row. The reason we allow the wrapping rows is that it facilitates the construction of an orthogonal matrix – if we truncate these rows, we would have many more conditions to satisfy in order to make W_N orthogonal. The major drawback of this construction is that we implicitly assume that our data are periodic. Note that dotting row four of H_5 in (7.32) with $\mathbf{v}\in\mathbb{R}^{10}$ uses components v_7,v_8,v_9,v_{10},v_1, and v_2. Now if $\mathbf{v}$ can be viewed as a repeating or periodic signal, say $v_9=v_1,v_8=v_2,\ldots$ this computation makes perfect sense. But in practice, digital images or audio signals can hardly be considered periodic. And even if the data are periodic, there is no special correlation of the filter coefficients $h_0,\ldots,h_5$ to exploit this.

In this section we develop techniques utilizing symmetric filters that minimize the problems caused by wrapping rows in the transformation. The material that appears in this section is outlined in Walnut [100].

Using Symmetry with a Biorthogonal Filter Pair

To motivate the process for handling the wrapping rows, let's consider an example.

Example 7.3 (Symmetry and the Biorthogonal Wavelet Transformation). *Suppose that we wish to transform the vector* $\mathbf{v} = [v_1, v_2, \ldots, v_8]^T$ *using the* $(5,3)$ *biorthogonal spline filter pair given in (7.2). The averages portion of the transformation* $\widetilde{W}_8$ *uses the symmetric filter* $\widetilde{\mathbf{h}} = \left(\widetilde{h}_{-1}, \widetilde{h}_0, \widetilde{h}_1\right) = \left(\widetilde{h}_1, \widetilde{h}_0, \widetilde{h}_1\right)$. *We let* $\mathbf{a}$ *denote the averages portion of the biorthogonal wavelet transformation so that*

$$\widetilde{\mathbf{a}} = \widetilde{H}_4\mathbf{v} = \begin{bmatrix} \widetilde{h}_0 & \widetilde{h}_1 & 0 & 0 & 0 & 0 & 0 & \widetilde{h}_1 \\ 0 & \widetilde{h}_1 & \widetilde{h}_0 & \widetilde{h}_1 & 0 & 0 & 0 & 0 \\ 0 & 0 & 0 & \widetilde{h}_1 & \widetilde{h}_0 & \widetilde{h}_1 & 0 & 0 \\ 0 & 0 & 0 & 0 & 0 & \widetilde{h}_1 & \widetilde{h}_0 & \widetilde{h}_1 \end{bmatrix} \begin{bmatrix} v_1 \\ v_2 \\ v_3 \\ v_4 \\ v_5 \\ v_6 \\ v_7 \\ v_8 \end{bmatrix}$$

$$= \begin{bmatrix} \widetilde{h}_1 v_8 + \widetilde{h}_0 v_1 + \widetilde{h}_1 v_2 \\ \widetilde{h}_1 v_2 + \widetilde{h}_0 v_3 + \widetilde{h}_1 v_4 \\ \widetilde{h}_1 v_4 + \widetilde{h}_0 v_5 + \widetilde{h}_1 v_6 \\ \widetilde{h}_1 v_6 + \widetilde{h}_0 v_7 + \widetilde{h}_1 v_8 \end{bmatrix}.$$

Notice that the first element of $\mathbf{y}$ *is* $\widetilde{h}_0 v_1 + \widetilde{h}_1 v_2 + \widetilde{h}_1 v_8$. *This computation uses elements from the top and bottom of* $\mathbf{v}$, *but unlike the filter in the orthogonal case (7.32), the filter coefficients are symmetric. A more natural combination of elements for* $\widetilde{a}_1$ *should use only the first few elements from* $\mathbf{v}$.

Since we can never assume that our input would satisfy such a symmetry condition, we try instead to periodize the input. Let's create the 14*-vector*

$$\mathbf{v}^p = [v_1, v_2, \ldots, v_7, v_8, v_7, v_6, v_5, v_4, v_3, v_2]^T \tag{7.33}$$

and look at the structure of $\mathbf{v}^p$. *We have basically removed the first and last elements of* $\mathbf{v}$, *reversed the result, and appended it to* $\mathbf{v}$. *We are guaranteed an even length vector with this process and also note that the last element of* $\mathbf{v}^p$ *is* v_2. *Now consider the computation*

$$\widetilde{\mathbf{a}}^p = \widetilde{H}_7\mathbf{v}^p$$

$$= \begin{bmatrix} \widetilde{h}_0 & \widetilde{h}_1 & 0 & 0 & 0 & 0 & 0 & 0 & 0 & 0 & 0 & 0 & 0 & \widetilde{h}_1 \\ 0 & \widetilde{h}_1 & \widetilde{h}_0 & \widetilde{h}_1 & 0 & 0 & 0 & 0 & 0 & 0 & 0 & 0 & 0 & 0 \\ 0 & 0 & 0 & \widetilde{h}_1 & \widetilde{h}_0 & \widetilde{h}_1 & 0 & 0 & 0 & 0 & 0 & 0 & 0 & 0 \\ 0 & 0 & 0 & 0 & 0 & \widetilde{h}_1 & \widetilde{h}_0 & \widetilde{h}_1 & 0 & 0 & 0 & 0 & 0 & 0 \\ 0 & 0 & 0 & 0 & 0 & 0 & 0 & \widetilde{h}_1 & \widetilde{h}_0 & \widetilde{h}_1 & 0 & 0 & 0 & 0 \\ 0 & 0 & 0 & 0 & 0 & 0 & 0 & 0 & 0 & \widetilde{h}_1 & \widetilde{h}_0 & \widetilde{h}_1 & 0 & 0 \\ 0 & 0 & 0 & 0 & 0 & 0 & 0 & 0 & 0 & 0 & 0 & \widetilde{h}_1 & \widetilde{h}_0 & \widetilde{h}_1 \end{bmatrix} \cdot \begin{bmatrix} v_1 \\ v_2 \\ v_3 \\ v_4 \\ v_5 \\ v_6 \\ v_7 \\ v_8 \\ v_7 \\ v_6 \\ v_5 \\ v_4 \\ v_3 \\ v_2 \end{bmatrix}$$

$$= \left[\begin{array}{c} \widetilde{h}_0 v_1 + 2\widetilde{h}_1 v_2 \\ \widetilde{h}_1 v_2 + \widetilde{h}_0 v_3 + \widetilde{h}_1 v_4 \\ \widetilde{h}_1 v_4 + \widetilde{h}_0 v_5 + \widetilde{h}_1 v_6 \\ \widetilde{h}_1 v_6 + \widetilde{h}_0 v_7 + \widetilde{h}_1 v_8 \\ \hline \widetilde{h}_1 v_8 + \widetilde{h}_0 v_7 + \widetilde{h}_1 v_6 \\ \widetilde{h}_1 v_6 + \widetilde{h}_0 v_5 + \widetilde{h}_1 v_4 \\ \widetilde{h}_1 v_4 + \widetilde{h}_0 v_3 + \widetilde{h}_1 v_2 \end{array}\right]. \tag{7.34}$$

Note that the elements of $\widetilde{\mathbf{a}}^p$ *exhibit some symmetry. In particular,* $\widetilde{a}^p_2 = \widetilde{a}^p_7$, $\widetilde{a}^p_3 = \widetilde{a}^p_6$, *and* $\widetilde{a}^p_4 = \widetilde{a}^p_5$. *So* $\widetilde{\mathbf{a}}^p$ *has the form*

$$\widetilde{\mathbf{a}}^p = \left[\widetilde{a}^p_1,\, \widetilde{a}^p_2,\, \widetilde{a}^p_3,\, \widetilde{a}^p_4 \middle| \widetilde{a}^p_4,\, \widetilde{a}^p_3,\, \widetilde{a}^p_2\right]^T. \tag{7.35}$$

More important, the first element $\widetilde{a}^p_1 = \widetilde{h}_0 v_1 + 2\widetilde{h}_1 v_2$ *uses a combination of only the first two elements of* $\mathbf{v}$.

The value $\widetilde{a}^p_1$ *combines only the first few elements of* $\mathbf{v}$, *where* $\widetilde{a}_1$ *uses* v_8. *So* $\widetilde{a}^p_1$ *provides a more accurate representation of the top portion of* $\mathbf{v}$ *than* $\widetilde{a}_1$ *does. Also note that* $\widetilde{a}^p_2 = \widetilde{a}_2$, $\widetilde{a}^p_3 = \widetilde{a}_3$, *and* $\widetilde{a}^p_4 = \widetilde{a}_4$. *Thus, it makes sense to use* $[\widetilde{a}^p_1, \widetilde{a}^p_1, \widetilde{a}^p_3, \widetilde{a}^p_4]^T$ *as the averages portion of our transformation.*

We now consider the differences portion for both $\widetilde{G}_4\mathbf{v}$ *and* $\widetilde{G}_7\mathbf{v}^p$. *The differences portion of the transformation uses the* 5*-term filter* $\widetilde{\mathbf{g}}$, *where* $\widetilde{g}_k = (-1)^k h_{1-k}$, $k =$

$-1,\ldots,3$ *given in* (7.14). *We have*

$$
\widetilde{\mathbf{d}} = \widetilde{G}_4\mathbf{v} = \begin{bmatrix} \widetilde{g}_0 & \widetilde{g}_1 & \widetilde{g}_2 & \widetilde{g}_3 & 0 & 0 & 0 & \widetilde{g}_{-1} \\ 0 & \widetilde{g}_{-1} & \widetilde{g}_0 & \widetilde{g}_1 & \widetilde{g}_2 & \widetilde{g}_3 & 0 & 0 \\ 0 & 0 & 0 & \widetilde{g}_{-1} & \widetilde{g}_0 & \widetilde{g}_1 & \widetilde{g}_2 & \widetilde{g}_3 \\ \widetilde{g}_2 & \widetilde{g}_3 & 0 & 0 & 0 & \widetilde{g}_{-1} & \widetilde{g}_0 & \widetilde{g}_1 \end{bmatrix} \cdot \begin{bmatrix} v_1 \\ v_2 \\ v_3 \\ v_4 \\ v_5 \\ v_6 \\ v_7 \\ v_8 \end{bmatrix}
$$

$$
= \begin{bmatrix} \widetilde{g}_{-1}v_8 + \widetilde{g}_0 v_1 + \widetilde{g}_1 v_2 + \widetilde{g}_2 v_3 + \widetilde{g}_3 v_4 \\ \widetilde{g}_{-1}v_2 + \widetilde{g}_0 v_3 + \widetilde{g}_1 v_4 + \widetilde{g}_2 v_5 + \widetilde{g}_3 v_6 \\ \widetilde{g}_{-1}v_4 + \widetilde{g}_0 v_5 + \widetilde{g}_1 v_6 + \widetilde{g}_2 v_7 + \widetilde{g}_3 v_8 \\ \widetilde{g}_{-1}v_6 + \widetilde{g}_0 v_7 + \widetilde{g}_1 v_8 + \widetilde{g}_2 v_1 + \widetilde{g}_3 v_2 \end{bmatrix}
$$

$$
= \begin{bmatrix} -h_2 v_8 + h_1 v_1 - h_0 v_2 + h_1 v_3 - h_2 v_4 \\ -h_2 v_2 + h_1 v_3 - h_0 v_4 + h_1 v_5 - h_2 v_6 \\ -h_2 v_4 + h_1 v_5 - h_0 v_6 + h_1 v_7 - h_2 v_8 \\ -h_2 v_6 + h_1 v_7 - h_0 v_8 + h_1 v_1 - h_2 v_2 \end{bmatrix}.
$$

We now multiply

$$
\widetilde{G}_7 = \begin{bmatrix} \widetilde{g}_0 & \widetilde{g}_1 & \widetilde{g}_2 & \widetilde{g}_3 & 0 & 0 & 0 & 0 & 0 & 0 & 0 & 0 & 0 & \widetilde{g}_{-1} \\ 0 & \widetilde{g}_{-1} & \widetilde{g}_0 & \widetilde{g}_1 & \widetilde{g}_2 & \widetilde{g}_3 & 0 & 0 & 0 & 0 & 0 & 0 & 0 & 0 \\ 0 & 0 & 0 & \widetilde{g}_{-1} & \widetilde{g}_0 & \widetilde{g}_1 & \widetilde{g}_2 & \widetilde{g}_3 & 0 & 0 & 0 & 0 & 0 & 0 \\ 0 & 0 & 0 & 0 & 0 & \widetilde{g}_{-1} & \widetilde{g}_0 & \widetilde{g}_1 & \widetilde{g}_2 & \widetilde{g}_3 & 0 & 0 & 0 & 0 \\ 0 & 0 & 0 & 0 & 0 & 0 & 0 & \widetilde{g}_{-1} & \widetilde{g}_0 & \widetilde{g}_1 & \widetilde{g}_2 & \widetilde{g}_3 & 0 & 0 \\ 0 & 0 & 0 & 0 & 0 & 0 & 0 & 0 & 0 & \widetilde{g}_{-1} & \widetilde{g}_0 & \widetilde{g}_1 & \widetilde{g}_2 & \widetilde{g}_3 \\ \widetilde{g}_2 & \widetilde{g}_3 & 0 & 0 & 0 & 0 & 0 & 0 & 0 & 0 & 0 & \widetilde{g}_{-1} & \widetilde{g}_0 & \widetilde{g}_1 \end{bmatrix}
$$

and the periodized 14-*vector* $\mathbf{v}^p$ *given in* (7.33) *to obtain*

$$
\widetilde{\mathbf{d}}^p = \widetilde{G}_7\mathbf{v}^p = \left[\begin{array}{c} \widetilde{g}_{-1}v_2 + \widetilde{g}_0 v_1 + \widetilde{g}_1 v_2 + \widetilde{g}_2 v_3 + \widetilde{g}_3 v_4 \\ \widetilde{g}_{-1}v_2 + \widetilde{g}_0 v_3 + \widetilde{g}_1 v_4 + \widetilde{g}_2 v_5 + \widetilde{g}_3 v_6 \\ \widetilde{g}_{-1}v_4 + \widetilde{g}_0 v_5 + \widetilde{g}_1 v_6 + \widetilde{g}_2 v_7 + \widetilde{g}_3 v_8 \\ \widetilde{g}_{-1}v_6 + \widetilde{g}_0 v_7 + \widetilde{g}_1 v_8 + \widetilde{g}_2 v_7 + \widetilde{g}_3 v_6 \\ \hline \widetilde{g}_{-1}v_8 + \widetilde{g}_0 v_7 + \widetilde{g}_1 v_6 + \widetilde{g}_2 v_5 + \widetilde{g}_3 v_4 \\ \widetilde{g}_{-1}v_6 + \widetilde{g}_0 v_5 + \widetilde{g}_1 v_4 + \widetilde{g}_2 v_3 + \widetilde{g}_3 v_2 \\ \widetilde{g}_{-1}v_4 + \widetilde{g}_0 v_3 + \widetilde{g}_1 v_2 + \widetilde{g}_2 v_1 + \widetilde{g}_3 v_2 \end{array}\right]. \tag{7.36}
$$

Note that $\widetilde{\mathbf{d}}^p$ *possesses some symmetry although the symmetry is a bit different than that possessed by the elements of* $\widetilde{\mathbf{a}}^p$. *We have* $\widetilde{d}^p_5 = \widetilde{d}^p_3$, $\widetilde{d}^p_6 = \widetilde{d}^p_2$, *and* $\widetilde{d}^p_7 = \widetilde{d}^p_1$, *so that* $\widetilde{\mathbf{d}}^p$ *has the form*

$$
\widetilde{\mathbf{d}}^p = \left[\, \widetilde{d}^p_1,\, \widetilde{d}^p_2,\, \widetilde{d}^p_3,\, \widetilde{d}^p_4 \,\middle|\, \widetilde{d}^p_3,\, \widetilde{d}^p_2,\, \widetilde{d}^p_1 \right]^T. \tag{7.37}
$$

We also observe that $\widetilde{d}^p_2 = \widetilde{d}_2$ *and* $\widetilde{d}^p_3 = \widetilde{d}_3$. *Unlike* $\widetilde{d}_1$, $\widetilde{d}^p_1$ *uses only the first four elements of* $\mathbf{v}$ *and in a similar way, we see that* $\widetilde{d}^p_4$ *uses only the last three elements of* $\mathbf{v}$. *Thus we will take the first four elements of* $\widetilde{\mathbf{d}}^p$ *as the differences portion of the wavelet transformation. Our "symmetric" wavelet transformation is*

$$
\left[\widetilde{a}^p_1, \widetilde{a}^p_2, \widetilde{a}^p_3, \widetilde{a}^p_4 \,|\, \widetilde{d}^p_1, \widetilde{d}^p_2, \widetilde{d}^p_3, \widetilde{d}^p_4 \right]^T. \tag{7.38}
$$

□

More on Symmetric Filter Lengths

Before developing general algorithms for symmetric biorthogonal wavelet transformations and their inverses, we need to know more about the relationship of the lengths of symmetric biorthogonal filter pairs. When attempting to find $\mathbf{h}$ to pair with the 4-term symmetric filter $\widetilde{\mathbf{h}}$ given by (7.19), we considered filter lengths that *didn't* work. In particular, Problem 7.18 in the previous section outlined a proof to show that the length of $\mathbf{h}$ cannot be odd. The following proposition generalizes this result.

Proposition 7.2 (Length Characteristics of Symmetric Biorthogonal Filter Pairs). *Suppose* $\left(\mathbf{h}, \widetilde{\mathbf{h}}\right)$ *is a biorthogonal filter pair with lengths* N *and* $\widetilde{N}$. *Further suppose each filter is symmetric in the sense of Definition 7.1.*[1] *Then* N, $\widetilde{N}$ *are either both even or both odd.* □

Proof. We consider the case where N is odd and leave the proof when N is even as Problem 7.34.

Since N is odd, we use Proposition 7.1 to conclude that $\mathbf{h} = (h_{-L}, \ldots, h_0, \ldots, h_L)$ for some positive integer L. Suppose that $\widetilde{N}$ is even so that again by Proposition 7.1 we have $\widetilde{\mathbf{h}} = \left(\widetilde{h}_{-\widetilde{L}+1}, \ldots, \widetilde{h}_0, \ldots, \widetilde{h}_{\widetilde{L}}\right)$ for some positive integer $\widetilde{L}$.

If $\widetilde{L}$ is even, then $-\widetilde{L}+1$ is odd and shifting $\widetilde{\mathbf{h}}$ to the right $L - \widetilde{L}$ (even) units, as we would do to consider the entries of $I = \widetilde{H}H^T$, results in the following inner product whose value is 0:

$$\begin{array}{ccccccccccc} \cdots & h_{-L} & \cdots & h_0 & \cdots & h_L & 0 & \cdots & 0 & \cdots \\ \cdots & 0 & 0 & 0 & \cdots & \widetilde{h}_{-\widetilde{L}+1} & \widetilde{h}_{-\widetilde{L}+2} & \cdots & \widetilde{h}_{\widetilde{L}} & \cdots. \end{array}$$

This implies either $h_L = 0$ or $\widetilde{h}_{\widetilde{L}} = 0$ but this contradicts the convention (page 263) we have made for lengths of symmetric filters.

If $\widetilde{L}$ is odd, then shifting $\widetilde{\mathbf{h}}$ to the left $\widetilde{L} - (-L) = \widetilde{L} + L$ (even) units results in the following dot product whose value is 0:

$$\begin{array}{ccccccccccc} \cdots & 0 & \cdots & 0 & h_{-L} & \cdots & h_0 & \cdots & h_L & \cdots \\ \cdots & \widetilde{h}_{-\widetilde{L}+1} & \cdots & \widetilde{h}_{\widetilde{L}-1} & \widetilde{h}_{\widetilde{L}} & 0 & 0 & \cdots & 0 & \cdots. \end{array}$$

This implies that either $h_{-L} = 0$ or $\widetilde{h}_{\widetilde{L}} = 0$ which again contradicts the convention on (page 263). Thus $\widetilde{N}$ cannot be odd. □

[1] This assumption can be relaxed so that the starting index of one filter is less than the stopping index of the other filter. Since we only consider biorthogonal filters in this book, we opt not to use this relaxed assumption.

A Symmetric Biorthogonal Wavelet Transformation

The algorithm for Example 7.3 is as follows:

1. Form $\mathbf{v}^p$ using (7.33).

2. Compute $\mathbf{w} = \widetilde{W}_{14}\mathbf{v}^p$.

3. If $\widetilde{\mathbf{a}}^p$ and $\widetilde{\mathbf{d}}^p$ denote the averages and differences portions of $\mathbf{w}$, respectively, then take as the symmetric transformation the 8-vector given by (7.38).

An algorithm for applying the symmetric biorthogonal transformation constructed from odd-length biorthogonal filter pairs applied to vectors of even length N is presented next without justification. The justification for this algorithm is presented at the end of the section.

Algorithm 7.1 (Symmetric BWT for Odd-Length Filter Pairs). *This algorithm takes as input the N-vector $\mathbf{v}$ (N even) and an odd-length biorthogonal filter pair $\mathbf{h}$ and $\widetilde{\mathbf{h}}$ and returns a symmetric version of the biorthogonal wavelet transformation. The vector $\mathbf{v}$ is periodized to form vector $\mathbf{v}^p \in \mathbb{R}^{2N-2}$. Then we compute the biorthogonal wavelet transformation of $\mathbf{v}^p$. If $\widetilde{\mathbf{a}}^p$ and $\widetilde{\mathbf{d}}^p$ are the averages and differences portions, respectively, of the biorthogonal wavelet transformation, then the algorithm returns the symmetric biorthogonal wavelet transformation*

$$\left[\widetilde{a}_1, \ldots, \widetilde{a}_{N/2} \,|\, \widetilde{d}_1, \ldots, \widetilde{d}_{N/2}\right]^T.$$

1. *Given $\mathbf{v} \in \mathbb{R}^N$, N even, form $\mathbf{v}^p \in \mathbb{R}^{2N-2}$ by the rule*

$$\mathbf{v}^p = \left[v_1, v_2, \ldots, v_{N-1}, v_N, v_{N-1}, \ldots, v_3, v_2\right]^T. \tag{7.39}$$

2. *Compute the biorthogonal wavelet transformation $\mathbf{w} = \widetilde{W}_{2N-2}\mathbf{v}^p$. Denote the averages portion of the transformation by $\widetilde{\mathbf{a}}^p$ and the differences portion of the transformation by $\widetilde{\mathbf{d}}^p$.*

3. *Return $\left[\widetilde{a}_1^p, \ldots, \widetilde{a}_{N/2}^p \,|\, \widetilde{d}_1^p, \ldots, \widetilde{d}_{N/2}^p\right]^T$.*

□

Let's look at an example that illustrates Algorithm 7.1.

Example 7.4 (Using Algorithm 7.1). *Consider the vector $\mathbf{v} \in \mathbb{R}^{24}$ whose components are given by the rule $v_k = k$, $k = 1, \ldots, 24$. The components of vector $\mathbf{v}$ are plotted in Figure 7.3.*

We next compute the biorthogonal wavelet transformation $\mathbf{w}$ and the symmetric biorthogonal wavelet transformation $\mathbf{w}^s$ (here the superscript s denotes a symmetric transformation) of $\mathbf{v}$ using the $\widetilde{\mathbf{h}}$ and $\widetilde{\mathbf{g}}$ given in (7.13) *and* (7.14), *respectively.*

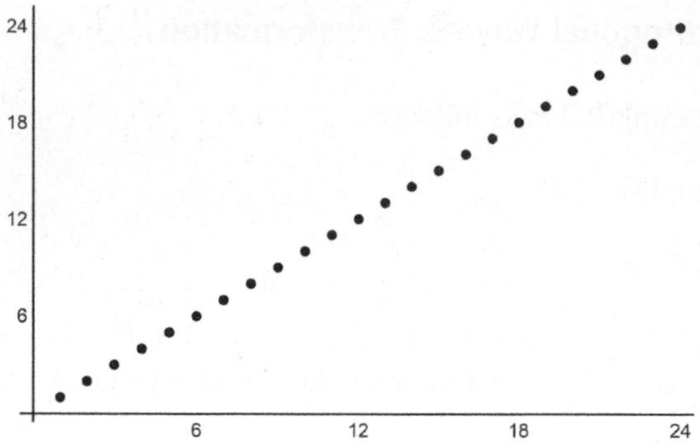

Figure 7.3 The components of vector $\mathbf{v}$.

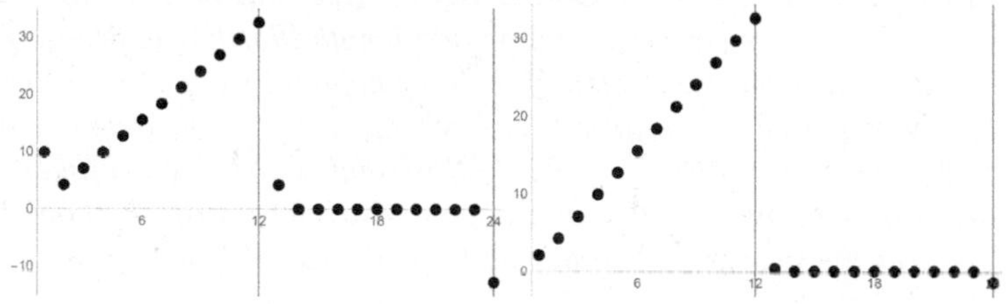

(a) $\mathbf{w}$ (b) $\mathbf{w}^s$

Figure 7.4 Plots of the biorthogonal wavelet transformations $\mathbf{w}$ and $\mathbf{w}^s$.

Denote these transformations by $\mathbf{w}$ *and* $\mathbf{w}^s$*, respectively. In Figure 7.4 we plot* $\mathbf{w}$ *and* $\mathbf{w}^s$.

Note that the averages portion of $\mathbf{w}^s$ *is a better approximation of* $\mathbf{v}$ *than is the averages portion of* $\mathbf{w}$.

(Live example: Visit `stthomas.edu/wavelets` *and click on Live Examples.)*

□

Inverting the Symmetric Transformation for Odd-Length Filter Pairs

We next consider the inverse process. There are a couple of ways we could proceed. The first is the most computationally efficient way to proceed but is difficult to implement. The second option is simple to implement but slows both the symmetric biorthogonal transformation and its inverse by roughly a factor of 2.

Finding the Matrix for the Inverse Symmetric Transformation

Let's first consider the computationally efficient method. The idea here is that the symmetric biorthogonal wavelet transformation maps the input vector $\mathbf{v} \in \mathbb{R}^N$ to an N-vector $\mathbf{w}$, so we should be able to represent this transformation with an $N \times N$ matrix $\widetilde{W}_N^s$. Once we have identified this matrix, then we compute its inverse and write an algorithm to implement its use. To better understand this method, let's again look at the symmetric transformation of Example 7.3.

Example 7.5 (Inverting the Symmetric Biorthogonal Transformation: Filter Lengths 3 and 5). *Consider the symmetric biorthogonal wavelet transformation from Example 7.3. Recall that the transformation worked on vectors of length eight and used the* $(5,3)$ *biorthogonal spline filter pair given in Definition 7.2.*

Using (7.34), (7.36), *and* (7.38), *we can write down the mapping*

$$\mathbf{v} = \begin{bmatrix} v_1 \\ v_2 \\ v_3 \\ v_4 \\ v_5 \\ v_6 \\ v_7 \\ v_8 \end{bmatrix} \rightarrow \left[\begin{array}{c} \widetilde{h}_0 v_1 + 2\widetilde{h}_1 v_2 \\ \widetilde{h}_1 v_2 + \widetilde{h}_0 v_3 + \widetilde{h}_1 v_4 \\ \widetilde{h}_1 v_4 + \widetilde{h}_0 v_5 + \widetilde{h}_1 v_6 \\ \widetilde{h}_1 v_6 + \widetilde{h}_0 v_7 + \widetilde{h}_1 v_8 \\ \hline h_1 v_1 - (h_0 + h_2) v_2 + h_1 v_3 - h_2 v_4 \\ -h_2 v_2 + h_1 v_3 - h_0 v_4 + h_1 v_5 - h_2 v_6 \\ -h_2 v_4 + h_1 v_5 - h_0 v_6 + h_1 v_7 - h_2 v_8 \\ -2h_2 v_6 + 2h_1 v_7 - h_0 v_8 \end{array}\right]$$

It is straightforward to identify the transformation matrix in this case. We have

$$\left[\begin{array}{c} \widetilde{\mathbf{a}}^s \\ \hline \widetilde{\mathbf{d}}^s \end{array}\right] = \widetilde{W}_8^s \mathbf{v}$$

where

$$\widetilde{W}_8^s = \left[\begin{array}{cccccccc} \widetilde{h}_0 & 2\widetilde{h}_1 & 0 & 0 & 0 & 0 & 0 & 0 \\ 0 & \widetilde{h}_1 & \widetilde{h}_0 & \widetilde{h}_1 & 0 & 0 & 0 & 0 \\ 0 & 0 & 0 & \widetilde{h}_1 & \widetilde{h}_0 & \widetilde{h}_1 & 0 & 0 \\ 0 & 0 & 0 & 0 & 0 & \widetilde{h}_1 & \widetilde{h}_0 & \widetilde{h}_1 \\ \hline h_1 & -h_0 - h_2 & h_1 & -h_2 & 0 & 0 & 0 & 0 \\ 0 & -h_2 & h_1 & -h_0 & h_1 & -h_2 & 0 & 0 \\ 0 & 0 & 0 & -h_2 & h_1 & -h_0 & h_1 & -h_2 \\ 0 & 0 & 0 & 0 & 0 & -2h_2 & 2h_1 & -h_0 \end{array}\right] \tag{7.40}$$

and $\widetilde{\mathbf{a}}^s$, $\widetilde{\mathbf{d}}^s$ *are nothing more than the first four elements in* (7.35) *and* (7.37), *respectively.*

This matrix is similar to $\widetilde{W}_8$, *but rows one, five, and eight are different. These changes will undoubtedly alter the inverse of* $\widetilde{W}_8^s$ *– we cannot expect the inverse to* *be the transpose of a wavelet matrix. There are other unanswered questions. How does the inverse change when* N *changes? We will need* W_8^s *for a two-dimensional transformation – what does its inverse look like? What happens if we use other odd-length filters?*

As you can see, this method has several disadvantages. It is possible, however, to analyze the inverse for particular filters. For example, if we use the $(5,3)$ *biorthogonal spline filter pair given in Definition 7.2 in* (7.40), *we have*

$$\widetilde{W}_8^s = \left[\begin{array}{cccccccc} \frac{\sqrt{2}}{2} & \frac{\sqrt{2}}{2} & 0 & 0 & 0 & 0 & 0 & 0 \\ 0 & \frac{\sqrt{2}}{4} & \frac{\sqrt{2}}{2} & \frac{\sqrt{2}}{4} & 0 & 0 & 0 & 0 \\ 0 & 0 & 0 & \frac{\sqrt{2}}{4} & \frac{\sqrt{2}}{2} & \frac{\sqrt{2}}{4} & 0 & 0 \\ 0 & 0 & 0 & 0 & 0 & \frac{\sqrt{2}}{4} & \frac{\sqrt{2}}{2} & \frac{\sqrt{2}}{4} \\ \hline \frac{\sqrt{2}}{4} & -\frac{5\sqrt{2}}{8} & \frac{\sqrt{2}}{4} & \frac{\sqrt{2}}{8} & 0 & 0 & 0 & 0 \\ 0 & \frac{\sqrt{2}}{8} & \frac{\sqrt{2}}{4} & -\frac{3\sqrt{2}}{4} & \frac{\sqrt{2}}{4} & \frac{\sqrt{2}}{8} & 0 & 0 \\ 0 & 0 & 0 & \frac{\sqrt{2}}{8} & \frac{\sqrt{2}}{4} & -\frac{3\sqrt{2}}{4} & \frac{\sqrt{2}}{4} & \frac{\sqrt{2}}{8} \\ 0 & 0 & 0 & 0 & 0 & \frac{\sqrt{2}}{4} & \frac{\sqrt{2}}{2} & -\frac{3\sqrt{2}}{4} \end{array}\right]$$

and we can use a ***CAS*** *to compute the inverse*

$$(\widetilde{W}_8^s)^{-1} = \left[\begin{array}{cccc|cccc} \frac{3\sqrt{2}}{4} & -\frac{\sqrt{2}}{4} & 0 & 0 & \frac{\sqrt{2}}{2} & 0 & 0 & 0 \\ \frac{\sqrt{2}}{4} & \frac{\sqrt{2}}{4} & 0 & 0 & -\frac{\sqrt{2}}{2} & 0 & 0 & 0 \\ -\frac{\sqrt{2}}{8} & \frac{3\sqrt{2}}{4} & -\frac{\sqrt{2}}{8} & 0 & \frac{\sqrt{2}}{4} & \frac{\sqrt{2}}{4} & 0 & 0 \\ 0 & \frac{\sqrt{2}}{4} & \frac{\sqrt{2}}{4} & 0 & 0 & -\frac{\sqrt{2}}{2} & 0 & 0 \\ 0 & -\frac{\sqrt{2}}{8} & \frac{3\sqrt{2}}{4} & -\frac{\sqrt{2}}{8} & 0 & \frac{\sqrt{2}}{4} & \frac{\sqrt{2}}{4} & 0 \\ 0 & 0 & \frac{\sqrt{2}}{4} & \frac{\sqrt{2}}{4} & 0 & 0 & -\frac{\sqrt{2}}{2} & 0 \\ 0 & 0 & -\frac{\sqrt{2}}{8} & \frac{5\sqrt{2}}{8} & 0 & 0 & \frac{\sqrt{2}}{4} & \frac{\sqrt{2}}{4} \\ 0 & 0 & 0 & \frac{\sqrt{2}}{2} & 0 & 0 & 0 & -\frac{\sqrt{2}}{2} \end{array}\right].$$

There are some similarities here between $(\widetilde{W}_8^s)^{-1}$ and $\widetilde{W}_8^{-1} = W_8^T$. We have

$$
W_8^T = \left[\begin{array}{cccc|cccc}
\frac{3\sqrt{2}}{4} & -\frac{\sqrt{2}}{8} & 0 & -\frac{\sqrt{2}}{8} & \frac{\sqrt{2}}{4} & 0 & 0 & \frac{\sqrt{2}}{4} \\
\frac{\sqrt{2}}{4} & \frac{\sqrt{2}}{4} & 0 & 0 & -\frac{\sqrt{2}}{2} & 0 & 0 & 0 \\
-\frac{\sqrt{2}}{8} & \frac{3\sqrt{2}}{4} & -\frac{\sqrt{2}}{8} & 0 & \frac{\sqrt{2}}{4} & \frac{\sqrt{2}}{4} & 0 & 0 \\
0 & \frac{\sqrt{2}}{4} & \frac{\sqrt{2}}{4} & 0 & 0 & -\frac{\sqrt{2}}{2} & 0 & 0 \\
0 & -\frac{\sqrt{2}}{8} & \frac{3\sqrt{2}}{4} & -\frac{\sqrt{2}}{8} & 0 & \frac{\sqrt{2}}{4} & \frac{\sqrt{2}}{4} & 0 \\
0 & 0 & \frac{\sqrt{2}}{4} & \frac{\sqrt{2}}{4} & 0 & 0 & -\frac{\sqrt{2}}{2} & 0 \\
-\frac{\sqrt{2}}{8} & 0 & -\frac{\sqrt{2}}{8} & \frac{3\sqrt{2}}{4} & 0 & 0 & \frac{\sqrt{2}}{4} & \frac{\sqrt{2}}{4} \\
\frac{\sqrt{2}}{4} & 0 & 0 & \frac{\sqrt{2}}{4} & 0 & 0 & 0 & -\frac{\sqrt{2}}{2}
\end{array}\right]
$$

and we can see that columns one, four, and eight are the only ones that differ between the two matrices.

We can also write down W_8^s and check $(W_8^s)^T$ against $(\widetilde{W}_8^s)^{-1}$ (see Problem 7.36). You will see that these two matrices are also similar and differ only in columns four and eight.

In Problems 7.37 and 7.38 you will further investigate the structure of $\widetilde{W}_N^m$ and W_N^m, and their inverses for general even N.

(Live example: Visit `stthomas.edu/wavelets` *and click on Live Examples.)*

□

The point of Example 7.5 is to see that it is difficult to analyze the inverse matrix for our symmetric biorthogonal wavelet transformation.

A Simple Algorithm for Inverting the Symmetric Transformation

We use an alternative method to invert the transformation.

Recall that Algorithm 7.1 takes an N-vector $\mathbf{v}$, periodizes it using (7.39) to obtain $\mathbf{v}^p$, and then computes the biorthogonal wavelet transformation to obtain

$$
\left[\begin{array}{c} \widetilde{\mathbf{a}}^p \\ \hline \widetilde{\mathbf{d}}^p \end{array}\right]
$$

where $\widetilde{\mathbf{a}}^p$ and $\widetilde{\mathbf{d}}^p$ are the averages and differences portions of the transformation, respectively.

The algorithm then returns

$$\left[\frac{\widetilde{\mathbf{a}}^s}{\widetilde{\mathbf{d}}^s}\right] = \left[\widetilde{a}_1^p \cdots \widetilde{a}_{N/2}^p \middle| \widetilde{d}_1^p \cdots \widetilde{d}_{N/2}^p\right]^T. \tag{7.41}$$

As we will verify in Propositions 7.4 and 7.6, $\widetilde{\mathbf{a}}^p$ and $\widetilde{\mathbf{d}}^p$ are always of the form

$$\widetilde{\mathbf{a}}^p = \left[\widetilde{a}_1^p, \widetilde{a}_2^p, \ldots, \widetilde{a}_{N/2}^p \middle| \widetilde{a}_{N/2}^p, \widetilde{a}_{N/2-1}^p, \cdots, \widetilde{a}_3^p, \widetilde{a}_2^p\right]^T$$
$$\widetilde{\mathbf{d}}^p = \left[\widetilde{d}_1^p, \widetilde{d}_2^p, \ldots, \widetilde{d}_{N/2}^p \middle| \widetilde{d}_{N/2-1}^p, \widetilde{d}_{N/2-2}^p, \cdots, \widetilde{d}_2^p, \widetilde{d}_1^p\right]^T$$

Our process for inverting is straightforward.

Algorithm 7.2 (Inverting the Symmetric Biorthogonal Transformation for Odd--Length Filters). *Given the N-vector $\mathbf{w} = \left[\widetilde{\mathbf{a}}^s \middle| \widetilde{\mathbf{d}}^s\right]^T$ where $\widetilde{\mathbf{a}}^s$ and $\widetilde{\mathbf{d}}^s \in \mathbb{R}^{\frac{N}{2}}$ are given by* (7.41)*, and a biorthogonal filter pair whose lengths are odd, perform the following steps:*

1. *Form the $(2N-2)$-vector*

$$\mathbf{w}^p = \left[\widetilde{a}_1^p, \ \ldots, \ \widetilde{a}_{N/2}^p, \ \widetilde{a}_{N/2}^p, \ \widetilde{a}_{N/2-1}^p, \ \ldots, \ \widetilde{a}_2^p \ \middle| \ \widetilde{d}_1^p, \ \ldots, \ \widetilde{d}_{N/2}^p, \ \widetilde{d}_{N/2-1}^p, \ \widetilde{d}_{N/2-2}^p, \ \ldots, \ \widetilde{d}_1^p\right]^T.$$

2. *Compute the inverse biorthogonal wavelet transformation of $\mathbf{w}^p$ to obtain $\mathbf{v}^p$ given by* (7.39).

3. *Return $[v_1, \ldots, v_N]^T$.*

□

Of course, the problem with Algorithms 7.1 and 7.2 is that we must process vectors of length $2N-2$ in order to obtain vectors of length N. These algorithms are bound to be slower, but considering the flexibility and ease of programming, they will work fine for our purposes.

The Symmetric Biorthogonal Wavelet Transformation for Even-Length Filter Pairs

We now turn our attention to developing an algorithm for computing the symmetric biorthogonal transformation for even-length filter pairs. The ideas are very similar to those discussed in detail for the odd-length filter pairs. As a result, we motivate the process with an example, state the algorithm, and leave the technical results that verify the process as problems. Let's consider an example.

Example 7.6 (Symmetry in the Even-Length Filter Case). *Suppose that* $\mathbf{v}$ *is a vector of length* $N = 12$ *and suppose that we wish to compute the biorthogonal wavelet transformation of* $\mathbf{v}$ *using* $\mathbf{h}$ *and* $\widetilde{\mathbf{g}}$ *given in* (7.19) *and* (7.27), *respectively.*

The averages portion of the transformation is

$$\widetilde{\mathbf{a}} = \widetilde{H}_6\mathbf{v} = \begin{bmatrix} \widetilde{h}_1 & \widetilde{h}_1 & \widetilde{h}_2 & 0 & 0 & 0 & 0 & 0 & 0 & 0 & 0 & \widetilde{h}_2 \\ 0 & \widetilde{h}_2 & \widetilde{h}_1 & \widetilde{h}_1 & \widetilde{h}_2 & 0 & 0 & 0 & 0 & 0 & 0 & 0 \\ 0 & 0 & 0 & \widetilde{h}_2 & \widetilde{h}_1 & \widetilde{h}_1 & \widetilde{h}_2 & 0 & 0 & 0 & 0 & 0 \\ 0 & 0 & 0 & 0 & 0 & \widetilde{h}_2 & \widetilde{h}_1 & \widetilde{h}_1 & \widetilde{h}_2 & 0 & 0 & 0 \\ 0 & 0 & 0 & 0 & 0 & 0 & 0 & \widetilde{h}_2 & \widetilde{h}_1 & \widetilde{h}_1 & \widetilde{h}_2 & 0 \\ \widetilde{h}_2 & 0 & 0 & 0 & 0 & 0 & 0 & 0 & 0 & \widetilde{h}_2 & \widetilde{h}_1 & \widetilde{h}_1 \end{bmatrix} \cdot \begin{bmatrix} v_1 \\ v_2 \\ v_3 \\ v_4 \\ v_5 \\ v_6 \\ v_7 \\ v_8 \\ v_9 \\ v_{10} \\ v_{11} \\ v_{12} \end{bmatrix}$$

$$= \begin{bmatrix} \widetilde{h}_2 v_{12} + \widetilde{h}_1 v_1 + \widetilde{h}_1 v_2 + \widetilde{h}_2 v_3 \\ \widetilde{h}_2 v_2 + \widetilde{h}_1 v_3 + \widetilde{h}_1 v_4 + \widetilde{h}_2 v_5 \\ \widetilde{h}_2 v_4 + \widetilde{h}_1 v_5 + \widetilde{h}_1 v_6 + \widetilde{h}_2 v_7 \\ \widetilde{h}_2 v_6 + \widetilde{h}_1 v_7 + \widetilde{h}_1 v_8 + \widetilde{h}_2 v_9 \\ \widetilde{h}_2 v_8 + \widetilde{h}_1 v_9 + \widetilde{h}_1 v_{10} + \widetilde{h}_2 v_{11} \\ \widetilde{h}_2 v_{10} + \widetilde{h}_1 v_{11} + \widetilde{h}_1 v_{12} + \widetilde{h}_2 v_1 \end{bmatrix}.$$

The differences portion of the transformation is

$$\widetilde{\mathbf{d}} = \widetilde{G}_6\mathbf{v} = \begin{bmatrix} h_1 & -h_1 & h_2 & -h_3 & h_4 & 0 & 0 & 0 & 0 & -h_4 & h_3 & -h_2 \\ h_3 & -h_2 & h_1 & -h_1 & h_2 & -h_3 & h_4 & 0 & 0 & 0 & 0 & -h_4 \\ 0 & -h_4 & h_3 & -h_2 & h_1 & -h_1 & h_2 & -h_3 & h_4 & 0 & 0 & 0 \\ 0 & 0 & 0 & -h_4 & h_3 & -h_2 & h_1 & -h_1 & h_2 & -h_3 & h_4 & 0 \\ h_4 & 0 & 0 & 0 & 0 & -h_4 & h_3 & -h_2 & h_1 & -h_1 & h_2 & -h_3 \\ h_2 & -h_3 & h_4 & 0 & 0 & 0 & 0 & -h_4 & h_3 & -h_2 & h_1 & -h_1 \end{bmatrix} \cdot \begin{bmatrix} v_1 \\ v_2 \\ v_3 \\ v_4 \\ v_5 \\ v_6 \\ v_7 \\ v_8 \\ v_9 \\ v_{10} \\ v_{11} \\ v_{12} \end{bmatrix}$$

$$= \begin{bmatrix} -h_4 v_{10} + h_3 v_{11} - h_2 v_{12} + h_1 v_1 - h_1 v_2 + h_2 v_3 - h_3 v_4 + h_4 v_5 \\ -h_4 v_{12} + h_3 v_1 - h_2 v_2 + h_1 v_3 - h_1 v_4 + h_2 v_5 - h_3 v_6 + h_4 v_7 \\ -h_4 v_2 + h_3 v_3 - h_2 v_4 + h_1 v_5 - h_1 v_6 + h_2 v_7 - h_3 v_8 + h_4 v_9 \\ -h_4 v_4 + h_3 v_5 - h_2 v_6 + h_1 v_7 - h_1 v_8 + h_2 v_9 - h_3 v_{10} + h_4 v_{11} \\ -h_4 v_6 + h_3 v_7 - h_2 v_8 + h_1 v_9 - h_1 v_{10} + h_2 v_{11} - h_3 v_{12} + h_4 v_1 \\ -h_4 v_8 + h_3 v_9 - h_2 v_{10} + h_1 v_{11} - h_1 v_{12} + h_2 v_1 - h_3 v_2 + h_4 v_3 \end{bmatrix}.$$

Note that there is wrapping in both portions of the transformation. For example, the first element in $\widetilde{\mathbf{a}}$ *is a linear combination of* v_1, v_2, v_3 *and* v_{12}. *We proceed in a manner similar to what was done in Example 7.3. We define the* 24*-vector*

$$\mathbf{v}^p = \left[\, v_1,\ v_2,\ \ldots,\ v_{11},\ v_{12} \,\middle|\, v_{12},\ v_{11},\ \ldots,\ v_2,\ v_1 \,\right]^T$$

and compute $\widetilde{\mathbf{a}}^p = \widetilde{W}_{24}\mathbf{v}^p$. *In Problem 7.39 you will verify that the averages portion of the transformation is*

$$\widetilde{\mathbf{a}}^p = \left[\begin{array}{c} (\widetilde{h}_1 + \widetilde{h}_2)v_1 + \widetilde{h}_1 v_2 + \widetilde{h}_2 v_3 \\ \widetilde{h}_2 v_2 + \widetilde{h}_1 v_3 + \widetilde{h}_1 v_4 + \widetilde{h}_2 v_5 \\ \widetilde{h}_2 v_4 + \widetilde{h}_1 v_5 + \widetilde{h}_1 v_6 + \widetilde{h}_2 v_7 \\ \widetilde{h}_2 v_6 + \widetilde{h}_1 v_7 + \widetilde{h}_1 v_8 + \widetilde{h}_2 v_9 \\ \widetilde{h}_2 v_8 + \widetilde{h}_1 v_9 + \widetilde{h}_1 v_{10} + \widetilde{h}_2 v_{11} \\ \widetilde{h}_2 v_{10} + \widetilde{h}_1 v_{11} + (\widetilde{h}_1 + \widetilde{h}_2) v_{12} \\ \hline \widetilde{h}_2 v_{10} + \widetilde{h}_1 v_{11} + (\widetilde{h}_1 + \widetilde{h}_2) v_{12} \\ \widetilde{h}_2 v_8 + \widetilde{h}_1 v_9 + \widetilde{h}_1 v_{10} + \widetilde{h}_2 v_{11} \\ \widetilde{h}_2 v_6 + \widetilde{h}_1 v_7 + \widetilde{h}_1 v_8 + \widetilde{h}_2 v_9 \\ \widetilde{h}_2 v_4 + \widetilde{h}_1 v_5 + \widetilde{h}_1 v_6 + \widetilde{h}_2 v_7 \\ \widetilde{h}_2 v_2 + \widetilde{h}_1 v_3 + \widetilde{h}_1 v_4 + \widetilde{h}_2 v_5 \\ (\widetilde{h}_1 + \widetilde{h}_2)v_1 + \widetilde{h}_1 v_2 + \widetilde{h}_2 v_3 \end{array}\right] \tag{7.42}$$

and the differences portion of the transformation is

$$\widetilde{\mathbf{d}}^p = \left[\begin{array}{c} (h_1-h_2)v_1+(h_3-h_1)v_2-(h_4-h_2)v_3-h_3v_4+h_4v_5 \\ (h_3-h_4)v_1-h_2v_2+h_1v_3-h_1v_4+h_2v_5-h_3v_6+h_4v_7 \\ -h_4v_2+h_3v_3-h_2v_4+h_1v_5-h_1v_6+h_2v_7-h_3v_8+h_4v_9 \\ -h_4v_4+h_3v_5-h_2v_6+h_1v_7-h_1v_8+h_2v_9-h_3v_{10}+h_4v_{11} \\ -h_4v_6+h_3v_7-h_2v_8+h_1v_9-h_1v_{10}+h_2v_{11}-(h_3-h_4)v_{12} \\ -h_4v_8+h_3v_9-(h_2-h_4)v_{10}+(h_1-h_3)v_{11}-(h_1-h_2)v_{12} \\ \hline h_4v_8-h_3v_9+(h_2-h_4)v_{10}-(h_1-h_3)v_{11}+(h_1-h_2)v_{12} \\ h_4v_6-h_3v_7+h_2v_8-h_1v_9+h_1v_{10}-h_2v_{11}+(h_3-h_4)v_{12} \\ h_4v_4-h_3v_5+h_2v_6-h_1v_7+h_1v_8-h_2v_9+h_3v_{10}-h_4v_{11} \\ h_4v_2-h_3v_3+h_2v_4-h_1v_5+h_1v_6-h_2v_7+h_3v_8-h_4v_9 \\ -(h_3-h_4)v_1+h_2v_2-h_1v_3+h_1v_4-h_2v_5+h_3v_6-h_4v_7 \\ -(h_1-h_2)v_1+(h_1-h_3)v_2+(h_4-h_2)v_3+h_3v_4-h_4v_5 \end{array}\right]. \tag{7.43}$$

There is some symmetry in $\widetilde{\mathbf{a}}^p$ *and* $\widetilde{\mathbf{d}}^p$. *In fact, the structures of these vectors are*

$$\widetilde{\mathbf{a}}^p = \left[\, \widetilde{a}_1^p,\, \widetilde{a}_2^p,\, \widetilde{a}_3^p,\, \widetilde{a}_4^p,\, \widetilde{a}_5^p,\, \widetilde{a}_6^p \,\middle|\, \widetilde{a}_6^p,\, \widetilde{a}_5^p,\, \widetilde{a}_4^p,\, \widetilde{a}_3^p,\, \widetilde{a}_2^p,\, \widetilde{a}_1^p \,\right]^T$$

$$\widetilde{\mathbf{d}}^p = \left[\, \widetilde{d}_1^p,\, \widetilde{d}_2^p,\, \widetilde{d}_3^p,\, \widetilde{d}_4^p,\, \widetilde{d}_5^p,\, \widetilde{d}_6^p \,\middle|\, -\widetilde{d}_6^p,\, -\widetilde{d}_5^p,\, -\widetilde{d}_4^p,\, -\widetilde{d}_3^p,\, -\widetilde{d}_2^p,\, -\widetilde{d}_1^p \,\right]^T.$$

We take as our symmetric transformation the 12*-vector*

$$\left[\begin{array}{c} \widetilde{\mathbf{a}}^s \\ \hline \widetilde{\mathbf{d}}^s \end{array}\right] = \left[\, \widetilde{a}_1^p,\, \ldots,\, \widetilde{a}_6^p \,\middle|\, \widetilde{d}_1^p,\, \ldots,\, \widetilde{d}_6^p \,\right]^T.$$

□

Algorithms for the Symmetric Biorthogonal Wavelet Transformation and its Inverse for Even-Length Filter Pairs

We are now ready to state the algorithms for computing the symmetric biorthogonal wavelet transformation for even-length filter pairs and its inverse. The algorithms are similar to Algorithms 7.1 and 7.2. In Problems 7.48 – 7.50 you will verify the validity of the following algorithm.

Algorithm 7.3 (Symmetric Biorthogonal Transformation for Even-Length Filter Pairs). *This algorithm takes as input the N-vector $\mathbf{v}$ (N even) and an even-length biorthogonal filter pair $\left(\mathbf{h}, \widetilde{\mathbf{h}}\right)$ and returns a symmetric biorthogonal wavelet transformation. The vector $\mathbf{v}$ is periodized to form vector $\mathbf{v}^p \in \mathbb{R}^{2N}$. We compute the biorthogonal wavelet transformation of $\mathbf{v}^p$. If $\widetilde{\mathbf{a}}^p$ and $\mathbf{z}^p$ are the averages and differences portions, respectively, of the biorthogonal wavelet transformation, then the algorithm returns the symmetric biorthogonal wavelet transformation*

$$\left[\widetilde{a}_1, \ldots, \widetilde{a}_{N/2} \,|\, \widetilde{d}_1, \ldots, \widetilde{d}_{N/2}\right]^T.$$

1. *Given $\mathbf{v} \in \mathbb{R}^N$, N even, form $\mathbf{v}^p \in \mathbb{R}^{2N}$ by the rule*

$$\mathbf{v}^p = [v_1, \ldots, v_N, v_N, \ldots, v_1]^T. \tag{7.44}$$

2. *Compute the biorthogonal wavelet transformation $\mathbf{w} = \widetilde{W}_{2N}\mathbf{v}^p$. Denote the averages portion of the transformation by $\widetilde{\mathbf{a}}^p$ and the differences portion of the transformation by $\widetilde{\mathbf{d}}^p$.*

3. *Return* $\left[\widetilde{a}_1^p, \ldots, \widetilde{a}_{N/2}^p \,|\, \widetilde{d}_1^p, \ldots, \widetilde{d}_{N/2}^p\right]^T.$

□

The algorithm for the inverse symmetric biorthogonal transformation is very similar to Algorithm 7.2.

Algorithm 7.4 (Symmetric Inverse Biorthogonal Transformation for Even-Length Filter Pairs). *Given the N-vector $\mathbf{w} = \left[\widetilde{\mathbf{a}}^s \,|\, \widetilde{\mathbf{d}}^s\right]^T$, where $\widetilde{\mathbf{a}}^s$ and $\widetilde{\mathbf{d}}^s \in \mathbb{R}^{\frac{N}{2}}$ are the averages and differences portions, respectively, of the symmetric biorthogonal wavelet transformation, and a biorthogonal filter pair whose lengths are even, perform the following steps:*

1. *Form the $2N$-vector*

$$\mathbf{w}^p = \left[\widetilde{a}_1^p, \ldots, \widetilde{a}_{N/2}^p, \widetilde{a}_{N/2}^p, \ldots, \widetilde{a}_1^p \,|\, \widetilde{d}_1^p, \ldots, \widetilde{d}_{N/2}^p, -\widetilde{d}_{N/2}^p, \ldots, -\widetilde{d}_1^p\right]^T.$$

2. *Compute the inverse biorthogonal wavelet transformation of $\mathbf{w}^p$ to obtain*

$$\mathbf{v}^p = [v_1, \ldots, v_N, v_N, \ldots, v_1]^T.$$

3. *Return* $[v_1, \ldots, v_N]^T$.

□

Symmetric Two-Dimensional Biorthogonal Wavelet Transformations

It is straightforward to use Algorithms 7.1 and 7.3 to compute symmetric biorthogonal wavelet transformations of matrices. Suppose A is an $M \times N$ matrix with M, N even. For the sake of illustration, assume we use a biorthogonal filter pair $\left(\mathbf{h}, \widetilde{\mathbf{h}}\right)$ whose lengths are odd. Then the two-dimensional biorthogonal wavelet transformation is $B = \widetilde{W}_M A W_N^T$. We let $U = \widetilde{W}_N A$ and observe that

$$\begin{aligned} U &= \widetilde{W}_M A \\ &= \widetilde{W}_M \begin{bmatrix} \mathbf{a}^1 & \mathbf{a}^2 & \cdots & \mathbf{a}^N \end{bmatrix} \\ &= \begin{bmatrix} \widetilde{W}_M \mathbf{a}^1 & \widetilde{W}_M \mathbf{a}^2 & \cdots & \widetilde{W}_M \mathbf{a}^N \end{bmatrix} \end{aligned}$$

where $\mathbf{a}^1, \ldots, \mathbf{a}^N$ are the columns of A. For each product $\widetilde{W}_M \mathbf{a}^k$, $k = 1, \ldots, N$, we use Algorithm 7.1. The remaining computation is $B = U W_N^T$. We consider instead $B^T = W_M U^T$ and apply Algorithm 7.1 to the columns of U^T. Transposing this result gives us a symmetric two-dimensional biorthogonal transformation of A.

Why Does Algorithm 7.1 Work?

We have seen that Algorithm 7.1 works for vectors of length eight and a biorthogonal filter pair whose lengths are 5 and 3. We conclude this section by showing that when $\mathbf{v}^p$ is formed from $\mathbf{v} \in \mathbb{R}^N$ with N even, Step 2 of Algorithm 7.1 maps $\mathbf{v}^p$ to $\begin{bmatrix} \widetilde{\mathbf{a}}^p \\ \hline \widetilde{\mathbf{d}}^p \end{bmatrix}$ where

$$\widetilde{\mathbf{a}}^p = [\widetilde{a}_1^p, \widetilde{a}_2^p, \ldots, \widetilde{a}_{N/2}^p, \widetilde{a}_{N/2}^p, \widetilde{a}_{N/2-1}^p, \ldots, \widetilde{a}_2^p]^T \tag{7.45}$$

and

$$\widetilde{\mathbf{d}}^p = [\widetilde{d}_1^p, \widetilde{d}_2^p, \ldots, \widetilde{d}_{N/2}^p, \widetilde{d}_{N/2-1}^p, \widetilde{d}_{N/2-2}^p, \ldots, \widetilde{d}_1^p]^T. \tag{7.46}$$

To prove these facts, we must return to the ideas of convolution and filters from Section 2.4.

Symmetry in the Lowpass Portion $\mathbf{y}^p$

Let's first analyze the averages portion of the transformation returned by Algorithm 7.1. The entire transformation process can be viewed as convolving two bi-infinite sequences $\widetilde{\mathbf{h}}$ and $\mathbf{x}$ and then *downsampling* the result. We learned in Section 2.4 that we can write the convolution process as an infinite-dimensional matrix times $\mathbf{x}$. We first write our symmetric filter as a bi-infinite sequence

$$\widetilde{\mathbf{h}} = (\ldots, 0, 0, \widetilde{h}_{\widetilde{L}}, \widetilde{h}_{\widetilde{L}-1}, \ldots, \widetilde{h}_1, \widetilde{h}_0, \widetilde{h}_1, \ldots, \widetilde{h}_{\widetilde{L}}, 0, 0, \ldots)$$

If $\mathbf{x} = (\ldots, x_{-2}, x_{-1}, x_0, x_1, x_2, \ldots)$, we can write $\mathbf{y} = \mathbf{h} * \mathbf{x} = \widetilde{H}\mathbf{x}$ using the matrix

$$\widetilde{H} = \begin{bmatrix} \ddots & \ddots & \ddots & \ddots & \ddots & & & & & & & & & \\ & \widetilde{h}_{\widetilde{L}} & \ldots & \widetilde{h}_2 & \widetilde{h}_1 & \widetilde{\mathbf{h}}_{\mathbf{0}} & \widetilde{h}_1 & \widetilde{h}_2 & \ldots & \widetilde{h}_{\widetilde{L}} & 0 & 0 & 0 & 0 & \\ & 0 & \widetilde{h}_{\widetilde{L}} & \ldots & \widetilde{h}_2 & \widetilde{h}_1 & \widetilde{\mathbf{h}}_{\mathbf{0}} & \widetilde{h}_1 & \widetilde{h}_2 & \ldots & \widetilde{h}_{\widetilde{L}} & 0 & 0 & 0 & \\ \ldots & 0 & 0 & \widetilde{h}_{\widetilde{L}} & \ldots & \widetilde{h}_2 & \widetilde{h}_1 & \widetilde{\mathbf{h}}_{\mathbf{0}} & \widetilde{h}_1 & \widetilde{h}_2 & \ldots & \widetilde{h}_{\widetilde{L}} & 0 & 0 & \ldots \\ & 0 & 0 & 0 & \widetilde{h}_{\widetilde{L}} & \ldots & \widetilde{h}_2 & \widetilde{h}_1 & \widetilde{\mathbf{h}}_{\mathbf{0}} & \widetilde{h}_1 & \widetilde{h}_2 & \ldots & \widetilde{h}_{\widetilde{L}} & 0 & \\ & 0 & 0 & 0 & 0 & \widetilde{h}_{\widetilde{L}} & \ldots & \widetilde{h}_2 & \widetilde{h}_1 & \widetilde{\mathbf{h}}_{\mathbf{0}} & \widetilde{h}_1 & \widetilde{h}_2 & \ldots & \widetilde{h}_{\widetilde{L}} & \\ & & & & & & \ddots & \ddots & \ddots & \ddots & \ddots & & & & \ddots \end{bmatrix}$$

where the boldface elements are on the main diagonal of $\widetilde{H}$. Recall that the elements of $\mathbf{y}$ are given by the formula[2]

$$y_n = \sum_k \widetilde{h}_k \, x_{n-k} = \sum_k \widetilde{h}_{n-k} \, x_k. \tag{7.47}$$

To form the averages portion of a wavelet transformation, we first *downsample* $\widetilde{H}\mathbf{x}$ by discarding every other row and then truncate the output. The truncation process typically involves the introduction of wrapping rows – let's hold off on truncation for the moment. We can represent the downsampling by slightly altering the convolution formula (7.47). We form the bi-infinite sequence $\mathbf{a}$ using the rule

$$a_n = y_{2n} = \sum_k \widetilde{h}_{2n-k} \, x_k. \tag{7.48}$$

We load $\mathbf{x}$ with repeated copies of the $\mathbf{v}^p$ given in (7.39). We start with $x_k = v_{k+1}$ for $k = 0, \ldots, 2N-3$ and then continue loading every $2N-2$ elements of $\mathbf{x}$ with the elements of $\mathbf{v}^p$. Abusing notation, we have

$$\mathbf{x} = (\ldots, \mathbf{v}^p, \mathbf{v}^p, \mathbf{v}^p, \ldots). \tag{7.49}$$

By construction, $\mathbf{x}$ is a $(2N-2)$-periodic sequence so that

$$x_k = x_{2N-2+k} \tag{7.50}$$

for all $k \in \mathbb{Z}$. In Problem 7.42, you will show that $\mathbf{x}$ is symmetric about zero. It might be helpful to work Problem 7.41 before trying the general proof.

Now we consider $\mathbf{a}$ where a_n is given by (7.48). The following proposition tells us that $\mathbf{a}$ is periodic and symmetric.

Proposition 7.3 (Sequence a Is Periodic and Symmetric). *Assume that* $\mathbf{v} \in \mathbb{R}^N$ *where N is even and that* $\mathbf{v}^p$ *is given by* (7.39). *Further suppose that*

$$\widetilde{\mathbf{h}} = (\ldots, 0, 0, \widetilde{h}_{\widetilde{L}}, \widetilde{h}_{\widetilde{L}-1}, \ldots, \widetilde{h}_1, \widetilde{h}_0, \widetilde{h}_1, \ldots, \widetilde{h}_{\widetilde{L}-1}, \widetilde{h}_{\widetilde{L}}, 0, 0, \ldots)$$

and $\mathbf{x}$ *is given by* (7.49). *If* $\mathbf{a}$ *is formed componentwise by the rule* (7.48), *then* $\mathbf{a}$ *is* $(N-1)$*-periodic and* $a_n = a_{-n}$ *for all* $n \in \mathbb{Z}$. □

[2]By Problem 2.53 in Section 2.4, we know that the convolution product is commutative.

Proof. Let's first show that $\mathbf{a}$ is an $(N-1)$-periodic sequence. That is, we must show that $a_n = a_{n+N-1}$ for all $n \in \mathbb{Z}$. Using (7.48), we have

$$a_{n+N-1} = \sum_k \widetilde{h}_{2(n+N-1)-k}\, x_k = \sum_k \widetilde{h}_{2n-(k+2-2N)}\, x_k.$$

Now make the substitution $m = k + 2 - 2N$. Then $k = m + 2N - 2$, and since k runs over all integers, so does m. We have

$$a_{n+N-1} = \sum_m \widetilde{h}_{2n-m}\, x_{m+2N-2}.$$

But $\mathbf{x}$ is $(2N-2)$-periodic, so by (7.50) we can rewrite the previous identity as

$$a_{n+N-1} = \sum_m \widetilde{h}_{2n-m}\, x_m.$$

But this is exactly a_n so the first part of the proof is complete.

To show that $a_n = a_{-n}$, we use (7.48) to write

$$a_{-n} = \sum_k \widetilde{h}_{2(-n)-k}\, x_k = \sum_k \widetilde{h}_{-(2n+k)}\, x_k = \sum_k \widetilde{h}_{2n+k}\, x_k$$

since we are given that $\widetilde{\mathbf{h}}$ satisfies $\widetilde{h}_m = \widetilde{h}_{-m}$. Now by Problem 7.42, we know that $x_k = x_{-k}$, so we can write

$$a_{-n} = \sum_k \widetilde{h}_{2n+k}\, x_{-k}.$$

Now make the substitution $m = -k$ in the previous identity. Then $k = -m$ and since k runs over all integers, so will m. We have

$$a_{-n} = \sum_m \widetilde{h}_{2n-m} x_m = a_n$$

and the proof is complete. □

The following corollary is almost immediate.

Corollary 7.1 (Characterizing the Elements of Sequence a). *Each period of* $\mathbf{a}$ *constructed using* (7.48) *has the form*

$$a_0, a_1, a_2, \ldots, a_{N/2-1}, a_{N/2-1}, \ldots, a_2, a_1.$$

□

Proof. We know from Proposition 7.3 that $\mathbf{a}$ is $(N-1)$-periodic, so the elements of each period of $\mathbf{a}$ are

$$a_0, a_1, \ldots, a_{N/2-1}, a_{N/2}, \ldots, a_{N-2}. \tag{7.51}$$

We must show that for $k = 1, \ldots, \frac{N}{2} - 1$, we have $a_k = a_{N-1-k}$.

Let $k = 1, \ldots, \frac{N}{2} - 1$. Since $\mathbf{a}$ is $(N-1)$-periodic, we know that

$$a_{-k} = a_{N-1-k}.$$

But from Proposition 7.3, we know that the elements of $\mathbf{a}$ satisfy $a_k = a_{-k}$. So

$$a_k = a_{-k} = a_{N-1-k}$$

and the proof is complete. □

Let's look at an example.

Example 7.7 (Using Proposition 7.3 and Corollary 7.1). *Let's analyze* $\mathbf{a}$ *in the case where* $N = 8$. *Then by Proposition 7.3,* $\mathbf{a}$ *is 7-periodic sequence, and from Corollary 7.1 we have*

$$[a_0, a_1, a_2, a_3, a_4, a_5, a_6]^T = [a_0, a_1, a_2, a_3, a_3, a_2, a_1]^T$$

This is exactly the structure (modulo shifting the index by one) of $\widetilde{\mathbf{a}}^p$ *in (7.35)! It is left for you in Problem 7.43 to show that* $\widetilde{a}^p_k = a_{k-1}$ *for* $k = 1, \ldots, 4$ *when we use the 3-term symmetric filter* $\widetilde{\mathbf{h}}$ *from Example 7.3.* □

Characterizing the $(N-1)$-Periodic Values of $\mathbf{a}$

Our next proposition confirms that the elements of the lowpass vector $\widetilde{\mathbf{a}}^p$ from Algorithm 7.1 are given by (7.51).

Proposition 7.4 (Connecting $\widetilde{\mathbf{a}}^p$ and the Elements of $\mathbf{a}$). *Assume that* $\mathbf{v} \in \mathbb{R}^N$ *where* N *is even and that* $\mathbf{v}^p \in \mathbb{R}^{2N-2}$ *is given by (7.39). Suppose that* $\widetilde{\mathbf{h}}$ *is an odd-length, symmetric FIR filter. Form the lowpass portion of the biorthogonal wavelet transformation* $\widetilde{\mathbf{a}}^p = \widetilde{H}_{N-1}\mathbf{v}^p \in \mathbb{R}^{N-1}$ *(see Step 2 of Algorithm 7.1) and naturally extend* $\widetilde{\mathbf{h}}$ *to a bi-infinite sequence (packing with zeros) to form* $\mathbf{a}$ *whose components are given by (7.48). Then* $\widetilde{\mathbf{a}}^p_k = a_{k-1}$, $k = 1, \ldots, \frac{N}{2}$. □

Proof. This proof is technical and is left as Problem 7.44. □

Symmetry in the Highpass Portion $\widetilde{\mathbf{d}}^p$

The procedure for verifying the symmetry (7.46) exhibited by $\widetilde{\mathbf{d}}^p$ from Algorithm 7.1 is quite similar to the steps to show the symmetry (7.45) for $\widetilde{\mathbf{a}}^p$. We use $\mathbf{x}$ defined by (7.49) but now use the bi-infinite sequence

$$\widetilde{\mathbf{g}} = (\ldots, 0, 0, \widetilde{g}_{-L+1}, \widetilde{g}_{-L+2}, \ldots, \widetilde{g}_{-1}, \widetilde{g}_0, \ldots, \widetilde{g}_L, \widetilde{g}_{L+1}, 0, 0, \ldots)$$

in our convolution product

$$z_n = \sum_k \widetilde{g}_k \, x_{n-k} = \sum_k \widetilde{g}_{n-k} \, x_k.$$

Recall that the filter $\widetilde{\mathbf{g}}$ is built by the component-wise rule $\widetilde{g}_k = (-1)^k h_{1-k}$, where $\mathbf{h}$ is a symmetric, FIR filter of odd length.

We define the bi-infinite sequence $\mathbf{d}$ of downsampled terms by

$$d_n = z_{2n} = \sum_k \widetilde{g}_{2n-k} \, x_k. \tag{7.52}$$

The following proposition is analogous to Proposition 7.3.

Proposition 7.5 (Sequence d Is Periodic and Symmetric). *Assume that* $\mathbf{v} \in \mathbb{R}^N$ *where* N *is even and that* $\mathbf{v}^p$ *is given by* (7.39)*. Further suppose that*

$$\widetilde{\mathbf{g}} = (\ldots, 0, 0, \widetilde{g}_{-L+1}, \widetilde{g}_{-L+2}, \ldots, \widetilde{g}_{-1}, \widetilde{g}_0, \widetilde{g}_1, \ldots, \widetilde{g}_L, \widetilde{g}_{L+1}, 0, 0, \ldots)$$

is constructed from $\mathbf{h} = (\ldots, 0, 0, h_L, h_{L-1}, \ldots, h_1, h_0, h_1, \ldots, h_{L-1}, h_L, 0, 0, \ldots)$ *using the rule* $\widetilde{g}_k = (-1)^k h_{1-k}$*. If* $\mathbf{x}$ *is given by* (7.49) *and* $\mathbf{d}$ *is formed component-wise by the rule* (7.52)*, then* $\mathbf{d}$ *is* $(N-1)$*-periodic and* $d_n = d_{1-n}$ *for all* $n \in \mathbb{Z}$.

□

Proof. The proof that $\mathbf{d}$ is $(N-1)$-periodic is quite similar to the proof that $\mathbf{a}$ is $(N-1)$-periodic given for Proposition 7.3, so we leave it as Problem 7.45.

To see that $d_n = d_{1-n}$, we use (7.52) and the fact that $x_k = x_{-k}$ to write

$$d_{1-n} = \sum_k g_{2(1-n)-k} \, x_k = \sum_k g_{2-2n-k} \, x_k = \sum_k g_{2-2n-k} \, x_{-k}.$$

Now we make the substitution $m = -k$. Since k runs through all the integers, so does m and we have

$$d_{1-n} = \sum_m g_{2-2n+m} \, x_m = \sum_m g_{2-(2n-m)} \, x_m.$$

In Problem 7.11 of Section 7.1, we showed that $\widetilde{g}_k = \widetilde{g}_{2-k}$. Thus we can write the previous identity as

$$d_{1-n} = \sum_m g_{2n-m} \, x_m = d_n$$

and the proof is complete. □

The following corollary characterizes the structure of each period of $\mathbf{d}$.

Corollary 7.2 (Characterizing the Elements of Sequence d). *Each period of* $\mathbf{d}$ *has the form* $d_1, d_2, \ldots, d_{N/2-1}, d_{N/2}, d_{N/2-1}, \ldots, d_2, d_1$. □

Proof. The proof is left as Problem 7.46. □

Note that the symmetry is a bit different for $\mathbf{d}$ than it was for $\mathbf{a}$. Let's again consider Proposition 7.5 and Corollary 7.2 in the $N = 8$ case.

Example 7.8 (Using Proposition 7.5 and Corollary 7.2). *Let's analyze $\mathbf{d}$ in the case where $N = 8$. Then by Proposition 7.5, $\mathbf{d}$ is a 7-periodic sequence and from Corollary 7.2 we have*

$$[d_1, d_2, d_3, d_4, d_5, d_6, d_7]^T = [d_1, d_2, d_3, d_4, d_3, d_2, d_1]^T .$$

The structure of the period is identical to that of $\widetilde{\mathbf{d}}^p$ in (7.37). Moreover, if we use the 5-term filter from Definition 7.2, we see that $\widetilde{d}^p_k = d_k$, $k = 1, \ldots, 4$ (see Problem 7.43). □

The next proposition is similar to Proposition 7.6. Again, the proof is quite technical and is left as a challenging problem.

Proposition 7.6 (Connecting $\widetilde{\mathbf{d}}^p$ and the Elements of $\mathbf{d}$). *Assume that $\mathbf{v} \in \mathbb{R}^N$ where N is even and that $\mathbf{v}^p \in \mathbb{R}^{2N-2}$ is given by (7.39). Suppose that $\widetilde{\mathbf{g}}$ is constructed from an odd-length, symmetric FIR filter $\mathbf{h}$ using the rule $\widetilde{g}_k = (-1)^k h_{1-k}$. Form the highpass portion of the biorthogonal wavelet transformation $\widetilde{\mathbf{d}}^p = \widetilde{G}_{N-1}\mathbf{v}^p \in \mathbb{R}^{N-1}$ (see Step 2 of Algorithm 7.1) and naturally extend $\widetilde{\mathbf{g}}$ to a bi-infinite sequence (packing with zeros) to form $\mathbf{d}$ whose components are given by (7.52). Then $\widetilde{\mathbf{d}}^p_k = d_k$, $k = 1, \ldots, \frac{N}{2}$.* □

Proof. The proof is left as Problem 7.47. □

PROBLEMS

7.34 Complete the proof of Proposition 7.2 in the case where N is even.

7.35 Let $\mathbf{v} = [5, 2, 8, -4, 6, 1, 1, 0]^T$. Use Algorithm 7.1 to compute the symmetric biorthogonal wavelet transformation using the $(5, 3)$ biorthogonal spline filter pair.

7.36 In Example 7.5 we built the matrix $\widetilde{W}^s_8$ that computes the symmetric biorthogonal wavelet transformation using the $\widetilde{\mathbf{h}}$ and $\widetilde{\mathbf{g}}$ given by (7.13) and (7.14).

(a) We could just as well have used $\mathbf{h}$ and $\mathbf{g}$ given in (7.12) and (7.15), respectively, to form the symmetric biorthogonal transformation matrix W^s_8. Use the ideas from Example 7.5 and (7.17) to write down W^s_8.

(b) How does W^s_8 differ from W_8?

(c) Using a **CAS** to compute $(W^s_8)^{-1}$. How does it compare to $W_8^{-1} = \widetilde{W}_8^T$?

7.37 Suppose that N is even and let $\widetilde{W}_N$ and W_N be matrices associated with the biorthogonal transformation and its inverse constructed from the $(5, 3)$ biorthogonal spline filter pair. If $\widetilde{\mathbf{w}}^k$ are the rows of $\widetilde{W}_N$ and $\mathbf{w}^k$ are the rows of W_N, $k = 1, \ldots, N$, show that

$$\widetilde{W}_N^s = \left[\begin{array}{ccccccccc}
\frac{\sqrt{2}}{2} & \frac{\sqrt{2}}{2} & 0 & 0 & \cdots & 0 & 0 & 0 & 0 \\
 & & & & \widetilde{\mathbf{w}}^2 & & & & \\
 & & & & \vdots & & & & \\
 & & & & \widetilde{\mathbf{w}}^{N/2} & & & & \\
\hline
\frac{\sqrt{2}}{4} & -\frac{5\sqrt{2}}{8} & \frac{\sqrt{2}}{4} & \frac{\sqrt{2}}{8} & 0 & \cdots & 0 & 0 & 0 \\
 & & & & \widetilde{\mathbf{w}}^{N/2+2} & & & & \\
 & & & & \vdots & & & & \\
 & & & & \widetilde{\mathbf{w}}^{N-1} & & & & \\
0 & 0 & 0 & 0 & \cdots & 0 & \frac{\sqrt{2}}{4} & \frac{\sqrt{2}}{2} & \frac{3\sqrt{2}}{4}
\end{array}\right]$$

and

$$W_N^s = \left[\begin{array}{ccccccccc}
\frac{3\sqrt{2}}{4} & \frac{\sqrt{2}}{2} & -\frac{\sqrt{2}}{4} & 0 & \cdots & 0 & 0 & 0 & 0 \\
 & & & & \mathbf{w}^2 & & & & \\
 & & & & \vdots & & & & \\
 & & & & \mathbf{w}^{N/2-1} & & & & \\
0 & 0 & 0 & 0 & \cdots & -\frac{\sqrt{2}}{8} & \frac{\sqrt{2}}{4} & \frac{5\sqrt{2}}{8} & \frac{\sqrt{2}}{4} \\
\hline
 & & & & \mathbf{w}^{N/2+1} & & & & \\
 & & & & \mathbf{w}^{N/2+2} & & & & \\
 & & & & \vdots & & & & \\
 & & & & \mathbf{w}^{N-1} & & & & \\
0 & 0 & 0 & 0 & \cdots & 0 & 0 & \frac{\sqrt{2}}{2} & -\frac{\sqrt{2}}{2}
\end{array}\right].$$

Hint: Look at the entries of $\widetilde{W}_{2N-2}\mathbf{v}^p$ and $W_{2N-2}\mathbf{v}^p$ where $\mathbf{v}^p$ is given by (7.39).

7.38 Under the assumptions of Problem 7.37, let

$$U = \left[\begin{array}{cccccccccc}
\frac{3\sqrt{2}}{4} & \frac{\sqrt{2}}{4} & -\frac{\sqrt{2}}{8} & 0 & 0 & 0 & \cdots & 0 & 0 \\
-\frac{\sqrt{2}}{4} & \frac{\sqrt{2}}{4} & \frac{3\sqrt{2}}{4} & \frac{\sqrt{2}}{4} & -\frac{\sqrt{2}}{8} & 0 & \cdots & 0 & 0 \\
 & & & & & \mathbf{w}^3 & & & \\
 & & & & & \vdots & & & \\
 & & & & & \mathbf{w}^{N/2-1} & & & \\
0 & 0 & 0 & \cdots & 0 & -\frac{\sqrt{2}}{8} & \frac{\sqrt{2}}{4} & \frac{11\sqrt{2}}{8} & -\sqrt{2} \\
\hline
 & & & & & \mathbf{w}^{N/2+1} & & & \\
 & & & & & \vdots & & & \\
 & & & & & \mathbf{w}^{N-1} & & & \\
0 & 0 & 0 & \cdots & 0 & 0 & 0 & -\frac{\sqrt{2}}{2} & \sqrt{2}
\end{array}\right]$$

and

$$V = \left[\begin{array}{cccccccc} \frac{\sqrt{2}}{2} & \frac{\sqrt{2}}{4} & 0 & 0 & \cdots & 0 & 0 & 0 \\ & & & \widetilde{\mathbf{w}}^2 & & & & \\ & & & \vdots & & & & \\ & & & \widetilde{\mathbf{w}}^{N/2} & & & & \\ \hline \frac{\sqrt{2}}{2} & -\frac{5\sqrt{2}}{8} & \frac{\sqrt{2}}{4} & \frac{\sqrt{2}}{8} & 0 & \cdots & 0 & 0 \\ & & & \widetilde{\mathbf{w}}^{N/2+2} & & & & \\ & & & \vdots & & & & \\ & & & \widetilde{\mathbf{w}}^{N-1} & & & & \\ 0 & 0 & 0 & \cdots & 0 & \frac{\sqrt{2}}{8} & \frac{\sqrt{2}}{4} & -\frac{3\sqrt{2}}{4} \end{array}\right].$$

Show that U^T is the inverse of $\widetilde{W}_N^s$ and V is the inverse of W_N^s. (*Hint:* You know that the inner product

$$\widetilde{\mathbf{w}}^k \cdot \mathbf{w}^\ell = \begin{cases} 1, & k = \ell \\ 0, & k \neq \ell \end{cases}$$

You need to check the "special rows" – for example, the inner product of the first row $\left(\frac{\sqrt{2}}{2}, \frac{\sqrt{2}}{4}, 0, \ldots, 0\right)$ of $\widetilde{W}_N^s$ and the ℓth row of U should be 1 if $\ell = 1$ and 0 otherwise. It is easy to check this condition as well as to verify that the other "special rows" satisfy the necessary orthogonality conditions.)

7.39 Verify the computations (7.42) and (7.43) in Example 7.6.

7.40 Let $\mathbf{v} = [0, 0, 8, 0, 2, 2, 0, 3, 5, 0, 1, 9]^T$. Use Algorithm 7.3 to compute the symmetric biorthogonal wavelet transformation using the $(8, 4)$ biorthogonal spline filter pair.

7.41 In this problem, you will investigate the symmetric nature of $\mathbf{x}$ given in (7.49) when $\mathbf{v} \in \mathbb{R}^6$. Suppose $\mathbf{v} \in \mathbb{R}^6$. Using (7.39), form $\mathbf{v} \in \mathbb{R}^{10}$ in this case and then write down a few periods of $\mathbf{x}$. We know by its periodic construction that $x_{-10} = x_0 = x_{10}$. You should be able to see that $x_k = x_{-k}$, $k = -9, \ldots, 9$. If

$$x_k = \begin{cases} v_{k+1}, & 1 \le k \le 5 \\ v_{12-k-1}, & 6 \le k \le 9, \end{cases}$$

how can we express the elements x_k for $-9 \le k \le -1$? This should give proof that $x_k = x_{-k}$ for $k = -9, \ldots, 9$. The fact that $\mathbf{x}$ is 10-periodic proves the result in general.

7.42 Let $\mathbf{v}^p$ be given by (7.39) and $\mathbf{x}$ be given by (7.49). Show that $\mathbf{x}$ is symmetric about zero. In other words, show that for all $k \in \mathbb{Z}$, we have $x_k = x_{-k}$. *Hint:* You should be able to generalize the result of Problem 7.41.

7.43 Consider the biorthogonal filter pair from Example 7.3. Suppose that we form

$$\widetilde{\mathbf{h}} = \left(\ldots, 0, 0, \widetilde{h}_1, \widetilde{h}_0, \widetilde{h}_1, 0, 0, \ldots\right)$$

and

$$\mathbf{h} = (\ldots, 0, 0, h_2, h_1, h_0, h_1, h_2, 0, 0, \ldots).$$

For $N = 8$, use (7.33) to form $\mathbf{v}^p$, and from there, use (7.49) to form $\mathbf{x}$. From Examples 7.7 and 7.8 we know that each period of $\mathbf{a}$ and $\mathbf{d}$ (defined in (7.48) and (7.52)) have the structures $(a_0, a_1, a_2, a_3, a_3, a_2, a_1)$ and $(d_1, d_2, d_3, d_4, d_3, d_2, d_1)$, respectively. Show that $\widetilde{a}_k^p = a_{k-1}$ and $\widetilde{d}_k^p = d_k$, $k = 1, 2, 3, 4$ where the values $\widetilde{a}_k^p$ and $\widetilde{d}_k^p$ are given by (7.34) and (7.36), respectively.

7.44 Prove Proposition 7.4.

7.45 Complete the proof of Proposition 7.5 and show that $\mathbf{d}$ is $(N-1)$-periodic.

7.46 Prove Corollary 7.2. (*Hint:* Show $d_n = d_{N-n}$.)

7.47 Prove Proposition 7.6.

7.48 Suppose that $\mathbf{v} = [v_1, \ldots, v_N]^T \in \mathbf{R}^N$, N even. Form the vector $\mathbf{v}^p \in \mathbb{R}^{2N}$ as

$$\mathbf{v}^p = [v_1, \ldots, v_N, v_N, \ldots, v_1]^T.$$

Build $\mathbf{x}$ out of infinitely many copies of $\mathbf{v}^p$. That is, $\mathbf{x}$ is the bi-infinite sequence

$$\mathbf{x} = (\ldots, \mathbf{v}^p, \mathbf{v}^p, \mathbf{v}^p, \ldots).$$

where $x_0 = v_1$. Show that

(a) $\mathbf{x}$ is $2N$-periodic

(b) $x_k = x_{1-k}$ for all $k \in \mathbb{Z}$

7.49 Let $\mathbf{v}, \mathbf{v}^p$, and $\mathbf{x}$ be as defined in Problem 7.48 and suppose that $\widetilde{\mathbf{h}}$ is the symmetric bi-infinite sequence given by

$$\begin{aligned}\widetilde{\mathbf{h}} &= (\ldots, 0, 0, \widetilde{h}_{-\widetilde{L}+1}, \ldots, \widetilde{h}_0, \widetilde{h}_1, \ldots, \widetilde{h}_{\widetilde{L}}, 0, 0, \ldots) \\ &= (\ldots, 0, 0, \widetilde{h}_{\widetilde{L}}, \ldots, \widetilde{h}_1, \widetilde{h}_1, \ldots, \widetilde{h}_{\widetilde{L}}, 0, 0, \ldots).\end{aligned}$$

Define the bi-infinite sequence $\mathbf{a}$ componentwise by

$$a_n = \sum_k \widetilde{h}_{2n-k} x_k.$$

(a) Show that $\mathbf{a}$ is $2N$-periodic. *Hint:* Recall that convolution is commutative (see Problem 2.53 in Section 2.4) and use Problem 7.48(a).

(b) Use the fact that $\widetilde{h}_k = \widetilde{h}_{1-k}$, $k \in \mathbb{Z}$ and Problem 7.48(b) to show that for all $n \in \mathbb{Z}$, $a_n = a_{1-n}$.

(c) Use (a) and (b) to show that $a_{N+1-n} = a_n$, $n \in \mathbb{Z}$.

(d) Use (a) and (c) to show that each period of $\mathbf{a}$ has the form

$$[a_1, a_2, \ldots, a_{N/2}, a_{N/2}, \ldots, a_2, a_1]^T.$$

(e) Let $\widetilde{\mathbf{a}}^p$ be the lowpass portion of the modified biorthogonal wavelet transformation returned by Algorithm 7.3 with lowpass filter

$$\widetilde{h} = \left(\widetilde{h}_{\widetilde{L}}, \ldots, \widetilde{h}_1, \widetilde{h}_1, \ldots, \widetilde{h}_{\widetilde{L}}\right).$$

Show that $\left(a_1, \ldots, a_{N/2}\right) = \left(\widetilde{a}_1^p, \ldots, \widetilde{a}_{N/2}^p\right)$.

7.50 Let $\mathbf{v}, \mathbf{v}^p$, and $\mathbf{x}$ be as defined in Problem 7.48 and suppose that $\widetilde{\mathbf{g}}$ is the symmetric bi-infinite sequence

$$\widetilde{\mathbf{g}} = (\ldots, \widetilde{g}_{-L+1}, \ldots, \widetilde{g}_0, \widetilde{g}_1, \ldots, \widetilde{g}_L, \ldots)$$

constructed from the symmetric bi-infinite sequence

$$\begin{aligned}\mathbf{h} &= (\ldots, 0, 0, h_{-L+1}, \ldots, h_0, h_1, \ldots, h_L, 0, 0, \ldots) \\ &= (\ldots, 0, 0, h_L, \ldots, h_1, h_1, \ldots, h_L, 0, 0, \ldots).\end{aligned}$$

Define the bi-infinite sequence $\mathbf{b}$ component-wise by

$$d_n = \sum_k \widetilde{g}_{2n-k} x_k.$$

(a) Show that $\mathbf{d}$ is a $2N$-periodic sequence. (*Hint:* The proof should be very similar to that given for Problem 7.49(a).)

(b) Use the fact that $\widetilde{g}_k = -\widetilde{g}_{1-k}$, $k \in \mathbb{Z}$ (see Problem 7.30 in Section 7.2) and Problem 7.48(b) to show that for all $n \in \mathbb{Z}$, $d_n = -d_{1-n}$.

(c) Use (a) and (b) to show that $d_{N+1-n} = -d_n$, $n \in \mathbb{Z}$.

(d) Use (a) and (c) to show that each period of $\mathbf{d}$ has the form

$$\left[d_1, d_2, \ldots, d_{N/2}, -d_{N/2}, \ldots, -d_2, -d_1\right]^T.$$

(e) Let $\widetilde{\mathbf{d}}^p$ be the highpass portion of the symmetric biorthogonal wavelet transformation returned by Algorithm 7.3 with highpass filter $\widetilde{\mathbf{g}} = (\widetilde{g}_{-L+1}, \ldots, \widetilde{g}_0, \widetilde{g}_1, \ldots, \widetilde{g}_L)$. Show that $\left[d_1, \ldots, d_{N/2}\right]^T = \left[\widetilde{d}_1^p, \ldots, \widetilde{d}_{N/2}^p\right]^T$.

Computer Lab

7.3 The Symmetric Biorthogonal Wavelet Transformation. From `stthomas.edu/wavele` access the file `SymmetricBiorth`. In this lab, you will develop and test efficient algorithms for computing the one- and two-dimensional symmetric biorthogonal wavelet transformations using the $(5, 3)$ and the $(8, 4)$ biorthogonal spline filter pairs.

7.4 Image Compression and Image Pansharpening

In this section we revisit the application of image compression and investigate the application of image *pansharpening* to which the biorthogonal wavelet transform is well-suited.

In Example 7.9, we see that the $(5, 3)$ biorthogonal spline filter pair can be modified and used in conjunction with the symmetric biorthogonal wavelet transform to produce a much lower bits per pixel value than can be acheived using the D6 filter. The modification allows us to map integer-valued intensities from the input image to integers that are divided by powers of two. The orthogonal/biorthogonal filters constructed so far contain irrational terms, so mapping integers to rational numbers is an improvement. In Chapter 11, we learn how to further modify the $(5, 3)$ biorthogonal spline filter pair so that it can map integers to integers.

Example 7.9 (Image Compression Using the Biorthogonal Wavelet Transform). *We consider the* 256×384 *image from Examples 4.15 and 5.3. In this example we perform image compression using the* $(5, 3)$ *biorthogonal spline filter pair. and compare our results with those obtained using the D6 orthogonal filter.*

Let A *denote the matrix of the image in Figure 4.20(a). We begin by computing three iterations of the symmetric biorthogonal wavelet transform and the orthogonal wavelet transform of the centered version of* A *(subtract* 128 *from each element of* A*). While we use the D6 filter for the orthogonal filter, we make a slight adjustment to the* $(5, 3)$ *biorthogonal spline filter pair before computing the transform.*

An iteration of the biorthogonal wavelet transform of A *is computed using the matrices* $\widetilde{W}_N$ *and* W_N^T *from (7.16) and (7.17), respectively. We can "rationalize" our filters in each matrix if we divide* $\widetilde{W}_N$ *by* $\sqrt{2}$ *and multiply* W_N *by* $\sqrt{2}$*. This is exactly the process described in Exercise 7.10. We use this problem with* $r = \dfrac{1}{\sqrt{2}}$ *so that the resulting diagonal matrix is*

$$D_N = \begin{bmatrix} \frac{1}{\sqrt{2}} & & & & & \\ & \ddots & & & & \\ & & \frac{1}{\sqrt{2}} & & & \\ & & & \sqrt{2} & & \\ & & & & \ddots & \\ & & & & & \sqrt{2} \end{bmatrix}.$$

Following Exercise 7.10, we define the new transform matrices $\widetilde{U}_N = D_N \widetilde{W}_N$ *and* $U_N = D_N^{-1} W_N$*.*

$$B = \widetilde{U}_{256} A U_{384}^T = \left(D_{256} \widetilde{W}_{256} \right) A \left(W_{384}^T D_{384}^{-1} \right). \tag{7.53}$$

As pointed out in Exercise 7.10, the effect of this step is to generate our wavelet matrices from the filters

$$\frac{1}{\sqrt{2}}\widetilde{\mathbf{h}} = \left(\frac{1}{4}, \frac{1}{2}, \frac{1}{4}\right) \quad \text{and} \quad \sqrt{2}\mathbf{h} = \left(-\frac{1}{4}, \frac{1}{2}, \frac{3}{2}, \frac{1}{2}, -\frac{1}{4}\right) \tag{7.54}$$

where $\widetilde{\mathbf{h}}$ *and* $\mathbf{h}$ *are given in Definition 7.2.*

The elements of B *will be integers and integer multiples of* $\frac{1}{2}$, $\frac{1}{4}$ *and* $\frac{1}{8}$. *These values are not integers, but at least they are no longer irrational numbers. The hope is removing irrational values from* B *will increase the effectiveness of the Huffman coding of the quantized transform.*

It is straightforward to recover A *from* B. *From Exercise 7.10 we know that* $\widetilde{U}_{256}^{-1} = U_{256}^T$ *and* $U_{384}^T = \widetilde{U}_{384}^{-1}$. *Thus we can multiply both sides of (7.53) by* U_{256}^T *on the left and* $\widetilde{U}_{384}$ *on the right to obtain*

$$U_{256}^T B \widetilde{U}_{384} = \left(U_{256}^T \widetilde{U}_{256}\right) A \left(U_{384}^T \widetilde{U}_{384}\right) = I_{256} A I_{384} = A.$$

The above equation can be rewritten as

$$\begin{aligned} A &= U_{256}^T B \widetilde{U}_{384} \\ &= \left(W_{256}^T D_{256}^{-1}\right) B \left(D_{384} \widetilde{W}_{384}\right) \\ &= W_{256}^T \left(D_{256}^{-1} B D_{384}\right) \widetilde{W}_{384}. \end{aligned}$$

The wavelet transforms are plotted in Figure 7.5.

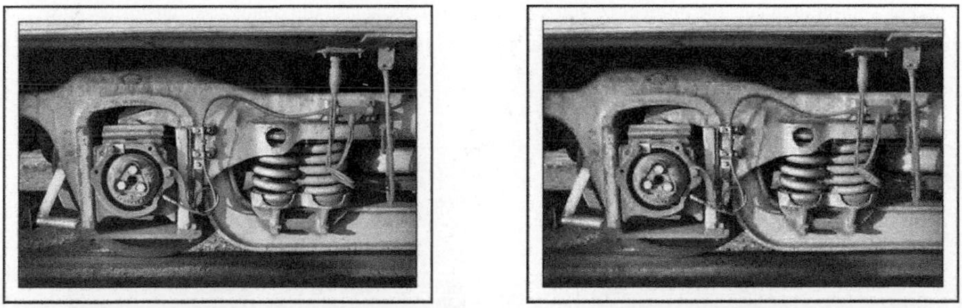

(a) D6 (b) BWT

Figure 7.5 Three iterations of the biorthogonal and orthogonal wavelet transforms of A using the filter pair in (7.54) and the D6 filters

We use cumulative energy to quantize the transforms. The cumulative energy vectors for each transform is plotted in Figure 7.6. The biorthogonal wavelet transform does a much better job conserving energy.

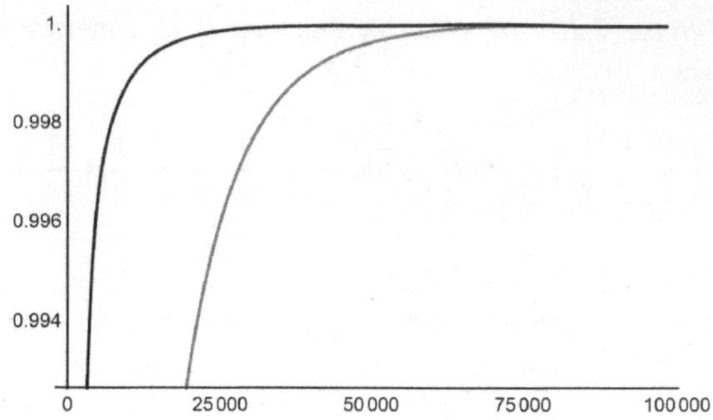

Figure 7.6 The cumulative energy vectors for the biorthogonal wavelet transform (black) and the orthogonal wavelet transform (gray).

We find the largest elements (in absolute value) in each wavelet transform that constitutes 99.9% *of the energy. These largest values are floored to the nearest integer. All other wavelet coefficients are set to zero. Table 7.1 summarizes the compression information for each wavelet transformation. The quantized transformations are plotted in Figure 7.7.*

Table 7.1 Data for the quantized wavelet transformations using the modified filter pair in (7.54) and the D6 filter.

Filter	Nonzero Terms	% Zero
D6	39,154	60.17
$(5,3)$	10,857	89.96

[a]The second column indicates the number of nonzero terms still present in the quantized transforms, and the third column lists the percentage of elements that are zero in each quantized transform.

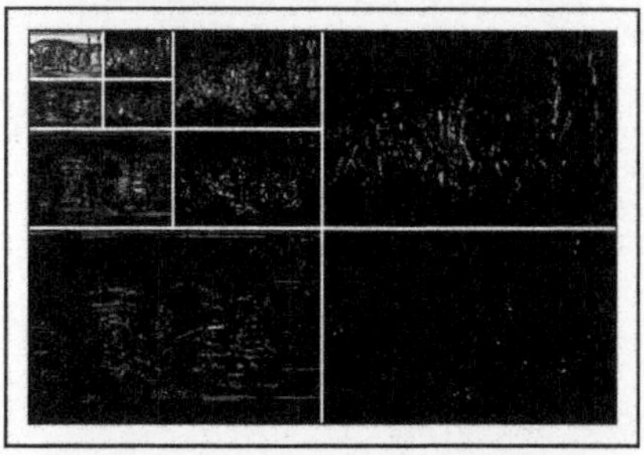

(a) D6

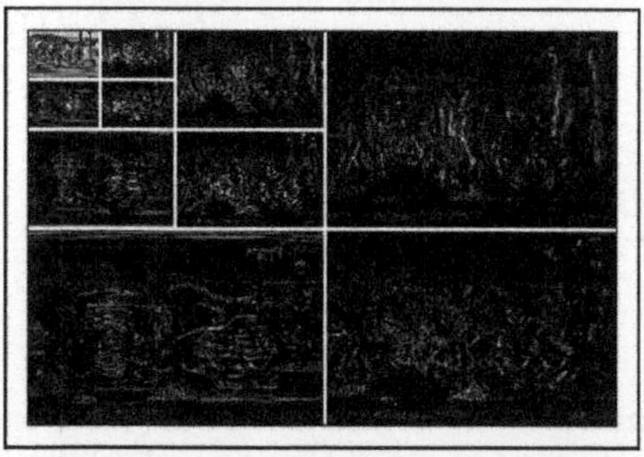

(b) BWT

Figure 7.7 The quantized wavelet transforms using the modified filter pair given in (7.54) and the D6 filter.

We next round all the elements in the two wavelet transforms and apply Huffman coding to the results. Coding information for each transform is given in Table 7.2.

Table 7.2 Huffman bit stream length, bpp, and entropy for each quantized transform.

Filter	Bit Stream Length	bpp	Entropy
D6	385,159	3.918	3.877
$(5, 3)$	205,781	2.093	1.590

Table 7.2 tells us that the modified $(5, 3)$ *biorthogonal spline filter pair does a better job improving the effectiveness of the coding than does the D6 filter. To check the quality of the approximation, we apply the inverse transforms to each of the quantized transforms and add* 128 *to the results to offset the original centering step. The approximate images are plotted in Figure 7.8. The captions of these figures contain information on PSNR.*

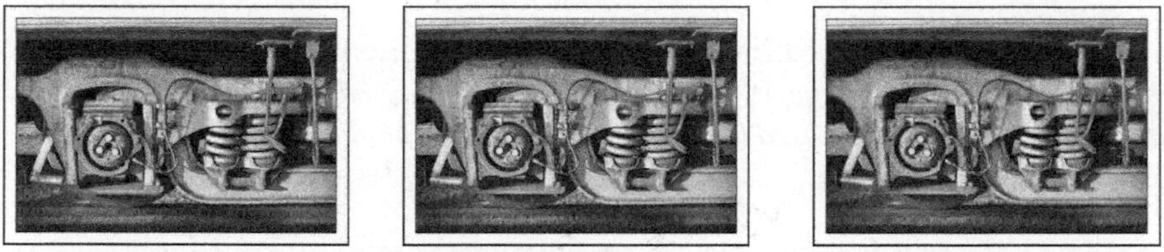

(a) Original Image (b) D6, PSNR = 41.21 (c) BWT, PSNR = 25.70

Figure 7.8 The approximate images using the modified filter pair given in (7.54) and the D6 filter.

The bits per pixel for the approximation built using the biorthogonal wavelet transform is quite low, but the PSNR value is low as well. Repeating the compression process using 99.99% *of the energy leads to the results summarized in Table 7.3.*

Table 7.3 Results of compression method when 99.99% of the energy is retained.

Filter	% Zero	Bit Stream Length	bpp	Entropy	PSNR
D6	36.89	516,313	5.252	5.186	49.76
$(5, 3)$	71.25	342,532	3.484	3.342	34.34

(Live example: Visit `stthomas.edu/wavelets` *and click on Live Examples.)*

□

Image Pansharpening

Companies such as Google Maps rely on satellite imagery to create maps. Satellites images typically consist of several multispectral bands and a panchromatic band. Among the multispectral bands are red, green and blue color channels and the panchromatic channel can be viewed as a grayscale channel. Landsat 8 was launched by NASA on February 11, 2013 and captures red, green, and blue channels at 30 m resolution and the panchromatic channel at 15 m resolution (see [89]). The Worldview-2 satellite was launched on October 8, 2009 by DigitalGlobe, Inc. Its red, green, and blue channels are acquired at 1.84 m resolution and the panchromatic channel is obtained at 0.46 m resolution (see [34]). Lower resolution is used for multispectral channels in order to maximize storage space.

A natural question is how might we combine the low resolution color channels and the high resolution grayscale channel to produce a high resolution color channel? In image processing, the procedure for creating a high resolution color image is called *pansharpening* ("pan" for panchromatic). There are many different methods for performing pansharpening of images, several of the proprietary nature. We conclude this section by investigating image pansharpening via wavelets.

A standard wavelet-based pansharpening algorithm can be found in [72]. The procedure is described in Algorithm 7.5.

Algorithm 7.5 (Wavelet-Based Image Pansharpening). *Given high resolution panchromatic image (grayscale) image P and low resolution color channels R, G, and B, this algorithm uses wavelet-based image pansharpening to produce a high resolution color image.*

1. *Enlarge R, G, and B so that their dimensions are the same as P.*

2. *Convert R, G, and B to HSI space producing channels H, S and I.*

3. *Perform histogram matching (see Section 3.1) with I as the reference channel and P as the target channel to obtain P'.*

4. *Compute j iterations of the symmetric biorthogonal wavelet transform of both I and P'.*

5. *Replace the blur portion of the transform of I with the corresponding portion of the transform of P'.*

6. *Compute j iterations of the inverse symmetric biorthogonal wavelet transform to obtain the new intensity channel I'.*

7. *Convert H, S, and I' to RGB space.*

□

The resolution panchromatic channel of the Landsat 8 data is a factor of two finer than the color channels. For Worldview-2 satellite imagery, the factor is 4. Thus we assume the panchromatic channel is a factor of 2^ℓ finer than the resolution of the color channels, for some ℓ.

The initial step in the pansharpening algorithm is new. To enlarge a color channel, we create a zero matrix whose dimensions are the same as the panchromatic channel and overwrite the upper left corner with the color channel. We consider this matrix as the ℓth iterative Haar wavelet transform of some matrix A and perform ℓ iterations of the inverse Haar wavelet transform to obtain A. Problem 4.30 provides more details about the elements of the resulting matrix and Figure 7.9 illustrates the enlarging process on an image matrix.

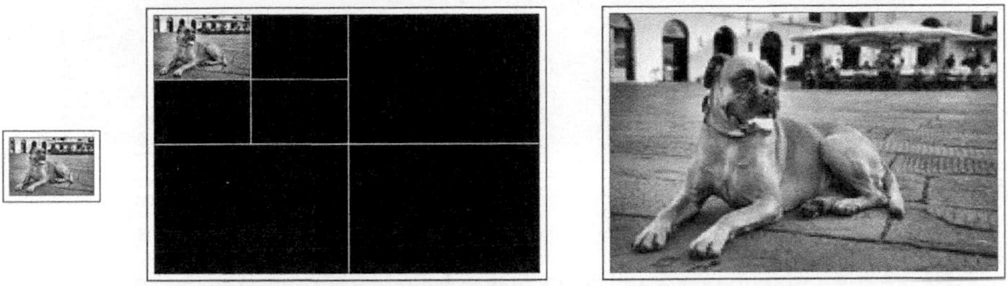

(a) A (b) Wavelet transform (c) Enlarged A

Figure 7.9 Enlarge an image using the inverse Haar wavelet transform.

We conclude this section with an example that illustrates how Algorithm 7.5 works.

Example 7.10 (Image Pansharpening). *We consider satellite images of Arcadia National Park plotted in Figure 7.10(a) and 7.10(b), respectively.*[3] *The dimensions of the color and grayscale images are* 200×200 *and* 800×800*, respectively.*

The first step in Algorithm 7.5 is to enlarge the red, green and blue channels of the low resolution color image. Since the dimensions of the color image are a factor of four smaller than the high resolution grayscale image, for each color channel, we create an 800×800 *zero matrix and overwrite the top left corner with a centered version of the color channel. Then we perform two iterations of the inverse Haar wavelet transform on this matrix and add 128 to each element to obtain the enlarged version of the color channel. The enlarged color channel is plotted in Figure 7.10(c).*

The enlarged color image is next converted to HSI space and we use the I channel as a reference in histogram matching for the high resolution grayscale channel P. The histogram matched channel P' is plotted next to P for comparative purposes in Figures 7.11(b) and 7.11(c).

Next we compute four iterations of the symmetric biorthogonal wavelet transform using the $(8,4)$ *biorthogonal spline filter pair of both I and P'. As was the case in*

[3]The images that appear in Figure 7.10 are constructed from the public domain Landsat imagery courtesy of NASA Goddard Space Flight Center and U.S. Geological Survey available at `landsat.visibleearth.nasa.gov/`.

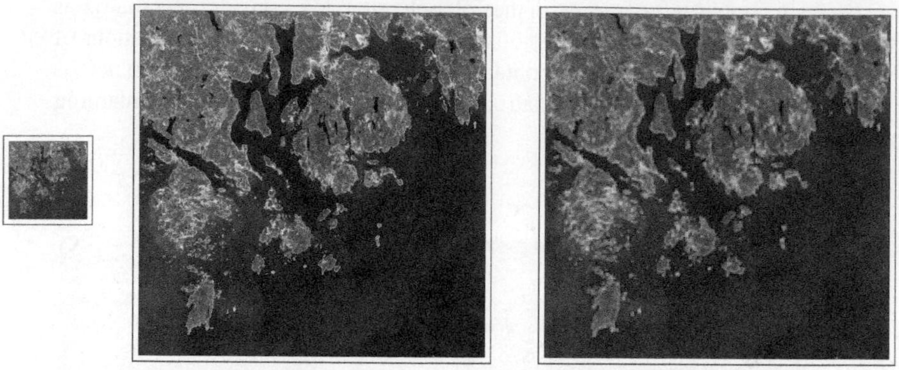

(a) Color (b) Grayscale (c) Enlarged color

Figure 7.10 A low resolution color image, its high resolution grayscale counterpart and an enlarged version of the color image. See the color plates for color versions of these images.

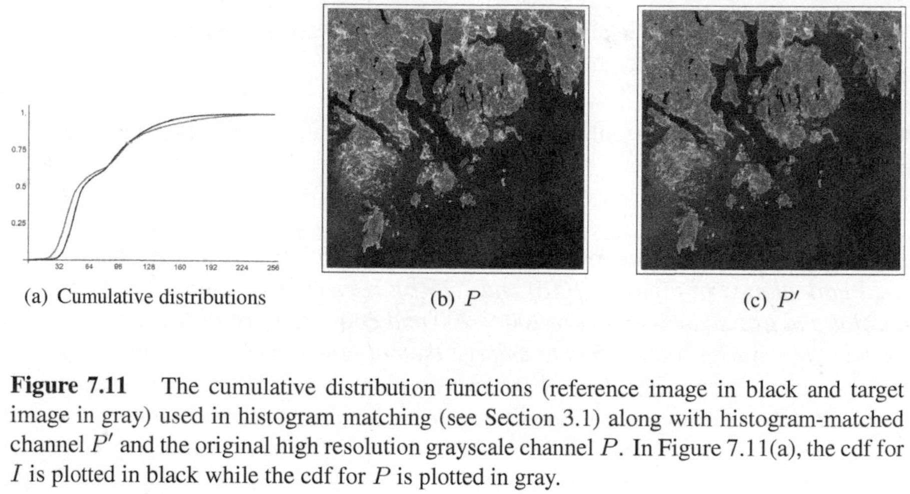

(a) Cumulative distributions (b) P (c) P'

Figure 7.11 The cumulative distribution functions (reference image in black and target image in gray) used in histogram matching (see Section 3.1) along with histogram-matched channel P' and the original high resolution grayscale channel P. In Figure 7.11(a), the cdf for I is plotted in black while the cdf for P is plotted in gray.

Example 7.9, we divide $\widetilde{\mathbf{h}}$ by $\sqrt{2}$ and multiply $\mathbf{h}$ by $\sqrt{2}$. In this case, our filters are

$$\widetilde{\mathbf{h}} = \left(\frac{1}{8}, \frac{3}{8}, \frac{3}{8}, \frac{1}{8}\right)$$

$$\mathbf{h} = \left(\frac{3}{32}, -\frac{9}{32}, \frac{7}{32}, \frac{45}{32}, \frac{45}{32}, \frac{7}{32}, -\frac{9}{32}, \frac{3}{32}\right).$$

Adjusting our transform in this way ensures that the values in the output are rational. We replace the blur portion of the wavelet transform of P' with the blur portion of the transform of I. The transform and its inverse, the new I' channel, are plotted in Figure 7.12.

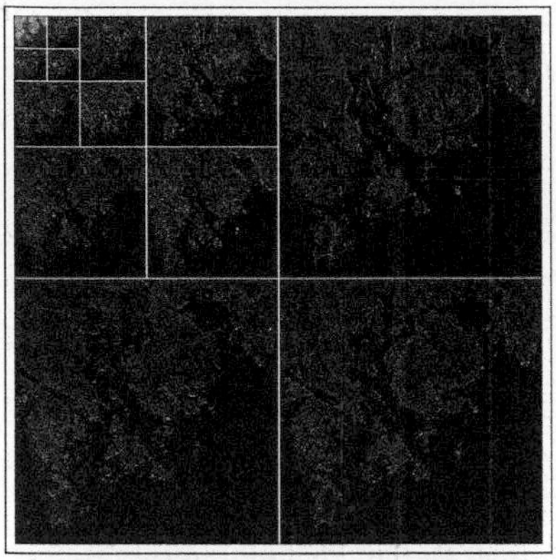

(a) Fused wavelet transform

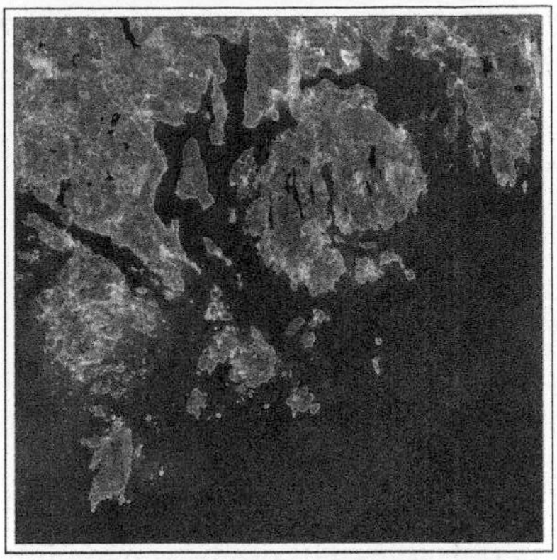

(b) Inverse wavelet transform I'

Figure 7.12 The fused wavelet transform (the blur portion from the transform of I replacing the blur portion of the transform of P') and its inverse transform I'.

In practice, some of the values of I' may land outside the interval $[0, 1]$. Before proceeding, we "clip" these values – any value greater than one is converted to one and any negative value is changed to zero. Finally, we join I' with the hue and saturation channels and convert these channels back to RGB space. The resulting pansharpened image is plotted in Figure 7.13(b). The enlarged color image from Figure 7.10(a) is plotted alongside in Figure 7.13(a) for comparative purposes.

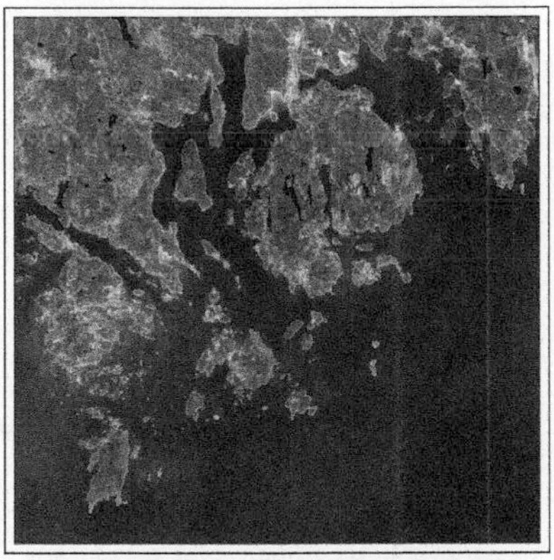

(a) Pansharpened image

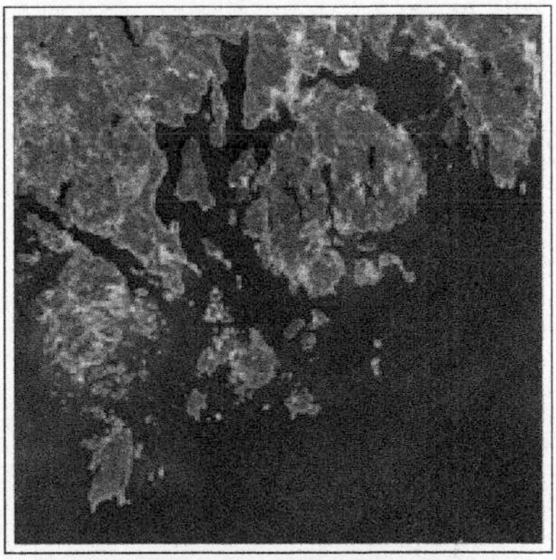

(b) Enlarged color image

Figure 7.13 The fused wavelet transform (the blur portion from the transform of I replacing the blur portion of the transform of P') and its inverse transform I'. See the color plates for color versions of these images.

(Live example: Visit `stthomas.edu/wavelets` *and click on Live Examples.)*

□

Computer Lab

7.4 Image Pansharpening. From `stthomas.edu/wavelets`, access the file `PanSharpen`. In their paper [72], Mitianoudis, Tzimiropoulos and Stathaki propose a couple of changes to Algorithm 7.5. The authors suggest first converting the low resolution color image to HSI space but then only enlarging the hue and saturation channels. Instead of histogram-matching the intensity channel I with the panchromatic channel P as described in Algorithm 7.5, I is histogram matched with $\widetilde{P} = \overline{x}P + v$, where $\overline{x}$, v are the mean and variance of I. The authors claim that this modification will produce a smoother histogram matched P'. Then j iterations of the symmetric biorthogonal wavelet transform of P' are computed compared to just $j - 2$ iterations of the wavelet transform of the intensity channel I. Since the dimensions of I are a factor of four smaller than those of P', we can still replace the blur portion of the transform of P' with the wavelet transform of I. The new intensity channel I' is obtained as before and, coupled with the hue and saturation channels, converted to RGB space. In this lab, you will investigate the effectiveness of these changes to Algorithm 7.5.

CHAPTER 8

COMPLEX NUMBERS AND FOURIER SERIES

Complex numbers and Fourier series play vital roles in digital and signal processing. This chapter begins with an introduction to complex numbers and complex arithmetic.

Fourier series are discussed in Section 8.2. The idea is similar to that of Maclaurin series – we rewrite the given function in terms of basic elements. While the family of functions x^n, $n = 0, 1, \ldots$ serve as building blocks for Maclaurin series, complex exponentials $e^{ik\omega}$, $k \in \mathbb{Z}$ are used to construct Fourier series. Also included in Section 8.2 are several useful properties obeyed by Fourier series. We connect Fourier series with convolution and filters in Section 8.3.

We have constructed (bi)orthogonal wavelet filters/filter pairs in Chapters 4, 5 and 7. The constructions were somewhat ad hoc and in the case of the biorthogonal filter pairs, limiting. Fourier series opens up a completely new and systematic way to look at wavelet filter construction. It is important you master the material in this chapter before using Fourier series to construct (bi)orthogonal wavelet filters/filter pairs in Chapter 9.

Discrete Wavelet Transformations: An Elementary Approach With Applications, Second Edition.
Patrick J. Van Fleet.

8.1 The Complex Plane and Arithmetic

In this section we review complex numbers and some of their basic properties. We will also discuss elementary complex arithmetic, modulus, and conjugates. Let's start with the *imaginary number*

$$i = \sqrt{-1}.$$

It immediately follows that

$$i^2 = (\sqrt{-1})^2 = -1 \qquad i^3 = i^2 \cdot i = -i \qquad i^4 = i^2 \cdot i^2 = (-1)(-1) = 1.$$

In Problem 8.2 you will compute i^n for any nonnegative integer n.

We define a *complex number* to be any number $z = a + bi$, where a and b are any real numbers. The number a is called the *real part* of z, and b is called the *imaginary part* of z.

It is quite easy to plot complex numbers. Since each complex number has a real and an imaginary part, we simply associate with each complex number $z = a + bi$ the ordered pair (a, b). We use the connection to two-dimensional space and plot complex points in a plane. Our horizontal axis will be used for the real part of z, and the vertical axis will represent the imaginary part of z. Some points are plotted in the complex plane in Figure 8.1.

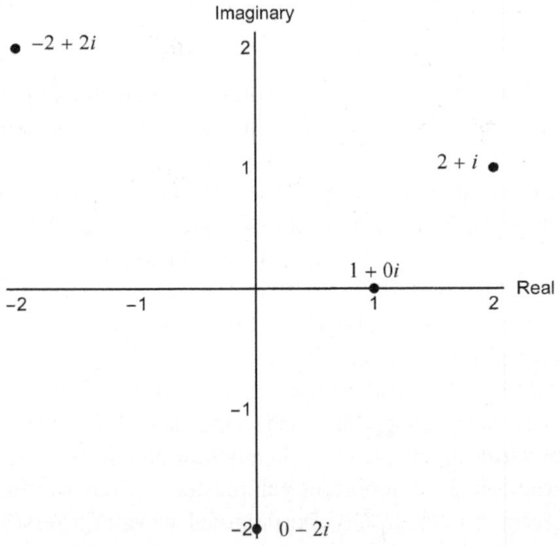

Figure 8.1 Some complex numbers.

Trigonometric Review

It's always good to sprinkle in a little review, especially if it will be of use later. Remember how the trigonometric functions cosine and sine are defined? We draw a circle of radius one centered at the origin and then pick any point on the circle. If we draw the right triangle pictured in Figure 8.2, we obtain an angle θ. The length of the horizontal side of this triangle is $\cos\theta$ and the length of the vertical side of this triangle is $\sin\theta$. (Here, we will allow lengths to be negative.) So any point on the circle can be determined by an angle θ, and its coordinates are $(\cos\theta, \sin\theta)$.

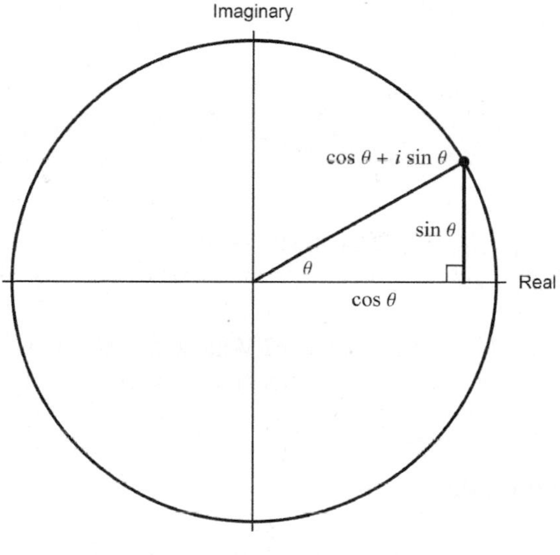

Figure 8.2 Circle and trig functions.

How does this trigonometric review relate to complex numbers? We have already associated complex numbers with two-dimensional space, so it is natural to see that just as any circle of radius one centered at the origin has points defined by $(\cos\theta, \sin\theta)$, we see that the function

$$f(\theta) = \cos\theta + i\sin\theta \tag{8.1}$$

defines a circle in the complex plane. We evaluate $f(\theta)$ at three points and plot the output in Figure 8.3. The points are

$$\begin{aligned} f\left(\frac{\pi}{6}\right) &= \cos\frac{\pi}{6} + i\sin\frac{\pi}{6} = \frac{\sqrt{3}}{2} + i\frac{1}{2} \\ f\left(\frac{3\pi}{4}\right) &= \cos\frac{3\pi}{4} + i\sin\frac{3\pi}{4} = -\frac{\sqrt{2}}{2} + i\frac{\sqrt{2}}{2} \\ f(\pi) &= \cos\pi + i\sin\pi = -1. \end{aligned}$$

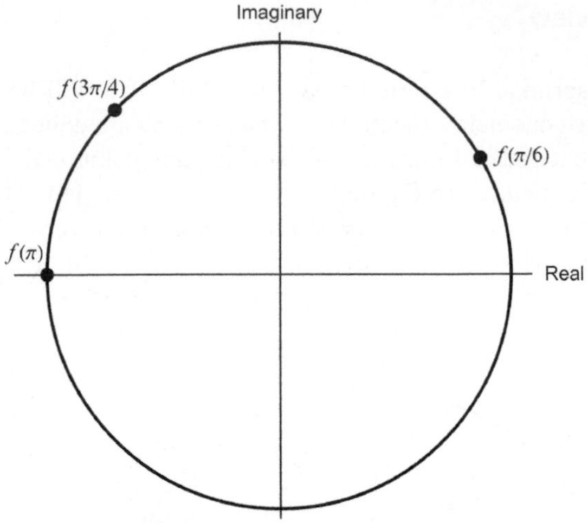

Figure 8.3 The function $f(\theta) = \cos\theta + i\sin\theta$.

For $0 \leq \theta \leq 2\pi$, what does $f(2\theta)$ look like? What about $f(4\theta)$ or $f(\frac{\theta}{2})$? See the problems at the end of this section for more functions to plot.

Basic Complex Arithmetic

Now that we have been introduced to complex numbers and the complex plane, let's discuss some basic arithmetic operations.

Just as with real numbers, we can add, subtract, multiply, and divide complex numbers. Addition and subtraction are very straightforward. To add (subtract) two complex numbers $z = a + bi$ and $w = c + di$, we simply add (subtract) the real parts and add (subtract) the imaginary parts. Thus,

$$z \pm w = (a + bi) \pm (c + di) = (a \pm c) + (b \pm d)i.$$

For example, the sum of $3 + 5i$ and $2 - i$ is $5 + 4i$, and the difference of $3 + 5i$ and $2 - i$ is $1 + 6i$.

Multiplication is also quite straightforward. To multiply $z = a + bi$ and $w = c + di$, we simply multiply the numbers just like we would two algebraic expressions

$$z \cdot w = (a + bi) \cdot (c + di) = ac + adi + bci + bdi^2 = (ac - bd) + (ad + bc)i.$$

For example, the product of $3 + 5i$ and $2 - i$ is $11 + 7i$.

Conjugates

Before we discuss division, we need to introduce the idea of the *conjugate* of a complex number.

Definition 8.1 (Conjugate of a Complex Number). *Let $z = a + bi \in \mathbb{C}$. Then the* conjugate *of z, denoted by $\overline{z}$, is given by $\overline{z} = a - bi$.* □

As you can see, the conjugate merely serves to negate the imaginary part of a complex number. Let's look at some simple examples and then we will look at the geometric implications of this operation.

Example 8.1 (Computing Conjugates). *Find the conjugates of $3 + i$, $-2i$, 5, and $\cos\frac{\pi}{4} + i\sin\frac{\pi}{4}$.*

Solution.
$\overline{3 + i} = 3 - i$ and the conjugate of $-2i$ is $2i$. Since 5 has no imaginary part, it is its own conjugate and the conjugate of $\cos\frac{\pi}{4} + i\sin\frac{\pi}{4}$ is $\cos\frac{\pi}{4} - i\sin\frac{\pi}{4}$.
(Live example: Visit `stthomas.edu/wavelets` and click on Live Examples.)

□

Some properties of the conjugate are immediate. For example, $\overline{\overline{z}} = z$ and a complex number is real if and only if $\overline{z} = z$. Now let's look at some of the geometry behind the conjugate operator.

To plot a complex number $z = a + bi$ and its conjugate $\overline{z} = a - bi$, we simply reflect the ordered pair (a, b) over the real (horizontal) axis to obtain $(a, -b)$. This is illustrated in Figure 8.4.

Now to get even more geometric insight into the conjugate operator and circles, recall the function $f(\theta) = \cos\theta + i\sin\theta$, $0 \le \theta \le 2\pi$. If we start at $\theta = 0$ and plot points in a continuous manner until we reach $\theta = 2\pi$, we not only graph a circle but we do so in a counterclockwise manner.

What happens if we perform the same exercise with $\overline{f(\theta)}$, $0 \le \theta \le 2\pi$? We get exactly the same picture, but now the graph is drawn in a clockwise manner.

One last remark about the conjugate. Consider the product of $z = a + bi$ and its conjugate $\overline{z} = a - bi$:

$$z \cdot \overline{z} = a^2 - b^2 i^2 = a^2 + b^2.$$

The Modulus of a Complex Number

Now let's plot $z = a + bi$ and connect it to the origin. We can form a right triangle whose horizontal side has length a and whose vertical side has length b (see Figure 8.5). By the Pythagorean theorem, the hypotenuse $h = \sqrt{a^2 + b^2}$, so that

$$h^2 = a^2 + b^2 = z \cdot \overline{z}.$$

Now the length of h also represents the distance from z to the origin, so we immediately see that this distance is related to the product of z and $\overline{z}$. We are ready for the following definition.

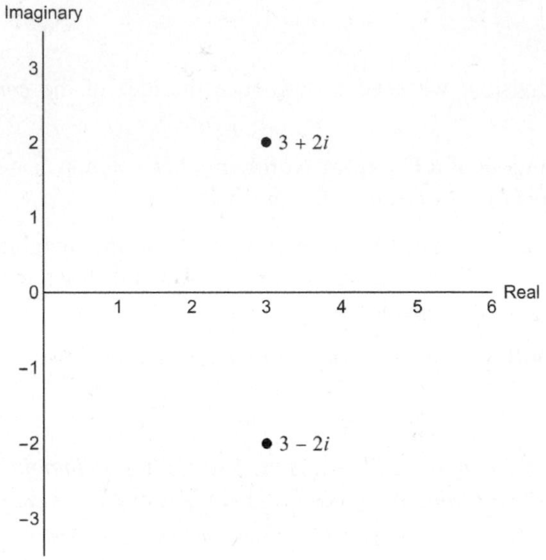

Figure 8.4 A complex number and its conjugate.

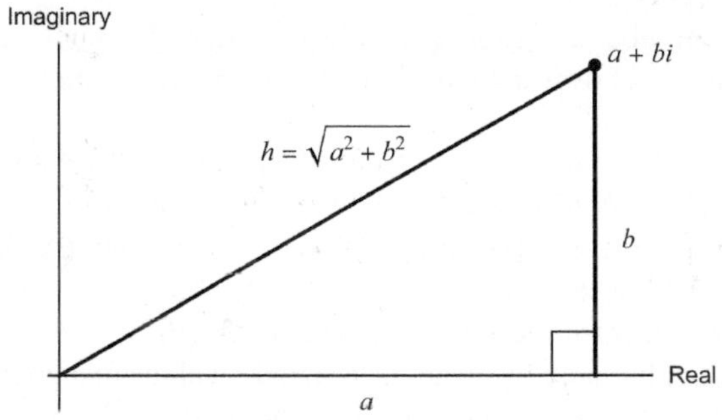

Figure 8.5 The right triangle formed from $z = a + bi$.

Definition 8.2 (Modulus of a Complex Number). *The* modulus *or the* absolute value *of a complex number $z = a + bi$ is denoted by $|z|$ and computed as*

$$|z| = \sqrt{a^2 + b^2}.$$

□

We immediately see that

$$\boxed{|z|^2 = z \cdot \overline{z}.}$$

Example 8.2 (Computing the Modulus of a Complex Number). *Find the modulus of* $3+5i$, $-2i$, 5 *and* $\cos\theta + i\sin\theta$.

Solution. $|3+5i| = \sqrt{34}$ *while* $|-2i| = 2$ *and* $|5| = 5$. *To compute* $|\cos\theta + i\sin\theta|$, *we can think about the point geometrically and since the point is on a circle of radius one centered at the origin, arrive at the answer of* 1, *or we can use a well-known trigonometric identity to compute*

$$|\cos\theta + i\sin\theta| = \sqrt{\cos^2\theta + \sin^2\theta} = 1.$$

(Live example: Visit `stthomas.edu/wavelets` *and click on Live Examples.)*

□

We conclude this section with a discussion of division of complex numbers. The quotient

$$\frac{a+bi}{c+di}$$

can be simplified by multiplying top and bottom by the conjugate of the denominator:

$$\frac{a+bi}{c+di} = \frac{a+bi}{c+di}\cdot\frac{c-di}{c-di} = \frac{(ac+bd)+(bc-ad)i}{c^2+d^2} = \frac{ac+bd}{c^2+d^2} + \frac{bc-ad}{c^2+d^2}i.$$

Maclaurin Series and Euler's Formula

To develop the basic theory going forward, we first introduce the concept of a complex exponential function. We recall the notion of a Maclaurin series from calculus. If you need to review Maclaurin series, it is strongly recommended that you consult a calculus text (e.g. Stewart [84]) before proceeding with the remainder of the material in this chapter.

You may recall that any function f with continuous derivatives of all order can be expanded as a Maclaurin series

$$f(t) = f(0) + f'(0)t + \frac{f''(0)}{2!}t^2 + \frac{f'''(0)}{3!}t^3 + \cdots = \sum_{n=0}^{\infty}\frac{f^{(n)}(0)}{n!}t^n.$$

Here, $f^{(n)}$ denotes the nth derivative of f for $n = 0, 1, 2, \ldots$ with $f^{(0)} = f$.

We are interested in the Maclaurin series for three well-known functions: $\cos t$, $\sin t$, and e^t. The Maclaurin series for these functions are

$$\begin{aligned}
\cos t &= 1 - \frac{t^2}{2!} + \frac{t^4}{4!} - \frac{t^6}{6!} + \frac{t^8}{8!} - + \cdots \\
\sin t &= \frac{t}{1!} - \frac{t^3}{3!} + \frac{t^5}{5!} - \frac{t^7}{7!} + - \cdots \\
e^t &= 1 + \frac{t}{1!} + \frac{t^2}{2!} + \frac{t^3}{3!} + \frac{t^4}{4!} + \frac{t^5}{5!} + \cdots
\end{aligned} \tag{8.2}$$

It is interesting to note the similarities in these series. The terms in e^t above look like a combination of the terms in $\cos t$ and $\sin t$ with some missing minus signs. Undoubtedly, Euler felt that there should be a connection when he discovered the famous result that bears his name.

To derive Euler's formula, we need the Maclaurin series above as well as a bit of knowledge about the powers of i. Let's begin by replacing t with it in (8.2). Using the fact that $i^2 = -1$, $i^3 = -i$, $i^4 = 1$, and so on, we have

$$\begin{aligned} e^{it} &= 1 + \frac{it}{1!} + \frac{(it)^2}{2!} + \frac{(it)^3}{3!} + \frac{(it)^4}{4!} + \frac{(it)^5}{5!} + \cdots \\ &= 1 + it - \frac{t^2}{2!} - i\frac{t^3}{3!} + \frac{t^4}{4!} + i\frac{t^5}{5!} - \cdots \\ &= \left(1 - \frac{t^2}{2!} + \frac{t^4}{4!} - \frac{t^6}{6!} + - \cdots\right) \\ &\quad + i\left(\frac{t}{1!} - \frac{t^3}{3!} + \frac{t^5}{5!} - \frac{t^7}{7!} + - \cdots\right) \\ &= \cos t + i \sin t. \end{aligned}$$

Thus we have *Euler's formula.*

$$\boxed{e^{it} = \cos t + i \sin t}. \tag{8.3}$$

There are several problems designed to help you become familiar with Euler's formula. It is perhaps one of the most important results we will use throughout the book. We will spend more time with Euler's formula before moving on to Fourier series.

At first it seems strange to see a connection between an exponential function and sines and cosines. But look a little closer at the right side of (8.3). Have you seen it before? Recall from (8.1) that $f(\theta) = \cos\theta + i\sin\theta$ has a graph that is simply a graph of a circle centered at the origin with radius 1 in the complex plane. In fact, we could let θ be any real number and the effect would simply be a retracing of our circle. So $e^{i\theta}$ is a graph of a circle in the complex plane (Figure 8.6)!

What does a picture of $e^{2i\theta}$ from $0 \le \theta \le 2\pi$ look like? How about $e^{i\theta/2}$, $e^{4i\theta}$?

Since cosine and sine are 2π-periodic functions, it follows that $e^{i\theta}$ is also 2π-periodic. What about the periodicity of $e^{i\theta/2}$, $e^{4i\theta}$? What about $\overline{e^{i\theta}}$? We can easily compute the conjugate as follows:

$$\overline{e^{i\theta}} = \overline{\cos\theta + i\sin\theta} = \cos\theta - i\sin\theta.$$

Let's take this result a step further. Recall that cosine is an even function so that $\cos\theta = \cos(-\theta)$, and sine is an odd function, so that $\sin(-\theta) = -\sin\theta$. We use these results to compute and simplify $e^{-i\theta}$:

$$e^{-i\theta} = \cos(-\theta) + i\sin(-\theta) = \cos\theta - i\sin\theta = \overline{e^{i\theta}}.$$

So we see that

$$\overline{e^{i\theta}} = e^{-i\theta} = \cos\theta - i\sin\theta. \tag{8.4}$$

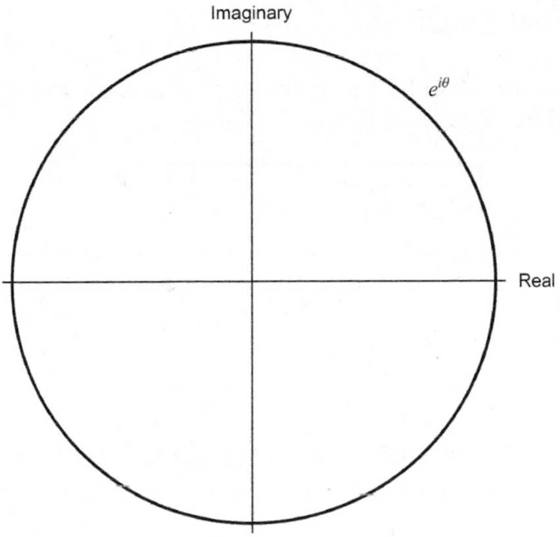

Figure 8.6 A graph of $e^{i\theta} = \cos\theta + i\sin\theta$.

Geometrically, we recall that as θ varies from 0 to 2π, the graph of the circle $\overline{e^{i\theta}}$ is drawn in a clockwise manner. So we see that the same is true for $e^{-i\theta}$.

It is very important that you are familiar with some special values of $e^{i\theta}$.

Example 8.3 (Computing Values of $\mathbf{e^{i\theta}}$). *Compute $e^{i\theta}$ for $\theta = 0, \frac{\pi}{2}, \pi$, and 2π.*

Solution. *We have*

$$\begin{aligned} e^{i\cdot 0} &= 1 \\ e^{\frac{i\pi}{2}} &= \cos\frac{\pi}{2} + i\sin\frac{\pi}{2} = i \\ e^{i\pi} &= \cos\pi + i\sin\pi = -1 \\ e^{2\pi i} &= \cos 2\pi + i\sin 2\pi = 1. \end{aligned}$$

For the last two values of θ we first need to note that at any integer multiple of π, the sine function is 0 while the cosine function is ± 1. In particular, if k is even, $\cos k\pi = 1$, and if k is odd, $\cos k\pi = -1$. We can combine these two results to write

$$\cos k\pi = (-1)^k.$$

Now we see that

$$\boxed{e^{\pi ik} = \cos k\pi + i\sin k\pi = (-1)^k}$$

and

$$e^{2\pi ik} = \cos 2k\pi + i\sin 2k\pi = 1.$$

(Live example: Visit `stthomas.edu/wavelets` *and click on Live Examples.)*

□

Complex Exponential Functions

Now that we are familiar with Euler's formula for $e^{i\omega}$, it is time to put it to use. To do so, we define a family of *complex exponential functions*:

$$\boxed{e_k(\omega) = e^{ik\omega}, \qquad k \in \mathbb{Z}.} \tag{8.5}$$

Note: Hereafter, we use ω as our independent variable when working with complex exponentials and Fourier series – this choice of notation is in keeping with the majority of the literature on wavelets.

Although it is clear that each function e_k is $\frac{2\pi}{k}$-periodic ($k \neq 0$), it is helpful to view each e_k as 2π-periodic with k copies of cosine and sine in each interval of length 2π. Figure 8.7 illustrates this observation as we graph the real and imaginary parts of $e_k(\omega)$ for different values of k.

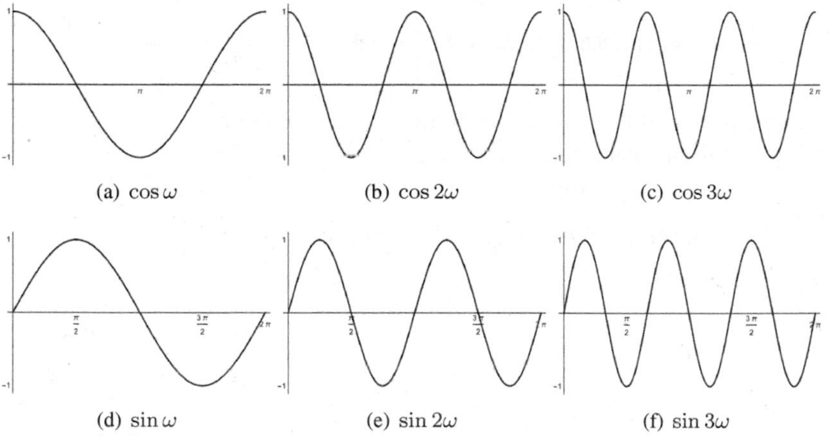

(a) $\cos\omega$ (b) $\cos 2\omega$ (c) $\cos 3\omega$

(d) $\sin\omega$ (e) $\sin 2\omega$ (f) $\sin 3\omega$

Figure 8.7 A graph of the real (top) and imaginary (bottom) parts of $e_k(\omega)$ for $k = 1, 2, 3$.

Let's now consider the integral

$$\int_{-\pi}^{\pi} e_k(\omega)\overline{e_j(\omega)}\,d\omega$$

for $j, k \in \mathbb{Z}$. We have

$$\begin{aligned}\int_{-\pi}^{\pi} e_k(\omega)\overline{e_j(\omega)}\,d\omega &= \int_{-\pi}^{\pi} e^{ik\omega}e^{(-ij\omega)}\,d\omega = \int_{-\pi}^{\pi} e^{i(k-j)\omega}\,d\omega \\ &= \int_{-\pi}^{\pi} \cos((k-j)\omega)\,d\omega + i\int_{-\pi}^{\pi} \sin((k-j)\omega)\,d\omega.\end{aligned}$$

Now if $k = j$, the right-hand side above reduces to

$$\int_{-\pi}^{\pi} e_k(\omega)\overline{e_j(\omega)}\,\mathrm{d}\omega = \int_{-\pi}^{\pi} e^{ik\omega}e^{-ik\omega}\,\mathrm{d}\omega = \int_{-\pi}^{\pi} 1\cdot \mathrm{d}\omega = 2\pi.$$

If $k \neq j$, we make the substitution $u = (k-j)\omega$ in each integral to obtain

$$\int_{-\pi}^{\pi} e_k(\omega)\overline{e_j(\omega)}\,\mathrm{d}\omega = \frac{1}{k-j}\int_{-(k-j)\pi}^{(k-j)\pi} \cos u\,\mathrm{d}u + \frac{i}{k-j}\int_{-(k-j)\pi}^{(k-j)\pi} \sin u\,\mathrm{d}u$$

Now sine is an odd function and we are integrating it over an interval $[-(k-j)\pi, (k-j)\pi]$ that is symmetric about the origin, so the value of the second integral on the right-hand side above is zero (see Problem 8.17).

We also utilize the fact that cosine is an even function and since we are also integrating it over an interval $[-(k-j)\pi, (k-j)\pi]$ that is symmetric about the origin, we find that

$$\frac{1}{k-j}\int_{-(k-j)\pi}^{(k-j)\pi} \cos u\,\mathrm{d}u = \frac{2}{k-j}\int_{0}^{(k-j)\pi} \cos u\,\mathrm{d}u = \frac{2}{k-j}\sin u\Big|_0^{(k-j)\pi} = 0$$

since sine at any integer multiple of π is zero.

So we have

$$\boxed{\int_{-\pi}^{\pi} e_k(\omega)\overline{e_j(\omega)}\,\mathrm{d}\omega = \begin{cases} 2\pi, & k = j \\ 0 & k \neq j. \end{cases}} \tag{8.6}$$

More generally, we can think of

$$\int_{-\pi}^{\pi} f(\omega)\overline{g(\omega)}\,\mathrm{d}\omega \tag{8.7}$$

as the definition of the *inner product* of functions $f, g\colon [-\pi, \pi] \to \mathbb{C}$. It is a quite natural definition – the inner product of two vectors $\mathbf{v}, \mathbf{w} \in \mathbb{R}^n$ is defined to be the sum of componentwise multiplication. That is,

$$\mathbf{v}\cdot\mathbf{w} = v_1w_1 + v_2w_2 + \cdots + v_nw_n = \sum_{k=1}^{n} v_kw_k.$$

But recall from elementary calculus that the definition of the integral in (8.7) is

$$\int_{-\pi}^{\pi} f(t)\overline{g(t)}\,\mathrm{d}t = \lim_{n\to\infty}\sum_{k=1}^{n} f(t_k^*)\overline{g(t_k^*)}\,\Delta t = \lim_{n\to\infty}\Delta t\left(\sum_{k=1}^{n} f(t_k^*)\overline{g(t_k^*)}\right).$$

We can think of the factors $f(t_k^*), \overline{g(t_k^*)}$ in each term of the summand as vector elements – the definition of the integral simply adds a multiplication by the differential Δt and the "length" of the vectors tends to ∞.

Thus, just as $\mathbf{v}, \mathbf{w} \in \mathbb{R}^n$ are called orthogonal vectors if $\mathbf{v} \cdot \mathbf{w} = 0$, we say that for $j \neq k$, the functions $e_j(\omega)$ and $e_k(\omega)$ are *orthogonal functions* on $[-\pi, \pi]$ since

$$\int_{-\pi}^{\pi} e_k(\omega)\overline{e_j(\omega)}\, d\omega = 0.$$

We can use Problem 8.18 to see that $e_k(\omega)$, $e_j(\omega)$ are orthogonal on any interval of length 2π.

PROBLEMS

8.1 Plot the following points in the complex plane.

(a) $2+i$, $3+2i$, 5, $-4i$

(b) $\cos\theta + i\sin\theta$ where $\theta = \frac{\pi}{3}, \frac{3\pi}{2}, \frac{7\pi}{6}$

8.2 Let n be a nonnegative integer. Find a closed formula for i^n.

8.3 Consider the function $f(\theta) = \cos\theta + i\sin\theta$. Describe the path traced out by the following functions as θ ranges from 0 to 2π.

(a) $f_1(\theta) = f(-\theta)$

(b) $f_2(\theta) = f(\frac{\theta}{3})$

(c) $f_3(\theta) = f(k\theta)$, where k is an integer

8.4 Plot the following numbers in the complex plane: $2+i$, $1-i$, $3i$, 5, $\cos\frac{\pi}{4} + i\sin\frac{\pi}{4}$, $\overline{3-2i}$, $\frac{1+i}{3+4i}$.

8.5 Find the modulus of each of the numbers in Problem 8.4.

8.6 Compute the following values.

(a) $(2+3i)+(1-i)$

(b) $(2-5i)-(2-2i)$

(c) $(2+3i)\overline{(4+i)}$

(d) $\overline{(2+3i)(4+i)}$

(e) $\overline{(2+3i)}(4+i)$

(f) $(1+i) \div (1-i)$

★**8.7** Let $z = a+bi$ and $y = c+di$. Show that $\overline{yz} = \overline{y} \cdot \overline{z}$.

8.8 Let $z = a+bi$. Show that $|z| = |\overline{z}|$.

★**8.9** Show that z is a real number if and only if $z = \overline{z}$.

★**8.10** Let $y = a + bi$ and $z = c + di$. Show that $|yz| = |y| \cdot |z|$. Note that in particular, if y is real, say $y = c$, $c \in \mathbb{R}$, we have $|cz| = |c||z|$.

8.11 Let $y = a+bi$ and $z = c+di$. Show that $\overline{y+z} = \overline{y} + \overline{z} = (a+c) - i(b+d)$.

★**8.12** This is a generalization of Problem 8.11. Suppose that you have n complex numbers $z_k = a_k + ib_k$, $k = 1, 2, \ldots, n$. Show that

$$\overline{\sum_{k=1}^{n} z_k} = \sum_{k=1}^{n} \overline{z_k}$$
$$= \sum_{k=1}^{n} a_k - i \sum_{k=1}^{n} b_k.$$

Note: If (a_k) and (b_k) are convergent sequences then the result holds for infinite sums as well.

8.13 Let $z = a + bi$. Find the real and imaginary parts of $z^{\;1} = \frac{1}{z}$.

★**8.14** Let $f(\omega) = e^{i\omega}$. Show that $f'(\omega) = ie^{i\omega}$. In this text, the only complex-valued functions we differentiate are those built from complex exponentials. Write $e^{i\omega} = \cos(\omega) + i\sin(\omega)$ and differentiate "normally" treating i as a constant.

★**8.15** Suppose $f : \mathbb{C} \to \mathbb{C}$. For a given $z \in \mathbb{C}$, describe all complex numbers that have the same modulus as $f(z)$.

★**8.16** Find $e^{i\theta}$ for $\theta = -\pi, 2\pi, \pm\frac{\pi}{4}, \frac{k\pi}{4}, \frac{\pi}{6}, k\pi, \frac{k\pi}{2}$, and $2k\pi$. Here k is any integer.

8.17 Suppose that $f(\omega)$ is an odd function. That is, $f(-\omega) = -f(\omega)$. Show that for any real number a,

$$\int_{-a}^{a} f(\omega)\, d\omega = 0.$$

8.18 Suppose that $f(\omega)$ is a 2π-periodic function and a is any real number. Show that

$$\int_{-\pi-a}^{\pi-a} f(\omega)\, d\omega = \int_{-\pi}^{\pi} f(\omega)\, d\omega.$$

(*Hint:* The integral on the left side can be written as

$$\int_{-\pi-a}^{\pi-a} f(\omega)\, d\omega = \int_{-\pi-a}^{-\pi} f(\omega)\, d\omega + \int_{-\pi}^{\pi-a} f(\omega)\, d\omega$$

and the integral on the right side can be written as

$$\int_{-\pi}^{\pi} f(\omega)\, d\omega = \int_{-\pi}^{\pi-a} f(\omega)\, d\omega + \int_{\pi-a}^{\pi} f(\omega)\, d\omega.$$

It suffices to show that

$$\int_{-\pi-a}^{-\pi} f(\omega)\, d\omega = \int_{\pi-a}^{\pi} f(\omega)\, d\omega.$$

Use the fact that $f(\omega) = f(\omega - 2\pi)$ in the integral on the right-hand side of the above equation and then make an appropriate u-substitution. It is also quite helpful to sketch a graph of the function from $-\pi - a$ to π to see how to solve the problem.

★8.19 Show that

$$\cos\omega = \frac{e^{i\omega} + e^{-i\omega}}{2} \qquad \text{and} \qquad \sin\omega = \frac{e^{i\omega} - e^{-i\omega}}{2i}.$$

We will see in the next section that these identities are the Fourier series for $\cos\omega$ and $\sin\omega$. *Hint:* Use Euler's formula (8.3) and (8.4).

★8.20 In a trigonometry class, you learned about half-angle formulas involving sine and cosine. Use the results from Problem 8.19 to prove that

$$\sin^2\omega = \frac{1 - \cos 2\omega}{2} \qquad \text{and} \qquad \cos^2\omega = \frac{1 + \cos 2\omega}{2}.$$

★8.21 Prove *DeMoivre's theorem*. That is, for any nonnegative integer n, show that

$$(\cos\omega + i\sin\omega)^n = \cos n\omega + i\sin n\omega.$$

8.2 Fourier Series

In this section we introduce Fourier series, which are some of the most useful tools in applied mathematics. Simply put, if f is a 2π-periodic function and *well behaved*, then we can represent it as a linear combination of complex exponential functions. What do we mean by a well behaved function? For our purposes, f will be at the very least a piecewise continuous function with a finite number of jump discontinuities in the interval $[-\pi, \pi]$. Recall that if $f(t)$ has a finite jump discontinuity at $t = a$, then $\lim\limits_{t\to a^+} f(t)$ and $\lim\limits_{t\to a^-} f(t)$ both exist but are not the same. Note that if f is piecewise continuous with a finite number of jump discontinuities on $[-\pi, \pi]$, then we are assured that the integral $\int\limits_{-\pi}^{\pi} f(t)\,dt$ exists.

Fourier Series Defined

As we learned in Section 8.1, the family $\{e_k(\omega)\}_{k\in\mathbb{Z}}$ forms an orthogonal set. Moreover, it can be shown that the family forms a *basis* for the space of all 2π-periodic square-integrable[1] functions. That is, any function f in the space can be written as a

[1]Recall that a real-valued function f is square integrable on $[-\pi, \pi]$ if $\int\limits_{-\pi}^{\pi} f^2(t)\,dt < \infty$. It can be shown (see e. g., Bartle and Sherbert [2]) that a piecewise smooth function with a finite number of jump discontinuities in $[-\pi, \pi]$ is square integrable.

linear combination of complex exponential functions. Such a linear combination is called a *Fourier series*.[2]

Definition 8.3 (Fourier Series). *Suppose that f is a 2π-periodic, absolutely integrable function. Then the* Fourier series *for f is given by*

$$\boxed{f(\omega) = \sum_{k=-\infty}^{\infty} c_k e^{ik\omega}.} \tag{8.8}$$

□

While we work with Fourier series in an informal manner in this text, some remarks are in order regarding Definition 8.3.

The right-hand side of (8.8) is an infinite series, and as you learned in calculus, some work is typically required to show convergence of such a series. A simple condition that guarantees convergence of the series is to require that the series of coefficients *converge absolutely* (see Stewart [84]). That is, if

$$\sum_{k=-\infty}^{\infty} |c_k e^{ik\omega}| = \sum_{k=-\infty}^{\infty} |c_k||e^{ik\omega}| = \sum_{k=-\infty}^{\infty} |c_k| < \infty$$

then the series converges absolutely for all $\omega \in \mathbb{R}$. Most of the Fourier series we study in this book satisfy this condition.

Even though the series on the right-hand side of (8.8) might converge for $\omega \in \mathbb{R}$, it is quite possible that the series value is not $f(\omega)$. This statement brings into question the use of the "=" in (8.8). Indeed, if we were to present a more rigorous treatment of Fourier series, we would learn exactly the context of the "=" in (8.8). This is a fascinating study, but one beyond the scope of this book. The reader new to the topic is referred to Kammler [62] or Boggess and Narcowich [7] and the reader with some background in analysis is referred to Körner [64]. For our purposes, some general statements will suffice.

As long as $f(\omega)$ is a continuous function and the derivative exists at ω, then the series converges to $f(\omega)$. If f is piecewise continuous and ω is a jump discontinuity, then the series converges to half the sum of the left and right limits of f at ω. This value may or may not be the value of $f(\omega)$. For example, the Fourier series for the function plotted in Figure 8.8(a) converges to $f(\omega)$ as long as ω is not an odd multiple of π. In the case that ω is an odd multiple of π, the Fourier series converges to zero.

Fourier Series Coefficients

To understand, use and analyze Fourier series, we need a way to compute c_k in (8.8). We first rewrite (8.8) using the index variable j and then multiply both sides of the resulting equation by $\overline{e_k(\omega)}$ to obtain

[2]Joseph Fourier first studied these series in his attempt to understand how heat disperses through a solid.

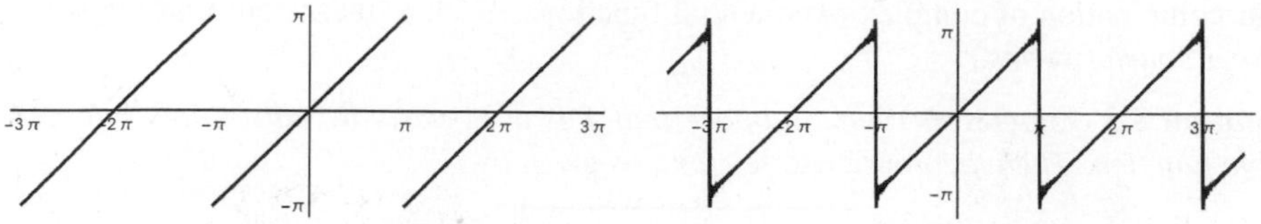

(a) Periodic linear function

(b) Fourier series approximate

Figure 8.8 A graph of a 2π-periodic linear function and its Fourier series approximate.

$$f(\omega)\overline{e_k(\omega)} = \sum_{j=-\infty}^{\infty} c_j e^{ij\omega}\overline{e_k(\omega)}.$$

Now integrate both sides of the above equation over the interval $[-\pi, \pi]$. Some analysis is required to justify passing the integral through the infinite summation, but we assume that the step is valid and write

$$\begin{aligned}\int_{-\pi}^{\pi} f(\omega)\overline{e_k(\omega)}\,d\omega &= \sum_{j=-\infty}^{\infty} c_j \int_{-\pi}^{\pi} e^{ij\omega}\overline{e_k(\omega)}\,d\omega \\ &= \sum_{j=-\infty}^{\infty} c_j \int_{-\pi}^{\pi} e_j(\omega)\overline{e_k(\omega)}\,d\omega \\ &= 2\pi c_k.\end{aligned}$$

The value $2\pi c_k$ appears by virtue of (8.6). The only integral on the right-hand side above that is not zero is the term where $j = k$. We can easily solve the last equation to obtain a formula for the *Fourier coefficients* c_k. We give this formula in the following definition.

Proposition 8.1 (Fourier Coefficient). *Let f be any integrable, 2π-periodic function. When f is expressed as a Fourier series* (8.8), *then the* Fourier coefficients c_k, $k \in \mathbb{Z}$, *satisfy*

$$\boxed{c_k = \frac{1}{2\pi}\int_{-\pi}^{\pi} f(\omega)\overline{e_k(\omega)}\,d\omega = \frac{1}{2\pi}\int_{-\pi}^{\pi} f(\omega)e^{-ik\omega}\,d\omega.} \tag{8.9}$$

□

These c_k are known as the *Fourier coefficients* for $f(\omega)$. Now let's see how the c_k were computed for the sawtooth function shown in Figure 8.8.

Example 8.4 (Fourier Series for the Sawtooth Function). *Find the Fourier series for the function plotted in Figure 8.8(a).*

Solution. *If we restrict the function shown in Figure 8.8(a) to $(-\pi, \pi)$, we have $f(\omega) = \omega$. So to compute the Fourier coefficients, we must compute the following integral:*

$$c_k = \frac{1}{2\pi}\int_{-\pi}^{\pi} \omega e^{-ik\omega}\, d\omega = \frac{1}{2\pi}\left(\int_{-\pi}^{\pi} \omega\cos(k\omega)\, d\omega - i\int_{-\pi}^{\pi} \omega\sin(k\omega)\, d\omega\right).$$

Now ω is an odd function and $\cos(k\omega)$ is an even function so that the product $\omega\cos(k\omega)$ is odd. Since we are integrating over an interval $(-\pi, \pi)$ that is symmetric about zero, this integral is zero. In a similar way, we see that since $\omega\sin(k\omega)$ is an even function, we have

$$c_k = \frac{-i}{2\pi}\int_{-\pi}^{\pi} \omega\sin(k\omega)\, d\omega = \frac{-i}{\pi}\int_{0}^{\pi} \omega\sin(k\omega)\, d\omega$$

We next integrate by parts. Let $u = \omega$ so that $du = d\omega$, and set $dv = \sin k\omega\, d\omega$ so that $v = -\frac{1}{k}\cos k\omega$.

Note that we can only integrate by parts if $k \neq 0$ (otherwise v is undefined), so let's assume that $k \neq 0$ and return to the $k = 0$ case later. We see that when $k \neq 0$, the integral simplifies to

$$\begin{aligned} c_k &= \frac{-i}{\pi}\left(-\frac{\omega}{k}\cos(k\omega)\Big|_0^{\pi} + \frac{1}{k}\int_0^{\pi} \cos(k\omega)\, d\omega\right) \\ &= \frac{-i}{\pi}\left(-\frac{\pi}{k}\cos(k\pi) + \frac{1}{k^2}\sin(k\omega)\Big|_0^{\pi}\right) \end{aligned}$$

Recall that for $k \in \mathbb{Z}$, we have $\sin(k\pi) = 0$ and $\cos(k\pi) = (-1)^k$. Using these facts and simplifying gives

$$c_k = (-1)^k \frac{i}{k}.$$

If $k = 0$, we use (8.9) to write

$$c_0 = \frac{1}{2\pi}\int_{-\pi}^{\pi} \omega e^{i \cdot 0 \cdot \omega}\, d\omega = \frac{1}{2\pi}\int_{-\pi}^{\pi} \omega\, d\omega = 0.$$

We have

$$c_k = \begin{cases} 0, & k = 0 \\ (-1)^k \frac{i}{k}, & k \neq 0 \end{cases} \tag{8.10}$$

which results in the following Fourier series for f:

$$f(\omega) = \sum_{k \neq 0} (-1)^k \frac{i}{k} e^{ik\omega}. \tag{8.11}$$

You might find it odd that the Fourier coefficients are purely imaginary for this function.[3] *If you work through Problems 8.25 and 8.27, you will see that if a function*

[3] If we check the absolute convergence of the Fourier coefficients (see page 335), we find that $\sum\limits_{k \neq 0} \left|\frac{(-1)^k i}{k}\right| = 2\sum\limits_{k=1}^{\infty} \frac{1}{k}$. This is the harmonic series and we know from calculus that it is a divergent series. Kammler [62, page 43] gives a nice argument that shows the series converges for $\omega \in \mathbb{R}$.

is odd (even), then its Fourier coefficients are imaginary (real) and the series can be reduced to one of sines (cosines). This should make sense to you – f is an odd function, so we should not need cosine functions to represent it. Indeed, you will show in Problem 8.30 that we can reduce (8.11) *to*

$$f(\omega) = -2\sum_{k=1}^{\infty} \frac{(-1)^k}{k} \sin(k\omega).$$

Let's get an idea of what this series looks like. We define the sequence of partial sums by

$$f_n(\omega) = -2\sum_{k=1}^{n} \frac{(-1)^k}{k} \sin(k\omega)$$

and plot f_n for various values of n. Figure 8.9 shows f, f_1, f_3, f_{10}, f_{20}, and f_{50}.

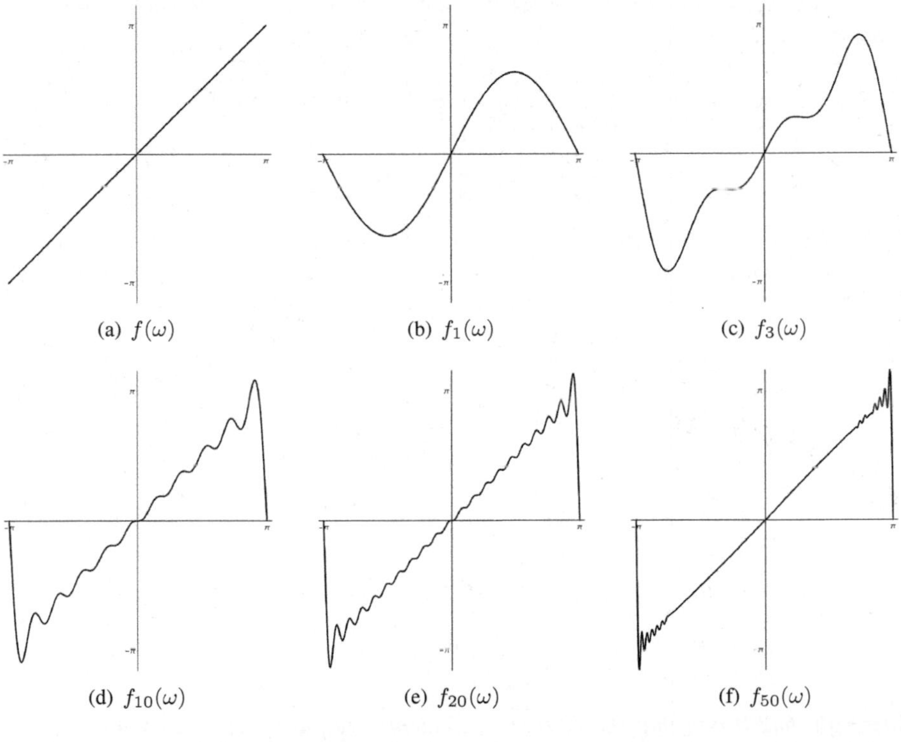

(a) $f(\omega)$ (b) $f_1(\omega)$ (c) $f_3(\omega)$

(d) $f_{10}(\omega)$ (e) $f_{20}(\omega)$ (f) $f_{50}(\omega)$

Figure 8.9 A graph of $f(\omega)$, $f_1(\omega)$, $f_3(\omega)$, $f_{10}(\omega)$, $f_{20}(\omega)$, and $f_{50}(\omega)$ on the interval $[-\pi, \pi]$.

(Live example: Visit `stthomas.edu/wavelets` *and click on Live Examples.)*

□

I still find it amazing that the periodic piecewise linear function can be represented with sine functions. As k gets larger, $\sin(k\omega)$ becomes more and more oscillatory. But the coefficient $\frac{(-1)^k}{k}$ multiplying it becomes smaller and smaller – just the right combination to produce f. You will notice a little undershoot and overshoot at the points of discontinuity of f. This is the famous Gibbs phenomenon. The topic is fascinating in itself but is beyond the scope of this book. The reader is referred to Kammler [62] for more details.

Let's look at another example.

Example 8.5 (Fourier Series for a Piecewise Constant Function). *For the 2π-periodic piecewise constant function $f(\omega)$ shown in Figure 8.10, find the Fourier series for $f(\omega)$.*

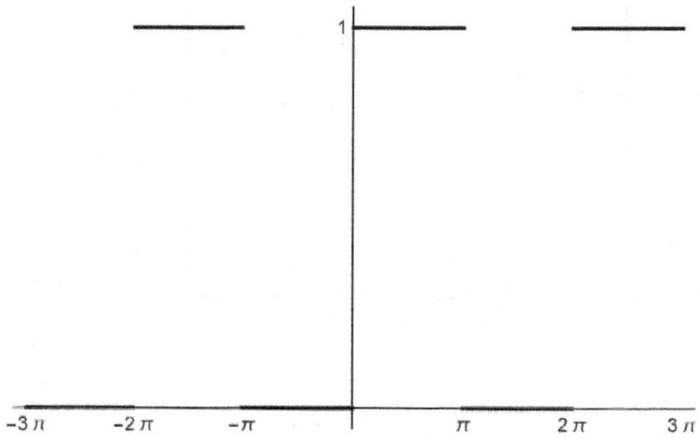

Figure 8.10 Three periods of a 2π-periodic piecewise constant function.

Solution. *When $k \neq 0$, the Fourier coefficients are given by*

$$
\begin{aligned}
c_k &= \frac{1}{2\pi}\int_{-\pi}^{\pi} f(\omega)e^{-ik\omega}\,\mathrm{d}\omega \\
&= \frac{1}{2\pi}\int_{0}^{\pi} 1\cdot e^{-ik\omega}\,\mathrm{d}\omega \\
&= \frac{1}{2\pi}\left(\int_0^{\pi}\cos(k\omega)\,\mathrm{d}\omega - i\int_0^{\pi}\sin(k\omega)\,\mathrm{d}\omega\right) \\
&= \frac{1}{2\pi}\left(\frac{1}{k}\sin(k\omega)\Big|_0^{\pi} + \frac{i}{k}\cos(k\omega)\Big|_0^{\pi}\right) \\
&= \frac{i}{2k\pi}\left((-1)^k - 1\right) \\
&= \begin{cases} 0, & k = \pm 2, \pm 4, \ldots \\ \frac{-i}{k\pi}, & k = \pm 1, \pm 3, \pm 5, \ldots \end{cases}
\end{aligned}
\tag{8.12}
$$

When $k = 0$, we have $c_0 = \frac{1}{2\pi}\int_0^{\pi} 1\, d\omega = \frac{1}{2}$. So the Fourier series is

$$f(\omega) = \frac{1}{2} - \sum_{k \text{ odd}} \frac{i}{k\pi} e^{ik\omega}. \tag{8.13}$$

In Problem 8.31 you will show that this series can be reduced to

$$f(\omega) = \frac{1}{2} + \frac{2}{\pi}\sum_{k=1}^{\infty} \frac{1}{2k-1}\sin((2k-1)\omega).$$

We graph this Fourier series in Figure 8.11.

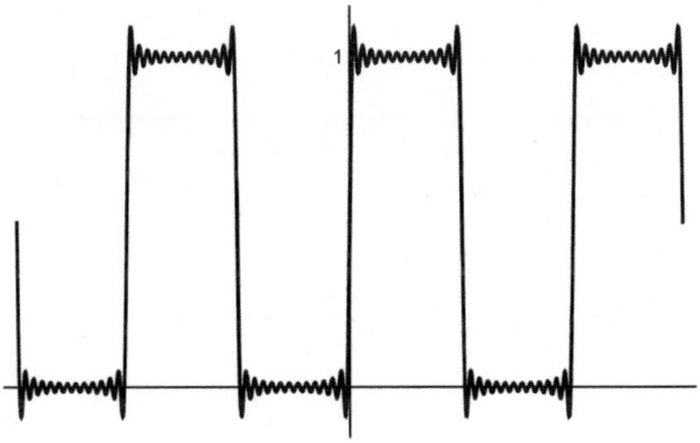

Figure 8.11 Three periods of the partial Fourier series ($n = 25$) for the function in Figure 8.10.

(Live example: Visit `stthomas.edu/wavelets` and click on Live Examples.)

□

As you can see, the computation of Fourier coefficients can be somewhat tedious. Fortunately, there exists an entire calculus for computing Fourier coefficients. The idea is to build Fourier series from a "library" of known Fourier series. Perhaps this concept is best illustrated with an example.

Example 8.6 (Computing One Fourier Series from Another). *Find the Fourier series for the 2π-periodic piecewise constant function $g(\omega)$ shown in Figure 8.12.*

Solution. *Let's denote the Fourier coefficients of $g(\omega)$ with d_k so that*

$$g(\omega) = \sum_{k=-\infty}^{\infty} d_k e^{ik\omega}.$$

If we compare $g(\omega)$ to $f(\omega)$ from Example 8.5, we see that $g(\omega) = f(\omega - \frac{\pi}{2})$. Instead of computing the Fourier coefficients for g directly, we consider

$$f\left(\omega - \frac{\pi}{2}\right) = \sum_{k=-\infty}^{\infty} c_k e^{ik(\omega-\pi/2)}$$

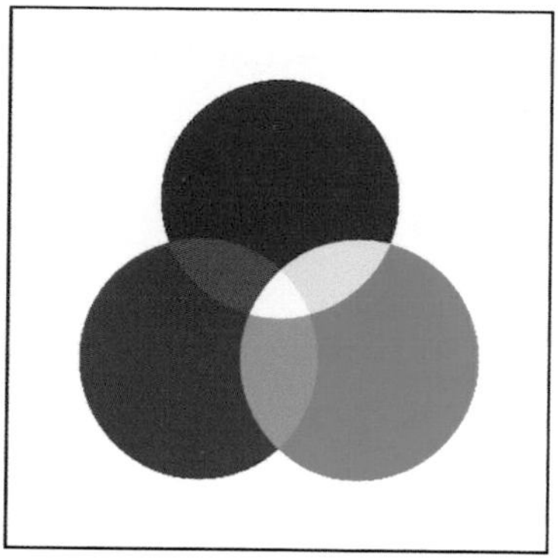

Figure 3.23 Using additive mixing to generate white and the secondary colors magenta, yellow, and cyan from the primary colors red, green, and blue.

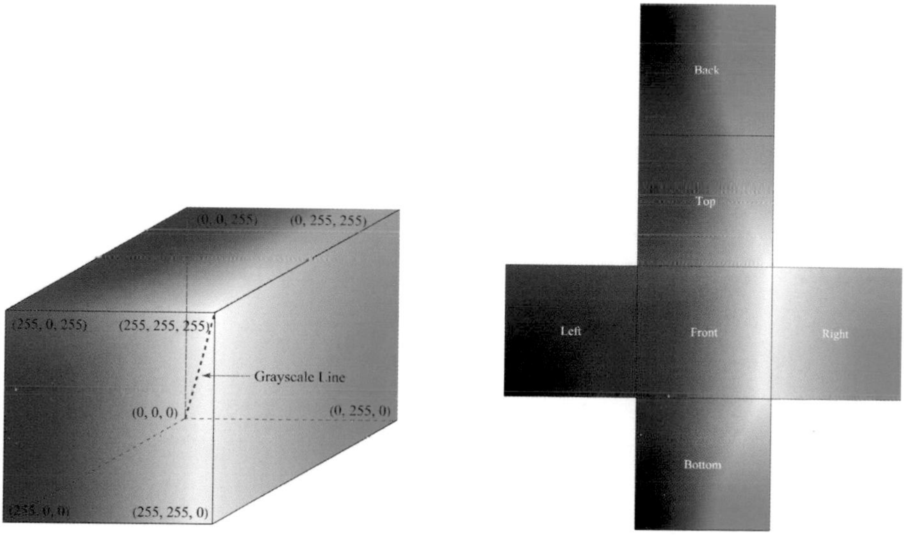

(a) RGB color cube

(b) Color cube schematic

Figure 3.24 The image at left is the RGB color cube. The image on the right is a schematic of the color cube.

Discrete Wavelet Transformations: An Elementary Approach With Applications, Second Edition.
Patrick J. Van Fleet.

(a) A color image

(b) The red channel

(c) The green channel

(d) The blue channel

Figure 3.25 A color image and the three primary color channels.

(a) A digital color image

(b) The negative image

Figure 3.26 Color image negation.

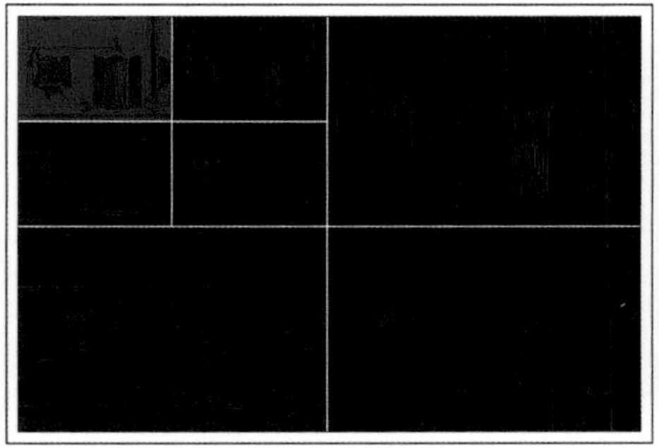

(a) Haar transform of R

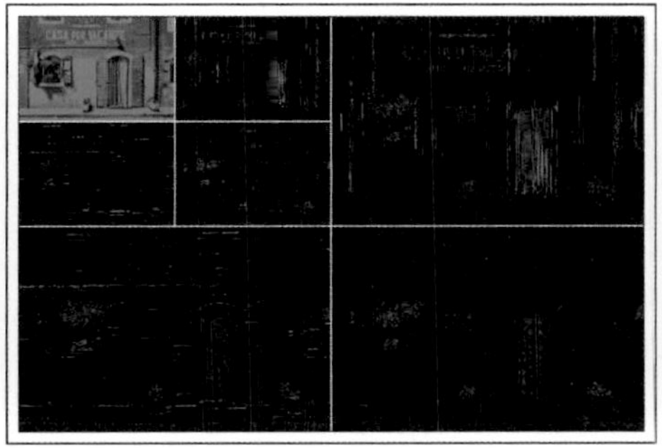

(b) Haar transform of G

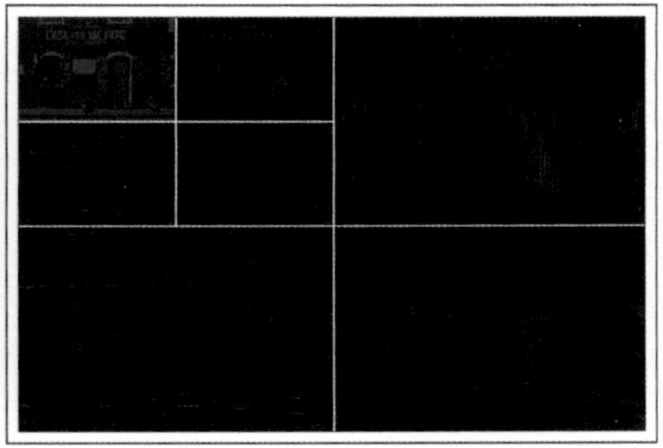

(c) Haar transform of B

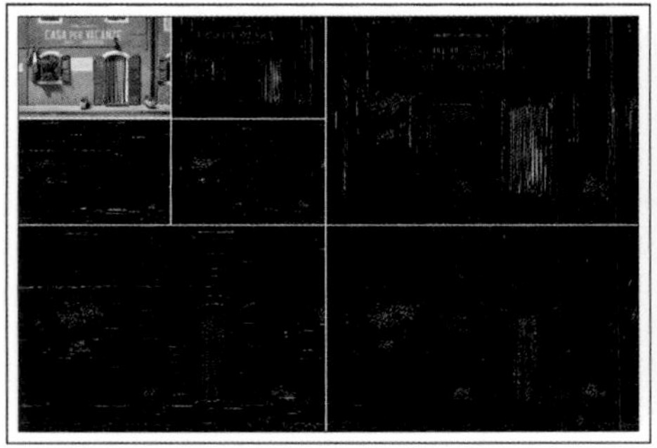

(d) Haar transform of Y

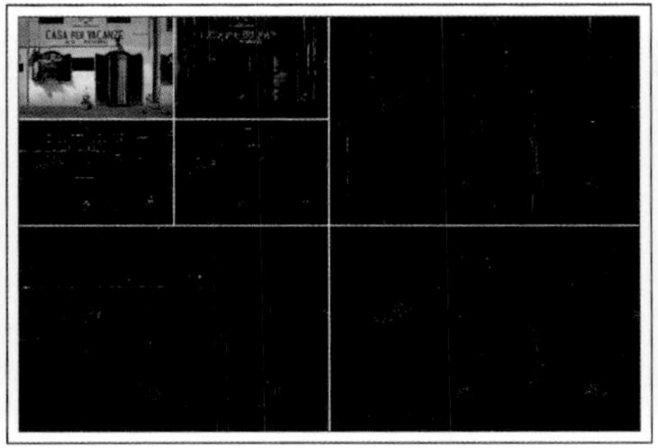

(e) Haar transform of Cb

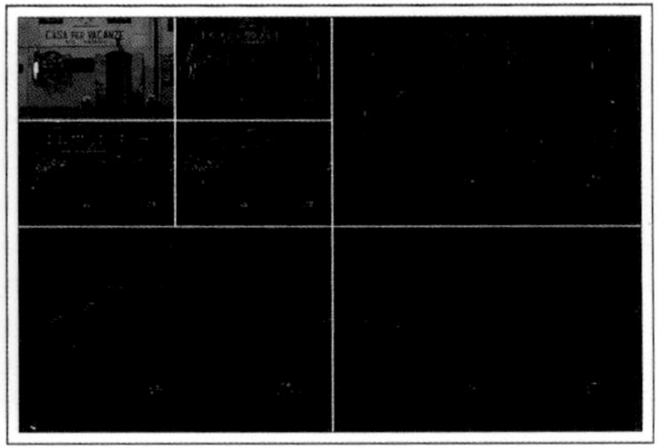

(f) Haar transform of Cr

Figure 4.19 The Haar wavelet transform of each of the color channels R, G, and B in Figure 3.25(b) – 3.25(d) and the Y, Cb, and Cr channels.

(a) Original image (b) 99.99% energy (c) Variable energy rates

Figure 4.29 The original and reconstructed images. The reconstructed image that uses 99.2% energy across each channel has PSNR = 35.68, while the reconstructed image that uses the energy rates given in Table 4.5 has PSNR = 3.88.

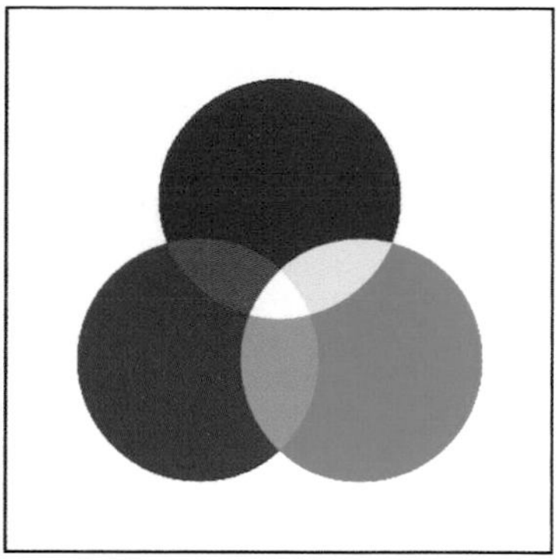

Figure 3.23 Using additive mixing to generate white and the secondary colors magenta, yellow, and cyan from the primary colors red, green, and blue.

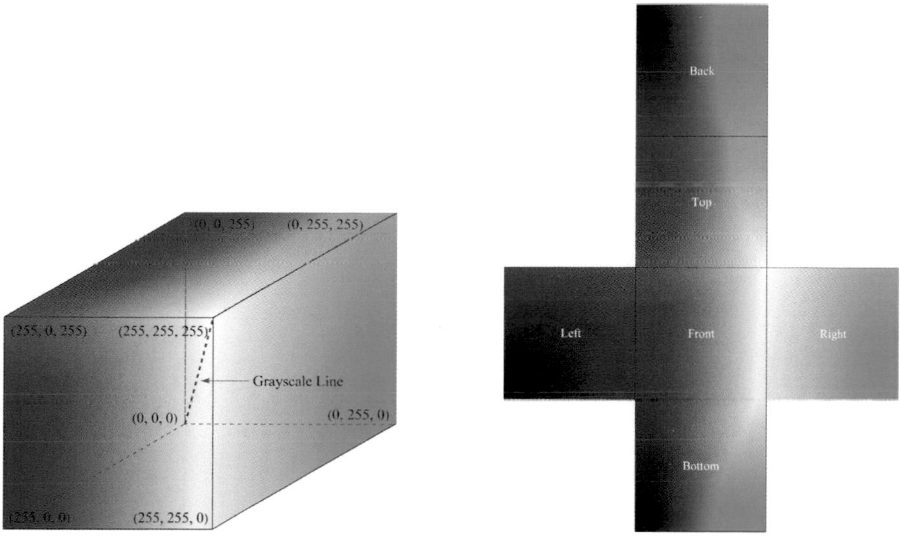

(a) RGB color cube

(b) Color cube schematic

Figure 3.24 The image at left is the RGB color cube. The image on the right is a schematic of the color cube.

Discrete Wavelet Transformations: An Elementary Approach With Applications, Second Edition.
Patrick J. Van Fleet.

(a) A color image

(b) The red channel

(c) The green channel

(d) The blue channel

Figure 3.25 A color image and the three primary color channels.

(a) A digital color image

(b) The negative image

Figure 3.26 Color image negation.

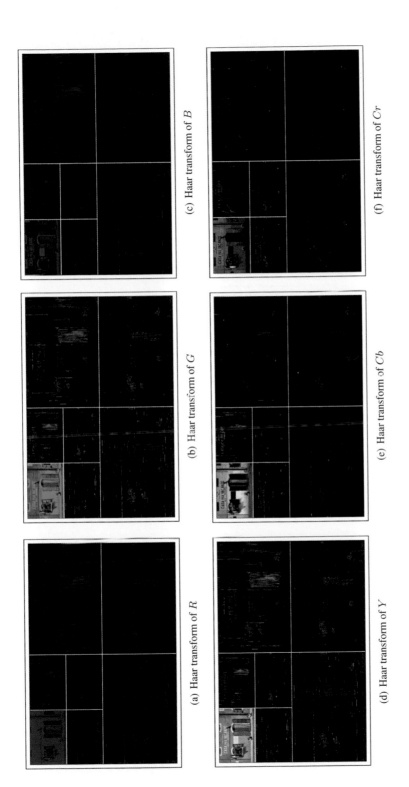

(a) Haar transform of R

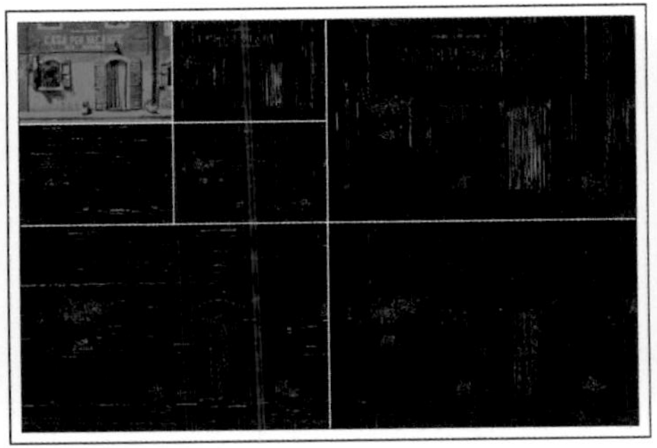

(b) Haar transform of G

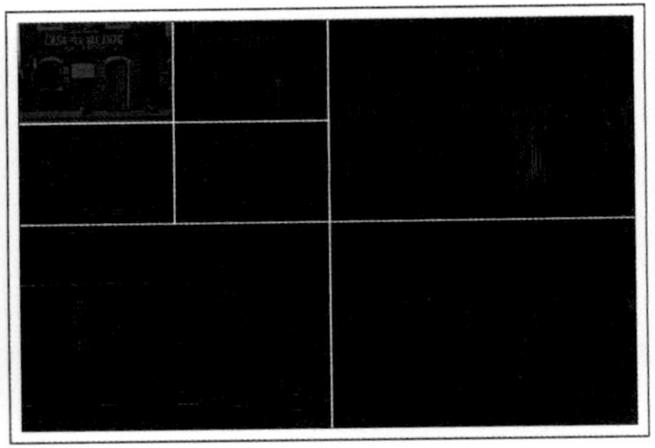

(c) Haar transform of B

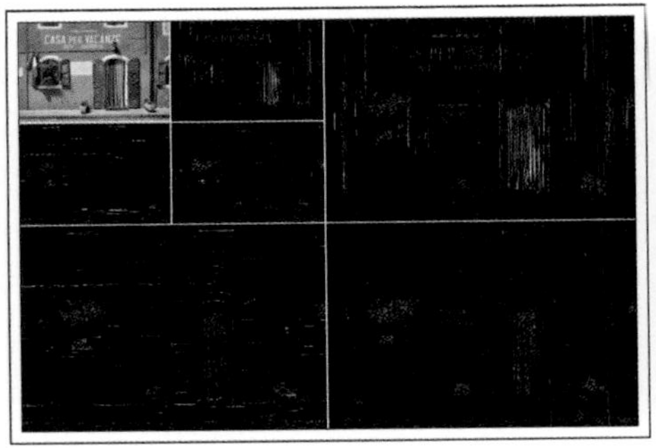

(d) Haar transform of Y

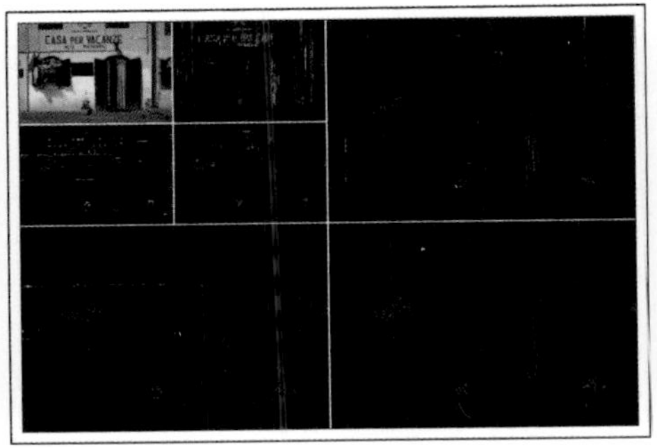

(e) Haar transform of Cb

(f) Haar transform of Cr

Figure 4.19 The Haar wavelet transform of each of the color channels R, G, and B in Figure 3.25(b) – 3.25(d) and the Y, Cb, and Cr channels.

(a) Original image (b) 99.99% energy (c) Variable energy rates

Figure 4.29 The original and reconstructed images. The reconstructed image that uses 99.2% energy across each channel has PSNR = 35.68, while the reconstructed image that uses the energy rates given in Table 4.5 has PSNR = 3.88.

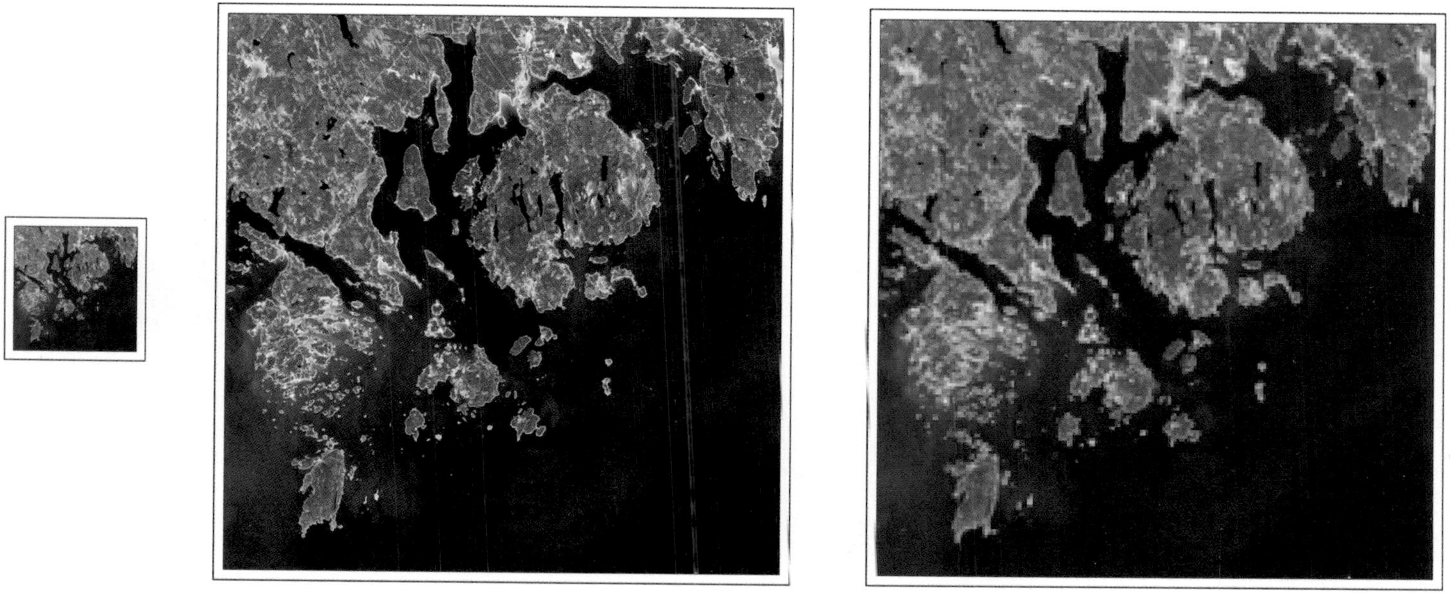

(a) Color (b) Grayscale (c) Enlarged color

Figure 7.10 A low resolution color image, its high resolution grayscale counterpart, and an enlarged version of the color image.

(a) Pansharpened image

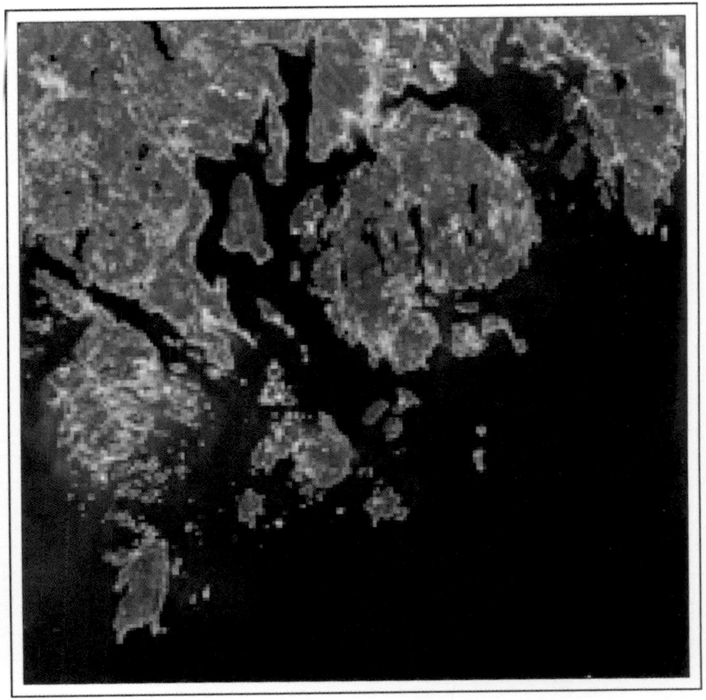

(b) Enlarged color image

Figure 7.13 The fused wavelet transform (the blur portion from the transform of I replacing the blur portion of the transform of P') and its inverse transform I'.

Figure 12.12 The original color image.

(a) JPEG2000 (b) JPEG

Figure 12.15 The compressed image using JPEG2000 and JPEG.

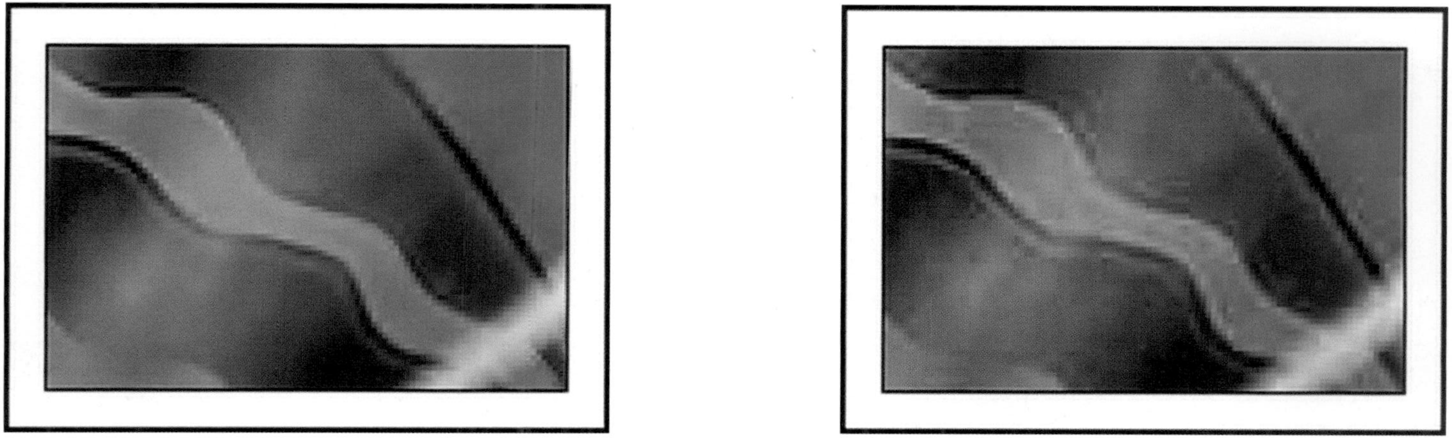

(a) JPEG2000

(b) JPEG

Figure 12.16 Corresponding 64×96 blocks of each compressed images in Figure 12.15.

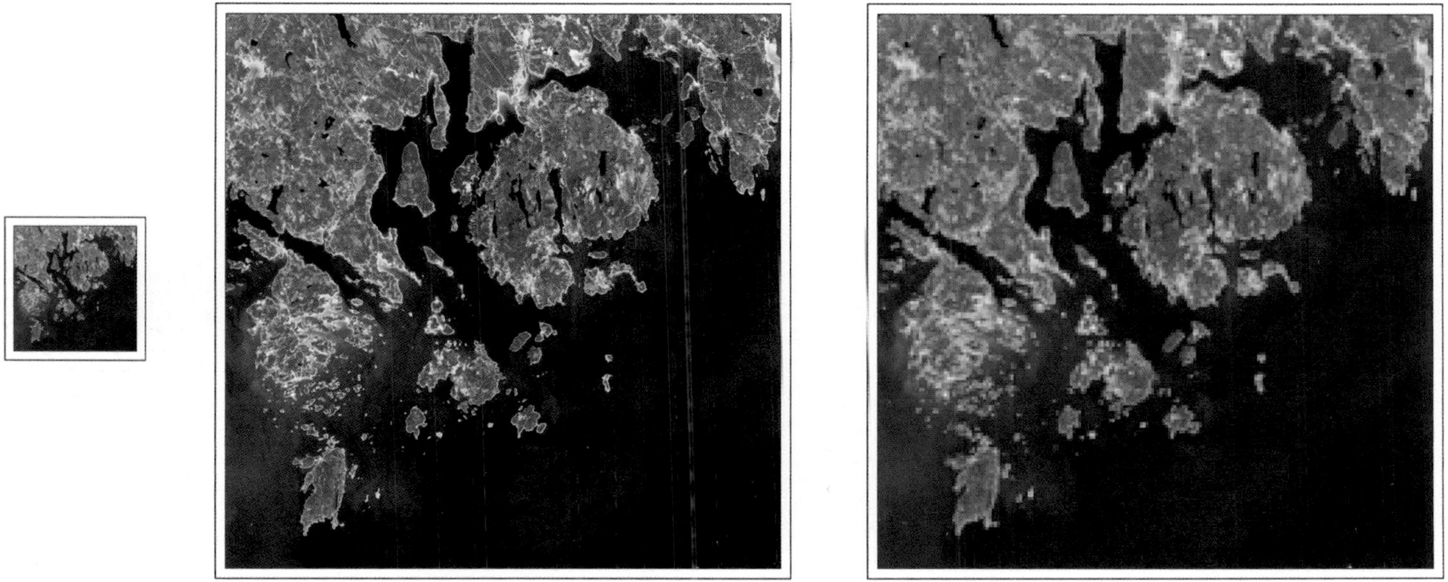

(a) Color (b) Grayscale (c) Enlarged color

Figure 7.10 A low resolution color image, its high resolution grayscale counterpart, and an enlarged version of the color image.

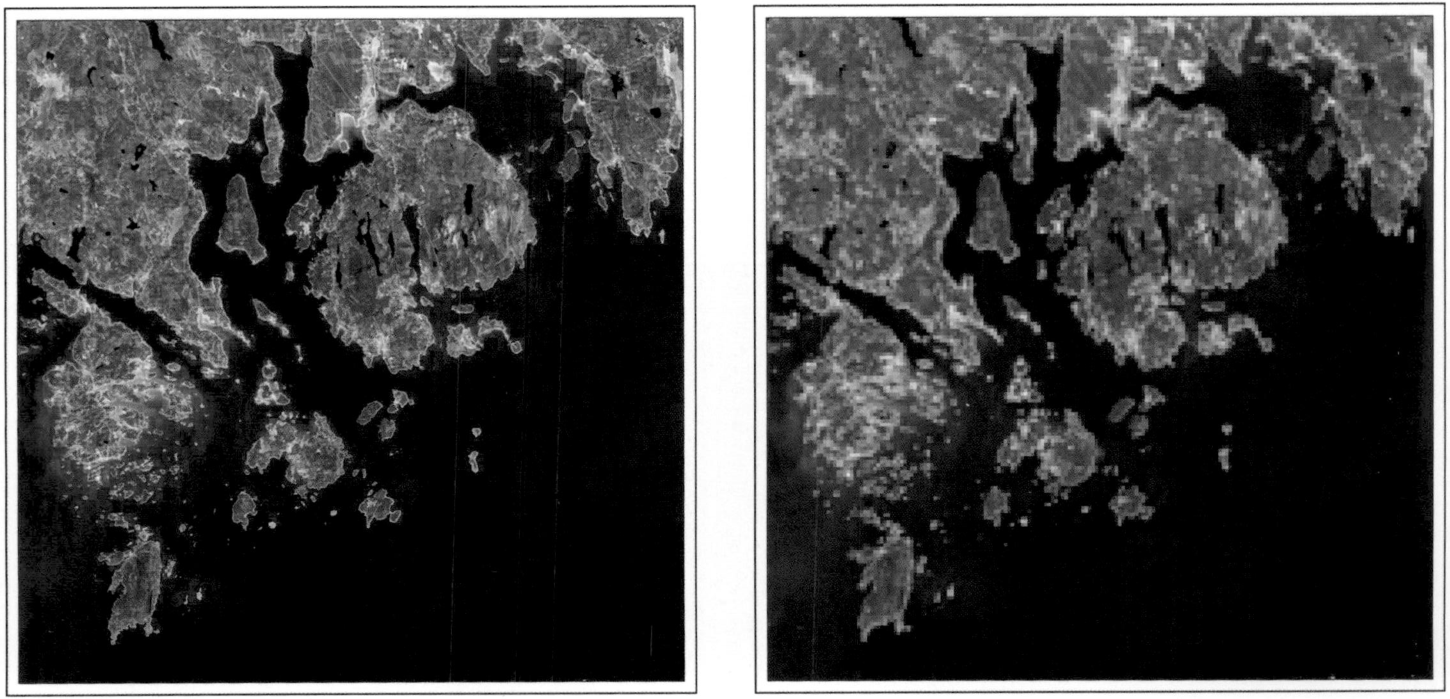

(a) Pansharpened image

(b) Enlarged color image

Figure 7.13 The fused wavelet transform (the blur portion from the transform of I replacing the blur portion of the transform of P') and its inverse transform I'.

Figure 12.12 The original color image.

(a) JPEG2000

(b) JPEG

Figure 12.15 The compressed image using JPEG2000 and JPEG.

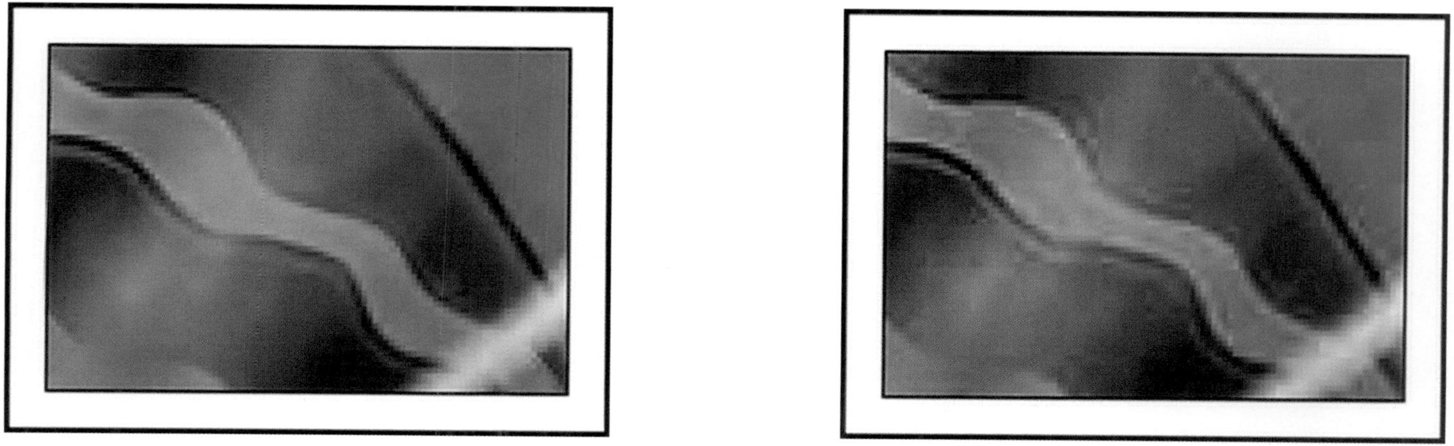

(a) JPEG2000 (b) JPEG

Figure 12.16 Corresponding 64×96 blocks of each compressed images in Figure 12.15.

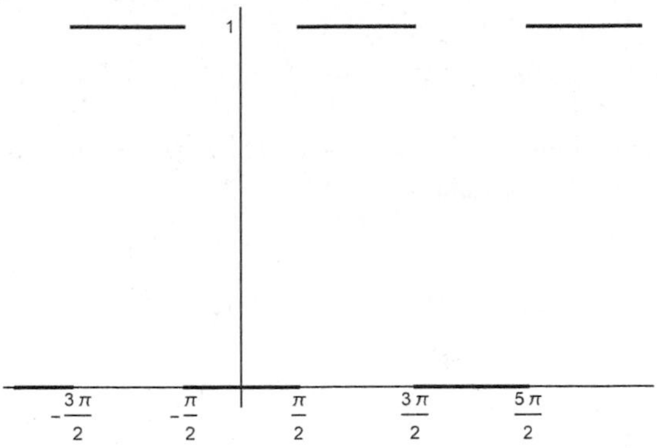

Figure 8.12 Three periods of a 2π-periodic piecewise constant function.

where $c_0 = \frac{1}{2}$ *and for* $k \neq 0$, c_k *is given by* (8.12). *It is to our advantage not to use the simplified series given by* (8.13). *Indeed, we have*

$$g(\omega) = f\left(\omega - \frac{\pi}{2}\right) = \sum_{k=-\infty}^{\infty} c_k e^{ik(\omega - \pi/2)} = \sum_{k=-\infty}^{\infty} \left(c_k e^{-ik\pi/2}\right) e^{ik\omega}.$$

Thus we see that $d_k = e^{-ik\pi/2} c_k$. *This is much easier than integrating! We can even simplify. Recall that when* $k = 0$, $c_0 = \frac{1}{2}$, *so that* $d_0 = e^0 c_0 = \frac{1}{2}$. *When* $k = \pm 2, \pm 4, \pm 6, \ldots$, $c_k = 0$ *so that* $d_k = 0$ *as well. Finally, for* k *odd, we have* $c_k = -\frac{i}{k\pi}$ *and* $e^{-ik\pi/2} = (-i)^k$, *so that* $d_k = \frac{(-i)^{k+1}}{k\pi}$. □

Rules for Computing Fourier Coefficients

In general we can always build the Fourier series of a function that is expressed as a translation of another function.

Proposition 8.2 (Translation Rule). *Suppose that* $f(\omega)$ *has the Fourier series representation*

$$f(\omega) = \sum_{k=-\infty}^{\infty} c_k e^{ik\omega}$$

and suppose that $g(\omega) = f(\omega - a)$. *If the Fourier series representation for* $g(\omega)$ *is*

$$g(\omega) = \sum_{k=-\infty}^{\infty} d_k e^{ik\omega}$$

then $d_k = e^{-ika} c_k$. □

Proof. We could prove the result by generalizing the argument given in Example 8.6. As an alternative, we prove the result using the definition of the Fourier coefficient (8.9) for $g(\omega)$. We have

$$d_k = \frac{1}{2\pi}\int_{-\pi}^{\pi} g(\omega)e^{-ik\omega}\,\mathrm{d}\omega = \frac{1}{2\pi}\int_{-\pi}^{\pi} f(\omega - a)e^{-ik\omega}\,\mathrm{d}\omega.$$

Now we do a u-substitution. Let $u = \omega - a$. Then $\mathrm{d}u = \mathrm{d}\omega$ and $\omega = u + a$. Changing endpoints gives the formula

$$\begin{aligned} d_k &= \frac{1}{2\pi}\int_{-\pi-a}^{\pi-a} f(u)e^{-ik(u+a)}\,\mathrm{d}u \\ &= \frac{1}{2\pi}e^{-ika}\int_{-\pi-a}^{\pi-a} f(u)e^{-iku}\,\mathrm{d}u \\ &= \frac{1}{2\pi}e^{-ika}\int_{-\pi}^{\pi} f(u)e^{-iku}\,\mathrm{d}u \\ &= e^{-ika}c_k. \end{aligned}$$

□

Note that Problem 8.18 from Section 8.1 allows us to change the limits of integration from $-\pi - a, \pi - a$ to $-\pi, \pi$.

Although there are several such rules for obtaining Fourier series coefficients, we only state one more. It will be useful later in the book. In the Problems section, other rules are stated and you are asked to provide proofs for them.

Proposition 8.3 (Modulation Rule). *Suppose $f(\omega)$ has the Fourier series representation*

$$f(\omega) = \sum_{k=-\infty}^{\infty} c_k e^{ik\omega}$$

and suppose that $g(\omega) = e^{im\omega}f(\omega)$ for some $m \in \mathbb{Z}$. If the Fourier series representation for $g(\omega)$ is

$$g(\omega) = \sum_{k=-\infty}^{\infty} d_k e^{ik\omega}$$

then $d_k = c_{k-m}$. □

Proof. The proof is left as Problem 8.34. □

We could easily spend an entire semester studying the fascinating topic of Fourier series, but we only attempt here to get a basic understanding of the concept so that we can use it in filter design.

Finite-Length Fourier Series

You might have decided that it is a little unnatural to construct a series for a function that is already known. There are several reasons for performing such a construction, but admittedly your point would be well taken. In many applications we typically do not know the function that generates a Fourier series – indeed, we usually have only the coefficients c_k or samples of some function.

In Chapters 4, 5, and 7 we constructed FIR filters that could be used in a wavelet transformation matrix to process data. The constructions were somewhat ad hoc, but we were ultimately able to tailor our filters so that they were lowpass, highpass, or annihilated polynomial data. A much cleaner way to construct these filters with the desired characteristics is to use Fourier series. Engineers, scientists, and mathematicians have long known about this technique. The idea is to use the elements h_k of an FIR filter $\mathbf{h}$ as the *Fourier coefficients* of the Fourier series of an unknown function $H(\omega)$. If the h_k are known, we can analyze $H(\omega)$ looking for special characteristics. If the h_k are unknown, then we can put constraints on $H(\omega)$ in an attempt to solve for the filter coefficients. Since $\mathbf{h}$ is an FIR filter, $H(\omega)$ will be a finite sum. In particular, if $\mathbf{h} = (h_\ell, \ldots, h_L)$, then

$$H(\omega) = \sum_{k=\ell}^{L} h_k e^{ik\omega}.$$

What do we stand to gain by forming such a Fourier series? Let's look at an example.

Example 8.7 (Finite-Length Fourier Series). *Let $h_{-1} = h_1 = \frac{1}{4}$ and $h_0 = \frac{1}{2}$. All other values for h_k are zero. Construct the Fourier series $H(\omega)$ for these h_k, and plot a graph of $|H(\omega)|$, for $-\pi \le \omega \le \pi$.*

Solution.

$$H(\omega) = \sum_{k=-1}^{1} h_k e^{ik\omega} = \frac{1}{4}e^{-i\omega} + \frac{1}{2} + \frac{1}{4}e^{i\omega}.$$

Using Problem 8.19, we can simplify $H(\omega)$ as

$$\begin{aligned} H(\omega) &= \frac{1}{4}e^{-i\omega} + \frac{1}{2} + \frac{1}{4}e^{i\omega} \\ &= \frac{1}{2} + \frac{1}{2}\left(\frac{e^{-i\omega} + e^{i\omega}}{2}\right) \\ &= \frac{1}{2} + \frac{1}{2}\cos\omega \\ &= \frac{1}{2}(1 + \cos\omega). \end{aligned}$$

Since $-1 \le \cos\omega \le 1$, we have $H(\omega) \ge 0$ so that $|H(\omega)| = H(\omega)$. We plot one period of $|H(\omega)|$ in Figure 8.13.

(Live example: Visit `stthomas.edu/wavelets` *and click on Live Examples.)* □

Notice that the function $|H(\omega)|$ plotted in Figure 8.13 is an even function. This is true in general for the modulus of a Fourier series as you will prove in Problem 8.33.

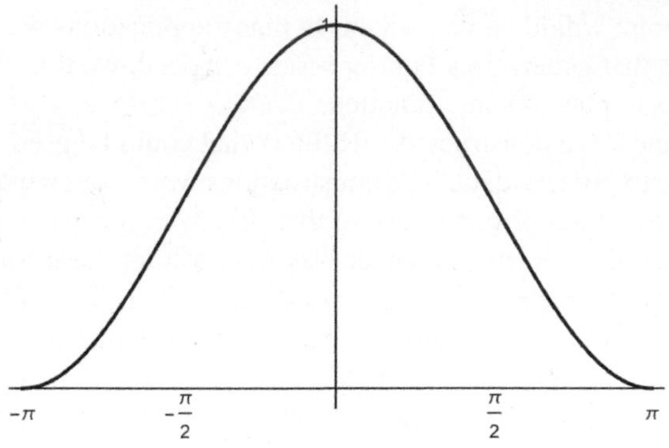

Figure 8.13 A graph of $|H(\omega)|$.

Thus we can restrict our attention to the interval $[0, \pi]$. In the following section, we will learn how Fourier series can be used to construct lowpass (highpass) filters and the important role the plot of $|H(\omega)|$ on $[0, \pi]$ plays in this process.

PROBLEMS

8.22 Use Problem 8.19 from Section 8.1 to write the Fourier coefficients for $\cos\omega$ and the Fourier coefficients for $\sin\omega$.

8.23 Suppose that $\sum\limits_{k=-\infty}^{\infty} |c_k| < \infty$. Show that the Fourier series (8.8) converges absolutely.

8.24 This problem is for students who have taken a real analysis course. Suppose that $\sum\limits_{k=-\infty}^{\infty} |c_k| < \infty$. Show that the Fourier series (8.8) converges uniformly for all $\omega \in \mathbb{R}$.

8.25 In this problem you will show that the Fourier series of an even function can be reduced to a series involving cosine functions. Assume that f is an even, 2π-periodic function.

(a) Show that the Fourier coefficient $c_k = \frac{1}{\pi}\int\limits_0^{\pi} f(\omega)\cos(k\omega)\,d\omega$. Thus, c_k is real whenever f is even.

(b) Using (a), show that $c_k = c_{-k}$.

(c) Using (b), show that $f(\omega) = c_0 + 2\sum\limits_{k=1}^{\infty} c_k\cos(k\omega)$. Organize your work using the following steps.

(i) Start with $f(\omega) = \sum_{k\in\mathbb{Z}} c_k e^{ik\omega}$ and write it as $f(\omega) = c_0 + \sum_{k=1}^{\infty} c_k e^{ik\omega} + \sum_{k=-\infty}^{-1} c_k e^{ik\omega}$.

(ii) Change the indices on the last sum to run from $k = 1$ to ∞.

(iii) Combine the two infinite sums into one and use (b) to replace c_{-k} with c_k.

(iv) Factor out a c_k from each term and then use Problem 8.19 from Section 8.1.

8.26 Assume that $H(\omega) = \sum_{k\in\mathbb{Z}} h_k e^{ik\omega}$ is a Fourier series whose coefficients satisfy $h_k = h_{-k}$ for all $k \in \mathbb{Z}$. Show that $H(\omega) = H(-\omega)$. That is, show that $H(\omega)$ is an even function. *Hint:* Use the ideas from Problem 8.25.

8.27 In this problem you will show that the Fourier series of an odd function can be reduced to a series involving sine functions. Assume that f is an odd, 2π-periodic function.

(a) Show that the Fourier coefficient $c_k = \frac{-i}{\pi}\int_0^{\pi} f(\omega)\sin(k\omega)\,d\omega$. Thus, c_k is imaginary whenever f is odd.

(b) Using (a), show that $c_k = -c_{-k}$. Note that in particular, $c_0 = 0$.

(c) Using (b), show that $f(\omega) = 2i\sum_{k=1}^{\infty} c_k \sin(k\omega)$. *Hint:* The ideas from Problem 8.25(c) will be helpful.

8.28 Assume that $H(\omega) = \sum_{k\in\mathbb{Z}} h_k e^{ik\omega}$ is a Fourier series whose coefficients satisfy $h_k = -h_{-k}$ for all $k \in \mathbb{Z}$. Show that $H(\omega) = -H(-\omega)$. That is, show that $H(\omega)$ is an odd function. *Hint:* Use the ideas from Problem 8.27.

8.29 Use the defining relation (8.9) to find the Fourier series for the function plotted in Figure 8.14. On the interval $[-\pi, \pi]$, the function is $f(\omega) = e^{-|\omega|}$.

8.30 In Example 8.4 the Fourier coefficients are given by (8.10) and the resulting Fourier series is given by (8.11). Show that this series can be reduced to the sine series

$$f(\omega) = -2\sum_{k=1}^{\infty} \frac{(-1)^k}{k}\sin(k\omega).$$

Hint: We know that $c_0 = 0$. Group the $k = \pm 1$ terms together, simplify, and use Problem 8.19 from Section 8.1. Do the same with the $k = \pm 2, \pm 3, \pm 4, \ldots$ terms.

8.31 In Example 8.5 the Fourier series is given by (8.13). Show that this series can be reduced to the sine series of odd terms

$$f(\omega) = \frac{1}{2} + \frac{2}{\pi}\sum_{k=1}^{\infty} \frac{1}{2k-1}\sin\left((2k-1)\omega\right).$$

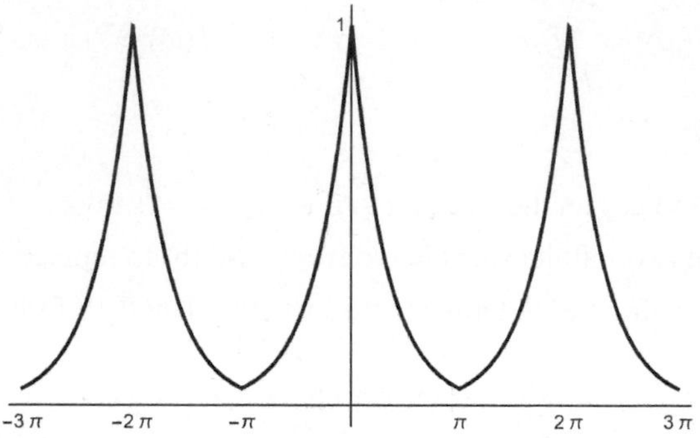

Figure 8.14 The graph of $f(\omega)$ for Problem 8.29.

Hint: We know that $c_0 = \frac{1}{2}$. Group the $k = \pm 1$ terms together, simplify, and use Problem 8.19 from Section 8.1. Do the same with the $k = \pm 3, \pm 5, \ldots$ terms.

8.32 Let

$$f_n(\omega) = \frac{1}{2} + \frac{2}{\pi} \sum_{k=1}^{n} \frac{1}{2k-1} \sin\left((2k-1)\omega\right)$$

be a sequence of partial sums for the Fourier series given in Problem 8.31. Use a **CAS** to plot f_n, $n = 1, 2, 5, 10, 50$.

8.33 Suppose that $H(\omega) = \sum\limits_{k=-\infty}^{\infty} h_k e^{ik\omega}$ with h_k a real number for $k \in \mathbb{Z}$.

(a) Show that $H(-\omega) = \overline{H(\omega)}$. (You can "pass" the conjugation operator through the infinite sum without justification – in a complex analysis class, you will learn the conditions necessary to legitimize this operation.)

(b) Show that $|H(\omega)|$ is an even function by verifying that $|H(\omega)| = |H(-\omega)|$. *Hint:* Use (a) along with Problem 8.8 from Section 8.1.

8.34 Prove the modulation rule given by Proposition 8.3. *Hint:* Start with the definition of d_k.

8.35 Let $f(\omega)$ have the Fourier series representation $f(\omega) = \sum\limits_{k=-\infty}^{\infty} c_k e^{ik\omega}$ and $g(\omega)$ have the Fourier series representation $g(\omega) = \sum\limits_{k=-\infty}^{\infty} d_k e^{ik\omega}$. Prove the following properties about Fourier series.

(a) If $g(\omega) = f(-\omega)$, show that $d_k = c_{-k}$. *Hint:* You can prove this either by using the definition of the Fourier coefficients (8.9) or by inserting $-\omega$ into the representation for f above.

(b) If $g(\omega) = \overline{f(-\omega)}$, show that $d_k = \overline{c_k}$. *Hint:* Problem 8.12 will be helpful.

(c) If $g(\omega) = f'(\omega)$, show that $d_k = ikc_k$. *Hint:* Use (8.9) and integrate by parts. Alternatively, you could differentiate the Fourier series for $f(\omega)$ *if* you can justify differentiating each term of an infinite series. In an analysis class, you will learn how to determine if such a step can be performed.

8.36 Let $f(\omega)$ be the function given in Example 8.4. The Fourier coefficients for this function are given in (8.10). Find the Fourier coefficients of each of the following functions.

(a) $f_1(\omega) = f(\omega - \pi)$

(b) $f_2(\omega) = f(-\omega)$

(c) $f_3(\omega) = e^{2i\omega} f(\omega)$

8.37 The graph for $f(\omega) = \sum\limits_{k=-\infty}^{\infty} c_k e^{ik\omega}$ is plotted in Figure 8.15. Sketch a graph of g_1, g_2, and g_3. Problem 8.35 and Propositions 8.2 and 8.3 will be helpful.

(a) $g_1(\omega) = \sum\limits_{k=-\infty}^{\infty} (-1)^k c_k e^{ik\omega}$

(b) $g_2(\omega) = \sum\limits_{k=-\infty}^{\infty} c_{-k} e^{ik\omega}$

(c) $g_3(\omega) = i \sum\limits_{k=-\infty}^{\infty} k c_k e^{ik\omega}$

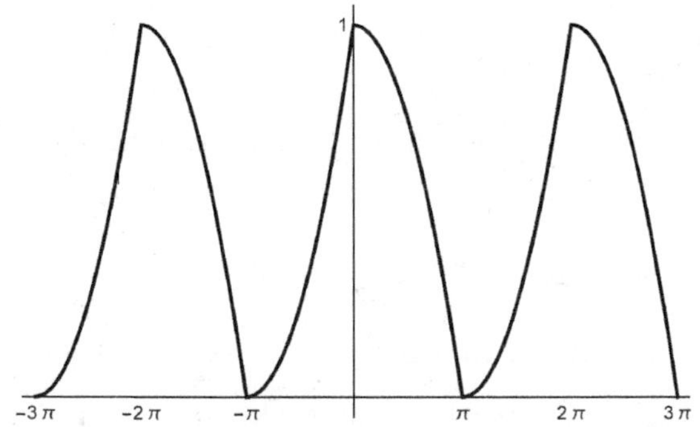

Figure 8.15 A graph of the $f(\omega)$ for Problem 8.37.

★**8.38** Let $A(\omega) = \sum\limits_{k=\ell}^{L} a_k e^{ik\omega}$ where $a_k \in \mathbb{R}$, $k = \ell, \ldots, L$. Show that $\overline{A'(\omega)} = \left(\overline{A(\omega)}\right)'$. *Hint:* Problem 8.14 will be helpful.

8.39 We developed the notion of Fourier series for 2π-periodic functions. It turns out that the Fourier series representation is valid for functions of period $2L$ ($L > 0$). In this problem you will learn about the complex exponential functions and the Fourier coefficients needed for this representation. The following steps will help you organize your work.

(a) Show that the complex exponential function $e_k^L(\omega) = e^{2\pi ik\omega/2L}$ is a $2L$-periodic function. Here $k \in \mathbb{Z}$ and we add the superscript L to distinguish these complex exponential functions from the ones defined in (8.5) for 2π-periodic functions.

(b) Show that for $j, k \in \mathbb{Z}$, we have

$$\int_0^{2L} e_j^L(\omega)\overline{e_k^L(\omega)}\,d\omega = \int_{-L}^{L} e^{2\pi ij\omega/2L}e^{-2\pi ik\omega/2L}\,d\omega = \begin{cases} 2L, & j = k \\ 0, & j \neq k. \end{cases}$$

(c) It can be shown (see e.g. Boggess and Narcowich [7]) that the set $\{e^{2\pi ik\omega/2L}\}_{k\in\mathbb{Z}}$ forms a basis for $2L$-periodic, integrable (over $[0, 2L]$) functions. Thus, for such a function, we can write

$$f(\omega) = \sum_{k\in\mathbb{Z}} c_k e^{2\pi ik\omega/(2L)}.$$

Using an argument similar to that preceding Proposition 8.1 show that

$$c_k = \frac{1}{2L}\int_{-L}^{L} f(\omega)e^{-2\pi ik\omega/(2L)}\,d\omega.$$

★8.40 Let $H(\omega) = \cos^2(\frac{\omega}{2})$. Show that the Fourier series for $H(\omega)$ is $H(\omega) = \frac{1}{4}(e^{i\omega} + 2 + e^{-i\omega}) = \frac{1}{2}(1 + \cos\omega)$.

★8.41 Let $G(\omega) = \sin^2(\frac{\omega}{2})$. Show that the Fourier series for $G(\omega)$ is $G(\omega) = \frac{1}{4}(-e^{i\omega} + 2 - e^{-i\omega}) = \frac{1}{2}(1 - \cos\omega)$.

★8.42 Consider the finite-length Fourier series $H(\omega) = \sum_{k=0}^{3} h_k e^{ik\omega}$. Let $G(\omega) = \sum_{k=0}^{3} g_k e^{ik\omega}$ where $g_k = (-1)^k h_{3-k}$ for $k = 0, 1, 2, 3$. Write $G(\omega)$ in terms of $H(\omega)$. The following steps will help you organize your work.

(a) Write down $\overline{G(\omega)}$ and factor $-e^{-3i\omega}$ from the series to obtain

$$\overline{G(\omega)} = -e^{-3i\omega}\left(h_0 - h_1e^{i\omega} + h_2e^{2i\omega} - h_3e^{3i\omega}\right).$$

(b) Chose a value a so that $H(\omega + a)$ gives the four-term series on the right in (a).

(c) Conjugate the identity in (b) to obtain the desired result.

★**8.43** Let $H(\omega) = \cos^N\left(\frac{\omega}{2}\right)$ where N is an even positive integer with $L = N/2$. Show that the Fourier series for $H(\omega)$ is

$$H(\omega) = \sum_{k=-L}^{L} \frac{1}{2^N}\binom{N}{k+L}e^{ik\omega}.$$

Compare your result to that of Problem 7.17(c). *Hint:* $\cos\frac{\omega}{2} = \frac{1}{2}(e^{i\omega/2} + e^{-i\omega/2}) = \frac{1}{2}e^{-i\omega/2}\left(e^{i\omega}+1\right)$. Apply the binomial theorem (7.30) to $\left(e^{i\omega}+1\right)$.

★**8.44** Let $H(\omega) = e^{i\omega/2}\cos^N\left(\frac{\omega}{2}\right)$ where N is an odd positive integer with $L = \frac{N-1}{2}$. Show that

$$H(\omega) = \sum_{k=-L}^{L+1} \frac{1}{2^N}\binom{N}{k+L}e^{ik\omega}.$$

Compare your result to that of Problem 7.17(c). *Hint:* The ideas of Problem 8.43 will be useful.

8.45 Let $G(\omega) = (\sin\frac{\omega}{2})^N$ where N is an even, positive integer with $L = N/2$. Find the Fourier series for $G(\omega)$. Show that

$$G(\omega) = \sum_{k=-L}^{L} \frac{(-1)^k}{(2i)^N}\binom{N}{k+L}e^{ik\omega}.$$

Hint: The ideas from Problem 8.43 should be helpful.

8.46 In Example 8.7 we created a Fourier series from the three nonzero coefficients $h_0 = h_2 = \frac{1}{4}$ and $h_1 = \frac{1}{2}$. The resulting Fourier series is

$$H(\omega) = e^{i\omega}\frac{1}{2}(1+\cos\omega).$$

This is counter to having a known 2π-periodic, integrable function $f(\omega)$ and using it to create Fourier coefficients (see Example 8.4). It is a good exercise to use (8.9) to directly compute the integral and show that

$$h_k = \frac{1}{2\pi}\int_{-\pi}^{\pi} H(\omega)e^{-ik\omega}\,d\omega = \begin{cases} \frac{1}{4}, & k = 0, 2 \\ \frac{1}{2}, & k = 1 \\ 0, & \text{otherwise} \end{cases}$$

8.3 Filters and Convolution in the Fourier Domain

In this section, we consider the notions of lowpass and highpass filters in terms of Fourier series. We also learn via the convolution theorem about an important connection between convolution and Fourier series. In Section 2.4 we talked about computing $\mathbf{y} = \mathbf{h} * \mathbf{v}$ but we have yet to talk about the possibility of recovering $\mathbf{v}$ from $\mathbf{y}$. The convolution theorem sheds some light on this problem and in some ways provides further motivation for the construction of discrete wavelet transformations.

Fourier Series and Lowpass Filters

In Example 8.7 from Section 8.2, we analyzed the Fourier series generated by $\mathbf{h} = (h_{-1}, h_0, h_1) = \left(\frac{1}{4}, \frac{1}{2}, \frac{1}{4}\right)$. It is easily verified using Definition 2.19 that $\mathbf{h}$ is a lowpass filter. We computed the Fourier series $H(\omega) = \frac{1}{2}(1 + \cos\omega)$ and plotted the modulus $|H(\omega)|$ from $[-\pi, \pi]$. In addition to being 2π-periodic, we learned in Problem 8.33 in Section 8.2 that $|H(\omega)|$ is an even function. So we have reproduced the plot in Figure 8.16 for the interval $[0, \pi]$.

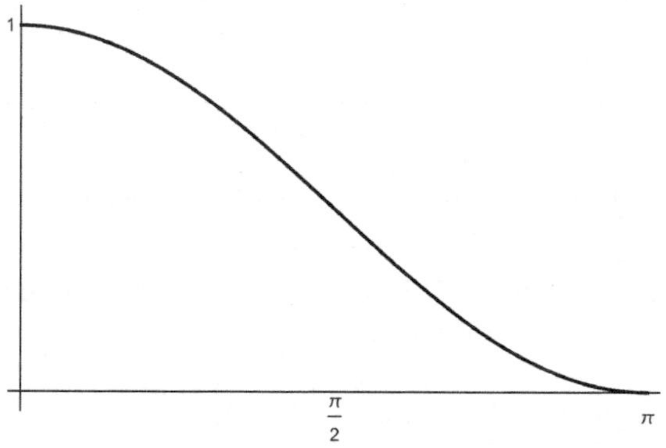

Figure 8.16 $|H(\omega)|$ on $[0, \pi]$.

Now $|H(\omega)|$ is constructed using a linear combination of complex exponential functions $e^{ik\omega}$, each of which by Euler's formula (8.3) is represented using $\cos(k\omega)$, $\sin(k\omega)$, $k = 0, 1, 2$. Both k and ω play a role in the oscillatory nature of $\cos(k\omega)$, $\sin(k\omega)$ as ω ranges from 0 to π. The larger the value of k, the more oscillatory the $\cos(k\omega)$ and $\sin(k\omega)$, and at the extreme right of the interval, $\omega = \pi$, we are at a point where $\cos(k\omega)$ and $\sin(k\omega)$ have oscillated the most over the interval $[0, \pi]$. Conversely, at $\omega = 0$, no oscillation has occurred in $\cos(k\omega)$ and $\sin(k\omega)$. Now look at the graph of $|H(\omega)|$. We have $|H(0)| = 1$ and $|H(\pi)| = 0$ and the way we interpret this is that when $\mathbf{h}$ is convolved with bi-infinite sequence $\mathbf{v}$, it preserves portions of $\mathbf{v}$ where the components do not vary much, and it annihilates or attenuates portions of $\mathbf{v}$ where the components do vary.

This is essentially the idea of a lowpass filter. You can see by the graph of $|H(\omega)|$ that the term *filter* might better apply in the Fourier domain. Near $\omega = 0$, $|H(\omega)| \approx 1$ and for $\omega \approx \pi$, we have $|H(\omega)| \approx 0$. The shape of $|H(\omega)|$ makes it easy to characterize lowpass filters in the Fourier domain.

An Ideal Lowpass Filter

The ideal lowpass filter $\mathbf{h}$ with Fourier series $H(\omega)$ occurs when the graph of $|H(\omega)|$ is of the form given in Figure 8.17. For this filter there is a range $[0, a]$ with $0 < a <$

π of low frequencies where the filter $\mathbf{h}$ preserves low oscillatory tendencies when convolved with $\mathbf{v}$ and a range of high frequencies $[a, \pi]$ where $\mathbf{h}$ annihilates high oscillatory tendencies when convolved with $\mathbf{v}$.

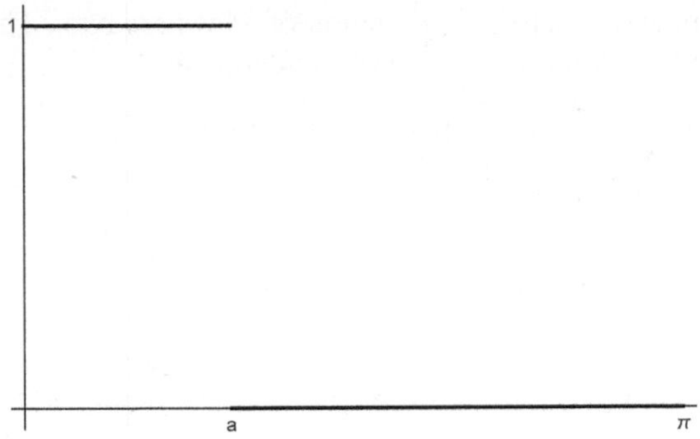

Figure 8.17 $|H(\omega)|$ of an ideal lowpass filter on $[0, \pi]$.

In Problem 8.47, you are asked to compute the values of a filter $\mathbf{h}$ that generates $|H(\omega)|$ and investigate the practical considerations of convolving with $\mathbf{h}$. Here you will see that the implementation of $\mathbf{h}$ is computationally intractable.

Lowpass Filter Defined (in the Sense of Fourier Series)

Below we define a lowpass filter in the sense of Fourier series. You will see that the definition allows for a wide variety of filters to be considered lowpass. What follows might be a definition found in a standard text on signal processing (see, e. g., Smith [82]). To develop some terminology, let's consider the $|H(\omega)|$ plotted in Figure 8.18.

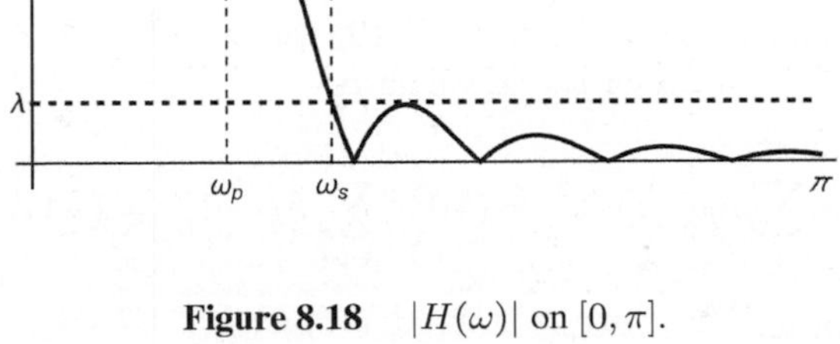

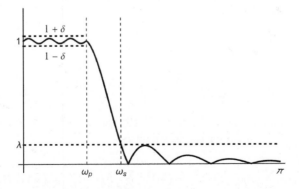

Figure 8.18 $|H(\omega)|$ on $[0, \pi]$.

Consider the vertical dashed line segment $\omega = \omega_p$ and note that $1-\delta \leq |H(\omega)| \leq 1+\delta$ for $0 \leq \omega \leq \omega_p$. The interval $[0, \omega_p]$ is called the *passband*. Next, note that for $\omega_s \leq \omega \leq \pi$, we have $|H(\omega)| \leq \lambda$. The interval $[\omega_s, \pi]$ is called the *stopband* and λ is referred to as the *stopband attenuation*. Finally, we call the interval $[\omega_p, \omega_s]$ the *transition band*. We now state a formal definition of a lowpass filter. This definition is a generalization of the Definition 2.19 given in Section 2.4.

Definition 8.4 (Lowpass Filter (in Terms of Fourier Series)). *Let* $\mathbf{h}$ *be a bi-infinite sequence with Fourier series* $H(\omega)$. *Let* $0 < \omega_p \leq \omega_s < \pi$ *and suppose that there exists* $0 < \delta < \frac{1}{2}$, *with* $1 - \delta \leq |H(\omega)| \leq 1 + \delta$ *for* $0 \leq \omega \leq \omega_p$ *and* λ *with* $0 < \lambda < \frac{1}{2}$ *so that for* $\omega_s \leq \omega \leq \pi$, $|H(\omega)| \leq \lambda$. *Then we call* $\mathbf{h}$ *a* lowpass filter. □

A good working definition is to say that if $\mathbf{h}$ is a lowpass filter, then $|H(\omega)|$ attains its maximum in the interval $[0, \omega_p]$ and $H(\omega) \approx 0$ for $\omega_s \leq \omega \leq \pi$ where we require that $\omega_p \leq \omega_s < \pi$. The following proposition connects Definitions 2.19 and 8.4.

Proposition 8.4 (Lowpass Filter Condition). *Suppose* $\mathbf{h} = (h_\ell, \ldots, h_L)$ *is an FIR filter and let*

$$H(\omega) = \sum_{k=\ell}^{L} h_k e^{ik\omega}$$

be its Fourier series. Then $\mathbf{h}$ *is a lowpass filter in the sense of Definition 2.19 if and only if* $H(\pi) = 0$ *and* $H(0) \neq 0$. □

Proof. We assume that $\mathbf{h}$ is a lowpass filter in the sense of Definition 2.19 and show that $H(\pi) = 0$ and $H(0) \neq 0$. The proof of the reverse implication is left to the reader as Problem 8.54. Note that $H(0) = h_\ell + \cdots + h_L$. We need to show that this sum is nonzero. Since $\mathbf{h}$ is lowpass in the sense of Definition 2.19, we know that $\mathbf{h} * \mathbf{u} = c\mathbf{u}$ where $c \neq 0$ and $\mathbf{u}$ is a bi-infinite sequence with each $u_k = 1$. Then for all $n \in \mathbb{Z}$, $(\mathbf{h} * \mathbf{u})_n = c$ and

$$(\mathbf{h} * \mathbf{u})_n = \sum_{k=\ell}^{L} h_k u_{n-k} = \sum_{k=\ell}^{L} h_k \cdot 1 = H(0).$$

So $H(0) = c \neq 0$ as desired.

We plug $\omega = \pi$ into $H(\omega)$ and use the fact that $e^{ik\pi} = (-1)^k$ (see Example 8.3) to write

$$H(\pi) = \sum_{k=\ell}^{L} h_k e^{ik\pi} = \sum_{k=\ell}^{L} h_k (-1)^k.$$

Since $\mathbf{h}$ is lowpass in the sense of Definition 2.19, we know $\mathbf{0} = \mathbf{h} * \mathbf{v}$ where the entries of $\mathbf{v}$ are $v_k = (-1)^k$. So for $n \in \mathbb{Z}$ we have

$$0 = (\mathbf{h} * \mathbf{v})_n = \sum_{k} h_k (-1)^{n-k} = (-1)^n \sum_{k-\ell}^{L} h_k (-1)^k = (-1)^n H(\pi)$$

and the proof is complete. □

Let's look at some examples of lowpass filters.

Example 8.8 (Lowpass Filters). *For each of the filters below, write down its Fourier series $H(\omega)$, show that it satisfies the conditions stated in Proposition 8.4 and plot $|H(\omega)|$ on $[0, \pi]$.*

(a) $\mathbf{h}_1 = (h_0, h_1) = \left(\frac{1}{2}, \frac{1}{2}\right)$

(b) $\mathbf{h}_2 = (h_{-1}, h_0, h_1) = \left(\frac{2}{3}, \frac{1}{2}, -\frac{1}{6}\right)$

(c) $\mathbf{h}_3 = (h_0, \ldots, h_3) = \left(\frac{1+\sqrt{3}}{4\sqrt{2}}, \frac{3+\sqrt{3}}{4\sqrt{2}}, \frac{3-\sqrt{3}}{4\sqrt{2}}, \frac{1-\sqrt{3}}{4\sqrt{2}}\right)$

Solution. *Note that $\mathbf{h}_1$ is the Haar filter. We use Problem 8.19 from Section 8.1 to write*

$$H_1(\omega) = \frac{1}{2} + \frac{1}{2}e^{i\omega} = e^{i\omega/2}\left(\frac{e^{-i\omega/2} + e^{i\omega/2}}{2}\right) = e^{i\omega/2}\cos\left(\frac{\omega}{2}\right).$$

Now we can easily verify that $H_1(\pi) = e^{i\pi/2}\cos\left(\frac{\pi}{2}\right) = 0$. It also follows that

$$|H_1(\omega)| = |e^{i\pi/2}| \cdot \left|\cos\left(\frac{\omega}{2}\right)\right| = \cos\left(\frac{\omega}{2}\right)$$

and we have plotted it in Figure 8.19(a).

For part (b), we compute

$$H_2(\omega) = \frac{2}{3}e^{-i\omega} + \frac{1}{2} - \frac{1}{6}e^{i\omega}$$

so that

$$H_2(\pi) = -\frac{2}{3} + \frac{1}{2} - \left(-\frac{1}{6}\right) = 0.$$

There is no nice simplification for $|H^2(\omega)|$ in this case – we have plotted it in Figure 8.19(b).

We recognize the filter $\mathbf{h}_3$ as the D4 filter. The Fourier series is

$$H_3(\omega) = \frac{1+\sqrt{3}}{4\sqrt{2}} + \frac{3+\sqrt{3}}{4\sqrt{2}}e^{i\omega} + \frac{3-\sqrt{3}}{4\sqrt{2}}e^{2i\omega} + \frac{1-\sqrt{3}}{4\sqrt{2}}e^{3i\omega}$$

and

$$H_3(\pi) = \frac{1+\sqrt{3}}{4\sqrt{2}} - \frac{3+\sqrt{3}}{4\sqrt{2}} + \frac{3-\sqrt{3}}{4\sqrt{2}} - \frac{1-\sqrt{3}}{4\sqrt{2}} = 0.$$

We have plotted $|H_3(\omega)|$ in Figure 8.19(c).

(Live example: Visit `stthomas.edu/wavelets` *and click on Live Examples.)*

□

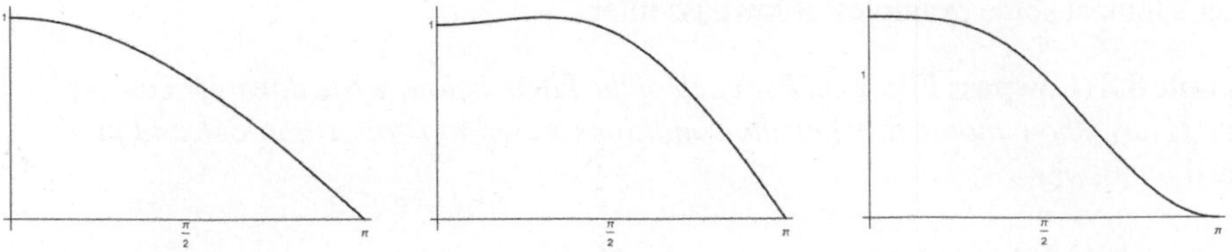

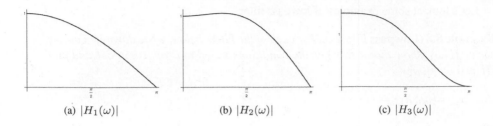

(a) $|H_1(\omega)|$ (b) $|H_2(\omega)|$ (c) $|H_3(\omega)|$

Figure 8.19 Moduli of the Fourier series from Example 2.17 plotted on $[0, \pi]$.

Highpass Filters in the Fourier Domain

As was the case with lowpass filters, we can determine the characteristics of a highpass filter $\mathbf{g}$ by examining $|G(\omega)|$. Let's start with an example.

Example 8.9 (Highpass Filter). *Let* $\mathbf{g} = (g_0, g_1) = \left(\frac{1}{2}, -\frac{1}{2}\right)$. *Compute* $G(\omega)$ *and plot* $|G(\omega)|$ *on the interval* $[0, \pi]$.

Solution. *We have, using Problem 8.19,*

$$\begin{aligned} G(\omega) &= \frac{1}{2} - \frac{1}{2}e^{i\omega} = e^{i\omega/2}\left(\frac{e^{-i\omega/2} - e^{i\omega/2}}{2}\right) \\ &= -ie^{i\omega/2}\left(\frac{e^{i\omega/2} - e^{-i\omega/2}}{2i}\right) \\ &= -ie^{i\omega/2}\sin\left(\frac{\omega}{2}\right). \end{aligned}$$

Since $|i| = \left|e^{i\omega/2}\right| = 1$, *we have* $|G(\omega)| = \left|\sin\left(\frac{\omega}{2}\right)\right|$. *The modulus of* $G(\omega)$ *is plotted in Figure 8.20.*

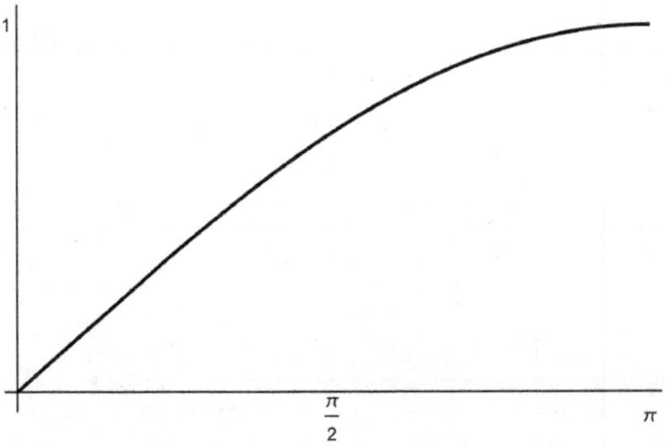

Figure 8.20 $|G(\omega)|$ on $[0, \pi]$.

(Live example: Visit `stthomas.edu/wavelets` *and click on Live Examples.)*

□

Highpass Filter Defined (in Terms of Fourier Series)

A good working definition is to say that if **g** is a highpass filter, then $|G(\omega)| \approx 0$ for $\omega \in [0, \omega_p]$ and $|G(\omega)| \approx 1$ for $\omega \in [\omega_s, \pi]$ where we require $\omega_p \leq \omega_s < \pi$. A good signal-processing book (see, e.g. Smith [82]) will contain more information on highpass filters. The terminology is exactly the same as that developed prior to Definition 8.4. Note that Definition 8.5 is a generalization of Definition 2.20 from Section 2.4.

Definition 8.5 (Highpass Filter (in Terms of Fourier Series)). *Let* **g** *be a bi-infinite sequence with Fourier series $G(\omega)$. Let $0 < \omega_p \leq \omega_s < \pi$ and suppose that there exists λ with $0 < \lambda < \frac{1}{2}$ so that $|G(\omega)| \leq \lambda$ for $0 \leq \omega \leq \omega_p$ and δ with $0 < \delta < \frac{1}{2}$ and $1 - \delta \leq |G(\omega)| \leq 1 + \delta$ for $\omega_s \leq \omega \leq \pi$. Then we call* **g** *a* highpass filter. □

The following proposition connects Definitions 8.5 and 2.20.

Proposition 8.5 (Highpass Filter Condition). *Suppose* $\mathbf{g} = (g_\ell, \ldots, g_L)$ *is an FIR filter with Fourier series*

$$G(\omega) = \sum_{k=\ell}^{L} g_k e^{ik\omega}.$$

Then **g** *is a highpass filter in the sense of Definition 2.20 if and only if $G(0) = 0$ and $G(\pi) \neq 0$.* □

Proof. The proof is straightforward and resembles that of Proposition 8.4. It is left as Problem 8.55. □

We include the following proposition at this time since we have typically required our highpass filters to annihilate polynomial data.

Proposition 8.6 (Annihilating Polynomial Data). *Suppose* $\mathbf{g} = (g_\ell, \ldots, g_L)$ *is an FIR filter with Fourier series*

$$G(\omega) = \sum_{k=\ell}^{L} g_k e^{ik\omega}.$$

If **g** *annihilates polynomial data of degree m then $G^{(m)}(0) = 0$.* □

Proof. The case where $m = 0$ is addressed in Proposition 8.5 so assume $m > 0$. Assume that **g** annihilates data of degree m. Then according to Proposition 2.3 with $n = 0$, we have

$$\sum_{k=\ell}^{L} g_k(-k)^m = (-1)^m \sum_{k=\ell}^{L} g_k k^m = 0. \tag{8.14}$$

We can easily compute the mth derivative as

$$G^{(m)}(\omega) = \sum_{k=\ell}^{L} g_k (ik)^m e^{ik\omega}$$

and evaluate it at $\omega = 0$ to obtain

$$G^{(m)}(0) = \sum_{k=\ell}^{L} g_k (ik)^m = i^m \sum_{k=\ell}^{L} g_k k^m.$$

Multiplying both sides of the above equation by i^m gives

$$i^m G^{(m)}(0) = i^{2m} \sum_{k=\ell}^{L} g_k k^m = (-1)^m \sum_{k=\ell}^{L} g_k k^m = 0$$

by (8.14). Since $i^m \neq 0$, we have $G^{(m)}(0) = 0$.

□

The Convolution Theorem

The convolution theorem is an important result that characterizes convolution in the Fourier domain. We state and prove it now.

Theorem 8.1 (Convolution Theorem). *Let* $\mathbf{h}, \mathbf{x}$, *and* $\mathbf{y}$ *be bi-infinite sequences with* $\mathbf{y} = \mathbf{h} * \mathbf{x}$. *Let* $H(\omega), X(\omega)$, *and* $Y(\omega)$ *denote the Fourier series of* $\mathbf{h}, \mathbf{x}$, *and* $\mathbf{y}$, *respectively. Then*

$$Y(\omega) = H(\omega)X(\omega).$$

□

Proof. We start with the Fourier series $Y(\omega)$. Our goal will be to manipulate this series so that ultimately we arrive at $H(\omega)X(\omega)$. We have

$$Y(\omega) = \sum_{n=-\infty}^{\infty} y_n e^{in\omega} = \sum_{n=-\infty}^{\infty} \left(\sum_{k=-\infty}^{\infty} h_k x_{n-k} \right) e^{in\omega}.$$

What we need to do next is interchange the sums. To interchange infinite sums we need to know that the terms in the series are suitably well behaved (i.e. decay rapidly to zero). We assume in this chapter that all bi-infinite sequence we convolve decay rapidly enough to ensure that the interchanging of sums is valid. The reader interested in the exact details of this part of the argument should review an analysis text such as Rudin [78]. We have

$$Y(\omega) = \sum_{k=-\infty}^{\infty} h_k \left(\sum_{n=-\infty}^{\infty} x_{n-k} e^{in\omega} \right).$$

The inner sum looks a bit like the Fourier series for $\mathbf{x}$ but the indices don't match up. So let's change the index on the inner sum and set $j = n - k$. Then $n = j + k$. We substitute these values and write

$$\begin{aligned} Y(\omega) &= \sum_{k=-\infty}^{\infty} h_k \left(\sum_{n=-\infty}^{\infty} x_{n-k} e^{in\omega} \right) \\ &= \sum_{k=-\infty}^{\infty} h_k \left(\sum_{j=-\infty}^{\infty} x_j e^{i(j+k)\omega} \right) \\ &= \sum_{k=-\infty}^{\infty} h_k \left(\sum_{j=-\infty}^{\infty} x_j e^{ij\omega} e^{ik\omega} \right) \\ &= \sum_{k=-\infty}^{\infty} h_k e^{ik\omega} \left(\sum_{j=-\infty}^{\infty} x_j e^{ij\omega} \right). \end{aligned}$$

But the inner sum is simply $X(\omega)$, so that

$$Y(\omega) = \sum_{k=-\infty}^{\infty} h_k e^{ik\omega} X(\omega)$$

and if we factor out the $X(\omega)$, we see that the remaining sum is $H(\omega)$ and the theorem is proved. $\square$

An immediate observation we make is that the complicated operation of convolution turns into simple multiplication in the Fourier domain. To get a better feel for the convolution theorem, let's look at the following example.

Example 8.10 (Using the Convolution Theorem). *Let* $\mathbf{h} = (h_0, h_1, h_2) = (1, 2, 1)$ *and* $\mathbf{g} = (g_0, g_1) = (1, 1)$. *Consider the convolution product* $\mathbf{y} = \mathbf{g} * \mathbf{h}$. *Find the components of the sequence* $\mathbf{y}$ *and then use them to find* $Y(\omega)$. *Then form* $H(\omega)$ *and* $G(\omega)$ *and verify the convolution theorem for these sequences by computing* $H(\omega)G(\omega)$.

Solution. *We have*

$$y_n = \sum_{k=0}^{2} h_k g_{n-k} = h_0 g_n + h_1 g_{n-1} + h_2 g_{n-2} = g_n + 2g_{n-1} + g_{n-2}$$

Since $g_n = 0$ *unless* $n = 0, 1$ *the only time* $y_n \neq 0$ *is when* $n = 0, 1, 2, 3$. *In these cases we have*

$$y_0 = g_0 + 2g_{-1} + g_{-2} = 1 \qquad y_1 = g_1 + 2g_0 + g_{-1} = 3$$
$$y_2 = g_2 + 2g_1 + g_0 = 3 \qquad y_3 = g_3 + 2g_2 + g_1 = 1.$$

Thus, the Fourier series for $\mathbf{y}$ *is*

$$Y(\omega) = 1 + 3e^{i\omega} + 3e^{2i\omega} + e^{3i\omega}.$$

The Fourier series for $\mathbf{h}$ *and* $\mathbf{g}$ *are*

$$H(\omega) = 1 + 2e^{i\omega} + e^{2i\omega} \qquad \textit{and} \qquad G(\omega) = 1 + e^{i\omega}.$$

We compute the product

$$\begin{aligned} H(\omega)G(\omega) &= (1 + 2e^{i\omega} + e^{2i\omega})(1 + e^{i\omega}) \\ &= (1 + 2e^{i\omega} + e^{2i\omega}) + (e^{i\omega} + 2e^{2i\omega} + e^{3i\omega}) \\ &= 1 + 3e^{i\omega} + 3e^{2i\omega} + e^{3i\omega} \end{aligned}$$

to see that, indeed, $Y(\omega) = H(\omega)G(\omega)$.

(Live example: Visit `stthomas.edu/wavelets` *and click on Live Examples.)*

□

De-convolution

We have talked about processing sequence $\mathbf{v}$ via filter $\mathbf{h}$ and the convolution operator to obtain $\mathbf{y} = \mathbf{h} * \mathbf{v}$. In previous chapters, we used a lowpass/highpass filter pair in a truncated, downsampled convolution matrix to compute the wavelet transformation of input data. In this case the transformation is invertible. Can the same be said about the convolution operator? That is, can we recover $\mathbf{v}$ from $\mathbf{y}$? We turn to Fourier series for help with this problem. Recall the convolution theorem:

$$\text{If } \mathbf{y} = \mathbf{h} * \mathbf{x}, \text{ then } Y(\omega) = H(\omega)X(\omega).$$

If $H(\omega) \neq 0$, we can divide both sides by it and arrive at

$$X(\omega) = \frac{1}{H(\omega)} Y(\omega). \tag{8.15}$$

Since we have $\mathbf{y}$ and know $\mathbf{h}$, we could form and compute $\frac{1}{H(\omega)} Y(\omega)$ to obtain $X(\omega)$. But then we would have to try to write $X(\omega)$ as a Fourier series. This might be problematic, but if we could do it, we would have as coefficients of the Fourier series for $X(\omega)$ the sequence $\mathbf{x}$.

Alternatively, we could try to find a Fourier series representation for the 2π-periodic function $\frac{1}{H(\omega)}$. Suppose that we could write down a Fourier series (with coefficient sequence $\mathbf{m}$) for this function:

$$\frac{1}{H(\omega)} = \sum_{k=-\infty}^{\infty} m_k e^{ik\omega}.$$

Then by the convolution theorem we would know that $\mathbf{x}$ could be obtained by convolving $\mathbf{m}$ with $\mathbf{y}$. This process is known as *de-convolution*.

Note that the convolution theorem also tells us *when* it is possible to perform a de-convolution. If $H(\omega) = 0$ for some ω, then we cannot perform the division and arrive at (8.15). In particular, note that by Proposition 8.4 if $\mathbf{h}$ is a lowpass filter, then $H(\pi) = 0$, and by Proposition 8.5 if $\mathbf{h}$ is a highpass filter, then $H(0) = 0$. In either case, we cannot solve for $X(\omega)$.

This makes some practical sense if we consider the averaging filter $\mathbf{h} = (h_0, h_1) = (\frac{1}{2}, \frac{1}{2})$. In this case (see Example 2.17), $\mathbf{y} = \mathbf{h} * \mathbf{x}$ has components $y_n = \frac{x_n}{2} + \frac{x_{n-1}}{2}$ so that y_n is the average of x_n and its previous component. Think about it: If you are told the average of two numbers is 12, could you determine what two numbers produced this average? The answer is no – there are an infinite number of possibilities. So it makes sense that we can't de-convolve with this sequence or any lowpass or highpass sequence.

Example 8.11 (De-convolution). *Consider the FIR filter* $\mathbf{h} = (h_0, h_1) = (2, 1)$. *If possible, find the Fourier series for* $\frac{1}{H(\omega)}$ *and write down the de-convolution filter* $\mathbf{m}$.

Solution. *It is quite simple to draw* $H(\omega) = 2 + e^{i\omega}$ *in the complex plane. We recognize from Chapter 8 that* $e^{i\omega}$ *is a circle centered at* 0 *with radius* 1. *By adding two to it, we're simply moving the circle two units right on the real axis.* $H(\omega)$ *is plotted in Figure 8.21.*

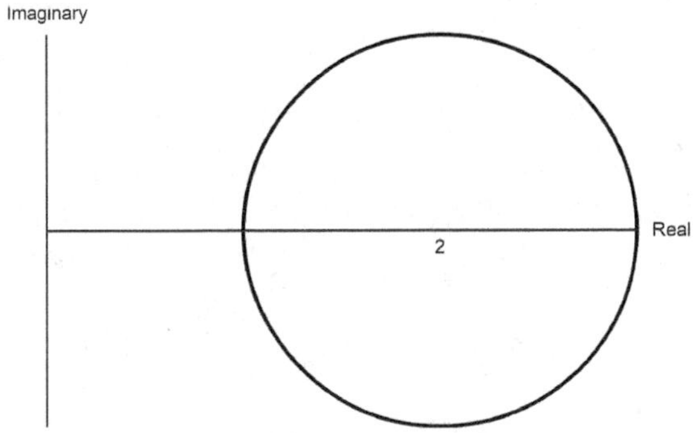

Figure 8.21 $H(\omega) = 2 + e^{i\omega}$.

Since $H(\omega) \neq 0$ *for all* $\omega \in \mathbb{R}$, *we can write*

$$X(\omega) = \frac{1}{H(\omega)} Y(\omega) = \frac{1}{2 + e^{i\omega}} Y(\omega).$$

To write $\frac{1}{2+e^{i\omega}}$ as a Fourier series, we recall the Maclaurin series from calculus for the function $\frac{1}{1+x}$:

$$\frac{1}{1+x} = 1 - x + x^2 - x^3 + x^4 - x^5 \pm \cdots = \sum_{k=0}^{\infty} (-1)^k x^k. \tag{8.16}$$

This series converges as long as $|x| < 1$. We can factor the two out of the denominator of $\frac{1}{2+e^{i\omega}}$ to obtain

$$\frac{1}{2+e^{i\omega}} = \frac{1}{2(1+\frac{1}{2}e^{i\omega})} = \frac{1}{2} \cdot \frac{1}{1+\frac{1}{2}e^{i\omega}}.$$

Let $t = \frac{1}{2}e^{i\omega}$. Then $|t| = \frac{1}{2} < 1$, so we use (8.16) with this choice of t to write

$$\frac{1}{2+e^{i\omega}} = \frac{1}{2} \cdot \frac{1}{1+\frac{1}{2}e^{i\omega}} = \frac{1}{2} \sum_{k=0}^{\infty} (-1)^k \left(\frac{1}{2}e^{i\omega}\right)^k = \sum_{k=0}^{\infty} (-1)^k \frac{1}{2^{k+1}} e^{ik\omega}.$$

So we see that

$$m_k = \begin{cases} (-1)^k/2^{k+1}, & k \geq 0 \\ 0, & k < 0. \end{cases}$$

or $\mathbf{m}$ is the infinite-length causal filter $\mathbf{m} = (\frac{1}{2}, -\frac{1}{4}, \frac{1}{8}, -\frac{1}{16}, \frac{1}{32}, -\frac{1}{64}, \ldots)$.

Although we can, indeed, write down the de-convolution filter $\mathbf{m}$, the fact that $\mathbf{m}$ is of infinite length means that we will have to truncate the convolution $\mathbf{m} * \mathbf{y}$, and thus we will not exactly recover $\mathbf{x}$. □

PROBLEMS

Note: *In the problems that follow, lowpass, highpass filters are in the sense of Definitions 2.19, 2.20, respectively.*

8.47 Consider the ideal lowpass filter that generates $|H(\omega)|$ from Figure 8.17. Certainly, $H(\omega)$ could be the following 2π-periodically extended box function:

$$H(\omega) = \sum_{k=-\infty}^{\infty} h_k e^{ik\omega} = \begin{cases} 1, & -a \leq 0 \leq a \\ 0, & \text{otherwise} \end{cases}$$

where $0 < a < \pi$. We know by (8.9) from Section 8.2 that we can compute the h_k by the formula

$$h_k = \frac{1}{2\pi} \int_{-a}^{a} 1 \cdot e^{-ik\omega} \, d\omega.$$

Simplify the integral above to find expressions for h_k. *Hint:* consider the $k = 0$ case separately). Is $\mathbf{h}$ an FIR filter? If your answer is no, observe then that such an ideal filter is not computationally tractable.

8.48 Let $\mathbf{h} = (h_{-1}, h_0, h_1, h_2) = (\frac{1}{4}, \frac{1}{4}, \frac{1}{4}, \frac{1}{4})$.

(a) Show that $H(\omega) = e^{i\omega/2}\cos\left(\frac{\omega}{2}\right)\cos\omega$.

(b) Use (a) to plot $|H(\omega)|$. Determine if $\mathbf{h}$ a lowpass filter, highpass filter or neither.

8.49 Suppose $\mathbf{h} = (h_\ell, \ldots, h_L)$ is a lowpass filter with Fourier series $H(\omega)$. We create a new filter $\mathbf{t}$ using $H(\omega)$. In particular, we take $T(\omega) = e^{ij\omega}H(\omega)$ where $j \in \mathbb{Z}$.

(a) Is $\mathbf{t}$ a lowpass filter? Explain your answer.

(b) Find the elements of $\mathbf{t}$ in terms of $\mathbf{h}$.

8.50 Find a causal filter $\mathbf{c}$ such that $|C(\omega)| = \cos^2(\frac{\omega}{2})$. Is your $\mathbf{c}$ lowpass, highpass, or neither? Is your choice of $\mathbf{c}$ unique? *Hint:* Use Problem 8.19 from Section 8.1 with $\cos(\frac{\omega}{2})$.

8.51 Suppose that $\mathbf{h} \neq \mathbf{0}$ is a causal filter that satisfies $h_k = h_{7-k}$.

(a) Show that

$$\begin{aligned}|H(\omega)| &= 2\left|e^{(7i\omega/2)}\right| \cdot \left|h_0\cos\left(\frac{7\omega}{2}\right) + h_1\cos\left(\frac{5\omega}{2}\right)\right. \\ &\quad \left. + h_2\cos\left(\frac{3\omega}{2}\right) + h_3\cos\left(\frac{\omega}{2}\right)\right| \\ &= 2\left|\sum_{k=0}^{3} h_k\cos\left(\frac{(7-2k)\omega}{2}\right)\right|.\end{aligned}$$

(b) Can $\mathbf{h}$ ever be a highpass filter? Explain.

8.52 Suppose that $\mathbf{h}$ is a (nonzero) causal filter that satisfies $h_k = h_{L-k}$ for some positive odd integer L. Can $\mathbf{h}$ ever be highpass? Explain. *Hint:* Try to generalize the factorization from the preceding problem.

★8.53 Suppose that $\mathbf{h} = (h_0, \ldots, h_L)$ is a lowpass filter, where L is an odd positive integer. Suppose that we construct the filter $\mathbf{g}$ by insisting that $G(\omega) = H(\omega + \pi)$.

(a) Is $\mathbf{g}$ lowpass, highpass, or neither?

(b) What are the coefficents g_k in terms of h_k?

(c) Use your results to find $\mathbf{g}$ when $\mathbf{h} = \left(\frac{1}{8}, \frac{3}{8}, \frac{3}{8}, \frac{1}{8}\right)$ so that $\mathbf{g}$ is highpass.

(d) Note that $\mathbf{h}\cdot\mathbf{g} = 0$. In general, will we always have $\mathbf{h}\cdot\mathbf{g} = 0$ when we construct $\mathbf{g}$ as suggested by (b)? Write down a condition for $\mathbf{h} = (h_0, h_1, \ldots, h_L)$ that guarantees $\mathbf{h}\cdot\mathbf{g} = 0$.

8.54 Complete the proof of Proposition 8.4.

8.55 Prove Proposition 8.5.

8.56 Consider the filter $\mathbf{h} = (h_0, h_1, h_2) = \left(\frac{1}{5}, \frac{3}{5}, \frac{1}{5}\right)$.

(a) Show that $|H(\omega)| = \frac{2}{5}\left(\frac{3}{2} + \cos\omega\right)$.

(b) Show that we can write $H(\omega)$ as $H(\omega) = e^{i\omega}|H(\omega)|$.

It turns out that we can write any complex-valued function $f(\omega)$ as

$$f(\omega) = e^{i\rho(\omega)}|f(\omega)|$$

This is analogous to using polar coordinates for representing points in $\mathbb{R}^2$. You can think of $|f(\omega)|$ as the "radius" and $\rho(\omega)$ is called the *phase angle*. The phase angle for this problem is $\rho(\omega) = \omega$, and this is a linear function. In image processing the phase angle plays an important role in filter design. It can be shown that for symmetric filters ($h_k = h_{N-k}$ as in this problem) or antisymmetric filters ($h_k = -h_{N-k}$), the phase angle is a (piecewise) linear function. For more information on phase angles, see Gonzalez and Woods [48] or Embree and Kimble [40].

8.57 Let $X(\omega) = 2 + e^{i\omega} + 2e^{2i\omega}$ and $Y(\omega) = 1 + 3e^{i\omega} + 2e^{2i\omega}$. Compute $Z(\omega) = X(\omega)Y(\omega)$ and write down the nonzero elements of $\mathbf{z}$ associated with $Z(\omega)$. Compare your results to Problem 2.50 in Section 2.4.

8.58 Let $\mathbf{h}$ be the bi-infinite sequence whose only nonzero elements are $h_0 = h_1 = 1$. In Example 2.13(c), we learned that the nonzero elements of $\mathbf{y} = \mathbf{h} * \mathbf{h}$ are $y_0 = y_2 = 1$ and $y_1 = 2$. Verify this result using the convolution theorem.

8.59 Use the convolution theorem to provide an easy proof that convolution is a commutative operator.

8.60 Let $\mathbf{x}$ and $\mathbf{y}$ be bi-infinite sequences with the components of $\mathbf{y}$ given by $y_k = x_{k+1}$.

(a) Find $Y(\omega)$ in terms of $X(\omega)$.

(b) Use the result in (a) to find a sequence $\mathbf{h}$ such that $\mathbf{y} = \mathbf{h} * \mathbf{x}$.

8.61 In Problem 2.52 in Section 2.4, you computed $\mathbf{y} = \mathbf{e}^m * \mathbf{x}$. Find the Fourier series $E_m(\omega)$ and verify the convolution theorem holds for $\mathbf{e}^m$, $\mathbf{x}$ and $\mathbf{y}$.

8.62 Suppose that $Y(\omega) = \sin(2\omega)$.

(a) Find $\mathbf{y}$.

(b) Write $\mathbf{y}$ as a convolution product of two sequences $\mathbf{h}$ and $\mathbf{g}$. *Hint:* Think about a trigonometric identity for $\sin(2\omega)$.

8.63 In this problem you will learn more about which FIR filters can be used as de-convolution filters.

(a) Let $\mathbf{h} = (h_0, h_1)$, where h_0 and h_1 are nonzero real numbers. Is it possible to construct an FIR filter $\mathbf{m}$ so that the associated Fourier series $M(\omega)$ satisfies $M(\omega) = \frac{1}{H(\omega)}$?

(b) Now suppose that $\mathbf{h} = (h_0, h_1, h_2)$, where at least two of h_0, h_1, and h_2 are nonzero. Explain why the filter $\mathbf{m}$ associated with Fourier series $M(\omega) = \frac{1}{H(\omega)}$ cannot be an FIR filter.

(c) Repeat (b) using $\mathbf{h} = (h_0, \ldots, h_L)$, where at least two of the elements of $\mathbf{h}$ are nonzero.

(d) Can you now characterize the FIR filters $\mathbf{h}$ for which the filter $\mathbf{m}$ associated with $M(\omega) = \frac{1}{H(\omega)}$ is an FIR filter?

CHAPTER 9

FILTER CONSTRUCTION IN THE FOURIER DOMAIN

We have constructed orthogonal filters in Chapters 4 and 5 and biorthogonal filter pairs in Chapter 7. The constructions were ad hoc and somewhat limiting in that it was difficult to produce a method for finding filters of arbitrary length. Equipped with Fourier series from Chapter 8, we can reproduce and indeed generalize the filter (pair) construction of previous chapters in a much more systematic way.

In Section 9.1 we reformulate the problem of finding filters (filter pairs) in the Fourier domain. The orthogonality relationships satisfied by the Daubechies filters of Chapter 5 can be reduced to three identities involving their Fourier series, and similar results hold true for biorthogonal spline filter pairs and their associated highpass filters. Conditions necessary to ensure $\mathbf{g}$ ($\widetilde{\mathbf{g}}$) are highpass filters that annihilate polynomial data are reformulated in the Fourier domain. We characterize the Daubechies system (5.65) in terms of Fourier series in Section 9.1. In the section that follows, we learn how to modify this system to produce a new family of orthogonal filters called *Coiflet filters*. The entire family of biorthogonal spline filter pairs introduced in Chapter 7 is developed in Section 9.4. The Cohen–Daubechies–Feauveau biorthogonal filter pair is utilized by the Federal Bureau of Investigation for compressing digitized images of fingerprints and by the JPEG2000 committee for the

Patrick J. Van Fleet.

JPEG2000 Image Compression Standard. This important filter pair is constructed in Section 9.5.

9.1 Filter Construction

In Chapters 4 and 5, we constructed orthogonal matrices

$$W_N = \left[\begin{array}{c} H_{N/2} \\ \hline G_{N/2} \end{array}\right]$$

from lowpass/highpass filter pairs $(\mathbf{h}, \mathbf{g})$. The fact that $W_N W_N^T = I_N$ gave the identities

$$\begin{aligned} H_{N/2}H_{N/2}^T &= G_{N/2}G_{N/2}^T = I_{N/2} \\ H_{N/2}G_{N/2}^T &= G_{N/2}H_{N/2}^T = 0_{N/2}. \end{aligned} \tag{9.1}$$

These identities in turn led to conditions on the elements of the filters. By nature of the construction of $H_{N/2}$, $G_{N/2}$, we required that the following identities hold (thinking now of our filters as bi-infinite sequences):

$$\begin{aligned} \sum_k h_k h_{k-2m} &= \sum_k g_k g_{k-2m} = \begin{cases} 1, & m = 0 \\ 0, & m \neq 0 \end{cases} \\ \sum_k h_k g_{k-2m} &= \sum_k h_{k-2m} g_k = 0, \quad m \in \mathbb{Z}. \end{aligned} \tag{9.2}$$

The first equation above says that $\mathbf{h}$ $(\mathbf{g})$ must dot with itself to produce value one and dot with any two-shift of itself with resulting value of zero. These conditions ensure the first set of matrix equations in (9.1) hold.

We constructed matrices $\widetilde{W}_N$ and W_N and highpass filter $\widetilde{\mathbf{g}}$, $\mathbf{g}$ from a biorthogonal spline filter pair $\left(\mathbf{h}, \widetilde{\mathbf{h}}\right)$ in Chapter 7 where $\widetilde{W}_N W_N^T = I_N$. In order for this matrix equation to hold we required

$$\begin{aligned} \sum_k \widetilde{h}_k h_{k-2m} &= \sum_k \widetilde{g}_k g_{k-2m} = \begin{cases} 1, & m = 0 \\ 0, & m \neq 0 \end{cases} \\ \sum_k \widetilde{h}_k g_{k-2m} &= \sum_k \widetilde{g}_k h_{k-2m} = 0. \end{aligned} \tag{9.3}$$

In this section, we are interested in characterizing (9.2), (9.3) in terms of the Fourier series $H(\omega)$ and $G(\omega)$. To accomplish this goal, we start first with some familiar filters.

The Orthogonal Haar Filter Revisited

Before looking at the Fourier series of the orthogonal Haar filter, we state a basic identity that is used repeatedly in this section. This identity is easily established (see Problem 9.1).

$$\boxed{e^{i\omega+k\pi} = (-1)^k e^{i\omega}.} \tag{9.4}$$

Consider the orthogonal Haar filter $\mathbf{h} = (h_0, h_1) = \left(\frac{\sqrt{2}}{2}, \frac{\sqrt{2}}{2}\right)$. Using (8.19) we see the corresponding Fourier series can be simplified as

$$H(\omega) = \frac{\sqrt{2}}{2} + \frac{\sqrt{2}}{2}e^{i\omega} = \sqrt{2}e^{i\omega/2}\left(\frac{e^{-i\omega/2} + e^{i\omega/2}}{2}\right) = \sqrt{2}e^{i\omega/2}\cos\left(\frac{\omega}{2}\right). \tag{9.5}$$

We plot the 2π-periodic function

$$|H(\omega)|^2 = H(\omega)\overline{H(\omega)} = 2\cos^2\left(\frac{\omega}{2}\right) \tag{9.6}$$

in Figure 9.1(a). The plot in Figure 9.1(b) shows a left-shift by π units of $|H(\omega)|^2$ along with $|H(\omega)|^2$ and Figure 9.1(c) shows the sum of the functions in the previous plots.

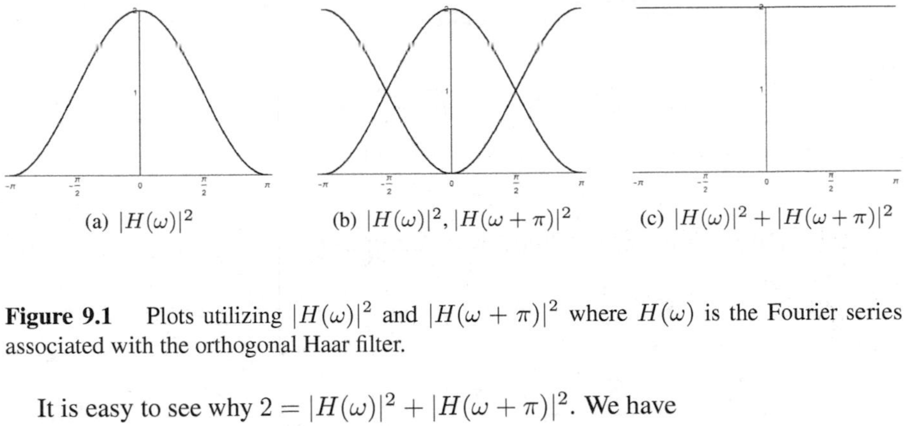

(a) $|H(\omega)|^2$ (b) $|H(\omega)|^2, |H(\omega+\pi)|^2$ (c) $|H(\omega)|^2 + |H(\omega+\pi)|^2$

Figure 9.1 Plots utilizing $|H(\omega)|^2$ and $|H(\omega+\pi)|^2$ where $H(\omega)$ is the Fourier series associated with the orthogonal Haar filter.

It is easy to see why $2 = |H(\omega)|^2 + |H(\omega+\pi)|^2$. We have

$$|H(\omega+\pi)|^2 = 2\cos^2(\omega+\pi) = 2\left(-\sin(\omega)\right)^2 = 2\sin^2(\omega)$$

so that

$$|H(\omega)|^2 + |H(\omega+\pi)|^2 = 2\cos^2(\omega) + 2\sin^2(\omega) = 2. \tag{9.7}$$

The Fourier series $G(\omega)$ is built from the filter $\mathbf{g} = (g_0, g_1) = \left(\frac{\sqrt{2}}{2}, -\frac{\sqrt{2}}{2}\right)$. We have

$$G(\omega) = \frac{\sqrt{2}}{2} - \frac{\sqrt{2}}{2}e^{i\omega} = \frac{\sqrt{2}}{2} + \frac{\sqrt{2}}{2}e^{i\pi}e^{i\omega} = \frac{\sqrt{2}}{2} + \frac{\sqrt{2}}{2}e^{i(\omega+\pi)} = H(\omega+\pi).$$

In this case, we use the above identity and the fact that $H(\omega)$ is 2π-periodic to write

$$\begin{aligned}|G(\omega)|^2 + |G(\omega+\pi)|^2 &= |H(\omega+\pi)|^2 + |H(\omega+2\pi)|^2\\ &= |H(\omega+\pi)|^2 + |H(\omega)|^2\\ &= 2.\end{aligned}$$

Finally, we use (9.5) and consider

$$\begin{aligned}H(\omega)\overline{G(\omega)} &= H(\omega)\overline{H(\omega+\pi)}\\ &= \sqrt{2}e^{i\omega/2}\cos\left(\frac{\omega}{2}\right)\cdot\sqrt{2}e^{-i(\omega+\pi)/2}\cos\left(\frac{\omega+\pi}{2}\right)\\ &= 2e^{-i\pi/2}\cos\left(\frac{\omega}{2}\right)\cos\left(\frac{\omega}{2}+\frac{\pi}{2}\right)\\ &= 2(-i)\cos\left(\frac{\omega}{2}\right)\left(-\sin\left(\frac{\omega}{2}\right)\right)\\ &= 2i\cos\left(\frac{\omega}{2}\right)\sin\left(\frac{\omega}{2}\right)\\ &= i\sin(\omega),\end{aligned} \tag{9.8}$$

where we have used the fact that $e^{-i\pi/2} = -i$ and the trigonometric identities $\cos\left(t+\frac{\pi}{2}\right) = -\sin(t)$ and $\sin(2t) = 2\cos(t)\sin(t)$. Using (9.8), we have

$$H(\omega+\pi)\overline{G(\omega+\pi)} = i\sin(\omega+\pi) = -i\sin(\omega)$$

so that

$$H(\omega)\overline{G(\omega)} + H(\omega+\pi)\overline{G(\omega+\pi)} = 0. \tag{9.9}$$

Let's try the same exercise on a biorthogonal filter pair.

The (5,3) Biorthogonal Spline Filter Pair Revisited

Let's look at the $(5,3)$ biorthogonal spline filter pair given in Definition 7.2. The associated Fourier series are

$$\widetilde{H}(\omega) = \frac{\sqrt{2}}{4}e^{-i\omega} + \frac{\sqrt{2}}{2} + \frac{\sqrt{2}}{4}e^{i\omega} \tag{9.10}$$

$$H(\omega) = -\frac{\sqrt{2}}{8}e^{-2i\omega} + \frac{\sqrt{2}}{4}e^{-i\omega} + \frac{3\sqrt{2}}{4} + \frac{\sqrt{2}}{4}e^{i\omega} - \frac{\sqrt{2}}{8}e^{2i\omega}. \tag{9.11}$$

For the orthogonal Haar filter, we established (9.6). We have two filters in this case, so we consider

$$\widetilde{H}(\omega)\overline{H(\omega)} + \widetilde{H}(\omega+\pi)\overline{H(\omega+\pi)}.$$

In Problem 9.2, you will show

$$\widetilde{H}(\omega)\overline{H(\omega)} = -\frac{1}{16}e^{-3i\omega} + \frac{9}{16}e^{-i\omega} + 1 + \frac{9}{16}e^{i\omega} - \frac{1}{16}e^{3i\omega}. \tag{9.12}$$

Plugging $w + \pi$ into (9.12) and making repeated use of (9.4) gives

$$\begin{aligned}\widetilde{H}(\omega+\pi)\overline{H(\omega+\pi)} &= -\frac{1}{16}e^{-3i(\omega+\pi)} + \frac{9}{16}e^{-i(\omega+\pi)} + 1 + \frac{9}{16}e^{i(\omega+\pi)} - \frac{1}{16}e^{3i(\omega+\pi)} \\ &= \frac{1}{16}e^{-3i\omega} - \frac{9}{16}e^{-i\omega} + 1 - \frac{9}{16}e^{i\omega} + \frac{1}{16}e^{3i\omega}.\end{aligned}$$

Using this last identity in conjunction with (9.12) we find that

$$\widetilde{H}(\omega)\overline{H(\omega)} + \widetilde{H}(\omega+\pi)\overline{H(\omega+\pi)} = 2. \tag{9.13}$$

Graphs consisting of the functions $\widetilde{H}(\omega)\overline{H(\omega)}$ and $\widetilde{H}(\omega+\pi)\overline{H(\omega+\pi)}$ are plotted in Figure 9.2.

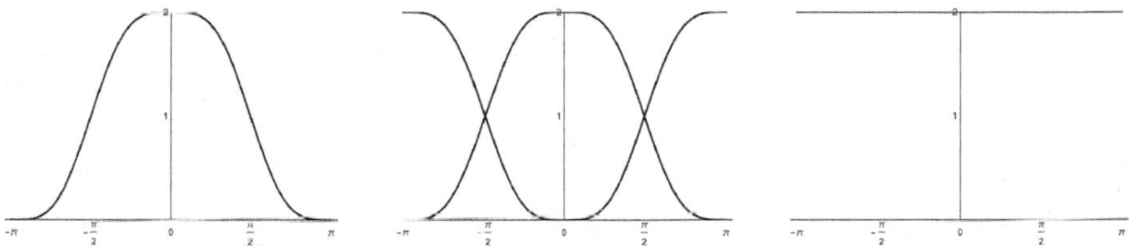

(a) $f(\omega) = \widetilde{H}(\omega)\overline{H(\omega)}$ (b) $f(\omega), f(\omega+\pi)$ (c) $f(\omega) + f(\omega+\pi)$

Figure 9.2 Plots utilizing $\widetilde{H}(\omega)$ and $H(\omega)$ where $\widetilde{H}(\omega)$ and $H(\omega)$ are the Fourier series associated with the $(5, 3)$ biorthogonal spline filter.

Recall that in order to construct the highpass filter $\widetilde{\mathbf{g}}$, we used the element-wise rule $\widetilde{g}_k = (-1)^k h_{1-k}$, $k = -1, \ldots, 3$. Then the Fourier series $\widetilde{G}(\omega)$ is

$$\widetilde{G}(\omega) = \sum_{k=-1}^{3}(-1)^k h_{1-k}e^{ik\omega} = \sum_{k=-1}^{3} e^{ik\pi}h_{1-k}e^{ik\omega} = \sum_{k=-1}^{3} h_{1-k}e^{ik(\omega+\pi)}.$$

Make the index substitution $j = 1 - k$ so that $k = 1 - j$. When $k = -1$, $j = 2$ and when $k = 3$, $j = -2$. The Fourier series becomes

$$\begin{aligned}\widetilde{G}(\omega) &= \sum_{j=2}^{-2} h_j e^{i(1-j)(\omega+\pi)} \\ &= e^{i(\omega+\pi)}\sum_{j=-2}^{2} h_j e^{-ij(\omega+\pi)} \\ &= -e^{i\omega}\overline{\sum_{j=-2}^{2} h_j e^{ij(\omega+\pi)}} \\ &= -e^{i\omega}\overline{H(\omega+\pi)}.\end{aligned} \tag{9.14}$$

In Problem 9.3, you will show that

$$G(\omega) = -e^{i\omega}\overline{\widetilde{H}(\omega+\pi)} \tag{9.15}$$

and in Problem 9.4, you will show that

$$\begin{aligned}\widetilde{G}(\omega)\overline{G(\omega)} + \widetilde{G}(\omega+\pi)\overline{G(\omega+\pi)} &= 2\\ \widetilde{H}(\omega)\overline{G(\omega)} + \widetilde{H}(\omega+\pi)\overline{G(\omega+\pi)} &= 0\\ \widetilde{G}(\omega)\overline{H(\omega)} + \widetilde{G}(\omega+\pi)\overline{H(\omega+\pi)} &= 0.\end{aligned} \tag{9.16}$$

Utilizing the Fourier Series Identities

For both the orthogonal Haar filter and the $(5,3)$ biorthogonal spline filter pair, we produced identities (9.7) and (9.13) that involve the Fourier series of the lowpass filters. Once Fourier series for the highpass filters were constructed, we created (9.9) for the Haar filter and the system (9.16) for the $(5,3)$ filter pair. What good are these identities?

It turns out the relationships completely characterize (bi)orthogonality in the Fourier domain! Let's look at an example.

Example 9.1 (Constructing the (5, 3) Spline Filter Pair Using Fourier Series). *Using* (9.13)*, construct the five-term symmetric lowpass filter* $\mathbf{h} = (h_{-2}, h_{-1}, h_0, h_1, h_2)$ *given that* $\widetilde{\mathbf{h}} = \left(\widetilde{h}_{-1}, \widetilde{h}_0, \widetilde{h}_1\right) = \left(\frac{\sqrt{2}}{4}, \frac{\sqrt{2}}{2}, \frac{\sqrt{2}}{4}\right)$.

Solution. *We start by writing the Fourier series associated with* $\mathbf{h}$.

$$\begin{aligned}H(\omega) &= h_{-2}e^{-2i\omega} + h_{-1}e^{-i\omega} + h_0 + h_1e^{i\omega} + h_2e^{2i\omega}\\ &= h_2e^{-2i\omega} + h_2e^{-i\omega} + h_0 + h_1e^{i\omega} + h_2e^{2i\omega}.\end{aligned} \tag{9.17}$$

Since $\mathbf{h}$ *is a lowpass filter we can use Proposition 8.4 to write*

$$0 = H(\pi) = h_2 - h_1 + h_0 - h_1 + h_2 = h_0 - 2h_1 + 2h_2. \tag{9.18}$$

Note the above identity is (7.7) from Section 7.1.

We next consider $\widetilde{H}(\omega)\overline{H(\omega)} = \widetilde{H}(\omega)H(\omega)$ *since* $\mathbf{h}$ *is odd-length and symmetric. In Problem 9.5 you will show that*

$$\widetilde{H}(\omega)H(\omega) = ae^{-3i\omega} + be^{-2i\omega} + ce^{-i\omega} + d + ce^{i\omega} + be^{2i\omega} + ae^{3i\omega} \tag{9.19}$$

where

$$\begin{aligned}a &= \frac{\sqrt{2}}{4}h_2\\ b &= \frac{\sqrt{2}}{4}h_1 + \frac{\sqrt{2}}{2}h_2\\ c &= \frac{\sqrt{2}}{4}h_0 + \frac{\sqrt{2}}{2}h_1 + \frac{\sqrt{2}}{4}h_2\\ d &= \frac{\sqrt{2}}{2}\left(h_0 + h_1\right).\end{aligned}$$

We use (9.19) *to write*

$$\widetilde{H}(\omega+\pi)H(\omega+\pi) = -ae^{-3i\omega}+be^{-2i\omega}-ce^{-i\omega}+\frac{\sqrt{2}}{2}(h_0+h_1)-ce^{i\omega}+be^{2i\omega}-ae^{3i\omega}.$$

Adding this result to (9.19) *and using* (9.13) *gives*

$$\begin{aligned} 2 &= \widetilde{H}(\omega)\overline{H(\omega)} + \widetilde{H}(\omega+\pi)\overline{H(\omega+\pi)} \\ &= \widetilde{H}(\omega)H(\omega) + \widetilde{H}(\omega+\pi)H(\omega+\pi) \\ &= 2be^{-2i\omega} + 2d + 2be^{2i\omega}. \end{aligned}$$

We can express the left side above as $0e^{-2i\omega} + 2 + 0e^{2i\omega}$ *and equating coefficients gives*

$$2 = 2d \qquad \text{and} \qquad 0 = 2b$$

or

$$1 = \frac{\sqrt{2}}{2}(h_0 + h_1) \qquad \text{and} \qquad 0 = \frac{\sqrt{2}}{4}h_1 + \frac{\sqrt{2}}{2}h_2.$$

These two equations are exactly (7.6) *and we can use them along with* (9.18) *to obtain the length 5 filter given in Definition 7.2.*

(Live example: Visit `stthomas.edu/wavelets` *and click on Live Examples.)*

□

Orthogonality in the Fourier Domain

We are ready for the main result of the section. It allows us to completely characterize (bi)orthogonality in the Fourier domain.

Theorem 9.1 (Orthogonality in the Fourier Domain). *Suppose* a, b *are bi-infinite sequences that satisfy*

$$\sum_{(j,k)\in\mathbb{Z}^2} \left|a_j b_k\right| < \infty.$$

Write the Fourier series series of each sequence as

$$A(\omega) = \sum_{k=-\infty}^{\infty} a_k e^{ik\omega} \qquad \text{and} \qquad B(\omega) = \sum_{k=-\infty}^{\infty} b_k e^{ik\omega}$$

and let

$$p(\omega) = A(\omega)\overline{B(\omega)} + A(\omega+\pi)\overline{B(\omega+\pi)}.$$

Then $p(\omega) = 2$ *if and only if*

$$\sum_k a_k b_{k-2m} = \begin{cases} 1, & m = 0 \\ 0, & m \neq 0 \end{cases} \tag{9.20}$$

and $p(\omega) = 0$ *if and only if*

$$\sum_k a_k b_{k-2m} = 0 \tag{9.21}$$

for all $m \in \mathbb{Z}$. □

Remark: The $\mathbb{Z}^2$ that appears in the hypothesis of Theorem 9.1 is the set of all ordered pairs whose elements are integers. The absolute convergence result is necessary to apply Fubini's theorem to interchange infinite sums in the proof that follows. We refer the interested reader to Rudin [78]. We apply the theorem to FIR filters so the Fourier series will actually be finite sums. The proof is cleaner though in the case where $\mathbf{a}$, $\mathbf{b}$ are bi-infinite sequences.

Proof. We start by expanding and then simplifying $p(\omega)$. For ease of notation, we use $\sum\limits_n$ for $\sum\limits_{n=-\infty}^{\infty}$. We have

$$\begin{aligned}
p(\omega) &= \sum_k a_k e^{ik\omega} \sum_j b_j e^{-ij\omega} + \sum_k a_k e^{ik(\omega+\pi)} \sum_j b_j e^{-ij(\omega+\pi)} \\
&= \sum_k a_k e^{ik\omega} \sum_j b_j e^{-ij\omega} + \sum_k a_k (-1)^k e^{ik\omega} \sum_j b_j (-1)^j e^{-ij\omega} \\
&= \sum_k \sum_j a_k b_j e^{i(k-j)\omega} + \sum_k \sum_j a_k b_j (-1)^{k+j} e^{i(k-j)\omega} \\
&= \sum_k \sum_j a_k b_j \left(1 + (-1)^{j+k}\right) e^{i(k-j)\omega}.
\end{aligned}$$

Now make the index substitution $m = k - j$ for the inner sum. Then $j = k - m$ and $j + k = k - m + k = 2k - m$. Note for fixed k, as j ranges over all integers, so will m. We have

$$\begin{aligned}
p(\omega) &= \sum_k \sum_j a_k b_j \left(1 + (-1)^{j+k}\right) e^{i(k-j)\omega} \\
&= \sum_k \sum_m a_k b_{k-m} \left(1 + (-1)^{2k-m}\right) e^{im\omega} \\
&= \sum_k \sum_m a_k b_{k-m} \left(1 + (-1)^m\right) e^{im\omega}.
\end{aligned}$$

When m is even, the term $(1 + (-1)^m) = 2$ and when m is odd $(1 + (-1)^m) = 0$. Consequently, we can just sum over even values of m. We can do this by letting m

run over all integers but changing m to $2m$ in the summand. Continuing we find that

$$\begin{aligned} p(\omega) &= \sum_k \sum_m a_k b_{k-m} \left(1 + (-1)^m\right) e^{im\omega} \\ &= \sum_k \sum_m 2 a_k b_{k-2m} e^{2im\omega} \\ &= \sum_m \left(2 \sum_k a_k b_{k-2m} \right) e^{2im\omega}. \end{aligned}$$

The fact that $\sum a_k b_j$ is absolutely convergent allows us to interchange the summations in the last line above. We see that $p(\omega)$ is a Fourier series with coefficients

$$p_{2m} = 2 \sum_k a_k b_{k-2m} \quad \text{and} \quad p_{2m+1} = 0.$$

The proof follows quickly from here. We have $p(\omega) = 0$ if and only if $p_m = 0$ for all m which gives (9.21). In addition, $p(\omega) = 2$ if and only if the constant term $p_0 = 2 \sum\limits_k a_k b_k = 2$ and all other $p_m = 0$. This gives (9.20) and the proof is complete. □

The following corollaries for lowpass (bi)orthogonal filters are immediate.

Corollary 9.1 (Orthogonal Filters in Terms of Fourier Series). *Suppose* $\mathbf{h} = (h_\ell, \ldots, h_L)$ *is an FIR filter of even length. Further assume the Fourier series* $H(\omega)$ *satisfies*

$$H(\omega)\overline{H(\omega)} + H(\omega + \pi)\overline{H(\omega + \pi)} = 2. \tag{9.22}$$

Then

$$\sum_{k=\ell}^{L} h_k^2 = 1$$

$$\sum_{k=\ell+2m}^{L} h_k h_{k-2m} = 0, \qquad m = 1, \ldots, \frac{L - \ell - 1}{2}.$$

□

Proof. Since $\widetilde{H}(\omega)$ satisfies (9.22), we can use Theorem 9.1 with $A(\omega) = B(\omega) = H(\omega)$ to write

$$\sum_{k=-\infty}^{\infty} h_k h_{k-2m} = \begin{cases} 1, & m = 0 \\ 0, & m \neq 0. \end{cases}$$

When $m = 0$, we can restrict the sum to $\ell, \ldots, L$ since those are the nonzero entries of $\mathbf{h}$. Thus $\sum\limits_{k=\ell}^{L} h_k^2 = 1$ as desired. To establish the result for $m = 1, \ldots, \frac{L-\ell-1}{2}$, we think of the sum as an inner product with fixed $\mathbf{h}$ and $2m$-shift of $\mathbf{h}$. When $m = 1$, the terms line up as displayed in Figure 9.3(a) and the result holds. We can continue shifting by two to the right (shifting to the left yields the same result). Since the length of $\mathbf{h}$ is even, it must be that ℓ and L cannot both be even or both be odd. Then $L - \ell - 1$ is an even number. The last shift aligns h_{L-1}, h_L on top with h_0, h_1 on the bottom (see Figure 9.3(b)). Since $L - 1 = \ell + (L - \ell - 1)$, we see that the last shift is $L - \ell + 1$ units. This occurs when $2m = L - \ell - 1$ or $m = \frac{L-\ell-1}{2}$ and the proof is complete.

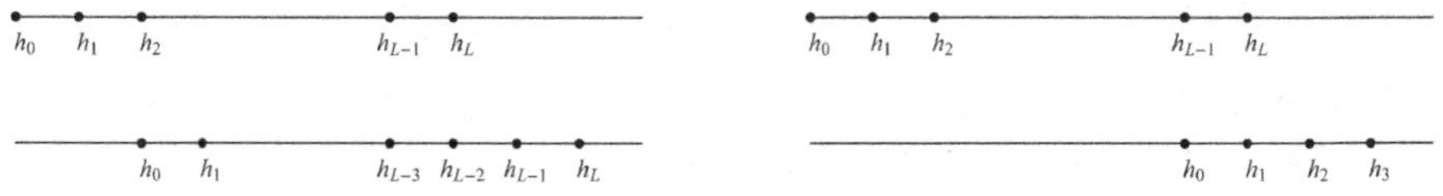

(a) $m = 1$ (b) $m = \frac{L-\ell-1}{2}$

Figure 9.3 Shifting inner products.

□

A similar result exists for biorthogonal filter pairs.

Corollary 9.2 (Symmetric Biorthogonal Filter Pairs in Terms of Fourier Series). *Suppose that* $\widetilde{\mathbf{h}} = \left(\widetilde{h}_{\widetilde{\ell}}, \ldots, \widetilde{h}_{\widetilde{L}}\right)$ *and* $\mathbf{h} = (h_\ell, \ldots, h_L)$ $(\widetilde{L} < L)$ *are FIR filters that satisfy*

$$\widetilde{H}(\omega)\overline{H(\omega)} + \widetilde{H}(\omega + \pi)\overline{H(\omega + \pi)} = 2. \tag{9.23}$$

Let $\widetilde{N}$, N *be the lengths of* $\widetilde{\mathbf{h}}$, $\mathbf{h}$, *respectively. If* $\widetilde{N}$, N *are both odd (even), then* $p = -\widetilde{L}$ $(p = -\widetilde{L} + 1)$ *and*

$$\sum_{k=p}^{\widetilde{L}} \widetilde{h}_k h_{k-2m} = \begin{cases} 1, & m = 0 \\ 0, & m \neq 0. \end{cases}$$

□

Proof. The proof is similar to that of Corollary 9.1 and left as Problem 9.8. □

Building $G(\omega)$ from $H(\omega)$

We conclude this section by demonstrating how to construct Fourier series for highpass filters once the lowpass filters are known. We start with the orthogonal filters from Chapter 5.

In Section 5.3, we defined our highpass filters $\mathbf{g}$ using the element-wise formula $g_k = (-1)^k h_{L-k}$ where $\mathbf{h}$ is an orthogonal lowpass filter and L is an odd integer. We can use this formula to construct the Fourier series for $\mathbf{g}$.

$$G(\omega) = \sum_{k=0}^{L} (-1)^k h_{L-k} e^{ik\omega} = \sum_{k=0}^{L} h_{L-k} e^{ik\pi} e^{ik\omega} = \sum_{k=0}^{L} h_{L-k} e^{ik(\omega+\pi)}.$$

Now make the index substitution $j = L - k$ so that $k = L - j$. When $k = 0$, $j = L$ and when $k = L$, $j = 0$. Inserting these values into the equation above gives

$$\begin{aligned} G(\omega) &= \sum_{k=0}^{L} h_{L-k} e^{ik(\omega+\pi)} \\ &= \sum_{j=L}^{0} h_j e^{i(L-j)(\omega+\pi)} \\ &= e^{iL(\omega+\pi)} \sum_{j=0}^{L} h_j e^{-ij(\omega+\pi)} \\ &= e^{iL\pi} e^{iL\omega} \overline{H(\omega+\pi)} \\ &= -e^{iL\omega} \overline{H(\omega+\pi)}. \end{aligned}$$

Similar representations exist for biorthogonal filter pairs. You can mimic the argument above and show that if $\mathbf{b}$ is defined element-wise as $b_k = (-1)^k a_{1-k}$, then $B(\omega) = -e^{i\omega} \overline{A(\omega+\pi)}$.

The above formulations of $G(\omega)$ lead to the following propositions. The first result tells that once we have an orthogonal lowpass filter $\mathbf{h}$ in hand, we can construct a highpass $\mathbf{g}$ so that all orthogonality conditions required of W_N are satisfied. The proposition that follows is analogous to the first but considers biorthogonal filter pairs.

Proposition 9.1 (Constructing $\mathbf{G(\omega)}$ From $\mathbf{H(\omega)}$). *Suppose* $\mathbf{h} = (h_\ell, \ldots, h_L)$ *is an even-length lowpass filter with Fourier series*

$$H(\omega) = \sum_{k=\ell}^{L} h_k e^{ik\omega}$$

that satisfies (9.22). *If* $G(\omega) = -e^{i\omega(L+\ell)} \overline{H(\omega+\pi)}$, *then*

$$G(\omega)\overline{G(\omega)} + G(\omega+\pi)\overline{G(\omega+\pi)} = 2 \tag{9.24}$$

$$H(\omega)\overline{G(\omega)} + H(\omega+\pi)\overline{G(\omega+\pi)} = 0. \tag{9.25}$$

Moreover, $\mathbf{g}$ *is a highpass filter whose are given element-wise by the formula*

$$g_k = (-1)^k h_{L+\ell-k}, \quad k = \ell, \ldots, L.$$

□

Proof. We start with

$$\begin{aligned} G(\omega)\overline{G(\omega)} &= \left(-e^{i\omega(L+\ell)}\overline{H(\omega+\pi)}\right)\overline{\left(-e^{i\omega(L+\ell)}\overline{H(\omega+\pi)}\right)} \\ &= \left(-e^{i\omega(L+\ell)}\overline{H(\omega+\pi)}\right)\left(-e^{-i\omega(L+\ell)}H(\omega+\pi)\right) \\ &= H(\omega+\pi)\overline{H(\omega+\pi)}. \end{aligned} \tag{9.26}$$

Plugging $\omega+\pi$ into (9.26), the last identity leads to

$$G(\omega+\pi)\overline{G(\omega+\pi)} = H(\omega+2\pi)\overline{H(\omega+2\pi)} = H(\omega)H(\omega+\pi).$$

Adding this result to (9.26) gives the desired result (9.24). To establish (9.25), we consider

$$H(\omega)\overline{G(\omega)} = H(\omega)\overline{\left(-e^{i\omega(L+\ell)}\overline{H(\omega+\pi)}\right)} = -e^{-i\omega(L+\ell)}H(\omega)H(\omega+\pi).$$

Since $\mathbf{h}$ is even-length, ℓ and L cannot both be even or both be odd. Since one is even and one is odd, their sum $L+\ell$ is odd. In this case $e^{-i\pi(L+\ell)} = -1$. We have

$$\begin{aligned} H(\omega+\pi)\overline{G(\omega+\pi)} &= -e^{-i(\omega+\pi)(L+\ell)}H(\omega+\pi)H(\omega+2\pi) \\ &= -e^{-i\pi(L+\ell)}e^{-i\omega(L+\ell)}H(\omega+\pi)H(\omega) \\ &= e^{-i\omega(L+\ell)}H(\omega)H(\omega+\pi) \\ &= -H(\omega)\overline{G(\omega)}. \end{aligned}$$

Thus $H(\omega)\overline{G(\omega)} + H(\omega+\pi)\overline{G(\omega+\pi)} = 0$ as desired.

Since $\mathbf{h}$ is lowpass, Proposition 8.4 tells us that $H(\pi) = 0$ and $H(0) \neq 0$. In this case

$$\begin{aligned} G(\pi) &= -e^{i\pi(L+\ell)}\overline{H(\pi+\pi)} = \overline{H(0)} \neq 0 \\ G(0) &= -e^{i\cdot 0(L+\ell)}\overline{H(\pi)} = -\overline{H(\pi)} = 0 \end{aligned}$$

so that by Proposition 8.5, $\mathbf{g}$ is a highpass filter. The derivation of the elements of $\mathbf{g}$ is similar to the argument that precedes this proposition and is thus left as Problem 9.9. □

Proposition 9.2 (Constructing $\mathbf{G}(\omega)$, $\widetilde{\mathbf{G}}(\omega)$ From $\mathbf{H}(\omega)$, $\widetilde{\mathbf{H}}(\omega)$). *Suppose* $\widetilde{\mathbf{h}} = \left(\widetilde{h}_{\widetilde{\ell}}, \ldots, \widetilde{h}_{\widetilde{L}}\right)$ *and* $\mathbf{h} = (h_\ell, \ldots, h_L)$ *are symmetric lowpass biorthogonal filters whose Fourier series* $\widetilde{H}(\omega)$ *and* $H(\omega)$*, respectively, satisfy* (9.23). *If*

$$\widetilde{G}(\omega) = -e^{i\omega}\overline{H(\omega+\pi)} \qquad \text{and} \qquad G(\omega) = -e^{i\omega}\overline{\widetilde{H}(\omega+\pi)}$$

then

$$\begin{aligned} \widetilde{G}(\omega)\overline{G(\omega)} + \widetilde{G}(\omega+\pi)\overline{G(\omega+\pi)} &= 2 \\ \widetilde{H}(\omega)\overline{G(\omega)} + \widetilde{H}(\omega+\pi)\overline{G(\omega+\pi)} &= 0 \\ \widetilde{G}(\omega)\overline{H(\omega)} + \widetilde{G}(\omega+\pi)\overline{H(\omega+\pi)} &= 0. \end{aligned}$$

Moreover, $\widetilde{\mathbf{g}}$ and $\mathbf{g}$ are highpass filters defined element-wise by

$$\widetilde{g}_k = (-1)^k h_{1-k}, \quad k = 1-L, \ldots, 1-\ell$$
$$g_k = (-1)\widetilde{h}_{1-k}, \quad k = 1-\widetilde{L}, \ldots, 1-\widetilde{\ell}.$$

□

Proof. The proof is similar to that of Proposition 9.1 and left as Problem 9.10. □

PROBLEMS

★9.1 Show that $e^{i(\omega+k\pi)} = (-1)^k e^{i\omega}$.

9.2 Use (9.10) and (9.11) to establish (9.12).

9.3 Let $\widetilde{\mathbf{h}} = \left(\widetilde{h}_{-1}, \widetilde{h}_0, \widetilde{h}_1\right) = \left(\frac{\sqrt{2}}{4}, \frac{\sqrt{2}}{2}, \frac{\sqrt{2}}{4}\right)$ and assume $\mathbf{g}$ is defined component-wise by $g_k = (-1)^k \widetilde{h}_{1-k}$, $k = 0, 1, 2$. Show that $G(\omega) = -e^{i\omega}\overline{\widetilde{H}(\omega+\pi)}$.

9.4 Use the Fourier series (9.10), (9.11), (9.14), and (9.15) to show that the equations (9.16) hold.

9.5 Using (9.10) and (9.17), establish the identity in (9.19).

9.6 Let $\mathbf{h} = (h_0, h_1)$. Then $H(\omega) = h_0 + h_1 e^{i\omega}$. Use (9.22) to build a system of equations that represent the orthogonality conditions satisfied by $\mathbf{h}$. Add to this system the condition that $H(\pi) = 0$ and then solve the system.

9.7 Repeat the previous problem using the D4 filter $\mathbf{h} = (h_0, h_1, h_2, h_3)$. *Hint:* The lowpass condition $H(\pi) = 0$ and the orthogonality relation (9.22) will only give you three of the four equations from the system (5.18). Any ideas how to express the fourth equation in terms of Fourier series? Proposition 8.6 might help.

9.8 Prove Corollary 9.2.

9.9 Complete the proof of Proposition 9.1. That is, for even-length filter $\mathbf{h} = (h_\ell, \ldots, h_L)$ and $G(\omega) = -e^{i\omega(L+\ell)}\overline{H(\omega+\pi)}$, show that the elements of $\mathbf{g}$ are given by the formula $g_k = (-1)^k h_{L+\ell-k}$, $k = \ell, \ldots, L$.

9.10 Prove Proposition 9.1.

9.11 Suppose $\widetilde{\mathbf{g}}$, $\mathbf{g}$ are the filters given in Proposition 9.2.

(a) Assume the lengths of $\widetilde{\mathbf{h}}$, $\mathbf{h}$ are even. Show that the range of indices for the elements of $\widetilde{\mathbf{g}}$ is $-L+1, \ldots, L$ and the range of indices for $\mathbf{g}$ is $-\widetilde{L}+1, \cdots, \widetilde{L}$.

(b) Assume the lengths of $\widetilde{\mathbf{h}}$, $\mathbf{h}$ are odd. Show that the range of indices for the elements of $\widetilde{\mathbf{g}}$ is $-L+1, \ldots, L+1$ and the range of indices for $\mathbf{g}$ is $-\widetilde{L}+1, \cdots, \widetilde{L}+1$.

Hint: Proposition 7.1 will be helpful.

9.12 Consider the filters[1]

$$\widetilde{\mathbf{h}} = (h_{-3}, h_{-2}, h_{-1}, h_0, h_1, h_2, h_3) = \left(-\frac{3\sqrt{2}}{280}, -\frac{3\sqrt{2}}{56}, \frac{73\sqrt{2}}{280}, \frac{17\sqrt{2}}{28}, \frac{73\sqrt{2}}{280}, -\frac{3\sqrt{2}}{56}, -\frac{3\sqrt{2}}{280}\right)$$

and

$$\mathbf{h} = (h_{-2}, h_{-1}, h_0, h_1, h_2) = \left(-\frac{\sqrt{2}}{20}, \frac{\sqrt{2}}{4}, \frac{3\sqrt{2}}{5}, \frac{\sqrt{2}}{4}, -\frac{\sqrt{2}}{20}\right).$$

(a) Show that $\widetilde{H}(0) = H(0) = \sqrt{2}$ and that $\widetilde{\mathbf{h}}$ and $\mathbf{h}$ are lowpass filters.

(b) Show that $\widetilde{H}(\omega)$ and $H(\omega)$ satisify the biorthogonality condition (9.23). A **CAS** might be helpful.

(c) Use Proposition 9.2 to find the highpass fiters $\widetilde{\mathbf{g}}$ and $\mathbf{g}$.

(d) Write down the wavelet matrices $\widetilde{W}_{10}$ and W_{10} and use a **CAS** to show that $\widetilde{W}_{10} W_{10}^T = I_N$. *Hint:* The matrices in (7.16) – (7.17), adjusted for the $(5, 7)$ filter pair, will be helpful.

9.13 In Proposition 9.2 we constructed the highpass filter $\mathbf{g}$ from the lowpass filter $\widetilde{\mathbf{h}}$ using the rule $g_k = (-1)^k \widetilde{h}_{1-k}$.

(a) Suppose that $\widetilde{\mathbf{h}}$ is a symmetric, odd-length filter. Show that $g_k = g_{2-k}$.

(b) Suppose that $\widetilde{\mathbf{h}}$ is a symmetric, even-length filter. Show that $g_k = -g_{1-k}$. We say this filter is *antisymmetric*.

9.14 Once an orthogonal lowpass filter $\mathbf{h}$ is determined, Proposition 9.1 tells us how to construct $G(\omega)$ and then $\mathbf{g}$. The choice for $G(\omega)$ is not unique.

(a) Suppose that n is an odd integer and b is a real number. Show that we can replace $G(\omega)$ in Proposition 9.1 with $G(\omega) = e^{i(n\omega+b)}\overline{H(\omega + \pi)}$ and the orthogonality relations (9.24), (9.25) still hold.

(b) What are the elements of $\mathbf{g}$ in this case?

(c) What are n and b for $G(\omega)$ as stated in Proposition 9.1?

9.2 Daubechies Filters

Our aim in this section is to rewrite the system (5.65) in terms of Fourier series. Most of the machinery to accomplish this task is in place.

[1]These filters are constructed in Daubechies [32]. The filter of length 5 is the well-known filter of Burt and Adelson [16] using in image encoding.

Daubechies Filters in the Fourier Domain

The system (5.65) for finding the Daubechies orthogonal filter $\mathbf{h} = (h_0, \ldots, h_L)$, L odd, consists of three parts. There are $\frac{L+1}{2}$ equations necessary to make the resulting wavelet matrix W_N orthogonal, a single condition to control the signs of the elements of $\mathbf{h}$ and finally $\frac{L+1}{2}$ equations to ensure the associated highpass filter $\mathbf{g}$ annihilates polynomial data of degree $0, \ldots, \frac{L-1}{2}$.

Corollary 9.1 describes how to rewrite the $\frac{L+1}{2}$ orthogonality conditions in terms of the Fourier series $H(\omega)$ of $\mathbf{h}$. We require

$$H(\omega)\overline{H(\omega)} + H(\omega+\pi)\overline{H(\omega+\pi)} = 2.$$

Given that

$$H(\omega) = \sum_{k=0}^{L} h_k e^{ik\omega}, \tag{9.27}$$

the redundancy condition in (5.65) can be obtained by evaluating $H(\omega)$ at $\omega = 0$ and setting the result equal to $\sqrt{2}$. Using (9.27) we have

$$H(0) = \sum_{k=0}^{L} h_k = \sqrt{2}.$$

Annihilating Polynomial Data and $\mathrm{H}(\omega)$

For the annihilation of constant data, the condition

$$\sum_{k=0}^{L}(-1)^k h_k = 0$$

is the lowpass condition on $H(\omega)$ from Proposition 8.4. Using (9.27), we see that

$$0 = H(\pi) = \sum_{k=0}^{L} h_k e^{ik\pi} = \sum_{k=0}^{L}(-1)^k h_k.$$

The remaining $\frac{L-1}{2}$ conditions are set so that the highpass filter $\mathbf{g}$ annihilates polynomial data of degree $1, \ldots, \frac{L-1}{2}$. Proposition 8.6 dealt with annihilating polynomial data in the Fourier domain so we start there. Assume $\mathbf{g}$ annihilates polynomial data of degree $m = 0, \ldots, \frac{L-1}{2}$ (we have already established this for $m = 0$ since $H(\pi) = G(0) = 0$). Then Proposition 8.6 says that

$$G^{(m)}(0) = 0, \quad m = 0, \ldots, \frac{L-1}{2}.$$

We want our new system in terms of the Fourier series $H(\omega)$, so we recall Proposition 9.1 and note that (with $\ell = 0$) $G(\omega) = -e^{iL\omega}\overline{H(\omega+\pi)}$. We solve this equation for $H(\omega)$ and start by replacing ω with $\omega+\pi$. We have

$$G(\omega+\pi) = -e^{iL(\omega+\pi)}\overline{H(\omega+2\pi)} = -e^{iL\pi}e^{iL\omega}\overline{H(\omega)} = e^{iL\omega}\overline{H(\omega)}.$$

Next we conjugate both sides and solve for $H(\omega)$.

$$\overline{G(\omega+\pi)} = e^{-iL\omega}H(\omega) \quad \Rightarrow \quad H(\omega) = e^{iL\omega}\overline{G(\omega+\pi)}.$$

We now compute the m-th derivative of $H(\omega)$ keeping in mind Problems 8.14 and 8.38, the latter saying the derivative of a conjugate is the conjugate of the derivative. We make use of the Leibniz rule for derivatives of products that is proved in Problem 5.30. We have

$$\begin{aligned} H^{(m)}(\omega) &= \sum_{n=0}^{m}\binom{m}{n}\left(e^{iL\omega}\right)^{(m-n)}\overline{G^{(n)}(\omega+\pi)} \\ &= \sum_{n=0}^{m}\binom{m}{n}(iL)^{m-n}e^{iL\omega}\overline{G^{(n)}(\omega+\pi)}. \end{aligned}$$

Evaluating the above identity at $\omega=\pi$ gives

$$\begin{aligned} H^{(m)}(\pi) &= \sum_{n=0}^{m}\binom{m}{n}(iL)^{m-n}e^{iL\pi}\overline{G^{(n)}(2\pi)} \\ &= -\sum_{n=0}^{m}\binom{m}{n}(iL)^{m-n}\overline{G^{(n)}(0)}. \end{aligned}$$

Since $G^{(n)}(0)=0$ for $n=0,\ldots,m$, we have $H^{(m)}(\pi)=0$. We can differentiate the Fourier series (9.27) m times to obtain

$$H^{(m)}(\omega) = \sum_{k=0}^{L} h_k(ik)^m e^{ik\omega}.$$

Plugging in $\omega=\pi$ gives

$$0 = H^{(m)}(\pi) = i^m\sum_{k=0}^{L} h_k k^m e^{ik\pi} = i^m\sum_{k=0}^{L}(-1)^k h_k k^m. \tag{9.28}$$

Note that the above identity is exactly the last set of equations in (5.65).

The Daubechies System in the Fourier Domain

Summarizing, we can write the system for finding Daubechies filter of length $L+1$, L odd, completely in terms of the Fourier series $H(\omega)$. We have

$$\boxed{\begin{array}{ll} H(\omega)\overline{H(\omega)} + H(\omega+\pi)\overline{H(\omega+\pi)} = 2 & \text{(orthogonality)} \\ H(0) = \sqrt{2} & \text{(redundancy condition)} \\ H(\pi) = 0 & \text{(constant data)} \\ H^{(m)}(\pi) = 0, \quad m = 1,\ldots,\frac{L-1}{2} & \text{(annihilate polynomial data)} \end{array}} \tag{9.29}$$

The system (9.29) is quadratic and must be solved on a computer for $L > 5$. In Section 5.3, we outlined a method for selecting a particular solution that was of interest to Daubechies. In [29], she describes a much more efficient way to obtain this particular solution.

Let's look at an example utilizing (9.29).

Example 9.2 (Using Fourier Series to Find the D6 Filter). *Set up the system necessary to find Daubechies D6 filter* $\mathbf{h} = (h_0, \ldots, h_5)$ *using* (9.29).

Solution.

Our goal is to reproduce the system (5.44). *Let* $H(\omega) = \sum_{k=0}^{5} h_k e^{ik\omega}$. *With a tedious amount of algebra, we find that*

$$\begin{aligned} p(\omega) &= H(\omega)\overline{H(\omega)} + H(\omega+\pi)\overline{H(\omega+\pi)} \\ &= ae^{-4i\omega} + be^{-2i\omega} + c + be^{2i\omega} + ae^{4i\omega} \end{aligned}$$

where

$$\begin{aligned} a &= 2h_0h_4 + 2h_1h_5 \\ b &= 2h_0h_2 + 2h_1h_3 + 2h_2h_4 + 2h_3h_5 \\ c &= 2h_0^2 + \cdots + 2h_5^2. \end{aligned}$$

The orthogonality condition says that $p(\omega) = 2$ *and equating coefficients leads to the desired orthogonality equations that are the first three equations in* (5.44). *Computing* $H(\pi) = 0$ *gives*

$$h_0 - h_1 + h_2 - h_3 + h_4 - h_5 = 0$$

which is the fourth equation in (5.44). *The last two equations in* (5.44) *can be represented in the Fourier domain by setting the first and second derivatives of* $H(\omega)$ *equal to zero. We have*

$$\begin{aligned} H'(\omega) &= i\sum_{k=1}^{5} kh_k e^{ik\omega} \\ H''(\omega) &= -\sum_{k=1}^{5} k^2 h_k e^{ik\omega}. \end{aligned}$$

Plugging in $\omega = \pi$ *above gives*

$$\begin{aligned} 0 = H'(\pi) &= i\sum_{k=1}^{5} (-1)^k kh_k \\ 0 = H''(\pi) &- \sum_{k=1}^{5} (-1)^k k^2 h_k. \end{aligned}$$

□

It is interesting to examine the effects of (9.28) on $|H(\omega)|$. We have plotted $|H(\omega)|$ in Figure 9.4 below for different filter lengths. As one would expect, the higher the order of derivatives set to zero at $\omega = \pi$, the flatter the curve $|H(\omega)|$ near $\omega = \pi$.

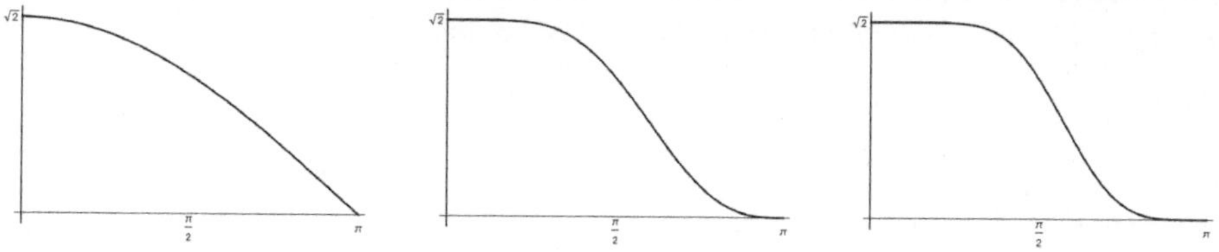

(a) Haar (b) D6 (c) D10

Figure 9.4 Modulus of the Fourier series constructed from the Haar, D6 and D10 filters.

PROBLEMS

9.15 Repeat Example 9.2 for the D4 filter.

9.16 Suppose $\mathbf{h}$ satisfies (9.29). Show that $-\mathbf{h}$ does as well.

9.17 In this problem, you will show that if $\mathbf{h} = (h_0, \ldots, h_L)$, L odd, solves (9.29), then so does its reflection $\mathbf{h}^r = (h_L, \ldots, h_0)$. The following steps will help you organize your work.

(a) Note that $h_k^r = h_{L-k}$ so that $H_r(\omega) = \sum\limits_{k=0}^{L} h_{L-k}e^{ik\omega}$. Make the index substitution $j = L - k$ to show that $H_r(\omega) = e^{iL\omega}\overline{H(\omega)}$.

(b) Use (a) to show the orthogonality condition and lowpass condition in (9.29) hold.

(c) Use Leibniz' rule (Problem 5.30) to differentiate $H_r^{(m)}(\omega)$ and then use this formula and the fact that $H^{(m)}(\pi) = 0$, $m = 1, \ldots, \frac{L-1}{2}$ to show that $H_r^{(m)}(\pi) = 0$.

9.18 Suppose $\mathbf{h}$ satisfies the orthogonality and constant data conditions of (9.29). Show that $H(0) = \pm\sqrt{2}$.

9.3 Coiflet Filters

In Section 9.2 we learned how to combine the orthogonality conditions and derivative conditions on the Fourier series $H(\omega)$ of the filter $\mathbf{h}$ to construct Daubechies' family of orthogonal filters.

For filter $\mathbf{h} = (h_0, \ldots, h_L)$, we insist that

$$H^{(m)}(\pi) = 0, \qquad m = 0, \ldots, \frac{L-1}{2}. \tag{9.30}$$

The derivative conditions in (9.30) "flatten" $|H(\omega)|$ near $\omega = \pi$. Such a filter will typically annihilate or attenuate high oscillations in the data.

In this section, we utilize the Fourier series to identify other desirable conditions satisfied by the lowpass filter $\mathbf{h}$ and in so doing, create a new family of orthogonal filters. The new conditions, suggested by Ronald Coifman, led Daubechies to develop the so-called Coiflet filters (see [32]). We give Daubechies' general formulation and explicitly construct two members of the Coiflet filter family.

Flattening the Fourier Series $H(\omega)$ at $\omega = 0$

It is worthwhile to think about modifying our filter further to better model the trends in our input vector. For example, is it possible to modify the filter so that the averages portion of the output does a better job approximating the original input? Daubechies considered this question [30] and her solution was to impose derivative conditions on $H(\omega)$ at $\omega = 0$. That is, we construct an orthogonal filter $\mathbf{h}$ that satisfies (9.30) and

$$H^{(m)}(0) = 0, \qquad m = 1, 2, \ldots \tag{9.31}$$

We will be more specific about the number of derivative conditions in the theorem that follows.

The conditions given in (9.31) will have the same effect on $|H(\omega)|$ at $\omega = 0$ as (9.30) has on $|H(\omega)|$ at $\omega = \pi$. As a result, the averages portion of the wavelet transformation should do a good job of modeling near-constant trends in the input data.

A Theorem for Constructing Coiflet Filters

Before we state the theorem for constructing Coiflet filters, we give a lemma that is used in the proof of both Theorems 9.2 and 9.3.

Lemma 9.1 (A Useful Trigonometric Identity). *For any nonnegative integer K, we have*

$$1 = \cos^{2K}\left(\frac{\omega}{2}\right) \sum_{j=0}^{K-1} \binom{K-1+j}{j} \sin^{2j}\left(\frac{\omega}{2}\right) + \sin^{2K}\left(\frac{\omega}{2}\right) \sum_{j=0}^{K-1} \binom{K-1+j}{j} \cos^{2j}\left(\frac{\omega}{2}\right). \tag{9.32}$$

□

Note: The binomial coefficient $\binom{n}{j}$ is defined in Problem 5.30.

Proof. We give one proof here[2] and outline a second proof in Problem 9.20. For convenience, we set $c = \cos\left(\frac{\omega}{2}\right)$ and $s = \sin\left(\frac{\omega}{2}\right)$. We then write the right hand side of (9.32) as

$$F_K = c^{2K} \sum_{j=0}^{K-1} \binom{K-1+j}{j} s^{2j} + s^{2K} \sum_{j=0}^{K-1} \binom{K-1+j}{j} c^{2j}. \tag{9.33}$$

We will show that $F_K = F_{K-1}$ for any positive integer K. Combining this with the fact that in the $K = 1$ case, we have $\cos^2 \omega + \sin^2 \omega = 1$, completes the proof. We compute

$$\begin{aligned} F_K - F_{K-1} = c^{2K-2} &\left(c^2 \sum_{j=0}^{K-1} \tbinom{K-1+j}{j} s^{2j} - \sum_{j=0}^{K-2} \tbinom{K-2+j}{j} s^{2j} \right) \\ + s^{2K-2} &\left(s^2 \sum_{j=0}^{K-1} \tbinom{K-1+j}{j} c^{2j} - \sum_{j=0}^{K-2} \tbinom{K-2+j}{j} c^{2j} \right). \end{aligned} \tag{9.34}$$

Let's consider the first term on the right side of (9.34). Let's call it P_K. We replace c^2 with $1 - s^2$, simplify, and write

$$P_K = c^{2K-2} \left(\sum_{j=0}^{K-1} \tbinom{K-1+j}{j} s^{2j} - \sum_{j=0}^{K-1} \tbinom{K-1+j}{j} s^{2j+2} - \sum_{j=0}^{K-2} \tbinom{K-2+j}{j} s^{2j} \right).$$

There are three summations in the identity above. We will group like terms for $j = 1$ to $K - 2$. For the first summation, we have

$$\sum_{j=0}^{K-1} \tbinom{K-1+j}{j} s^{2j} = 1 + \sum_{j=1}^{K-2} \tbinom{K-1+j}{j} s^{2j} + \tbinom{2K-2}{K-1} s^{2K-2}. \tag{9.35}$$

For the second summation, we replace j with $j - 1$ and write

$$\begin{aligned} \sum_{j=0}^{K-1} \tbinom{K-1+j}{j} s^{2j+2} &= \sum_{j=1}^{K} \tbinom{K-2+j}{j-1} s^{2j} \\ &= \sum_{j=1}^{K-2} \tbinom{K-2+j}{j-1} s^{2j} + \tbinom{2K-3}{K-2} s^{2K-2} + \tbinom{2K-2}{K-1} s^{2K}. \end{aligned} \tag{9.36}$$

[2]This proof was shown to me by Yongzhi Yang.

The third summation is

$$\sum_{j=0}^{K-2} \binom{K-2+j}{j} s^{2j} = 1 + \sum_{j=1}^{K-2} \binom{K-2+j}{j} s^{2j}. \tag{9.37}$$

Subtracting (9.36) and (9.37) from (9.35) gives

$$P_K = c^{2K-2} \left(\sum_{j=1}^{K-2} \left(\binom{K-1+j}{j} - \binom{K-2+j}{j-1} - \binom{K-2+j}{j} \right) s^{2j} \right.$$
$$\left. + \left(\binom{2K-2}{K-1} - \binom{2K-3}{K-2} \right) s^{2K-2} - \binom{2K-2}{K-1} s^{2K} \right). \tag{9.38}$$

Now by Pascal's identity (7.31),

$$\binom{K-1+j}{j} - \binom{K-2+j}{j-1} - \binom{K-2+j}{j} = 0$$

so (9.38) reduces to

$$P_K = c^{2K-2} \left(\left[\binom{2K-2}{K-1} - \binom{2K-3}{K-2} \right] s^{2K-2} - \binom{2K-2}{K-1} s^{2K} \right). \tag{9.39}$$

Using Problem 9.19, we see that

$$\binom{2K-2}{K-1} - \binom{2K-3}{K-2} = \frac{1}{2} \binom{2K-2}{K-1}.$$

Inserting this identity into (9.39) gives

$$\begin{aligned} P_K &= \frac{1}{2} \binom{2K-2}{K-1} c^{2K-2} s^{2K-2} - \binom{2K-2}{K-1} c^{2K-2} s^{2K} \\ &= \binom{2K-2}{K-1} (cs)^{2K-2} \left(\frac{1}{2} - s^2 \right). \end{aligned}$$

In a similar manner, if we let Q_K denote the second term in (9.34), we have

$$Q_K = \binom{2K-2}{K-1} (cs)^{2K-2} \left(\frac{1}{2} - c^2 \right).$$

Finally, we add P_K and Q_K to obtain

$$\begin{aligned} F_K - F_{K-1} &= P_K + Q_K \\ &= \binom{2K-1}{K-1} (cs)^{2K-2} \left(\frac{1}{2} - s^2 + \frac{1}{2} - c^2 \right) \\ &= \binom{2K-1}{K-1} (cs)^{2K-2} \left(1 - s^2 - c^2 \right) \\ &= 0. \end{aligned}$$

□

The result that follows is due to Daubechies [30] and suggests a way to construct filters $\mathbf{h}$ whose Fourier series $H(\omega)$ are "flattened" at $\omega = 0$. We do not give a complete proof of the result, as part of the proof is beyond the scope of this book.

Theorem 9.2 (Coiflet Filters). *Let $K = 1, 2, \ldots$ and define the 2π-periodic function $H(\omega)$ as*

$$H(\omega) = \sqrt{2}\cos^{2K}\left(\frac{\omega}{2}\right) \cdot \left(\sum_{j=0}^{K-1} \binom{K-1+j}{j} \sin^{2j}\left(\frac{\omega}{2}\right) + \sin^{2K}\left(\frac{\omega}{2}\right) \sum_{\ell=0}^{2K-1} a_\ell\, e^{i\ell\omega} \right). \tag{9.40}$$

Then

$$\begin{aligned} H(0) &= \sqrt{2} \\ H^{(m)}(0) &= 0, \qquad m = 1, \ldots, 2K-1, \\ H^{(m)}(\pi) &= 0, \qquad m = 0, \ldots, 2K-1. \end{aligned} \tag{9.41}$$

Furthermore, real numbers $a_0, \ldots, a_{2K-1}$ exist so that

$$|H(\omega)|^2 + |H(\omega + \pi)|^2 = 2. \tag{9.42}$$

□

Discussion of the Proof of Theorem 9.2. There are two parts to the proof. The first part is to show that the Fourier series $H(\omega)$ given in (9.40) satisfies the conditions (9.41). We provide this proof below. The second part of the proof shows the existence of real number $a_0, \ldots, a_{2K-1}$ so that (9.40) satisfies (9.42). Recall from Corollary 9.1 that if $H(\omega)$ satisfies (9.42), then the corresponding filter $\mathbf{h}$ satisfies

$$\sum_{k=\ell+2m}^{L} h_k h_{k-2m} = \begin{cases} 1, & m = 0 \\ 0, & m = 1, \ldots, \frac{L-\ell-1}{2}. \end{cases}$$

The proof that $H(\omega)$ satisfies (9.42) is quite technical and requires some advanced mathematics that we do not cover in this book. The reader with some background in analysis is encouraged to look at Daubechies [30].

To show that $H^{(m)}(\pi) = 0$ for $m = 0, \ldots, 2K-1$, we note that

$$\cos\left(\frac{\omega}{2}\right) = \frac{e^{i\omega/2} + e^{-i\omega/2}}{2} = \frac{1}{2} e^{-i\omega/2}\left(e^{i\omega} + 1\right). \tag{9.43}$$

If we insert (9.43) into (9.40), we have

$$H(\omega) = \frac{\sqrt{2}}{2^{2K} e^{iK\omega}} (1 + e^{i\omega})^{2K} \cdot \left(\sum_{j=0}^{K-1} \binom{K-1+j}{j} \sin^{2j}(\frac{\omega}{2}) + \sin^{2K}(\frac{\omega}{2}) \sum_{\ell-0}^{2K-1} a_\ell\, e^{i\ell\omega} \right).$$

Since $H(\omega)$ contains the factor $(1+e^{i\omega})^{2K}$, we see that it has a root of multiplicity at least $2K$ at $\omega=\pi$. Using the ideas from Problem 5.31 from Section 5.3, we infer that $H^{(m)}(\pi)=0$, $m=0,\ldots,2K-1$.

Using Lemma 9.1 and Problem 9.21, we rewrite (9.40) as

$$\begin{aligned} H(\omega) &= \sqrt{2}+\sqrt{2}\sin^{2K}\left(\frac{\omega}{2}\right) \\ &\cdot\left(-\sum_{j=0}^{K-1}\binom{K-1+j}{j}\cos^{2j}\left(\frac{\omega}{2}\right)+\cos^{2K}\left(\frac{\omega}{2}\right)\sum_{\ell=0}^{2K-1}a_\ell\, e^{i\ell\omega}\right). \end{aligned} \tag{9.44}$$

We rewrite $\sin\left(\frac{\omega}{2}\right)$ as

$$\sin\left(\frac{\omega}{2}\right)=\frac{e^{i\omega/2}-e^{-i\omega/2}}{2i}=\frac{1}{2i}e^{-i\omega/2}\left(e^{i\omega}-1\right)=\frac{i}{2}e^{-i\omega/2}\left(1-e^{i\omega}\right). \tag{9.45}$$

If we insert (9.45) into (9.44), we have

$$\begin{aligned} H(\omega) &= \sqrt{2}+\sqrt{2}\left(\frac{i}{2}\right)^{2K}e^{-iK\omega}\left(1-e^{i\omega}\right)^{2K} \\ &\cdot\left(-\sum_{j=0}^{K-1}\binom{K-1+j}{j}\cos^{2j}\left(\frac{\omega}{2}\right)+\cos^{2K}\left(\frac{\omega}{2}\right)\sum_{\ell=0}^{2K-1}a_\ell\, e^{i\ell\omega}\right). \end{aligned}$$

Now if we let $Q(\omega)=H(\omega)-\sqrt{2}$, we see that $Q(\omega)$ contains the factor $(1-e^{i\omega})^{2K}$ and thus has a root of multiplicity at least $2K$ at $\omega=0$. Thus, $Q^{(m)}(0)=0$ for $m=0,\ldots,2K-1$.

For $m=0$, we have $Q(0)=H(0)-\sqrt{2}=0$ or $H(0)=\sqrt{2}$. It is easy to see that $Q^{(m)}(\omega)=H^{(m)}(\omega)$ for $m=1,\ldots,2K-1$, so we can infer that $H^{(m)}(0)=0$ for $m=1,\ldots,2K-1$. □

The Coiflet Filter for $K=1$

Let's use Theorem 9.2 to find a Coiflet filter for the case $K=1$. Using (9.40), we see that

$$\begin{aligned} H(\omega) &= \sqrt{2}\cos^2\left(\frac{\omega}{2}\right)\left(1+\sin^2\left(\frac{\omega}{2}\right)\sum_{\ell=0}^{1}a_\ell e^{i\ell\omega}\right) \\ &= \sqrt{2}\Big(\cos^2\left(\frac{\omega}{2}\right)+\cos^2\left(\frac{\omega}{2}\right)\sin^2\left(\frac{\omega}{2}\right)\left(a_0+a_1e^{i\omega}\right)\Big) \\ &= \sqrt{2}\Big(\cos^2\left(\frac{\omega}{2}\right)+\frac{\sin^2(\omega)}{4}\left(a_0+a_1e^{i\omega}\right)\Big) \end{aligned} \tag{9.46}$$

where we use the double-angle formula $\sin(2\omega)=2\sin(\omega)\cos(\omega)$ in the last step.

We use Problem 8.40 to write $\cos^2\left(\frac{\omega}{2}\right) = \frac{1}{4}\left(e^{i\omega} + 2 + e^{-i\omega}\right)$ and Problem 8.41 to write $\sin^2(\omega) = \frac{1}{4}\left(-e^{2i\omega} + 2 - e^{-2i\omega}\right)$. We insert these identities into (9.46) and expand to obtain a Fourier series for $H(\omega)$:

$$\begin{aligned}H(\omega) &= \sqrt{2}\Big(\frac{1}{4}(e^{i\omega} + 2 + e^{-i\omega}) + \frac{1}{16}(-e^{2i\omega} + 2 - e^{-2i\omega})(a_0 + a_1 e^{i\omega})\Big)\\ &= \frac{\sqrt{2}}{16}\Big(- a_0 e^{-2i\omega} + (4 - a_1)e^{-i\omega} + (8 + 2a_0) + (4 + 2a_1)e^{i\omega}\\ &\qquad - a_0 e^{2i\omega} - a_1 e^{3i\omega}\Big).\end{aligned}$$

If we gather like terms in the identity above, we have the following for $H(\omega)$:

$$\begin{aligned}H(\omega) = &-\frac{\sqrt{2}\,a_0}{16}\,e^{-2i\omega} + \frac{\sqrt{2}(4 - a_1)}{16}\,e^{-i\omega} + \frac{\sqrt{2}(8 + 2a_0)}{16}\\ &+ \frac{\sqrt{2}(4 + 2a_1)}{16}\,e^{i\omega} - \frac{\sqrt{2}\,a_0}{16}\,e^{2i\omega} - \frac{\sqrt{2}\,a_1}{16}\,e^{3i\omega}.\end{aligned} \tag{9.47}$$

We can now read our filter coefficients from (9.47). We have

$$\begin{aligned}h_{-2} &= \frac{-\sqrt{2}\,a_0}{16} & h_{-1} &= \frac{\sqrt{2}(4 - a_1)}{16} & h_0 &= \frac{\sqrt{2}(8 + 2a_0)}{16}\\ h_1 &= \frac{\sqrt{2}(4 + 2a_1)}{16} & h_2 &= \frac{-\sqrt{2}\,a_0}{16} & h_3 &= \frac{-\sqrt{2}\,a_1}{16}.\end{aligned} \tag{9.48}$$

Now we must find a_0 and a_1 so that the (noncausal) filter

$$\mathbf{h} = (h_{-2}, h_{-1}, h_0, h_1, h_2, h_3)$$

satisfies the orthogonality condition (9.42). At this point, we can use Corollary 9.1 (with $\ell = -2$ and $L = 3$) to see that the filter must satisfy

$$\begin{aligned}h_{-2}^2 + h_{-1}^2 + h_0^2 + h_1^2 + h_2^2 + h_3^2 &= 1\\ \sum_{k=-2+2m}^{3} h_k h_{k-2m} &= 0, \qquad m = 1, 2,\end{aligned}$$

or

$$\begin{aligned}h_{-2}^2 + h_{-1}^2 + h_0^2 + h_1^2 + h_2^2 + h_3^2 &= 1\\ h_{-2}h_0 + h_{-1}h_1 + h_0 h_2 + h_1 h_3 &= 0\\ h_{-2}h_2 + h_{-1}h_3 &= 0.\end{aligned} \tag{9.49}$$

If we insert the values from (9.48) into (9.49) and simplify, we obtain the following system (you are asked to verify the algebra in Problem 9.22) with unknowns a_0, a_1:

$$\begin{aligned}16a_0 + 3a_0^2 + 4a_1 + 3a_1^2 &= 16\\ 4 - 4a_0 - a_0^2 - a_1^2 &= 0\\ -4a_1 + a_0^2 + a_1^2 &= 0.\end{aligned} \tag{9.50}$$

Adding the last two equations of (9.50) together and solving for a_0 gives $a_0 = 1 - a_1$. Putting $a_0 = 1 - a_1$ into the first equation of (9.50), we obtain

$$
\begin{aligned}
16(1-a_1) + 3(1-a_1)^2 + 4a_1 + 3a_1^2 &= 16 \\
2a_1^2 - 6a_1 + 1 &= 0
\end{aligned}
$$

so that $a_1 = \frac{1}{2}(3 \pm \sqrt{7})$. This leads to the following two solutions to (9.50):

$$
\begin{aligned}
a_0 &= \frac{1}{2}(\sqrt{7}-1), \quad a_1 = \frac{1}{2}(3-\sqrt{7}) \\
a_0 &= -\frac{1}{2}(\sqrt{7}+1), \quad a_1 = \frac{1}{2}(3+\sqrt{7}).
\end{aligned}
\tag{9.51}
$$

Coiflet Length 6 Filter

We can plug either of these solutions into (9.48) to find the filter coefficients. For the first pair in (9.51), we have

$$
\boxed{
\begin{aligned}
h_{-2} &= \tfrac{1}{16\sqrt{2}}(1-\sqrt{7}), & h_{-1} &= \tfrac{1}{16\sqrt{2}}(5+\sqrt{7}) \\
h_0 &= \tfrac{1}{8\sqrt{2}}(7+\sqrt{7}), & h_1 &= \tfrac{1}{8\sqrt{2}}(7-\sqrt{7}) \\
h_2 &= \tfrac{1}{16\sqrt{2}}(1-\sqrt{7}), & h_3 &= \tfrac{1}{16\sqrt{2}}(-3+\sqrt{7}).
\end{aligned}
}
\tag{9.52}
$$

In Figure 9.5 we plot $|H(\omega)|$ for the Coiflet filter given in (9.52). You are asked to find the second solution in Problem 9.23. We use (9.52) to formulate our next definition.

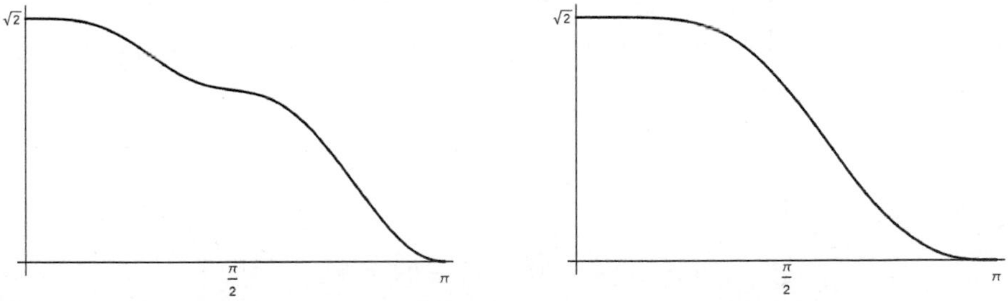

(a) C6 Filter (b) D6 Filter

Figure 9.5 Plots of $|H(\omega)|$ for the C6 and D6 filters.

Definition 9.1 (Coiflet Length 6 Filter). *We define the* Coiflet six-term orthogonal filter *by the values given in* (9.52). *We call this filter as the* C6 filter. □

An Alternative Method for Finding the C6 Filter

You can probably imagine that to find $\mathbf{h}$ for $K = 2, 3, \ldots$, the steps (analogous to those used to find the Coiflet six-term filter) are quite tedious. The number of unknowns a_ℓ in the Fourier series grows as K increases and once we express the filter in terms of the a_ℓ, we must insert these identities into the orthogonality condition (9.42) and solve the resulting quadratic system. We present an alternative form of solution that at least cuts out the need to solve for the a_ℓ. We use the $K = 1$ case to illustrate.

Expanding $H(\omega)$ in the $K = 1$ case reveals that the Fourier series has the form given by (9.47). Rather than worrying about finding the values of a_0 and a_1, let's instead note that the Fourier series starts at $\ell = -2$ and terminates at $L = 3$ and try to find numbers $h_{-2}, \ldots, h_3$, that satisfy the orthogonality conditions (9.49) and the derivative conditions (9.41). We have written the orthogonality conditions in (9.49), so let's write down the equations for the derivative conditions. We have

$$H(\omega) = h_{-2}\, e^{-2i\omega} + h_{-1}\, e^{-i\omega} + h_0 + h_1\, e^{i\omega} + h_2\, e^{2i\omega} + h_3\, e^{3i\omega}. \tag{9.53}$$

If we plug in $\omega = 0$ into (9.53) and set the result equal to $\sqrt{2}$, we have

$$h_{-2} + h_{-1} + h_0 + h_1 + h_2 + h_3 = \sqrt{2}.$$

Inserting $\omega = \pi$ into (9.53) and setting the result equal zero gives

$$h_{-2} - h_{-1} + h_0 - h_1 + h_2 - h_3 = 0.$$

Differentiating (9.53) gives

$$H'(\omega) = -2ih_{-2}\, e^{-2i\omega} - ih_{-1}\, e^{-i\omega} + ih_1\, e^{i\omega} + 2ih_2\, e^{2i\omega} + 3ih_3\, e^{3i\omega}.$$

The derivative conditions (9.41) are $H'(0) = H'(\pi) = 0$. Plugging 0 and π into $H'(\omega)$, simplifying, and setting the results to zero gives

$$\begin{aligned} -2h_{-2} - h_{-1} + h_1 + 2h_2 + 3h_3 &= 0 \\ -2h_{-2} + h_{-1} - h_1 + 2h_2 - 3h_3 &= 0. \end{aligned}$$

Thus, the system we seek to solve is

$$\begin{aligned} h_{-2}^2 + h_{-1}^2 + h_0^2 + h_1^2 + h_2^2 + h_3^2 &= 1 \\ h_{-2}h_0 + h_{-1}h_1 + h_0h_2 + h_1h_3 &= 0 \\ h_{-2}h_2 + h_{-1}h_3 &= 0 \\ h_{-2} + h_{-1} + h_0 + h_1 + h_2 + h_3 &= \sqrt{2} \\ h_{-2} - h_{-1} + h_0 - h_1 + h_2 - h_3 &= 0 \\ -2h_{-2} - h_{-1} + h_1 + 2h_2 + 3h_3 &= 0 \\ -2h_{-2} + h_{-1} - h_1 + 2h_2 - 3h_3 &= 0. \end{aligned} \tag{9.54}$$

Table 9.1 Two Coiflet filters for the case $K = 1$.

	Solution 1[a]	**Solution 2**
h_{-2}	-0.072733	0.161121
h_{-1}	0.337898	0.104044
h_0	0.852572	0.384865
h_1	0.384865	0.852572
h_2	-0.072733	0.161121
h_3	-0.015656	-0.249509

[a]This is the Daubechies C6 filter.

If we use a **CAS** to solve (9.54), we obtain the two solutions (rounded to six digits) listed in Table 9.1.

It is easy to check that the first system is the C6 filter given in (9.52) and that the second solution comes from plugging the second solution pair from (9.51) into (9.48) (see Problem 9.23). In Problem 9.25 you learn how to solve the system (9.54) by hand.

A General Method for Finding Coiflet Filters

We use the ideas outlined in the preceding subsection to suggest a general method for finding Coiflet filters. The key is knowing the starting and stopping indices ℓ and L, respectively, of the finite-length filter $\mathbf{h}$ associated with the Fourier series $H(\omega)$ given by (9.40). Once we have ℓ and L, we can write down the orthogonality conditions and the derivative conditions and then use a **CAS** to solve the system.

In Problem 9.26(b) you will show that when $K = 2$, the associated finite-length filter is $\mathbf{h} = (h_{-4}, h_{-3}, \ldots, h_6, h_7)$, so that $\ell = -4$ and $L = 7$. In Problem 9.26(c), you will show that for $K = 3$, the starting index is $\ell = -6$ and the stopping index is $L = 11$. Finally, in Problem 9.27, you will provide a proof for the following proposition.

Proposition 9.3 (Finite-Length Coiflet Filter). *Suppose that* $\mathbf{h}$ *is a finite-length filter associated with the Fourier series* $H(\omega)$ *given by Theorem 9.2. Then the length of* $\mathbf{h}$ *is* $6K$ *and*

$$\mathbf{h} = (h_\ell, h_{\ell+1}, \ldots, h_{L-1}, h_L)$$

where $\ell = -2K$ *and* $L = 4K - 1$. □

Proof. The proof is left as Problem 9.27. □

Coiflets for K = 2

When $K = 2$, we have $\ell = -4$, $L = 7$, and $(L - \ell - 1)/2 = 5$. We use Corollary 9.1 to write down the orthogonality conditions that $\mathbf{h} = (h_{-4}, \ldots, h_7)$ must satisfy

$$\sum_{k=-4}^{11} h_k^2 = 1 \qquad \text{and} \qquad \sum_{k=-4+2m}^{11} h_k h_{k-2m} = 0, \qquad m = 1, \ldots, 5. \tag{9.55}$$

Since $2K - 1 = 3$, the derivative conditions (9.41) are

$$\begin{aligned} H(0) &= \sqrt{2} \\ H(\pi) &= 0 \\ H^{(m)}(0) &= 0 \qquad m = 1, \ldots, 3 \\ H^{(m)}(\pi) &= 0 \qquad m = 1, \ldots, 3. \end{aligned} \tag{9.56}$$

In Problem 9.28 you are asked to explicitly write down the system for finding Coiflet filters when $K = 2$. Solving this system numerically yields four solutions. They are listed (rounded to six digits) in Table 9.2.

Table 9.2 Four Coiflet filters for $K = 2$.

	Solution 1[a]	Solution 2	Solution 3	Solution 4
h_{-4}	0.016387	−0.001359	−0.021684	−0.028811
h_{-3}	−0.041465	−0.014612	−0.047599	0.009542
h_{-2}	−0.067373	−0.007410	0.163254	0.113165
h_{-1}	0.386110	0.280612	0.376511	0.176527
h_0	0.812724	0.750330	0.270927	0.542555
h_1	0.417005	0.570465	0.516748	0.745265
h_2	−0.07649	−0.071638	0.545852	0.102774
h_3	−0.059434	−0.155357	−0.239721	−0.296788
h_4	0.023680	0.050024	−0.327760	−0.020498
h_5	0.005611	0.024804	0.136027	0.078835
h_6	−0.001823	−0.012846	0.076520	−0.002078
h_7	−0.000721	0.001195	−0.034858	−0.006275

[a] This is the Coiflet C12 filter.

Definition 9.2 (Coiflet Length 12 Filter). *We define the* Coiflet 12-term orthogonal filter *by the first solution in Table 9.2. We call this filter the* C12 *filter.* □

We will not cover the method by which Daubechies chooses a particular filter. The interested reader is referred to Daubechies [30].

A Highpass Filter for the Coiflet Filter

Now that we have the means to construct a Coiflet filter $\mathbf{h}$, we need to build the associated wavelet filter $\mathbf{g}$. For the causal filters from Chapters 4 and 5, we used the rule $g_k = (-1)^k h_{L-k}$. In this case, the elements g_k are obtained by reversing the elements in $\mathbf{h}$ and multiplying by $(-1)^k$. You could also think of this operation as reflecting the filter about some center. In particular, the operation $h_k \mapsto h_{L-k}$ has the effect of reflecting the numbers about the line $k = L/2$. As an example, consider the D6 filter $h_0, h_1, h_2, h_3, h_4, h_5$. Here $L = 5$ and $L/2 = 5/2 = 2.5$. If you think about constructing $\mathbf{g}$ by the rule $g_k = (-1)^k h_{5-k}$, we could instead reflect the elements of $\mathbf{h}$ about the midpoint of the list $0, \ldots, 5$ in order to reverse the filter and then multiply by ± 1.

Our convention for Coiflet filters will be the same. However, these filters are noncausal, so we will have to think about the line of reflection.

In the case $K = 1$, the reflection line would be $k = \frac{1}{2}$, since three of the filter coefficient indices are to the left of $\frac{1}{2}$ and the other three are to the right. It is easy to see where the $\frac{1}{2}$ comes from – we simply find the midpoint of the list $\ell, \ldots, L$. In order to reflect about the line $k = \frac{1}{2}$, we use the rule

$$g_k = (-1)^k h_{1-k}$$

so that $g_{-2} = h_3$, $g_{-1} = -h_2$, $g_0 = h_1$, $g_1 = -h_0$, $g_2 = -h_{-1}$, and $g_3 = h_{-2}$.

In the case $K = 2$, the reflection line would be $k = \frac{3}{2}$ since six of the filter coefficient indices are to the left of $\frac{3}{2}$ and six are to the right of $\frac{3}{2}$. The rule we would then use in the $K = 2$ case is

$$g_k = (-1)^k h_{3-k}.$$

For general K, Proposition 9.3 tells us that $\ell = -2K$ and $L = 4K - 1$, so that the midpoint of the list $-2K, \ldots, 4K - 1$ is $\frac{-2K+4K-1}{2} = \frac{2K-1}{2}$. Thus we will reflect about the line $k = \frac{2K-1}{2}$. The rule

$$g_k = (-1)^k h_{2K-1-k}. \tag{9.57}$$

will incorporate the desired reflection. For some applications, it may be convenient to multiply the highpass filter by -1. In this case the elements of the highpass filter are

$$g_k = (-1)^{k+1} h_{2K-1-k}. \tag{9.58}$$

In Problem 9.30 you will find the Fourier series $G(\omega)$ associated with $\mathbf{g}$.

We are free to pick any odd index (in our case $2K - 1$) that we like in (9.57) – reflecting about the midpoint of the indices simply keeps the wavelets associated with Coiflet filters consistent with the highpass filters we constructed for Daubechies filters.

The Coiflet Filter in the Wavelet Transformation Matrix

Proposition 9.3 tells us that the Coiflet filters are not causal filters. How do we implement such filters in our wavelet transformamtion algorithms? One answer is simply to treat these filters just like causal filters. That is, the associated wavelet matrix W_N is

$$
W_N = \left[\begin{array}{cccccccccccccc}
h_3 & h_2 & h_1 & h_0 & h_{-1} & h_{-2} & 0 & 0 & \cdots & 0 & 0 & 0 & 0 & 0 & 0 \\
0 & 0 & h_3 & h_2 & h_1 & h_0 & h_{-1} & h_{-2} & & 0 & 0 & 0 & 0 & 0 & 0 \\
\vdots & & & & & & & & \ddots & & & & & & \vdots \\
0 & 0 & 0 & 0 & & & & & & h_3 & h_2 & h_1 & h_0 & h_{-1} & h_{-2} \\
h_{-1} & h_{-2} & 0 & 0 & & & & \cdots & & 0 & 0 & h_3 & h_2 & h_1 & h_0 \\
h_1 & h_0 & h_{-1} & h_{-2} & & & & & & & & 0 & 0 & h_3 & h_2 \\
\hline
-h_{-2} & h_{-1} & -h_0 & h_1 & -h_2 & h_3 & 0 & 0 & \cdots & 0 & 0 & 0 & 0 & 0 & 0 \\
0 & 0 & -h_{-2} & h_{-1} & -h_0 & h_1 & -h_2 & h_3 & & 0 & 0 & 0 & 0 & 0 & 0 \\
\vdots & & & & & & & & \ddots & & & & & & \vdots \\
0 & 0 & 0 & 0 & & & & & & -h_{-2} & h_{-1} & -h_0 & h_1 & -h_2 & h_3 \\
-h_2 & h_3 & 0 & 0 & & & & \cdots & & 0 & 0 & -h_{-2} & h_{-1} & -h_0 & h_1 \\
-h_0 & h_1 & -h_2 & h_3 & & & & & & & & 0 & 0 & -h_{-2} & h_{-1}
\end{array}\right]. \qquad (9.59)
$$

An Alternative Form of the Wavelet Transformation Matrix

If you look at other texts or journal articles about discrete wavelet transformations, you will often see the filters in reverse order as well as shifted so that terms with negative indices appear at the far right of the first row. For example, with the C6 filter, W_N might be constructed as follows:

$$
W_N = \left[\begin{array}{cccccccccccccc}
h_0 & h_1 & h_2 & h_3 & 0 & 0 & 0 & 0 & & 0 & 0 & 0 & 0 & h_{-2} & h_{-1} \\
h_{-2} & h_{-1} & h_0 & h_1 & h_2 & h_3 & 0 & 0 & \cdots & 0 & 0 & 0 & 0 & 0 & 0 \\
0 & 0 & h_{-2} & h_{-1} & h_0 & h_1 & h_2 & h_3 & & & & 0 & 0 & 0 & 0 \\
\vdots & & & & & & & & \ddots & & & & & & \vdots \\
0 & 0 & 0 & 0 & & & \cdots & & & h_{-2} & h_{-1} & h_0 & h_1 & h_2 & h_3 \\
h_2 & h_3 & 0 & 0 & & & & & & 0 & 0 & h_{-2} & h_{-1} & h_0 & h_1 \\
\hline
h_1 & -h_0 & h_{-1} & -h_{-2} & 0 & 0 & 0 & 0 & & 0 & 0 & 0 & 0 & h_3 & -h_2 \\
h_3 & -h_2 & h_1 & -h_0 & h_{-1} & -h_{-2} & 0 & 0 & \cdots & 0 & 0 & 0 & 0 & 0 & 0 \\
0 & 0 & h_3 & -h_2 & h_1 & -h_0 & h_{-1} & -h_{-2} & & & & 0 & 0 & 0 & 0 \\
\vdots & & & & & & & & \ddots & & & & & & \vdots \\
0 & 0 & 0 & 0 & & & \cdots & & & h_3 & -h_2 & h_1 & -h_0 & h_{-1} & -h_{-2} \\
h_{-1} & -h_{-2} & 0 & 0 & & & & & & 0 & 0 & h_3 & -h_2 & h_1 & -h_0
\end{array}\right] \qquad (9.60)
$$

This form of the transformation is especially useful for application of symmetric biorthogonal wavelet filters in applications such as image compression.

It is easy to see that the rows in (9.60) are orthogonal to each other if we assume that W_N in (9.59) is an orthogonal matrix. Each row in (9.60) is obtained by reversing the elements of the corresponding row in (9.59) and cyclically translating the result four units right. These operations do not disrupt the orthogonality of the rows of (9.60).

We learned in Problem 5.28 that if $\mathbf{h} = (h_0, \ldots, h_L)$ solves (5.52), then its reflection $(h_L, h_{L-1}, \ldots, h_1, h_0)$ solves (5.52) as well. Note that none of the solutions in Table 9.1 or 9.2 are reflections of other solutions, yet we are proposing the use of reflections in (9.60). In Problem 9.32 you will show that reflections of Coiflet filters also satisfy the orthogonality condition (9.42) and the derivative conditions (9.41) and see why these reflections did not appear as solutions in the $K = 1, 2$ cases.

Example: Signal Compression

We conclude this section by returning to Example 5.2.

Example 9.3 (Signal Compression Using C6). *We perform six iterations of the wavelet transformation on the audio file consisting of 23,616 samples from Example 5.2 using first the D6 filter and next the C6 filter. In Figure 9.6 we have plotted the cumulative energy for each transformation. Note that the energy in the transformation using the C6 filter is comparable to that of the energy of the transformation using the D6 filter.*

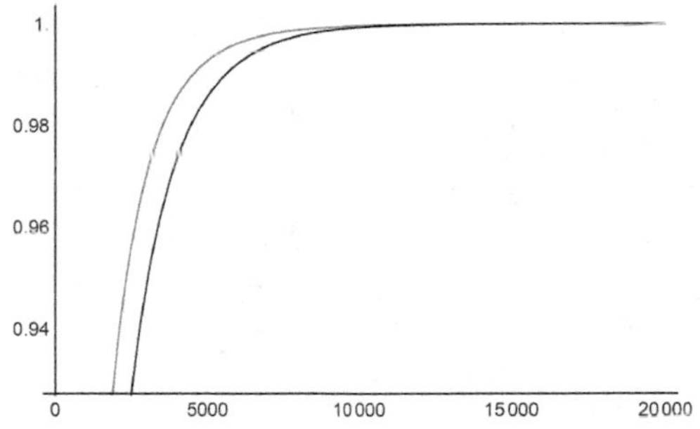

Figure 9.6 Plot of the cumulative energies of wavelet transformations using the D6 filter (dark gray), the C6 filter (black).

(Live example: Visit `stthomas.edu/wavelets` *and click on Live Examples.)*

□

PROBLEMS

9.19 For integers $K \geq 2$, use the definition of the binomial coefficient (5.70) to show that

$$\binom{2K-2}{K-1} = 2\binom{2K-3}{K-2}.$$

The following steps will help you organize your work.

(a) Show that

$$\binom{2K-2}{K-1} = \frac{K\cdot(K+1)\cdots(2K-2)}{1\cdot 2\cdots(K-1)}$$

and

$$\binom{2K-3}{K-2} = \frac{K\cdot(K+1)\cdots(2K-3)}{1\cdot 2\cdots(K-2)}.$$

(b) Multiply the top and bottom of the second identity in (a) by $K-1$ and observe that the denominators for both identities are now the same.

(c) Show that $2(K-1)\cdot K\cdots(2K-3) = K\cdots(2K-2)$ to complete the proof.

9.20 In this problem you will provide an alternative proof to Lemma 9.1. This identity plays an important role in the proof of Theorems 9.2 and 9.3. The argument is a bit tedious and recommended to those students who are familiar with proof by induction.[3] The following steps will help you organize your work.

(a) Show directly that the result holds when ω is any integer multiple of π. *Hint:* You might want to consider separately the cases $\omega = 2k\pi$ and $\omega = (2k+1)\pi$, where $k \in \mathbb{Z}$.

For the remainder of the proof, we assume that ω is not an integer multiple of π.

(b) Show that the result holds for $K = 1$.

The induction part of the proof relies on the identity

$$\sum_{\ell=0}^{m}\binom{n+\ell}{\ell} = \binom{n+m+1}{m}. \tag{9.61}$$

You can either prove (9.61) yourself by repeated use Pascal's identity (7.31) or consult an elementary text on combinatorics (e.g. Brualdi [12]).

To better understand the induction step, it is helpful to derive the result for $K = 4$ assuming that it holds for $K = 3$.

(c) Let $c = \cos(\frac{\omega}{2})$ and $s = \sin(\frac{\omega}{2})$. Write down (9.32) for $K = 3$.

(d) Write down the right-hand side of (9.32) for $K = 4$. Use (9.61) to show that in this case we have

$$c^8\sum_{j=0}^{3}\sum_{\ell=0}^{j}\binom{2+\ell}{\ell}s^{2j} + s^8\sum_{j=0}^{3}\sum_{\ell=0}^{j}\binom{2+\ell}{\ell}c^{2j}.$$

[3]This argument was shown to me by Bo Green.

(e) Consider the first term in (d). Show that we can rewrite this term as

$$c^8 \sum_{j=0}^{3} \sum_{\ell=0}^{j} \binom{2+\ell}{\ell} s^{2j} = c^8 \sum_{\ell=0}^{3} \binom{2+\ell}{\ell} \sum_{j=\ell}^{3} s^{2j}$$
$$= c^8 \sum_{\ell=0}^{3} \binom{2+\ell}{\ell} s^{2\ell} \sum_{j=0}^{3-\ell} s^{2j}.$$

(f) Separate out the $\ell = 3$ term in (e) to show that

$$c^8 \sum_{j=0}^{3} \sum_{\ell=0}^{j} \binom{2+\ell}{\ell} s^{2j} = \binom{5}{3} c^8 s^6 + c^8 \sum_{\ell=0}^{2} \binom{2+\ell}{\ell} s^{2\ell} \sum_{j=0}^{3-\ell} (s^2)^j.$$

(g) Now use the identity

$$1 + t + t^2 + \cdots + t^n = \frac{1 - t^{n+1}}{1 - t}$$

for $t \neq 1$ to show that

$$c^8 \sum_{j=0}^{3} \sum_{\ell=0}^{j} \binom{2+\ell}{\ell} s^{2j} = \binom{5}{3} c^8 s^6 + c^8 \sum_{\ell=0}^{2} \binom{2+\ell}{\ell} s^{2\ell} \frac{1 - s^{8-2\ell}}{1 - s^2}$$
$$= \binom{5}{3} c^8 s^6 + c^6 \sum_{\ell=0}^{2} \binom{2+\ell}{\ell} s^{2\ell} (1 - s^{8-2\ell})$$
$$= \binom{5}{3} c^8 s^6 + c^6 \sum_{\ell=0}^{2} \binom{2+\ell}{\ell} s^{2\ell} - c^6 s^8 \sum_{\ell=0}^{2} \binom{2+\ell}{\ell}.$$

Note that since ω is not an odd-integer multiple of π, $1 - s^2 \neq 0$.

(h) Using the ideas from (e) – (g) show that the second term in (d) can be written as

$$s^8 \sum_{j=0}^{3} \sum_{\ell=0}^{j} \binom{2+\ell}{\ell} c^{2j} = \binom{5}{3} s^8 c^6 + s^6 \sum_{\ell=0}^{2} \binom{2+\ell}{\ell} c^{2\ell} - s^6 c^8 \sum_{\ell=0}^{2} \binom{2+\ell}{\ell}.$$

(i) Now use (9.61) and (c) to show that the terms in (g) and (h) add to 1.

(j) For the general induction step, assume that (9.32) holds and mimic steps (d) – (i) to show that

$$1 = \cos^{2K+2}\left(\frac{\omega}{2}\right) \sum_{j=0}^{K} \binom{K+j}{j} \sin^{2j}\left(\frac{\omega}{2}\right)$$
$$+ \sin^{2K+2}\left(\frac{\omega}{2}\right) \sum_{j=0}^{K} \binom{K+j}{j} \cos^{2j}\left(\frac{\omega}{2}\right)$$

9.21 Use Lemma 9.1 to verify (9.44).

9.22 Verify the algebra necessary to show that when we insert the filter coefficients given by (9.48) into (9.49), we obtain the system (9.50).

9.23 Find the filter coefficients $\mathbf{h} = (h_{-2}, h_{-1}, h_0, h_1, h_2, h_3)$ for the second pair of a_0, a_1 in (9.51).

9.24 Let $\mathbf{h} = (h_0, h_1, h_2)$ where $h_0 = h_2 = \frac{\sqrt{2}}{4}$ and $h_1 = \frac{\sqrt{2}}{2}$.

(a) Verify that $H(0) = \sqrt{2}$ and that $H(\pi) = H'(\pi) = 0$.

(b) Show that $H(\omega) = \sqrt{2}\, e^{i\omega} \cos^2(\frac{\omega}{2})$. *Hint:* Write down $H(\omega)$ and then factor out e^{iw}.

(c) Use (b) to show that $|H(\omega)|^2 = 2\cos^4\left(\frac{\omega}{2}\right)$.

(d) Use (c) to show that $|H(\omega+\pi)|^2 = 2\sin^4\left(\frac{\omega}{2}\right)$. *Hint:* The identity $\cos\left(t + \frac{\pi}{2}\right) = -\sin(t)$ will be useful.

(e) Use (c) and (d) to show that

$$|H(\omega)|^2 + |H(\omega+\pi)|^2 \neq 2.$$

so that by the converse of Corollary 9.1, $\mathbf{h}$ does not satisfy the orthogonality conditions (9.22).

9.25 In this problem you will solve the system (9.54) algebraically. The following steps will help you organize your work.

(a) The last four equations of (9.54) are linear. Row reduce these linear equations to show that $h_{-2}, \ldots, h_1$ can be expressed in terms of h_2 and h_3. In particular show that

$$h_0 = \frac{\sqrt{2}}{2} - 2h_2, \quad h_{-1} = \frac{\sqrt{2}}{4}, \quad h_1 = \frac{\sqrt{2}}{4} - 2h_3, \quad h_{-2} = h_2.$$

(b) Insert the values for $h_{-2}, \ldots, h_1$ you found in (a) into the third equation of (9.54) and simplify to show that

$$h_2^2 + \frac{\sqrt{2}}{4}h_3 + h_3^2 = 0.$$

(c) Insert the values for $h_{-2}, \ldots, h_1$ you found in (a) into the second equation of (9.54) and simplify to show that

$$\sqrt{2}h_2^2 - 4h_2^2 + 4h_3^2 = -\frac{1}{8}.$$

(d) Insert the values for $h_{-2},\ldots,h_1$ you found in (a) into the first equation of (9.54) and simplify to show that

$$-2\sqrt{2}h_2 + 6h_2^2 - \frac{\sqrt{2}}{2}h_3 + 6h_3^2 = -\frac{1}{4}.$$

(e) Combine the results of (b)– (d) in an appropriate way to show that $h_2 = -\left(h_3 + \frac{\sqrt{2}}{16}\right)$.

(f) Using (a) and (e), write $h_{-2},\ldots,h_2$ in terms of h_3.

(g) Insert the results from (f) into the first equation in (9.54) and simplify to obtain

$$12h_3^2 + \frac{9\sqrt{2}}{4}h_3 + \frac{3}{64} = 0.$$

(h) Use the quadratic formula to solve for h_3 in (g) and show that $h_3 = \frac{\sqrt{2}}{32}(-3 \pm \sqrt{7})$.

(i) Insert the values from (h) into the identities you obtained in (f) to find the two solutions listed in Table 9.1. You will need a calculator or a **CAS** to verify your solutions match those given in the table.

9.26 In this problem you will analyze the finite-length filter $\mathbf{h}$ associated with Fourier series $H(\omega)$ from Theorem 9.2 in the cases $K = 1, 2$.

(a) Let $K = 2$. Use (9.40) and the identities given by (9.43) and (9.45) to write $H(\omega)$ as a linear combination of complex exponentials of the form $e^{ik\omega}$.

(b) Use the result in (a) to show that the finite filter $\mathbf{h}$ has starting index $\ell = -4$ and stopping index $L = 7$, so that the length of $\mathbf{h}$ is 12.

(c) Repeat (a) and (b) in the case $K = 3$.

9.27 Prove Proposition 9.3.

9.28 Write down the explicit system (9.55) – (9.56) you must solve to find Coiflet filters in the case $K = 2$.

9.29 Let $K = 1, 2, \ldots$ Write down the explicit system satisfied by the Coiflet filter $\mathbf{h} = (h_{-2K}, h_{-2K+1}, \ldots, h_{4K-2}, h_{4K-1})$. That is, use Corollary 9.1 to write down the general orthogonality conditions and compute the mth derivative of $H(\omega)$, evaluate it at zero and π, and simplify to obtain explicit forms of the derivative conditions (9.41).

9.30 Let $\mathbf{g}$ be given element-wise by (9.57). Show that the associated Fourier series is

$$G(\omega) = -e^{i(2K-1)\omega}\overline{H(\omega + \pi)}.$$

9.31 The C6 filter satisfied $H(0) = \sqrt{2}$, $H'(0) = H(\pi) = H'(\pi) = 0$ plus the orthogonality condition (9.42). Suppose that $\mathbf{h} = (h_{-2}, h_{-1}, h_0, h_1, h_2, h_3)$ is *any* filter that satisfies the orthogonality condition (9.42).

(a) Can we find $\mathbf{h}$ that satisfies

$$H(0) = \sqrt{2}, \quad H'(0) = 0, \quad H''(0) = 0, \quad H(\pi) = 0?$$

(b) Why might someone want to use this filter?

9.32 Suppose that $\mathbf{h} = (h_{-2K}, \ldots, h_{4K-1})$ is a Coiflet filter whose Fourier transformation is given by (9.40). Let $\mathbf{p} = (h_{4K-1}, \ldots, h_{-2K})$ be the reflection of $\mathbf{h}$. Show that $\mathbf{p}$ satisfies (9.42) and (9.41). *Hint:* Problem 9.17 will be useful. After working this problem, you should be able to explain why $\mathbf{p}$ for $K = 1, 2$ did not show up in Table 9.1 or 9.2.

9.4 Biorthogonal Spline Filter Pairs

In Chapter 7 we constructed the $(5, 3)$ and $(8, 4)$ biorthogonal spline filter pairs. The construction was ad hoc but greatly aided by the facts that the filters in both pairs are symmetric (reducing the number of distinct filter elements) and that one filter in the pair was fixed from the start. In the case of orthogonal filters, we have the system (9.29) in Section 9.2 that consists of the orthogonality relation (9.22) from Corollary 9.1 and conditions on $H(\omega)$ and its derivatives evaluated at $\omega = \pi$, but (9.22) leads to a quadratic system of equations and such systems can be difficult to solve. Corollary 9.2 gives us a similar biorthogonality relation

$$\widetilde{H}(\omega)\overline{H(\omega)} + \widetilde{H}(\omega + \pi)\overline{H(\omega + \pi)} = 2 \tag{9.62}$$

and with $\widetilde{\mathbf{h}}$ known, this system is linear and thus easily solvable. But for filters $\mathbf{h}$ of longer length (relative to the length of $\widetilde{\mathbf{h}}$), the system (9.62) is underdetermined and could possibly possess an infinite amount of solutions. We can remedy the discrepancy between the number of equations and unknowns in (9.62) by adding derivative conditions on $\widetilde{H}(\omega)$ and $H(\omega)$ at $\omega = \pi$. Recall we added the equation

$$h_1 - 3h_2 + 5h_3 - 7h_4 = 0 \tag{9.63}$$

to the system we developed to find $\mathbf{h}$ for the $(8, 4)$ biorthogonal spline filter pair. In Problem 9.33, you will show that this condition is equivalent to $H'(\pi) = 0$.

We could continue the construction introduced in Chapter 7, but it is not clear how to choose the lengths of each filter in the pair, how to choose $\widetilde{\mathbf{h}}$ in general or which derivative condition(s) to impose on $H(\omega)$. In some cases, derivative conditions on $H(\omega)$ are built in. Suppose that $\mathbf{h}$ is an symmetric, even-length filter of length N. In the case when N is even, it turns out $H(\pi) = 0$ (see Problem 9.35) and when N is odd, $H^{(m)}(\pi) = 0$ when m is odd (see Problem 9.34).

In [30], Daubechies provides answers to the above questions. In this section, we learn how she chooses the symmetric, lowpass filter $\widetilde{\mathbf{h}}$ and then uses the length of this filter to construct the Fourier series $H(\omega)$ of the companion filter (also symmetric and lowpass) $\mathbf{h}$.

The Filter $\widetilde{\mathrm{H}}(\omega) = \sqrt{2}\cos^{\widetilde{\mathrm{N}}}\left(\frac{\omega}{2}\right)$

The question of how to select $\widetilde{\mathbf{h}}$ has in large part already been answered. If you worked Problem 8.43 in Section 8.2, then you learned that if $\widetilde{N}$ is an even integer,

$$\cos^{\widetilde{N}}\left(\frac{\omega}{2}\right) = \sum_{k=-\widetilde{N}/2}^{\widetilde{N}/2} a_k e^{ik\omega}$$

where

$$a_k = \frac{1}{2^{\widetilde{N}}}\binom{\widetilde{N}}{k+\frac{\widetilde{N}}{2}} = \frac{\widetilde{N}!}{2^{\widetilde{N}}(k+\frac{\widetilde{N}}{2})!\,(\frac{\widetilde{N}}{2}-k)!}.$$

Let's take

$$\widetilde{H}(\omega) = \sqrt{2}\cos^{\widetilde{N}}\left(\frac{\omega}{2}\right) = \sqrt{2}\sum_{k=-\widetilde{N}/2}^{\widetilde{N}/2} \frac{1}{2^{\widetilde{N}}}\binom{\widetilde{N}}{k+\frac{\widetilde{N}}{2}} e^{ik\omega}. \tag{9.64}$$

In this form, we see immediately that $\widetilde{H}(0) = \sqrt{2}\cos^{\widetilde{N}}(0) = \sqrt{2}$ and $\widetilde{H}(\pi) = \sqrt{2}\cos^{\widetilde{N}}(\frac{\pi}{2}) = 0$. Moreover, we can use (9.64) to write $\widetilde{h}_k = \sqrt{2}\,a_k$, and it is readily evident that $a_k = a_{-k}$, so that $\widetilde{h}_k = \widetilde{h}_{-k}$. Thus, $\widetilde{\mathbf{h}}$ is a symmetric filter with odd length $\widetilde{N}+1$. Let's look at some examples.

Example 9.4 (Even Powers of $\cos\left(\frac{\omega}{2}\right)$**).** *Write down the Fourier series for*

(a) $\widetilde{H}(\omega) = \sqrt{2}\cos^2\left(\frac{\omega}{2}\right)$

(b) $\widetilde{H}(\omega) = \sqrt{2}\cos^4\left(\frac{\omega}{2}\right)$

(c) $\widetilde{H}(\omega) = \sqrt{2}\cos^6\left(\frac{\omega}{2}\right)$

Solution. *For part (a) we have* $\widetilde{N} = 2$ *and*

$$a_k = \frac{2!}{2^2(1-k)!\,(1+k)!}, \qquad k = -1, 0, 1$$

so that $a_0 = \frac{1}{2}$ *and* $a_1 = a_{-1} = \frac{1}{4}$. *We have* $\widetilde{\mathbf{h}} = \left(\widetilde{h}_{-1}, \widetilde{h}_0, \widetilde{h}_1\right) = \left(\frac{\sqrt{2}}{4}, \frac{\sqrt{2}}{2}, \frac{\sqrt{2}}{4}\right)$ *and the Fourier series is*

$$\widetilde{H}(\omega) = \frac{\sqrt{2}}{4}e^{-i\omega} + \frac{\sqrt{2}}{2} + \frac{\sqrt{2}}{4}e^{i\omega}.$$

For part (b) we have $\widetilde{N} = 4$ and

$$a_k = \frac{4!}{2^4(2-k)!\,(2+k)!}, \qquad k = -2, -1, 0, 1, 2,$$

so that $a_0 = \frac{3}{8}$, $a_1 = a_{-1} = \frac{1}{4}$, and $a_2 = a_{-2} = \frac{1}{16}$. Thus, the Fourier series for part (b) is

$$\widetilde{H}(\omega) = \frac{\sqrt{2}}{16}\,e^{-2i\omega} + \frac{\sqrt{2}}{4}\,e^{-i\omega} + \frac{3\sqrt{2}}{8} + \frac{\sqrt{2}}{4}\,e^{i\omega} + \frac{\sqrt{2}}{16}\,e^{2i\omega}.$$

Finally, for part (c), $\widetilde{N} = 6$ and

$$a_k = \frac{6!}{2^6(3-k)!\,(3+k)!}, \qquad k = 0, \pm 1, \pm 2, \pm 3.$$

We have $a_0 = \frac{5}{16}$, $a_1 = a_{-1} = \frac{15}{64}$, $a_2 = a_{-2} = \frac{3}{32}$, and $a_3 = a_{-3} = \frac{1}{64}$. The Fourier series for part (c) is

$$\begin{aligned}\widetilde{H}(\omega) = &\frac{\sqrt{2}}{64}\,e^{-3i\omega} + \frac{3\sqrt{2}}{32}\,e^{-2i\omega} + \frac{15\sqrt{2}}{64}\,e^{-i\omega} + \frac{5\sqrt{2}}{16} \\ &+ \frac{15\sqrt{2}}{64}\,e^{i\omega} + \frac{3\sqrt{2}}{32}\,e^{2i\omega} + \frac{\sqrt{2}}{64}\,e^{3i\omega}.\end{aligned}$$

(Live example: Visit `stthomas.edu/wavelets` *and click on Live Examples.)*

□

When $\widetilde{N}$ is even, we have seen that $\widetilde{H}(\omega) = \sqrt{2}\cos^{\widetilde{N}}\left(\frac{\omega}{2}\right)$ produces a symmetric filter $\widetilde{\mathbf{h}} = \left(\widetilde{h}_{-\widetilde{N}/2}, \ldots, \widetilde{h}_{\widetilde{N}/2}\right)$ of odd length $\widetilde{N}+1$. How can we generate even-length symmetric filters? Let's try raising $\cos\left(\frac{\omega}{2}\right)$ to an odd power. For example, for $\widetilde{N} = 3$, we have

$$\begin{aligned}\cos^3\left(\frac{\omega}{2}\right) &= \left(\frac{e^{i\omega/2} + e^{-i\omega/2}}{2}\right)^3 \\ &= \frac{1}{8}\left(e^{-i\omega/2}(e^{i\omega}+1)\right)^3 \\ &= \frac{e^{-3i\omega/2}}{8}\left(1 + 3\,e^{i\omega} + 3\,e^{2i\omega} + e^{3i\omega}\right) \\ &= \frac{1}{8}\,e^{-3i\omega/2} + \frac{3}{8}\,e^{-i\omega/2} + \frac{3}{8}\,e^{i\omega/2} + \frac{1}{8}\,e^{3i\omega/2}\end{aligned}$$

Unfortunately, this is not a Fourier series, due to the division by 2 in each of the complex exponentials. It is promising, though, since the length of the series is even and the coefficients are symmetric. Perhaps we can modify the expression so that it would work. In particular, if we multiply the entire expression by $e^{i\omega/2}$, we would shift the complex exponentials so that they are of the proper form. We have

$$e^{i\omega/2}\cos^3\left(\frac{\omega}{2}\right) = \frac{1}{8}\,e^{-i\omega} + \frac{3}{8} + \frac{3}{8}\,e^{i\omega} + \frac{1}{8}\,e^{2i\omega}.$$

Thus, if we take

$$\widetilde{H}(\omega) = \sqrt{2}\, e^{i\omega/2}\cos^3\left(\frac{\omega}{2}\right) = \frac{\sqrt{2}}{8}\, e^{-i\omega} + \frac{3\sqrt{2}}{8} + \frac{3\sqrt{2}}{8}\, e^{i\omega} + \frac{\sqrt{2}}{8}\, e^{2i\omega},$$

we see immediately that

$$\widetilde{H}(0) = \sqrt{2}\, e^{0}\cos^3(0) = \sqrt{2} \qquad \text{and} \qquad \widetilde{H}(\pi) = \sqrt{2}\, e^{i\pi/2}\cos^3(\pi/2) = 0$$

so that $\widetilde{H}(\omega)$ produces a lowpass filter. Moreover, the length of the filter is 4 and the filter $\widetilde{\mathbf{h}} = \left(\widetilde{h}_{-1}, \widetilde{h}_0, \widetilde{h}_1, \widetilde{h}_2\right) = \left(\frac{\sqrt{2}}{8}, \frac{3\sqrt{2}}{8}, \frac{3\sqrt{2}}{8}, \frac{\sqrt{2}}{8}\right)$ satisfies $\widetilde{h}_k = \widetilde{h}_{1-k}$, so that it is symmetric about $k = \frac{1}{2}$. In Problem 9.38 you will show that for $\widetilde{N}$ odd,

$$\widetilde{H}(\omega) = \sqrt{2}\, e^{i\omega/2}\cos^{\widetilde{N}}\left(\frac{\omega}{2}\right) = \sqrt{2}\sum_{k=-\frac{\widetilde{N}-1}{2}}^{\frac{\widetilde{N}+1}{2}} \frac{1}{2^{\widetilde{N}}}\binom{\widetilde{N}}{\frac{\widetilde{N}-1}{2}+k} e^{ik\omega} \tag{9.65}$$

produces an even-length filter $\widetilde{\mathbf{h}}$ that is symmetric about $k = \frac{1}{2}$.

Spline Filters

We can combine (9.64) and (9.65) to form the following proposition.

Definition 9.3 (Spline Filter of Length $\widetilde{\mathbf{N}} + 1$). *Let $\widetilde{N}$ be a positive integer. Then $\widetilde{\mathbf{h}}$, given element-wise by*

$$\boxed{\widetilde{h}_k = \frac{\sqrt{2}}{2^{\widetilde{N}}}\binom{\widetilde{N}}{\frac{\widetilde{N}}{2} - k}, \qquad k = 0, \pm 1, \ldots, \pm\widetilde{N}/2} \tag{9.66}$$

when $\widetilde{N}$ is even, and

$$\boxed{\widetilde{h}_k = \frac{\sqrt{2}}{2^{\widetilde{N}}}\binom{\widetilde{N}}{\frac{\widetilde{N}-1}{2} + k}, \qquad k = -\frac{\widetilde{N}-1}{2}, \ldots, \frac{\widetilde{N}+1}{2}} \tag{9.67}$$

when $\widetilde{N}$ is odd, is called the spline filter of length $\widetilde{N}+1$. *We will call*

$$\widetilde{H}(\omega) = \begin{cases} \sqrt{2}\cos^{\widetilde{N}}\left(\frac{\omega}{2}\right), & \widetilde{N} = 2\widetilde{\ell} \\ \sqrt{2}e^{i\omega/2}\cos^{\widetilde{N}}\left(\frac{\omega}{2}\right), & \widetilde{N} = 2\widetilde{\ell} + 1 \end{cases} \tag{9.68}$$

the Fourier series associated with the spline filter. *Here $\widetilde{\ell}$ is any positive integer.* □

In Problem 9.40 you will show that for a given positive integer $\widetilde{N}$, we can write the spline filter given in (9.66) – (9.67) as

$$\boxed{\widetilde{h}_{k-m} = \frac{\sqrt{2}}{2^{\widetilde{N}}}\binom{\widetilde{N}}{k}, \quad k = 0, \ldots, \widetilde{N}, \quad m = \begin{cases} \frac{\widetilde{N}}{2}, & \widetilde{N} \text{ even}, \\ \frac{\widetilde{N}-1}{2}, & \widetilde{N} \text{ odd}. \end{cases}} \tag{9.69}$$

Those readers familiar with Pascal's triangle will undoubtedly recognize the spline filter of length $\widetilde{N}+1$ as the $\widetilde{N}$th row of Pascal's identity multiplied by $\sqrt{2}/2^{\widetilde{N}}$.

Why the Name "Spline Filters"?

Let's digress for just a moment and discuss the term "spline filter." In mathematics, *spline functions*[4] are piecewise polynomials that are used in many areas of applied mathematics. The fundamental (B-spline) piecewise constant spline is the *box function*

$$B_0(t) = \begin{cases} 1, & 0 \le t < 1 \\ 0, & \text{otherwise.} \end{cases}$$

Higher-degree splines $B_{\widetilde{N}}(t)$ are defined recursively by the formula[5]

$$B_{\widetilde{N}+1}(t) = \int_0^1 B_{\widetilde{N}}(t-u)\,du. \tag{9.70}$$

The spline $B_{\widetilde{N}}(t)$ is a piecewise polynomial of degree $\widetilde{N}$. It is straightforward to show that

$$B_1(t) = \begin{cases} t, & 0 \le t < 1 \\ 2-t, & 1 \le t < 2 \\ 0, & \text{otherwise} \end{cases} \tag{9.71}$$

is the *triangle function* (see Problem 9.41). The splines $B_0(t)$ and $B_1(t)$ are plotted in Figure 9.7.

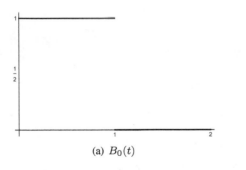

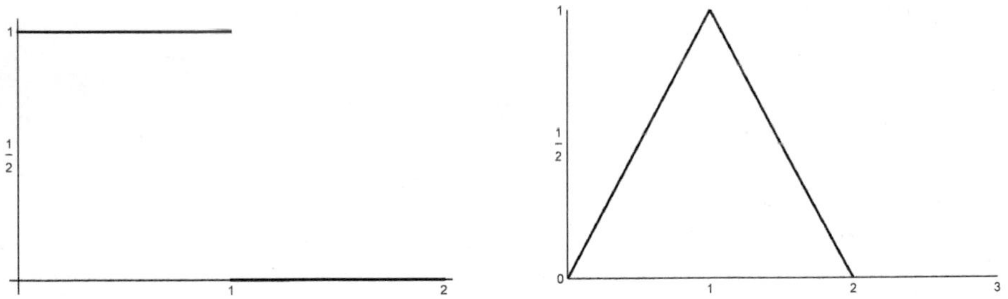

(a) $B_0(t)$ (b) $B_1(t)$

Figure 9.7 The piecewise constant spline $B_0(t)$ and the piecewise linear spline $B_1(t)$.

[4]The word *spline* comes from architecture. A spline is a flexible piece of hard rubber that is used in mechanical drawing to render curves.

[5]Those readers with some background in analysis will recognize that $B_{\widetilde{N}}(t)$ is simply the $\widetilde{N}$-fold convolution product of $B_0(t)$ with itself.

One important property of splines is that they satisfy a *dilation equation*. A function $f(t)$ satisfies a dilation equation if it can be written in the form

$$f(t) = \sum_{k \in \mathbb{Z}} h_k f(2t - k).$$

So $f(t)$ is formed by a linear combination of two-dilates and translates of itself. We have seen the dilation equation in Chapter 1 and Section 4.1. The dilation equation is of fundamental importance in the classical derivation of wavelet theory. Indeed, the filters used for the discrete transformations we developed are the coefficients h_k in the dilation equation for some function $f(t)$.

It is easy to check that both $B_0(t)$ and $B_1(t)$ satisfy dilation equations. We have

$$B_0(t) = \mathbf{1} \cdot B_0(2t) + \mathbf{1} \cdot B_0(2t - 1)$$

and (see Figure 9.8)

$$B_1(t) = \frac{\mathbf{1}}{\mathbf{2}} B_1(2t) + \mathbf{1} \cdot B_1(2t - 1) + \frac{\mathbf{1}}{\mathbf{2}} B_1(2t - 2).$$

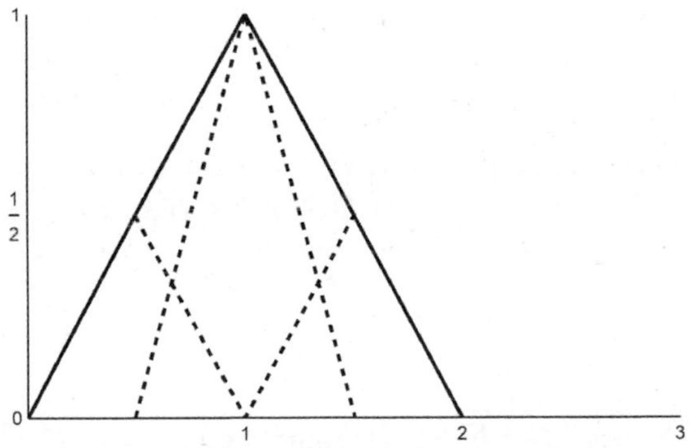

Figure 9.8 The two-dilates and translates $\frac{1}{2} B_1(2t)$, $B_1(2t-1)$, and $\frac{1}{2} B_1(2t-2)$ of $B_1(t)$ used to construct $B_1(t)$.

Note that the coefficient sets $(1, 1)$ and $\left(\frac{1}{2}, 1, \frac{1}{2}\right)$ above are $\frac{\sqrt{2}}{2}$ times the filters from Definition 9.3 with $\widetilde{N} = 1, 2$, respectively. In fact, it can be shown (see Chui

[21]) that for $\widetilde{N} = 1, 2, \ldots$

$$\begin{aligned}
B_{\widetilde{N}-1}(t) &= \sum_{k=0}^{\widetilde{N}} 2^{-\widetilde{N}+1} \binom{\widetilde{N}}{k} B_{\widetilde{N}-1}(2t-k) \\
&= \sum_{k=0}^{\widetilde{N}} \frac{2}{2^{\widetilde{N}}} \binom{\widetilde{N}}{k} B_{\widetilde{N}-1}(2t-k) \\
&= \frac{\sqrt{2}}{2} \sum_{k=0}^{\widetilde{N}} \widetilde{h}_{k-m} B_{\widetilde{N}-1}(2t-k)
\end{aligned}$$

where $\widetilde{h}_{k-m}$ is given by (9.69).

Traditionally, the approach for obtaining wavelet filters consists of a search for a *scaling function* (such as $B_{\widetilde{N}}(t)$) that satisfies a dilation equation. This is a fascinating topic but not the approach we adopt in this book. The interested reader is strongly encouraged to view the books by Boggess and Narcowich [7], Frazier [42], Ruch and Van Fleet [77], and Walnut [100]. The advanced reader would enjoy the development of Daubechies [32].

Daubechies Formulation of $\mathrm{H}(\omega)$

As you can see from Section 7.2, sometimes we have to add extra conditions to the orthogonality constraints in order to obtain our solution for $\mathbf{h}$. Rather than trying to write down derivative conditions for $H(\omega)$ at $\omega = \pi$ for every possible pair of filter lengths, or trying to determine conditions on the lengths of the filters which ensure that a solution for $\mathbf{h}$ exists, we instead appeal to a result due to Daubechies [30]. This result tells us exactly when, given a spline filter $\widetilde{\mathbf{h}}$, we can find a lowpass, symmetric companion filter $\mathbf{h}$. Moreover, her result gives an explicit form for $H(\omega)$.

Theorem 9.3 (Daubechies Solution for $\mathbf{H}(\omega)$). *Suppose that $\widetilde{\mathbf{h}}$ is the spline filter whose Fourier series is given by (9.68) in Definition 9.3. If N and $\widetilde{N}$ are both even or both odd and $\mathbf{h}$ is the filter whose Fourier series is*

$$H(\omega) = \begin{cases} \sqrt{2}\cos^N\left(\frac{\omega}{2}\right) \sum\limits_{j=0}^{\ell+\widetilde{\ell}-1} \binom{\ell+\widetilde{\ell}-1+j}{j} \sin^{2j}\left(\frac{\omega}{2}\right), & N = 2\ell \\ \sqrt{2}e^{i\omega/2}\cos^N\left(\frac{\omega}{2}\right) \sum\limits_{j=0}^{\ell+\widetilde{\ell}} \binom{\ell+\widetilde{\ell}+j}{j} \sin^{2j}\left(\frac{\omega}{2}\right), & N = 2\ell+1 \end{cases} \tag{9.72}$$

then $\mathbf{h}$ is a symmetric finite-length filter (with $h_k = h_{-k}$ when N is even and $h_k = h_{1-k}$ when N is odd) with

$$H(0) = \sqrt{2} \qquad \textit{and} \qquad H(\pi) = 0$$

and $\widetilde{H}(\omega)$ and $H(\omega)$ satisfy the bi-orthogonality condition (9.62). Here $\widetilde{N} = 2\widetilde{\ell}$ when N is even and $\widetilde{N} = 2\widetilde{\ell}+1$ when N is odd. □

Proof. Since $H(\omega)$ is a finite sum, it is clear that $\mathbf{h}$ will be a finite length filter. It is also clear from (9.72) that $H(0) = \sqrt{2}$ and $H(\pi) = 0$.

To show that $\mathbf{h}$ is a symmetric filter, consider the case first where N is even. Since $\cos\left(\frac{\omega}{2}\right) = \cos(-\frac{\omega}{2})$ and $\sin^{2j}(-\frac{\omega}{2}) = \left(-\sin\left(\frac{\omega}{2}\right)\right)^{2j} = \sin^{2j}\left(\frac{\omega}{2}\right)$, we have $H(\omega) = H(-\omega)$. We can thus use Problem 8.25 from Section 8.2 to infer that $h_k = h_{-k}$. Thus when N is even, $\mathbf{h}$ must be an odd length filter symmetric about zero.

In the case when N is odd, note that the function

$$P(\omega) = \sqrt{2}\cos^N\left(\frac{\omega}{2}\right)\sum_{j=0}^{\ell+\widetilde{\ell}}\binom{\ell+\widetilde{\ell}+j}{j}\sin^{2j}\left(\frac{\omega}{2}\right),$$

by the preceding argument, will be even. If we expand $P(\omega)$ as a linear combination of complex exponentials, the coefficients will be symmetric about zero. However, since N is odd, the complex exponentials coming from $\cos^N\left(\frac{\omega}{2}\right)$ will be of the form $e^{(2m+1)i\omega/2}$ for integer m (see the discussion on page 402) so that multiplying $P(\omega)$ by $e^{i\omega/2}$ not only gives $H(\omega) = e^{i\omega/2}P(\omega)$, but it shifts the complex exponentials $\frac{1}{2}$ unit right. Thus we see that $H(\omega)$ is a Fourier series with the coefficients of $P(\omega)$ shifted 1 unit right. This gives us $h_k = h_{1-k}$, so that $\mathbf{h}$ is an even-length filter that is symmetric about $k = \frac{1}{2}$.

To complete the proof, we need to show that $\widetilde{H}(\omega)$ and $H(\omega)$ satisfy the orthogonality relation (9.62). We will give the proof for the case when $\widetilde{N}$ and N are both even and leave the odd case as Problem 9.43.

Suppose that $\widetilde{N}$ and N are even. Then both $\widetilde{H}(\omega)$ and $H(\omega)$ are real-valued, so

$$\begin{aligned}
\widetilde{H}(\omega)\overline{H(\omega)} &= \widetilde{H}(\omega)H(\omega)\\
&= \sqrt{2}\cos^{\widetilde{N}}\left(\frac{\omega}{2}\right)\sqrt{2}\cos^{N}\left(\frac{\omega}{2}\right)\sum_{j=0}^{\ell+\widetilde{\ell}-1}\binom{\ell+\widetilde{\ell}-1+j}{j}\sin^{2j}\left(\frac{\omega}{2}\right)\\
&= 2\cos^{\widetilde{N}+N}\left(\frac{\omega}{2}\right)\sum_{j=0}^{\ell+\widetilde{\ell}-1}\binom{\ell+\widetilde{\ell}-1+j}{j}\sin^{2j}\left(\frac{\omega}{2}\right).
\end{aligned} \tag{9.73}$$

Using the trigonometric identities $\cos(t + \frac{\pi}{2}) = -\sin t$ and $\sin(t + \frac{\pi}{2}) = \cos t$, we replace ω by $\omega + \pi$ in (9.73) to write

$$\begin{aligned}
\widetilde{H}(\omega+\pi)\overline{H(\omega+\pi)} &= \widetilde{H}(\omega+\pi)H(\omega+\pi)\\
&= 2\cos^{\widetilde{N}+N}\left(\frac{\omega}{2}+\frac{\pi}{2}\right)\sum_{j=0}^{\ell+\widetilde{\ell}-1}\binom{\ell+\widetilde{\ell}-1+j}{j}\sin^{2j}\left(\frac{\omega}{2}+\frac{\pi}{2}\right)\\
&= 2\left(-\sin\left(\frac{\omega}{2}\right)\right)^{\widetilde{N}+N}\sum_{j=0}^{\ell+\widetilde{\ell}-1}\binom{\ell+\widetilde{\ell}-1+j}{j}\cos^{2j}\left(\frac{\omega}{2}\right)\\
&= 2\sin^{\widetilde{N}+N}\left(\frac{\omega}{2}\right)\sum_{j=0}^{\ell+\widetilde{\ell}-1}\binom{\ell+\widetilde{\ell}-1+j}{j}\cos^{2j}\left(\frac{\omega}{2}\right)
\end{aligned}\tag{9.74}$$

where the last identity in (9.74) results from the fact that $\widetilde{N}+N$ is an even integer. Combining (9.73) and (9.74) yields

$$\begin{aligned}
&\widetilde{H}(\omega)\overline{H(\omega)}+\widetilde{H}(\omega+\pi)\overline{H(\omega+\pi)}\\
&\quad= 2\left(\cos^{\widetilde{N}+N}\left(\frac{\omega}{2}\right)\sum_{j=0}^{\ell+\widetilde{\ell}-1}\binom{\ell+\widetilde{\ell}-1+j}{j}\sin^{2j}\left(\frac{\omega}{2}\right)\right.\\
&\qquad\left.+\sin^{\widetilde{N}+N}\left(\frac{\omega}{2}\right)\sum_{j=0}^{\ell+\widetilde{\ell}-1}\binom{\ell+\widetilde{\ell}-1+j}{j}\cos^{2j}\left(\frac{\omega}{2}\right)\right)
\end{aligned}$$

Now let $K=\ell+\widetilde{\ell}$ so that $\widetilde{N}+N=2(\widetilde{\ell}+\ell)=2K$. Rewriting the last identity gives

$$\begin{aligned}
&\widetilde{H}(\omega)\overline{H(\omega)}+\widetilde{H}(\omega+\pi)\overline{H(\omega+\pi)}\\
&\quad= 2\left(\cos^{2K}\left(\frac{\omega}{2}\right)\sum_{j=0}^{K-1}\binom{K-1+j}{j}\sin^{2j}\left(\frac{\omega}{2}\right)\right.\\
&\qquad\left.+\sin^{2K}\left(\frac{\omega}{2}\right)\sum_{j=0}^{K-1}\binom{K-1+j}{j}\cos^{2j}\left(\frac{\omega}{2}\right)\right).
\end{aligned}$$

Using Lemma 9.1, we see that the right side of this identity is two and we have proved (9.62). □

In Problem 9.49 you will show that the length of $\mathbf{h}$ in Theorem 9.3 is $2N+\widetilde{N}-1$. We now use Theorem 9.3 to state the following definition.

Definition 9.4 (Biorthogonal Spline Filter Pair). *Suppose that N and $\widetilde{N}$ are either both even or both odd. Any biorthogonal filter pair formed using Theorem 9.3 is called the* $(2N+\widetilde{N}-1,\widetilde{N})$ biorthogonal spline filter pair *or a* biorthogonal spline filter pair. □

Using the Daubechies Theorem

Let's look at an example that implements Theorem 9.3.

Example 9.5 (Creating Biorthogonal Spline Filter Pairs Using Theorem 9.3). *Let's create a biorthogonal spline filter pair (and the corresponding highpass filters) using Theorem 9.3 for $\widetilde{N}=2$ and $N=4$. In this case, $\widetilde{\ell}=1$ and $\ell=2$. For $\widetilde{\mathbf{h}}$ we have*

$$\widetilde{H}(\omega)=\sqrt{2}\cos^2\left(\frac{\omega}{2}\right)=\frac{\sqrt{2}}{4}e^{-i\omega}+\frac{\sqrt{2}}{2}+\frac{\sqrt{2}}{4}e^{i\omega}.$$

For this Fourier series, we have $\widetilde{\mathbf{h}}=(h_{-1},h_0,h_1)=\left(\frac{\sqrt{2}}{4},\frac{\sqrt{2}}{2},\frac{\sqrt{2}}{4}\right)$. We now use (9.72) to write

$$\begin{aligned}H(\omega)&=\sqrt{2}\cos^4\left(\frac{\omega}{2}\right)\sum_{j=0}^{2}\binom{2+j}{j}\sin^{2j}\left(\frac{\omega}{2}\right)\\&=\sqrt{2}\cos^4\left(\frac{\omega}{2}\right)\left(1+3\sin^2\left(\frac{\omega}{2}\right)+6\sin^2\left(\frac{\omega}{2}\right)\right)\\&=\sqrt{2}\left(\frac{e^{i\omega/2}+e^{-i\omega/2}}{2}\right)^4\cdot\\&\qquad\left(1+3\left(\frac{e^{i\omega/2}-e^{-i\omega/2}}{2i}\right)^2+6\left(\frac{e^{i\omega/2}-e^{-i\omega/2}}{2i}\right)^4\right).\end{aligned}$$

After some tedious algebra computations, we find

$$\begin{aligned}H(\omega)=&\frac{3\sqrt{2}}{128}e^{-4i\omega}-\frac{3\sqrt{2}}{64}e^{-3i\omega}-\frac{\sqrt{2}}{8}e^{-2i\omega}+\frac{19\sqrt{2}}{64}e^{-i\omega}\\&+\frac{45\sqrt{2}}{64}+\frac{19\sqrt{2}}{64}e^{i\omega}+\frac{\sqrt{2}}{8}e^{2i\omega}-\frac{3\sqrt{2}}{64}e^{3i\omega}+\frac{3\sqrt{2}}{128}e^{4i\omega}.\end{aligned}$$

We can read the symmetric filter $\mathbf{h}$ from the Fourier series. We have

$$\begin{aligned}\mathbf{h}&=(h_{-4},h_{-3},h_{-2},h_{-1},h_0,h_1,h_2,h_3,h_4)\\&=\left(\frac{3\sqrt{2}}{128},-\frac{3\sqrt{2}}{64},-\frac{\sqrt{2}}{8},\frac{19\sqrt{2}}{64},\frac{45\sqrt{2}}{64},\frac{19\sqrt{2}}{64},-\frac{\sqrt{2}}{8},-\frac{3\sqrt{2}}{64},\frac{3\sqrt{2}}{128}\right).\end{aligned}$$

We use Proposition 9.2 to note that $\widetilde{g}_k=(-1)^k h_{1-k}$, $k=-3,\ldots,5$ and $g_k=(-1)^k\widetilde{h}_{1-k}$, $k=0,1,2$. We have

$$\begin{aligned}\widetilde{\mathbf{g}}&=(\widetilde{g}_{-3},\widetilde{g}_{-2},\widetilde{g}_{-1},\widetilde{g}_0,\widetilde{g}_1,\widetilde{g}_2,\widetilde{g}_3,\widetilde{g}_4,\widetilde{g}_5)\\&=(-h_4,h_3,-h_2,h_1,-h_0,h_{-1},-h_{-2},h_{-3},-h_{-4})\\&=\left(-\frac{3\sqrt{2}}{128},-\frac{3\sqrt{2}}{64},\frac{\sqrt{2}}{8},\frac{19\sqrt{2}}{64},-\frac{45\sqrt{2}}{64},\frac{19\sqrt{2}}{64},\frac{\sqrt{2}}{8},-\frac{3\sqrt{2}}{64},-\frac{3\sqrt{2}}{128}\right).\end{aligned}$$

and

$$\begin{aligned}\mathbf{g} &= (g_0, g_1, g_2)\\ &= \left(\widetilde{h}_1, -\widetilde{h}_0, \widetilde{h}_{-1}\right)\\ &= \left(\frac{\sqrt{2}}{4}, -\frac{\sqrt{2}}{2}, \frac{\sqrt{2}}{4}\right).\end{aligned}$$

(Live example: Visit `stthomas.edu/wavelets` *and click on Live Examples.)*

□

In Problem 9.45, you will show that the biorthogonal filter pair obtained in Section 7.2 can be constructed using Theorem 9.3 with $\widetilde{N} = N = 3$.

Which Biorthogonal Filter Pair to Use?

Using Definition 9.3 and Theorem 9.3, we can easily generate many biorthogonal filter pairs for use in applications. How do we know which filter pair to choose?

As a rule of thumb, longer filters $\widetilde{\mathbf{h}}$ whose Fourier series $\widetilde{H}(\omega)$ satisfies $\widetilde{H}^{(m)}(\pi) = 0$ for a relatively large m return a more accurate approximation of the original data in the averages portion of the transformation. The classical treatment of wavelet theory develops the lowpass filters by creating *scaling functions* that are building blocks used to represent the space of square-integrable functions. Here, the designer has control of features such as orthogonality, symmetry, and smoothness. The length of the resulting lowpass filter is directly proportional to the smoothness of the scaling functions. We touched on this fact in our earlier discussion of the spline scaling functions $B_N(t)$ on page 404. Note that as N increases, the spline function $B_N(t)$ becomes smoother and the resulting spline filter grows in length.

One disadvantage of using a long filter is that the computation speed is slower. Another problem is that biorthogonal spline filter pairs often have lengths where one filter is quite a bit longer than the other (see Problem 9.49). This can be problematic since $\widetilde{\mathbf{g}}$ is constructed from $\mathbf{h}$, and if $\mathbf{h}$ is substantially longer say than $\widetilde{h}$, then the resulting wavelet transformation matrix $\widetilde{W}$ is built with filters where the approximation ability of the averages portion of the transformation is not as good as that of the differences portion of the transformation. In applications such as image compression, we prefer filter pairs whose lengths are similar. In Section 9.5 we address this problem by constructing a symmetric biorthogonal filter pair whose lengths differ by 2.

PROBLEMS

9.33 Let

$$\mathbf{h} = (h_{-3}, h_{-2}, h_{-1}, h_0, h_1, h_2, h_3, h_4) = (h_4, h_3, h_2, h_1, h_1, h_2, h_3, h_4)$$

and suppose $H(\omega)$ is the associated Fourier series. Show that (9.63) is equivalent to $H'(\pi) = 0$.

9.34 Suppose $\mathbf{h}$ is a symmetric filter of length $2L+1$.

(a) Show that

$$H(\omega) = h_0 + \sum_{k=1}^{L} h_k e^{ik\omega} + \sum_{k=1}^{L} h_k e^{-ik\omega}.$$

(b) Use (a) and Problem 8.19 to show that

$$H(\omega) = h_0 + 2\sum_{k=1}^{L} h_k \cos(k\omega).$$

(c) Show that $H'(\pi) = 0$.

(d) Explain why $H^{(m)}(\pi) = 0$ whenever m is odd.

9.35 Suppose $\mathbf{h}$ is a symmetric, even-length filter with associated Fourier series $H(\omega)$. Show that $H(\pi) = 0$.

9.36 Suppose that

$$\mathbf{h} = (h_{-L}, h_{-L+1}, \ldots, h_0, \ldots, h_{L-1}, h_L) = (h_L, h_{L-1}, \ldots, h_0, \ldots, h_{L-1}, h_L)$$

is a symmetric, odd-length lowpass filter with L even and associated Fourier series $H(\omega)$ that satisfies $H(0) = \sqrt{2}$. Show that

$$h_0 + 2\left(h_2 + h_4 + \cdots + h_L\right) = \frac{\sqrt{2}}{2}$$

and

$$h_1 + h_3 + \cdots + h_{L-1} = \frac{\sqrt{2}}{4}.$$

What can you infer if L is odd?

9.37 Suppose that $\mathbf{h}$ and $\widetilde{\mathbf{h}}$ are filters whose Fourier series $H(\omega)$ and $\widetilde{H}(\omega)$, respectively, satisfy $\widetilde{H}(0) = \sqrt{2}$, $\widetilde{H}(\pi) = 0$, and (9.62). Show that $H(0) = \sqrt{2}$.

9.38 Show that the filter given by

$$\widetilde{H}(\omega) = \sqrt{2}\, e^{i\omega/2} \cos^{\widetilde{N}}\left(\frac{\omega}{2}\right)$$

has the form given by (9.65) and show that $\widetilde{h}_k = \widetilde{h}_{1-k}$ for $k = -\frac{\widetilde{N}-1}{2}, \ldots, \frac{\widetilde{N}+1}{2}$.

9.39 Write down the spline filter of length $\widetilde{N}+1$ where

(a) $\widetilde{N} = 1$

(b) $\widetilde{N} = 3$

(c) $\widetilde{N} = 5$

9.40 Use Definition 9.3 to show the relation (9.69) holds. Thus, the spline filter of length $\widetilde{N} + 1$ can be written using the $\widetilde{N} + 1$ row of Pascal's triangle.

9.41 Use the defining relation (9.70) to verify the formula for the triangle function (9.71).

9.42 Use (9.70) and (9.71) to find an explicit representation for the piecewise quadratic B-spline $B_2(t)$.

9.43 Complete the proof that $\widetilde{H}(\omega)$ and $H(\omega)$ satisfy (9.62) in Theorem 9.3 in the case where both $\widetilde{N}$ and N are odd. *Hint:* Take $K = \ell + \widetilde{\ell} + 1$ and mimic the proof when $\widetilde{N}$ N are both even.

9.44 What "biorthogonal" spline filter pair do you get if you take $N = \widetilde{N} = 1$ in Theorem 9.3 and Definition 9.3?

9.45 Consider the spline filter $\widetilde{\mathbf{h}}$ given by (9.67) for $\widetilde{N} = 3$. Show that if we take this filter and $N = 3$, Theorem 9.3 produces the eight-term filter $\mathbf{h}$ constructed in Definition 7.3.

9.46 Use Theorem 9.3 and Definition 9.3 to find the biorthogonal spline filter pairs for the given values of $\widetilde{N}$ and N. You might wish to use a **CAS** for some of the computations.

(a) $\widetilde{N} = 3$, $N = 1$

(b) $\widetilde{N} = 1$, $N = 3$

(c) $\widetilde{N} = 4$, $N = 2$

9.47 For each of the filter pairs you found in Problem 9.46, find m and $\widetilde{m}$ such that

$$H^{(k)}(\pi) = 0,\ k = 0, \ldots, m-1 \qquad \text{and} \qquad \widetilde{H}^{(k)}(\pi) = 0,\ k = 0, \ldots, \widetilde{m}-1$$

but $H^{(m)}(\pi) \neq 0$ and $\widetilde{H}^{(\widetilde{m})}(\pi) \neq 0$.

9.48 Let $\widetilde{H}(\omega)$ be the Fourier series given by (9.64). Show that $\widetilde{H}^{(k)}(\pi) = 0$ for $k = 0, \ldots, \widetilde{N} - 1$ and $\widetilde{H}^{(\widetilde{N})}(\pi) \neq 0$.

9.49 Suppose that N and $\widetilde{N}$ are either both even or both odd. In this problem you will show that the length of the filter $\mathbf{h}$ from Theorem 9.3 is $2N + \widetilde{N} - 1$. The following steps will help you organize your work.

(a) Consider the case where both N and $\widetilde{N}$ are even. Write $N = 2\ell$ and $\widetilde{N} = 2\widetilde{\ell}$. Argue that

$$\cos\left(\frac{\omega}{2}\right)^N = \cos\left(\frac{\omega}{2}\right)^{2\ell} = \sum_{k=-\ell}^{\ell} a_k e^{ik\omega}$$

and

$$\sum_{j=0}^{\ell+\tilde{\ell}-1} \binom{\ell+\tilde{\ell}-1+j}{j} \sin^{2j}\left(\frac{\omega}{2}\right) = \sum_{m=-\ell-\tilde{\ell}+1}^{\ell+\tilde{\ell}-1} b_k e^{im\omega}.$$

for numbers a_k, $k = -\ell, \ldots, \ell$ and b_m, $m = -\ell - \tilde{\ell} + 1, \ldots, \ell + \tilde{\ell} - 1$. Use Problem 8.19 from Section 8.1 to expand $\cos\left(\frac{\omega}{2}\right)$ and $\sin\left(\frac{\omega}{2}\right)$. We do not want to find a_k and b_m explicitly – we just want to know where these series start and stop.

(b) Multiply the results from (a) to show that $H(\omega)$ can be expanded as a Fourier series of the form

$$H(\omega) = \sum_{k=-2\ell+\tilde{\ell}-1}^{2\ell+\tilde{\ell}-1} c_k e^{ik\omega}$$

for some numbers c_k.

(c) Use (b) to show that the length of the Fourier series $H(\omega)$ is $4\ell + 2\tilde{\ell} - 1 = 2N + \tilde{N} - 1$, as desired.

(d) Repeat steps (a) – (c) for the case when both N and $\tilde{N}$ are odd.

9.50 Suppose that $\tilde{\mathbf{h}}$ and $\mathbf{h}$ are a biorthogonal spline filter pair constructed using Theorem 9.3. Then the length of $\tilde{\mathbf{h}}$ is $\tilde{N} + 1$, and from Problem 9.49 we know that the length of $\mathbf{h}$ is $2N + \tilde{N} - 1$.

(a) Show that the length of $\tilde{\mathbf{h}}$ is the same as the length of $\mathbf{h}$ if and only if $N = 1$.

(b) From (a) we see that odd-length biorthogonal filter pairs cannot have the same length. Is it possible to construct an odd-length biorthogonal filter pair using Theorem 9.3 so that the difference of the lengths of the filters is 2? If so, explain how you would choose $\tilde{N}$ and N.

9.51 Suppose that N and $\tilde{N}$ are both even. Then from Problem 9.49 we know that the length of $\mathbf{h} = (h_{-L}, \ldots, h_L)$ is $2N + \tilde{N} - 1$. We also know from Definition 9.3 that the length of $\tilde{\mathbf{h}} = (\tilde{h}_{-L'}, \ldots, \tilde{h}_{L'})$ is $\tilde{N} + 1$. Find L and L' in terms of N and $\tilde{N}$. Repeat the problem in the case where N and $\tilde{N}$ are both odd.

Computer Lab

9.1 Biorthogonal Spline Filter Pairs From `stthomas.edu/wavelets`, access the file `SplineFilters`. In this lab, you will construct code for creating biorthogonal spline filter pairs.

9.5 The Cohen–Daubechies–Feauveau 9/7 Filter

The CDF $9/7$ Filter Pair

The biorthogonal filter pair discussed in this section is involved in two of the most important applications of discrete wavelet transformations. The Cohen–Daubechies–Feauveau 9/7 biorthogonal filter pair (hereafter the CDF97 filter pair) is constructed in [23]. A group of scientists at Los Alamos National Laboratory used the CDF97 filter pair in their Wavelet Scalar Quantization Method [11] that was adopted by the Federal Bureau of Investigation for use in compressing digitized images of fingerprints. The filter pair was also chosen by the JPEG2000 committee for the lossy version of the JPEG2000 compression standard (see, for example [95]). We discuss the FBI fingerprint standard in Section 10.3 and JPEG2000 in Chapter 12. Although this filter pair is not a member of the biorthogonal spline filter pair family that we developed in Section 9.4, the filters are symmetric and their Fourier transformations satisfy the conditions necessary to classify them as lowpass filters.

The filter lengths of the CDF97 filter pair are relatively short and similar in length. Although it is possible to develop a biorthogonal spline filter pair with lengths 9 and 7 (see Problem 9.52), Unser and Blu [96] cite several reasons why the CDF97 filter pair is preferred over the $(9,7)$ biorthogonal spline filter pair. Unser and Blu report that the CDF97 filter pair produces a wavelet transformation matrix that is closer to orthogonal than that generated by the $(9,7)$ biorthogonal spline filter pair.[6]

Smoothness of CDF97 Fourier Series at $\omega = \pi$

A key advantage of the CDF97 over the $(9,7)$ biorthogonal spline pair is recognized when we analyze the Fourier series of each filter. Suppose that $\widetilde{\mathbf{h}}$ and $\mathbf{h}$ are the biorthogonal spline filter pair whose lengths are 7 and 9, respectively. These filters were constructed in Example 9.5. Let

$$\widetilde{H}(\omega) = \sum_{k=-3}^{3} \widetilde{h}_k e^{ik\omega} \qquad \text{and} \qquad H(\omega) = \sum_{k=-4}^{4} h_k e^{ik\omega}$$

denote the Fourier series associated with each filter. Then we can show (see Problem 9.53) that

$$H(\pi) = H'(\pi) = 0 \qquad \text{and} \qquad \widetilde{H}^{(m)}(\pi) = 0, \qquad m = 0, \ldots, 5 \tag{9.75}$$

The moduli of the Fourier series $H(\omega)$ and $\widetilde{H}(\omega)$ are plotted in Figure 9.9.

It turns out that the Fourier series $H(\omega)$ and $\widetilde{H}(\omega)$ for the CDF97 filters $\mathbf{h}$ and $\widetilde{\mathbf{h}}$, respectively, satisfy

$$H^{(m)}(\pi) = \widetilde{H}^{(m)}(\pi) = 0, \qquad m = 0, 1, 2, 3. \tag{9.76}$$

[6] The mathematics behind this argument is beyond the scope of this book.

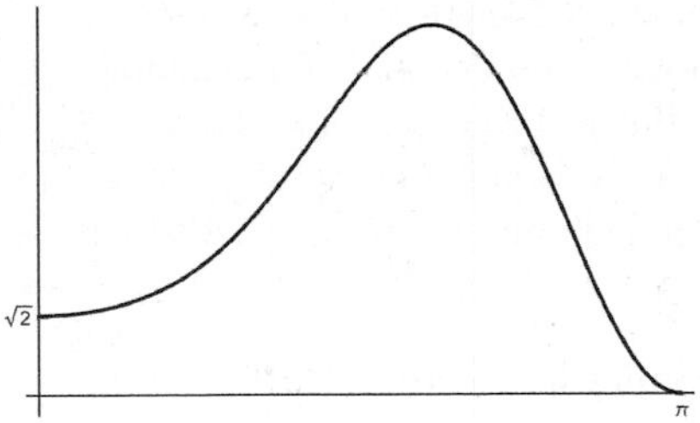

(a) $|H(\omega)|$

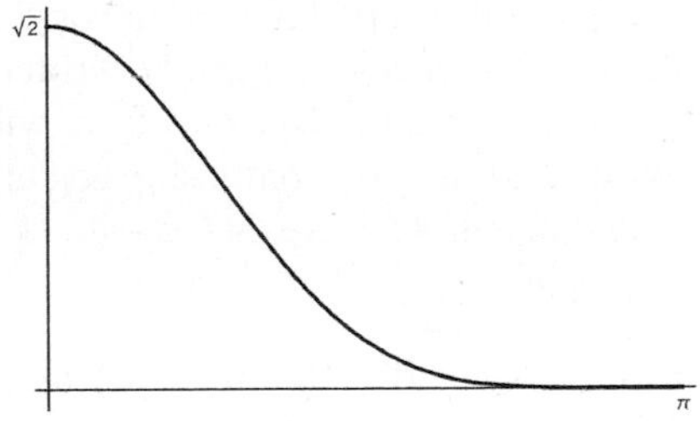

(b) $|\widetilde{H}(\omega)|$

Figure 9.9 $|H(\omega)|$ and $|\widetilde{H}(\omega)|$ for the $(9,7)$ biorthogonal spline filter pair. Note that, as indicated by (9.75), $H(\omega)$ is not very "flat" at $\omega = \pi$.

Thus, we see that while the combined number of zeros at $\omega = \pi$ of $\widetilde{H}(\omega)$ and $H(\omega)$ for each filter pair is eight, the zeros at $\omega = \pi$ of $\widetilde{H}(\omega)$ and $H(\omega)$ for the CDF97 filter pair are equally balanced.

As we work through the construction of the CDF97 filter pair, we will have a better understanding of why the zeros are equally balanced. The process for constructing the CDF97 filter pair requires us to better understand the form of the solutions to the biorthogonality condition

$$\widetilde{H}(\omega)\overline{H(\omega)} + \widetilde{H}(\omega + \pi)\overline{H(\omega + \pi)} = 2 \tag{9.77}$$

that is fundamental for constructing biorthogonal filter pairs.

Note: The results that follow are from Cohen et al. [23]. Whereas we consider only odd-length symmetric filters in this section, the results given in Cohen et al. [23] consider even-length symmetric filters as well.

$H(\omega)$ for Symmetric Odd-Length Filters

Suppose that $\mathbf{h} = (h_{-L}, \ldots, h_L)$ is an odd-length symmetric filter. In Problem 8.26 from Section 8.2 we showed that if $h_k = h_{-k}$, then the Fourier series $H(\omega) = \sum\limits_{k=-L}^{L} h_k e^{ik\omega}$ is an even function. We use part (c) of Problem 8.25 in Section 8.2 to write $H(\omega)$ as a linear combinations of cosine functions:

$$H(\omega) = h_0 + 2\sum_{k=1}^{L} h_k \cos(k\omega). \tag{9.78}$$

It turns out that we can write $H(\omega)$ in (9.78) as a linear combination of powers of $\cos(\omega)$. In other words, when we start with a symmetric, odd-length filter $\mathbf{h}$, $H(\omega)$

can be expressed as a polynomial in $\cos(\omega)$. In Problem 9.54, you will show that for $k \in \mathbf{Z}$, the function $\cos(k\omega)$ can be written as a polynomial in $\cos(\omega)$. For example, $\cos(3\omega) = 4\cos^3(\omega) - 3\cos(\omega)$ (you will verify this in Problem 9.55). Here, we have written $\cos(3\omega)$ by composing $\cos(\omega)$ and the polynomial $P(t) = 4t^3 - 3t$.

Using Problem 9.54, we can rewrite (9.78) as a polynomial in $\cos(\omega)$. Let's look at an example.

Example 9.6 (Writing Even $\mathbf{H(\omega)}$ as a Polynomial in Cosine). *Consider the length 7 symmetric filter*

$$\mathbf{h} = (h_{-3}, \ldots, h_3) = \left(\frac{3\sqrt{2}}{32}, -\frac{3\sqrt{2}}{8}, \frac{5\sqrt{2}}{32}, \frac{5\sqrt{2}}{4}, \frac{5\sqrt{2}}{32}, -\frac{3\sqrt{2}}{8}, \frac{3\sqrt{2}}{32}\right)$$

Using part (c) of Problem 8.25 in Section 8.2, we can write the associated Fourier series $H(\omega)$ as

$$H(\omega) = \frac{5\sqrt{2}}{4} + 2\left(\frac{5\sqrt{2}}{32}\cos(\omega) - \frac{3\sqrt{2}}{8}\cos(2\omega) + \frac{3\sqrt{2}}{32}\cos(3\omega)\right).$$

Since $\cos(2\omega) = 2\cos^2(\omega) - 1$ *and* $\cos(3\omega) = 4\cos^3(\omega) - 3\cos(\omega)$, *we have*

$$\begin{aligned} H(\omega) &= \frac{5\sqrt{2}}{4} + \frac{5\sqrt{2}}{16}\cos(\omega) - \frac{3\sqrt{2}}{4}(2\cos^2(\omega) - 1) + \frac{3\sqrt{2}}{16}(4\cos^3(\omega) - 3\cos(\omega)) \\ &= 2\sqrt{2} - \frac{\sqrt{2}}{4}\cos(\omega) - \frac{3\sqrt{2}}{2}\cos^2(\omega) + \frac{3\sqrt{2}}{4}\cos^3(\omega) \end{aligned}$$

Thus, we see $H(\omega)$ can be written as a cubic polynomial P in $\cos(\omega)$ *where*

$$P(t) = 2\sqrt{2} - \frac{\sqrt{2}}{4}t - \frac{3\sqrt{2}}{2}t^2 + \frac{3\sqrt{2}}{4}t^3. \tag{9.79}$$

□

Lowpass Conditions and $\mathrm{H}(\omega)$

For $\mathbf{h}$ to be a lowpass filter, we require $H(0) = \sqrt{2}$ and $H(\pi) = 0$. Moreover, if $H(\omega)$ must also satisfy $H^{(\ell)}(\pi) = 0$ for some $\ell = 0, 1, \ldots$, then $(1 + \cos(\omega))^\ell$ must be a factor of $H(\omega)$. Thus, we can write

$$H(\omega) = \sqrt{2}\,(1 + \cos(\omega))^\ell\, q(\cos(\omega))$$

for some polynomial q where $q(\cos(\pi)) = q(-1) \neq 0$. Since

$$\sqrt{2} = H(0) = \sqrt{2}\,(1 + \cos(0))^\ell\, q(\cos(0)) = \sqrt{2} \cdot 2^\ell \cdot q(1)$$

we also must have $q(1) = 2^{-\ell}$. Using the half-angle formula

$$\cos^2\left(\frac{\omega}{2}\right) = \frac{1}{2}\left(1 + \cos(\omega)\right) \qquad \text{or} \qquad 2\cos^2\left(\frac{\omega}{2}\right) = 1 + \cos(\omega)$$

we can write

$$H(\omega) = \sqrt{2}\cos^{2\ell}\left(\frac{\omega}{2}\right) 2^\ell q(\cos(\omega)).$$

If we let $p(\omega) = 2^\ell q(\omega)$, we have $p(-1) = 2^\ell q(-1) \neq 0$ and $p(1) = 2^\ell q(1) = 2^\ell \cdot 2^{-\ell} = 1$, so that

$$H(0) = \sqrt{2}\cos^{2\ell}(0)\, p(\cos(0)) = \sqrt{2}.$$

We summarize this discussion in Proposition 9.4.

Proposition 9.4 (H(ω) for Odd-Length Symmetric Filters). *Consider the filter* $\mathbf{h} = (h_{-L}, \ldots, h_L)$ *with* $h_k = h_{-k}$ *for* $k = 1, \ldots L$ *whose Fourier series* $H(\omega)$ *satisfies* $H(0) = \sqrt{2}$. *Then*

$$H(\omega) = \sum_{k=-L}^{L} h_k e^{ik\omega}$$

is an even function (see Problem 8.26 from Section 8.2) and $H(\omega)$ *can be written as a power of* $\cos\left(\frac{\omega}{2}\right)$ *times a polynomial in* $\cos(\omega)$. *That is,*

$$H(\omega) = \sqrt{2}\cos^{2\ell}\left(\frac{\omega}{2}\right) p(\cos(\omega)) \tag{9.80}$$

where $p(1) = 1$ *and* $p(-1) \neq 0$.

□

Let's apply Proposition 9.4 to the filter from Example 9.6.

Example 9.7 (Expressing H(ω) Using Proposition 9.4). *For the filter* $\mathbf{h}$ *in Example 9.6, we found that* $H(\omega)$ *could be expressed in terms of* $P(\cos(\omega))$ *where* P *is given by* (9.79). *We factor out* $\sqrt{2}/4(1+t)$ *to obtain*

$$P(t) = \frac{\sqrt{2}}{4}(1+t)(3t^2 - 9t + 8).$$

Our goal is to write $H(\omega)$ *in the form given by* (9.80). *Toward this end, we substitute* $t = \cos(w)$ *and use the identity* $1 + \cos(\omega) = 2\cos^2\left(\frac{\omega}{2}\right)$ *to write*

$$\begin{aligned}
H(\omega) &= P(\cos(\omega)) \\
&= \frac{\sqrt{2}}{4}(1 + \cos(\omega))(3\cos^2(\omega) - 9\cos(\omega) + 8) \\
&= \frac{\sqrt{2}}{2}\cos^2\left(\frac{\omega}{2}\right)(3\cos^2(\omega) - 9\cos(\omega) + 8) \\
&= \sqrt{2}\cos^2\left(\frac{\omega}{2}\right)\left(\frac{3}{2}\cos^2(\omega) - \frac{9}{2}\cos(\omega) + 4\right)
\end{aligned}$$

so we see that $\ell = 1$ *and* $p(t)$ *from Proposition 9.4 is*

$$p(t) = \frac{3}{2}t^2 - \frac{9}{2}t + 4.$$

It is easy to check that $p(-1) \neq 0$ *and* $p(1) = 1$. □

Existence of Even Solutions to the Orthogonality Condition

The next result says that for a given *even* $H(\omega)$, if we find a solution $\widetilde{H}(\omega)$ to the orthogonality condition (9.77), then we can construct an even solution. This result was stated and proved in Cohen et al. [23].

Proposition 9.5 (Even Solutions to the Orthogonality Condition). *Suppose that* $H(\omega)$ *is given and* $H(\omega) = H(-\omega)$. *If there exists a solution* $\widetilde{H}(\omega)$ *to (9.77), then there exists an even solution.*

□

Proof. If $\widetilde{H}(\omega) = \widetilde{H}(-\omega)$, then the proof is complete. Otherwise, consider the Fourier series

$$E(\omega) = \frac{1}{2}\left(\widetilde{H}(\omega) + \widetilde{H}(-\omega)\right).$$

It is easy to check that $E(\omega) = E(-\omega)$. We need to verify that $E(\omega)$ satisfies (9.77). Toward this end, we first compute

$$E(\omega)\overline{H(\omega)} = \frac{1}{2}(\widetilde{H}(\omega)+\widetilde{H}(-\omega))\overline{H(\omega)} = \frac{1}{2}\widetilde{H}(\omega)\overline{H(\omega)}+\frac{1}{2}\widetilde{H}(-\omega)\overline{H(\omega)} \tag{9.81}$$

and next compute

$$\begin{aligned} E(\omega+\pi)\overline{H(\omega+\pi)} &= \frac{1}{2}(\widetilde{H}(\omega+\pi) + \widetilde{H}(-(\omega+\pi)))\overline{H(\omega+\pi)} \\ &= \frac{1}{2}\widetilde{H}(\omega+\pi)\overline{H(\omega+\pi)} + \frac{1}{2}\widetilde{H}(-(\omega+\pi))\overline{H(\omega+\pi)} \end{aligned} \tag{9.82}$$

We next add $E(\omega)\overline{H}(\omega)$ and $E(\omega+\pi)\overline{H(\omega+\pi)}$. Adding the first terms of (9.81) and (9.82) gives

$$\frac{1}{2}\widetilde{H}(\omega)\overline{H(\omega)} + \frac{1}{2}\widetilde{H}(\omega+\pi)\overline{H(\omega+\pi)} = \frac{1}{2}(\widetilde{H}(\omega)\overline{H(\omega)} + \widetilde{H}(\omega+\pi)\overline{H(\omega+\pi)})$$

But $\widetilde{H}(\omega)$ is a solution to (9.77) so this sum is 1. Thus we have

$$\begin{aligned} &E(\omega)\overline{H(\omega)} + E(\omega+\pi)\overline{H(\omega+\pi)} \\ &\qquad = 1 + \frac{1}{2}\left(\widetilde{H}(-\omega)\overline{H(\omega)} + \widetilde{H}(-\omega-\pi)\overline{H(\omega+\pi)}\right) \end{aligned} \tag{9.83}$$

We need to replace the $-\omega$ that appears as an argument in the functions in (9.83). In Problem 9.56 you will verify that replacing ω by $-\omega$ leaves the left side of (9.83) unchanged. Let's work on the right side of (9.83). In particular, let

$$g(\omega) = \widetilde{H}(-\omega)\overline{H(\omega)} + \widetilde{H}(-\omega-\pi)\overline{H(\omega+\pi)}.$$

We replace ω by $-\omega$ and use the fact that $H(\omega)$ is even and both $H(\omega)$ and $\widetilde{H}(\omega)$ are 2π-periodic functions in the second term to write

$$\begin{aligned} g(-\omega) &= \widetilde{H}(\omega)\overline{H(-\omega)} + \widetilde{H}(\omega-\pi)\overline{H(-(\omega-\pi))} \\ &= \widetilde{H}(\omega)\overline{H(\omega)} + \widetilde{H}(\omega-\pi)\overline{H(\omega-\pi)} \\ &= \widetilde{H}(\omega)\overline{H(\omega)} + \widetilde{H}(\omega-\pi+2\pi)\overline{H(\omega-\pi+2\pi)} \\ &= \widetilde{H}(\omega)\overline{H(\omega)} + \widetilde{H}(\omega+\pi)\overline{H(\omega+\pi)} = 2 \end{aligned}$$

since $\widetilde{H}(\omega)$ is a solution to (9.77). Inserting this result into the second term on the right side of (9.83) gives the desired result. □

To summarize, if $\mathbf{h}$ is a symmetric, odd-length filter, then we can write its Fourier series in the form

$$H(\omega) = \sqrt{2}\cos^{2\ell}\left(\frac{\omega}{2}\right)p(\cos(\omega)) \tag{9.84}$$

where ℓ is a nonnegative integer and p is a polynomial with $p(-1) \neq 0$ and $p(1) = 1$.

Moreover, Proposition 9.5 tells us that for this even $H(\omega)$, if we can find a solution to (9.77), then we can find an even solution to (9.77). Combining this result with Proposition 9.4 tells us that the even solution $\widetilde{H}(\omega)$ to (9.77) can be expressed as

$$\widetilde{H}(\omega) = \sqrt{2}\cos^{2\widetilde{\ell}}\left(\frac{\omega}{2}\right)\widetilde{p}(\cos(\omega)) \tag{9.85}$$

where $\widetilde{\ell}$ is a nonnegative integer and $\widetilde{p}$ is a polynomial with $\widetilde{p}(-1) \neq 0$ and $\widetilde{p}(1) = 1$.

Characterizing Solutions to (9.77)

We are now ready for the main result of this section. Theorem 9.4, due to Cohen et al. [23],[7] allows us to characterize the polynomials p and $\widetilde{p}$ given in (9.84) and (9.85), respectively, and thus develop many different solutions to the orthogonality condition (9.77).

[7]We only state and prove the result for odd-length symmetric filters. The result is valid for even-length symmetric filters as well. Moreover, we have only given a special case of the result. The result, Proposition 6.4 in [23], gives an even more general constraint that the product of p and $\widetilde{p}$ must satisfy. This general constraint leads to even more possible formulations of $\widetilde{\mathbf{h}}$ and $\mathbf{h}$.

Theorem 9.4 (Characterizing Solutions to (9.77)**).** *Suppose that* $\widetilde{\mathbf{h}}$ *and* $\mathbf{h}$ *are odd-length, symmetric filters whose Fourier series given by* (9.85) *and* (9.84)*, respectively, solve* (9.77)*. Then the polynomials* p *and* $\widetilde{p}$ *must satisfy*

$$p(\cos(\omega))\,\widetilde{p}(\cos(\omega)) = \sum_{j=0}^{K-1} \binom{K-1+j}{j} \sin^{2j}\left(\frac{\omega}{2}\right) \tag{9.86}$$

where $K = \ell + \widetilde{\ell}$ *and the degree of* $p(t)\widetilde{p}(t)$ *is less than* K. □

Proof. The proof of Theorem 9.4 is quite technical. For this reason, we leave it as (challenging!) Problem 9.57. □

You should recognize the right-hand side of (9.86) – it appeared as part of $H(\omega)$ for odd-length biorthogonal spline filter pairs in Theorem 9.3.[8] That is, the Fourier series for the biorthogonal spline filter pair $\widetilde{\mathbf{h}}$ and $\mathbf{h}$ are

$$\widetilde{H}(\omega) = \sqrt{2}\cos^{\widetilde{N}}\left(\frac{\omega}{2}\right)$$

and

$$H(\omega) = \sqrt{2}\cos^{N}\left(\frac{\omega}{2}\right) \sum_{j=0}^{K-1} \binom{K-1+j}{j} \sin^{2j}\left(\frac{\omega}{2}\right)$$

where $\widetilde{N} = 2\widetilde{\ell}$ and $N = 2\ell$ are nonnegative, even integers. So we see now that this filter pair is simply a special case of Theorem 9.4 with $\widetilde{p}(t) = 1$ and $p(t) = \sum_{j=0}^{K-1} \binom{K-j+1}{j} t^j$.

The Construction of Cohen, Daubechies, and Feauveau

In developing the CDF97 filter pair, Cohen et al. decided to factor the right-hand side of (9.86) and spread the factors as equally as possibly between $\widetilde{p}$ and p. This differs from the spline case where all the factors of the right-hand side of (9.86) are assigned to $p(t)$. Thus, the idea is to choose a K and then factor

$$P(t) = \sum_{j=0}^{K-1} \binom{K-1+j}{j} t^j.$$

Recall that the roots of any polynomial with real coefficients are either real or occur in complex conjugate pairs. Cohen et al. sought to build one of the polynomials (say, p) by multiplying the linear factors built from the complex roots of $P(t)$ and the other (say, $\widetilde{p}$) as the product of the linear factors from the real roots of $P(t)$. In

[8] The right side of (9.86) also appears in the Fourier series for the Coiflet filter of Theorem 9.2.

Problem 9.58 you will find the roots of $P(t)$ where $K = 2$ or 3. When $K = 1$ there are no roots, and in the other cases there are only real roots or only complex roots. The first case, where there is a combination of roots, is when $K = 4$. The CDF97 filter pair is built from this case. When $K = 4$, we take $\ell = \widetilde{\ell} = 2$ and

$$P(t) = \sum_{j=0}^{3} \binom{3+j}{j} t^j = 1 + 4t + 10t^2 + 20t^3.$$

Using a **CAS**, we can find that the roots (to six digits) of $P(t)$ are

$$r_1 = -0.342484 \qquad r_2 = -0.078808 - .373391i \qquad r_3 = -0.078808 + .373391i.$$

Since the leading coefficient of $P(t)$ is 20, we have

$$P(t) = 20(t - r_1)(t - r_2)(t - r_3).$$

We build $\widetilde{p}$ using r_1 and p from r_2 and r_3. The question is how to distribute the leading coefficient 20 of $P(t)$. For now, we write

$$\widetilde{p}(t) = a(t - r_1) \qquad \text{and} \qquad p(t) = \frac{20}{a}(t - r_2)(t - r_3).$$

We now use (9.84) and (9.85) with $\widetilde{\ell} = \ell = 2$ and $\widetilde{p}(t) = a(t + 0.342484)$. If we make the substitution $t = \sin^2\left(\frac{\omega}{2}\right)$, we can write $\widetilde{H}(\omega)$ as

$$\widetilde{H}(\omega) = \sqrt{2}\cos^4\left(\frac{\omega}{2}\right) a\left(\sin^2\left(\frac{\omega}{2}\right) + 0.342484\right) \tag{9.87}$$

and

$$H(\omega) = \sqrt{2}\cos^4\left(\frac{\omega}{2}\right) \frac{20}{a}\left(\sin^2\left(\frac{\omega}{2}\right) + 0.078808 + 0.373391i\right) \cdot \left(\sin^2\left(\frac{\omega}{2}\right) + 0.078808 - 0.373391i\right). \tag{9.88}$$

We can find a by remembering that both $\widetilde{H}(\omega)$ and $H(\omega)$ must satisfy the lowpass condition $\widetilde{H}(0) = H(0) = \sqrt{2}$. So we can plug $\omega = 0$ into, say, (9.87), set the result equal to $\sqrt{2}$, and solve for a. We can then plug the value for a into (9.88). We have

$$\sqrt{2} = \widetilde{H}(0) = \sqrt{2}\cos^4(0)a\left(\sin^2(0) + 0.342484\right) = \sqrt{2}\,(0.342484)a$$

so that $a = 2.920696$.

The last step in the construction is easy but tedious. We first insert a into (9.87) and (9.88). Next we must expand each of (9.87) and (9.88) into Fourier series. The coefficients of these Fourier series will be our filter coefficients.

To expand (9.87) and (9.88) into Fourier series, we replace $\cos\left(\frac{\omega}{2}\right)$ and $\sin\left(\frac{\omega}{2}\right)$ by $\frac{e^{i\omega/2}+e^{-i\omega/2}}{2}$, $\frac{e^{i\omega/2}-e^{-i\omega/2}}{2i}$, respectively, and then use a **CAS** to expand both series. In so doing (see Problem 9.60), we see that $\widetilde{H}(\omega)$ consists of seven terms and $H(\omega)$ has nine terms. That is, to six digits,

$$H(\omega) = \sum_{k=-4}^{4} h_k e^{ik\omega} \quad \text{and} \quad \widetilde{H}(\omega) = \sum_{k=-3}^{3} \widetilde{h}_k e^{ik\omega}$$

where

$$\boxed{\begin{aligned} h_0 &= 0.852699 \\ h_{-1} = h_1 &= 0.377403 \\ h_{-2} = h_2 &= -0.110624 \\ h_{-3} = h_3 &= -0.023850 \\ h_{-4} = h_4 &= 0.037829 \end{aligned}} \quad \text{and} \quad \boxed{\begin{aligned} \widetilde{h}_0 &= 0.788486 \\ \widetilde{h}_{-1} = \widetilde{h}_1 &= 0.418092 \\ \widetilde{h}_{-2} = \widetilde{h}_2 &= -0.040689 \\ \widetilde{h}_{-3} = \widetilde{h}_3 &= -0.064539 \end{aligned}}. \tag{9.89}$$

We have the following definition.

Definition 9.5 (Cohen–Daubechies–Feauveau 9/7 Filter Pair). *We define the* Cohen–Daubechies–Feauveau 9/7 biorthogonal filter pair *by the values given in* (9.89). *We call this filter pair the* CDF97 filter pair. □

In Problem 9.61 you will show that the Fourier series $\widetilde{H}(\omega)$, $H(\omega)$ for the CDF97 filters $\widetilde{\mathbf{h}}$ and $\mathbf{h}$, respectively, satisfy

$$\widetilde{H}^{(m)}(\pi) = H^{(m)}(\pi) = 0, \qquad m = 0, 1, 2, 3.$$

In Figure 9.10 we have plotted $|H(\omega)|$ and $|\widetilde{H}(\omega)|$ for the CDF97 filter pair. You can see both filters are equally "flat" at $\omega = \pi$. Compare these functions to those plotted in Figure 9.9.

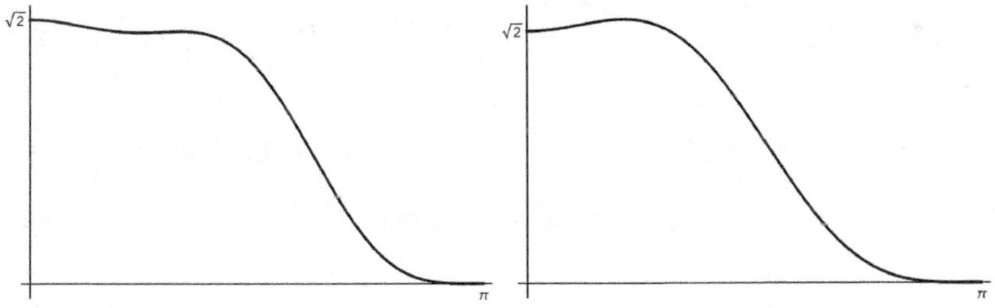

(a) $|H(\omega)|$ (b) $|\widetilde{H}(\omega)|$

Figure 9.10 Plots of $|H(\omega)|$ and $|\widetilde{H}(\omega)|$ for the CDF97 filter.

The Wavelet Filters $\widetilde{\mathbf{g}}$ and $\mathbf{g}$

We use the element-wise formulas from Proposition 9.2 to compute the associated wavelet filters $\widetilde{\mathbf{g}}$ and $\mathbf{g}$ for the CDF97 filter pair. The results are displayed in Table 9.3.

Table 9.3 The highpass filter pair $\mathbf{g}$ and $\widetilde{\mathbf{g}}$ for the CDF97 filter pair.

$\widetilde{\mathbf{g}}_\mathbf{k}$	$(-1)^\mathbf{k}\mathbf{h}_{1-\mathbf{k}}$		$\mathbf{g}_\mathbf{k}$	$(-1)^\mathbf{k}\widetilde{\mathbf{h}}_{1-\mathbf{k}}$	
$\widetilde{g}_{-3}$	$-h_4$	-0.037829			
$\widetilde{g}_{-2}$	h_3	-0.023850	g_{-2}	$\widetilde{h}_3$	-0.064539
$\widetilde{g}_{-1}$	$-h_2$	0.110624	g_{-1}	$-\widetilde{h}_2$	0.040689
$\widetilde{g}_0$	h_1	0.377403	g_0	$\widetilde{h}_1$	0.418092
$\widetilde{g}_1$	$-h_0$	-0.852699	g_1	$-\widetilde{h}_0$	-0.788486
$\widetilde{g}_2$	h_1	0.377403	g_2	$\widetilde{h}_1$	0.418092
$\widetilde{g}_3$	$-h_2$	0.110624	g_3	$-\widetilde{h}_2$	0.040689
$\widetilde{g}_4$	h_3	-0.023850	g_4	$\widetilde{h}_3$	-0.064539
$\widetilde{g}_5$	$-h_4$	-0.037829			

PROBLEMS

9.52 What values of $\widetilde{N}$ and N produce the $(9,7)$ biorthogonal spline filter pair?

9.53 Use the software you wrote in Computer Lab 9.1 from Section 9.4 to find the $(9,7)$ biorthogonal spline filter pair. Form the Fourier series $H(\omega)$ and $\widetilde{H}(\omega)$ for each filter and show that the formulas (9.75) hold. Show also that $H''(\pi) \neq 0$ and $\widetilde{H}^{(6)}(\pi) \neq 0$. *Hint:* A **CAS** is useful for this problem.

9.54 In this problem you show that for any nonnegative integer k, $\cos(k\omega)$ can be written as a polynomial in $\cos(\omega)$.

(a) By DeMoivre's theorem (Problem 8.21 from Section 8.1), we know that

$$(\cos(\omega) + i\sin(\omega))^k = \cos(k\omega) + i\sin(k\omega).$$

Use the binomial theorem (Problem 8.43 from Section 8.2) to show that for any nonnegative integer k, we have

$$\cos(k\omega) + i\sin(k\omega) = \sum_{j=0}^{k} \binom{k}{j} \cos^{k-j}(\omega) i^j \sin^j(\omega).$$

(b) Identify the real portion of the right-hand side of the identity in (a). Can you give an explicit formula for this real part?

(c) Use (b) to equate the real parts of the identity in (a) to complete the proof.

9.55 Use the ideas from Problem 9.54 to verify the following trigonometric identities:

(a) $\cos(2\omega) = 2\cos^2(\omega) - 1$

(b) $\cos(3\omega) = 4\cos^3(\omega) - 3\cos(\omega)$

(c) $\cos(4\omega) = 8\cos^4(\omega) - 8\cos^2(\omega) + 1$

9.56 Let $g(\omega) = E(\omega)\overline{H(\omega)} + E(\omega+\pi)\overline{H(\omega+\pi)}$ and assume that $E(\omega)$ and $H(\omega)$ are even, 2π-periodic functions. Show that $g(-\omega) = g(\omega)$.

9.57 In this problem you will prove Theorem 9.4. This is a challenging problem, so we have divided the proof into the following steps.

(a) Insert (9.85) and (9.84) into (9.77) and use trigonometric identities for $\cos\left(t+\frac{\pi}{2}\right)$ and $\cos(t+\pi)$ to show that

$$\cos^{2K}\left(\frac{\omega}{2}\right)p(\cos(\omega))\widetilde{p}(\cos(\omega)) + \sin^{2K}\left(\frac{\omega}{2}\right)p(-\cos(\omega))\widetilde{p}(-\cos(\omega)) = 1$$

where $K = \ell + \widetilde{\ell}$.

(b) Explain why we can always rewrite $p(\cos(\omega))$ and $\widetilde{p}(\cos(\omega))$ as polynomials whose arguments are $\frac{1-\cos(\omega)}{2}$.

(c) Let $P(t) = p(t)\widetilde{p}(t)$. Use the trigonometric identities $\sin^2\left(\frac{\omega}{2}\right) = \frac{1-\cos(\omega)}{2}$ and $\cos^2\left(\frac{\omega}{2}\right) = \frac{1+\cos(\omega)}{2}$ to rewrite the identity in (a) as

$$\cos^{2K}\left(\frac{\omega}{2}\right)P\left(\sin^2\left(\frac{\omega}{2}\right)\right) + \sin^{2K}\left(\frac{\omega}{2}\right)P\left(\cos^2\left(\frac{\omega}{2}\right)\right) = 1$$

or

$$(1-t)^K P(t) + t^K P(1-t) = 1 \tag{9.90}$$

where $t = \sin^2\left(\frac{\omega}{2}\right)$.

(d) The proof now relies on a result known as *Bezout's Theorem.* We state below the version of this theorem given in [32].

Bezout's Theorem: If p_1 and p_2 are polynomials of degree n_1 and n_2, respectively, and if p_1 and p_2 have no common zeros, then there exist unique polynomials q_1 and q_2 of degree at most $n_2 - 1$ and $n_1 - 1$, respectively, so that

$$p_1(t)q_1(t) + p_2(t)q_2(t) = 1. \tag{9.91}$$

Note that the polynomials $p_1(t) = (1-t)^K$ and $p_2(t) = t^K$ share no common zeros and both are degree K polynomials. Thus, by Bezout's theorem, there exist *unique* polynomials $q_1(t)$ and $q_2(t)$, both of degree at most $K - 1$, such that

$$(1-t)^K q_1(t) + t^K q_2(t) = 1.$$

Replace t by $1-t$ and use Bezout's theorem to explain why $q_2(t) = q_1(1-t)$.

(e) Use (d) to show that

$$(1-t)^K q_1(t) + t^K q_1(1-t) = 1.$$

so that $P(t) = q_1(t)$ solves (9.90) and must be a polynomial of at most $K-1$.

(f) Use (9.90) to show that

$$P(t) = (1-t)^{-K} - t^K(1-t)^{-K}P(1-t).$$

(g) Show that the Maclaurin series for $f(t) = (1-t)^{-K}$ is

$$f(t) = \sum_{j=0}^{\infty} \binom{K+j-1}{j} t^j.$$

Find the interval of convergence for this series.

(h) Insert the series you found in (g) into the identity from (f). Explain why

$$P(t) = \sum_{j=0}^{K-1} \binom{K+j-1}{j} t^j.$$

Hint: What is the degree of $P(t)$? Replacing t by $\sin^2\left(\frac{\omega}{2}\right)$ completes the proof.

9.58 Consider the polynomial given in (h). Write down the polynomial $P(t)$ for $K=1$, $K=2$, and $K=3$. For $K=2,3$ find the roots of $P(t)$.

9.59 The proof of Bezout's theorem is broken into two parts. The first part of the proof establishes the existence of a solution to the identity (9.91). Existence is established using the Euclidean algorithm and proof by induction. Although this part of the proof is a bit advanced for the scope of this book (the interested reader is referred to Daubechies [32]), the second part of proof, showing that the solution is unique, is straightforward. In this exercise you will establish the uniqueness of solutions to (9.91) in Bezout's theorem. The following steps will help you organize your work.

(a) To establish the uniqueness of solution to (9.91), assume that there are two solution pairs, $q_1(t), q_2(t)$ and $r_1(t), r_2(t)$, with degrees at most $n_2 - 1$ and $n_1 - 1$, respectively. Insert these solutions into (9.91) and subtract to show that

$$p_1(t)(q_1(t) - r_1(t)) = p_2(t)(r_2(t) - q_2(t)).$$

(b) Why must the set of zeros of $p_2(t)$ be contained in the set of zeros for $q_1(t) - r_1(t)$?

(c) What is the only polynomial $q_1(t) - r_1(t)$ for which the set of zeros of $p_2(t)$ is contained in the set of zeros for $q_1(t) - r_1(t)$? Explain your answer and why this establishes the uniqueness of the solution to (9.91).

9.60 Use a **CAS** with $\widetilde{\ell} = 2$, $\widetilde{p}(\cos(\omega)) = 2.920696(\sin^2\left(\frac{\omega}{2}\right) + 0.342484)$ and the identities for $\cos\left(\frac{\omega}{2}\right)$ and $\sin\left(\frac{\omega}{2}\right)$ from Problem 8.19 in Section 8.1 to verify that the coefficients of the Fourier series for (9.87) are listed in (9.89). Repeat the problem for the Fourier series given by (9.88).

9.61 Show that (9.76) holds and that neither $\widetilde{H}^{(4)}(\pi)$ nor $H^{(4)}(\pi)$ equals zero.

Computer Lab

9.2 The CDF97 Biorthogonal Filter Pair. From `stthomas.edu/wavelets`, access the file `CDF97Filters`. In this you will write a function that produces the CDF97 filter pair. Your code will expand the Fourier series given by (9.87) and (9.88) to find the filters. You will also write functions to compute the (inverse) wavelet transformation of a vector or matrix using the CDF97 filter pair.

CHAPTER 10

WAVELET PACKETS

Suppose we compute $j = 1, \ldots, i$ iterations of the wavelet transform of matrix A using a prescribed filter (or biorthogonal filter pair). Then counting the matrix itself, we have $i + 1$ representations of A. In previous chapters, we have seen that the number of iterations of the wavelet transform varies between applications. In image compression, we learn that increasing the number of iterations concentrates the energy of the transformed image to relatively few elements. In an application such as edge detection, using a high iteration transform might make it difficult to detect changes in the image and in the signal de-noising, we learn that the differences portion of the first iteration is comprised essentially of noise.

Since a variety of representations of the data are desirable, it is natural to ask if we can create even more representations of the data via the wavelet transformation. This question was considered by Coifman, Meyer and Wickerhauser in [25]. They decided to perform a *full decomposition* of the data. That is, instead of iteratively applying the wavelet transformation to the preceding averages portion, they suggested applying it to all preceding portions.

Section 10.1 describes the wavelet packet transformation and the large number or redundant representations that result. In Section 10.2, we introduce the notion of a *cost function* that is used in conjunction with an algorithm for selecting the "best"

Discrete Wavelet Transformations: An Elementary Approach With Applications, Second Edition.
Patrick J. Van Fleet.

representation. The algorithm is stated and examples are provided to illustrate its implementation. Probably the most notable application of wavelet packet transformation is the FBI Wavelet Scalar Quantization (WSQ)Specification developed by Bradley and Christopher Brislawn at Los Alamos National Laboratory and Hopper from the Federal Bureau of Investigation [8, 9, 10, 11]. The WSQ method is discussed in Section 10.3.

10.1 The Wavelet Packet Transform

We motivate the wavelet packet transformation by looking at a schematic representation of an iterated wavelet transformation. Suppose we compute three iterations of an orthogonal (or biorthogonal) wavelet transformation of vector $\mathbf{v}$. The graphic in Figure 10.1 describes how the iterative transformation is computed.

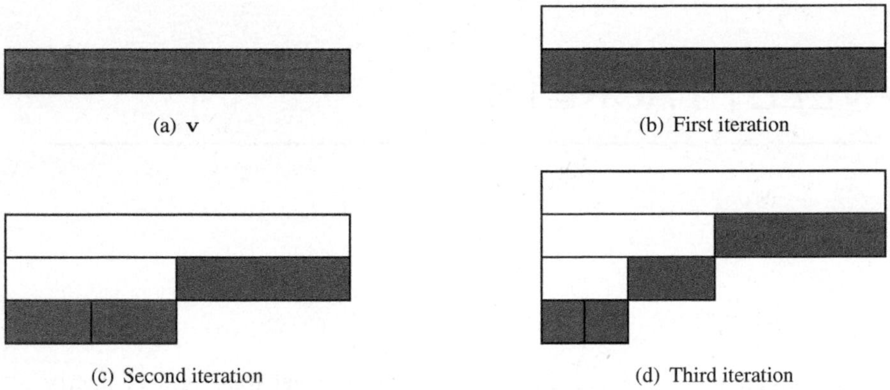

Figure 10.1 Three iterations of a wavelet transformation plotted schematically.

We can see from Figure 10.1 that iteration i is obtained by applying the wavelet transformation to the data represented by the left-most rectangle in iteration $i-1$. Rather than just transforming the data represented by the left-most rectangle in the preceding iteration, Coifman, Meyer and Wickerhauser [24, 25] suggest instead applying the wavelet transformation to data represented by *each* rectangle in the preceding iteration. The result is the *one-dimensional wavelet packet transformation.*

One-Dimensional Wavelet Packet Transformation

We compute i iterations of the *wavelet packet transformation* by applying the wavelet transformation to each portion of the transformation at the subsequent level as the following definition suggests.

Definition 10.1 (One-Dimensional Wavelet Packet Transformation). *Suppose i is a nonnegative integer and assume that $\mathbf{v} \in \mathbb{R}^N$ where N is divisible by 2^i. Set $\mathbf{w}_0^0 = \mathbf{v}$ and for $j = 1, \ldots, i-1$ and $k = 0, \ldots, 2^j - 1$, compute the wavelet transformation $\left[\frac{\mathbf{a}}{\mathbf{d}}\right]$ of $\mathbf{w}_k^j$ and set*

$$\mathbf{w}_{2k}^{j+1} = \mathbf{a} \quad \textit{and} \quad \mathbf{w}_{2k+1}^{j+1} = \mathbf{d}.$$

The collection of vectors $\mathcal{W} = \left\{\mathbf{w}_k^j\right\}$, $j = 0, \ldots, i$, $k = 0, \ldots, 2^j - 1$ constitutes the one-dimensional wavelet packet transformation *of* $\mathbf{v}$. *We refer to* $\mathbf{w}_{2k}^j$ *and* $\mathbf{w}_{2k+1}^j$ *as the* descendants *of the* parent $\mathbf{w}_k^j$. □

Any filter or biorthogonal filter pair can be used in Definition 10.1.

Note: We can recover a parent $\mathbf{w}_k^j$ from its descendants $\mathbf{w}_{2k}^j$, $\mathbf{w}_{2k+1}^j$ via the inverse wavelet transformation. That is,

$$\left[\frac{\mathbf{w}_{2k}^{j+1}}{\mathbf{w}_{2k+1}^{j+1}}\right] \xrightarrow{\text{IWT}} \mathbf{w}_k^j. \tag{10.1}$$

The notion of averages and differences portions is lost when computing wavelet packet transformations although $\mathbf{w}_0^i$ is the averages portion $\mathbf{a}^i$ and $\mathbf{w}_1^j$ are the differences portion $\mathbf{d}^j$ of the jth iteration of a regular wavelet transformation, $j = 1, \ldots, i$. Figure 10.2 illustrates the process for $i = 3$ iterations.

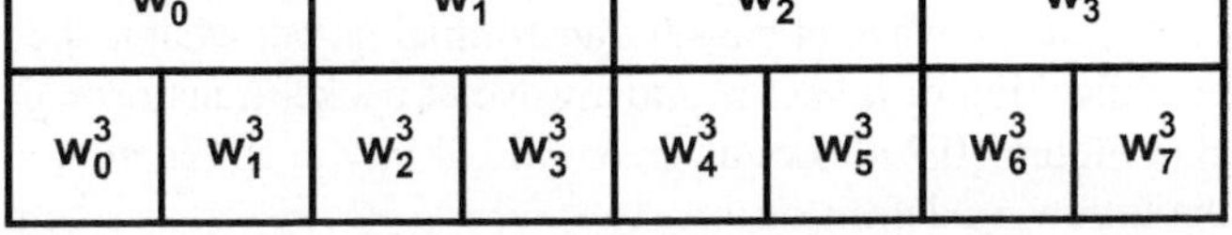

Figure 10.2 A schematic of the wavelet packet transformation for $i = 3$ iterations.

Let's look at a couple of examples.

Example 10.1 (Computing a Wavelet Packet Transformation). *Compute three iterations of the wavelet packet transformation using the Haar wavelet transformation* (4.2) *of the vector* $\mathbf{v} = [64, 24, 0, 64, 72, 8, 24, 32]$.

Solution. *We start with* $\mathbf{w}_0^0 = \mathbf{v}$ *and compute its Haar wavelet transformation. The averages and differences portions are stored as* $\mathbf{w}_0^1 = \mathbf{a}^1$ *and* $\mathbf{w}_1^1 = \mathbf{d}^1$*. We then apply the Haar wavelet transformation to each of* $\mathbf{w}_0^1$ *and* $\mathbf{w}_1^1$*. Application of the Haar wavelet transformation to* $\mathbf{w}_0^1$ *results in the usual averages and differences portions* $\mathbf{a}^2$, $\mathbf{d}^2$ *that are stored in* $\mathbf{w}_0^2$, $\mathbf{w}_1^2$*, respectively. Application of the Haar wavelet transformation to the differences portion* $\mathbf{w}_0^1 = \mathbf{d}^1$ *results in two new pieces* $\mathbf{w}_2^2 = [6, -14]^T$ *and* $\mathbf{w}_3^2 = [26, 18]^T$*. The Haar wavelet transformation is then applied to each of* $\mathbf{w}_k^2$, $k = 0, 1, 2, 3$ *with the results listed in the last row of Table 10.1.*

Table 10.1 The results of computing the wavelet packet transformation of $\mathbf{v}$.

64	24	0	64	72	8	24	32
44	32	40	28	-20	32	-32	4
38	34	-6	-6	6	-14	26	18
36	-2	-6	0	-4	-10	22	-4

(Live example: Visit `stthomas.edu/wavelets` *and click on Live Examples.)*

□

Example 10.2 (The Wavelet Packet Transformation of a Signal). *We return to Example 6.6 from Section 6.3. The vector* $\mathbf{v}$ *in this example contains recorded* 4304 *boat speeds at equally spaced time intervals. We compute three iterations of the wavelet packet transformation using the Haar filter. The results are plotted in Figure 10.3.*

(Live example: Visit `stthomas.edu/wavelets` *and click on Live Examples.)*

□

Basis Representations

We can certainly compute the wavelet packet transformation, but what is the motivation? The full decomposition of a vector into a wavelet packet transformation such as that diagramed in Figure 10.2 allows us greater flexibility in applications.

We consider the following definition.

Definition 10.2 (One-Dimensional Basis Representation). *Let* $\mathcal{S}$ *be nonempty subset of the wavelet packet transformation* $\mathcal{W} = \left\{\mathbf{w}_k^j\right\}$, $j = 0, \ldots, i$, $k = 0, \ldots, 2^j - 1$ *of vector* $\mathbf{v}$*. We say* $\mathcal{S}$ *is a* basis representation *of* $\mathbf{v}$ *if*

(a) $\mathbf{w}_k^j \in \mathcal{S}$, $j < i$, *then* $\mathbf{w}_{2^m k+n}^{j+m} \notin \mathcal{S}$, $m = 1, \ldots, i - j$, $n = 0, \ldots, 2^m - 1$, *and*

(b) we start at level i *and recursively determine, via* (10.1), *parents from the descendants in* $\mathcal{S}$*, we ultimately obtain* $\mathbf{w}_0^0 = \mathbf{v}$.

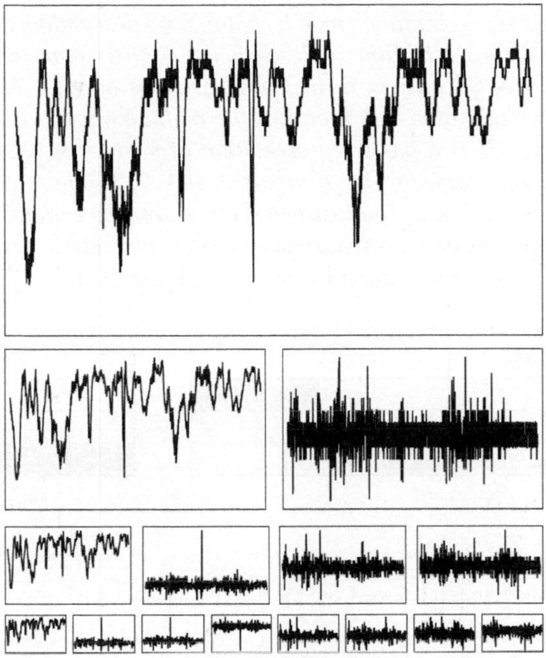

Figure 10.3 Three iterations of the wavelet packet transformation of the boat speeds from Example 6.6.

□

Condition (a) above says that of $\mathbf{w}_k^j \in \mathcal{S}$ and $\mathbf{w}_k^j$ is a parent, then none of its descendants can be in $\mathcal{S}$, while condition (b) ensures that there are no "holes" in the representation. Note that (b) also says $\mathbf{w}_{2k}^j \in \mathcal{S}$ if and only if $\mathbf{w}_{2k+1}^j \in \mathcal{S}$.

The following example lists collections of vectors of a wavelet packet transformation $\mathcal{W}$. Some of the collections are basis representations, some are not.

Example 10.3 (Examples of Basis Representations). *Suppose $i = 3$ and let $\mathcal{W}$ denote the wavelet packet transformation of vector $\mathbf{v}$. Determine whether or not the collections below are basis representations of $\mathbf{v}$. Figure 10.2 is helpful for this example.*

(a) $\mathcal{S}_1 = \{\mathbf{w}_0^0\}$

(b) $\mathcal{S}_2 = \{\mathbf{w}_0^1, \mathbf{w}_1^1, \mathbf{w}_7^3\}$

(c) $\mathcal{S}_3 = \{\mathbf{w}_1^1, \mathbf{w}_1^2, \mathbf{w}_1^3, \mathbf{w}_0^3\}$

(d) $\mathcal{S}_4 = \{\mathbf{w}_k^3\}_{k=0,\ldots,7}$

(e) $\mathcal{S}_5 = \{\mathbf{w}_0^2, \mathbf{w}_1^2, \mathbf{w}_2^2, \mathbf{w}_6^3\}$

Solution. *$\mathcal{S}_1$ is a basis representation since it contains no descendants and trivially satisfies condition (b) from Definition 10.2. $\mathcal{S}_2$ is not a basis representation since $\mathbf{w}_7^3$ is a descendant of $\mathbf{w}_3^2$ which in turn is a descendant of $\mathbf{w}_1^1$. The collection $\mathcal{S}_3$ is a basis representation of $\mathbf{v}$ – in fact, it is the regular wavelet transformation (three iterations) of $\mathbf{v}$. $\mathcal{S}_4$ is a basis representation of $\mathbf{v}$ since $\mathbf{w}_{2k}^3$, $\mathbf{w}_{2k+1}^3$, $k = 0, 1, 2, 3$ are descendants, respectively, of $\mathbf{w}_k^2$, $k = 0, 1, 2, 3$. These vectors are in turn descendants of $\mathbf{w}_0^1$ and $\mathbf{w}_1^1$. The last two vectors are descendants of $\mathbf{w}_0^0$. The last collection $\mathcal{S}_5$ is not a basis representation of $\mathbf{v}$ since it is missing $\mathbf{w}_7^3$.*

Schematics of the basis representations appear in Figure 10.4.

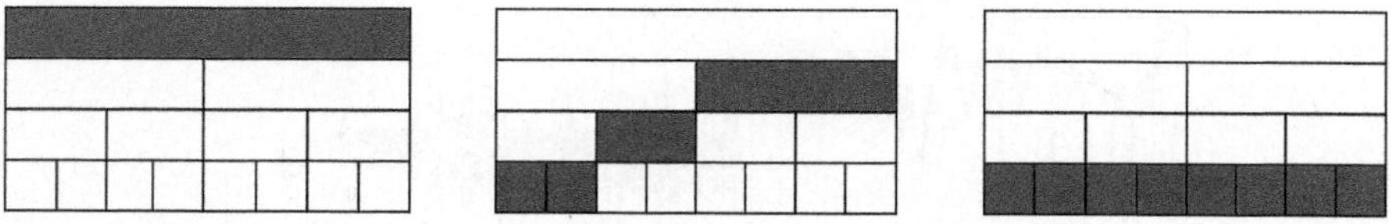

(a) $\mathcal{S}_1$ (b) $\mathcal{S}_3$ (c) $\mathcal{S}_4$

Figure 10.4 Basis representations from the wavelet packet transformation of $\mathbf{v}$.

□

The number of basis representations of a vector grows rapidly as the number of iterations increases. The proposition below gives a formula for counting the number of basis representations in a wavelet packet transformation of i iterations.

Proposition 10.1 (Number of One-Dimensional Basis Representations). *Suppose $\mathcal{W}$ is a wavelet packet transformation with i iterations of vector $\mathbf{v}$. If $c_{\mathbf{v}}(i)$ represents the number of possible basis representations of A, then $c_{\mathbf{v}}(0) = 1$, $c_{\mathbf{v}}(1) = 2$, and for $i > 1$ we have*

$$c_{\mathbf{v}}(i) = c_{\mathbf{v}}(i-1)^2 + 1. \tag{10.2}$$

□

Proof. When $i = 0$, the only basis representation of $\mathbf{v}$ is $\mathbf{v}$ itself so that $c_{\mathbf{v}}(0) = 1$. When $i = 1$, $\mathcal{W} = \left\{\mathbf{w}_0^0, \mathbf{w}_0^1, \mathbf{w}_1^1\right\}$ and the basis representations are $\mathcal{S}_1 = \left\{\mathbf{w}_0^0\right\}$ and $\mathcal{S}_2 = \left\{\mathbf{w}_0^1, \mathbf{w}_1^1\right\}$. For $i > 1$, we start the wavelet packet transformation construction by computing $\mathbf{w}_0^1$ and $\mathbf{w}_1^1$ from $\mathbf{w}_0^0$ via a wavelet transformation (see Figure 10.5).

A wavelet packet decomposition of $\mathbf{w}_0^1$ consists of $i - 1$ iterations and there are $c(i)$ possible basis representations for this decomposition. Similarly, we see that there are $c_{\mathbf{v}}(i-1)$ basis representations of $\mathbf{w}_1^1$. Since for every basis representation of $\mathbf{w}_0^1$, there are $c_{\mathbf{v}}(i-1)$ basis representations for $\mathbf{w}_1^1$, and there are a total of $c_{\mathbf{v}}(i-1)^2$ basis representations of the vector $\left[\mathbf{w}_0^1, \mathbf{w}_1^1\right]^T$. Add one more to this number to reflect that $\mathbf{w}_0^0 = \mathbf{v}$ itself is a possible representation and we see there are $c_{\mathbf{v}}(i) = c_{\mathbf{v}}(i-1)^2 + 1$ basis representations of a wavelet packet transformation with i iterations of $\mathbf{v}$ and the proof is complete. □

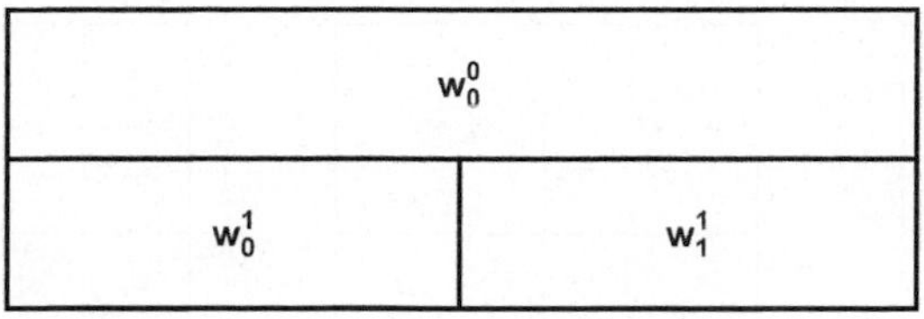

Figure 10.5 The first step in a wavelet packet transformation.

Two-Dimensional Wavelet Packet Transformations

It is straightforward to generalize the concept of the wavelet packet transformation to matrices. We start with a definition similar to Definition 10.1.

Definition 10.3 (Two-Dimensional Wavelet Packet Transformation). *Suppose i is a nonnegative integer and assume that A is an $M \times N$ matrix where M, N are divisible by 2^i. Set $W_0^0 = A$ and for $j = 1, \ldots, i-1$ and $k = 0, \ldots, 4^j - 1$, compute the wavelet transformation*

$$\left[\begin{array}{c|c} \mathcal{B} & \mathcal{V} \\ \hline \mathcal{H} & \mathcal{D} \end{array}\right]$$

of W_k^j and set

$$\begin{aligned} W_{4k}^{j+1} &= \mathcal{B} \\ W_{4k+1}^{j+1} &= \mathcal{V} \\ W_{4k+2}^{j+1} &= \mathcal{H} \\ W_{4k+3}^{j+1} &= \mathcal{D}. \end{aligned}$$

The collection of matrices $\mathcal{W} = \left\{W_k^j\right\}$, $j = 0, \ldots, i$, $k = 0, \ldots, 4^j - 1$ constitutes the two-dimensional wavelet packet transformation *of A. We refer to W_{4k}^j, W_{4k+1}^j, W_{4k+2}^j, and W_{4k+3}^j as the* descendants *of the* parent *W_k^j.* □

As is the case with the one-dimensional wavelet packet transformation, any filter or biorthogonal filter pair can be used in Definition 10.3. We can recover a parent from its descendants via the inverse wavelet transformation. That is,

$$\left[\begin{array}{c|c} W_{4k}^{j+1} & W_{4k+1}^{j+1} \\ \hline W_{4k+2}^{j+1} & W_{4k+3}^{j+1} \end{array}\right] \stackrel{\text{IWT}}{\longrightarrow} W_k^j. \tag{10.3}$$

Figure 10.6 is a schematic of the two-dimensional wavelet packet transformation with $i = 2$ and the regions labeled and Figure 10.7 shows a wavelet packet transformation with $i = 3$ iterations of an image matrix.

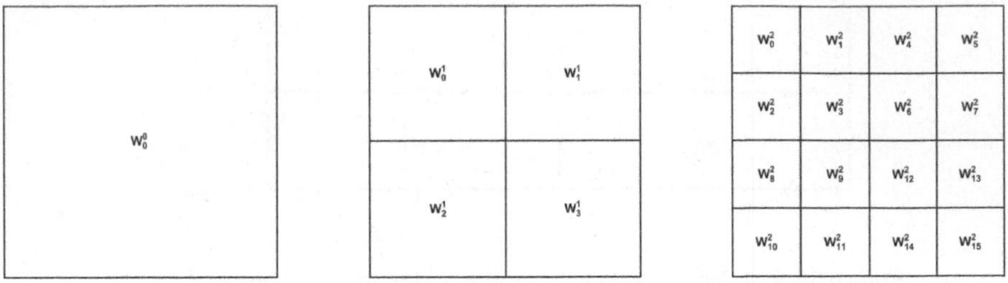

(a) $i = 0$ (b) $i = 1$ (c) $i = 2$

Figure 10.6 Notation for portions of the two-dimensional wavelet packet transformation with $i = 2$ iterations.

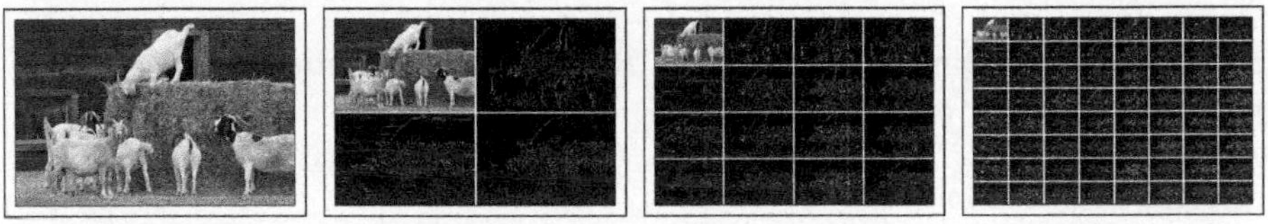

(a) A (b) $i = 1$ (c) $i = 2$ (d) $i = 3$

Figure 10.7 Three iterations of a two-dimensional wavelet transformation.

The definition of basis representations extends to two-dimensional transformations.

Definition 10.4 (Two-Dimensional Basis Representation). *Let $\mathcal{S}$ be nonempty subset of the wavelet packet transformation $\mathcal{W} = \left\{W_k^j\right\}$, $j = 0, \ldots, i$, $k = 0, \ldots, 4^j - 1$ of matrix A. We say $\mathcal{S}$ is a* basis representation *of A if*

(a) $W_k^j \in \mathcal{S}$, $j < i$, then $W_{4^m k+n}^{j+m} \notin \mathcal{S}$, $m = 1, \ldots, i-j$, $n = 0, \ldots, 4^m - 1$, and

(b) we start at level i and recursively determine parents, via (10.3), *from the descendants in $\mathcal{S}$, we ultimately obtain $W_0^0 = A$.* □

Figure 10.8 shows three different basis representations for the two-dimensional wavelet packet transformation with $i = 3$ iterations of an image matrix.

The final proposition of the section gives a count of the number of basis representations possible for iterated two-dimensional wavelet packet transformations.

Proposition 10.2 (Number of Two-Dimensional Basis Representations). *Suppose $\mathcal{W}$ is a wavelet packet transformation with i iterations of matrix A. If $c_{\mathcal{W}}(i)$ represents the number of possible basis representations of A, then $c_A(0) = 1$, $c_A(1) = 2$, and for $i > 1$ we have*

$$c_A(i) = c_A(i-1)^4 + 1. \tag{10.4}$$

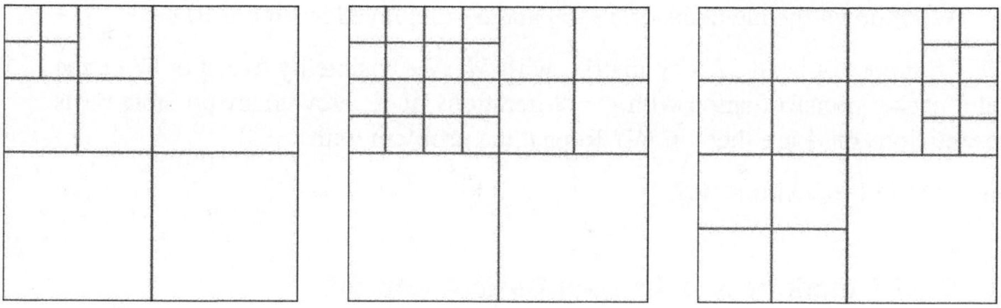
(a) $\mathcal{S}_1$ (b) $\mathcal{S}_2$ (c) $\mathcal{S}_3$

Figure 10.8 Schematic plots of three possible basis representations for a two-dimensional wavelet packet transformation.

□

Proof. The proof is similar to that of Proposition 10.1 and left as Problem 10.11. □

PROBLEMS

10.1 Let $\mathbf{v} \in \mathbb{R}^{16}$ with $v_k = k - 1$, $k = 1, \ldots, 16$. For $i = 3$ iterations, calculate by hand wavelet packet transformation of $\mathbf{v}$ using the Haar filter (4.2).

10.2 Let $\mathbf{v} \in \mathbb{R}^N$ where N is divisible by four and suppose $\mathcal{W}$ is the wavelet packet transformation of $\mathbf{v}$ with $i = 2$ iterations. Find all basis representations of $\mathbf{v}$ and sketch schematics of them.

10.3 Suppose $\mathcal{W}$ is a wavelet packet transformation of three iterations of $\mathbf{v}$ and assume $\mathbf{w}_3^1 \in \mathcal{S}$ where $\mathcal{S}$ is a basis representation of $\mathbf{w}$. Write down all the elements of $\mathcal{W}$ guaranteed *not* to be in $\mathcal{S}$.

10.4 Suppose $\mathcal{W}$ is a wavelet packet transformation of i iterations of $\mathbf{v}$ and assume $\mathcal{S}$ is a basis representation of $\mathbf{v}$ with $\mathbf{w}_{2k}^i \in \mathbf{S}$. Explain why $\mathbf{w}_{2k+1}^i$ is also a member of $\mathcal{S}$.

10.5 Use Proposition10.1 to determine the number of basis representations in $\mathcal{W}$ for $i = 3, 4, 5$ iterations.

10.6 Use mathematical induction to show that $c_{\mathbf{v}}(i) > 2^{2^i}$ whenever $i \geq 2$.

10.7 Suppose $\mathcal{W}$ is a wavelet packet transformation of three iterations of matrix A and assume $W_2^1 \in \mathcal{S}$ where $\mathcal{S}$ is a basis representation of A. Write down all the elements of $\mathcal{W}$ guaranteed *not* to be in $\mathcal{S}$.

10.8 Suppose $\mathcal{W}$ is a wavelet packet transformation of three iterations of matrix A and assume $W_1^1, W_2^1 \in \mathcal{S}$ where $\mathcal{S}$ is a basis representation of A. Write down all the possible sets $\mathcal{S}$ in this case.

10.9 Write down the elements of $\mathcal{S}_1, \mathcal{S}_2$ and $\mathcal{S}_3$ displayed in Figure 10.8.

10.10 Suppose A is an $M \times N$ matrix, with M, N divisible by four. Let $\mathcal{W}$ be the wavelet packet transformation with $i = 2$ iterations of A. How many possible basis representations of A are there in $\mathcal{W}$? Repeat the problem with $i = 3$.

10.11 Prove Proposition 10.2.

10.2 Cost Functions and the Best Basis Algorithm

Portions of the material that follows is adapted from Section 9.3 of Wavelet Theory: An Elementary Approach with Applications, *by David K. Ruch and Patrick J. Van Fleet [77] with permission from John Wiley & Sons, Inc.*

The Cost Function

As we have seen from Propositions 10.1 and 10.2, the number of possible basis representations for the wavelet packet transformation of a given vector or matrix grows rapidly as the number of iterations of the transformation increases. The large number of basis representations in a wavelet packet transformation is an advantage over an ordinary wavelet transformation since we have more options for decomposing the data. On the other hand, the large number of basis representations presents a problem: how to choose a *best* basis representation?

It is natural to assume that the choice of the best basis representation is influenced by the application as is the case in Section 10.3, but is it possible to construct a general method for determining the best representation from a wavelet packet transformation? Coifman and Wickerhauser [26] employed a *cost function* for this task. This nonnegative real-valued function is typically used to compare two representations of the same data and assign a smaller number to the one that is deemed more efficient. The advantage here is the fact that the cost function can be application-dependent – we can determine what comprises an efficient representation based on a particular application and design our cost function accordingly.

In applications such as image compression, we typically use cost functions to "reward" those representations that use relatively few elements of the transformed data. For the purposes of fast computation in the best-basis algorithm, we impose an additivity condition on the cost function. We have the following definition.

Definition 10.5 (Cost Function). *Suppose that* $\mathbf{v} \in \mathbb{R}^M$. *Then a* cost function $\mathcal{C}(\mathbf{v})$ *is any nonnegative real-valued function that satisfies* $\mathcal{C}(\mathbf{0}) = 0$, *where* $\mathbf{0}$ *is the length* M *zero vector, and for any partition*

$$\mathbf{v} = [\mathbf{v}_\ell \mid \mathbf{v}_r]^T = [v_1, \ldots, v_m \mid v_{m+1}, \ldots, v_M]^T .$$

where $1 \le m \le M$, *we have*

$$\mathcal{C}(\mathbf{v}) = \mathcal{C}\left(\left[\mathbf{v}_\ell \mid \mathbf{v}_r\right]^T\right) = \mathcal{C}\left(\mathbf{v}_\ell\right) + \mathcal{C}\left(\mathbf{v}_r\right). \tag{10.5}$$

□

There are several different cost functions described in the literature. We provide four in the example that follows.

Example 10.4 (Examples of Cost Functions). *In this example we list and describe four cost functions. The proofs that each are cost functions are left as exercises.*

Shannon Entropy. *This function is closely related to the entropy function* (3.16) *given in Definition 3.2. For* $\mathbf{v} \in \mathbb{R}^M$ *with* $\|\mathbf{v}\| = 1$, *we have*

$$\mathcal{C}_s(\mathbf{v}) = -\sum_{k=1}^{M} v_k^2 \ln\left(v_k^2\right). \tag{10.6}$$

where we define $v_k^2 \ln\left(v_k^2\right) = 0$ *in the case that* $v_k = 0$ *(see Problem 10.12). It is clear that* $\mathcal{C}_s(\mathbf{0}) = 0$, *and in Problem 10.13 you will show that* $\mathcal{C}_s(\mathbf{v})$ *satisfies Definition 10.5. Note that if* v_k^2 *is close to zero or one, the value of* $v_k^2 \ln\left(v_k^2\right)$ *will be small, so this function "rewards" portions of a basis representation comprised largely of zeros or elements of maximal size.*

Number Above Threshold. *For* $\mathbf{v} \in \mathbb{R}^M$ *and* $t > 0$, *we define*

$$\mathcal{C}_t(\mathbf{v}) = \#\{\, v_k \;:\; |v_k| \ge t\}. \tag{10.7}$$

For example, if $\mathbf{v} = [3, -4, -1, 0, 3, 1, 5]^T$, *then* $\mathcal{C}_2(\mathbf{v}) = 4$. *This function counts the number of "essential" coefficients in a given representation.*

Number of Nonzero Elements. *Let* $\mathbf{v} \in \mathbb{R}^M$ *and define*

$$\mathcal{C}_z(\mathbf{v}) = \#\{\, v_k \;:\; v_k \ne 0\}.$$

For example, if $\mathbf{v} = [2, 0, -13, 0, 0, 9, 0, 0, 1, 5]^T$, *then* $\mathcal{C}_z(\mathbf{v}) = 5$. *This function counts the number of elements contributing to the energy of* $\mathbf{v}$ *and does not take into account the magnitude of the coefficients.*

Sum of Powers. *For* $\mathbf{v} \in \mathbb{R}^M$ *and* $p > 0$, *we define*

$$\mathcal{C}_p(\mathbf{v}) = \sum_{k=1}^{M} |v_k|^p. \tag{10.8}$$

Suppose that $p > 1$. *Then* $|v_k|^p < |v_k|$ *whenever* $|v_k| < 1$. *In such cases this function "rewards" those representations that use coefficients that are small in magnitude.*

(Live example: Visit `stthomas.edu/wavelets` *and click on Live Examples.)*

□

Let's look at an example that serves as good motivation for how the cost function is used in the best basis algorithm.

Example 10.5 (The Cost Function and the Wavelet Packet Transformation). *Consider the wavelet packet transformation from Example 10.2. Using $\mathcal{C}_p$ with $p = 1/2$ and the number above threshold cost function $\mathcal{C}_t$ with $t = 0.85$, compute the cost of each component of the wavelet packet transformation plotted in Figure 10.3.*

Solution.

The results of the computations are plotted in Figure 10.9.

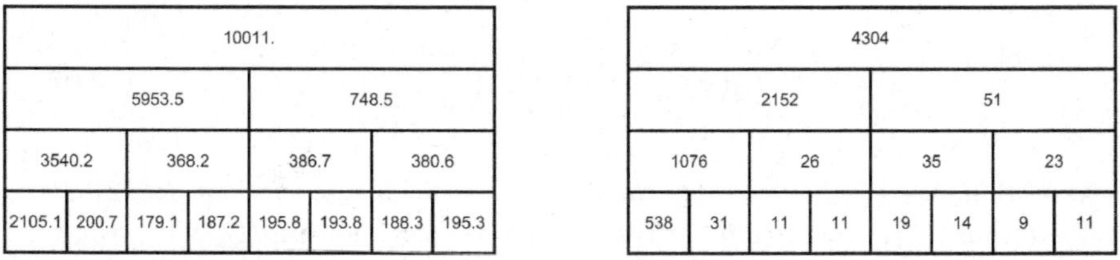

10011.
5953.5
748.5
3540.2
368.2
386.7
380.6
2105.1
200.7
179.1
187.2
195.8
193.8
188.3
195.3
4304
2152
51
1076
26
35
23
538
31
11
11
19
14
9
11

(a) Sum of Powers with $p = \frac{1}{2}$ (b) Number Above Threshold $t = 0.85$

Figure 10.9 The cost function values for each component of the wavelet packet transformation of Example 10.2.

(Live example: Visit `stthomas.edu/wavelets` *and click on Live Examples.)*

□

The Best Basis Algorithm

The idea of the best basis algorithm of Coifman and Wickerhauser [26] is quite straightforward. We use the fact that in a wavelet packet transformation, $\mathbf{w}_{2k}^{j+1}$ and $\mathbf{w}_{2k+1}^{j+1}$ are constructed from $\mathbf{w}_k^j$ along with the additivity of the cost function to make determinations about which of the portions of the packet transformation we should use.

The algorithm for the one-dimensional transformation is as follows. Compute $\mathcal{C}\left(\mathbf{w}_{2k}^i\right)$ and $\mathcal{C}\left(\mathbf{w}_{2k+1}^i\right)$ for $k = 0, \ldots, 2^{i-1} - 1$, and compare their sum (the cost function is additive) to $\mathcal{C}\left(\mathbf{w}_k^{i-1}\right)$. If the sum of the costs of the descendants is less than the cost of the parent, we set $\mathcal{C}\left(\mathbf{w}_j^{i-1}\right) = \mathcal{C}\left(\mathbf{w}_{2k}^i\right) + \mathcal{C}\left(\mathbf{w}_{2k+1}^i\right)$ and note that $\mathbf{w}_{2k}^i$ and $\mathbf{w}_{2k+1}^i$ are, at least at this stage, part of the best basis representation. We repeat this process with levels $i - 1$ and $i - 2$ continuing recursively updating cost function values as well as the components of the best basis representation until we ultimately compare levels one and zero.

Let's look at an example that illustrates the process.

Example 10.6 (Best Basis Algorithm). *Consider the wavelet packet transformation from Example 10.2 and the cost function values plotted in Figure 10.9(b) using* $\mathcal{C}_t$ *with* $t = 0.85$. *To start, we compare*

$$\mathcal{C}_t\left(\left[\frac{\mathbf{w}_0^3}{\mathbf{w}_1^3}\right]\right) = \mathcal{C}_t\left(\mathbf{w}_0^3\right) + \mathcal{C}_t\left(\mathbf{w}_1^3\right) = 538 + 31 = 569$$

to that of its parent $\mathcal{C}_t\left(\mathbf{w}_0^2\right) = 1076$. *Since the former value is smaller, we retain the descendants instead of the parent in the best basis representation. Continuing on comparing third iteration descendants to second iteration parents, we see that*

$$\mathcal{C}_t\left(\left[\frac{\mathbf{w}_2^3}{\mathbf{w}_3^3}\right]\right) = \mathcal{C}_t\left(\mathbf{w}_2^3\right) + \mathcal{C}_t\left(\mathbf{w}_3^3\right) = 11 + 11 = 22 < 26 = \mathcal{C}_t\left(\left[\mathbf{w}_1^2\right]\right)$$

$$\mathcal{C}_t\left(\left[\frac{\mathbf{w}_4^3}{\mathbf{w}_5^3}\right]\right) = \mathcal{C}_t\left(\mathbf{w}_4^3\right) + \mathcal{C}_t\left(\mathbf{w}_5^3\right) = 19 + 14 = 33 < 35 = \mathcal{C}_t\left(\left[\mathbf{w}_2^2\right]\right)$$

$$\mathcal{C}_t\left(\left[\frac{\mathbf{w}_6^3}{\mathbf{w}_7^3}\right]\right) = \mathcal{C}_t\left(\mathbf{w}_6^3\right) + \mathcal{C}_t\left(\mathbf{w}_7^3\right) = 9 + 11 = 20 < 23 = \mathcal{C}_t\left(\left[\mathbf{w}_3^2\right]\right).$$

Thus, all descendants at level three replace parents at level two. The schematic of our best basis representation thus far is plotted in Figure 10.10(a) and the updated cost function values are plotted in Figure 10.10(b).

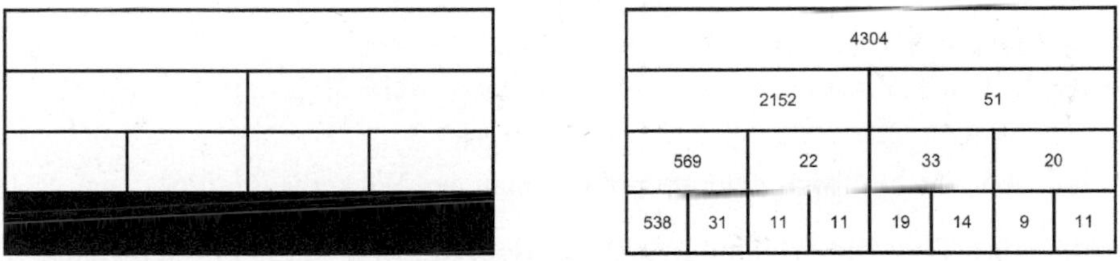

(a) Best basis

(b) Updated cost function values

Figure 10.10 The best basis schematic and updated cost function values after comparison of iterations two (parents) and three (descendants).

We repeat the process using the costs from levels two and one. We have

$$\mathcal{C}_t(\left(\left[\frac{\mathbf{w}_0^2}{\mathbf{w}_1^2}\right]\right) = \mathcal{C}_t\left(\mathbf{w}_0^2\right) + \mathcal{C}_t\left(\mathbf{w}_1^2\right) = 569 + 22 = 591 < 2152 = \mathcal{C}_t\left(\left[\mathbf{w}_0^1\right]\right)$$

$$\mathcal{C}_t(\left(\left[\frac{\mathbf{w}_2^2}{\mathbf{w}_3^2}\right]\right) = \mathcal{C}_t\left(\mathbf{w}_2^2\right) + \mathcal{C}_t\left(\mathbf{w}_3^2\right) = 33 + 20 = 53 > 51 = \mathcal{C}_t\left(\left[\mathbf{w}_1^1\right]\right).$$

In this case, we retain the components $\mathbf{w}_0^3, \ldots, \mathbf{w}_3^3$ *since their cost is smaller than that of* $\mathbf{w}_0^1$*, but we replace* $\mathbf{w}_4^3, \ldots, \mathbf{w}_7^3$ *with* $\mathbf{w}_1^1$ *since the latter vector has a smaller cost. Our best basis schematic and updated cost function values to this point are plotted in Figure 10.11(b).*

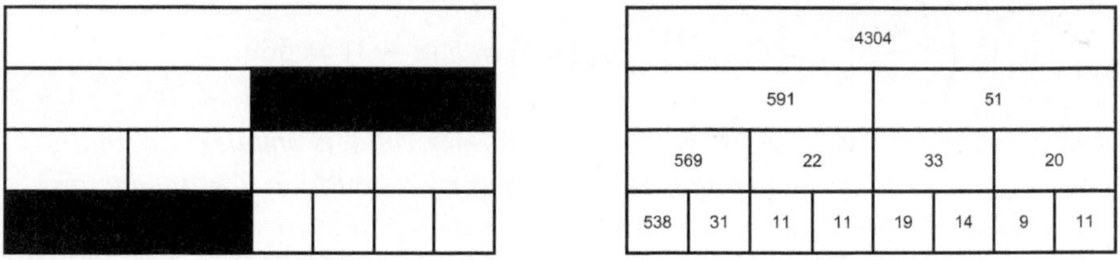

(a) Best basis (b) Updated cost function values

Figure 10.11 The best basis schematic and updated cost function values after comparison of iterations one (parents) and two (descendants).

The last step is to compare $\mathcal{C}_t\left(\mathbf{w}_0^0\right) = 4304$ *to that of*

$$\mathcal{C}_t\left(\mathbf{w}_0^3\right) + \mathcal{C}_t\left(\mathbf{w}_1^3\right) + \mathcal{C}_t\left(\mathbf{w}_2^3\right) + \mathcal{C}_t\left(\mathbf{w}_3^3\right) + \mathcal{C}_t\left(\mathbf{w}_1^1\right) = 591 + 51 = 642.$$

Since the latter value is smaller, we do not replace these components with $\mathbf{w}_0^0$*. Our best basis representation is*

$$\mathcal{S} = \left\{\mathbf{w}_0^3, \mathbf{w}_1^3, \mathbf{w}_2^3, \mathbf{w}_3^3, \mathbf{w}_1^1\right\}$$

and the schematic of $\mathcal{S}$ *appears in Figure 10.11(a).*
(Live example: Visit `stthomas.edu/wavelets` *and click on Live Examples.)*

□

We now state the best basis algorithm of Coifman and Wickerhauser [26].

Algorithm 10.1 (One-Dimensional Best Basis Algorithm). *Suppose* $\mathcal{W}$ *is a wavelet packet transformation with* i *iterations of vector* $\mathbf{v}$*. The following algorithm returns the best basis representation* $\mathcal{S}$ *along with a best basis tree* $\mathbf{b} = \left\{b_k^j\right\}$, $j = 0, \ldots, i$, $k = 0, \ldots, 2^j - 1$.

1. *Initialize the basis tree* $\mathbf{b}$ *by setting* $b_k^i = 1$, $k = 0, \ldots, 2^i - 1$ *and* $b_k^j = 0$, $j = 0, \ldots, i-1$, $k = 0, \ldots, 2^j - 1$.

2. *Compute the cost values* $c_k^j = \mathcal{C}\left(\mathbf{w}_k^j\right)$, $j = 0, \ldots, i$, $k = 0, \ldots, 2^j - 1$.

3. *For* $j = i, \ldots, 1$, $k = 0, \ldots, 2^{i-1} - 1$,

   If $c_k^{j-1} \leq c_{2k}^j + c_{2k+1}^j$,

   $b_k^{j-1} = 1$,

For $m = 1, \ldots, i - j + 1$

For $n = 0, \ldots, 2^m - 1,$

$$b^{j-1+m}_{2^m k+n} = 0,$$

else,

$$\text{set } c^{j-1}_k = c^j_{2k} + c^j_{2k+1}.$$

4. *Return* $\mathbf{b}$ *and* $\mathcal{S} = \left\{ \mathbf{w}^j_k \right\}$ *where* $b^j_k = 1$.

□

Some comments are in order regarding Algorithm 10.1.

1. In Step 3, if we choose the element corresponding to c^j_k, then we need to convert all values of $\mathbf{b}$ descending from b^j_k to zero.

2. The basis tree array $\mathbf{b}$ tells us which elements $\mathbf{w}^j_k \in \mathcal{W}$ we must choose for the best basis representation $\mathcal{S}$.

3. The algorithm is not as efficient as it could be – the innermost For loops (on m and n) convert *all* values b_{mn} to zero that lie below $b_{j+1,k}$. In fact, many of these values are already possibly zero, so these assignments are redundant. We have chosen readability of the pseudocode over efficiency. The interested reader is encouraged to perform the steps necessary to make the code more efficient.

Note that we can use this same algorithm to find the best basis representation of a two-dimensional wavelet packet transformation. In this case, we simply convert W^j_k to a vector and proceed in a manner similar to Example 10.6. The difference is that we are comparing the costs of descendants $W^j_{4k}, W^j_{4k+1}, W^j_{4k+2}, W^j_{4k+3}$ to the cost of the parent W^{j-1}_k.

Best Basis and Image Compression

We conclude this section by using the best basis algorithm in conjunction with the two-dimensional wavelet packet transformation to perform image compression.

Example 10.7 (Wavelet Packets and Image Compression). *We consider the problem of compressing the* 512×768 *image plotted in Figure 10.12(a).*

We compute the wavelet packet transformation of the centered image matrix A *using the Haar filter with* $i = 3$ *iterations. We find the best basis representation using the number above threshold cost function* $\mathcal{C}_t$ *and the number of nonzero terms cost function* $\mathcal{C}_z$ *given in Example 10.4. For comparative purposes, we also include* $i = 3$ *iterations of the Haar wavelet transformation. In fact, we use the histogram in Figure 10.12(b) that shows the distribution of the elements in three iterations of the Haar wavelet transformation as an informal guide for choosing* $t = 30$ *in the*

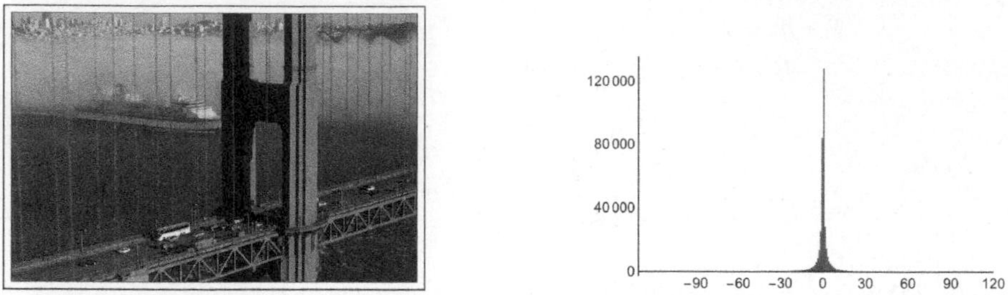

(a) Test image (b) Distribution of HWT elements

Figure 10.12 A test image and distribution of the elements of three iterations of its Haar wavelet transformation

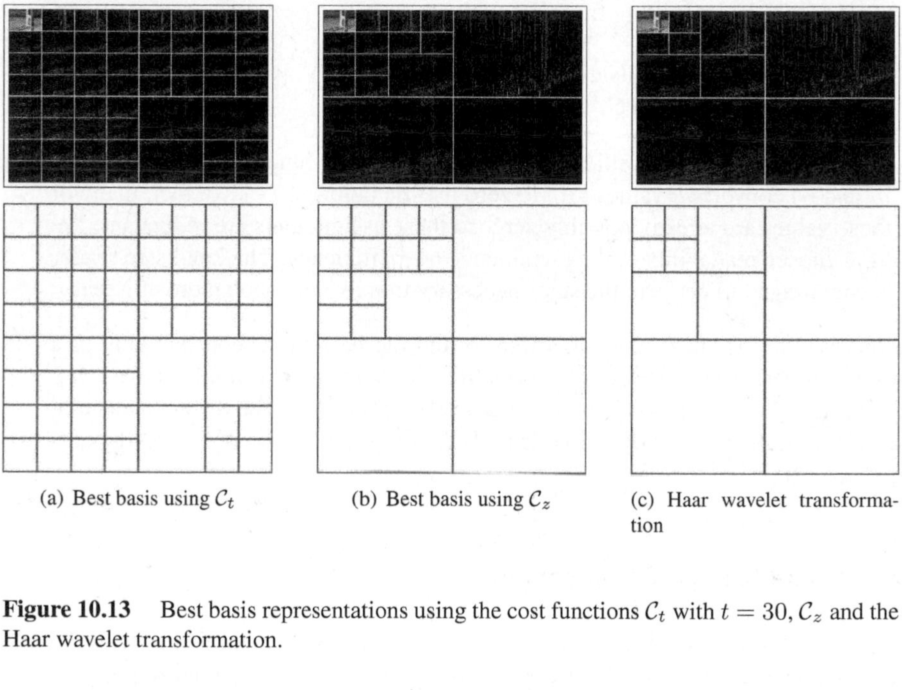

(a) Best basis using $\mathcal{C}_t$ (b) Best basis using $\mathcal{C}_z$ (c) Haar wavelet transformation

Figure 10.13 Best basis representations using the cost functions $\mathcal{C}_t$ with $t = 30$, $\mathcal{C}_z$ and the Haar wavelet transformation.

number above threshold function $\mathcal{C}_t$. *The best basis representations for each cost function are plotted in Figure 10.13.*

For each transformation, we compute the cumulative energy vector and use it to determine the number of terms needed to comprise 99.5% *of the energy. The results are plotted in Figure 10.14.*

We keep those largest elements (in absolute value) in each transformation that comprise 99.5% *of the energy in the transformation and convert the remaining terms to zero. Table 10.2 summarizes the quantization process for each transformation. Note that each transformation consists of* $512 \cdot 768 = 393{,}216$ *intensity values. We*

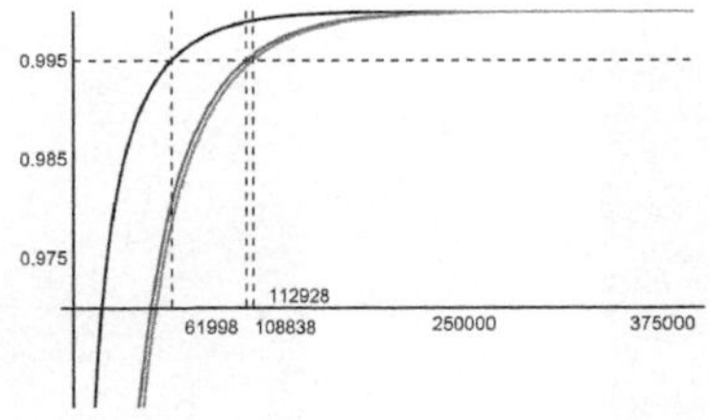

Figure 10.14 Cumulative energies for the transformations. The packet transformation/best basis using $\mathcal{C}_t$, $\mathcal{C}_z$ are plotted in black and dark gray, respectively, and the HWT is plotted in gray. The packet transformation using $\mathcal{C}_t$ outperforms the other two transformations, which are comparable.

Table 10.2 Quantization information for the two packet transformations and the Haar transformation from Example 10.7.[a]

Transformation	Nonzero Terms	% Zero
Packet with $\mathcal{C}_t$	62,439	84.12
Packet with $\mathcal{C}_z$	114,176	70.97
Haar	113,280	71.19

[a]Note that the numbers in the nonzero terms column do not match the values in Figure 10.14. The quantization process finds the k largest values (in absolute value) that constitute 99.5% of the energy in the transformation. However, there might be multiple copies of the smallest of these values. Since it is impossible to know which of these values to keep or discard, the selection routine keeps all of them, thus creating the discrepancies we see in this table and Figure 10.14.

next round the elements in each transformation to complete the quantization process and then apply Huffman coding to the quantized transformation. Table 10.3 summarizes the result. Note that there are $512 \cdot 768 \cdot 8 = 3,145,728$ *total bits that comprise each quantized transformation. Table 10.3 also includes information about optimal*

Table 10.3 Coding information for the Haar, D4, D6, and D4 packet transformations.

Transformation	Bitstream Length	bpp	Entropy	PSNR
Packet with $\mathcal{C}_t$	697,768	1.77	1.60	35.50
Packet with $\mathcal{C}_z$	1,007,684	2.56	2.36	39.76
Haar	1,021,454	2.60	2.35	41.25

compression rates (entropy) and accuracy of the approximate images after reconstruction (PSNR). To recover the approximate images, we decode the Huffman codes and apply the inverse transformation. Note that we need the trees returned by Algorithm 10.1 to perform the inverse packet transformation. The approximate images, along with the original, are plotted in Figure 10.15.

(Live example: Visit `stthomas.edu/wavelets` *and click on Live Examples.)*

□

(a) Original

(b) $\mathcal{C}_t$

(c) $\mathcal{C}_z$

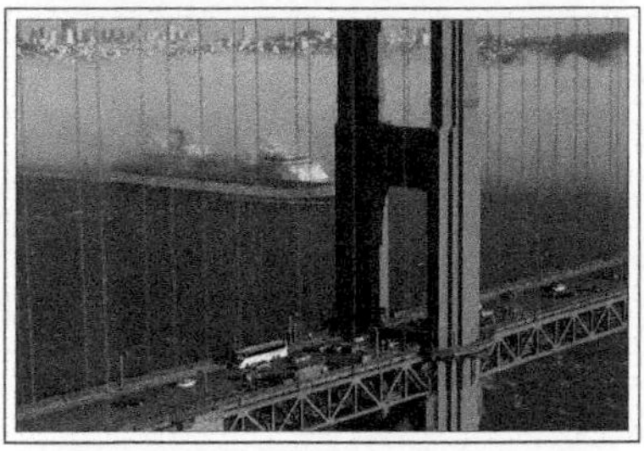

(d) HWT

Figure 10.15 The original image and approximates using wavelet packet transformations and the Haar wavelet transformation.

PROBLEMS

Note: *A computer algebra system and the software on the course web site will be useful for some of these problems. See the Preface for more details.*

10.12 Use L'Hôpital's rule to prove that $\lim\limits_{t\to 0} t\ln(t) = 0$. In this way we justify the assignment $v_k^2 \ln\left(v_k^2\right) = 0$ when $v_k = 0$ in (10.6).

10.13 Show that $\mathcal{C}_s(\mathbf{v})$ defined in (10.6) is a nonnegative real-valued function with $\mathcal{C}_s(\mathbf{0}) = 0$ and also satisfies (10.5).

10.14 We know that $\mathcal{C}_s(\mathbf{0}) = 0$. Describe all other vectors $\mathbf{v}$ for which $\mathcal{C}_s(\mathbf{v}) = 0$.

10.15 Find $\mathbf{C}_s(\mathbf{v})$ for each of the following vectors.

(a) $\mathbf{v} \in \mathbf{R}^M$ where $v_k = \frac{1}{\sqrt{M}}$, for $k = 1, \ldots, M$

(b) $\mathbf{v} \in \mathbf{R}^M$ where M is an even integer; half of the $v_k = 0$ and the other half take the value $\pm\sqrt{2/M}$

10.16 Show that the function $\mathcal{C}_t(\mathbf{v})$ given by (10.7) satisfies Definition 10.5.

10.17 Let $\mathcal{C}_t(\mathbf{v})$ be the cost function defined by (10.7). Show that $\mathbf{C}_t(\mathbf{v}) = 0$ if and only if $\mathbf{v}$ is the zero vector.

10.18 Show that $\mathcal{C}_p(\mathbf{v})$ given by (10.8) satisfies Definition 10.5.

10.19 For the vector $\mathbf{v} = [0, 3, 4, 3, 0, -5, -6, -5, -2, 1, 2, 1, 0, -3, -4, -3]^T$, compute $\mathcal{C}_s(\mathbf{v})$, $\mathcal{C}_t(\mathbf{v})$. $t = 2$, and $\mathcal{C}_z(\mathbf{v})$. Note that for $\mathcal{C}_s(\mathbf{v})$ you will have to normalize $\mathbf{v}$ so that it has unit length.

10.20 Let $\mathbf{v} \in \mathbb{R}^M$ and consider the function $f(\mathbf{v}) = \max\{|v_k| : 1 \leq k \leq M\}$. Is $f(\mathbf{v})$ a cost function?

10.21 A wavelet packet transformation of $i = 3$ iterations is applied to a matrix and then the cost is computed for each region of the packet transformation. The values for each iterations are shown in Figure 10.16. Use the two-dimensional analog of Algorithm 10.1 to determine the best basis representation and sketch a schematic using Figure 10.13 as a guide.

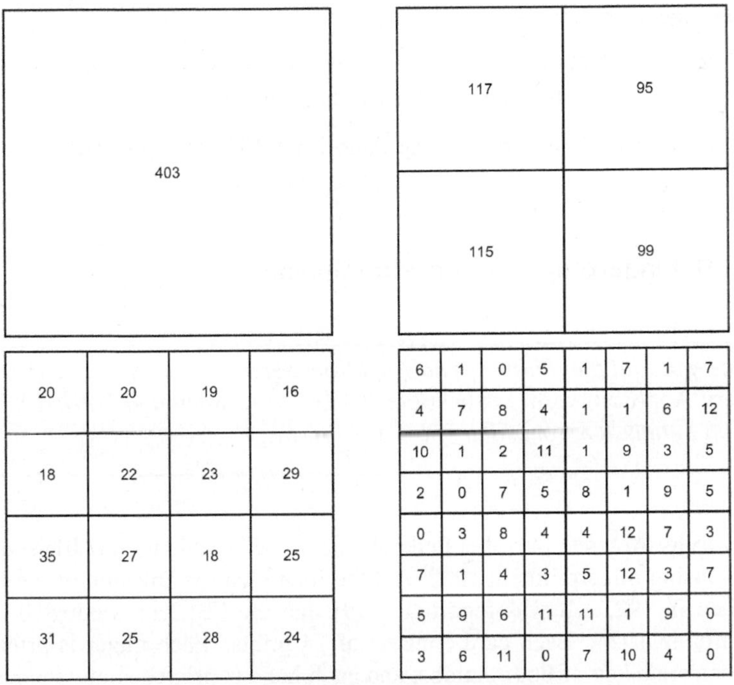

Figure 10.16 Cost function values for Problem 10.21.

10.22 Apply the best basis algorithm to the cost function data given in Figure 10.9(a). Sketch a schematic of the best basis representation using Figure 10.11(a).

10.23 Let $\mathbf{v} \in \mathbf{R}^M$ with $\|\mathbf{v}\|^2 = 1$ and consider $\mathcal{C}(\mathbf{v}) = -\sum_{k=1}^{M} f(v_k^2)$ where

$$f(t) = \begin{cases} \ln(t), & t \neq 0 \\ 0, & t = 0. \end{cases}$$

Show that $\mathcal{C}(\mathbf{v})$ is a cost function.

10.24 Let $\mathbf{v} \in \mathbb{R}^M$ and consider the *nonnormalized Shannon entropy function*

$$\mathcal{C}(\mathbf{v}) = \sum_{k=1}^{M} s(v_k)\, v_k^2 \ln\left(v_k^2\right)$$

with the convention that $0\ln(0) = 0$. Here $s(t)\colon [0,\infty) \to \{-1, 1\}$ is given by

$$s(t) = \begin{cases} -1, & 0 \le t < 1 \\ 1, & t \ge 1. \end{cases}$$

Show that $\mathcal{C}(\mathbf{v})$ is a cost function.

10.25 Is the entropy function (3.16) from Definition 3.2 a cost function? Either prove it is or provide a counterexample to show that it is not.

10.26 Write a best basis algorithm (use Algorithm 10.1 as a guide) for matrix input.

10.3 The FBI Fingerprint Compression Specification

Portions of the material that follows is adapted from Section 9.4 of Wavelet Theory: An Elementary Approach with Applications, *by David K. Ruch and Patrick J. Van Fleet [77] with permission from John Wiley & Sons, Inc.*

According to its Web site [41], the Federal Bureau of Investigation (FBI) started collecting and using fingerprints in 1902, and the Identification Division of the FBI was established in 1921. It was reported in [9] that the FBI had about 810,000 fingerprint cards in 1924. Each card consists of 14 prints. Each finger is printed, flat impressions are taken of both thumbs, and each hand is printed simultaneously. In 1992, the Identification Division was renamed the Criminal Justice Information Services (CJIS) Division. In 1996, the CJIS fingerprint collection consisted of over 200 million print cards (see [10]) and in 2008, totaled over 250 million cards. The CJIS Web site [41] reported that about 80 million of these cards had been digitized and that they add or digitally convert about 7000 new cards daily to their electronic database. Each digitized print (an example is plotted in Figure 10.17) from a fingerprint card is scanned at 500 dots per inch (dpi) and each print is of varying size

(e. g., a flat thumbprint measuring 0.875×1.875 inches is stored in a matrix of dimensions 455×975 pixels). According to Bradley, Brislawn, and Hopper [8], about 10.7 megabytes is necessary to store each fingerprint card. Thus over 1000 terabytes of space are needed to store the digitized cards as raw data. Given the daily growth rate of the digitized card database, it is clear why the CJIS Division decided to use image compression to store the fingerprint cards.

Figure 10.17 A digital fingerprint of size 768×768 pixels.

The FBI first considered the JPEG image compression standard (see Section 12.1) but ultimately decided not to use it for fingerprint compression. The JPEG standard starts by partitioning the image into 8×8 blocks and then applying the *discrete cosine transformation* to each block. Quantization is applied individually to each block, which is then encoded using a version of Huffman coding. The compressed image is recovered by unencoding the image data and then applying the inverse discrete cosine transformation to each 8×8 block. The method is highly effective, but the decoupling of the image into 8×8 blocks often gives the compressed image a "blocky" look. The original fingerprint image in Figure 10.17 was compressed using JPEG, and the center 96×96 portion is enlarged in Figure 10.18. The blocky structure of the compression method is evident in this figure.

In 1993 the FBI adopted the *Wavelet Scalar Quantization (WSQ) Specification* [8]. This specification was developed by Thomas Hopper of the FBI in conjunction with researchers Jonathan Bradley and Christopher Brislawn at Los Alamos National Laboratory [9, 10, 11].

Our interest in this specification is the fact that it uses the CDF97 filter pair developed in Section 9.5 and the wavelet packet transformation from Section 10.1. Unlike

Figure 10.18 The enlarged 96×96 center of the JPEG-compressed fingerprint from Figure 10.17. The decoupling of 8×8 blocks is evident in this image.

the image compression schemes described earlier in the book, this specification allows the user to set the compression rate *before* the quantization step. Although we can set the compression rate to whatever value we desire, we must be aware of the fact that there is an objective test for the effectiveness of this specification. In a US court of law, experts use ten points on a fingerprint to uniquely identify the person whose finger produced the print (see Hawthorne [53]). If the compressed version of a digitized fingerprint does not preserve these ten points accurately, then the compression scheme is useless in a court of law. Remarkably, a typical compression rate used by the FBI is 0.75 bit per pixel, or a 10.7:1 ratio if we consider eight bits per pixel a 1:1 compression ratio.

The Basic Wavelet/Scalar Quantization Specification Algorithm

We now present the basic algorithm for implementing the FBI WSQ specification.

Algorithm 10.2 (FBI Wavelet/Scalar Quantization Specification). *This algorithm gives a basic description of the FBI WSQ specification developed by Bradley, Brislawn and Hopper [8, 9, 10, 11] for compressing digitized fingerprints. Suppose that matrix A holds a grayscale version of a digitized fingerprint image.*

1. *Normalize A.*

2. *Apply the wavelet packet transformation using the CDF97 filter pair. At each iteration of the packet transformation, use a two-dimensional version of the symmetric biorthogonal wavelet transformation (see Algorithm 7.1) to better handle boundary effects. Rather than determining a best basis via Algorithm 10.1, use a prescribed basis representation for the packet transformation.*

3. *Perform quantization on each portion of the packet transformation.*

4. *Apply adaptive Huffman coding.*

□

Note: The material that follows is based on the work [8, 9, 10, 11] of the designers Bradley, Brislawn and Hopper of the WSQ specification.

We look at each step of Algorithm 10.2 in more detail using the fingerprint plotted in Figure 10.17 as a running example. Each step contains some computations that we do not attempt to justify – to do so would be beyond the scope of this book. The specification designers have done much empirical work and analysis in developing the first three steps of the algorithm. For more details the interested reader is encouraged to see the references cited in the note above.

Normalizing the Image

The first step in Algorithm 10.2 is to normalize the digitized fingerprint image A. If μ is the mean value of the elements of A and m_1 and M_1 are the minimum and maximum elements of A, respectively, then we normalize A to obtain $\widetilde{A}$, where the elements of $\widetilde{A}$ are

$$\widetilde{A}_{i,j} = \frac{A_{i,j} - \mu}{R}.$$

Here

$$R = \frac{1}{128}\max\left\{M_1 - \mu, \mu - m_1\right\}. \tag{10.9}$$

Bradley, Brislawn, and Hopper [8] remark that this normalization produces a mean of approximately zero in the averages portion of the iterated packet transformation (Region 0 in Figure 10.20) that is computed in the second step of the algorithm. In this way the distribution of values in this portion of the transformation is similar to the distribution of other portions of the transformation, making the Huffman coding in the final step more efficient. For the fingerprint plotted in Figure 10.17, we have $\mu = 186.499$, $m_1 = 0$, and $M_1 = 255$. The mean of $\widetilde{A}$ is -9.55808×10^{-12}. The normalized image is plotted in Figure 10.19.

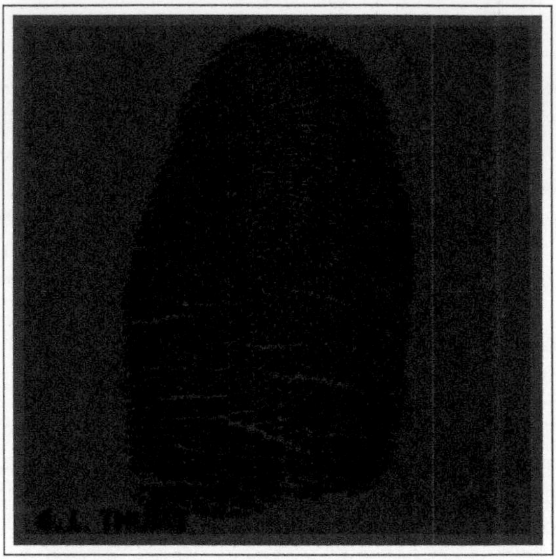

Figure 10.19 The normalized image of the fingerprint.

Applying the Discrete Wavelet Packet Transformation

The second step in Algorithm 10.2 is the application of the wavelet packet transformation to $\widetilde{A}$. The packet transformation is constructed from the CDF97 filter pair from Section 9.5 and is computed via a two-dimensional version of Algorithm 7.1 to exploit the symmetry of the filter pair and better handle the boundary effects produced by the transformation matrix. For this application the filters are *reversed* so that $\widetilde{W}$ in Algorithm 7.1 is constructed from the length 9 filter $\mathbf{h}$ given in (9.89). The similar lengths of each filter in the pair and the balanced smoothness of the corresponding scaling functions were desirable features for the designers of the specification.

The number of possible representations of the two-dimensional transformation grows quickly as the number of iterations increase,[1] and it is thus interesting to learn how the specification designers choose the "best" packet representation. We introduced the best basis algorithm in Section 10.2 and it would seem natural to expect it to be used in conjunction with some cost function that measures the effectiveness of each portion of the transformation with regard to image compression. But the designers of the specification did not use the best basis algorithm – undoubtedly, the cost of computing a best basis for each fingerprint weighed in their decision not to use it. Unlike the compression of generic images, the fingerprint compression con-

[1]Walnut remarks [100] that if A is $M \times M$ with $M = 2^i$, then there are more than $2^{M^2/2}$ possible wavelet packet representations.

siders a class of images with many similar traits and tendencies. The authors of the specification [8] used empirical studies and analysis of the *power spectral density*[2] of fingerprint images to determine a static packet representation that could be used for *all* digitized fingerprints. The power spectral density analysis gives information about the frequencies at which values in the packet transformation occur. This information is useful when designing a quantization method that will be most effective with the static wavelet packet representation selected by the specification designers.

For fingerprint compression, the static representation used in the WSQ specification is displayed in Figure 10.20. It is interesting to note that the schematic in Figure 10.20 implies that five iterations of the discrete wavelet packet transformation are computed for the specification. The 64 portions of this discrete wavelet packet transformation representation are numbered and referred to as the *bands* of the transformation. Note that no portions of the first and third iterations are used, and only four bands of the fifth iteration are computed. We recognize bands 0 through 3 as the fifth iteration of the symmetric biorthogonal wavelet transformation from Algorithm 7.1 applied to $\widetilde{A}$. The transformation of the fingerprint image from Figure 10.17 is plotted in Figure 10.21.

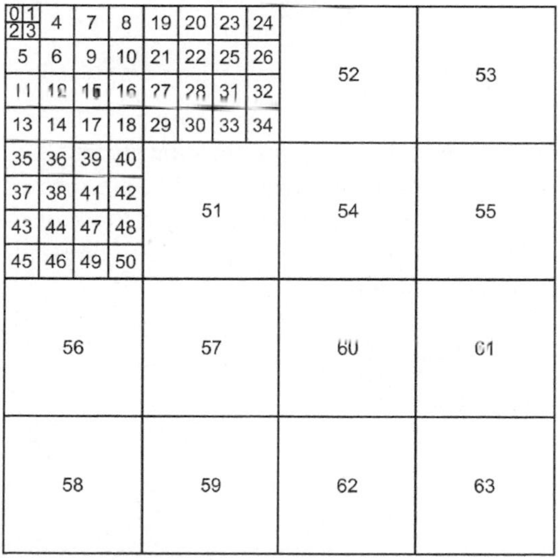

Figure 10.20 The wavelet packet representation used for the second step of Algorithm 10.2.

[2]The power spectral density of a continuous signal can be viewed as the norm squared of its Fourier transformation. If the signal is discrete, then the power spectral density is in terms of the norm squared of the Fourier series built from the signal.

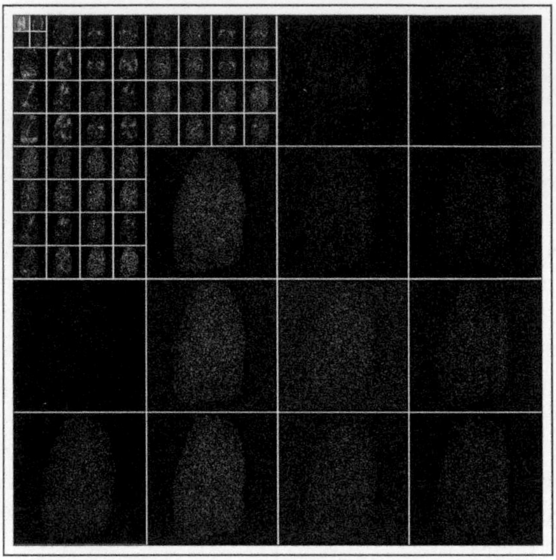

Figure 10.21 The wavelet packet transformation of the fingerprint from Figure 10.17.

Performing Quantization

Perhaps the most important step in the WSQ specification algorithm is quantization. Remarkably, bands 60 through 63 of the transformation are discarded (i.e., their elements are converted to zero). Mathematically, we can think of the elements in these bands as quantized to zero. Quantization is performed on bands zero through 59 using the piecewise constant function

$$f(t,Q,Z) = \begin{cases} \left\lfloor \dfrac{t - Z/2}{Q} \right\rfloor + 1, & t > Z/2 \\ 0, & -Z/2 \le t \le Z/2 \\ \left\lceil \dfrac{t + Z/2}{Q} \right\rceil - 1, & t < -Z/2. \end{cases} \tag{10.10}$$

Here Z and Q are positive numbers. The breakpoints of this function are $\pm\left(\frac{Z}{2} + kQ\right)$, $k \in \mathbb{Z}$. The value Z is called the *zero bin width* since it determines the range of values quantized to zero. The value Q is called the *bin width*. The quantization function is plotted in Figure 10.22.

For $k = 0, \ldots, 59$ we compute the bin widths Q_k and the zero bin width Z_k for band k and use these values to compute the quantized transformation elements for each band. Indeed, if $w^k_{i,j}$ are the elements in band k, then we compute the quantized values

$$\widehat{w}^k_{i,j} = f\left(w^k_{i,j}, Q_k, Z_k\right).$$

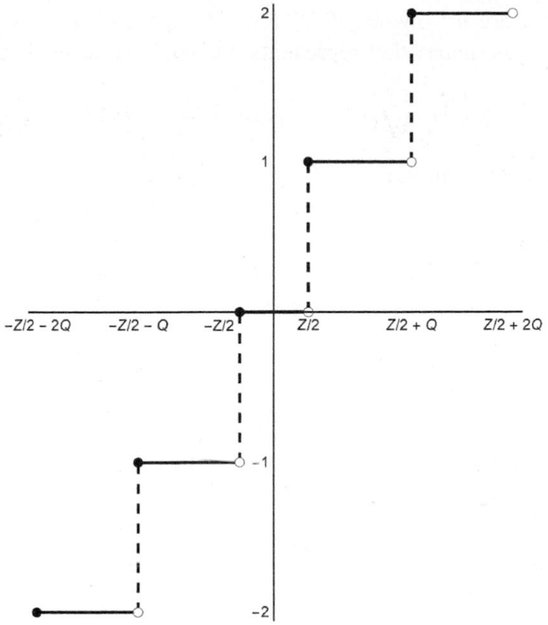

Figure 10.22 The quantization function for the FBI WSQ algorithm.

The zero bin width Z_k is assigned the value

$$Z_k = 1.2Q_k$$

for $k = 0, \ldots, 59$ and Q_k is computed using

$$Q_k = \begin{cases} \dfrac{1}{q}, & 0 \le k \le 3 \\ \dfrac{10}{qA_k \ln\left(\sigma_k^2\right)}, & 4 \le k \le 59. \end{cases} \tag{10.11}$$

The values Q_k depend on values q, A_k, and σ_k. The constants $A_k \approx 1$ chosen by the designers are listed in Table 10.4. The value σ_k^2 is the variance of a modified

Table 10.4 The weights A_k used to determine the bin width (10.11).

Band	A_k
$4, \ldots, 51$	1.00
52, 56	1.32
53, 55, 58, 59	1.08
54, 57	1.42

version of band k. In particular, we compute σ_k^2 from a subregion of band k. Recall the elements of band k are $w_{i,j}^k$, where $0 \le i < M_k$ and $0 \le j < N_k$. Instead of using the entire $M_k \times N_k$ matrix that represents the band, we instead set

$$\widetilde{M}_k = \left\lfloor \frac{7M_k}{16} \right\rfloor \qquad \widetilde{N}_k = \left\lfloor \frac{3N_k}{4} \right\rfloor \tag{10.12}$$

and consider the subregion with indices

$$\begin{aligned} i_{0,k} &= \left\lfloor \frac{9M_k}{32} \right\rfloor \le i \le i_{0,k} + \widetilde{M}_k - 1 = i_{1,k} \\ j_{0,k} &= \left\lfloor \frac{N_k}{8} \right\rfloor \le j \le j_{0,k} + \widetilde{N}_k - 1 = j_{1,k}. \end{aligned}$$

We compute σ_k^2 using the formula

$$\sigma_k^2 = \frac{1}{\widetilde{M}_k \widetilde{N}_k - 1} \sum_{i=i_{0,k}}^{i_{1,k}} \sum_{j=j_{0,k}}^{j_{1,k}} \left(w_{i,j}^k - \mu_k \right)^2 \tag{10.13}$$

where μ_k is the mean of band k.

For example, Band 39 is a square matrix with $M_{39} = N_{39} = 48$. Using (10.12) we see that $\widetilde{M}_{39} = 28$ and $\widetilde{N}_{39} = 36$. We compute the subregion boundary indices to be

$$i_{0,39} = \left\lfloor \frac{9 \cdot 48}{32} \right\rfloor = 13, \quad i_{1,39} = 13 + 28 - 1 = 40$$

and

$$j_{0,39} = \left\lfloor \frac{48}{8} \right\rfloor = 6, \quad j_{1,39} = 6 + 36 - 1 = 41.$$

Using (10.13) we compute $\sigma_{39}^2 = 592.505$. If the variance $\sigma_k^2 < 1.01$ for any band k, then that band has all elements set to zero.

Derivation of the Bin Widths Q_k

The last value needed to determine the bin widths Q_k given in (10.11) is the parameter q. This value chosen so that we are guaranteed a compression rate of r bits per pixel.

Suppose L_k denotes the number of distinct elements in band k. Then we will assume the bit rate r_k for band k is

$$r_k = \log_2 (L_k).$$

If μ_k and σ_k^2 are the mean and variance, respectively, of band k, then it is natural to assume that the values L_k are contained in an interval $[\mu_k - \gamma\sigma_k, \mu_k + \gamma\sigma_k]$, where γ is called the *loading factor*. Statistically, we can think of γ as the number of standard deviations we go left and right of the mean μ_k to capture (most if not all) the distinct elements of band k. The bin width can be expressed as

$$Q_k = \frac{2\gamma\sigma_k}{L_k}. \tag{10.14}$$

For $k = 0, \ldots, 59$, set

$$P_k = qQ_k = \begin{cases} 1, & 0 \le k \le 3 \\ \dfrac{10}{A_k \ln(\sigma_k^2)}, & 4 \le k \le 59. \end{cases}$$

and by (10.14) we have

$$L_k = \frac{2\gamma\sigma_k}{Q_k} = \frac{2\gamma\sigma_k q}{P_k}. \tag{10.15}$$

Our next task is to identify an alternate expression for the bit rate r. To this end, if the dimensions of A are $M \times N$, set

$$m_k = \frac{MN}{M_k N_k}$$

where the dimensions of the matrix in band k are $M_k \times N_k$. In engineering terms we can think of m_k as the *downsampling* rate for band k. For example, the dimensions of the fingerprint in Figure 10.17 are 832×832 and the dimensions of band 39 are $M_{39} \times N_{39} = 48 \times 48$. Thus $m_k = \frac{768^2}{48^2} = 256$. The values of m_k for the fingerprint in Figure 10.17 are given in Table 10.5. Since we have downsampled the elements

Table 10.5 The values m_k for the fingerprint in Figure 10.17.

Band	m_k
$0, \ldots, 3$	1024
$4, \ldots, 50$	256
$51, \ldots 59$	16

in band k but m_k, we consider the modified rate $\frac{r_k}{m_k}$. Summing up these values produces an alternative form for the bit rate r. We have

$$r = \sum_{k=0}^{59} \frac{r_k}{m_k}. \tag{10.16}$$

This formulation of r is certainly motivated by the definition of entropy (Definition 3.2) that appears in Section 3.4. For our computation of q, we will need the fraction of non-discarded wavelet coefficients. Denote this value by S and note that it can be obtained using the m_k values. We have

$$S = \sum_{k=0}^{59} \frac{1}{m_k} = \frac{3}{4}.$$

We are ready to compute q. We start by inserting (10.3) into (10.16) to obtain

$$r = \sum_{k=0}^{59} \frac{\log_2(L_k)}{m_k} = \sum_{k=0}^{59} \log_2(L_k)^{\frac{1}{m_k}}.$$

We use the identity (10.15) to replace L_k above and simplify:

$$\begin{aligned}
r &= \sum_{k=0}^{59} \log_2 (L_k)^{\frac{1}{m_k}} \\
&= \sum_{k=0}^{59} \log_2 \left(\frac{2\gamma\sigma_k q}{P_k}\right)^{\frac{1}{m_k}} \\
&= \log_2 \left(\prod_{k=0}^{59} \left(\frac{2\gamma\sigma_k q}{P_k}\right)^{\frac{1}{m_k}}\right) \\
&= \log_2 \left(\prod_{k=0}^{59} (2\gamma q)^{\frac{1}{m_k}} \prod_{k=0}^{59} \left(\frac{\sigma_k}{P_k}\right)^{\frac{1}{m_k}}\right) \\
&= \log_2 \left((2\gamma q)^S \prod_{k=0}^{59} \left(\frac{\sigma_k}{P_k}\right)^{\frac{1}{m_k}}\right).
\end{aligned}$$

We use the last identity above to write

$$\begin{aligned}
2^r &= (2\gamma q)^S \prod_{k=0}^{59} \left(\frac{\sigma_k}{P_k}\right)^{\frac{1}{m_k}} \\
2^{r-S} &= \gamma^S q^S \left(\prod_{k=0}^{59} \left(\frac{\sigma_k}{P_k}\right)^{\frac{1}{m_k}}\right) \\
2^{r/S-1} &= \gamma q \left(\prod_{k=0}^{59} \left(\frac{\sigma_k}{P_k}\right)^{\frac{1}{m_k}}\right)^{\frac{1}{S}}
\end{aligned}$$

and finally infer that

$$q = \frac{1}{\gamma} \cdot 2^{r/S-1} \left(\prod_{k=0}^{59} \left(\frac{\sigma_k}{P_k}\right)^{\frac{1}{m_k}}\right)^{-1/S}. \tag{10.17}$$

The designers of the WSQ specification use $\gamma = 2.5$ as a loading factor so that $\frac{1}{\gamma} = 0.4$. Plugging $r = 2$ into (10.17) gives $q = 0.178598$. So that we can get a sense of the size of the bin widths Q_k, we have listed them in Table 10.6 for the fingerprint in Figure 10.17. Using the values of Q_k from Table 10.6 and $Z_k = 1.2Q_k$ in the quantization function (10.10), we quantize the first 60 bands of the discrete packet transformation and replace the elements in the remaining four bands with zeros. The quantized packet transformation is plotted in Figure 10.23.

Encoding the Quantized Transformation

The final step in Algorithm 10.2 is to encode the values in the quantized transformation. The WSQ specification uses an adaptive Huffman coding method that is

Table 10.6 The bin widths Q_k (with bit rate $r = 2$) from (10.11) for the fingerprint in Figure 10.17.

Band	Q_k	Band	Q_k	Band	Q_k	Band	Q_k
0	5.599	15	6.764	30	8.377	45	7.392
1	5.599	16	6.302	31	7.478	46	7.309
2	5.599	17	6.215	32	8.115	47	7.411
3	5.599	18	5.779	33	7.292	48	7.325
4	6.137	19	8.939	34	7.844	49	7.951
5	5.873	20	8.048	35	8.512	50	7.944
6	5.830	21	9.114	36	8.983	51	9.604
7	5.769	22	8.047	37	7.802	52	11.926
8	6.171	23	6.433	38	7.839	53	10.395
9	5.729	24	7.439	39	9.112	54	14.076
10	5.672	25	6.574	40	9.345	55	12.212
11	5.712	26	7.402	41	8.112	56	11.246
12	5.797	27	9.250	42	8.354	57	12.867
13	5.880	28	8.344	43	6.514	58	9.923
14	5.641	29	9.712	44	6.558	59	11.988

much like the one used by the JPEG group (see Pennebaker and Mitchell [74]) for its compression standard for this task. This coding scheme is a bit more sophisticated than the basic Huffman coding method detailed in Section 3.3. The interested reader is referred to Gersho and Gray [46]. Recall that the last four bands are not even included in the quantized packet transformation. The image is coded at 0.75 bit per pixel or at a compression ratio of 10.7 : 1.

Recovering the Compressed Image

To recover the compressed image, we perform the following steps:

1. Decode the Huffman codes.
2. Apply a *dequantization function.*
3. Compute the inverse wavelet packet transformation to obtain the normalized approximation $\widetilde{B}$ to $\widetilde{A}$.
4. For μ the mean of A and R given by (10.9), we compute

$$B = R\widetilde{B} + \mu. \tag{10.18}$$

The dequantization function is interesting and we describe it now. The quantization function given in (10.10) maps values larger than $Z/2$ to positive numbers, values between and including $-Z/2$ and $Z/2$ to zero, and values smaller than $-Z/2$ to negative numbers. Then it makes sense that the dequantization function $d(y, Q, Z)$

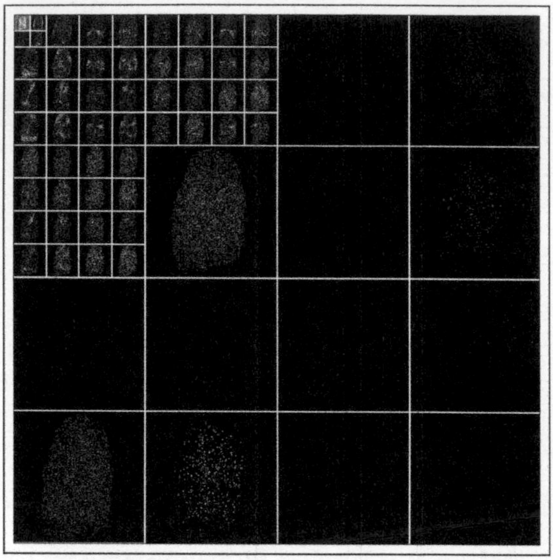

Figure 10.23 The quantized wavelet packet transformation of the fingerprint from Figure 10.17.

should be piecewise defined with different formulas for positive, negative, and zero values. The easiest case is $y = 0$. Here we define $d(0, Q, Z) = 0$, since we have no way of knowing which $t \in [-Z/2, Z/2]$ yields $f(t, Q, Z) = 0$. If $t > Z/2$, we have

$$y = f(t, Q, Z) = \left\lfloor \frac{t - Z/2}{Q} \right\rfloor + 1 \qquad \text{or} \qquad y - 1 = \left\lfloor \frac{t - Z/2}{Q} \right\rfloor. \tag{10.19}$$

Exact inversion is impossible due to the floor function. Instead of subtracting one from y, the designers of the specification subtracted a value $0 < C < 1$ instead to account for the floor function. So for $y > 0$, we define

$$d(y, Q, Z) = (y - C)Q + Z/2$$

solving (10.19) for t as if the floor function was not present. We can add the same "fudge factor" in the case when $y < 0$ and thus define our dequantization function as

$$d(y, Q, Z) = \begin{cases} (y - C)Q + Z/2, & y > 0 \\ 0, & y = 0 \\ (y + C)Q - Z/2, & y < 0. \end{cases} \tag{10.20}$$

We apply $f(y, Q_k, Z_k)$ to each element in band k, $k = 0, \ldots, 59$. A typical value for C used by the designers is $C = 0.44$. We have applied $d(y, Q_k, Z_k)$ to the packet transformation in Figure 10.23. The result is plotted in Figure 10.24. Note that we have added zero matrices in the place of bands 60 through 63.

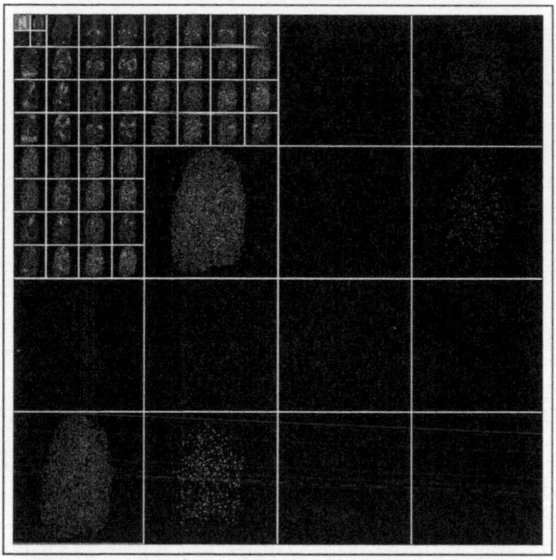

Figure 10.24 The dequantized wavelet packet transformation constructed by applying $d(y, Q_k, Z_k)$ to the matrix in Figure 10.23.

We next apply the inverse wavelet packet transformation to obtain an approximation of the normalized image $\widetilde{A}$. The result is plotted in Figure 10.25(a). The last step in recovering the compressed fingerprint image is to apply (10.18) to each element in the image matrix in Figure 10.24. The result is plotted in Figure 10.25(b).

The images in Figures 10.17 and 10.25(b) are difficult to distinguish. The PSNR of the compressed and original images is 29.8781. In Figures 10.26(a) and 10.26(b) we have plotted the fingerprint of Figure 10.17 compressed at one and two bits per pixel, respectively. Table 10.7 summarizes the results.

Table 10.7 Different compression results for the fingerprint in Figure 10.17.

Bits per Pixel	Compression Ratio	q	PSNR
0.75	10.7 : 1	0.056255	29.8781
1.00	8 : 1	0.070877	30.9036
2.00	4 : 1	0.178598	35.4621

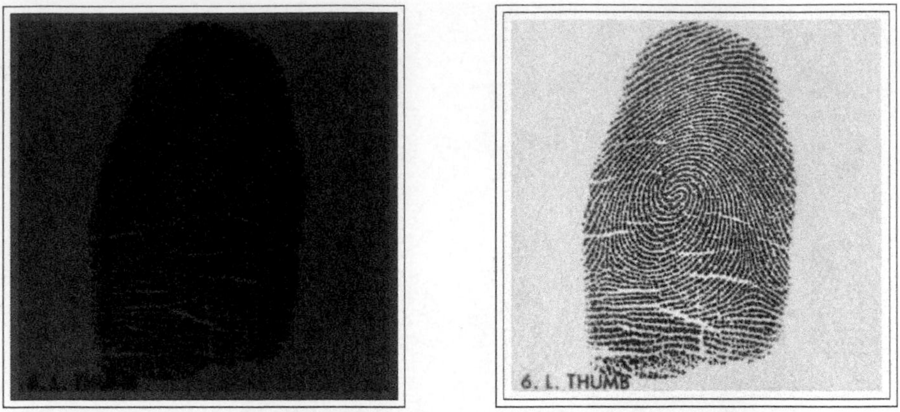

(a) Approximation of the normalized image (b) Compressed image

Figure 10.25 An approximate to the normalized image that appears in Figure 10.19 and the compressed image.

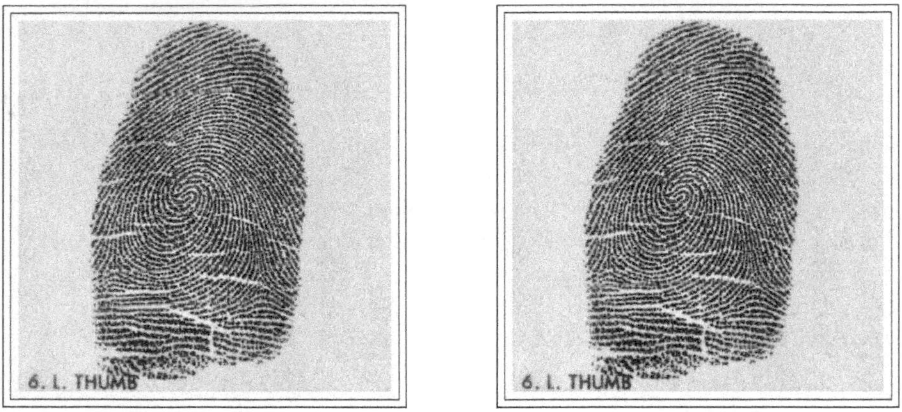

(a) $r = 1$ or $8 : 1$ ratio (b) $r = 2$ or $4 : 1$ ratio

Figure 10.26 Compressed images at ratios $8 : 1$ and $4 : 1$.

Computer Lab

10.1 The FBI WSQ Specification. From `stthomas.edu/wavelets`, access the file `FBIFingerprint`. In this lab, you will implement the FBI Wavelet Scalar Quantization method outlined in this section.

CHAPTER 11

LIFTING

If you worked Problem 4.36 in Section 4.4, then you have had a brief introduction to lifting and how to use it in lossless compression.

The lifting method for computing wavelet transforms was introduced by Wim Sweldens[1] in [92]. For a quick introduction to the topic, see Sweldens' tutorial paper [91]. Another good reference on lifting is the book by Jensen and la Cour-Harbo [61].

This chapter begins with a section that introduces lifting via the LeGall biorthogonal filter pair. This filter pair is a modification of the $(5, 3)$ biorthogonal spline filter pair introduced in Section 7.1. Section 11.2 introduces Z-transforms and Laurent polynomials. We introduce and consider several properties of Z-transforms and Laurent polynomials. In particular, we discuss the identification of a greatest common divisor of two Laurent polynomials obtained via the Euclidean algorithm. The matrix formulation of this greatest common divisor is a critical component in the derivation of a lifting method for a lowpass/highpass filter pair. In Section 11.3, we

[1]The term *lifting* was introduced by Sweldens to name the method he devised for increasing the number of derivatives m and n for which the Fourier series $\widetilde{H}(\omega)$ and $H(\omega)$ associated with filters $\widetilde{\mathbf{h}}, \mathbf{h}$, respectively, satisfy $\widetilde{H}^{(m)}(\pi) = H^{(n)}(\pi) = 0$.

introduce the concept of a polyphase matrix. The factorization of this matrix, due to Daubechies and Sweldens [31], leads to a lifting scheme for a wavelet transform. We state this result and provide a constructive proof in this section. The final section contains several examples of lifting methods for lowpass/highpass filter pairs.

11.1 The LeGall Wavelet Transform

One of the goals of the JPEG2000 standard was to include lossless compression as a feature. The designers decided to use a biorthogonal wavelet transform for lossless compression. The key was choosing the correct filter pair. The desirable filter pair would consist of filters of similar and relatively short lengths and since an image is a matrix (or three matrices for color) with integer-valued elements, the transform needs to produce integer output. None of the biorthogonal filter pairs we have developed (Sections 7.1, 7.2, 9.4, and 9.5) possess this feature. In Example 7.9, we scaled the $(5, 3)$ biorthogonal spline filter pair (7.54) so that the output was rational. We could have further scaled the filter by a factor of eight but then the range of the output, for one iteration of the transform, increases by a factor of 8.

In order to map integers to integers, the designers of the JPEG2000 standard adopted a scaled version of the $(5, 3)$ filter pair (7.54), but used a computation method known as *lifting* instead of the in-place computational method employed thus far (see for example Section 4.3). The filter pair is known as the *LeGall filter pair* and the lifting scheme allows us to round the output of an iteration of the wavelet transform and still recover it exactly via an inverse wavelet transform. In this section, we develop the lifting method for the LeGall filter pair and then illustrate how it can be modified so that it maps integers to integers.

The LeGall Filter Pair

Recall from Definition 7.2, the $(5, 3)$ biorthogonal spline filter pair is

$$\widetilde{\mathbf{h}} = \left(\widetilde{h}_{-1}, \widetilde{h}_0, \widetilde{h}_1\right) = \left(\frac{\sqrt{2}}{4}, \frac{\sqrt{2}}{2}, \frac{\sqrt{2}}{4}\right)$$

and

$$\mathbf{h} = (h_{-2}, h_{-1}, h_0, h_1, h_2) = \left(-\frac{\sqrt{2}}{8}, \frac{\sqrt{2}}{4}, \frac{3\sqrt{2}}{4}, \frac{\sqrt{2}}{4}, -\frac{\sqrt{2}}{8}\right).$$

As is the case in lossy compression, the JPEG2000 reverses the roles of these filters. We use the five-term filter in the lowpass portion of the transform and the three-term filter in the differences portion of the transform.

The filter pair is further modified by multiplying the five-term filter by $\frac{\sqrt{2}}{2}$ and multiplying the three-term filter by $\frac{2}{\sqrt{2}} = \sqrt{2}$. The resulting filter is often called the *LeGall filter pair* (see LeGall and Tabatabai [66]).

Definition 11.1 (The LeGall Filter Pair). *We define the* LeGall filter pair *as*

$$\boxed{\begin{aligned}\widetilde{\mathbf{h}} &= \left(\widetilde{h}_{-2}, \widetilde{h}_{-1}, \widetilde{h}_0, \widetilde{h}_1, \widetilde{h}_2\right) = \left(-\tfrac{1}{8}, \tfrac{1}{4}, \tfrac{3}{4}, \tfrac{1}{4}, -\tfrac{1}{8}\right)\\ \mathbf{h} &= (h_{-1}, h_0, h_1) = \left(\tfrac{1}{2}, 1, \tfrac{1}{2}\right).\end{aligned}} \tag{11.1}$$

□

The highpass filters for the LeGall filter pair are obtained in the usual way, with the exception that all filter elements are multiplied by -1.

$$\begin{aligned}\widetilde{g}_k &= (-1)\cdot(-1)^k h_{1-k} = (-1)^{k+1} h_{1-k}\\ g_k &= (-1)\cdot(-1)^k \widetilde{h}_{1-k} = (-1)^{k+1}\widetilde{h}_{1-k}\end{aligned} \tag{11.2}$$

Using (11.1) and (11.2), we see the highpass filters associated with the LeGall filters are

$$\boxed{\begin{aligned}\widetilde{\mathbf{g}} &= (\widetilde{g}_0, \widetilde{g}_1, \widetilde{g}_2) = \left(-\tfrac{1}{2}, 1, -\tfrac{1}{2}\right)\\ \mathbf{g} &= (g_{-1}, g_0, g_1, g_2, g_3) = \left(-\tfrac{1}{8}, -\tfrac{1}{4}, \tfrac{3}{4}, -\tfrac{1}{4}, -\tfrac{1}{8}\right).\end{aligned}} \tag{11.3}$$

The wavelet matrix, in the 10×10 case is

$$\widetilde{W}_{10} = \left[\begin{array}{cccccccccc} \frac{3}{4} & \frac{1}{4} & -\frac{1}{8} & 0 & 0 & 0 & 0 & 0 & -\frac{1}{8} & \frac{1}{4}\\ -\frac{1}{8} & \frac{1}{4} & \frac{3}{4} & \frac{1}{4} & -\frac{1}{8} & 0 & 0 & 0 & 0 & 0\\ 0 & 0 & -\frac{1}{8} & \frac{1}{4} & \frac{3}{4} & \frac{1}{4} & -\frac{1}{8} & 0 & 0 & 0\\ 0 & 0 & 0 & 0 & -\frac{1}{8} & \frac{1}{4} & \frac{3}{4} & \frac{1}{4} & -\frac{1}{8} & 0\\ -\frac{1}{8} & 0 & 0 & 0 & 0 & 0 & -\frac{1}{8} & \frac{1}{4} & \frac{3}{4} & \frac{1}{4}\\ \hline -\frac{1}{2} & 1 & -\frac{1}{2} & 0 & 0 & 0 & 0 & 0 & 0 & 0\\ 0 & 0 & -\frac{1}{2} & 1 & -\frac{1}{2} & 0 & 0 & 0 & 0 & 0\\ 0 & 0 & 0 & 0 & -\frac{1}{2} & 1 & -\frac{1}{2} & 0 & 0 & 0\\ 0 & 0 & 0 & 0 & 0 & 0 & -\frac{1}{2} & 1 & -\frac{1}{2} & 0\\ -\frac{1}{2} & 0 & 0 & 0 & 0 & 0 & 0 & 0 & -\frac{1}{2} & 1 \end{array}\right] \tag{11.4}$$

and its inverse is

$$W_{10}^{T} = \left[\begin{array}{ccccc|ccccc}
1 & 0 & 0 & 0 & 0 & -\frac{1}{4} & 0 & 0 & 0 & -\frac{1}{4} \\
\frac{1}{2} & \frac{1}{2} & 0 & 0 & 0 & \frac{3}{4} & -\frac{1}{8} & 0 & 0 & -\frac{1}{8} \\
0 & 1 & 0 & 0 & 0 & -\frac{1}{4} & -\frac{1}{4} & 0 & 0 & 0 \\
0 & \frac{1}{2} & \frac{1}{2} & 0 & 0 & -\frac{1}{8} & \frac{3}{4} & -\frac{1}{8} & 0 & 0 \\
0 & 0 & 1 & 0 & 0 & 0 & -\frac{1}{4} & -\frac{1}{4} & 0 & 0 \\
0 & 0 & \frac{1}{2} & \frac{1}{2} & 0 & 0 & -\frac{1}{8} & \frac{3}{4} & -\frac{1}{8} & 0 \\
0 & 0 & 0 & 1 & 0 & 0 & 0 & -\frac{1}{4} & -\frac{1}{4} & 0 \\
0 & 0 & 0 & \frac{1}{2} & \frac{1}{2} & 0 & 0 & -\frac{1}{8} & \frac{3}{4} & -\frac{1}{8} \\
0 & 0 & 0 & 0 & 1 & 0 & 0 & 0 & -\frac{1}{4} & -\frac{1}{4} \\
\frac{1}{2} & 0 & 0 & 0 & \frac{1}{2} & -\frac{1}{8} & 0 & 0 & -\frac{1}{8} & \frac{3}{4}
\end{array}\right].$$

The Lifting Method and the LeGall Filter Pair

The lifting method produces the same results as any efficient algorithm (see Computer Lab 7.3) we could write for computing the symmetric biorthogonal wavelet transform using the LeGall filter pair. Let's motivate the lifting scheme using a length-10 vector. We alter our normal indexing practice so that the derivation that follows is in line with the construction of the general lifting method in Section 11.3. For $\mathbf{v} = [v_0, v_1, \ldots, v_9]^T$, we use Algorithm 7.1 to compute the symmetric transform. We first create the length-18 vector

$$\mathbf{v}^p = [v_0, v_1, \ldots, v_9, v_8, \ldots, v_1]^T$$

and then form $\widetilde{W}_{18}$ to obtain

$$\left[\begin{array}{c} \widetilde{\mathbf{a}}^p \\ \hline \widetilde{\mathbf{d}}^p \end{array}\right] = \widetilde{W}_{18}\mathbf{v}^p.$$

The transform is completed by joining the first five values of $\widetilde{\mathbf{a}}^p$ with the first five values of $\widetilde{\mathbf{d}}^p$. Specifically, the averages portion of the transform is

$$\widetilde{\mathbf{a}} = \begin{bmatrix} \widetilde{a}_0 \\ \widetilde{a}_1 \\ \widetilde{a}_2 \\ \widetilde{a}_3 \\ \widetilde{a}_4 \end{bmatrix} = \begin{bmatrix} -\frac{1}{8}v_2 + \frac{1}{4}v_1 + \frac{3}{4}v_0 + \frac{1}{4}v_1 - \frac{1}{8}v_2 \\ -\frac{1}{8}v_0 + \frac{1}{4}v_1 + \frac{3}{4}v_2 + \frac{1}{4}v_3 - \frac{1}{8}v_4 \\ -\frac{1}{8}v_2 + \frac{1}{4}v_3 + \frac{3}{4}v_4 + \frac{1}{4}v_5 - \frac{1}{8}v_6 \\ -\frac{1}{8}v_4 + \frac{1}{4}v_5 + \frac{3}{4}v_6 + \frac{1}{4}v_7 - \frac{1}{8}v_8 \\ -\frac{1}{8}v_6 + \frac{1}{4}v_7 + \frac{3}{4}v_8 + \frac{1}{4}v_9 - \frac{1}{8}v_8 \end{bmatrix}. \tag{11.5}$$

while the differences portion of the transform is

$$\widetilde{\mathbf{d}} = \begin{bmatrix} \widetilde{d}_0 \\ \widetilde{d}_1 \\ \widetilde{d}_2 \\ \widetilde{d}_3 \\ \widetilde{d}_4 \end{bmatrix} = \begin{bmatrix} -\frac{1}{2}v_0 + v_1 - \frac{1}{2}v_2 \\ -\frac{1}{2}v_2 + v_3 - \frac{1}{2}v_4 \\ -\frac{1}{2}v_4 + v_5 - \frac{1}{2}v_6 \\ -\frac{1}{2}v_6 + v_7 - \frac{1}{2}v_8 \\ -\frac{1}{2}v_8 + v_9 - \frac{1}{2}v_8 \end{bmatrix}. \tag{11.6}$$

We could certainly simplify the elements $\widetilde{a}_0$ and $\widetilde{a}_4$ in (11.5) and $\widetilde{d}_4$ in (11.6). To better illustrate the lifting method, we choose not to do so. We now show how the same results can be obtained via the lifting scheme.

Note that the center terms in the elements of $\widetilde{\mathbf{a}}$ in (11.5) are built from the even elements of $\mathbf{v}$, and the center terms of the elements of $\widetilde{\mathbf{d}}$ in (11.6) are built from the odd elements of $\mathbf{v}$. Thus, the first step in lifting is to partition $\mathbf{v}$ into "even" and "odd" vectors:

$$\mathbf{e} = [e_0, e_1, e_2, e_3, e_4]^T = [v_0, v_2, v_4, v_6, v_8]^T.$$

and

$$\mathbf{o} = [o_0, o_1, o_2, o_3, o_4]^T = [v_1, v_3, v_5, v_7, v_9]^T$$

We can easily compute $\widetilde{\mathbf{d}}$ from $\mathbf{e}$ and $\mathbf{o}$. We have

$$\begin{aligned}
\widetilde{d}_0 &= o_0 - \frac{1}{2}(e_0 + e_1) \\
\widetilde{d}_1 &= o_1 - \frac{1}{2}(e_1 + e_2) \\
\widetilde{d}_2 &= o_2 - \frac{1}{2}(e_2 + e_3) \\
\widetilde{d}_3 &= o_3 - \frac{1}{2}(e_3 + e_4) \\
\widetilde{d}_4 &= o_4 - \frac{1}{2}(e_4 + e_4).
\end{aligned}$$

If we introduce $e_5 = e_4$, we can write the elements of $\widetilde{\mathbf{d}}$ as

$$\widetilde{d}_k = o_k - \frac{1}{2}(e_k + e_{k+1}), \qquad k = 0, \ldots, 4. \tag{11.7}$$

The so-called *lifting step* is evident when we make the observation that the sum of $\widetilde{d}_0$ and $\widetilde{d}_1$ is related to $\widetilde{a}_1$. For example, if we compute $\widetilde{d}_0 + \widetilde{d}_1$, we have

$$\begin{aligned}
\widetilde{d}_0 + \widetilde{d}_1 &= o_0 - \frac{1}{2}(e_0 + e_1) + o_1 - \frac{1}{2}(e_1 + e_2) \\
&= v_1 - \frac{1}{2}(v_0 + v_2) + v_3 - \frac{1}{2}(v_2 + v_4) \\
&= -\frac{1}{2}v_0 + v_1 - v_2 + v_3 - \frac{1}{2}v_4.
\end{aligned}$$

The result is "close" to

$$\widetilde{a}_1 = -\frac{1}{8}v_0 + \frac{1}{4}v_1 + \frac{3}{4}v_2 + \frac{1}{4}v_3 - \frac{1}{8}v_4.$$

Indeed, if we divide the sum by 4 and manipulate the term containing v_2, we see that

$$\begin{aligned}\frac{\widetilde{d}_0 + \widetilde{d}_1}{4} &= -\frac{1}{8}v_0 + \frac{1}{4}v_1 - \frac{1}{4}v_2 + \frac{1}{4}v_3 - \frac{1}{8}v_4 \\ &= \left(-\frac{1}{8}v_0 + \frac{1}{4}v_1 + \frac{3}{4}v_2 + \frac{1}{4}v_3 - \frac{1}{8}v_4\right) - v_2 \\ &= \widetilde{a}_1 - v_2 \\ &= \widetilde{a}_1 - e_1.\end{aligned}$$

Solving for $\widetilde{a}_1$ gives

$$\widetilde{a}_1 = e_1 + \frac{1}{4}\left(\widetilde{d}_0 + \widetilde{d}_1\right).$$

In a similar manner, we can show that

$$\widetilde{a}_k = e_k + \frac{1}{4}(\widetilde{d}_k + \widetilde{d}_{k-1}), \qquad k = 1, 2, 3, 4. \tag{11.8}$$

Some special consideration is needed for $\widetilde{a}_0$. We can write

$$\begin{aligned}\widetilde{a}_0 &= -\frac{1}{8}v_2 + \frac{1}{4}v_1 + \frac{3}{4}v_0 + \frac{1}{4}v_1 - \frac{1}{8}v_2 \\ &= -\frac{1}{4}v_2 + \frac{1}{2}v_1 + \frac{3}{4}v_0 \\ &= \left(-\frac{1}{4}v_2 + \frac{1}{2}v_1 - \frac{1}{4}v_0\right) + v_0 \\ &= \frac{1}{4}\left(-v_2 + 2v_1 - v_0\right) + v_0 \\ &= \frac{1}{4}\left(\widetilde{d}_0 + \widetilde{d}_0\right) + v_0 \\ &= \frac{1}{4}\left(\widetilde{d}_0 + \widetilde{d}_0\right) + e_0.\end{aligned}$$

Thus, we have constructed the averages portion of the transform from the differences portion of the transform.

An Algorithm for Performing Lifting with the LeGall Filter

We can easily generalize (11.7) and (11.8) to create a lifting process for the symmetric biorthogonal transform using the LeGall filter pair. For an N-vector $\mathbf{v}$, where N is even, we first partition $\mathbf{v}$ into even and odd vectors $\mathbf{e}$ and $\mathbf{o}$ each of length $N/2$.

We then compute the differences portion $\widetilde{\mathbf{d}}$ of the transform by first introducing the value $e_{N/2} = e_{N/2-1}$ and using

$$\boxed{\widetilde{d}_k = o_k - \frac{1}{2}\left(e_k + e_{k+1}\right), \qquad k = 0, \ldots, \frac{N}{2} - 1.} \tag{11.9}$$

Once we have $\widetilde{\mathbf{d}}$, we can compute the averages portion $\widetilde{\mathbf{a}}$ of the transformation. We set $\widetilde{d}_{-1} = \widetilde{d}_0$ and then use

$$\boxed{\widetilde{a}_k = e_k + \frac{1}{4}\left(\widetilde{d}_k + \widetilde{d}_{k-1}\right), \qquad k = 0, \ldots, \frac{N}{2} - 1.} \tag{11.10}$$

This process is summarized in the following algorithm.

Algorithm 11.1 (Lifting with the LeGall Filter Pair). *This algorithm takes a vector* $\mathbf{v}$ *of even length* N *and uses the LeGall filter pair* (11.1) *to compute the symmetric biorthogonal wavelet transform.*

1. *Form vectors* $\mathbf{e}$ *and* $\mathbf{o}$ *using* $e_k = v_{2k}$ *and* $o_k = v_{2k+1}$, $k = 0, \ldots, \frac{N}{2} - 1$.
2. *Append* $\mathbf{e}$ *with the element* $e_{N/2} = e_{N/2-1}$.
3. *Compute the difference portion* $\widetilde{\mathbf{d}}$ *via the formula* $\widetilde{d}_k = o_k - \frac{1}{2}\left(e_k + e_{k+1}\right)$, $k = 0, \ldots, \frac{N}{2} - 1$.
4. *Prepend* $\mathbf{d}$ *with the element* $\widetilde{d}_{-1} = \widetilde{d}_0$.
5. *Compute the averages portion* $\widetilde{\mathbf{a}}$ *via the formula* $\widetilde{a}_k = e_k + \frac{1}{4}\left(\widetilde{d}_k + \widetilde{d}_{k-1}\right)$, $k = 0, \ldots, \frac{N}{2} - 1$.
6. *Drop* $\widetilde{d}_{-1}$ *from* $\widetilde{\mathbf{d}}$.
7. *Return the vector* $\left[\begin{array}{c}\widetilde{\mathbf{a}} \\ \hline \widetilde{\mathbf{d}}\end{array}\right]$.

□

Let's look at an example.

Example 11.1 (Using the Lifting Scheme). *Let* $\mathbf{v} = [-5, 5, 0, 1, -3, 4, -1, -2]^T$. *Compute the symmetric biorthogonal transform with the LeGall filter pair using the lifting method.*

Solution. *We begin by partitioning* $\mathbf{v}$ *into even and odd vectors each of length* 4. *We have*

$$\mathbf{e} = [-5, 0, -3, -1]^T \qquad \textit{and} \qquad \mathbf{o} = [5, 1, 4, -2]^T.$$

We next set $e_4 = e_3 = -1$ *and use* (11.9) *to compute the differences portion* $\widetilde{\mathbf{d}}$. *We have*

$$\begin{aligned}
\widetilde{d}_0 &= o_0 - \tfrac{1}{2}\left(e_0 + e_1\right) = 5 - \tfrac{1}{2}(-5+0) = \tfrac{15}{2}\\
\widetilde{d}_1 &= o_1 - \tfrac{1}{2}\left(e_1 + e_2\right) = 1 - \tfrac{1}{2}(0-3) = \tfrac{5}{2}\\
\widetilde{d}_2 &= o_2 - \tfrac{1}{2}\left(e_2 + e_3\right) = 4 - \tfrac{1}{2}(-3-1) = 6\\
\widetilde{d}_3 &= o_3 - \tfrac{1}{2}\left(e_3 + e_4\right) = -2 - \tfrac{1}{2}(-1-1) = -1.
\end{aligned}$$

Next we set $\widetilde{d}_{-1} = \widetilde{d}_0 = \frac{15}{2}$ *and use* (11.10) *to compute the averages portion* $\widetilde{\mathbf{a}}$ *of the transform.*

$$\begin{aligned}
\widetilde{a}_0 &= e_0 + \tfrac{1}{4}\left(\widetilde{d}_0 + \widetilde{d}_{-1}\right) = -5 + \tfrac{1}{4}\left(\tfrac{15}{2} + \tfrac{15}{2}\right) = -\tfrac{5}{4}\\
\widetilde{a}_1 &= e_1 + \tfrac{1}{4}\left(\widetilde{d}_1 + \widetilde{d}_0\right) = 0 + \tfrac{1}{4}\left(\tfrac{5}{2} + \tfrac{15}{2}\right) = \tfrac{5}{2}\\
\widetilde{a}_2 &= e_2 + \tfrac{1}{4}\left(\widetilde{d}_2 + \widetilde{d}_1\right) = -3 + \tfrac{1}{4}\left(6 + \tfrac{5}{2}\right) = -\tfrac{7}{8}\\
\widetilde{a}_3 &= e_3 + \tfrac{1}{4}\left(\widetilde{d}_3 + \widetilde{d}_2\right) = -1 + \tfrac{1}{4}(-1+6) = \tfrac{1}{4}.
\end{aligned}$$

(Live example: Visit `stthomas.edu/wavelets` *and click on Live Examples.)*

□

Inverting the Lifting Process

It is a straightforward process to invert the lifting process. We assume that $\widetilde{\mathbf{a}}$ and $\widetilde{\mathbf{d}}$ are given and follow the notation in Algorithm 11.1 to set $\widetilde{d}_{-1} = \widetilde{d}_0$. Then solving for e_k in (11.10) gives

$$e_k = \widetilde{a}_k - \frac{1}{4}\left(\widetilde{d}_k + \widetilde{d}_{k-1}\right), \qquad k = 0, \ldots, \frac{N}{2} - 1. \tag{11.11}$$

Now we have the even components $\mathbf{e}$ of $\mathbf{v}$! If we set $e_{N/2} = e_{N/2-1}$ we can solve (11.9) for o_k:

$$o_k = \widetilde{d}_k + \frac{1}{2}\left(e_k + e_{k+1}\right), \qquad k = 0, \ldots, \frac{N}{2} - 1. \tag{11.12}$$

Finally, to obtain $\mathbf{v}$, we simply intertwine the elements of $\mathbf{e}$ and $\mathbf{o}$. In Problem 11.2 you are asked to write an algorithm to perform inverse lifting.

Lifting and Integer to Integer Transforms

Suppose we are given vector $\mathbf{v}$ of even length whose elements are integers. If we apply the symmetric biorthogonal transform via Algorithm 7.1 using the LeGall filter pair, integers or rational numbers with denominators two, four, or eight are returned (see Problem 11.4). Rounding or truncating these values to the nearest integer, applying the inverse transform, and rounding or truncating the result does not necessarily return the original vector $\mathbf{v}$ (see Problem 11.3). As you will see in Problem 11.5, it is possible to modify the LeGall filter pair and then use Algorithm 7.1 to map integers to integers. The process is reversible, but it is not desirable because of its adverse effect on the quantization step in applications such as image compression.

The big advantage to the lifting process is it can easily be adjusted to map integers to integers in a such a way as to make the inverse transform exact. Let $\mathbf{v}$ be a vector of even length N and suppose that it has been partitioned into even and odd parts $\mathbf{e}$ and $\mathbf{o}$. Recall that for $e_{N/2} = e_{N/2-1}$, we built the differences portion $\widetilde{\mathbf{d}}$ of the transform via the formula (11.9)

$$\widetilde{d}_k = o_k - \frac{1}{2}\left(e_k + e_{k+1}\right), \qquad k = 0, \ldots, \frac{N}{2} - 1$$

and then constructed the averages portion by setting $\widetilde{d}_{-1} = \widetilde{d}_0$ and using (11.10) to compute

$$\widetilde{a}_k = e_k + \frac{1}{4}\left(\widetilde{d}_k + \widetilde{d}_{k-1}\right), \qquad k = 0, \ldots, \frac{N}{2} - 1.$$

To alter these formulas so that they map integers to integers, we simply take

$$\boxed{\widetilde{d}^*_k = o_k - \left\lfloor \frac{1}{2}\left(e_k + e_{k+1}\right)\right\rfloor, \qquad k = 0, \ldots, \frac{N}{2} - 1.} \tag{11.13}$$

where $e_{N/2} = e_{N/2-1}$ to obtain the integer-valued differences portion $\widetilde{\mathbf{d}}^*$ of the transform. Here, $\lfloor \cdot \rfloor$ is the floor function defined by (3.2) in Section 3.1. Once the integer-valued $\widetilde{d}^*_k$ are known, we set $\widetilde{d}^*_{-1} = \widetilde{d}^*_0$ and compute

$$\boxed{\widetilde{a}^*_k = e_k + \left\lfloor \frac{1}{4}\left(\widetilde{d}^*_k + \widetilde{d}^*_{k-1}\right) + \frac{1}{2}\right\rfloor, \qquad k = 0, \ldots, \frac{N}{2} - 1.} \tag{11.14}$$

to obtain the integer-valued averages portion $\widetilde{\mathbf{a}}^*$ of the transform.

As the authors point out in [17], $\frac{1}{2}$ is added to the floored part of $\widetilde{a}^*_k$ to eliminate bias. Note as well that $\lfloor n + \frac{1}{2} \rfloor = \text{Round}(n)$. In Problem 11.7 you will continue to explore this step.

Let's look at an example.

Example 11.2 (Mapping Integers to Integers Using Lifting). *We return to Example 11.1. If we apply* (11.13) *and* (11.14) *to* $\mathbf{v}$, *we obtain*

$$\begin{aligned}
\widetilde{d}^*_0 &= o_0 - \left\lfloor \tfrac{1}{2}\left(e_0 + e_1\right)\right\rfloor = 5 - \left\lfloor \tfrac{1}{2}\left(-5 + 0\right)\right\rfloor = 8\\
\widetilde{d}^*_1 &= o_1 - \left\lfloor \tfrac{1}{2}\left(e_1 + e_2\right)\right\rfloor = 1 - \left\lfloor \tfrac{1}{2}\left(0 - 3\right)\right\rfloor = 3\\
\widetilde{d}^*_2 &= o_2 - \left\lfloor \tfrac{1}{2}\left(e_2 + e_3\right)\right\rfloor = 4 - \left\lfloor \tfrac{1}{2}\left(-3 - 1\right)\right\rfloor = 6\\
\widetilde{d}^*_3 &= o_3 - \left\lfloor \tfrac{1}{2}\left(e_3 + e_4\right)\right\rfloor = -2 - \left\lfloor \tfrac{1}{2}\left(-1 - 1\right)\right\rfloor = -1.
\end{aligned}$$

and

$$\begin{aligned}
\widetilde{a}_0^* &= e_0 + \left\lfloor \tfrac{1}{4}\left(\widetilde{d}_0^* + \widetilde{d}_{-1}^*\right) + \tfrac{1}{2} \right\rfloor &&= -5 + \left\lfloor \tfrac{1}{4}(8+8) + \tfrac{1}{2} \right\rfloor &&= -1 \\
\widetilde{a}_1^* &= e_1 + \left\lfloor \tfrac{1}{4}\left(\widetilde{d}_1^* + \widetilde{d}_0^*\right) + \tfrac{1}{2} \right\rfloor &&= 0 + \left\lfloor \tfrac{1}{4}(3+8) + \tfrac{1}{2} \right\rfloor &&= 3 \\
\widetilde{a}_2^* &= e_2 + \left\lfloor \tfrac{1}{4}\left(\widetilde{d}_2^* + \widetilde{d}_1^*\right) + \tfrac{1}{2} \right\rfloor &&= -3 + \left\lfloor \tfrac{1}{4}(6+3) + \tfrac{1}{2} \right\rfloor &&= -1 \\
\widetilde{a}_3^* &= e_3 + \left\lfloor \tfrac{1}{4}\left(\widetilde{d}_3^* + \widetilde{d}_2^*\right) + \tfrac{1}{2} \right\rfloor &&= -1 + \left\lfloor \tfrac{1}{4}(-1+6) + \tfrac{1}{2} \right\rfloor &&= 0.
\end{aligned}$$

(Live example: Visit `stthomas.edu/wavelets` *and click on Live Examples.)*

□

Inverting the Integer Lifting Process

To invert this process, we simply first solve (11.14) for e_k to obtain the identity

$$e_k = \widetilde{a}_k^* - \left\lfloor \frac{1}{4}\left(\widetilde{d}_k^* + \widetilde{d}_{k-1}^*\right) + \frac{1}{2} \right\rfloor, \qquad k = 0, \ldots, \frac{N}{2} - 1 \tag{11.15}$$

where $\widetilde{d}_{-1}^* = \widetilde{d}_0^*$. Note that this part of the inversion process is exact – $\widetilde{a}_k^*$ was built by adding some number to e_k, so e_k can be recovered by subtracting that value from $\widetilde{a}_k^*$.

Once we have the even vector $\mathbf{e}$, we solve (11.13) for o_k to write

$$o_k^* = \widetilde{d}_k^* + \left\lfloor \frac{1}{2}\left(e_k + e_{k+1}\right) \right\rfloor, \qquad k = 0, \ldots, \frac{N}{2} - 1 \tag{11.16}$$

where $e_{N/2} = e_{N/2-1}$. For now we have denoted the odd vector in (11.16) by $\mathbf{o}^*$. The reason we do this is that we have not yet verified that (11.16) returns $\mathbf{o}$.

Why does this process work? It turns out that the answer relies on a simple proposition.

Proposition 11.1 (Flooring Half-Integers). *Suppose that a and b are integers. Then*

$$\left\lfloor \frac{1}{2}(a+b) \right\rfloor = \begin{cases} \frac{1}{2}(a+b), & a+b \text{ even} \\ \frac{1}{2}(a+b) - \frac{1}{2}, & a+b \text{ odd.} \end{cases}$$

□

Proof. The proof is left as Problem 11.9. □

If we apply Proposition 11.1 to (11.13), simplify, and compare the result with (11.9), we see that

$$\begin{aligned}
\widetilde{d}_k^* &= o_k - \left\lfloor \frac{1}{2}\left(e_k + e_{k+1}\right)\right\rfloor \\
&= o_k - \begin{cases} \frac{1}{2}\left(e_k + e_{k+1}\right), & \text{if } e_k + e_{k+1} \text{ is even} \\ \frac{1}{2}\left(e_k + e_{k+1}\right) - \frac{1}{2}, & \text{if } e_k + e_{k+1} \text{ is odd} \end{cases} \\
&= \begin{cases} o_k - \frac{1}{2}\left(e_k + e_{k+1}\right), & \text{if } e_k + e_{k+1} \text{ is even} \\ o_k - \frac{1}{2}\left(e_k + e_{k+1}\right) + \frac{1}{2}, & \text{if } e_k + e_{k+1} \text{ is odd} \end{cases} \\
&= \begin{cases} \widetilde{d}_k, & \text{if } e_k + e_{k+1} \text{ is even} \\ \widetilde{d}_k + \frac{1}{2}, & \text{if } e_k + e_{k+1} \text{ is odd.} \end{cases}
\end{aligned} \tag{11.17}$$

Let's consider (11.16) in the case where $e_k + e_{k+1}$ is even. By (11.17) we know that $\widetilde{d}_k^* = \widetilde{d}_k$ and Proposition 11.1 gives $\lfloor \frac{1}{2}\left(e_k + e_{k+1}\right)\rfloor = \frac{1}{2}\left(e_k + e_{k+1}\right)$. Inserting these results into (11.16) and comparing with (11.12) gives

$$\begin{aligned}
o_k^* &= \widetilde{d}_k^* + \left\lfloor \frac{1}{2}\left(e_k + e_{k+1}\right)\right\rfloor \\
&= \widetilde{d}_k + \frac{1}{2}\left(e_k + e_{k+1}\right) \\
&= o_k.
\end{aligned}$$

Now if $e_k + e_{k+1}$ is odd, (11.17) tells us that $\widetilde{d}_k^* = \widetilde{d}_k + \frac{1}{2}$ and Proposition 11.1 gives $\lfloor \frac{1}{2}\left(e_k + e_{k+1}\right)\rfloor = \frac{1}{2}\left(e_k + e_{k+1}\right) - \frac{1}{2}$. Inserting these results into (11.16) and comparing with (11.12) gives

$$\begin{aligned}
o_k^* &= \widetilde{d}_k^* + \left\lfloor \frac{1}{2}\left(e_k + e_{k+1}\right)\right\rfloor \\
&= \widetilde{d}_k + \frac{1}{2} + \frac{1}{2}\left(e_k + e_{k+1}\right) - \frac{1}{2} \\
&= o_k.
\end{aligned}$$

In both cases we see that $o_k^* = o_k$, so (11.16) exactly returns the even vector $\mathbf{o}$.

PROBLEMS

11.1 Let $\mathbf{v} = [3, 1, 0, 5, 10, 20, 0, 17]^T$. Compute the symmetric biorthogonal transform of $\mathbf{v}$ by hand first using the lifting method (Algorithm 11.1) and then Algorithm 7.1. Which algorithm worked better as far as ease of computation and speed?

11.2 Write an algorithm that implements (11.11) and (11.12) and inverts the lifting scheme.

11.3 For $\mathbf{v}$ from Problem 11.1, compute the symmetric biorthogonal transform (Algorithm 7.1) using the LeGall filter pair and use $\lfloor \cdot \rfloor$ to truncate the elements of the transform to integers. Next, compute the inverse biorthogonal transform using Algorithm 7.2. Do you recover $\mathbf{v}$?

11.4 Suppose that $\mathbf{v}$ is a vector of even length whose values are integers from the set $\{-128, \ldots, 0, \ldots, 127\}$. If we apply Algorithm 7.1 to $\mathbf{v}$ using the LeGall filter pair, explain why elements of the averages portion of the data are integers or half-integers from the set $\{-256, \ldots, 254\}$ and the elements of the differences portion of the data are rational numbers with denominators one, two, four, or eight from the set $\{-765/4, \ldots, 765/4\}$.

11.5 Let $\mathbf{v}$ be a vector of even length N whose elements are integers in the range $0, \ldots, 255$. Suppose we create $\widetilde{W}_N$ using (11.4) as a template but use scaled versions of the LeGall filters to build the matrix in order to produce integer-valued output.

(a) How can we scale each of the LeGall filters in Definition 11.1 so $\mathbf{y} = \widetilde{W}_N \mathbf{v}$ is integer-valued?

(b) What is the range of values in $\mathbf{y} = \widetilde{W}_N \mathbf{v}$?

(c) Why is this range not desirable for coding purposes?

(d) How would the filters in the inverse transform W_N^T need to be scaled so that we can recover $\mathbf{v}$ from the product $W_N^T \mathbf{y}$?

11.6 Let $\mathbf{v}$ be as given in Problem 11.1. Use (11.13) and (11.14) to compute the integer-valued biorthogonal wavelet transform of $\mathbf{v}$. Now use (11.15) and (11.16) to exactly recover $\mathbf{v}$.

11.7 In this problem we investigate further the reason for adding $\frac{1}{2}$ to the floored part of (11.14). Suppose n is an integer. Then we can write n as $n = 4m + j$ where m and j are integers with $j = 0, 1, 2$, or 3. Dividing both sides of this equation by four gives $\frac{n}{4} = m + \frac{j}{4}$. Note that $m \le \frac{n}{4} < m + 1$ What is the value of $\left\lfloor \frac{n}{4} \right\rfloor$ regardless the value of remainder j? For each remainder j, what are the values of $\left\lfloor \frac{n}{4} + \frac{1}{2} \right\rfloor$? In this way, we see that adding $\frac{1}{2}$ to $\frac{n}{4}$ before flooring creates a more even distribution of possible values.

11.8 Consider $\mathbf{v}$ as given in Example 11.1. In Example 11.2 we used (11.13) and (11.14) to create an integer-valued transform. Use (11.15) and then (11.16) on the vectors $\widetilde{\mathbf{d}}^*$ and $\widetilde{\mathbf{a}}^*$ from Example 11.2 to show that we recover $\mathbf{v}$ exactly.

11.9 Prove Proposition 11.1.

11.10 Suppose that we replace the floor function $\lfloor \cdot \rfloor$ by the *ceiling function* $\lceil \cdot \rceil$ in (11.13) – (11.16). Is the integer-to-integer transformation process still reversible? If your answer is yes, provide a proof. Here, the ceiling function $\lceil t \rceil$ returns the smallest integer greater than or equal to t.

11.11 Repeat Problem 11.10, but in this case use normal rounding instead of the ceiling function.

11.12 A floating point operation (FLOP) is defined to be an arithmetic operation (usually either addition or multiplication) performed on two real numbers using floating-point arithmetic (see Burden and Faires [15] for more information on floating-point arithmetic). FLOPs (floating point operations per second) are often used to measure the speed and efficiency of an algorithm.

Suppose that our task is to compute one iteration of the symmetric biorthogonal wavelet transformation (stored in $\mathbf{y}$) of N-vector $\mathbf{v}$ (N even) using the LeGall filter pair. In this problem you calculate FLOP counts necessary to compute $\mathbf{y}$ using Algorithm 7.1 and Algorithm 11.1.

(a) Algorithm 7.1, modulo increasing the length of $\mathbf{v}$ and then dropping some terms, computes an arbitrary element of the averages portion of the transform as

$$-\frac{1}{8}v_{j-2} + \frac{1}{4}v_{j-1} + \frac{3}{4}v_j + \frac{1}{4}v_{j+1} - \frac{1}{8}v_{j+2}.$$

Assuming that we do not count the operations necessary to form $\frac{1}{8}$, $\frac{1}{4}$, and $\frac{3}{4}$, how many additions are performed in this computation? How many multiplications are performed in this computation?

(b) Using (a) determine the FLOP count needed to compute the averages portion of the transform.

(c) An arbitrary element of the differences portion of the transform is given by

$$-\frac{1}{2}v_{j-1} + v_j - \frac{1}{2}v_{j+1}.$$

What is the FLOP count to compute the differences portion of the transform? Combine your answer with that obtained in (b) to determine the FLOP count needed to compute the wavelet transform of $\mathbf{v}$ using Algorithm 7.1.

(d) Using (11.9), verify that the FLOP count needed to compute the differences portion of the transform using lifting is the same as that needed for Algorithm 7.1.

(e) Using (11.10), find the FLOP count needed to compute the averages portion of the wavelet transform via lifting.

(f) Combine your answers in (d) and (e) to determine the FLOP count needed to compute the transform using Algorithm 11.1. Which algorithm has a smaller FLOP count?

Computer Lab

11.1 The LeGall Filter Pair and Lifting. From `stthomas.edu/wavelets`, access the file `Lifting`. In this lab, you develop and test an efficient algorithm for computing the one- and two-dimensional symmetric biorthogonal wavelet transforms using the LeGall filter pair via lifting.

11.2 Z–Transforms and Laurent Polynomials

In this section, we introduce the Z-transform. This transform is related to Fourier series and it is a valuable tool for the development of the lifting method presented in Section 11.3.

The Z-Transform Defined

We begin by stating the definition of a Z-transform.

Definition 11.2 (The Z-Transform). *Suppose* $\mathbf{x} = (\ldots, x_{-1}, x_0, x_1, \ldots)$ *is a bi-infinite sequence. We define the* Z-transform *of* $\mathbf{x}$ *as*

$$X(z) = \sum_{k=-\infty}^{\infty} x_k z^{-k}$$

where $z \in \mathbb{C}$ *is any value for which the series converges.* □

We have seen series like the one in Definition 11.2 in calculus. For example, if $\mathbf{x}$ is the bi-infinite sequence whose elements are given by

$$x_k = \begin{cases} 1, & k \leq 0 \\ 0, & k > 0 \end{cases}$$

then the Z-transform is

$$X(z) = \sum_{k=-\infty}^{0} z^{-k} = \sum_{k=0}^{\infty} z^k.$$

When $|z| < 1$, we know from calculus (see, e.g. [84]) that the series converges. We have

$$X(z) = \sum_{k=0}^{\infty} z^k = \frac{1}{1-z}. \tag{11.18}$$

We can write the Fourier series constructed from bi-infinite sequence $\mathbf{x}$ as a Z-transform. We have

$$\sum_{k=-\infty}^{\infty} x_k e^{ik\omega} = X\left(e^{-i\omega}\right).$$

Let's look at some examples.

Example 11.3 (Examples of Z-Transforms). *Find the* Z*-transforms of the following bi-infinite sequences. Find the region of convergence for each transform. Here* $\mathbf{x}$ *is a bi-infinite sequence whose elements are given below.*

(a) $x_k = \begin{cases} 1, & k \geq 0 \\ 0, & k < 0 \end{cases}$

$$(b)\ x_k = \begin{cases} 2^k, & k \geq 0 \\ 0, & k < 0 \end{cases}$$

$$(c)\ x_k = \begin{cases} \frac{1}{4}, & k = \pm 1 \\ \frac{1}{2}, & k = 0 \\ 0, & \text{otherwise} \end{cases}$$

Solution. *For (a) we have*

$$X(z) = \sum_{k=0}^{\infty} z^{-k} = \sum_{k=0}^{\infty} \left(z^{-1}\right)^k.$$

Since the Z-transform in (11.18) *converges for $|z| < 1$, we note that $X(z)$ here converges when $\left|z^{-1}\right| < 1$ or $|z| > 1$. In this case we have*

$$X(z) = \frac{1}{1 - z^{-1}} = \frac{z}{z - 1}.$$

For (b), we write

$$X(z) = \sum_{k=0}^{\infty} 2^k z^{-k} = \sum_{k=0}^{\infty} \left(\frac{2}{z}\right)^k.$$

and note that it converges whenever $\left|\frac{2}{z}\right| < 1$ or $|z| > 2$. For $|z| > 2$, we have

$$X(z) = \frac{1}{1 - \frac{2}{z}} = \frac{z}{z - 2}.$$

For the final Z-transform we have

$$X(z) = \sum_{k=-1}^{1} h_k z^{-k} = \frac{1}{4} z + \frac{1}{2} + \frac{1}{4} z^{-1}.$$

Since this Z-transform is a finite sum, it converges for any value of $z \neq 0$. □

Rules for Computing Z-Transforms

As is the case for Fourier series, rules can be derived that allow us to build one Z-transform from another. While there are many such rules, we concentrate on three that will be useful in Section 11.3. Some other rules are given as problems at the end of the section. The first rule tells us that the Z-transform is a linear transformation.

Proposition 11.2 (Linearity of the Z-Transform). *Suppose $\mathbf{u}$ and $\mathbf{v}$ are two bi-infinite sequences with Z-transforms $U(z)$ and $V(z)$, respectively, and assume $c_1, c_2 \neq 0$. If $Y(z)$ is the Z-transform of $c_1\mathbf{u} + c_2\mathbf{v}$, then $Y(z) = c_1 U(z) + c_2 V(z)$.* □

Proof. We start by writing the Z-transform for $c_1\mathbf{u} + c_2\mathbf{v}$. We have

$$\begin{aligned} Y(z) &= \sum_{k=-\infty}^{\infty} (c_1 u_k + c_2 v_k) z^{-k} \\ &= c_1 \sum_{k=-\infty}^{\infty} u_k z^{-k} + c_2 \sum_{k=-\infty}^{\infty} v_k z^{-k} \\ &= c_1 U(z) + c_2 V(z). \end{aligned}$$

□

A common action on terms of a bi-infinite sequence is translation left or right by a certain number of units. The next rule describes what happens to the Z-transform in this case.

Proposition 11.3 (The Translation Rule for Z-Transforms). *Suppose* $\mathbf{x}$ *is a bi-infinite sequence with* Z*-transform* $X(z)$*. If* $\mathbf{y}$ *is formed element-wise by the formula* $y_k = x_{k-n}$*, then its* Z*-transform is* $Y(z) = z^{-n}X(z)$*.* □

Proof. We start with

$$Y(z) = \sum_{k=-\infty}^{\infty} y_k z^{-k} = \sum_{k=-\infty}^{\infty} x_{k-n} z^{-k}$$

and make the index substitution $j = k - n$. Then $k = j + n$ and since k runs over all integers, so does j. We have

$$Y(z) = \sum_{j=-\infty}^{\infty} x_j z^{-j-n} = z^{-n} \sum_{j=-\infty}^{\infty} x_j z^{-j} = z^{-n}X(z).$$

□

The following example illustrates the effect of the translation rule.

Example 11.4 (Example of the Translation Rule). *We again consider the* Z*-transform from Example 11.3(c),* $X(z) = \frac{1}{4}z + \frac{1}{2} + \frac{1}{4}z^{-1}$*, constructed from the sequence* $\mathbf{x}$ *whose only nonzero terms were* $x_{-1} = x_1 = \frac{1}{4}$ *and* $x_0 = \frac{1}{2}$*. We create the sequence* $\mathbf{y}$ *by shifting the elements of* $\mathbf{x}$ *two units left. That is, the nonzero elements of* $\mathbf{y}$ *are* $y_{-3} = x_{-1} = x_1 = y_{-1} = \frac{1}{4}$ *and* $y_{-2} = x_0 = \frac{1}{2}$*. The resulting* Z*-transform is*

$$Y(z) = z^2X(z) = \frac{1}{4}z^3 + \frac{1}{2}z^2 + \frac{1}{4}z.$$

□

The final rule we present is the convolution rule. It turns out the result is exactly the same as that of Theorem 8.1 for Fourier series.

Proposition 11.4 (The Convolution Rule for Z-Transforms). *Suppose* $\mathbf{u}$ *and* $\mathbf{v}$ *are bi-infinite sequences with* Z*-transforms* $U(z)$ *and* $V(z)$*, respectively. Then* $\mathbf{y} = \mathbf{u} * \mathbf{v}$ *if and only if its* Z*-transform is* $Y(z) = U(z)V(z)$. □

Proof. The proof is very similar to that of Theorem 8.1 and left as Problem 11.14. □

Z-Transforms of Even and Odd Components of a Sequence

The lifting method presented in Section 11.3 used to efficiently compute a wavelet transform starts by splitting a bi-infinite sequence $\mathbf{x}$ into bi-infinite sequences $\mathbf{e}$ and $\mathbf{o}$ of even- and odd-indexed elements, respectively. Since the method is constructed in the Z-transform domain, we need to understand how the Z-transform $X(z)$ can be expressed in terms of the "even" and "odd" Z-transforms $E(z)$ and $O(z)$ and in turn, how $E(z)$ and $O(z)$ can each be expressed in terms of $X(z)$.

Given bi-infinite sequence $\mathbf{x} = (\ldots, x_{-1}, x_0, x_1, \ldots)$, we can define the associated even and odd bi-infinite sequences. We have

$$\begin{aligned}\mathbf{e} &= (\ldots, x_{-4}, x_{-2}, x_0, x_2, x_4, \ldots)\\ \mathbf{o} &= (\ldots, x_{-5}, x_{-3}, x_{-1}, x_1, x_3, x_5, \ldots).\end{aligned}$$

Then the corresponding Z-transforms are

$$\begin{aligned}E(z) &= \sum_{k=-\infty}^{\infty} e_k z^{-k} = \sum_{k=-\infty}^{\infty} x_{2k} z^{-k}\\ O(z) &= \sum_{k=-\infty}^{\infty} o_k z^{-k} = \sum_{k=-\infty}^{\infty} x_{2k+1} z^{-k}.\end{aligned} \tag{11.19}$$

Notation: In the following sections, we find it convenient to emphasize that the transforms in (11.19) are constructed from bi-infinite sequence $\mathbf{x}$. For this reason we will often refer to the even and odd transforms as

$$X_e(z) = E(z) \qquad \text{and} \qquad X_o(z) = O(z),$$

respectively.

Note that if we evaluate each Z-transform at z^2, we have

$$\begin{aligned}E\left(z^2\right) &= \sum_{k=-\infty}^{\infty} x_{2k} z^{-2k}\\ O\left(z^2\right) &= \sum_{k=-\infty}^{\infty} x_{2k+1} z^{-2k}.\end{aligned}$$

For $E\left(z^2\right)$, the power of z matches the index of the coefficient, but this is not the case for $O\left(z^2\right)$. If we multiply $O\left(z^2\right)$ by z^{-1} and add the result to $E\left(z^2\right)$ we obtain

$$E\left(z^2\right) + z^{-1}O\left(z^2\right) = \sum_{k=-\infty}^{\infty} x_{2k}z^{-2k} + \sum_{k=-\infty}^{\infty} x_{2k+1}z^{-2k-1} = X(z). \quad (11.20)$$

In Problem 11.19, you will show that we can construct $E(z)$ and $O(z)$ using the following formulas:

$$\boxed{\begin{aligned} E\left(z^2\right) &= \frac{1}{2}\left(X(z) + X(-z)\right) \\ O\left(z^2\right) &= \frac{1}{2}z\left(X(z) - X(-z)\right). \end{aligned}} \quad (11.21)$$

Connecting Z-Transforms and (Bi)Orthogonality Conditions

Let $\mathbf{x}$ be a bi-infinite sequence. Then we can form the associated Z-transform $X(z)$ and a Fourier series $X(\omega)$ constructed from $\mathbf{x}$. There is a natural connection between $X(\omega)$ and $X(z)$. We "abuse notation" and use X to denote both the Z-transform and the Fourier series built from $\mathbf{z}$. If $z = e^{i\omega}$ we have

$$X(\omega) = \sum_{k=-\infty}^{\infty} x_k e^{ik\omega} = \sum_{k=-\infty}^{\infty} x_k z^k = \sum_{k=-\infty}^{\infty} x_k\left(z^{-1}\right)^{-k} = X\left(z^{-1}\right). \quad (11.22)$$

In Problem 11.20 you will derive the formulas

$$\begin{aligned} \overline{X(\omega)} &= X(z) \\ X(\omega+\pi) &= X\left(-z^{-1}\right) \\ \overline{X(\omega+\pi)} &= X(-z). \end{aligned} \quad (11.23)$$

Given an orthogonal filter $\mathbf{h} = (h_\ell, \ldots, h_L)$ and its Fourier series $H(\omega)$, Proposition 9.1 told us to write the Fourier series of the associated highpass filter $\mathbf{g}$ as $G(\omega) = -e^{i\omega(\ell+L)}\overline{H(\omega+\pi)}$ Setting $z = e^{i\omega}$ and using the last identity in (11.23), we see that the corresponding Z-transform is

$$G(z) = -z^{\ell+L}H(-z). \quad (11.24)$$

In the case of a biorthogonal filter pair $\left(\widetilde{\mathbf{h}}, \mathbf{h}\right)$ with Fourier series $\widetilde{H}(\omega)$ and $H(\omega)$, we use Proposition 9.2 to see that the Fourier series for the corresponding highpass filters $(\widetilde{\mathbf{g}}, \mathbf{g})$ are $\widetilde{G}(\omega) = -e^{i\omega}\overline{H(\omega+\pi)}$ and $G(\omega) = -e^{i\omega}\overline{\widetilde{H}(\omega+\pi)}$. The related Z-transforms are

$$\widetilde{G}(z) = -zH(-z) \quad \text{and} \quad G(z) = -z\widetilde{H}(-z). \quad (11.25)$$

We seek biorthogonal filter pairs $\left(\widetilde{\mathbf{h}}, \mathbf{h}\right)$ whose Fourier series satisfy $\widetilde{H}(\omega)\overline{H(\omega)} + \widetilde{H}(\omega+\pi)\overline{H(\omega+\pi)} = 2$. In terms of Z-transforms, this relation becomes

$$\widetilde{H}\left(z^{-1}\right) H(z) + \widetilde{H}\left(-z^{-1}\right) H(-z) = 2. \tag{11.26}$$

Assuming the Z-transforms $\widetilde{H}(z)$ and $H(z)$ of the filter pair $\left(\widetilde{\mathbf{h}}, \mathbf{h}\right)$ satisfy (11.26) and the Z-transforms of the corresponding highpass filter pair $(\widetilde{\mathbf{g}}, \mathbf{g})$ satisfy (11.25), the conclusions of Proposition 9.2 become

$$\begin{aligned} \widetilde{G}\left(z^{-1}\right) G(z) + \widetilde{G}\left(-z^{-1}\right) G(-z) &= 2 \\ \widetilde{H}\left(z^{-1}\right) G(z) + \widetilde{H}\left(-z^{-1}\right) G(-z) &= 0 \\ \widetilde{G}\left(z^{-1}\right) H(z) + \widetilde{G}\left(-z^{-1}\right) H(-z) &= 0. \end{aligned} \tag{11.27}$$

In Problem 11.21, you will rewrite the orthogonality condition and conclusions from Proposition 9.1 in terms of Z-transforms.

Laurent Polynomials

The Z-transform in Example 11.3(c) is an example of a *Laurent polynomial.* These polynomials are generalizations of polynomials we encounter in a calculus course.

Definition 11.3 (Laurent Polynomial). *Suppose* $\mathbf{c} = (c_m, \ldots, c_M)$ *where* $m \le M$ *and* $c_m, c_M \ne 0$. *Then*

$$p(z) = \sum_{k=m}^{M} c_k z^k$$

is called a Laurent polynomial. *We denote the* degree *of* $p(z)$ *as* deg$(p) = M - m$. *We say the degree of the zero Laurent polynomial is* $-\infty$. □

Notice that standard polynomials with real-valued coefficients are a special case of Laurent polynomials. While our definition allows coefficients of z^k to be complex-valued, we only consider real-valued coefficients c_k. Any Z-transform whose coefficients are taken from an FIR filter is a Laurent polynomial as well.

Let's look at some examples.

Example 11.5 (Examples of Laurent Polynomials). *Find*

(a) the Laurent polynomial associated with $\mathbf{c} = (c_{-2}, c_{-1}, c_0, c_1) = (-3, 1, 5, 6)$

(b) the Laurent polynomial of the Haar filter $\mathbf{h} = (h_0, h_1) = \left(\frac{1}{2}, \frac{1}{2}\right)$

(c) all Laurent polynomials that are of degree zero

Solution. *For (a) we have*

$$p(z) = -3z^{-2} + z^{-1} + 5 + 6z$$

and for (b) the Laurent polynomial is

$$p(z) = \frac{1}{2} + \frac{1}{2}z.$$

Note that the Z-transform of $\mathbf{h}$ is $H(z) = \frac{1}{2} + \frac{1}{2}z^{-1}$ is also a Laurent polynomial. If a Laurent polynomial is of degree zero, then $M = m$ in Definition 11.3. In this case we have

$$p(z) = c_m z^m, \quad m \in \mathbb{Z}.$$

Thus monomials are Laurent polynomials of degree zero. □

Recall for standard polynomials the only invertible ones are of the form $p(t) = c$, $c \neq 0$ since $q(t) = \frac{1}{c}$ satisfies $p(t)q(t) = 1$. For Laurent polynomials, it turns out the only invertible polynomials are the monomials.

Proposition 11.5 (Inverses of Laurent Polynomials). *Let $p(z)$ be a Laurent polynomial. Then $p(z)$ is invertible if and only if $p(z) = cz^m$, where $c \neq 0$ and $m \in \mathbb{Z}$.* □

Proof. Suppose $p(z) = cz^m$ where $c \neq 0$ and $m \in \mathbb{Z}$. Then $q(z) = \frac{1}{c}z^{-m}$ satisfies $p(z)q(z) = 1$ so that $p(z)$ is invertible. Now assume $p(z)$ is invertible and can be expressed as

$$p(z) = \sum_{k=m}^{M} c_k z^k$$

where $m < M$ (we have already considered the monomial case $m = M$). We assume as well that $c_m, c_M \neq 0$. Since $p(z)$ is invertible, we know there is a Laurent polynomial $q(z)$ of the form

$$q(z) = \sum_{k=n}^{N} d_k z^k$$

with $n \leq N$ and $c_n, c_N \neq 0$, such that $p(z)q(z) = 1$. But

$$1 = p(z)q(z) = c_m d_n z^{n+m} + \cdots c_M d_N z^{N+M}.$$

The Laurent polynomial on the left is degree zero but due to the fact that $m < M$, the Laurent polynomial on the right has positive degree. Thus our assumption that $m < M$ leads to a contradiction and it must be that $m = M$ or $p(z)$ is a monomial. □

Laurent Polynomials and the Euclidean Algorithm

In Section 11.3 we learn that application of the Euclidean algorithm leads to factorizations of the wavelet transform that improve computational efficiency (see [31]) and allows us to modify the transform so that it maps integer input to integer output while maintaining invertibility.

You may recall the Euclidean algorithm as a way to find the greatest common divisor of two integers $n > m > 0$. The steps are as follows.

1. Divide n by m, finding the quotient q and remainder r. This produces $n = qm + r$.

2. If $r = 0$, then the greatest common divisor is m. Otherwise, set $n = m$ and $m = r$. Return to the first step.

For example, if $n = 84$ and $m = 18$, we have $84 = 4 \cdot 18 + 12$. We then set $n = 18$ and $m = 12$. We have $18 = 1 \cdot 12 + 6$. Repeating the process again gives $12 = 2 \cdot 6 + 0$ so that the greatest common divisor is six. We previously made use of the Euclidean algorithm in Example 3.1 from Section 3.1 when we converted an integer to base 2.

The Euclidean algorithm can be used to find the greatest common divisor of two standard polynomials with the result being unique up to a nonzero constant factor. The process is finite since the degree of the remainder at each step is at least one degree less than the previous step. The process terminates when the remainder is the zero polynomial. The Euclidean algorithm also applies to Laurent polynomials although the division is more interesting due to flexibility allowed by negative powers.

Algorithm 11.2 (Euclidean Algorithm for Laurent Polynomials). *Suppose $a(z)$ and $b(z)$ are Laurent polynomials with $\deg(a) \geq \deg(b)$. This algorithm produces a greatest common divisor of $a(z)$ and $b(z)$.*

1. *Set $a_0(z) = a(z)$ and $b_0(z) = b(z)$.*

2. *Find Laurent polynomials $q_1(z)$ and $r_1(z)$ such that $a_0(z) = q_1(z)b_0(z) + r_1(z)$.*

3. *If $\deg(r_1) = 0$, a greatest common divisor is $b_0(z)$.*

4. *Set $n = 1$,*

5. *While $\deg(r_n) > 0$,*

   $a_n(z) = b_{n-1}(z)$, $b_n(z) = r_n(z)$.

   Find $q_{n+1}(z)$, $r_{n+1}(z)$ such that $a_n(z) = q_{n+1}(z)b_n(z) + r_{n+1}(z)$.

   If $r_{n+1}(z) = 0$, stop. Return $b_n(z)$ as a greatest common divisor.

   $n = n + 1$.

□

Since monomials are degree zero Laurent polynomials, we see that if $d(z)$ divides $a(z)$ and $b(z)$, then so does $cz^m d(z)$ for any $c \neq 0$ and $m \in \mathbb{Z}$. The Euclidean algorithm is a finite process since the degree of the remainder $r_n(z)$ decreases by at least one at each step.

Let's look at an example. We will see that greatest common divisors can take different monomial forms.

Example 11.6 (Applying the Euclidean Algorithm to Laurent Polynomials). *Suppose $a(z) = a_0(z) = z^{-1} + 5 + 2z$ and $b(z) = b_0(z) = z^{-1} + 2$. Then $\deg(a) = 2$ and $\deg(b) = 1$. To compute a greatest common divisor, we divide $a(z)$ by $b(z)$ looking for quotient and remainder polynomials. We start with $q_1(z)$ to be of the form $cz + d$ since $b_0(z)(cz + d) = dz^{-1} + (c + 2d) + 2cz$ and this is the basic form of $a_0(z)$. We have*

$$\begin{aligned} a_0(z) &= q_1(z)b_0(z) + r_1(z) \\ z^{-1} + 5 + 2z &= dz^{-1} + (c + 2d) + 2cz + r_1(z). \end{aligned} \tag{11.28}$$

Now we have some choices. Let's line up the z^{-1} and constant terms. This gives $d = 1$ and $c + 2 \cdot 1 = 5$ or $c = 3$. We compute

$$z^{-1} + 5 + 2z = z^{-1} + 5 + 6z + r_1(z)$$

so that $q_1(z) = 3z + 1$ and $r_1(z) = -4z$. Since $r_1(z) \neq 0$, we continue the Euclidean algorithm, setting $a_1(z) = b_0(z) = z^{-1} + 2$ and $b_1(z) = r_0(z) = -4z$. Since $b_1(z)$ is a monomial, the next step will produce a zero remainder. We have

$$\begin{aligned} a_1(z) &= q_2(z)b_1(z) + r_2(z) \\ z^{-1} + 2 &= \left(-\frac{1}{4}z^{-2} - \frac{1}{2}z^{-1}\right)(-4z) + 0 \end{aligned}$$

so that $q_2(z) = -\frac{1}{4}z^{-2} - \frac{1}{2}z^{-1}$ and a greatest common divisor is $a_2(z) = b_1(z) = -4z$.

What happens if we instead line up the constant and z terms in (11.28)? Then $c = 1$ and $1 + 2d = 5$ or $d = 2$. In this case we have

$$z^{-1} + 5 + 2z = 2z^{-1} + 5 + 2z + r_1(z)$$

so that $q_1(z) = z + 2$ and $r_1(z) = -z^{-1}$. Continuing the Euclidean algorithm, we set $a_1(z) = b_0(z) = z^{-1} + 2$ and $b_1(z) = r_1(z) = -z^{-1}$. The next step is

$$\begin{aligned} a_1(z) &= q_2(z)b_1(z) + r_2(z) \\ z^{-1} + 2 &= (-1 - 2z)\left(-z^{-1}\right) + 0 \end{aligned}$$

so that $q_2(z) = -1 - 2z$ and a greatest common divisor in this case is $a_2(z) = b_1(z) = -z^{-1}$. There are certainly other choices for $q_1(z)$. In Problem 11.23, you find a greatest common divisor in the case where the z^{-1} and z terms are aligned in (11.28).

(Live example: Visit `stthomas.edu/wavelets` and click on Live Examples.)

□

The Euclidean Algorithm as a Matrix Product

Suppose the last step in the Euclidean algorithm yields $a_N(z)$ and $b_N(z) = r_N(z) = 0$. Then it is possible to write the vector $\begin{bmatrix} a_0(z) \\ b_0(z) \end{bmatrix}$ as a product of matrices times $\begin{bmatrix} a_N(z) \\ 0 \end{bmatrix}$. This factorization proves to be extremely useful in Section 11.3 when we develop lifting steps for computing wavelet transforms.

The first step of the Euclidean algorithm (assuming $r_1(z) \neq 0$) is to set $a_1(z) = b_0(z)$ and $b_1(z) = r_1(z)$ where $a_0(z) = q_1(z)b_0(z) + r_1(z)$ or $b_1(z) = r_1(z) = a_0(z) - q_1(z)b_0(z)$. In matrix form we have

$$\begin{bmatrix} a_1(z) \\ b_1(z) \end{bmatrix} = \begin{bmatrix} 0 & 1 \\ 1 & -q_1(z) \end{bmatrix} \begin{bmatrix} a_0(z) \\ b_0(z) \end{bmatrix}.$$

In a similar fashion, we can write

$$\begin{aligned} \begin{bmatrix} a_2(z) \\ b_2(z) \end{bmatrix} &= \begin{bmatrix} 0 & 1 \\ 1 & -q_2(z) \end{bmatrix} \begin{bmatrix} a_1(z) \\ b_1(z) \end{bmatrix} \\ &= \begin{bmatrix} 0 & 1 \\ 1 & -q_2(z) \end{bmatrix} \begin{bmatrix} 0 & 1 \\ 1 & -q_1(z) \end{bmatrix} \begin{bmatrix} a_0(z) \\ b_0(z) \end{bmatrix}. \end{aligned}$$

Continuing to the end we have

$$\begin{aligned} \begin{bmatrix} a_N(z) \\ 0 \end{bmatrix} &= \begin{bmatrix} 0 & 1 \\ 1 & -q_N(z) \end{bmatrix} \cdots \begin{bmatrix} 0 & 1 \\ 1 & -q_1(z) \end{bmatrix} \begin{bmatrix} a_0(z) \\ b_0(z) \end{bmatrix} \\ &= \left(\prod_{n=N}^{1} \begin{bmatrix} 0 & 1 \\ 1 & -q_n(z) \end{bmatrix} \right) \begin{bmatrix} a_0(z) \\ b_0(z) \end{bmatrix}. \end{aligned} \tag{11.29}$$

We can use Problem 2.21 to write

$$\begin{bmatrix} 0 & 1 \\ 1 & -q_n(z) \end{bmatrix}^{-1} = \begin{bmatrix} q_n(z) & 1 \\ 1 & 0 \end{bmatrix}$$

and making repeated use of Proposition 2.1(c) gives

$$\left(\prod_{n=N}^{1} \begin{bmatrix} 0 & 1 \\ 1 & -q_n(z) \end{bmatrix} \right)^{-1} = \prod_{n=1}^{N} \begin{bmatrix} q_n(z) & 1 \\ 1 & 0 \end{bmatrix}.$$

Thus we can write (11.29) as

$$\begin{bmatrix} a_0(z) \\ b_0(z) \end{bmatrix} = \left(\prod_{n=1}^{N} \begin{bmatrix} q_n(z) & 1 \\ 1 & 0 \end{bmatrix} \right) \begin{bmatrix} a_N(z) \\ 0 \end{bmatrix}. \tag{11.30}$$

The following example illustrates the factorization identity given in (11.30).

Example 11.7 (Euclidean Algorithm as a Matrix Product). *We consider the first instance of the Euclidean algorithm applied to $a(z) = z^{-1} + 5 + 2z$ and $b(z) = z^{-1} + 2$ in Example 11.6. In this case we had*

$$\begin{aligned} a_0(z) &= a(z) = z^{-1} + 5 + 2z \\ a_1(z) &= b_0(z) = z^{-1} + 2 \\ a_2(z) &= b_1(z) = -4z \\ q_1(z) &= 3z + 1 \\ q_2(z) &= -\frac{1}{4}z^{-2} - \frac{1}{2}z^{-1}. \end{aligned}$$

Applying (11.30)*, we see that*

$$\begin{aligned} \begin{bmatrix} a_0(z) \\ b_0(z) \end{bmatrix} &= \begin{bmatrix} q_1(z) & 1 \\ 1 & 0 \end{bmatrix} \begin{bmatrix} q_2(z) & 1 \\ 1 & 0 \end{bmatrix} \begin{bmatrix} a_2(z) \\ 0 \end{bmatrix} \\ &= \begin{bmatrix} 3z+1 & 1 \\ 1 & 0 \end{bmatrix} \begin{bmatrix} -\frac{1}{4}z^{-2} - \frac{1}{2}z^{-1} & 1 \\ 1 & 0 \end{bmatrix} \begin{bmatrix} -4z \\ 0 \end{bmatrix}. \end{aligned}$$

If we consider the second instance of the Euclidean algorithm applied to $a(z)$ and $b(z)$, we find that

$$\begin{aligned} \begin{bmatrix} a_0(z) \\ b_0(z) \end{bmatrix} &= \begin{bmatrix} q_1(z) & 1 \\ 1 & 0 \end{bmatrix} \begin{bmatrix} q_2(z) & 1 \\ 1 & 0 \end{bmatrix} \begin{bmatrix} a_2(z) \\ 0 \end{bmatrix} \\ &= \begin{bmatrix} z+2 & 1 \\ 1 & 0 \end{bmatrix} \begin{bmatrix} -1-2z & 1 \\ 1 & 0 \end{bmatrix} \begin{bmatrix} -z^{-1} \\ 0 \end{bmatrix}. \end{aligned}$$

(Live example: Visit `stthomas.edu/wavelets` *and click on Live Examples.)*

□

PROBLEMS

11.13 What are the Z-transforms of $\mathbf{s}$ and $\mathbf{t}$ from Problem 2.47?

11.14 Prove Proposition 11.4.

11.15 Suppose $\mathbf{x}$ has Z-transform $X(z)$ and the elements of $\mathbf{y}$ are $y_k = x_{-k}$. If $Y(z)$ is the Z-transform of $\mathbf{y}$, show that $Y(z) = X\left(z^{-1}\right)$.

11.16 Suppose $\mathbf{x}$ has Z-transform $X(z)$ and the elements of $\mathbf{y}$ are $y_k = a^k x_k$, $a \neq 0$. If $Y(z)$ is the Z-transform of $\mathbf{y}$, show that $Y(z) = X\left(\frac{z}{a}\right)$.

11.17 Suppose $\mathbf{x}$ has Z-transform $X(z)$ and the elements of $\mathbf{y}$ are $y_k = kx_k$. If $Y(z)$ is the Z-transform of $\mathbf{y}$, show that $Y(z) = -zX'(z)$.

11.18 Suppose $\mathbf{x}$ has Z-transform $X(z)$ and the elements of $\mathbf{y}$ are $y_k = e^{-ka}x_k$. If $Y(z)$ is the Z-transform of $\mathbf{y}$, show that $Y(z) = X\left(e^a z\right)$.

11.19 In this problem you derive the formulas given in (11.21). The following steps will help you organize your work.

(a) Replace z by $-z$ in (11.20) and simplify to obtain an identity for $X(-z)$.

(b) Solve the equations in (a) and (11.20) first for $E\left(z^2\right)$ and next for $O\left(z^2\right)$.

★11.20 Use (11.22) to establish the identities in (11.23).

★11.21 A condition satisfied by the Fourier series $H(\omega)$ of orthogonal filter h is $H(\omega)\overline{H(\omega)} + H(\omega+\pi)\overline{H(\omega+\pi)} = 2$.

(a) Rewrite the above condition in terms of Z-transforms.

(b) Use (a) and (11.24) to write the conclusions (9.24) and (9.25) from Proposition 9.1 in terms of Z-transforms.

11.22 Find a general form for all Laurent polynomials of degree one.

11.23 Find a greatest common divisor for the polynomials in Example 11.6 by aligning the z^{-1} and z terms in (11.28).

11.24 Find two monomials, each with different powers of z, that are greatest common divisors of $a(z) = 2 + 3z + 6z^2$ and $b(z) = z^{-1} + 1$.

★11.25 Let $a(z) = a_0(z) = \frac{1+\sqrt{3}}{4\sqrt{2}} + \frac{3-\sqrt{3}}{4\sqrt{2}}z^{-1}$ and $b(z) = b_0(z) = \frac{3+\sqrt{3}}{4\sqrt{2}} + \frac{1-\sqrt{3}}{4\sqrt{2}}z^{-1}$.

(a) Use the Euclidean algorithm (there are two steps) to find a greatest common divisor of $a(z)$ and $b(z)$. Start by aligning the z^{-1} terms in $a_0(z)$ and $b_0(z)$. Write down $q_1(z)$, $q_2(z)$ and $r_1(z)$ for this factorization.

(b) Use (11.30) to express the Euclidean algorithm in product form.

★11.26 Suppose we have used the Euclidean algorithm with Laurent polynomials $a_0(z)$ and $b_0(z)$ with $\deg(a_0) \geq \deg(b_0)$ to produce the factorization (11.30). We would like a factorization for $\begin{bmatrix} b_0(z) \\ a_0(z) \end{bmatrix}$. Find a Laurent polynomial $q_0(x)$ such that

$$\begin{bmatrix} b_0(z) \\ a_0(z) \end{bmatrix} = \begin{bmatrix} q_0(z) & 1 \\ 1 & 0 \end{bmatrix} \cdot \begin{bmatrix} a_0(z) \\ b_0(z) \end{bmatrix}.$$

Use this result to show that

$$\begin{bmatrix} b_0(z) \\ a_0(z) \end{bmatrix} = \left(\prod_{k=0}^{N} \begin{bmatrix} q_k(z) & 1 \\ 1 & 0 \end{bmatrix} \right) \begin{bmatrix} a_N(z) \\ 0 \end{bmatrix}.$$

Thus the Euclidean Algorithm can produce a factorization when $\deg(b_0) > \deg(a_0)$.

★11.27 In this problem, we consider the results of Corollary 9.1 and Proposition 9.1 in terms of Z-transforms. In so doing, we "abuse notation" by letting the same upper case letter denote both the Z-transform and the Fourier series and assume $z = e^{i\omega}$. Problem 11.20 will be helpful for establishing the results that follow.

(a) Suppose the Fourier series $H(\omega)$ of $\mathbf{h}$ satisfies (9.22). Show that

$$H\left(z^{-1}\right)H(z)+H\left(-z^{-1}\right)H(-z)=2.$$

(b) Suppose Fourier series $G(\omega)$ is constructed from $H(\omega)$ by the formula

$$G(\omega)=-e^{i\omega(L+\ell)}\overline{H(\omega+\pi)}$$

as is the case in Proposition 9.1. Show that $G(z)=-z^{(L+\ell)}H(-z)$.

(c) Suppose that $H(\omega)$ satisfies (9.22), $G(z)$ is given in (b) and the Fourier series $H(\omega)$, $G(\omega)$ satisfy (9.24) and (9.25). Show that

$$G\left(z^{-1}\right)G(z)+G\left(-z^{-1}\right)G(-z)=2$$
$$H\left(z^{-1}\right)G(z)+H\left(-z^{-1}\right)G\left(-z\right)=0.$$

11.3 A General Construction of the Lifting Method

Lifting in the Z-Transform Domain

To motivate the general lifting method, let's revisit lifting with the LeGall filter pair introduced in Section 11.1. We started by splitting the input $\mathbf{v}$ into "even" and "odd" sequences $\mathbf{e}$, where $e_k=v_{2k}$ and $\mathbf{o}$ where $o_k=v_{2k+1}$. Then the lifting steps are

$$\begin{aligned}\widetilde{d}_k&=o_k-\frac{1}{2}\left(e_k+e_{k+1}\right)\\ \widetilde{a}_k&=e_k+\frac{1}{4}\left(\widetilde{d}_k+\widetilde{d}_{k-1}\right).\end{aligned}$$

We use Problem 2.47 from Section 2.4 to rewrite the above relations in sequence form as

$$\begin{aligned}\widetilde{\mathbf{d}}&=\mathbf{o}-\mathbf{t}*\mathbf{e}\\ \widetilde{\mathbf{a}}&=\mathbf{e}+\mathbf{s}*\widetilde{\mathbf{d}}\end{aligned}\tag{11.31}$$

where $\mathbf{t}$ is the bi-infinite sequence whose nonzero entries are $t_{-1}=t_0=\frac{1}{2}$ and $\mathbf{s}$ is the bi-infinite sequence with nonzero entries $s_0=s_1=\frac{1}{4}$. Using Propositions 11.2 and 11.4, we compute the Z-transforms of sequences in (11.31) to be

$$\begin{aligned}\widetilde{D}(z)&=O(z)-T(z)E(z)\\ \widetilde{A}(z)&=E(z)+S(z)\widetilde{D}(z)\end{aligned}\tag{11.32}$$

where $T(z)=\frac{1}{2}+\frac{1}{2}z$ and $S(z)=\frac{1}{4}+\frac{1}{4}z^{-1}$.

The equations in (11.32) can be expressed in matrix/vector form. We start with the first equation of (11.32) and write

$$\begin{bmatrix}E(z)\\ \widetilde{D}(z)\end{bmatrix}=\begin{bmatrix}1&0\\ -T(z)&1\end{bmatrix}\begin{bmatrix}E(z)\\ O(z)\end{bmatrix}\tag{11.33}$$

and continuing with second equation of (11.32), we have

$$\begin{bmatrix} \widetilde{A}(z) \\ \widetilde{D}(z) \end{bmatrix} = \begin{bmatrix} 1 & S(z) \\ 0 & 1 \end{bmatrix} \begin{bmatrix} E(z) \\ \widetilde{D}(z) \end{bmatrix}. \tag{11.34}$$

Combining (11.33) and (11.34) gives

$$\begin{aligned} \begin{bmatrix} \widetilde{A}(z) \\ \widetilde{D}(z) \end{bmatrix} &= \begin{bmatrix} 1 & S(z) \\ 0 & 1 \end{bmatrix} \begin{bmatrix} 1 & 0 \\ -T(z) & 1 \end{bmatrix} \begin{bmatrix} E(z) \\ O(z) \end{bmatrix} \\ &= \begin{bmatrix} 1 & \frac{1}{4} + \frac{1}{4}z^{-1} \\ 0 & 1 \end{bmatrix} \begin{bmatrix} 1 & 0 \\ -\frac{1}{2} - \frac{1}{2}z & 1 \end{bmatrix} \begin{bmatrix} E(z) \\ O(z) \end{bmatrix}. \end{aligned} \tag{11.35}$$

We can easily derive a matrix product for recovering $E(z)$ and $O(z)$ from $\widetilde{A}(z)$ and $\widetilde{D}(z)$. Using Problem 2.20, we see that

$$\begin{bmatrix} 1 & 0 \\ -\frac{1}{2} - \frac{1}{2}z & 1 \end{bmatrix}^{-1} = \begin{bmatrix} 1 & 0 \\ \frac{1}{2} + \frac{1}{2}z & 1 \end{bmatrix}$$

and

$$\begin{bmatrix} 1 & \frac{1}{4} + \frac{1}{4}z^{-1} \\ 0 & 1 \end{bmatrix}^{-1} = \begin{bmatrix} 1 & -\frac{1}{4} - \frac{1}{4}z^{-1} \\ 0 & 1 \end{bmatrix}.$$

Multiplying both sides of (11.35) by these matrices gives

$$\begin{bmatrix} E(z) \\ O(z) \end{bmatrix} = \begin{bmatrix} 1 & 0 \\ \frac{1}{2} + \frac{1}{2}z & 1 \end{bmatrix} \begin{bmatrix} 1 & -\frac{1}{4} - \frac{1}{4}z^{-1} \\ 0 & 1 \end{bmatrix} \begin{bmatrix} \widetilde{A}(z) \\ \widetilde{D}(z) \end{bmatrix}. \tag{11.36}$$

Reading the Lifting Steps from the Factorization

The lower triangular matrix in the product (11.35) corresponds to the computation of $\widetilde{\mathbf{d}}$ and the upper triangular matrix corresponds to the computation of $\widetilde{\mathbf{a}}$. Sweldens [92] calls these steps the *prediction* and *update* steps, respectively. Given the Z-transforms $T(z)$, $E(z)$ and $O(z)$, we can determine the first lifting step from the product $\begin{bmatrix} 1 & 0 \\ -T(z) & 1 \end{bmatrix} \begin{bmatrix} E(z) \\ O(z) \end{bmatrix}$. The second component of this product is

$$\widetilde{D}(z) = O(z) - T(z)E(z)$$

and expanding each side in series form gives

$$\begin{aligned} \sum_{k=-\infty}^{\infty} \widetilde{d}_k z^{-k} &= \sum_{k=-\infty}^{\infty} o_k z^{-k} - \left(\frac{1}{2} + \frac{1}{2}z\right)\left(\sum_{k=-\infty}^{\infty} e_k z^{-k}\right) \\ &= \sum_{k=-\infty}^{\infty} o_k z^{-k} - \sum_{k=-\infty}^{\infty} \left(\frac{1}{2}e_k\right) z^{-k} - \sum_{k=-\infty}^{\infty} \left(\frac{1}{2}e_k\right) z^{-k+1}. \end{aligned}$$

We make the index substitution $j = k - 1$ in the last sum above so that $k = j + 1$. Note that j also runs over all integers so that we can write

$$\begin{aligned}\sum_{k=-\infty}^{\infty} \widetilde{d}_k z^{-k} &= \sum_{k=-\infty}^{\infty} o_k z^{-k} - \sum_{k=-\infty}^{\infty} \left(\frac{1}{2}e_k\right) z^{-k} - \sum_{j=-\infty}^{\infty} \left(\frac{1}{2}e_{j+1}\right) z^{-j} \\ &= \sum_{k=-\infty}^{\infty} \left(o_k - \frac{1}{2}e_k - \frac{1}{2}e_{k+1}\right) z^{-k}\end{aligned}$$

and equating coefficients gives

$$\widetilde{d}_k = o_k - \frac{1}{2}\left(e_k + e_{k+1}\right) = -\frac{1}{2}v_{2k} + v_{2k+1} - \frac{1}{2}v_{2k+2}. \tag{11.37}$$

The formula (11.37) is identical to (11.6). In a similar manner, we can start with the first component of $\begin{bmatrix}1 & S(z)\\ 0 & 1\end{bmatrix}\begin{bmatrix}E(z)\\ \widetilde{D}(z)\end{bmatrix}$ and expand

$$\widetilde{A}(z) = E(z) + S(z)\widetilde{D}(z)$$

as series and, using (11.37), equate coefficients to obtain

$$\begin{aligned}\widetilde{a}_k &= e_k + \frac{1}{4}\left(\widetilde{d}_k + \widetilde{d}_{k-1}\right) \\ &= v_{2k} + \frac{1}{4}\left(-\frac{1}{2}v_{2k} + v_{2k+1} - \frac{1}{2}v_{2k+2} + -\frac{1}{2}v_{2k-2} + v_{2k-1} - \frac{1}{2}v_{2k}\right) \\ &= -\frac{1}{8}v_{2k-2} + \frac{1}{4}v_{2k-1} + \frac{3}{4}v_{2k} + \frac{1}{4}v_{2k+1} - \frac{1}{8}v_{2k+2}.\end{aligned} \tag{11.38}$$

The preceding identity also appears in (11.5).

In Problem 11.28, you will "read" the lifting steps for recovering $\mathbf{v}$ from $\widetilde{\mathbf{a}}$ and $\widetilde{\mathbf{d}}$ from (11.36).

The Polyphase Matrix

If we multiply the matrix factors in (11.35) we obtain an interesting result. The product is

$$\begin{aligned}\begin{bmatrix}1 & S(z)\\ 0 & 1\end{bmatrix}\begin{bmatrix}1 & 0\\ -T(z) & 1\end{bmatrix} &= \begin{bmatrix}1 & \frac{1}{4}+\frac{1}{4}z^{-1}\\ 0 & 1\end{bmatrix}\begin{bmatrix}1 & 0\\ -\frac{1}{2}-\frac{1}{2}z & 1\end{bmatrix} \\ &= \begin{bmatrix}-\frac{1}{8}z+\frac{3}{4}-\frac{1}{8}z^{-1} & \frac{1}{4}+\frac{1}{4}z^{-1}\\ -\frac{1}{2}-\frac{1}{2}z & 1\end{bmatrix} \\ &= \begin{bmatrix}\widetilde{H}_e(z) & z^{-1}\widetilde{H}_o(z)\\ z\widetilde{G}_e(z) & \widetilde{G}_o(z)\end{bmatrix}\end{aligned} \tag{11.39}$$

where $\widetilde{H}_e(z)$, $\widetilde{H}_o(z)$, $\widetilde{G}_e(z)$ and $\widetilde{G}_o(z)$ are constructed from the LeGall filters

$$\widetilde{\mathbf{h}} = \left(\widetilde{h}_{-2}, \widetilde{h}_{-1}, \widetilde{h}_0, \widetilde{h}_1, \widetilde{h}_2\right) = \left(-\frac{1}{8}, \frac{1}{4}, \frac{3}{4}, \frac{1}{4}, -\frac{1}{8}\right)$$
$$\widetilde{\mathbf{g}} = \left(\widetilde{g}_0, \widetilde{g}_1, \widetilde{g}_2\right) = \left(-\frac{1}{2}, 1, -\frac{1}{2}\right)$$

given in (11.1) and (11.3). Indeed $\widetilde{H}_e(z)$ and $\widetilde{H}_o(z)$ are the Z-transforms of the "even" and "odd" filters (see (11.19))

$$\widetilde{\mathbf{h}}_e = \left(\widetilde{h}_{-2}, \widetilde{h}_0, \widetilde{h}_2\right) = \left(-\frac{1}{8}, \frac{3}{4}, -\frac{1}{8}\right) \quad \text{and} \quad \widetilde{\mathbf{h}}_o = \left(\widetilde{h}_{-1}, \widetilde{h}_1\right) = \left(\frac{1}{4}, \frac{1}{4}\right)$$

and

$$\widetilde{\mathbf{g}}_e = \left(\widetilde{g}_0, \widetilde{g}_2\right) = \left(-\frac{1}{2}, -\frac{1}{2}\right) \quad \text{and} \quad \widetilde{\mathbf{g}}_o = \left(\widetilde{g}_1\right) = (1)$$

constructed from $\widetilde{\mathbf{h}}$ and $\widetilde{\mathbf{g}}$, respectively.

The Laurent polynomials in $\widetilde{H}_e(z)$ and $\widetilde{H}_o(z)$ in (11.39) give rise to the *polyphase* representation of $\widetilde{\mathbf{h}}$. We have the following general definition.

Definition 11.4 (Polyphase Representation of a Filter). *Suppose* $\mathbf{h}$ *is an FIR filter with* $H(z)$ *the associated* Z*-transform. Then for Laurent polynomials*

$$H_e(z) = \sum_k h_{2k} z^{-k} \quad \textit{and} \quad H_o(z) = \sum_k h_{2k+1} z^{-k}$$

the polyphase representation of $\mathbf{h}$ *is given by*

$$H(z) = H_e\left(z^2\right) + z^{-1} H_o\left(z^2\right).$$

□

The polyphase representation above is nothing more than the decomposition of Z-transform $X(z)$ into its "even" and "odd" parts as given in (11.20). The representation given by Definition 11.4 allows us to define the general *polyphase matrix*.

Definition 11.5 (The Polyphase Matrix). *Suppose* $\mathbf{h}$ *and* $\mathbf{g}$ *are FIR filters with polyphase representations*

$$H(z) = H_e\left(z^2\right) + z^{-1} H_o\left(z^2\right) \quad \textit{and} \quad G(z) = G_e\left(z^2\right) + z^{-1} G_o\left(z^2\right),$$

respectively. Then the polyphase matrix associated with $\mathbf{h}$ and $\mathbf{g}$ *is given by*

$$\mathbf{P}(z) = \begin{bmatrix} H_e(z) & H_o(z) \\ G_e(z) & G_o(z) \end{bmatrix}. \tag{11.40}$$

□

Note that the matrix in (11.39) is not quite in the form given in (11.40). But by Proposition 11.3 we see that multiplying a Z-transform by powers of z simply translates the coefficients of the transform. Thus we consider the matrix in (11.40) to be a polyphase matrix.

The fact that $\mathbf{h}$ and $\mathbf{g}$ are FIR filters ensures that the elements of $\mathbf{P}(z)$ are Laurent polynomials.

Example 11.8 (Constructing a Polyphase Matrix). *Find the polyphase matrix $\mathbf{P}(z)$ for the C6 filter $\mathbf{h}$ given by (9.52) and the associated highpass filter $\mathbf{g}$ whose elements are given by (9.58). Verify that $\det(\mathbf{P}(z)) = 1$.*

Solution. *We have*

$$\begin{aligned} H_e(z) &= \frac{1}{16\sqrt{2}}(1-\sqrt{7})z + \frac{1}{8\sqrt{2}}(7+\sqrt{7}) + \frac{1}{16\sqrt{2}}(1-\sqrt{7})z^{-1} \\ H_o(z) &= \frac{1}{16\sqrt{2}}(5+\sqrt{7})z + \frac{1}{8\sqrt{2}}(7-\sqrt{7}) + \frac{1}{16\sqrt{2}}(-3+\sqrt{7})z^{-1} \end{aligned}$$

and

$$\begin{aligned} G_e(z) &= -\frac{1}{16\sqrt{2}}(-3+\sqrt{7})z - \frac{1}{8\sqrt{2}}(7-\sqrt{7}) - \frac{1}{16\sqrt{2}}(5+\sqrt{7})z^{-1} \\ G_o(z) &= \frac{1}{16\sqrt{2}}(1-\sqrt{7})z + \frac{1}{8\sqrt{2}}(7+\sqrt{7}) + \frac{1}{16\sqrt{2}}(1-\sqrt{7})z^{-1}. \end{aligned}$$

Note that $G_e(z) = -H_o\left(z^{-1}\right)$ and $G_o(z) = H_e\left(z^{-1}\right)$ so that

$$\mathbf{P}(z) = \begin{bmatrix} H_e(z) & H_o(z) \\ G_e(z) & G_o(z) \end{bmatrix} = \begin{bmatrix} H_e(z) & H_o(z) \\ -H_o\left(z^{-1}\right) & H_e\left(z^{-1}\right) \end{bmatrix}$$

and

$$\det(\mathbf{P}(z)) = H_e(z)H_e\left(z^{-1}\right) + H_o(z)H_o\left(z^{-1}\right).$$

*Using a **CAS** we compute*

$$\begin{aligned} H_e(z)H_e\left(z^{-1}\right) &= \frac{15}{32} + \frac{13\sqrt{7}}{128} + \left(\frac{1}{64} - \frac{\sqrt{7}}{256}\right)z^{-2} - \frac{3\sqrt{7}}{64}z^{-1} - \frac{3\sqrt{7}}{64}z \\ &\quad + \left(\frac{1}{64} - \frac{\sqrt{7}}{256}\right)z^2 \\ H_o(z)H_o\left(z^{-1}\right) &= \frac{17}{32} - \frac{13\sqrt{7}}{128} - \left(\frac{1}{64} - \frac{\sqrt{7}}{256}\right)z^{-2} + \frac{3\sqrt{7}}{64}z^{-1} + \frac{3\sqrt{7}}{64}z \\ &\quad - \left(\frac{1}{64} - \frac{\sqrt{7}}{256}\right)z^2. \end{aligned}$$

from which we see that

$$\det(\mathbf{P}(z)) = H_e(z)H_e\left(z^{-1}\right) + H_o(z)H_o\left(z^{-1}\right) = 1.$$

(Live example: Visit `stthomas.edu/wavelets` *and click on Live Examples.)*

□

Properties of Polyphase Matrices

Our goal is to factor the polyphase matrix $\mathbf{P}(z)$ for given filters $\mathbf{h}$ and $\mathbf{g}$ into a form similar to one given in (11.35). If $\mathbf{h}$ is an orthogonal filter and $\mathbf{g}$ is the associated highpass filter, then we can "read" the lifting steps from this factorization. If $\mathbf{h}$ is part of a biorthogonal filter pair, then we need to construct $\widetilde{\mathbf{P}}(z)$ as well.

The main result of this section is a theorem due to Daubechies and Sweldens [31]. This theorem says that under modest conditions, $\mathbf{P}(z)$ can *always* be factored in a form that leads to a lifting algorithm. Before stating and proving this result, we need some more information about polyphase matrices. The first result considers the invertibility of a polyphase matrix.

Proposition 11.6 (Invertibility of a Polyphase Matrix). *Let* $\mathbf{P}(z)$ *be the polyphase matrix given in* (11.40) *associated with filters* $\mathbf{h}$ *and* $\mathbf{g}$. *Then* $\mathbf{P}^{-1}(z)$ *exists and has Laurent polynomials as elements if and only if* $\det(\mathbf{P}(z))$ *is a nonzero monomial.*

□

Proof. Let $d(z) = \det(\mathbf{P}(z))$. We assume first that $\mathbf{P}(z)$ is invertible with the elements of $\mathbf{P}^{-1}(z)$ being Laurent polynomials. Then using Definition 2.13 we know that the determinant of $\mathbf{P}^{-1}(z)$ is the difference of two products of Laurent polynomials which is itself a Laurent polynomial. Then Problem 2.27 says that $q(z) = \det\left(\mathbf{P}^{-1}\right) = \dfrac{1}{d(z)}$. But $q(z)$ and $d(z)$ are inverses since

$$1 = \det\left(I_2\right) = \det\left(\mathbf{P}(z)\right) \cdot \det\left(\mathbf{P}^{-1}(z)\right) = q(z)d(z)$$

and by Proposition 11.5, this can only happen if both $q(z)$ and $d(z)$ are nonzero monomials.

Conversely, if $d(z) = cz^m$ is a nonzero monomial, then $\frac{1}{d(z)} = \frac{1}{c}z^{-m}$ and Problem 2.20 allows us to write the inverse as

$$\mathbf{P}^{-1}(z) = \frac{1}{c}z^{-m}\begin{bmatrix} G_o(z) & -H_o(z) \\ -G_e(z) & H_e(z) \end{bmatrix}.$$

Multiplying Laurent polynomials $H_e(z), H_o(z), G_e(z)$ and $G_o(z)$ by Laurent polynomial $\frac{1}{c}z^{-m}$ results in elements of $\mathbf{P}^{-1}(z)$ that are Laurent polynomials and the proof is complete.

□

We certainly need $\mathbf{P}(z)$ to be invertible so that we can compute inverse lifting steps. In this case, we need $\det(\mathbf{P}(z)) \neq 0$. Using the following argument, we can assume that $\det(\mathbf{P}(z)) = 1$.

By Proposition 11.6, we know that $\det(\mathbf{P}) = d(z)$, where $d(z)$ is a monomial. Thus we modify $\mathbf{P}(z)$ by dividing the bottom row $\begin{bmatrix} G_e(z) & G_o(z) \end{bmatrix}$ by $d(z)$. Then by Proposition 2.2(d), the determinant of the new polyphase matrix is one. We have altered the highpass filter $\mathbf{g}$ by doing this, but recall (see Proposition 11.3) that multiplying $G(z)$ by a monomial only serves to translate the elements of $\mathbf{g}$ – any formula we construct for lifting would still work modulo this shift. Thus going forward, we assume that $\det(\mathbf{P}(z)) = 1$.

Suppose we modify the bottom row of $\mathbf{P}(z)$ to create a new polyphase matrix $\mathbf{Q}(z)$ where $\det(\mathbf{Q}(z)) = 1$. The following proposal, due to Daubechies and Sweldens [31], tells us how $\mathbf{P}(z)$ and $\mathbf{Q}(z)$ are related.

Proposition 11.7 (Creating a Polyphase Matrix from an Existing One). *Suppose*

$$\mathbf{P}(z) = \begin{bmatrix} H_e(z) & H_o(z) \\ G_e(z) & G_o(z) \end{bmatrix} \quad \textit{and} \quad \mathbf{Q}(z) = \begin{bmatrix} H_e(z) & H_o(z) \\ G'_e(z) & G'_o(z) \end{bmatrix}$$

with $\det(\mathbf{P}(z)) = 1$. *Then* $\det(\mathbf{Q}(z)) = 1$ *if and only if*

$$G'(z) = G(z) + H(z)S\left(z^2\right) \tag{11.41}$$

where $S(z)$ *is a Laurent polynomial.*

□

Proof. We start by first finding $G'_e(z)$ and $G'_o(z)$ from (11.41). We compute $G'_e(z)$ and leave the computation of $G'_o(z)$ as an exercise. Since all our Z-transforms are Laurent polynomials, we can freely interchange the order of the summation. So that we don't have to worry about the limits of sums, we still allow them to range from $-\infty$ to ∞ and adopt the notation $\sum\limits_{k=-\infty}^{\infty} = \sum\limits_{k}$. We have

$$\begin{aligned} G'(z) &= G(z) + H(z)S\left(z^2\right) \\ &= \sum_k g_k z^{-k} + \left(\sum_k h_k z^{-k}\right)\left(\sum_j s_j z^{-2j}\right) \\ &= \sum_k g_k z^{-k} + \sum_j \sum_k h_k s_j z^{-(k+2j)}. \end{aligned}$$

In the double sum above, we make the substitution $m = k + 2j$ so that $k = m - 2j$. Since k, j run over all integers, so does m. We continue by writing

$$G'(z) = \sum_k g_k z^{-k} + \sum_m \left(\sum_j h_{m-2j} s_j\right) z^{-m} = \sum_k g_k z^{-k} + \sum_m u_m z^{-m}$$

where $u_m = \sum\limits_j h_{m-2j} s_j$. We can now identify $G'_e(z)$. We have

$$G'_e(z) = G_e(z) + \sum_m u_{2m} z^{-m} = G_e(z) + \sum_m \sum_j h_{2m-2j} s_j z^{-m}.$$

We next make the index substition $k = m - j$ and note that k, like m and j, range over all integers. Continuing we find that

$$\begin{aligned} G'_e(z) &= G_e(z) + \sum_k \sum_j h_{2k} s_j z^{-(k+j)} \\ &= G_e(z) + \left(\sum_k h_{2k} z^{-k}\right)\left(\sum_j s_j z^{-j}\right) \\ &= G_e(z) + H_e(z)S(z). \end{aligned} \tag{11.42}$$

In Problem 11.30, you will show that

$$G'_o(z) = G_o(z) + H_o(z)S(z). \tag{11.43}$$

Thus we can restate the proposition by replacing (11.41) with (11.42) and (11.43).

Let's assume that $\det(\mathbf{Q}(z)) = 1 = \det(\mathbf{P}(z))$. Then by Problem 2.28 we know that $G'_e(z) = G_e(z) + H_e(z)\alpha$ and $G'_o(z) = G_o(z) + H_o(z)\alpha$. Since $G'_e(z)$ and $G'_o(z)$ are Laurent polynomials, then α can be a Laurent polynomial. So $G'_e(z) = G_e(z) + H_e(z)S(z)$ and $G'_o(z) = G_o(z) + H_o(z)S(z)$, where $S(z)$ is a Laurent polynomial and the result holds.

Next assume that (11.42) and (11.43) are true. In this case we can write

$$\begin{aligned} \mathbf{Q}(z) &= \begin{bmatrix} H_e(z) & H_o(z) \\ G'_e(z) & G'_o(z) \end{bmatrix} \\ &= \begin{bmatrix} H_e(z) & H_o(z) \\ G_e(z) + H_e(z)S(z) & G_o(z) + H_o(z)S(z) \end{bmatrix} \\ &= \begin{bmatrix} 1 & 0 \\ S(z) & 1 \end{bmatrix} \begin{bmatrix} H_e(z) & H_o(z) \\ G_e(z) & G_o(z) \end{bmatrix} \\ &= \begin{bmatrix} 1 & 0 \\ S(z) & 1 \end{bmatrix} \mathbf{P}(z). \end{aligned} \tag{11.44}$$

We have $\det\left(\begin{bmatrix} 1 & 0 \\ S(z) & 1 \end{bmatrix}\right) = 1$ by Definition 2.13 and since $\det(\mathbf{P}(z)) = 1$, we can infer by Proposition 2.2(a) that $\det(\mathbf{Q}(z)) = 1$ as well. □

The corollary below follows immediately from (11.44) in Proposition 11.7.

Corollary 11.1 (Solving for $\mathbf{S(z)}$). *Suppose $\mathbf{P}(z)$ and $\mathbf{Q}(z)$ are as given in Proposition 11.7 and assume $S(z)$ satisfiies* (11.41). *Then*

$$S(z) = G'_e(z)G_o(z) - G'_o(z)G_e(z). \tag{11.45}$$

□

Proof. The proof is left as Problem 11.31. □

The Polyphase Matrix and Biorthogonality Conditions

There is a natural connection between the polyphase matrix $\mathbf{P}(z)$ and the biorthogonality conditions (11.26) and (11.27). We can use matrices to write these as

$$\begin{bmatrix} H(z) & H(-z) \\ G(z) & G(-z) \end{bmatrix} \begin{bmatrix} \widetilde{H}\left(z^{-1}\right) & \widetilde{G}\left(z^{-1}\right) \\ \widetilde{H}\left(-z^{-1}\right) & \widetilde{G}\left(-z^{-1}\right) \end{bmatrix} = 2\begin{bmatrix} 1 & 0 \\ 0 & 1 \end{bmatrix} = 2I_2. \tag{11.46}$$

We define

$$M(z) = \begin{bmatrix} H(z) & H(-z) \\ G(z) & G(-z) \end{bmatrix} \quad \text{and} \quad \widetilde{M}(z) = \begin{bmatrix} \widetilde{H}(z) & \widetilde{H}(-z) \\ \widetilde{G}(z) & \widetilde{G}(-z) \end{bmatrix} \tag{11.47}$$

and note that (11.46) becomes

$$M(z)\widetilde{M}^T\left(z^{-1}\right) = 2I_2. \tag{11.48}$$

We can use $M(z)$ and $\widetilde{M}(z)$ to develop a connection between the biorthogonality conditions given in (11.46) and corresponding polyphase matrices.

Proposition 11.8 (Biorthogonality Conditions and Polyphase Matrices). *Let* $\mathbf{P}(z)$ *and* $\widetilde{\mathbf{P}}(z)$ *be polyphase matrices constructed from* $\mathbf{h}$, $\mathbf{g}$ *and* $\widetilde{\mathbf{h}}$, $\widetilde{\mathbf{g}}$, *respectively. Then*

$$\mathbf{P}(z)\widetilde{\mathbf{P}}^T\left(z^{-1}\right) = I_2. \tag{11.49}$$

□

Proof. The proof depends on the identity

$$\mathbf{P}\left(z^2\right) = \frac{1}{2}M(z)\begin{bmatrix} 1 & z \\ 1 & -z \end{bmatrix} \tag{11.50}$$

that you will establish in Problem 11.33. A similar result holds for $\widetilde{\mathbf{P}}(z)$. Using this result, we compute

$$\begin{aligned}
\mathbf{P}\left(z^2\right)\widetilde{\mathbf{P}}^T\left(z^{-2}\right) &= \left(\frac{1}{2}M(z)\begin{bmatrix} 1 & z \\ 1 & -z \end{bmatrix}\right)\left(\frac{1}{2}\widetilde{M}\left(z^{-1}\right)\begin{bmatrix} 1 & z^{-1} \\ 1 & -z^{-1} \end{bmatrix}\right)^T \\
&= \frac{1}{4}M(z)\left(\begin{bmatrix} 1 & z \\ 1 & -z \end{bmatrix}\begin{bmatrix} 1 & 1 \\ z^{-1} & -z^{-1} \end{bmatrix}\right)\widetilde{M}^T\left(z^{-1}\right) \\
&= \frac{1}{4}M(z)\begin{bmatrix} 2 & 0 \\ 0 & 2 \end{bmatrix}\widetilde{M}^T\left(z^{-1}\right) \\
&= \frac{1}{2}M(z)\widetilde{M}^T\left(z^{-1}\right) \\
&= \frac{1}{2}\cdot 2\,I_2 \\
&= I_2.
\end{aligned}$$

The identity (11.48) was used in the next to last step. We replace z^2 by z in the identity $\mathbf{P}\left(z^2\right)\widetilde{\mathbf{P}}^T\left(z^{-2}\right) = I_2$ to complete the proof.

□

The result (11.49) says that $\mathbf{P}^{-1}(z) = \widetilde{\mathbf{P}}^T\left(z^{-1}\right)$. In Problem 11.40, you will develop results similar to those in Proposition 11.8 for orthogonal filters.

Suppose we have constructed $\mathbf{P}(z)$ so that $\det(\mathbf{P}(z)) = 1$. Then we can use (11.49) to find $\widetilde{\mathbf{P}}(z)$. Using Problem 2.20 we compute

$$\begin{aligned}\widetilde{\mathbf{P}}^T\left(z^{-1}\right) &= \begin{bmatrix}\widetilde{H}_e\left(z^{-1}\right) & \widetilde{G}_e\left(z^{-1}\right)\\ \widetilde{H}_o\left(z^{-1}\right) & \widetilde{G}_o\left(z^{-1}\right)\end{bmatrix}\\ &= \mathbf{P}^{-1}(z)\\ &= \begin{bmatrix}G_o(z) & -H_o(z)\\ -G_e(z) & H_e(z)\end{bmatrix}\end{aligned}$$

so that, replacing z by z^{-1}, we obtain

$$\begin{aligned}\widetilde{H}_e(z) &= G_o\left(z^{-1}\right), \quad & \widetilde{H}_o(z) &= -G_e\left(z^{-1}\right)\\ \widetilde{G}_e(z) &= -H_o\left(z^{-1}\right), \quad & \widetilde{G}_o(z) &= H_e\left(z^{-1}\right).\end{aligned} \tag{11.51}$$

In Problem 11.39, you will consider these relationships and lifting in the Z-transform domain via the filters introduced in Problem 4.36.

Factoring the Polyphase Matrix

We are now ready for the main result of this chapter. It is due to Daubechies and Sweldens [31]. The result says that any polyphase matrix $\mathbf{P}(z)$ with $\det(\mathbf{P}(z)) = 1$ can be factored into matrices from which a lifting method can be obtained. The proof is constructive and can be used with any of the filters we have studied in this book.

Theorem 11.1 (Factoring the Polyphase Matrix). *Suppose* $\mathbf{P}(z)$ *is a polyphase matrix with* $\det(\mathbf{P}(z)) = 1$. *Then there exists a constant* $K \neq 0$ *and Laurent polynomials* $S_k(z)$ *and* $T_k(z)$, $k = 1, \ldots, N$ *such that*

$$\mathbf{P}(z) = \begin{bmatrix}K & 0\\ 0 & K^{-1}\end{bmatrix}\prod_{k=1}^{N}\begin{bmatrix}1 & S_k(z)\\ 0 & 1\end{bmatrix}\begin{bmatrix}1 & 0\\ -T_k(z) & 1\end{bmatrix}.$$

□

Proof. Suppose $\mathbf{P}(z) = \begin{bmatrix}H_e(z) & H_o(z)\\ G_e(z) & G_o(z)\end{bmatrix}$. We can use the matrix factorization form (11.30) of the Euclidean Algorithm in Section 11.2 to write

$$\begin{bmatrix}H_e(z)\\ H_o(z)\end{bmatrix} = \left(\prod_{n=1}^{L}\begin{bmatrix}q_n(z) & 1\\ 1 & 0\end{bmatrix}\right)\begin{bmatrix}a(z)\\ 0\end{bmatrix} \tag{11.52}$$

where $a(z)$ is a monomial of the form Kz^m, $K \neq 0$. Note by Problem 11.26 if $\deg(H_o) > \deg(H_e)$, the first matrix factor in (11.52) is $\begin{bmatrix} 0 & 1 \\ 1 & 0 \end{bmatrix}$ which corresponds to $q_1(z) = 0$.

We can assume L is even based on the following argument. If L is odd, we start with Z-transforms $zH(z)$ and $-z^{-1}G(z)$ instead of $H(z)$ and $G(z)$. By Propositions 11.2 and 11.3 these operations translate the elements of $\mathbf{h}$ one unit left and negate and translate the elements of $\mathbf{g}$ one unit right. These operations do not hinder our ability to create a lifting method for the filters. This change also results in the new polyphase matrix

$$\widehat{\mathbf{P}}(z) = \begin{bmatrix} H_o(z) & zH_e(z) \\ -z^{-1}G_o(z) & -G_e(z) \end{bmatrix}$$

that satisfies $\det\left(\widehat{\mathbf{P}}(z)\right) = 1$ (see Problem 11.38). Note also the elements that were used to create the "even" Z-transforms are exchanged with those used to create the odd Z-transforms. Thus we perform the Euclidean algorithm on $H_o(z)$, $zH_e(z)$ instead of $H_e(z)$, $H_o(z)$. We consider two cases. If $\deg(H_e) > \deg(H_o)$, then the first step in the "new" Euclidean algorithm flips the Laurent polynomials $H_o(z)$ and $zH_e(z)$. After that, there are L factors in the factorization. Thus the total number of factors in this case is $L + 1$ which is even. Finally, consider the case where $\deg(H_o) \geq \deg(H_e)$. Then the first matrix in the original factorization (11.52) flips $H_e(z)$ and $H_o(z)$. But this step is not necessary when the Euclidean algorithm is applied to $H_o(z)$ and $zH_e(z)$. In this case, we would need only $L - 1$ even factors. Thus we assume going forward that L is always even.

We can rewrite (11.52) as

$$\begin{bmatrix} z^{-m}H_e(z) \\ z^{-m}H_o(z) \end{bmatrix} = \left(\prod_{n=1}^{L} \begin{bmatrix} q_n(z) & 1 \\ 1 & 0 \end{bmatrix}\right) \begin{bmatrix} K \\ 0 \end{bmatrix}$$

by recalling from Proposition 11.3 that multiplying the $H_e(z)$, $H_o(z)$ by z^{-m} is equivalent to translating the filter elements in $\mathbf{h}$ by m units (right or left, depending on the sign of m). We can remove this multiplication by starting with the elements of $\mathbf{h}$ translated m units (right or left, depending on the sign of m). Then multiplication of the Z-transforms by z^{-m} produces the Z-transforms we desire. Thus we write

$$\begin{bmatrix} H_e(z) \\ H_o(z) \end{bmatrix} = \left(\prod_{n=1}^{L} \begin{bmatrix} q_n(z) & 1 \\ 1 & 0 \end{bmatrix}\right) \begin{bmatrix} K \\ 0 \end{bmatrix}. \tag{11.53}$$

We can convert the vector $\begin{bmatrix} H_e(z) \\ H_o(z) \end{bmatrix}$ into a 2×2 polyphase matrix where the second column contains two Laurent polynomials provided the resulting matrix has determinant one. We have

$$\det\left(\prod_{n=1}^{L} \begin{bmatrix} q_n(z) & 1 \\ 1 & 0 \end{bmatrix}\right) = (-1)^L = 1$$

since L is even. We can convert $\begin{bmatrix} K \\ 0 \end{bmatrix}$ into a 2×2 matrix with determinant one by adding the column $\begin{bmatrix} 0 \\ K^{-1} \end{bmatrix}$. We add columns $\begin{bmatrix} G'_e(z) \\ G'_o(z) \end{bmatrix}$ and $\begin{bmatrix} 0 \\ K^{-1} \end{bmatrix}$ to the left and right sides, respectively, to (11.53) and transpose the result to obtain

$$\begin{bmatrix} H_e(z) & G'_e(z) \\ H_o(z) & G'_o(z) \end{bmatrix} = \left(\prod_{n=1}^{L} \begin{bmatrix} q_n(z) & 1 \\ 1 & 0 \end{bmatrix} \right) \begin{bmatrix} K & 0 \\ 0 & K^{-1} \end{bmatrix}$$

$$\begin{bmatrix} H_e(z) & H_o(z) \\ G'_e(z) & G'_o(z) \end{bmatrix} = \begin{bmatrix} K & 0 \\ 0 & K^{-1} \end{bmatrix} \left(\prod_{k=1}^{L} \begin{bmatrix} q_{L+1-k}(z) & 1 \\ 1 & 0 \end{bmatrix} \right). \tag{11.54}$$

We have made repeated use of Problem 2.23 above and notice in so doing, we needed to reverse the order of multiplication in the L-fold product. From the proof of Proposition 11.7, we know that

$$\begin{bmatrix} H_e(z) & H_o(z) \\ G'_e(z) & G'_o(z) \end{bmatrix} = \begin{bmatrix} 1 & 0 \\ S(z) & 1 \end{bmatrix} \mathbf{P}(z).$$

Plugging this rcsult into (11.54) gives

$$\begin{bmatrix} 1 & 0 \\ S(z) & 1 \end{bmatrix} \mathbf{P}(z) = \begin{bmatrix} H_e(z) & H_o(z) \\ G'_e(z) & G'_o(z) \end{bmatrix} = \begin{bmatrix} K & 0 \\ 0 & K^{-1} \end{bmatrix} \left(\prod_{k=1}^{L} \begin{bmatrix} q_{L-k+1}(z) & 1 \\ 1 & 0 \end{bmatrix} \right)$$

and noting that $\begin{bmatrix} 1 & 0 \\ S(z) & 1 \end{bmatrix}^{-1} = \begin{bmatrix} 1 & 0 \\ -S(z) & 1 \end{bmatrix}$ we find that

$$\mathbf{P}(z) = \begin{bmatrix} 1 & 0 \\ -S(z) & 1 \end{bmatrix} \begin{bmatrix} K & 0 \\ 0 & K^{-1} \end{bmatrix} \left(\prod_{k=1}^{L} \begin{bmatrix} q_{L-k+1}(z) & 1 \\ 1 & 0 \end{bmatrix} \right). \tag{11.55}$$

In Problem 11.32, you will verify that

$$\begin{bmatrix} 1 & 0 \\ -S(z) & 1 \end{bmatrix} \begin{bmatrix} K & 0 \\ 0 & K^{-1} \end{bmatrix} = \begin{bmatrix} K & 0 \\ 0 & K^{-1} \end{bmatrix} \begin{bmatrix} 1 & 0 \\ -K^2 S(z) & 1 \end{bmatrix}. \tag{11.56}$$

Plugging (11.56) into (11.55) results in

$$\mathbf{P}(z) = \begin{bmatrix} K & 0 \\ 0 & K^{-1} \end{bmatrix} \begin{bmatrix} 1 & 0 \\ -K^2 S(z) & 1 \end{bmatrix} \left(\prod_{k=1}^{L} \begin{bmatrix} q_{L-k+1}(z) & 1 \\ 1 & 0 \end{bmatrix} \right).$$

Since L is even, we can rewrite the above identity as

$$\mathbf{P}(z) = \begin{bmatrix} K & 0 \\ 0 & K^{-1} \end{bmatrix} \begin{bmatrix} 1 & 0 \\ -K^2 S(z) & 1 \end{bmatrix} \times \left(\prod_{k=1}^{L/2} \begin{bmatrix} q_{L-2(k-1)}(z) & 1 \\ 1 & 0 \end{bmatrix} \begin{bmatrix} q_{L-2k+1}(z) & 1 \\ 1 & 0 \end{bmatrix} \right). \tag{11.57}$$

In Problem 11.58, you will establish the identities

$$\begin{aligned}\begin{bmatrix} x & 1 \\ 1 & 0 \end{bmatrix} &= \begin{bmatrix} 1 & x \\ 0 & 1 \end{bmatrix} \begin{bmatrix} 0 & 1 \\ 1 & 0 \end{bmatrix} \\ &= \begin{bmatrix} 0 & 1 \\ 1 & 0 \end{bmatrix} \begin{bmatrix} 1 & 0 \\ x & 1 \end{bmatrix}.\end{aligned} \tag{11.58}$$

We substitute the first identity in (11.58) for $\begin{bmatrix} q_{L-2(k-1)}(z) & 1 \\ 1 & 0 \end{bmatrix}$ and the second identity in (11.58) for $\begin{bmatrix} q_{L-2-2k+1}(z) & 1 \\ 1 & 0 \end{bmatrix}$ in the $L/2$-fold product in (11.57) to obtain

$$\begin{aligned}&\prod_{k=1}^{L/2} \begin{bmatrix} q_{L-2(k-1)}(z) & 1 \\ 1 & 0 \end{bmatrix} \begin{bmatrix} q_{L-2k+1}(z) & 1 \\ 1 & 0 \end{bmatrix} \\ &\quad = \prod_{k=1}^{L/2} \begin{bmatrix} 1 & q_{L-2(k-1)}(z) \\ 0 & 1 \end{bmatrix} \begin{bmatrix} 0 & 1 \\ 1 & 0 \end{bmatrix} \begin{bmatrix} 0 & 1 \\ 1 & 0 \end{bmatrix} \begin{bmatrix} 1 & 0 \\ q_{L-2k+1}(z) & 1 \end{bmatrix} \\ &\quad = \prod_{k=1}^{L/2} \begin{bmatrix} 1 & q_{L-2(k-1)}(z) \\ 0 & 1 \end{bmatrix} \begin{bmatrix} 1 & 0 \\ q_{L-2k+1}(z) & 1 \end{bmatrix}.\end{aligned} \tag{11.59}$$

We have used the fact that

$$\begin{bmatrix} 0 & 1 \\ 1 & 0 \end{bmatrix} \begin{bmatrix} 0 & 1 \\ 1 & 0 \end{bmatrix} = I_2$$

in the above computations. We can use (11.59) to write (11.57) as

$$\begin{aligned}\mathbf{P}(z) = &\begin{bmatrix} K & 0 \\ 0 & K^{-1} \end{bmatrix} \begin{bmatrix} 1 & 0 \\ -K^2 S(z) & 1 \end{bmatrix} \\ &\cdot \left(\prod_{k=1}^{L/2} \begin{bmatrix} 1 & q_{L-2(k-1)}(z) \\ 0 & 1 \end{bmatrix} \begin{bmatrix} 1 & 0 \\ q_{L-2k+1}(z) & 1 \end{bmatrix} \right).\end{aligned} \tag{11.60}$$

The last step is some simple indexing. Note that the second factor $\begin{bmatrix} 1 & 0 \\ -K^2 S(z) & 1 \end{bmatrix}$ in (11.60) fits the form of the second matrix in the $L/2$-fold product. It lacks a "companion" matrix. Inserting I_2 (so the Laurent polynomial in the lower left corner is zero) between the first two matrix factors in (11.60) does the job. We set $N = \frac{L}{2}+1$ and

$$\begin{aligned} S_1(z) &= 0 \\ T_1(z) &= K^2 S(z) \\ S_{k+1}(z) &= q_{L-2(k-1)}(z), \quad k = 1, \ldots, \frac{L}{2} \\ T_{k+1}(z) &= -q_{L-2k+1}(z), \quad k = 1, \ldots, \frac{L}{2} \end{aligned}$$

to arrive at

$$\mathbf{P}(z) = \begin{bmatrix} K & 0 \\ 0 & K^{-1} \end{bmatrix} \left(\prod_{k=1}^{N} \begin{bmatrix} 1 & S_k(z) \\ 0 & 1 \end{bmatrix} \begin{bmatrix} 1 & 0 \\ -T_k(z) & 1 \end{bmatrix} \right). \tag{11.61}$$

It is easily verified via Definition 2.13 and Proposition 2.2(a) that $\det(\mathbf{P}(z)) = 1$ as desired. □

PROBLEMS

11.28 In this problem, you mimic the computations that produced (11.37) and (11.38) from (11.35) to recover $\mathbf{v}$ from $\widetilde{\mathbf{a}}$ and $\widetilde{\mathbf{d}}$. Start with the right-most product on the right side of (11.36). This product gives

$$\begin{bmatrix} E(z) \\ \widetilde{D}(z) \end{bmatrix} = \begin{bmatrix} 1 & -\frac{1}{4} - \frac{1}{4}z^{-1} \\ 0 & 1 \end{bmatrix} \begin{bmatrix} \widetilde{A}(z) \\ \widetilde{D}(z) \end{bmatrix}$$

from which we obtain $E(z) = \widetilde{A}(z) - \left(\frac{1}{4} + \frac{1}{4}z^{-1}\right)\widetilde{D}(z)$. Expand the Z-transforms of both sides of this identity to recover the "even" values v_{2k}. Then consider

$$\begin{bmatrix} E(z) \\ O(z) \end{bmatrix} = \begin{bmatrix} 1 & 0 \\ \frac{1}{2} + \frac{1}{2}z & 1 \end{bmatrix} \begin{bmatrix} E(z) \\ \widetilde{D}(z) \end{bmatrix}.$$

This product yields the equation $O(z) = \widetilde{D}(z) + \left(\frac{1}{2} + \frac{1}{2}z\right)E(z)$. Expand the Z-transforms on both sides of this identity to recover the "odd" values v_{2k+1}.

11.29 Verify that the polyphase matrix given by (11.39) has determinant 1.

11.30 Mimic the argument used to obtain (11.42) to establish the formula (11.43) for $G_o'(z)$.

11.31 Prove Corollary 11.1.

11.32 Verify the identity given in (11.56).

11.33 In this problem, you establish the identity given in (11.50). The following steps will help you organize your work.

(a) Plug z^2 into the polyphase matrix given in (11.40) and then rewrite this matrix using (11.21).

(b) Expand the right side of (11.50) and compare this to (a).

11.34 Results similar to those given in Proposition 11.8 exist for orthogonal filters. You will establish these results in the steps below.

(a) In Problem 11.21 you showed that the orthogonality conditions (9.22), (9.24), and (9.25) are equivalent in the Z-transform domain to

$$H\left(z^{-1}\right)H(z)+H\left(-z^{-1}\right)H(-z)=2$$
$$G\left(z^{-1}\right)G(z)+G\left(-z^{-1}\right)G(-z)=2$$
$$H\left(z^{-1}\right)G(z)+H\left(-z^{-1}\right)G(-z)=0.$$

For $M(z)$ given in (11.47), show the above conditions are equivalent to

$$M(z)M^T\left(z^{-1}\right)=2I_2.$$

(b) Consider the polyphase matrix $\mathbf{P}(z)=\begin{bmatrix}H_e(z) & H_o(z)\\ G_e(z) & G_o(z)\end{bmatrix}$. Using the result from (a) and the identity (11.50) established in Problem 11.33, show that

$$\mathbf{P}(z)\mathbf{P}^T\left(z^{-1}\right)=I_2.$$

(c) Use (b) to show that

$$H_e(z)H_e\left(z^{-1}\right)+H_o(z)H_o\left(z^{-1}\right)=1=G_e(z)G_e\left(z^{-1}\right)+G_o(z)G_o\left(z^{-1}\right).$$

★11.35 Consider the D4 orthogonal filter $\mathbf{h}$ given in (5.27) and the associated highpass filter $\mathbf{g}$ given by (5.28). We will see that the polyphase matrix $\mathbf{P}(z)$ formed from these filters does not satisfy $\det(\mathbf{P}(z))=1$. There are different options for how $\mathbf{g}$ is defined and we will learn how to modify this filter so that we can create a polyphase matrix $\mathbf{P}(z)$ with determinant 1.

(a) Verify that the "even" and "odd" Z-transforms built from $\mathbf{h}$ and $\mathbf{g}$ are

$$\begin{aligned}
H_e(z) &= \frac{1+\sqrt{3}}{4\sqrt{2}}+\frac{3-\sqrt{3}}{4\sqrt{2}}z^{-1}\\
H_o(z) &= \frac{3+\sqrt{3}}{4\sqrt{2}}+\frac{1-\sqrt{3}}{4\sqrt{2}}z^{-1}\\
G_e(z) &= \frac{1-\sqrt{3}}{4\sqrt{2}}+\frac{3+\sqrt{3}}{4\sqrt{2}}z^{-1}\\
G_o(z) &= -\frac{3-\sqrt{3}}{4\sqrt{2}}-\frac{1+\sqrt{3}}{4\sqrt{2}}z^{-1}.
\end{aligned}$$

(b) Using (a) show that $G_e(z)=z^{-1}H_o\left(z^{-1}\right)$ and $G_o(z)=-z^{-1}H_e\left(z^{-1}\right)$.

(c) Using (b) we can rewrite the polyphase matrix $\mathbf{P}(z)$ as

$$\mathbf{P}(z)=\begin{bmatrix}H_e(z) & H_o(z)\\ G_e(z) & G_o(z)\end{bmatrix}=\begin{bmatrix}H_e(z) & H_o(z)\\ z^{-1}H_o\left(z^{-1}\right) & -z^{-1}H_e\left(z^{-1}\right)\end{bmatrix}.$$

Use this formulation of the polyphase matrix along with Problem 11.34(c) to show that $\det(\mathbf{P}(z)) = -z^{-1}$.

(d) Note that if we had started with $\widehat{G}_e(z) = -zG_e(z)$ and $\widehat{G}_o(z) = -zG_o(z)$ then the resulting polyphase matrix

$$\widehat{\mathbf{P}}(z) = \begin{bmatrix} H_e(z) & H_o(z) \\ \widehat{G}_e(z) & \widehat{G}_o(z) \end{bmatrix}$$

satisfies $\det\left(\widehat{\mathbf{P}}(z)\right) = 1$. Using (11.20), find $\widehat{G}(z)$ in terms of $G(z)$.

(e) Recall that the elements of g are defined as $g_k = (-1)^k h_{3-k}$, $k = 0, 1, 2, 3$. Using Proposition 11.2 with $c_1 = -1, c_2 = 0$ and Proposition 11.3 with $n = -2$, show that the elements of $\widehat{\mathbf{g}}$ are defined as $\widehat{g}_{k-2} = -g_k = (-1)^{k+1} h_{3-k}$, $k = 0, 1, 2, 3$. That is, we need to negate our filter elements and shift them two units left in order to create a polyphase matrix with determinant one. Proposition 2.4 guarantees that these actions still result in a highpass filter. Neither shifting the elements of **g** two units left or multiplying them by -1 will affect its orthogonality with **h**.

(f) Suppose we *start* then with highpass filter $\widetilde{\mathbf{g}}$ where $\widetilde{g}_k$ is given in (e). Verify that the resulting Laurent polynomials $\widetilde{G}_e(z)$ and $\widetilde{G}_o(z)$ are

$$\begin{aligned} \widetilde{G}_e(z) &= -h_1 - h_3 z = -H_o\left(z^{-1}\right) \\ \widetilde{G}_o(z) &= h_0 + h_2 z = H_e\left(z^{-1}\right). \end{aligned}$$

and that the determinant of the resulting polyphase matrix is one.

★11.36 Suppose $H_e(z)$ and $H_o(z)$ are formed from a Daubechies orthogonal filter **h** of even length $L+1$. Suppose the elements of the associated highpass filter g are given by (5.59). Show that $G_e(z) = z^{\frac{1-L}{2}} H_o\left(z^{-1}\right)$ and $G_o(z) = -z^{\frac{1-L}{2}} H_e\left(z^{-1}\right)$. The following steps will help you organize your work for the identity involving $G_e(z)$. The proof is similar for the identity involving $G_o(z)$.

(a) Use the defining relation

$$G_e(z) = \sum_{k=-\infty}^{\infty} g_{2k} z^{-k}$$

and the formula for g_k to show that

$$G_e(z) = \sum_{k=0}^{\frac{L-1}{2}} h_{L-2k} z^{-k}.$$

(b) Noting that $L-2k$ is odd, make the substitution $2m+1 = L-2k$ in the above sum and simplify to obtain

$$G_e(z) = z^{\frac{1-L}{2}} \sum_{m=0}^{\frac{L-1}{2}} h_{2m+1} z^m.$$

The result follows directly from this identity.

★11.37 Assume that the polyphase matrix $\mathbf{P}(z)$ is constructed from the Laurent polynomials in Problem 11.36. Use Problem 11.34(c) to show that

$$\det(\mathbf{P}(z)) = -z^{\frac{1-L}{2}}.$$

In order to create a polyphase matrix with determinant one, we need to multiply $G_e(z)$ and $G_o(z)$ by $-z^{\frac{L-1}{2}}$. This has the effect (see Problem 11.35) of negating the elements of $\mathbf{g}$ and also shifting them $L-1$ units left.

★11.38 Suppose $H(z)$ and $G(z)$ are Z-transforms that generate the polyphase matrix $\mathbf{P}(z)$ with $\det(\mathbf{P}(z)) = 1$. Create new Z-transforms $\widehat{H}(z) = zH(z)$ and $\widehat{G}(z) = -z^{-1}G(z)$.

(a) Show that the coefficients of $\widehat{H}(z)$ and $\widehat{G}(z)$ can be expressed as

$$\widehat{h}_k = h_{k+1} \qquad \text{and} \qquad \widehat{g}_k = -g_{k-1}.$$

(b) Use (a) to show that

$$\begin{aligned}\widehat{H}_e(z) &= H_o(z)\\ \widehat{H}_o(z) &= zH_e(z)\\ \widehat{G}_e(z) &= -z^{-1}G_o(z)\\ \widehat{G}_o(z) &= -G_e(z).\end{aligned}$$

(c) Use (b) to show that the polyphase matrix $\widehat{\mathbf{P}}(z)$ constructed from $\widehat{H}(z)$ and $\widehat{G}(z)$ satisfies $\det\left(\widehat{\mathbf{P}}(z)\right) = 1$.

11.39 In this problem we develop the lifting method for the modified Haar filter given in Problem 4.36 in the Z-transform domain. Recall from Problem 4.36, for filters $\mathbf{h} = (h_0, h_1) = \left(\frac{1}{2}, \frac{1}{2}\right)$ and $\mathbf{g} = (g_0, g_1) = (-1, 1)$ we developed the following lifting algorith:

1. Split the input vector $\mathbf{v}$ into "even" and "odd" vectors $\mathbf{e}$ and $\mathbf{o}$.
2. Compute the differences portion $\mathbf{d}$ element-wise by $d_k = o_k - e_k$.
3. Compute the averages portion $\mathbf{a}$ element-wise by $a_k = e_k + \frac{1}{2}d_k$.

This lifting method is simple enough that we can perform the factorization of the polyphase matrix without using Theorem 11.1.

(a) Write the Z-transform analog of Step 2 above and then find matrix C_1 such that

$$\begin{bmatrix} E(z) \\ D(z) \end{bmatrix} = C_1 \begin{bmatrix} E(z) \\ O(z) \end{bmatrix}.$$

(b) Write the Z-transform analog of Step 3 above and then find matrix C_2 such that

$$\begin{bmatrix} A(z) \\ D(z) \end{bmatrix} = C_2 \begin{bmatrix} E(z) \\ D(z) \end{bmatrix}.$$

(c) Combine (a) and (b) to produce the polyphase matrix $\mathbf{P}(z)$ so that

$$\begin{bmatrix} A(z) \\ D(z) \end{bmatrix} = \mathbf{P}(z) \begin{bmatrix} E(z) \\ O(z) \end{bmatrix}.$$

(d) Use (c) to write down the Laurent polynomials $H_e(z)$, $H_o(z)$, $G_e(z)$ and $G_o(z)$.

11.40 This problem is a continuation of Problem 11.39. In this exercise, you find the polyphase matrix $\widetilde{\mathbf{P}}(z)$ three different ways.

(a) Use Problem 11.39(d) and (11.51) to write down the Laurent polynomials $\widetilde{H}_e(z)$, $\widetilde{H}_o(z)$, $\widetilde{G}_e(z)$ and $\widetilde{G}_o(z)$.

(b) Use (a) to write down $\widetilde{\mathbf{P}}(z)$.

(c) Use Problem 2.20 to write $\mathbf{P}^{-1}(z)$ for $\mathbf{P}(z)$ from Problem 11.39(c) and then use (11.49) to obtain $\widetilde{\mathbf{P}}(z)$.

(d) We can rewrite Step 2 of the lifting method outlined in Problem 11.39 to obtain

$$e_k = a_k - \frac{1}{2}d_k \qquad \text{and} \qquad o_k = e_k + d_k.$$

The lifting here represents the inverse transform. Mimic Problem 11.39(a)–(c) to produce a matrix C such that

$$\begin{bmatrix} E(z) \\ O(z) \end{bmatrix} = C \begin{bmatrix} A(z) \\ D(z) \end{bmatrix}.$$

Compare C with $\widetilde{\mathbf{P}}(z)$ obtained in (b) and (c).

★11.41 Consider the orthogonal matrix $U = \begin{bmatrix} \cos\theta & \sin\theta \\ -\sin\theta & \cos\theta \end{bmatrix}$ investigated in Problem 2.29. It is straightforward to ascertain that $\det(U) = 1$ so by Theorem 11.1 we should be able to produce a factorization of the form given in (11.61). Show that for $\theta \neq \frac{\pi}{2}$,

$$U = \begin{bmatrix} \cos\theta & 0 \\ 0 & \frac{1}{\cos\theta} \end{bmatrix} \begin{bmatrix} 1 & 0 \\ -\sin\theta\cos\theta & 1 \end{bmatrix} \begin{bmatrix} 1 & \frac{\sin\theta}{\cos\theta} \\ 0 & 1 \end{bmatrix}.$$

The following steps will help you organize your work.

(a) Start with $a_0(z) = \cos\theta$ and $b_0(z) = \sin\theta$. Perform two steps of the Euclidean algorithm first with $q_1(z) = 0$ and next with $q_2(z) = \frac{\sin\theta}{\cos\theta}$ to obtain

$$\begin{bmatrix} \cos\theta \\ \sin\theta \end{bmatrix} = \begin{bmatrix} 0 & 1 \\ 1 & 0 \end{bmatrix} \begin{bmatrix} \frac{\sin\theta}{\cos\theta} & 1 \\ 1 & 0 \end{bmatrix} \begin{bmatrix} \cos\theta \\ 0 \end{bmatrix}.$$

(b) Mimic the proof of Theorem 11.1 showing that

$$\begin{bmatrix} \cos\theta & \sin\theta \\ G_e'(z) & G_o'(z) \end{bmatrix} = \begin{bmatrix} \cos\theta & \sin\theta \\ 0 & \frac{1}{\cos\theta} \end{bmatrix}.$$

(c) Use (b) and (11.45) from Corollary 11.1 to show that $S(z) = \frac{\sin\theta}{\cos\theta}$.

(d) Use (11.60) to complete the factorization.

11.42 Verify the matrix identities given in (11.58).

11.4 The Lifting Method – Examples

We conclude this chapter by illustrating the ideas of Section 11.3 with three different filter pairs. You will have a chance to consider other factorizations in the Problems section and also verify that factorizations are not unique.

Applying Theorem 11.1

We derived (11.37) and (11.38) for implementing the LeGall filter pair by "reading" them from the factorization (11.35) of the polyphase matrix. The lower triangular matrix in (11.35) *predicted* the differences portion $\widetilde{\mathbf{d}}$ while the upper triangular matrix *updated* the averages portion $\widetilde{\mathbf{a}}$. We can use the same ideas to "read" the lifting steps from the factorization (11.61) in Theorem 11.1.

We start by setting $\begin{bmatrix} A^{(0)}(z) \\ D^{(0)}(z) \end{bmatrix} = \begin{bmatrix} E(z) \\ O(z) \end{bmatrix}$. Our goal is to compute

$$\begin{bmatrix} A(z) \\ D(z) \end{bmatrix} = \mathbf{P}(z) \begin{bmatrix} A^{(0)}(z) \\ D^{(0)}(z) \end{bmatrix}.$$

We accomplish this task by successively applying the factors in the N-fold product to $\begin{bmatrix} A^{(0)}(z) \\ D^{(0)}(z) \end{bmatrix}$. The first product gives us the new predictor $D^{(1)}(z)$:

$$\begin{bmatrix} A^{(0)}(z) \\ D^{(1)}(z) \end{bmatrix} = \begin{bmatrix} 1 & 0 \\ -T_N(z) & 1 \end{bmatrix} \begin{bmatrix} A^{(0)}(z) \\ D^{(0)}(z) \end{bmatrix} = \begin{bmatrix} A^{(0)}(z) \\ D^{(0)}(z) - T_N(z)A^{(0)}(z) \end{bmatrix}.$$

The Z-transform of the second component above comes from the lifting step

$$\mathbf{d}^{(1)} = \mathbf{d}^{(0)} - \mathbf{t}_N * \mathbf{a}^{(0)}.$$

Next we multiply $\begin{bmatrix} A^{(0)}(z) \\ D^{(1)}(z) \end{bmatrix}$ by $\begin{bmatrix} 1 & S_N(z) \\ 0 & 1 \end{bmatrix}$ to obtain

$$\begin{bmatrix} A^{(1)}(z) \\ D^{(1)}(z) \end{bmatrix} = \begin{bmatrix} 1 & S_N(z) \\ 0 & 1 \end{bmatrix} \begin{bmatrix} A^{(0)}(z) \\ D^{(1)}(z) \end{bmatrix} = \begin{bmatrix} A^{(0)}(z) + S_N(z) D^{(1)}(z) \\ D^{(1)}(z) \end{bmatrix}.$$

The Z-transform of the first component above comes from the lifting step

$$\mathbf{a}^{(1)} = \mathbf{a}^{(0)} + \mathbf{s}_N * \mathbf{d}^{(1)}.$$

We continue this process for all the factors in the N-fold product generating

$$\begin{aligned} \mathbf{d}^{(k)} &= \mathbf{d}^{(k-1)} - \mathbf{t}_{N-k+1} * \mathbf{a}^{(k-1)} \\ \mathbf{a}^{(k)} &= \mathbf{a}^{(k-1)} + \mathbf{s}_{N-k+1} * \mathbf{d}^{(k)} \end{aligned}$$

for $k = 1, \ldots, N$. The final step is to "normalize" the last iterates. That is,

$$\mathbf{a} = K\mathbf{a}^{(N)} \qquad \text{and} \qquad \mathbf{d} = K^{-1}\mathbf{d}^{(N)}.$$

The First Example – The D4 Orthogonal Filter

Example 11.9 (The D4 Orthogonal Filter). *We construct a lifting method for the D4 filter* $\mathbf{h}$ *given by (5.27). In order to construct a polyphase matrix* $\mathbf{P}(z)$ *with* $\det(\mathbf{P}(z)) = 1$*, we use the ideas from Problem 11.35 and use the associated highpass filter* $\mathbf{g}$ *where the elements of* $\mathbf{g}$ *are given by* $g_{k-2} = (-1)^{k+1} h_{3-k}$*. We have*

$$\begin{aligned} H_e(z) &= \frac{1+\sqrt{3}}{4\sqrt{2}} + \frac{3-\sqrt{3}}{4\sqrt{2}} z^{-1} \\ H_o(z) &= \frac{3+\sqrt{3}}{4\sqrt{2}} + \frac{1-\sqrt{3}}{4\sqrt{2}} z^{-1} \\ G_e(z) &= -\frac{1-\sqrt{3}}{4\sqrt{2}} z - \frac{3+\sqrt{3}}{4\sqrt{2}} \\ G_o(z) &= \frac{3-\sqrt{3}}{4\sqrt{2}} z + \frac{1+\sqrt{3}}{4\sqrt{2}}. \end{aligned} \tag{11.62}$$

Factoring the Polyphase Matrix.

The first task is to use the Euclidean algorithm (Algorithm 11.2) with $a_0(z) = H_e(z)$ *and* $b_0(z) = H_o(z)$*. In order to find* $q_1(z)$*, we match up the coefficients of* z^{-1} *in* $a_0(z)$ *and* $b_0(z)$*. That is, we seek* $q_1(z)$ *such that*

$$q_1(z) = \frac{3-\sqrt{3}}{4\sqrt{2}} \div \frac{1-\sqrt{3}}{4\sqrt{2}} = \frac{3-\sqrt{3}}{1-\sqrt{3}}.$$

Multiplying the numerator and denominator above by $1+\sqrt{3}$ *gives* $q_1(z) = -\sqrt{3}$. *Thus* $a_0(z) = b_0(z)q_1(z) + r_1(z)$ *or*

$$\begin{aligned} r_1(z) &= a_0(z) - b_0(z)q_1(z) \\ &= \frac{1+\sqrt{3}}{4\sqrt{2}} + \frac{3-\sqrt{3}}{4\sqrt{2}}z^{-1} + \sqrt{3}\left(\frac{3+\sqrt{3}}{4\sqrt{2}} + \frac{1-\sqrt{3}}{4\sqrt{2}}z^{-1}\right) \\ &= \frac{1+\sqrt{3}}{\sqrt{2}}. \end{aligned}$$

Updating, we have

$$\begin{aligned} a_1(z) &= b_0(z) = \frac{3+\sqrt{3}}{4\sqrt{2}} + \frac{1-\sqrt{3}}{4\sqrt{2}}z^{-1} \\ b_1(z) &= r_1(z) = a_0(z) - b_0(z)q_1(z) = \frac{1+\sqrt{3}}{\sqrt{2}} \end{aligned}$$

or in matrix form

$$\begin{bmatrix} a_1(z) \\ b_1(z) \end{bmatrix} = \begin{bmatrix} 0 & 1 \\ 1 & -q_1(z) \end{bmatrix} \begin{bmatrix} a_0(z) \\ b_0(z) \end{bmatrix} = \begin{bmatrix} 0 & 1 \\ 1 & \sqrt{3} \end{bmatrix} \begin{bmatrix} a_0(z) \\ b_0(z) \end{bmatrix}. \tag{11.63}$$

Now $b_1(z)$ *is a monomial so we know that the next remainder will be* $r_2(z) = 0$. *Thus we seek* $q_2(z) = a_1(z)/b_1(z)$. *We have*

$$\begin{aligned} q_2(z) &= \frac{a_1(z)}{b_1(z)} \\ &= \frac{\sqrt{2}}{1+\sqrt{3}}\left(\frac{3+\sqrt{3}}{4\sqrt{2}} + \frac{1-\sqrt{3}}{4\sqrt{2}}z^{-1}\right) \\ &= \frac{1}{4}\left(\frac{3+\sqrt{3}}{1+\sqrt{3}} + \frac{1-\sqrt{3}}{1+\sqrt{3}}z^{-1}\right) \\ &= \frac{\sqrt{3}}{4} + \frac{\sqrt{3}-2}{4}z^{-1}. \end{aligned}$$

The final update is

$$\begin{aligned} a_2(z) &= b_1(z) = \frac{1+\sqrt{3}}{\sqrt{2}} \\ b_2(z) &= r_2(z) = a_1(z) - b_1(z)q_2(z) = 0 \end{aligned}$$

or in matrix form

$$\begin{bmatrix} a_2(z) \\ 0 \end{bmatrix} = \begin{bmatrix} 0 & 1 \\ 1 & -q_2(z) \end{bmatrix} \begin{bmatrix} a_1(z) \\ b_1(z) \end{bmatrix} = \begin{bmatrix} 0 & 1 \\ 1 & -\frac{\sqrt{3}}{4} - \frac{\sqrt{3}-2}{4}z^{-1} \end{bmatrix} \begin{bmatrix} a_1(z) \\ b_1(z) \end{bmatrix}. \tag{11.64}$$

Combining (11.63) *and* (11.64) *gives*

$$\begin{bmatrix} a_2(z) \\ 0 \end{bmatrix} = \begin{bmatrix} 0 & 1 \\ 1 & -\frac{\sqrt{3}}{4} - \frac{\sqrt{3}-2}{4} z^{-1} \end{bmatrix} \begin{bmatrix} 0 & 1 \\ 1 & \sqrt{3} \end{bmatrix} \begin{bmatrix} a_0(z) \\ b_0(z) \end{bmatrix}$$

$$\begin{bmatrix} \frac{1+\sqrt{3}}{\sqrt{2}} \\ 0 \end{bmatrix} = \begin{bmatrix} 0 & 1 \\ 1 & -\frac{\sqrt{3}}{4} - \frac{\sqrt{3}-2}{4} z^{-1} \end{bmatrix} \begin{bmatrix} 0 & 1 \\ 1 & \sqrt{3} \end{bmatrix} \begin{bmatrix} H_e(z) \\ H_o(z) \end{bmatrix}.$$

Using (11.53) *we can multiply by the inverses of the matrices on the right of the above identity to obtain*

$$\begin{bmatrix} H_e(z) \\ H_o(z) \end{bmatrix} = \begin{bmatrix} -\sqrt{3} & 1 \\ 1 & 0 \end{bmatrix} \begin{bmatrix} \frac{\sqrt{3}}{4} + \frac{\sqrt{3}-2}{4} z^{-1} & 1 \\ 1 & 0 \end{bmatrix} \begin{bmatrix} \frac{1+\sqrt{3}}{\sqrt{2}} \\ 0 \end{bmatrix}.$$

We next mimic the proof of Theorem 11.1 and add second columns to the vectors in above identity. Transposing the resulting equation gives

$$\begin{bmatrix} H_e(z) & H_o(z) \\ G_e'(z) & G_o'(z) \end{bmatrix} = \begin{bmatrix} \frac{1+\sqrt{3}}{\sqrt{2}} & 0 \\ 0 & \frac{\sqrt{2}}{1+\sqrt{3}} \end{bmatrix} \begin{bmatrix} \frac{\sqrt{3}}{4} + \frac{\sqrt{3}-2}{4} z^{-1} & 1 \\ 1 & 0 \end{bmatrix} \begin{bmatrix} -\sqrt{3} & 1 \\ 1 & 0 \end{bmatrix}. \tag{11.65}$$

We can expand (11.65) *to obtain formulas for* $G_e'(z)$ *and* $G_o'(z)$. *Using a* **CAS**, *and knowing that the top row, by construction, contains* $H_e(z)$ *and* $H_o(z)$, *we obtain*

$$\begin{bmatrix} H_e(z) & H_o(z) \\ G_e'(z) & G_o'(z) \end{bmatrix} = \begin{bmatrix} H_e(z) & H_o(z) \\ \frac{\sqrt{3}-3}{\sqrt{2}} & \frac{-1+\sqrt{3}}{\sqrt{2}} \end{bmatrix}$$

from which we see that

$$G_e'(z) = \frac{\sqrt{3}-3}{\sqrt{2}} \qquad \textit{and} \qquad G_o'(z) = \frac{-1+\sqrt{3}}{\sqrt{2}}.$$

Using the formulas given in (11.62) *for* $G_e(z)$ *and* $G_o(z)$, *the Laurent polynomials* $G_e'(z), G_o'(z)$ *above and Corollary 11.1, we compute the Laurent polynomial* $S(z)$ *that appears in the proof of Theorem 11.1. Using a* **CAS** *to do the computations we have*

$$\begin{aligned} S(z) &= G_e'(z)G_o(z) - G_o'(z)G_e(z) \\ &= \frac{\sqrt{3}-3}{\sqrt{2}} \left(\frac{3-\sqrt{3}}{4\sqrt{2}} z + \frac{1+\sqrt{3}}{4\sqrt{2}} \right) \\ &\qquad - \frac{-1+\sqrt{3}}{\sqrt{2}} \left(-\frac{1-\sqrt{3}}{4\sqrt{2}} z - \frac{3+\sqrt{3}}{4\sqrt{2}} \right) \\ &= (-2+\sqrt{3})z. \end{aligned}$$

Next we use (11.55) *to write*

$$\mathbf{P}(z) = \begin{bmatrix} 1 & 0 \\ (2-\sqrt{3})z & 1 \end{bmatrix} \begin{bmatrix} \frac{1+\sqrt{3}}{\sqrt{2}} & 0 \\ 0 & \frac{\sqrt{2}}{1+\sqrt{3}} \end{bmatrix} \begin{bmatrix} \frac{\sqrt{3}}{4} + \frac{\sqrt{3}-2}{4} z^{-1} & 1 \\ 1 & 0 \end{bmatrix} \begin{bmatrix} -\sqrt{3} & 1 \\ 1 & 0 \end{bmatrix}. \tag{11.66}$$

For our factorization, we need $-K^2S(z)$, where $K = \frac{1+\sqrt{3}}{\sqrt{2}}$. We have

$$-K^2S(z) = -\frac{z}{2}(1+\sqrt{3})^2(-2+\sqrt{3}) = -\frac{z}{2}(4+2\sqrt{3})(-2+\sqrt{3}) = z.$$

Combining this result in conjunction with Problem 11.32, the identity (11.66) *becomes*

$$\mathbf{P}(z) = \begin{bmatrix} \frac{1+\sqrt{3}}{\sqrt{2}} & 0 \\ 0 & \frac{\sqrt{2}}{1+\sqrt{3}} \end{bmatrix} \begin{bmatrix} 1 & 0 \\ z & 1 \end{bmatrix} \begin{bmatrix} \frac{\sqrt{3}}{4} + \frac{\sqrt{3}-2}{4}z^{-1} & 1 \\ 1 & 0 \end{bmatrix} \begin{bmatrix} -\sqrt{3} & 1 \\ 1 & 0 \end{bmatrix}.$$

The final step in the factorization is to use (11.58) *on the two right-most factors above and write*

$$\mathbf{P}(z) = \begin{bmatrix} \frac{1+\sqrt{3}}{\sqrt{2}} & 0 \\ 0 & \frac{\sqrt{2}}{1+\sqrt{3}} \end{bmatrix} \begin{bmatrix} 1 & 0 \\ z & 1 \end{bmatrix} \begin{bmatrix} 1 & \frac{\sqrt{3}}{4} + \frac{\sqrt{3}-2}{4}z^{-1} \\ 0 & 1 \end{bmatrix} \begin{bmatrix} 1 & 0 \\ -\sqrt{3} & 1 \end{bmatrix}. \tag{11.67}$$

Reading the Lifting Steps from the Polyphase Matrix.

Suppose we wish to apply the wavelet transform to $\mathbf{v} = (\ldots, v_{-1}, v_0, v_1, \ldots)$. We set

$$\begin{bmatrix} A^{(0)}(z) \\ D^{(0)}(z) \end{bmatrix} = \begin{bmatrix} E(z) \\ O(z) \end{bmatrix}$$

where

$$E(z) = \sum_n e_n z^{-n} = \sum_n v_{2n} z^{-n} \qquad \text{and} \qquad O(z) = \sum_n o_n z^{-n} = \sum_n v_{2n+1} z^{-n}$$

and compute

$$\begin{bmatrix} A^{(0)}(z) \\ D^{(1)}(z) \end{bmatrix} = \begin{bmatrix} 1 & 0 \\ -\sqrt{3} & 1 \end{bmatrix} \begin{bmatrix} A^{(0)}(z) \\ D^{(0)}(z) \end{bmatrix} = \begin{bmatrix} A^{(0)}(z) \\ D^{(0)}(z) - \sqrt{3}A^{(0)}(z) \end{bmatrix}.$$

The Z-transform in the second component above comes from the lifting step

$$\mathbf{d}^{(1)} = \mathbf{d}^{(0)} - \sqrt{3}\mathbf{a}^{(0)}.$$

Writing this identity element-wise gives

$$d_n^{(1)} = d_n^{(0)} - \sqrt{3}a_n^{(0)} = o_n - \sqrt{3}e_n = v_{2n+1} - \sqrt{3}v_{2n}.$$

We next compute

$$\begin{aligned} \begin{bmatrix} A^{(1)}(z) \\ D^{(1)}(z) \end{bmatrix} &= \begin{bmatrix} 1 & \frac{\sqrt{3}}{4} + \frac{\sqrt{3}-2}{4}z^{-1} \\ 0 & 1 \end{bmatrix} \begin{bmatrix} A^{(0)}(z) \\ D^{(1)}(z) \end{bmatrix} \\ &= \begin{bmatrix} A^{(0)}(z) + \left(\frac{\sqrt{3}}{4} + \frac{\sqrt{3}-2}{4}z^{-1}\right) D^{(1)}(z) \\ D^{(1)}(z) \end{bmatrix}. \end{aligned}$$

The Z-transform in the first component $A^{(1)}(z)$ above comes from the identity

$$\mathbf{a}^{(1)} = \mathbf{a}^{(0)} + \mathbf{s} * \mathbf{d}^{(1)}$$

where $S(z) = \frac{\sqrt{3}}{4} + \frac{\sqrt{3}-2}{4}z^{-1}$. In this case, the only nonzero elements of $\mathbf{s}$ are $s_0 = \frac{\sqrt{3}}{4}$ and $s_1 = \frac{\sqrt{3}-2}{4}$. The n-th element of the convolution product is

$$\left(\mathbf{s} * \mathbf{d}^{(1)}\right)_n = \sum_k s_k d^{(1)}_{n-k} = \frac{\sqrt{3}}{4} d^{(1)}_n + \frac{\sqrt{3}-2}{4} d^{(1)}_{n-1}$$

so that

$$a^{(1)}_n = a^{(0)}_n + \frac{\sqrt{3}}{4} d^{(1)}_n + \frac{\sqrt{3}-2}{4} d^{(1)}_{n-1} = v_{2n} + \frac{\sqrt{3}}{4} d^{(1)}_n + \frac{\sqrt{3}-2}{4} d^{(1)}_{n-1}.$$

Our next computation produces the predictor $\mathbf{d}^{(2)}$. We have

$$\begin{bmatrix} A^{(1)}(z) \\ D^{(2)}(z) \end{bmatrix} = \begin{bmatrix} 1 & 0 \\ z & 1 \end{bmatrix} \begin{bmatrix} A^{(1)}(z) \\ D^{(1)}(z) \end{bmatrix} = \begin{bmatrix} A^{(1)}(z) \\ D^{(1)}(z) + zA^{(1)}(z) \end{bmatrix}$$

so that

$$\mathbf{d}^{(2)} = \mathbf{d}^{(1)} + \mathbf{t} * \mathbf{a}^{(1)}.$$

Here $T(z) = z$ so that the only nonzero component of $\mathbf{t}$ is $t_{-1} = 1$. In this case the n-th element of the convolution product is $a^{(1)}_{n+1}$. We have

$$d^{(2)}_n = d^{(1)}_n + a^{(1)}_{n+1}.$$

The final step is to normalize $\mathbf{a}^{(1)}$ and $\mathbf{d}^{(2)}$. Element-wise we have $a_n = \frac{1+\sqrt{3}}{\sqrt{2}} a^{(1)}_n$ and $d_n = \frac{\sqrt{2}}{1+\sqrt{3}} d^{(2)}_n$.

To summarize, a lifting method for computing the wavelet transform using the D4 orthogonal filter is

$$d^{(1)}_n = v_{2n+1} - \sqrt{3} v_{2n}$$

$$a^{(1)}_n = v_{2n} + \frac{\sqrt{3}}{4} d^{(1)}_n + \frac{\sqrt{3}-2}{4} d^{(1)}_{n-1}$$

$$d^{(2)}_n = d^{(1)}_n + a^{(1)}_{n+1}$$

$$a_n = \frac{1+\sqrt{3}}{\sqrt{2}} a^{(1)}_n$$

$$d_n = \frac{\sqrt{2}}{1+\sqrt{3}} d^{(2)}_n.$$

(Live example: Visit `stthomas.edu/wavelets` and click on Live Examples.)

□

In Problem 11.45 you will learn that the factorization (11.67) is not unique and in Problem 11.46 you construct a lifting method for the inverse wavelet transform using the D4 orthogonal filter.

The Second Example – The (8, 4) Biorthogonal Spline Filter Pair

Example 11.10 (The (8,4) Biorthogonal Spline Filter Pair). *We develop a lifting method that utilizes the* $(8,4)$ *biorthogonal spline filter pair. We take as* $\mathbf{h}$ *the filter given in* (7.25) *and for the associated highpass filter* $\mathbf{g}$, *we use the negative values of the elements given in* (7.26). *Using the negative values ensures that the resulting polyphase matrix* $\mathbf{P}(z)$ *satisfies* $\det(\mathbf{P}(z)) = 1$. *We have*

$$\begin{aligned} H_e(z) &= -\frac{9\sqrt{2}}{64}z + \frac{45\sqrt{2}}{64} - \frac{7\sqrt{2}}{64}z^{-1} + \frac{3\sqrt{2}}{64}z^{-2} \\ H_o(z) &= \frac{3\sqrt{2}}{64}z^2 - \frac{7\sqrt{2}}{64}z + \frac{45\sqrt{2}}{64} - \frac{9\sqrt{2}}{64}z^{-1} \\ G_e(z) &= -\frac{3\sqrt{2}}{8} - \frac{\sqrt{2}}{8}z^{-1} \\ G_o(z) &= \frac{\sqrt{2}}{8}z + \frac{3\sqrt{2}}{8}. \end{aligned}$$

Factoring the Polyphase Matrix

We apply the Euclidean algorithm to the degree three Laurent polynomials $a_0(z) = H_e(z)$ *and* $b_0(z) = H_o(z)$. *We start by eliminating the* z^{-2} *term in* $a_0(z)$. *Towards this end, we take* $q_1(z) = -\frac{1}{3}z^{-1}$. *Then*

$$\begin{aligned} r_1(z) &= a_0(z) - b_0(z)q_1(z) \\ &= -\frac{9\sqrt{2}}{64}z + \frac{45\sqrt{2}}{64} - \frac{7\sqrt{2}}{64}z^{-1} + \frac{3\sqrt{2}}{64}z^{-2} \\ &\qquad - \left(-\frac{1}{3}z^{-1}\right)\left(\frac{3\sqrt{2}}{64}z^2 - \frac{7\sqrt{2}}{64}z + \frac{45\sqrt{2}}{64} - \frac{9\sqrt{2}}{64}z^{-1}\right) \\ &= -\frac{\sqrt{2}}{8}z + \frac{2\sqrt{2}}{3} + \frac{\sqrt{2}}{8}z^{-1}. \end{aligned}$$

We next update setting

$$\begin{aligned} a_1(z) = b_0(z) &= \frac{3\sqrt{2}}{64}z^2 - \frac{7\sqrt{2}}{64}z + \frac{45\sqrt{2}}{64} - \frac{9\sqrt{2}}{64}z^{-1} \\ b_1(z) = r_1(z) &= -\frac{\sqrt{2}}{8}z + \frac{2\sqrt{2}}{3} + \frac{\sqrt{2}}{8}z^{-1}. \end{aligned}$$

We select $q_2(z) = -\frac{3}{8}z$ *in order to eliminate quadratic term in* $a_1(z)$. *We have*

$$\begin{aligned} a_1(z) &= b_1(z)q_2(z) + r_2(z) \\ \frac{3\sqrt{2}}{64}z^2 - \frac{7\sqrt{2}}{64}z + \frac{45\sqrt{2}}{64} - \frac{9\sqrt{2}}{64}z^{-1} &= \left(-\frac{3}{8}z\right)\left(-\frac{\sqrt{2}}{8}z + \frac{2\sqrt{2}}{3} + \frac{\sqrt{2}}{8}z^{-1}\right) \\ &= \frac{3\sqrt{2}}{64}z^2 - \frac{\sqrt{2}}{4}z - \frac{3\sqrt{2}}{64} + r_2(z) \end{aligned}$$

so that

$$r_2(z) = \frac{9\sqrt{2}}{64}z + \frac{3\sqrt{2}}{4} - \frac{9\sqrt{2}}{64}z^{-1}.$$

The next update gives

$$\begin{aligned} a_2(z) = b_1(z) &= -\frac{\sqrt{2}}{8}z + \frac{2\sqrt{2}}{3} + \frac{\sqrt{2}}{8}z^{-1} \\ b_2(z) = r_2(z) &= \frac{9\sqrt{2}}{64}z + \frac{3\sqrt{2}}{4} - \frac{9\sqrt{2}}{64}z^{-1}. \end{aligned}$$

We choose $q_3(z) = -\frac{8}{9}$ so that the linear term in $a_2(z)$ is eliminated:

$$\begin{aligned} a_2(z) &= b_2(z)q_3(z) + r_3(z) \\ -\frac{\sqrt{2}}{8}z + \frac{2\sqrt{2}}{3} + \frac{\sqrt{2}}{8}z^{-1} &= -\frac{8}{9}\left(\frac{9\sqrt{2}}{64}z + \frac{3\sqrt{2}}{4} - \frac{9\sqrt{2}}{64}z^{-1}\right) + r_3(z) \\ &= -\frac{\sqrt{2}}{8}z - \frac{2\sqrt{2}}{3} + \frac{\sqrt{2}}{8}z^{-1} + r_3(z). \end{aligned}$$

Solving for $r_3(z)$, we have

$$r_3(z) = \frac{4\sqrt{2}}{3}.$$

Since $r_3(z)$ is a monomial, we know the next remainder is $r_4(z) = 0$. Updating again gives

$$\begin{aligned} a_3(z) = b_2(z) &= \frac{9\sqrt{2}}{64}z + \frac{3\sqrt{2}}{4} - \frac{9\sqrt{2}}{64}z^{-1} \\ b_3(z) = r_3(z) &= \frac{4\sqrt{2}}{3}. \end{aligned}$$

We take

$$q_4(z) = \frac{a_3(z)}{b_3(z)} = \frac{3\sqrt{2}}{8}\left(\frac{9\sqrt{2}}{64}z + \frac{3\sqrt{2}}{4} - \frac{9\sqrt{2}}{64}z^{-1}\right) = \frac{27}{256}z + \frac{9}{16} - \frac{27}{256}z^{-1}.$$

Using (11.54) *and a* ***CAS****, we see that*

$$\begin{aligned} \begin{bmatrix} H_e(z) & H_o(z) \\ G'_e(z) & G'_o(z) \end{bmatrix} &= \begin{bmatrix} K & 0 \\ 0 & K^{-1} \end{bmatrix} \begin{bmatrix} q_4(z) & 1 \\ 1 & 0 \end{bmatrix} \begin{bmatrix} q_3(z) & 1 \\ 1 & 0 \end{bmatrix} \begin{bmatrix} q_2(z) & 1 \\ 1 & 0 \end{bmatrix} \begin{bmatrix} q_1(z) & 1 \\ 1 & 0 \end{bmatrix} \\ &= \begin{bmatrix} \frac{4\sqrt{2}}{3} & 0 \\ 0 & \frac{3\sqrt{2}}{8} \end{bmatrix} \begin{bmatrix} \frac{27}{256}z + \frac{9}{16} - \frac{27}{256z} & 1 \\ 1 & 0 \end{bmatrix} \begin{bmatrix} -\frac{8}{9} & 1 \\ 1 & 0 \end{bmatrix} \begin{bmatrix} -\frac{3z}{8} & 1 \\ 1 & 0 \end{bmatrix} \begin{bmatrix} -\frac{1}{3z} & 1 \\ 1 & 0 \end{bmatrix} \\ &= \begin{bmatrix} H_e(z) & H_o(z) \\ -\frac{3\sqrt{2}}{8} - \frac{\sqrt{2}}{8}z^{-1} & \frac{\sqrt{2}}{8}z + \frac{3\sqrt{2}}{8} \end{bmatrix}. \end{aligned}$$

To put the factorization in the form of (11.61)*, we need* $S(z)$ *from Corollary 11.1. Using* (11.45) *and a* **CAS***, we have*

$$\begin{aligned} S(z) &= G_e'(z)G_o(z) - G_o'(z)G_e(z) \\ &= \left(-\frac{3\sqrt{2}}{8} - \frac{\sqrt{2}}{8}z^{-1}\right)\left(\frac{\sqrt{2}}{8}z + \frac{3\sqrt{2}}{8}\right) \\ &\qquad - \left(\frac{\sqrt{2}}{8}z + \frac{3\sqrt{2}}{8}\right)\left(-\frac{3\sqrt{2}}{8} - \frac{\sqrt{2}}{8}z^{-1}\right) \\ &= 0. \end{aligned}$$

Thus $-K^2S(z) = 0$ *so that* $G_e(z) = G_e'(z)$ *and* $G_o(z) = G_o'(z)$*. Furthermore the matrix* $\begin{bmatrix} 1 & 0 \\ -K^2S(z) & 1 \end{bmatrix}$ *is the identity and can be omitted from the factorization. The final factorization of the polyphase matrix is*

$$\begin{aligned} \mathbf{P}(z) &= \begin{bmatrix} \frac{4\sqrt{2}}{3} & 0 \\ 0 & \frac{3\sqrt{2}}{8} \end{bmatrix} \begin{bmatrix} 1 & \frac{27}{256}z + \frac{9}{16} - \frac{27}{256}z^{-1} \\ 0 & 1 \end{bmatrix} \\ &\qquad \cdot \begin{bmatrix} 1 & 0 \\ -\frac{8}{9} & 1 \end{bmatrix} \begin{bmatrix} 1 & -\frac{3}{8}z \\ 0 & 1 \end{bmatrix} \begin{bmatrix} 1 & 0 \\ -\frac{1}{3}z^{-1} & 1 \end{bmatrix}. \end{aligned} \tag{11.68}$$

Reading the Lifting Steps from the Polyphase Matrix.

Suppose we wish to apply the wavelet transform to $\mathbf{v} = (\ldots, v_{-1}, v_0, v_1, \ldots)$*. Mimicking the approach in Example 11.9, we set*

$$\begin{bmatrix} A^{(0)}(z) \\ D^{(0)}(z) \end{bmatrix} = \begin{bmatrix} E(z) \\ O(z) \end{bmatrix}$$

where

$$E(z) = \sum_n e_n z^{-n} = \sum_n v_{2n} z^{-n} \qquad \textit{and} \qquad O(z) = \sum_n o_n z^{-n} = \sum_n v_{2n+1} z^{-n}$$

and first compute

$$\begin{bmatrix} A^{(0)}(z) \\ D^{(1)}(z) \end{bmatrix} = \begin{bmatrix} 1 & 0 \\ -\frac{1}{3}z^{-1} & 1 \end{bmatrix} \begin{bmatrix} A^{(0)}(z) \\ D^{(0)}(z) \end{bmatrix} = \begin{bmatrix} A^{(0)}(z) \\ D^{(0)}(z) - \frac{1}{3}z^{-1}A^{(0)}(z) \end{bmatrix}.$$

Expanding the second component $D^{(1)}(z)$ gives us our first lifting step. We have

$$
\begin{aligned}
D^{(1)}(z) &= D^{(0)}(z) - \frac{1}{3}z^{-1}A^{(0)}(z) \\
\sum_n d_n^{(1)} z^{-n} &= \sum_n d_n^{(0)} z^{-n} - \frac{1}{3}\sum_n a_n^{(0)} z^{-n-1} \\
&= \sum_n d_n^{(0)} z^{-n} - \frac{1}{3}\sum_n a_{n-1}^{(0)} z^{-n} \\
&= \sum_n \left(d_n^{(0)} - \frac{1}{3}a_{n-1}^{(0)} \right) z^{-n} \\
&= \sum_n \left(v_{2n+1} - \frac{1}{3}v_{2n-2} \right) z^{-n}
\end{aligned}
$$

so that

$$d_n^{(1)} = v_{2n+1} - \frac{1}{3}v_{2n-2}.$$

We next update $A^{(1)}(z)$ using

$$\begin{bmatrix} A^{(1)}(z) \\ D^{(1)}(z) \end{bmatrix} = \begin{bmatrix} 1 & -\frac{3}{8}z \\ 0 & 1 \end{bmatrix} \begin{bmatrix} A^{(0)}(z) \\ D^{(1)}(z) \end{bmatrix} = \begin{bmatrix} A^{(0)}(z) - \frac{3}{8}zD^{(1)}(z) \\ D^{(1)}(z) \end{bmatrix}.$$

The third lifting step is obtained by expanding the first component of the vector on the right side of the above equation. We have

$$
\begin{aligned}
A^{(1)}(z) &= A^{(0)}(z) - \frac{3}{8}zD^{(1)}(z) \\
\sum_n a_n^{(1)} z^{-n} &= \sum_n a_n^{(0)} z^{-n} - \frac{3}{8}\sum_n d_n^{(1)} z^{-(n-1)} \\
&= \sum_n a_n^{(0)} z^{-n} - \frac{3}{8}\sum_n d_{n+1}^{(1)} z^{-n} \\
&= \sum_n \left(a_n^{(0)} - \frac{3}{8}d_{n+1}^{(1)} \right) z^{-n} \\
&= \sum_n \left(v_{2n} - \frac{3}{8}d_{n+1}^{(1)} \right) z^{-n}
\end{aligned}
$$

so that

$$a_n^{(1)} = v_{2n} - \frac{3}{8}d_{n+1}^{(1)}.$$

The lifting step comes from $D^{(2)}(z) = D^{(1)}(z) - \frac{8}{9}A^{(1)}(z)$ is easily verified to be

$$d_n^{(2)} = d_n^{(1)} - \frac{8}{9}a_n^{(1)}.$$

The matrix equation

$$\begin{bmatrix} A^{(2)}(z) \\ D^{(2)}(z) \end{bmatrix} = \begin{bmatrix} 1 & \frac{27}{256}z + \frac{9}{16} - \frac{27}{256}z^{-1} \\ 0 & 1 \end{bmatrix} \begin{bmatrix} A^{(1)}(z) \\ D^{(2)}(z) \end{bmatrix}$$
$$= \begin{bmatrix} A^{(1)}(z) + \left(\frac{27}{256}z + \frac{9}{16} - \frac{27}{256}z^{-1}\right) D^{(2)}(z) \\ D^{(2)}(z) \end{bmatrix}.$$

yields the Z*-transform equation for the fourth lifting step. We have*

$$A^{(2)}(z) = A^{(1)}(z) + \left(\frac{27}{256}z + \frac{9}{16} - \frac{27}{256}z^{-1}\right) D^{(2)}(z)$$
$$\mathbf{a}^{(2)} = \mathbf{a}^{(1)} + \mathbf{t} * \mathbf{d}^{(2)}$$

where $\mathbf{t}$ *is the bi-infinite sequence with nonzero elements* $t_{-1} = \frac{27}{256}$, $t_0 = \frac{9}{16}$ *and* $t_1 = -\frac{27}{256}$. *In Problem 11.47, you will find a formula for* $\left(\mathbf{d}^{(2)} * \mathbf{t}\right)_n$. *Using this result we see that the last lifting step is*

$$a_n^{(2)} = a_n^{(1)} + \frac{27}{256}d_{n+1}^{(2)} + \frac{9}{16}d_n^{(2)} - \frac{27}{256}d_{n-1}^{(2)}.$$

The final step is to normalize $\mathbf{a}^{(2)}$ *and* $\mathbf{d}^{(2)}$. *Element-wise we have* $a_n = \frac{4\sqrt{2}}{3}a_n^{(2)}$ *and* $d_n = \frac{3\sqrt{2}}{8}d_n^{(2)}$.

To summarize, a lifting method for computing the wavelet transform using the (8,4) biorthogonal spline filter pair is

$$\begin{aligned} d_n^{(1)} &= v_{2n+1} - \frac{1}{3}v_{2n-2} \\ a_n^{(1)} &= v_{2n} - \frac{3}{8}d_{n-1}^{(1)} \\ d_n^{(2)} &= d_n^{(1)} - \frac{8}{9}a_n^{(1)} \\ a_n^{(2)} &= a_n^{(1)} + \frac{27}{256}d_{n+1}^{(2)} + \frac{9}{16}d_n^{(2)} - \frac{27}{256}d_{n-1}^{(2)} \\ a_n &= \frac{4\sqrt{2}}{3}a_n^{(2)} \\ d_n &= \frac{3\sqrt{2}}{8}d_n^{(2)}. \end{aligned} \tag{11.69}$$

(Live example: Visit `stthomas.edu/wavelets` *and click on Live Examples.)*

□

In Problem 11.48, you construct a lifting scheme for computing the inverse wavelet transform associated with the $(8,4)$ biorthogonal spline filter pair.

The Third Example – The CDF97 Biorthogonal Filter Pair

The CDF97 biorthogonal filter pair is used in the JPEG2000 image compression standard to perform lossy compression. The filter $\widetilde{\mathbf{h}}$ whose elements are given in (9.89) is used to compute the averages portion of the transform and the associated highpass filter $\widetilde{\mathbf{g}}$ is constructed from $\mathbf{h}$ and is thus of length 9. The elements of $\widetilde{\mathbf{g}}$ are given in Table 9.3. In order to create a polyphase matrix $\widetilde{\mathbf{P}}(z)$ with $\det\left(\widetilde{\mathbf{P}}(z)\right) = 1$, we negate the elements of the values given in Table 9.3.

Example 11.11 (The CDF97 Biorthogonal Filter Pair). *The Laurent polynomials that are elements of $\widetilde{\mathbf{P}}(z)$ are*

$$\begin{aligned}
\widetilde{H}_e(z) &= -0.0406894z + 0.788486 - 0.0406894z^{-1} \\
\widetilde{H}_o(z) &= -0.0645389z^2 + 0.418092z + 0.418092 - 0.0645389z^{-1} \\
\widetilde{G}_e(z) &= 0.0238495z - 0.377403 - 0.377403z^{-1} + 0.0238495z^{-2} \\
\widetilde{G}_o(z) &= 0.0378285\left(z^{-2} + z^2\right) - 0.110624\left(z^{-1} + z\right) + 0.852699.
\end{aligned} \tag{11.70}$$

We construct a factorization of the polyphase matrix $\widetilde{\mathbf{P}}(z) = \begin{bmatrix} \widetilde{H}_e(z) & \widetilde{H}_o(z) \\ \widetilde{G}_e(z) & \widetilde{G}_o(z) \end{bmatrix}$ *and leave the exercise of creating the lifting steps from this factorization as Problem 11.56.*

We start the Euclidean algorithm by setting $a_0(z) = \widetilde{H}_e(z)$ and $b_0(z) = \widetilde{H}_o(z)$. Since $\deg(a_0(z)) = 2 < 3 = \deg(b_0(z))$ the first step is to reverse the positions of $a_0(z)$ and $b_0(z)$. This is accomplished by setting $q_1(z) = 0$ so that $r_1(z) = a_0(z)$. Updating, we have $a_1(z) = b_0(z)$ and $b_1(z) = r_1(z) = a_0(z)$. In matrix form, we have

$$\begin{bmatrix} a_1(z) \\ b_1(z) \end{bmatrix} = \begin{bmatrix} 0 & 1 \\ 1 & 0 \end{bmatrix} \begin{bmatrix} a_0(z) \\ b_0(z) \end{bmatrix}.$$

We seek to eliminate the quadratic and linear terms in $a_1(z)$. Towards this end, we set $q_2(z) = a + bz$ and compute

$$\begin{aligned}
(a + bz)b_1(z) &= -0.0406894az + 0.788486a - 0.0406894az^{-1} \\
&\qquad - 0.0406894bz^2 + 0.788486bz - 0.0406894b \\
&= -0.0406894bz^2 + (-0.0406894a + 0.788486b)\,z \\
&\qquad + (0.788486a - 0.0406894b) - 0.0406894az^{-1}.
\end{aligned}$$

We find a and b by solving the system

$$\begin{aligned}
-0.0645389 &= -0.0406894b \\
0.418092 &= -0.0406894a + 0.788486b.
\end{aligned}$$

*Using a **CAS**, we find then that* $q_2(z) = 20.4611 + 1.586z$ *which gives*

$$\begin{aligned} r_2(z) &= a_1(z) - q_2(z)b_1(z) \\ &= -0.0645389z^2 + 0.418092z + 0.418092 - 0.0645389z^{-1} \\ &\quad - (20.4611 + 1.58613z)\left(-0.0406894z + 0.788486 - 0.0406894z^{-1}\right) \\ &= -15.6507 + 0.768013z^{-1}. \end{aligned}$$

Updating again gives

$$\begin{aligned} a_2(z) &= b_1(z) = -0.0406894z + 0.788486 - 0.0406894z^{-1} \\ b_2(z) &= r_2(z) = -15.6507 + 0.768013z^{-1}. \end{aligned}$$

We again choose $q_3(z)$ *of the form* $a+bz$ *with the hopes of eliminating the* z *and* z^{-1} *terms in* $a_2(z)$*. The result ensures that the remainder* $r_3(z)$ *is constant. We compute*

$$\begin{aligned} (a+bz)b_2(z) &= -15.6507a + 0.768013az^{-1} - 15.6507bz + 0.768013b \\ &= -15.6507bz + (-15.6507a + 0.768013b) + 0.768013az^{-1} \end{aligned}$$

and solve

$$\begin{aligned} -0.0406894 &= -15.6507b \\ -0.0406894 &= 0.768013a \end{aligned}$$

to obtain $q_3(z) = -0.05298 + 0.0026z$*. The constant remainder is found to be*

$$r_3(z) = a_2(z) - q_3(z)b_2(z) = -0.0426861.$$

We update to find

$$\begin{aligned} a_3(z) = b_2(z) &= -15.6507 + 0.768013z^{-1} \\ b_3(z) = r_3(z) &= -0.0426861. \end{aligned}$$

For the final step, we set $q_4(z) = a_3(z)/b_3(z) = 366.646 - 17.9921z^{-1}$ *so that* $r_4(z) = 0$*. The final update is gives*

$$a_4(z) = b_3(z) = -0.0426861.$$

Gathering up $q_1(z), \ldots, q_4(z)$ *and* $a_4(z)$*, we use* (11.54) *and a **CAS** to write*

$$\begin{aligned} \begin{bmatrix} \widetilde{H}_e(z) & \widetilde{H}_o(z) \\ \widetilde{G}'_e(z) & \widetilde{G}'_o(z) \end{bmatrix} &= \begin{bmatrix} -0.0426861 & 0 \\ 0 & -23.4268 \end{bmatrix} \\ &\quad \cdot \begin{bmatrix} 366.646 - 17.9921z^{-1} & 1 \\ 1 & 0 \end{bmatrix} \begin{bmatrix} -0.05298 + 0.0026z & 1 \\ 1 & 0 \end{bmatrix} \\ &\quad \cdot \begin{bmatrix} 20.4611 + 1.586z & 1 \\ 1 & 0 \end{bmatrix} \begin{bmatrix} 0 & 1 \\ 1 & 0 \end{bmatrix} \\ &= \begin{bmatrix} \widetilde{H}_e(z) & \widetilde{H}_o(z) \\ 1.24116 - 0.0609z & 1.96864 + 0.722429z - 0.0966054z^2 \end{bmatrix}. \end{aligned} \tag{11.71}$$

We read

$$\widetilde{G}'_e(z) = 1.24116 - 0.0609z$$
$$\widetilde{G}'_o(z) = 1.96864 + 0.722429z - 0.0966054z^2$$

from (11.71) *and then use* (11.70) *and* (11.45) *from Corollary 11.1 to write*

$$S(z) = \widetilde{G}'_e(z)\widetilde{G}_o(z) - \widetilde{G}'_o(z)\widetilde{G}_e(z) = 2.08299 + 0.586134z^{-1}.$$

We compute

$$-K^2S(z) = -a_4(z)S(z) = -0.0038 - 0.001z^{-1}$$

and use (11.60) *to obtain the final factorization*

$$\widetilde{\mathbf{P}}(z) = \begin{bmatrix} -0.0426861 & 0 \\ 0 & -23.4268 \end{bmatrix} \begin{bmatrix} 1 & 0 \\ -0.0038 - 0.001z^{-1} & 1 \end{bmatrix} \cdot \begin{bmatrix} 1 & 366.646 - 17.9921z^{-1} \\ 0 & 1 \end{bmatrix} \begin{bmatrix} 1 & 0 \\ -0.05298 + 0.0026z & 1 \end{bmatrix} \cdot \begin{bmatrix} 1 & 20.4611 + 1.586z \\ 0 & 1 \end{bmatrix}. \tag{11.72}$$

Note that $\begin{bmatrix} 1 & 0 \\ q_1(z) & 1 \end{bmatrix}$ *is the identity and thus omitted from* (11.72).

(Live example: Visit `stthomas.edu/wavelets` *and click on Live Examples.)*

□

PROBLEMS

11.43 Let's consider the computational efficiency of computing the wavelet transform using the D4 filter via lifting compared to a more standard implementation. Suppose the input sequence is $\mathbf{v}$.

(a) Suppose we wish to compute a_n, d_n utilizing the matrix (5.29). Ignoring the wrapping rows and zero multiplies, we have

$$a_n = h_3v_n + h_2v_{n-1} + h_1v_{n-2} + h_0v_{n-3}$$
$$d_n = -h_0v_n + h_1v_{n-1} - h_2v_{n-2} - h_3v_{n-3}.$$

What is the FLOP count for computing a_n and d_n in this manner?

(b) What is the FLOP count for the lifting method given in (11.69)? Which method is more computationally efficient?

11.44 In this problem we look at the effects of choosing different divisors/remainders in the Euclidean algorithm for the Laurent polynomials given in (11.62) from Example 11.9.

(a) In Example 11.9, we started the Euclidean algorithm by selecting $q_1(z) = -\sqrt{3}$. This choice "eliminated" the z^{-1} term after the first step. Suppose instead we "eliminate" the constant term in $a_0(z)$ in our first step. Show that $q_1(z) = \frac{\sqrt{3}}{3}$ works in this case.

(b) Show that the updated steps are

$$a_1(z) = b_0(z) = \frac{3+\sqrt{3}}{4\sqrt{2}} + \frac{1-\sqrt{3}}{4\sqrt{2}} z^{-1}$$
$$b_1(z) = r_1(z) = \frac{2}{3+\sqrt{3}} z^{-1}.$$

(c) We can produce a remainder of $r_2(z) = 0$ in the next step by choosing $q_2(z) = c + dz$. Show that $c = -\frac{\sqrt{3}}{4}$ and $d = \frac{6+3\sqrt{3}}{4}$.

(d) Updating gives $a_2(z) = b_1(z) = r_1(z)$ and $b_2(z) = 0$. Note that $a_2(z)$ is not constant. In the proof of Theorem 11.1, we suggest that in such a case, we can multiply $H_e(z)$ and $H_o(z)$ by an appropriate power of z (for this example, the power is one) and then the Euclidean algorithm produces a nonzero constant polynomial $a_2(z)$. Show that

$$\widehat{H}_e(z) = zH_e(z) = \frac{1+\sqrt{3}}{4\sqrt{2}} z + \frac{3-\sqrt{3}}{4\sqrt{2}}$$
$$\widehat{H}_o(z) = zH_o(z) = \frac{3+\sqrt{3}}{4\sqrt{2}} z + \frac{1-\sqrt{3}}{4\sqrt{2}}.$$

(e) We see from (11.62) that $G_e(z) = -H_o\left(z^{-1}\right)$ and $G_o(z) = H_e\left(z^{-1}\right)$. Use these identities to show that

$$\widehat{G}_e(z) = -\widehat{H}_o\left(z^{-1}\right) = -z^{-1}H_o\left(z^{-1}\right) = -\frac{1-\sqrt{3}}{4\sqrt{2}} - \frac{3+\sqrt{3}}{4\sqrt{2}} z^{-1}$$
$$\widehat{G}_o(z) = \widehat{H}_e\left(z^{-1}\right) = -z^{-1}H_e\left(z^{-1}\right) = \frac{3-\sqrt{3}}{4\sqrt{2}} + \frac{1+\sqrt{3}}{4\sqrt{2}} z^{-1}.$$

(f) Show that

$$\det\left(\widehat{P}(z)\right) = \det\left(\begin{bmatrix} \widehat{H}_e(z) & \widehat{H}_o(z) \\ \widehat{G}_e(z) & \widehat{G}_o(z) \end{bmatrix}\right) = 1.$$

11.45 Consider $\widehat{P}(z)$ given in Problem 11.44(f). We use this polyphase matrix to construct a second lifting method for the wavelet transform that uses the D4 filter.

(a) Start the Euclidean algorithm using $a_0(z) = \widehat{H}_e(z)$ and $b_0(z) = \widehat{H}_o(z)$ and "eliminate" the z term. Show that $q_1(z) = \frac{\sqrt{3}}{3}$. This is the $q_1(z)$ we used in Problem 11.44(a). Is the resulting $r_1(z)$ the same in this case?

(b) Update to obtain $a_1(z) = b_0(z)$ and $b_1(z) = r_1(z)$. Choose $q_2(z) = c + dz$ and solve for c and d. Verify that $r_2(z) = 0$. Is $q_2(z)$ different from those given in Problem 11.44(c)?

(c) Construct the matrix $\begin{bmatrix} \widehat{H}_e(z) & \widehat{H}_o(z) \\ G'_e(z) & G'_o(z) \end{bmatrix}$ using (11.65) as a guide. Use a **CAS** and simplify this factorization to show that $G'_e(z) = \frac{1+\sqrt{3}}{\sqrt{2}}$ and $G'_o(z) = \frac{3+\sqrt{3}}{\sqrt{2}}$.

(d) Use (11.45), Problem 11.44(e) and (c) to show that $S(z) = (2+\sqrt{3})z^{-1}$ and $-K^2S(z) = -\frac{1}{3}z^{-1}$, where $K = a_2(z) = \frac{\sqrt{2}}{3+\sqrt{3}}$.

(e) Now that you have $q_1(z), q_2(z)$ and $-K^2S(z)$, write down a factorization of $\mathbf{P}(z)$ of the form given in (11.60).

(f) Read the lifting steps from your factorization in (e) using Example 11.9 as a guide.

11.46 In this problem, you develop a lifting method for computing the inverse wavelet transform using the D4 filter.

(a) Construct a matrix factorization for $\mathbf{P}^{-1}(z)$. The factorization of $\mathbf{P}(z)$ given in (11.60) in the proof of Theorem 11.1 is helpful.

(b) Use the factorization from (a) to develop a lifting method for the inverse wavelet transform. Start with $\begin{bmatrix} E^{(0)}(z) \\ O^{(0)}(z) \end{bmatrix} = \begin{bmatrix} A(z) \\ D(z) \end{bmatrix}$. The factorization should, in turn, produce formulas for $E^{(1)}(z), O^{(1)}(z), O^{(2)}(z), E^{(2)}(z)$ and $O^{(3)}(z)$. From these Z-transforms, you should be able to write down formulas for the elements $e_n^{(1)}, o_n^{(1)}, o_n^{(2)}, e_n^{(2)}$ and $o_n^{(3)}$. The last two formulas are e_n and o_n, respectively.

11.47 Suppose $\mathbf{d}^{(2)}$ and $\mathbf{t}$ are bi-infinite sequences with $t_{-1} = \frac{27}{256}$, $t_0 = \frac{9}{16}$ and $t_1 = -\frac{27}{256}$ the only nonzero elements of $\mathbf{t}$. Show that

$$\left(\mathbf{t} * \mathbf{d}^{(2)}\right)_n = \frac{27}{256}d_{n+1}^{(2)} + \frac{9}{16}d_n^{(2)} - \frac{27}{256}d_{n-1}^{(2)}.$$

11.48 Use the ideas of Problem 11.46 to devise a lifting method for the inverse biorthogonal wavelet transform using the $(8, 4)$ biorthogonal spline filter. That is, use (11.68) to construct a factorization of $\mathbf{P}^{-1}(z)$ and then use this factorization to read off the lifting steps.

11.49 In the process of factoring $\mathbf{P}(z)$ in Example 11.10, we took $q_2(z) = -\frac{3}{8}z$. What happens if we instead use $q_2(z) = -\frac{9}{8} - \frac{3}{8}z$? Complete the factorization and discuss why we do not use this factorization of $\mathbf{P}(z)$.

11.50 Consider the orthogonal Haar filter $\mathbf{h} = (h_0, h_1) = \left(\frac{\sqrt{2}}{2}, \frac{\sqrt{2}}{2}\right)$ and the associated highpass filter $\mathbf{g} = (g_0, g_1) = \left(-\frac{\sqrt{2}}{2}, \frac{\sqrt{2}}{2}\right)$.

(a) Show that the polyphase matrix associated with these filters is

$$\mathbf{P}(z) = \begin{bmatrix} \frac{\sqrt{2}}{2} & \frac{\sqrt{2}}{2} \\ -\frac{\sqrt{2}}{2} & \frac{\sqrt{2}}{2} \end{bmatrix}.$$

(b) Construct a factorization of $\mathbf{P}(z)$ of the form given in (11.61) in the proof of Theorem 11.1. *Hint:* Problem 11.41 should be extremely useful for this construction.

(c) Write down the lifting steps for computing a_n and d_n based on the factorization from (a). Compare your result to the lifting steps found in Problem 4.36.

11.51 In this problem, you develop a lifting method for the $(5,3)$ biorthogonal spline filter pair. Here

$$\mathbf{h} = (h_{-2}, h_{-1}, h_0, h_1, h_2) = \left(-\frac{\sqrt{2}}{8}, \frac{\sqrt{2}}{4}, \frac{3\sqrt{2}}{4}, \frac{\sqrt{2}}{4}, -\frac{\sqrt{2}}{8}\right)$$

$$\mathbf{g} = (g_0, g_1, g_2) = \left(-\frac{\sqrt{2}}{4}, \frac{\sqrt{2}}{2}, -\frac{\sqrt{2}}{4}\right).$$

and

$$\mathbf{P}(z) = \begin{bmatrix} H_e(z) & H_o(z) \\ G_e(z) & G_o(z) \end{bmatrix} = \begin{bmatrix} -\frac{\sqrt{2}}{8}z + \frac{3\sqrt{2}}{4} - \frac{\sqrt{2}}{8}z^{-1} & \frac{\sqrt{2}}{4} + \frac{\sqrt{2}}{4}z \\ -\frac{\sqrt{2}}{4} - \frac{\sqrt{2}}{4}z^{-1} & \frac{\sqrt{2}}{2} \end{bmatrix}.$$

(a) Factor $\mathbf{P}(z)$ mimicking the proof of Theorem 11.1. Use Laurent polynomials $q_1(z) = -\frac{1}{2} - \frac{1}{2}z^{-1}$ and $q_2(z) = \frac{1}{4} + \frac{1}{4}z$.

(b) Read the lifting steps from the factorization in (a). How do your lifting steps compare to (11.9) and (11.10)?

11.52 Consider the D6 orthogonal filter elements given by (5.45) in Definition 5.2 and the associated highpass filter elements are given by (5.46). We make some minor modifications to these elements to create a filter pair $(\mathbf{h}, \mathbf{g})$. We shift the indices of the elements in (5.45) two units left so that the indices run from -2 to 3 instead of 0 to 5. We make the same shift for the elements in (5.46) and also negate them. Thus the filters we consider are $\mathbf{h}$ and $\mathbf{g}$ with

$$h_{-2} = \tfrac{\sqrt{2}}{32}\left(1 + \sqrt{10} + \sqrt{5+2\sqrt{10}}\right) \approx 0.332671$$

$$h_{-1} = \tfrac{\sqrt{2}}{32}\left(5 + \sqrt{10} + 3\sqrt{5+2\sqrt{10}}\right) \approx 0.806892$$

$$h_{0} = \tfrac{\sqrt{2}}{32}\left(10 - 2\sqrt{10} + 2\sqrt{5+2\sqrt{10}}\right) \approx 0.459878$$

$$h_{1} = \tfrac{\sqrt{2}}{32}\left(10 - 2\sqrt{10} - 2\sqrt{5+2\sqrt{10}}\right) \approx -0.135011$$

$$h_{2} = \tfrac{\sqrt{2}}{32}\left(5 + \sqrt{10} - 3\sqrt{5+2\sqrt{10}}\right) \approx -0.085441$$

$$h_{3} = \tfrac{\sqrt{2}}{32}\left(1 + \sqrt{10} - \sqrt{5+2\sqrt{10}}\right) \approx 0.035226.$$

and

$$\mathbf{g} = (g_{-2}, g_{-1}, g_0, g_1, g_2, g_3) = (-h_3, h_2, -h_1, h_0, -h_{-1}, h_{-2}).$$

so that

$$\begin{aligned} H_e(z) &= h_{-2}z + h_0 + h_2 z^{-1} \\ H_o(z) &= h_{-1}z + h_1 + h_3 z^{-1} \\ G_e(z) &= -h_3 z - h_1 - h_{-1}z^{-1} \\ G_o(z) &= h_2 z + h_0 + h_{-2}z^{-1}. \end{aligned}$$

Our task is to factor the polyphase matrix $\mathbf{P}(z) = \begin{bmatrix} H_e(z) & H_o(z) \\ G_e(z) & G_o(z) \end{bmatrix}$ into a form given by (11.60).

(a) Start the Euclidean algorithm by setting $a_0(z) = H_e(z)$, $b_0(z) = H_o(z)$ and $q_1(z) = h_{-2}/h_{-1}$. Show that the resulting $r_1(z) = 0.515540735853 - 0.0999646018199z^{-1}$.

(b) Update again setting $a_1(z) = b_0(z)$ and $b_1(z) = r_1(z)$. Choose $q_2(z)$ constant so as to "eliminate" the z^{-1} term in $a_1(z)$. Show that $r_2(z) = 0.04665917233295 + 0.8068915093111z$.

(c) Set $a_2(z) = b_1(z)$ and $b_2(z) = r_2(z)$. For the third step, select $q_3(z) = c + dz^{-1}$, picking c and d so that the z and z^{-1} term are eliminated in $a_2(z)$. Show that $q_2(z) = 2.14244267143z^{-1}$ and $r_2(z) = 2.24425953662$.

(d) Since $r_2(z)$ is constant, the next step produces $r_4(z) = 0$. Set $a_3(z) = b_2(z)$, $b_3(z) = r_2(z)$ and $q_4(z) = a_3(z)/b_3(z) = 0.0207904529631 + 0.359535738245z$. The final update then is $a_4(z) = b_3(z) = r_2(z) = 2.24425953662$ and $b_4(z) = r_4(z) = 0$.

(e) Form the matrix $\mathbf{P}'(z) = \begin{bmatrix} H_e(z) & H_o(z) \\ G'_e(z) & G'_o(z) \end{bmatrix}$ using (11.54) from the proof of Theorem 11.1. Here, $L = 4$.

(f) Using a **CAS**, simplify $\mathbf{P}'(z)$ to show that

$$G'_e(z) = 0.183707181957 - 0.81593883624z^{-1}$$
$$G'_o(z) = 0.445581263523 + 0.336400644565z^{-1}.$$

(g) Use (11.45) from Corollary 11.1 and a **CAS** to show that $S(z) = 0.1058894199481$.

(h) Use your formulas for $q_1(z), \ldots, q_4(z)$ and $S(z)$ to produce a factored form of $\mathbf{P}(z)$ in a form given by (11.60).

11.53 Use Problem 11.52(h) to read the lifting steps for the wavelet transform from the factors in the factorization.

11.54 Use Problem 11.52(h) to find a factorization of $\mathbf{P}^{-1}(z)$ and read the lifting steps for the inverse wavelet transform from this factorization.

11.55 Suppose the lifting method below is applied to $\mathbf{v} = (\ldots, v_{-1}, v_0, v_1, \ldots)$.

$$\begin{aligned} d_n^{(1)} &= v_{2n+1} - v_{2n-2} \\ a_n^{(1)} &= v_{2n} + \frac{9}{8}d_n^{(1)} - \frac{3}{8}d_{n+1}^{(1)} \\ d_n^{(2)} &= d_n^{(1)} - \frac{4}{9}a_n^{(1)} \\ a_n &= \frac{2\sqrt{2}}{3}a_n^{(1)} \\ d_n &= \frac{3}{2\sqrt{2}}d_n^{(2)}. \end{aligned}$$

(a) Use the lifting steps to write down a factorization of the polyphase matrix $\mathbf{P}(z)$.

(b) Use (a) to find $H_e(z)$, $H_o(z)$, $G_e(z)$ and $G_o(z)$.

(c) Can you identify the biorthogonal filter pair that generated $\mathbf{P}(z)$? The elements of the highpass filter $\mathbf{g}$ were built using the identity $g_k = (-1)^{k+1}\widetilde{h}_{1-k}$.

★**11.56** Suppose we wish to compute the wavelet transform of $\mathbf{v} = (\ldots, v_{-1}, v_0, v_1, \ldots)$ using the CDF97 filters described prior to Example 11.11. Use (11.72) to show that lifting steps are

$$a_n^{(1)} = v_{2n} + 20.461v_{2n+1} + 1.586v_{2n-1}$$

$$d_n^{(1)} = v_{2n+1} - 0.05298a_n^{(1)} + 0.0026a_{n-1}^{(1)}$$

$$a_n^{(2)} = a_n^{(1)} + 366.646d_n^{(1)} - 17.9921d_{n+1}^{(1)}$$

$$d_n^{(2)} = d_n^{(1)} - 0.0038a_n^{(2)} - 0.001a_{n+1}^{(2)}$$

$$a_n = -0.0426861a_n^{(2)}$$

$$d_n = -23.4268d_n^{(2)}.$$

11.57 In their paper [31], Daubechies and Sweldens gave a factorization for the dual polyphase matrix $\mathbf{P}(z)$ of $\widetilde{\mathbf{P}}(z)$ given in Example 11.11. They showed that

$$\mathbf{P}(z) = \begin{bmatrix} K & 0 \\ 0 & K^{-1} \end{bmatrix} \begin{bmatrix} 1 & 0 \\ -\delta\left(1+z^{-1}\right) & 1 \end{bmatrix} \begin{bmatrix} 1 & -\gamma(1+z) \\ 0 & 1 \end{bmatrix} \cdot \begin{bmatrix} 1 & 0 \\ -\beta\left(1+z^{-1}\right) & 1 \end{bmatrix} \begin{bmatrix} 1 & -\alpha(1+z) \\ 0 & 1 \end{bmatrix} \tag{11.73}$$

where

$$\alpha = -1.58613, \quad \beta = -0.05298, \quad \gamma = 0.88291, \quad \delta = 0.44351, \quad K = 1.14960.$$

(a) Use (11.49) from Proposition 11.8 to show that $\widetilde{\mathbf{P}}^{-1}(z) = \mathbf{P}^T\left(z^{-1}\right)$.

(b) Use the result from (a) and (11.73) to find a factorization for $\widetilde{\mathbf{P}}^{-1}(z)$.

(c) Use (b) to write down a lifting method for computing the inverse wavelet transform from Example 11.11.

★**11.58** This is a continuation of Problem 11.57.

(a) Use (b) from Problem 11.57 to write a factorization of $\widetilde{\mathbf{P}}(z)$.

(b) Write a lifting method for the filters from Example 11.11 using (a). Note that this lifting method is different than the one given in Example 11.11.

CHAPTER 12

THE JPEG2000 IMAGE COMPRESSION STANDARD

In 1992, JPEG became an international standard for compressing digital still images. JPEG is the acronym for the Joint Photographic Experts Group formed in the 1980s by members of the International Organization for Standardization (ISO) and the International Telecommunication Union (ITU). The JPEG compression standard remains very popular. According to a survey posted on `www.w3techs.com` [90] in March 2017, 73.9% of all websites use JPEG images[1] are stored using it. Despite the popularity of the JPEG standard, members of JPEG decided the algorithm could be improved in several areas and should be enhanced so as to meet the growing needs of applications using digital images. In 1997, work began on the development of the improved JPEG2000 compression standard. The core system of the JPEG2000 compression method is now a published standard [60] by the ISO (ISO 15444). Unfortunately, JPEG2000 is not supported by web browsers such as Mozilla Firefox or Google Chrome although some software programs, such as Adobe Photoshop allow users to save images in the JPEG2000 format. While JPEG2000 is a marked improvement over JPEG, it has not caught on as a popular way to perform image

[1]Digital images using the JPEG standard typically have the suffix `.jpg` attached to their file name.

Discrete Wavelet Transformations: An Elementary Approach With Applications, Second Edition.
Patrick J. Van Fleet.

compression. Nonetheless, it is an important application of the biorthogonal wavelet transformation and thus covered in this chapter.

One of the biggest changes in the JPEG2000 standard is that the biorthogonal wavelet transformation replaces the discrete cosine transformation as the preferred method for organizing data in a way that can best be utilized by encoders. The JPEG2000 standard also offers lossless and lossy compression. For lossy compression, JPEG2000 uses the CDF97 filter pair (Section 9.5), and for lossless compression, JPEG2000 uses the LeGall filter pair from Section 11.1. Both forms of compression also utilize the symmetric biorthogonal wavelet transformation developed in Section 7.3 to handle boundary issues in a more effective manner.

We begin this chapter with a brief overview of the JPEG standard. In Section 12.2, we describe the basic JPEG2000 standard. Both standards involve many more options than we present here. Our aim is to show the reader how the biorthogonal wavelet transformation is used in the JPEG2000 standard. Toward this end, we do not utilize any of the sophisticated encoders used by either standard. Instead, we use the basic Huffman coding method described in Section 3.3. In the final section of the chapter, examples are given to illustrate the implementation and effectiveness of the JPEG2000 standard.

12.1 An Overview of JPEG

Although it is beyond the scope of this book to present the JPEG standard in its complete form, an overview of the basic algorithm will help us understand the need for improvements and enhancements included in the JPEG2000 standard. For the sake of simplicity, let's assume that the grayscale image we wish to compress is of size $N \times M$ where both N and M are divisible by 8.

The first step in the JPEG method is to substract 128 from each of the pixel values in the image. This step has the effect of centering the intensity values about zero. We have seen this centering step in Section 4.4. The next step is to partition the image into nonoverlapping 8×8 blocks. The *discrete cosine transformation* (DCT) is then applied to each 8×8 matrix. The purpose of this step is similar to that of applying a wavelet transformation – the DCT will concentrate the energy of each block into just a few elements.

If A is an 8×8 matrix, then the DCT is computed by $C = UAU^T$ where $U =$

$$\frac{1}{2}\begin{bmatrix} \frac{\sqrt{2}}{2} & \frac{\sqrt{2}}{2} & \frac{\sqrt{2}}{2} & \frac{\sqrt{2}}{2} & \frac{\sqrt{2}}{2} & \frac{\sqrt{2}}{2} & \frac{\sqrt{2}}{2} & \frac{\sqrt{2}}{2} \\ \cos(\frac{\pi}{16}) & \cos(\frac{3\pi}{16}) & \cos(\frac{5\pi}{16}) & \cos(\frac{7\pi}{16}) & \cos(\frac{9\pi}{16}) & \cos(\frac{11\pi}{16}) & \cos(\frac{13\pi}{16}) & \cos(\frac{15\pi}{16}) \\ \cos(\frac{\pi}{8}) & \cos(\frac{3\pi}{8}) & \cos(\frac{5\pi}{8}) & \cos(\frac{7\pi}{8}) & \cos(\frac{9\pi}{8}) & \cos(\frac{11\pi}{8}) & \cos(\frac{13\pi}{8}) & \cos(\frac{15\pi}{8}) \\ \cos(\frac{3\pi}{16}) & \cos(\frac{9\pi}{16}) & \cos(\frac{15\pi}{16}) & \cos(\frac{21\pi}{16}) & \cos(\frac{27\pi}{16}) & \cos(\frac{33\pi}{16}) & \cos(\frac{39\pi}{16}) & \cos(\frac{45\pi}{16}) \\ \cos(\frac{\pi}{4}) & \cos(\frac{3\pi}{4}) & \cos(\frac{5\pi}{4}) & \cos(\frac{7\pi}{4}) & \cos(\frac{9\pi}{4}) & \cos(\frac{11\pi}{4}) & \cos(\frac{13\pi}{4}) & \cos(\frac{15\pi}{4}) \\ \cos(\frac{5\pi}{16}) & \cos(\frac{15\pi}{16}) & \cos(\frac{25\pi}{16}) & \cos(\frac{35\pi}{16}) & \cos(\frac{45\pi}{16}) & \cos(\frac{55\pi}{16}) & \cos(\frac{65\pi}{16}) & \cos(\frac{75\pi}{16}) \\ \cos(\frac{3\pi}{8}) & \cos(\frac{9\pi}{8}) & \cos(\frac{15\pi}{8}) & \cos(\frac{21\pi}{8}) & \cos(\frac{27\pi}{8}) & \cos(\frac{33\pi}{8}) & \cos(\frac{39\pi}{8}) & \cos(\frac{45\pi}{8}) \\ \cos(\frac{7\pi}{16}) & \cos(\frac{21\pi}{16}) & \cos(\frac{35\pi}{16}) & \cos(\frac{49\pi}{16}) & \cos(\frac{63\pi}{16}) & \cos(\frac{77\pi}{16}) & \cos(\frac{91\pi}{16}) & \cos(\frac{105\pi}{16}) \end{bmatrix}. \quad (12.1)$$

Element-wise, we have

$$u_{jk} = \begin{cases} \frac{\sqrt{2}}{4}, & j = 1 \\ \frac{1}{2}\cos((j-1)(2k-1)\pi/16), & j = 2, \ldots, 8. \end{cases}$$

It can be shown that U is an orthogonal matrix (a nice proof is given by Strang in [85]). Moreover, fast algorithms exist for computing the DCT (see e.g. Rao and Yip [76] or Wickerhauser [102]).

The elements in each 8×8 block of the DCT are normalized and the results are rounded to the nearest integer. As stated by Gonzalez and Woods [48], the normalization matrix is

$$Z = \begin{bmatrix} 16 & 11 & 10 & 16 & 24 & 40 & 51 & 61 \\ 12 & 12 & 14 & 19 & 26 & 58 & 60 & 55 \\ 14 & 13 & 16 & 24 & 40 & 57 & 69 & 56 \\ 14 & 17 & 22 & 29 & 51 & 87 & 80 & 62 \\ 18 & 22 & 37 & 56 & 68 & 109 & 103 & 77 \\ 24 & 35 & 55 & 64 & 81 & 104 & 113 & 92 \\ 49 & 64 & 78 & 87 & 103 & 121 & 120 & 101 \\ 72 & 92 & 95 & 98 & 112 & 100 & 103 & 99 \end{bmatrix}. \tag{12.2}$$

Thus, the quantization step renders an 8×8 matrix $\widehat{C}$ whose elements are given by

$$\widehat{c}_{jk} = \text{Round}\left(\frac{c_{jk}}{z_{jk}}\right).$$

The elements of $\widehat{C}$ are next loaded into the vector

$$\mathbf{d} = [d_0, \ldots, d_{63}]^T$$

using the following *zigzag* array

0	1	5	6	14	15	27	28
2	4	7	13	16	26	29	42
3	8	12	17	25	30	41	43
9	11	18	24	31	40	44	53
10	19	23	32	39	45	52	54
20	22	33	38	46	51	55	60
21	34	37	47	50	56	59	61
35	36	48	49	57	58	62	63

(12.3)

That is,

$$[d_0, d_1, d_2, d_3, d_4, \ldots, d_{61}, d_{62}, d_{63}]^T = [\widehat{c}_{11}, \widehat{c}_{12}, \widehat{c}_{21}, \widehat{c}_{31}, \ldots, \widehat{c}_{78}, \widehat{c}_{87}, \widehat{c}_{88}]^T. \tag{12.4}$$

Vectors for each 8×8 block are concatenated left to right, top to bottom to form a vector of length NM. Huffman coding[2] is applied to the length NM vector to complete the process.

[2]The JPG standard provides default *AC and DC Huffman tables*, but we can also use the Huffman coding scheme developed in Section 3.3.

The JPEG compression standard can also handle color digital images. The red, green, and blue channels of a color image are first converted to YCbCr space, and the process described above is then applied to each of the Y, Cb, and Cr channels. Let's look at an example.

Example 12.1 (The JPEG Compression Method). *Let's consider the* 256×256 *grayscale image plotted in Figure 12.1(a). Figure 12.1(b) shows the image partitioned into* 8×8 *blocks, resulting in a* 32×32 *block matrix representation of the original image.*

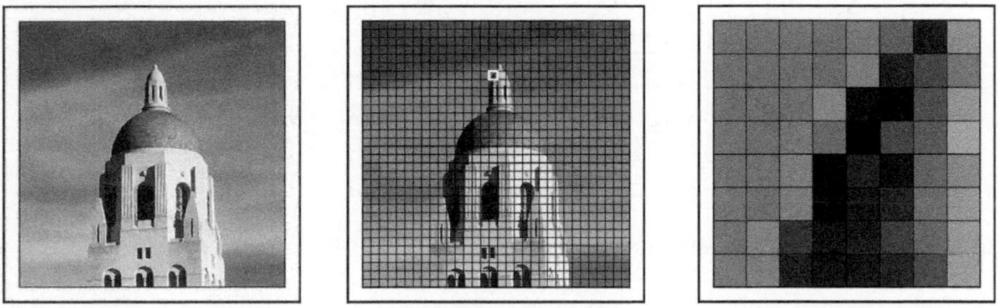

(a) Original image (b) Partitioned image (c) $(7, 16)$ block

Figure 12.1 An image, a partition of the image into a 32×32 block matrix where each block is 8×8, and the $(7, 16)$ block. In the middle image, we have highlighted the $(7, 16)$ block. We use this block to illustrate the JPEG algorithm throughout this example.

It is helpful to follow a single block through the JPEG algorithm. We consider the $(7, 16)$ *block*

$$A_{7,16} = \begin{bmatrix} 125 & 124 & 124 & 122 & 132 & 107 & 32 & 150 \\ 125 & 126 & 126 & 123 & 141 & 43 & 85 & 127 \\ 126 & 127 & 126 & 150 & 26 & 15 & 95 & 161 \\ 128 & 127 & 135 & 121 & 5 & 87 & 107 & 173 \\ 127 & 130 & 140 & 14 & 50 & 27 & 117 & 149 \\ 129 & 126 & 138 & 15 & 34 & 51 & 125 & 161 \\ 129 & 147 & 64 & 35 & 40 & 79 & 117 & 163 \\ 132 & 139 & 39 & 33 & 34 & 48 & 89 & 164 \end{bmatrix}.$$

that is highlighted in Figure 12.1(c).

The first step in the JPEG algorithm is to subtract 128 *from each element of* A*. In the case of the* $(7,16)$ *block, we obtain*

$$\widetilde{A}_{7,16} = \begin{bmatrix} -3 & -4 & -4 & -6 & 4 & -21 & -96 & 22 \\ -3 & -2 & -2 & -5 & 13 & -85 & -43 & -1 \\ -2 & -1 & -2 & 22 & -102 & -113 & -33 & 33 \\ 0 & -1 & 7 & -7 & -123 & -41 & -21 & 45 \\ -1 & 2 & 12 & -114 & -78 & -101 & -11 & 21 \\ 1 & -2 & 10 & -113 & -94 & -77 & -3 & 33 \\ 1 & 19 & -64 & -93 & -88 & -49 & -11 & 35 \\ 4 & 11 & -89 & -95 & -94 & -80 & -39 & 36 \end{bmatrix}.$$

We next compute the DCT of each block. The DCT of $\widetilde{A}_{7,16}$*, rounded to three decimal places, is*

$$T_{7,16} = U\widetilde{A}_{7,16}U^T$$

$$= \begin{bmatrix} -210.75 & 49.581 & 226.394 & -133.238 & 22.75 & -33.47 & 4.993 & 2.528 \\ 68.261 & 40.34 & -136.257 & -36.501 & 63.46 & 10.74 & 19.395 & -7.803 \\ -2.626 & -3.902 & -43.793 & 64.104 & 54.415 & 62.305 & -13.711 & -22.076 \\ 9.957 & -5.671 & -9.306 & 31.454 & -31.522 & -44.504 & 35.393 & 60.398 \\ -5.5 & -3.18 & 3.054 & 12.559 & -1.[illegible] & -19.620 & [illegible] & [illegible] \\ 22.804 & -16.797 & 3.099 & -1.379 & -7.737 & 14.748 & 27.086 & -72.661 \\ -13.525 & 7.907 & 16.539 & -32.161 & 3.979 & 15.279 & 14.543 & -9.79 \\ -19.232 & 8.822 & 11.488 & -5.251 & 7.698 & -8.103 & -18.037 & 10.458 \end{bmatrix}$$

where U *is given by* (12.1). *We next normalize the elements of each block* T_{jk}*,* $j,k = 1,\ldots,32$ *using* (12.2). *That is, if* t_{mn}*,* $m,n = 1,\ldots,8$*, are the elements of block* T_{jk}*, we compute* $\widetilde{t}_{mn} = t_{mn}/z_{mn}$*. We will call the resulting block* $\widetilde{T}_{jk}$*. In the case of the* $(7,16)$ *block, we have, rounded to three decimal places,*

$$\widetilde{T}_{7,16} = \begin{bmatrix} -13.172 & 4.507 & 22.639 & -8.327 & 0.948 & -0.837 & 0.098 & 0.041 \\ 5.688 & 3.362 & -9.733 & -1.921 & 2.441 & 0.185 & 0.323 & -0.142 \\ -0.188 & -0.3 & -2.737 & 2.671 & 1.36 & -1.093 & -0.199 & -0.394 \\ 0.711 & -0.334 & -0.423 & 1.085 & -0.618 & -0.512 & 0.442 & 0.974 \\ -0.306 & -0.145 & 0.083 & 0.224 & -0.015 & -0.18 & 0.287 & -0.256 \\ 0.95 & -0.48 & 0.056 & -0.022 & -0.096 & 0.142 & 0.24 & -0.79 \\ -0.276 & 0.124 & 0.212 & -0.37 & 0.039 & 0.126 & 0.121 & -0.097 \\ -0.267 & 0.096 & 0.121 & -0.054 & 0.069 & -0.074 & -0.175 & 0.106 \end{bmatrix}.$$

We quantize the elements of each block by rounding them to the nearest integer. We will call the resulting blocks $\widehat{C}_{jk}$*,* $j,k = 1,\ldots,32$*. In the case of the* $(7,16)$ *block,* *we have*

$$\widehat{C}_{7,16} = \begin{bmatrix} -13 & 5 & 23 & -8 & 1 & -1 & 0 & 0 \\ 6 & 3 & -10 & -2 & 2 & 0 & 0 & 0 \\ 0 & 0 & -3 & 3 & 1 & -1 & 0 & 0 \\ 1 & 0 & 0 & 1 & -1 & -1 & 0 & 1 \\ 0 & 0 & 0 & 0 & 0 & 0 & 0 & 0 \\ 1 & 0 & 0 & 0 & 0 & 0 & 0 & -1 \\ 0 & 0 & 0 & 0 & 0 & 0 & 0 & 0 \\ 0 & 0 & 0 & 0 & 0 & 0 & 0 & 0 \end{bmatrix}.$$

In $\widehat{C}_{7,16}$, we can see the effects of the DCT. This transformation "pushes" the large elements of the block toward the upper left-hand corner. As a result, the energy of each block is contained in only a few of the elements. We have plotted the cumulative energies of $A_{7,16}$ and $\widehat{C}_{7,16}$ in Figure 12.2. In Figure 12.3 we have plotted the quantized DCT for the image A.

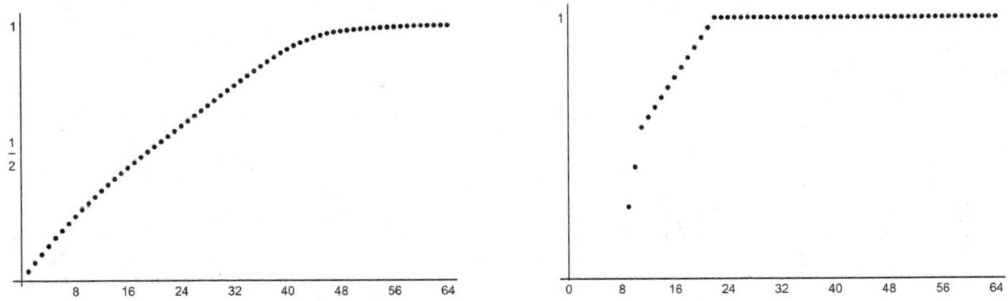

(a) Cumulative energy for $A_{7,16}$

(b) Cumulative energy for $\widehat{C}_{7,16}$

Figure 12.2 The cumulative energy vectors for the block $A_{7,16}$ and the quantized DCT block $\widehat{C}_{7,16}$.

We now load the elements of $\widehat{C}_{jk}$ into vector $\mathbf{d}_{jk}$ using (12.3). *In the case of $\widehat{C}_{7,16}$, we have*

$$\mathbf{d}_{7,16} = [-13, 5, 6, 0, 3, 23, -8, -10, 0, 1, 0, 0, -3, -2, -1, \ldots, 0, 0, -1, 0, 0, 0]\,.$$

The last step is to perform the Huffman coding. We create the vector $\mathbf{v}$ of length $256^2 = 65,536$ by chaining the $\mathbf{d}_{jk}$ together left to right, top to bottom. Huffman coding is then performed on $\mathbf{v}$. Using Algorithm 3.1 from Section 3.3, we find that the image, originally consisting of $8 \times 256^2 = 524{,}288$ bits, can be encoded using a total of $95,395$ bits or at a rate of 1.426 bpp.

To view the compressed image, we reverse the process. That is, we decode the Huffman codes and then rebuild the blocks $\widehat{C}_{jk}$. We multiply each of the elements $\widehat{c}_{mn}$ of $\widehat{C}_{jk}$ by z_{mn} (12.2) *to obtain an approximate $\widetilde{T}_{jk}$ of T_{jk}. The fact that we have rounded data in our quantization step means that we will never be able to exactly recover the original image. The inversion process is completed by applying*

Figure 12.3 The quantized DCT for image A.

the inverse DCT to each block $\widetilde{T}_{jk}$. That is, we obtain approximations $\widetilde{A}_{jk}$ to A_{jk} by computing

$$\widetilde{A}_{jk} = U^T \widetilde{T}_{jk} U$$

where $j, k = 1, \ldots, 32$. In Figure 12.4, we plot both $A_{7,16}$ and $\widetilde{A}_{7,16}$.

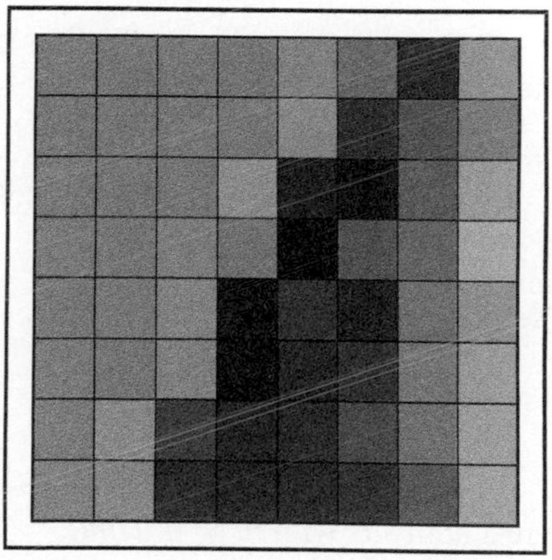

(a) $A_{7,16}$

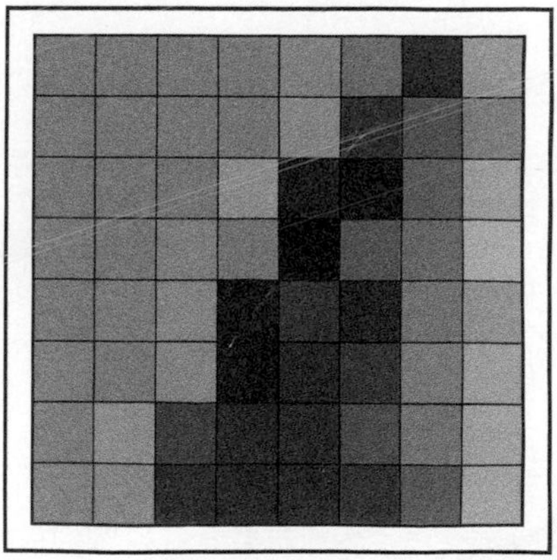

(b) $\widetilde{A}_{7,16}$

Figure 12.4 The block $A_{7,16}$ from the original image in Figure 12.1(a) and the JPEG-compressed $\widetilde{A}_{7,16}$.

The approximate $\widetilde{A}$ of the original image A is formed by constructing the 32×32 block matrix whose entries are $\widetilde{A}_{jk}$ and adding 128 to each element in $\widetilde{A}_{jk}$. We have plotted the image in JPEG format in Figure 12.5. The PSNR is computed to be 33.01.

Figure 12.5 The JPEG-compressed version of the original image.

(Live example: Visit `stthomas.edu/wavelets` *and click on Live Examples.)*

□

PROBLEMS

12.1 For each eight-vector $\mathbf{v}$, compute the one-dimensional DCT $\mathbf{y} = U\mathbf{v}$, where U is given by (12.1).

(a) $v_k = c$, $k = 1, \ldots, 8$ where c is any nonzero number

(b) $v_k = k$, $k = 1, \ldots, 8$

(c) $v_k = (-1)^k$, $k = 1, \ldots, 8$

(d) $v_k = mk + b$, $k = 1, \ldots, 8$, where m and b are real numbers with $m \neq 0$. *Hint:* Use (a) and (b).

12.2 Suppose that $\mathbf{v} = [v_1, v_2, v_3, v_4, v_4, v_3, v_2, v_1]^T$ and let $\mathbf{y} = U\mathbf{v}$ where U is given by (12.1). Show that the even elements of $\mathbf{y}$ are zero. (*Hint:* Use the trigonometric identity $\cos\theta = -\cos(\pi - \theta)$.)

12.3 Let $\mathbf{u}^1$ be the first row of U, where U is given by (12.1). Show that $\| \mathbf{u}^1 \| = 1$ and explain why $\mathbf{u}^1$ is orthogonal to the other seven rows of U.

12.4 Suppose that A is an 8×8 matrix whose entries are each $a_{jk} = c$, where $k = 1, \ldots, 8$ and c is any real nonzero number. Compute UAU^T where U is given by (12.1). *Hint:* Use (a) from Problem 12.1.

12.5 Suppose that, for a given 8×8 matrix A, we compute $B = UAU^T$ and then create the matrix C whose elements are b_{jk}/z_{jk}, $j, k = 1, \ldots, 8$, where Z is given by (12.2). The elements of C are rounded to give

$$\widehat{C} = \begin{bmatrix} 3 & 0 & 2 & -1 & -4 & -1 & 0 & -2 \\ 7 & -1 & - & 1 & 2 & 1 & 1 & 0 \\ 0 & 10 & -1 & 2 & 0 & -3 & 1 & -1 \\ 0 & 1 & -2 & -3 & 1 & 0 & 1 & 0 \\ 3 & -4 & 0 & -4 & 0 & -1 & 1 & -1 \\ -4 & -1 & 0 & -3 & -1 & 1 & -1 & 0 \\ 0 & 0 & -1 & 1 & 0 & 1 & 1 & 0 \\ -1 & -1 & 0 & 0 & 0 & 0 & 0 & 0 \end{bmatrix}$$

Use the zigzag array (12.3) to create the zigzag vector $\mathbf{d}$ (12.4) that will be sent to the Huffman coder.

12.2 The Basic JPEG2000 Algorithm

Example 12.1 illustrates a couple of problems with the current JPEG standard. The first problem is that the compression standard is *lossy*. That is, due to the rounding that occurs in the quantization step, it is generally impossible to exactly recover the original image from the compressed version.

In Figure 12.6 we have plotted an enlarged version of the near bottom left-hand corner (64×64 submatrix) of both the original image and the JPEG compressed image.

If you look closely at the image in Figure 12.6(b), you will notice that some pixels appear in "blocks." Partitioning the image into 8×8 blocks and processing these blocks independently leads to the artifacts in the compressed image.

JPEG2000 – Improving on the JPEG Standard

The JPEG2000 standard aspires to resolve many of the issues faced by the current standard. Christopoulos et al. [20] identify several features of JPEG2000. Some of these features are

1. **Lossy and lossless compression.** The JPEG2000 standard allows the user to perform lossless compression as well as lossy compression.

(a) Original image

(b) JPEG compressed version

Figure 12.6 A plot of the near bottom left corner of the original image A and the same region of the JPEG compressed version.

2. **Better compression.** The JEPG2000 standard produces higher-quality images for very low bit rates (below 0.25 bpp according to JPEG [60]).

3. **Progressive signal transmission.** JPEG2000 can reconstruct the digital image (at increasingly higher resolution) as it is received via the internet.

4. **Tiling.** JPEG2000 allows the user to *tile* the image. If desired, the image can be divided into non-overlapping tiles, and each tile can be compressed independently. This is a more general approach to the 8×8 block partitioning of JPEG. If tiling is used, the resulting uncompressed image is much smoother than that produced by the current JPEG standard.

5. **Regions of Interest.** The JPEG2000 standard allows users to define Regions of Interest (ROI) and encode these ROIs so that they can be uncompressed at a higher resolution than the rest of the image.

6. **Larger image size.** The current JPEG standard can handle images of size $64,000 \times 64,000$ or smaller. JPEG2000 can handle images of size $2^{32} - 1 \times 2^{32} - 1$ or smaller (see JPEG [60]).

7. **Multiple channels.** The current JPEG standard can only support three channels (these channels are used for color images). The new JPEG2000 standard can support up to 256 channels. JPEG2000 can also handle color models other than YCbCr.

The Basic JPEG2000 Algorithm

We now present a basic JPEG2000 algorithm. This algorithm appears in Gonzalez and Woods [48]. There are many features we omit at this time. The interested reader is encouraged to consult Taubman [95] for a more detailed description of the compression standard.

Algorithm 12.1 (Basic JPEG2000 Compression Standard). *This algorithm takes a grayscale image A whose dimensions are even and whose intensities are integers ranging from 0 to 255 and returns a compressed image using the JPEG2000 compression format.*

1. *Subtract 128 from each element of A.*

2. *Apply i iterations of the biorthogonal wavelet transformation. If the dimensions of A are $N \times M$ with $N = 2^p$ and $M = 2^q$, then the largest i can be is the minimum of p and q. For lossless compression, use the LeGall filter pair and for lossy compression, use the CDF97 biorthogonal filter pair.*

3. *For lossy compression, apply a quantizer similar to the shrinkage function* (6.3) *defined in Section 6.1.*

4. *Use an arithmetic coding algorithm known as Embedded Block Coding with Optimized Truncation (EBCOT) [94] to encode the quantized transformation data.*

□

To compress color images using the JPEG2000 standard, we first convert from RGB color space to YCbCr color space (see Section 3.2) and apply Algorithm 12.1 to each of the Y, Cb, and Cr channels. Next we look at Step 2 of Algorithm 12.1 in a bit more detail.

The Biorthogonal Wavelet Transformation for Lossy Compression

Perhaps the biggest change in the JPEG2000 compression standard is the use of the biorthogonal wavelet transformation instead of the discrete cosine transformation. The biorthogonal wavelet transformation is one reason the compression rates for JPEG2000 are superior to those of JPEG. The iterative process of the transformation makes it easy to see how it can be implemented to perform progressive image transmission. In a more advanced implementation of JPEG2000, the user can opt to reconstruct only a portion of the compressed image. The local nature of the biorthogonal wavelet transformation can be exploited to develop fast routines to perform this task.

For lossy compression, the CDF97 filter pair developed in Section 9.5 is used. The transformation is performed using the symmetric biorthogonal wavelet transformation from Section 7.3. The computation can be performed using Algorithm 7.1 or via lifting (see Problems 11.56 and 11.58 in Section 11.4 for possible lifting methods). In this way, the transformation is better suited to handle pixels at the right and bottom boundaries of the image.

In our derivation of the CDF97 filter pair, $\widetilde{\mathbf{h}}$ is the seven-term filter and it is used to construct the averages portion of the transformation matrix $\widetilde{W}$. The nine-term filter $\mathbf{h}$ gives rise to $\widetilde{\mathbf{g}}$ (see Table 9.3) and this filter is used to construct the differences portion of $\widetilde{W}$. We then use $\mathbf{h}$ and $\widetilde{\mathbf{h}}$ to form the averages and differences portions of W, respectively. Thus our two-dimensional transformation is $B = \widetilde{W}AW^T$.

JPEG2000 reverses the roles of the filters (see Christopoulos [20]). That is, the biorthogonal transformation is implemented as $B = WA\widetilde{W}^T$. The inverse transformation is then $A = \widetilde{W}^T BW$. It is easy to implement this change in Algorithm 7.1 – we simply call the algorithm with the filters reversed.

The Biorthogonal Wavelet Transformation for Lossless Compression

For lossless compression, the members of the JPEG settled on the LeGall filter pair. These filters are symmetric and close in length. The best feature of this filter pair is that it can easily be adapted to map integers to rational numbers as demonstrated in Section 11.1. This technique is reversible, so no information is lost.

The Quantization Process

Suppose that we have computed i iterations of the biorthogonal wavelet transformation. We denote by $\mathcal{V}^n$, $\mathcal{H}^n$, $\mathcal{D}^n$ the vertical, horizontal, and diagonal portions, respectively, of the nth iteration of the wavelet transformation. Here $n = 1, \ldots, i$. We denote by $\mathcal{B}^i$ the blur portion of the transformation. An example of this notation is given in Figure 12.7.

We quantize each of $\mathcal{V}^n, \mathcal{H}^n, \mathcal{D}^n$, and $\mathcal{B}^i$, $n = 1, \ldots, i$ separately. Suppose that t is any element in $\mathcal{V}^n, \mathcal{H}^n, \mathcal{D}^n$, or B^i, $i = 1, \ldots, n$. Then the quantization function is given by

$$q(t) = \text{sgn}(t)\lfloor |\,t\,|/d\,\rfloor. \tag{12.5}$$

Here $\text{sgn}(t)$ is the sign function given by (6.2) in Section 6.1 and $\lfloor\cdot\rfloor$ is the floor function defined by (3.2) in Section 3.1. The value d is called the *quantization step size*. We discuss the computation of d in the next subsection. A quantization step size must be provided for each of $\mathcal{V}^n, \mathcal{H}^n, \mathcal{D}^n$, and $\mathcal{B}^i$, $n = 1, \ldots, i$.

It is easy to understand how the quantization process operates. We first compute the absolute value of t. We divide this number by the quantization step size. The floor function truncates the result to an integer and the sgn function reattaches the original sign of t.

Note that if $|\,t|$ is smaller than the quantization step size, the floor function converts the result to 0. In this regard, (12.5) acts much like the shrinkage function (6.3) used in denoising in Chapter 6. If $d \le |t| < 2d$, the floor function returns one and the

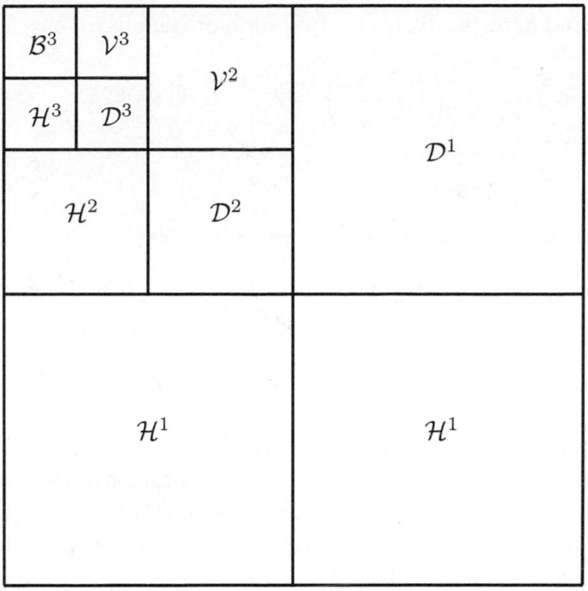

Figure 12.7 Three iterations of the biorthogonal wavelet transformation.

sgn function reattaches the original sign of t. In general, if $\ell d \leq |t| < (\ell + 1)d$, then the floor function returns ℓ. In Problem 12.6 you are asked to plot $q(t)$ for various values of d.

Creating the Quantization Step Sizes

To create the quantization step sizes, we form three-vectors $\mathbf{d}^n = [d_V^n, d_H^n, d_D^n]^T$, for $n = 1, \ldots, i-1$ and a four-vector $\mathbf{d}^i = \left[d_V^i, d_H^i, d_D^i, d_B^i\right]^T$. These vectors contain step sizes for the quantization. For $n = 1, \ldots, i-1$, d_V^n, d_H^n, and d_D^n are used to quantize $\mathcal{V}^n, \mathcal{H}^n$, and $\mathcal{D}^n$, respectively. The same pattern holds for $\mathbf{d}^i$ except that d_B^i will be used to quantize the blur portion of the transformation.

To form the vectors $\mathbf{d}^n$, $n = 1, \ldots, i$, we first need a *base step size*. We take

$$\tau = 2^{R-c+i}\left(1 + \frac{f}{2^{11}}\right)$$

where R is the number of bits needed to represent the values of the original image, and c and f are the number of bits used to represent the exponent and mantissa, respectively, of the elements in the blur portion of the transformation. Since our pixels range in intensity from 0 to $255 = 2^8 - 1$, we will take $R = 8$. Typical values for c and f are 8.5 and 8, respectively (see [48] for more details). Thus

$$\tau = 2^{8-8.5+i}\left(1+\frac{8}{2^{11}}\right) = 2^{i-1/2}1.00390625. \tag{12.6}$$

We use τ to write

$$\mathbf{d}^k = \left[d_V^k, d_H^k, d_D^k\right]^T = \frac{1}{2^{k-1}}\left[\tau, \tau, 2\tau\right]^T \tag{12.7}$$

for $k = 1, \ldots, i-1$ and for $\mathbf{d}^i$, we have

$$\mathbf{d}^i = \left[d_V^i, d_H^i, d_D^i, d_B^i\right]^T = \frac{1}{2^{i-1}}\left[\tau, \tau, 2\tau, \frac{\tau}{2}\right]^T.$$

Although we do not discuss the motivation for the derivation of $\mathbf{d}^k$ at this time, the formulation depends on the number of *analysis gain bits* for each level of the transformation. For more information, see Gonzalez and Woods [48] or Christopoulos [20]. This method for determining the step size is used in *implicit quantization*. In the case of *explicit quantization*, the user would supply all vectors $\mathbf{d}^n$, $n = 1, \ldots, i$. Let's look at an example.

Example 12.2 (Computing Quantization Step Sizes). *Suppose that we performed $i = 3$ iterations of the biorthogonal wavelet transformation. Using* (12.6)*, we see that*

$$\tau = 2^{3-1/2}1.00390625 \approx 5.678951$$

The step size vectors for the first and second iterations are

$$\mathbf{d}^1 = [\,d_V^1, d_H^1, d_D^1\,]^T = [\,\tau, \tau, 2\tau\,]^T = [\,5.678951, 5.678951, 11.357903\,]^T$$

$$\mathbf{d}^2 = [\,d_V^2, d_H^2, d_D^2\,]^T = \left[\frac{1}{2}\tau, \frac{1}{2}\tau, \tau\right]^T = [\,2.839476, 2.839476, 5.678951\,]^T$$

and the step size vector for the third iteration is

$$\begin{aligned}\mathbf{d}^3 &= [\,d_V^3, d_H^3, d_D^3, d_B^3\,]^T \\ &= \left[\frac{1}{4}\tau, \frac{1}{4}\tau, \frac{1}{2}\tau, \frac{1}{8}\tau\right]^T \\ &= [\,1.419738, 1.419738, 2.839476, 0.709869\,]^T\end{aligned}$$

Note that the step sizes get smaller as the iteration number increases. □

Encoding in JPEG2000

Rather than Huffman coding, the JPEG2000 standard uses a form of *arithmetic coding* to encode quantized transformation values. Arithmetic coding is different than

Huffman coding in that it takes a long string of integer values and stores them as a single rational number between 0 and 1. Naturally, this number has a binary representation, and the primary difference between arithmetic coding and Huffman coding is that Huffman coding assigns a single character to a list of bits, whereas arithmetic coding might use some combination of characters to define a list of bits. According to Wayner [101], it is possible for arithmetic coding methods to be 10% more efficient than Huffman coding.

The JPEG2000 uses a method of arithmetic coding called *Embedded Block Coding with Optimized Truncation* (EBCOT). This form of arithmetic coding was introduced by Taubman [94]. This coding method utilizes the wavelet transformation and facilitates many of the desired features of the JPEG2000 standard. The development of arithmetic coding and EBCOT are outside the scope of this book. The interested reader is referred to Sayood [80] or Taubman [95].

Inverting the Process

To invert the process, we begin by decoding the encoded quantized transformation. Next, we multiply the averages and the differences portions by the associated quantization step size. We then apply the inverse biorthogonal wavelet transformation using the CDF97 filter pair. Recall that Algorithm 12.1 stipulates that the transformation is computed using $B = WA\tilde{W}^T$ so the inverse transformation uses $\tilde{W}^T BW$. We then add 128 to the result and obtain a compressed version of the original image.

For color images, we decode the quantized transformations of the Y, Cb, and Cr channels and then apply the inversion process to each channel. As a last step, we convert from YCbCr color space to RGB color space.

PROBLEMS

12.6 Plot $q(t)$ given by (12.5) for $d = 1, 2, 5, 10$.

12.7 Suppose that

$$\mathbf{c} = [5, 103/4, 43, 21/2, 33/4, 11/4, -29, -121/4, 11/4, 25/2, -17/2, -1]^T.$$

Apply the quantization function $q(t)$ given by (12.5) to $\mathbf{c}$ using

(a) $d = 1$

(b) $d = 5$

(c) $d = 10$

(d) $d = 15$

12.8 Suppose that we compute i iterations of the symmetric biorthogonal wavelet transformation and for $k = 1, \ldots, i-1$. Let $\mathbf{d}^k$ be given by (12.7).

(a) Show that for $k = 1, \ldots, i-2$, $\mathbf{d}^{k+1} = \frac{1}{2}\mathbf{d}^k$.

(b) Why do you think the step sizes decrease by a factor of 2 for each successive iteration?

(c) Why do you think the step size for the diagonal block is twice as large as that used for the vertical and horizontal blocks?

12.3 Examples

We conclude this chapter with examples of how the naive JPEG2000 algorithm works. We perform lossy compression and lossless compression of the image in Figure 12.1(a) and then compress a color image.

Example 12.3 (Lossy JPEG2000 Compression). *We first subtract* 128 *from each element of the* 256×256 *image in Figure 12.1 of Example 12.1 and then perform three iterations of the symmetric biorthogonal wavelet transformation (Algorithm 7.1) using the reversed CDF97 filter pair. The result is plotted in Figure 12.8(a).*

(a) Biorthogonal wavelet transformation

(b) Quantized biorthogonal wavelet transformation

Figure 12.8 Three iterations of the biorthogonal wavelet transformation using the reversed CDF97 filter pair and the quantized version of the transformation.

The next step is to quantize the wavelet transformation. The quantization step size vectors are given in Example 12.2. We use these step sizes with the quantization function (12.5) *and quantize each portion of the wavelet transformation. The quantized transformation is plotted in Figure 12.8(b).*

At this point, the quantized transformation is encoded. Since the development of arithmetic coding (not to mention the EBCOT coding method) is beyond the scope of this book, we are left only with Huffman coding as a way to encode the quantized transformation. Using Algorithm 3.1 on the quantized transformation reduces the data from $256^2 \times 8 = 524,288$ *bits to* $128,295$ *bits. Thus, we would need about* 1.95763 *bpp to encode the data. Rest assured that the EBCOT method produces much better results.*

To recover the compressed image, the steps are reversed. The data are decoded, and then each portion of the transformation is multiplied by the appropriate quantization step size. The inverse biorthogonal wavelet transformation is applied and we add 128 *to the result. The compressed image is plotted in Figure 12.9.*

Figure 12.9 The JPEG2000 compressed version of the original image.

The PSNR is computed to be 42.82 *and this value is higher than the PSNR value* 33.01 *we achieved using JPEG compression in Example 12.1. Note also that the block artifacts that appear using the JPEG standard (see Figure 12.5) are no longer present.*

In Figure 12.10 we have plotted the near bottom left corner of the image in Figure 12.9. We have also plotted the same region for the JPEG compressed version. Note that the block artifacts that are present in the JPEG version no longer appear in the JPEG2000 version. You can also compare the results with the plot of the near bottom left corner of the original image in Figure 12.6(a).

(Live example: Visit `stthomas.edu/wavelets` *and click on Live Examples.)*

□

Example 12.4 (Lossless JPEG2000 Compression). *In this example we apply lossless compression to the image plotted in Figure 12.1(a). Let* A *be the* 200×200 *matrix that holds the grayscale intensity values of the image.*

(a) JPEG2000 compressed version.

(b) JPEG compressed version.

Figure 12.10 A plot of the near lower left corner of the JPEG2000 and the JPEG compressed versions of the original image A.

We begin by subtracting 128 *from each element in* A*. We now apply three iterations of the symmetric biorthogonal wavelet transformation with the LeGall filter using the lifting method (Algorithm 11.1). We utilize* (11.14) *and* (11.13) *so that the transformation values are integer-valued. The biorthogonal wavelet transformation is plotted in Figure 12.11.*

At this point we would use the EBCOT method to encode the data. Since we do not develop EBCOT in this text, we instead use the basic Huffman coding developed in Section 3.3. Using Huffman encoding, the transformation can be stored using $244,195$ *bits or* 3.726 *bpp.*

To invert the process and recover image A *exactly, we decode and then lift in conjunction with* (11.15) *and* (11.16) *to compute the inverse biorthogonal wavelet transformation.*

(Live example: Visit `stthomas.edu/wavelets` *and click on Live Examples.)*

□

Example 12.5 (Color JPEG2000 Compression). *In this example we use Algorithm 12.1 to compress the image plotted in Figure 12.12. The dimensions of this image are* 512×768.

The original image is given in RGB space. The first step is to convert the R, G, and B matrices to YCbCr space. Here, we use (3.7) *so that the elements of matrices* Y*,* Cb *and* Cr *are integer-valued and reside in the intervals* $[16, 235]$ *(for* Y*) and* $[16, 240]$ *(for* Cb *and* Cr*). These channels are plotted in Figure 12.13.*

We then subtract 128 *from each of* Y*,* Cb *and* Cr *and next compute three iterations of the biorthogonal wavelet transformation using the reversed CDF97 filter for each matrix. For quantization, we use the vectors* $\mathbf{d}^1, \mathbf{d}^2$*, and* $\mathbf{d}^3$ *in conjunction with the quantization function* (12.5) *to obtain the quantized transformations. The process here is exactly the same as in Example 12.3 except that in this case, we*

Figure 12.11 The biorthogonal wavelet transformation of the image.

Figure 12.12 The original color image. See the color plates for color versions of these images.

must perform it on three different matrices. The wavelet transformations and their quantized versions are plotted in Figure 12.14.

At this point we would encode using EBCOT. As in Example 12.3, we instead use basic Huffman encoding to obtain a basic understanding of the savings. Table 12.1

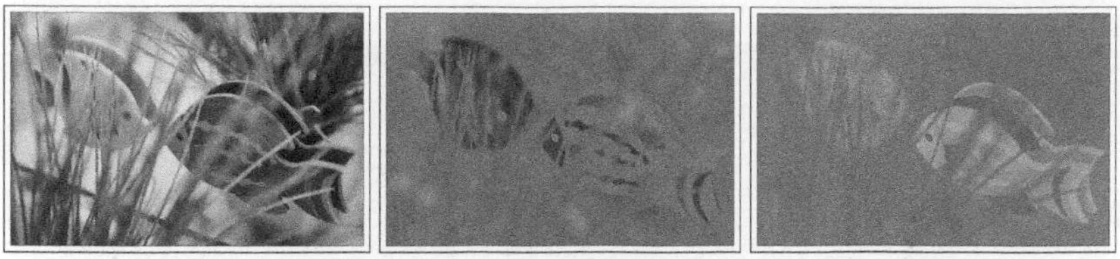

(a) Y (b) Cb (c) Cr

Figure 12.13 The Y, Cb, and Cr channels for the image in Figure 12.12.

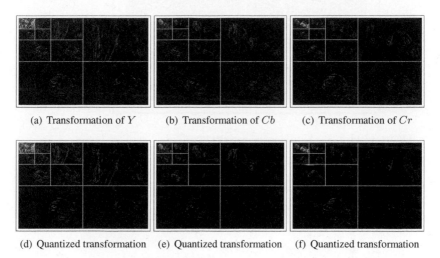

(a) Transformation of Y (b) Transformation of Cb (c) Transformation of Cr

(d) Quantized transformation (e) Quantized transformation (f) Quantized transformation

Figure 12.14 Three iterations of the biorthogonal wavelet transformation using the reversed CDF97 filter on each of the Y, Cb, nd Cr channels of the original image. The quantized transformations appear below each transformation.

gives the results. Recall that each of the R, G, *and* B *channels of the original image consists of* $512 \times 768 \times 8 = 3{,}145{,}@728$ *bits.*

Table 12.1 Bits needed to encode the Y, Cb, and Cr channels using basic Huffman encoding.

Channel	Bits Needed	bpp
Y	720,690	1.833
Cb	513,516	1.306
Cr	494,195	1.257

To invert the process, we multiply each portion of the transformation of the quantized transformation of Y *by the appropriate quantization step size and then apply three iterations of the inverse biorthogonal wavelet transformation. We repeat this process for the quantized transformations of the* Cb *and* Cr *channels. We now have* *compressed versions of Y, Cb, and Cr. We next add* 128 *to the elements in each of the compressed versions of Y, Cb and Cr and if necessary, "clip" the elements in these matrices so that values reside in the proper intervals* ($[16, 235]$ *for Y and* $[16, 240]$ *for Cb and Cr). The compressed versions are finally converted back to RGB space and the result is plotted in Figure 12.15(a). For comparative purposes we have plotted a compressed version of the original image using the JPEG standard in Figure 12.15(b).*

(a) JPEG2000 (b) JPEG

Figure 12.15 The compressed image using JPEG2000 and JPEG. See the color plates for color versions of these images.

To measure the PSNR for color images, we use only the Y channels. The PSNR for the JPEG2000 compressed image is 43.345 *while the PSNR for the JPEG compressed image is* 37.32. *The images in Figure 12.16 are obtained by partitioning the JPEG2000- and JPEG-compressed images into blocks of size* 64×96. *The* $(5, 6)$ *block for each compressed image is plotted below. Note that the JPEG version has block artifacts that are not present in the JPEG2000 image.*

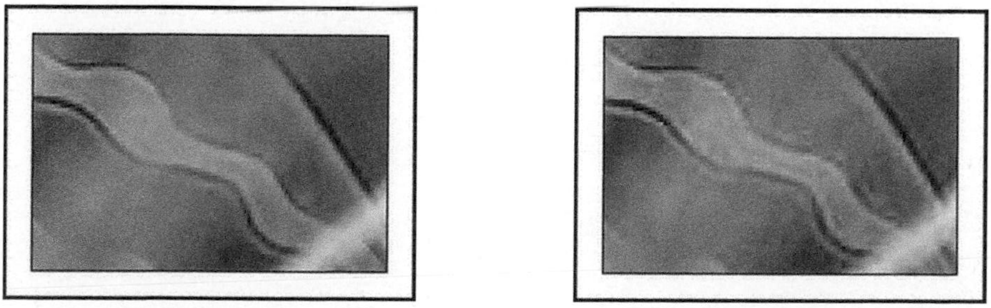

(a) JPEG2000 (b) JPEG

Figure 12.16 Corresponding 64×96 blocks of each compressed images in Figure 12.15. See the color plates for color versions of these images.

(Live example: Visit `stthomas.edu/wavelets` *and click on Live Examples.)* □

APPENDIX A

BASIC STATISTICS

The material in Chapter 6 makes use of several ideas from statistics. We also use some simple descriptive statistics in the edge detection application in Section 4.4. In this appendix we review some basic statistical concepts.

A.1 Descriptive Statistics

We start with the definition of the mean.

Definition A.1 (Mean). *Let* $\mathbf{x} = (x_1, x_2, \ldots, x_n)$ *be a list of numbers. Then the* mean $\overline{x}$ *is defined as*

$$\boxed{\overline{x} = \frac{x_1 + x_2 + \cdots + x_n}{n} = \frac{1}{n}\sum_{k=1}^{n} x_k.}$$

□

Patrick J. Van Fleet.

A second statistic that is used to measure the center of the data is the median. The median is a number x_{med} such that the number of data below x_{med} is equal to the number of data above x_{med}.

Definition A.2 (Median). *Let* $\mathbf{x} = (x_1, x_2, \ldots, x_n)$ *be a list of numbers. The median* x_{med} *is constructed by first sorting the values* $x_1, x_2, \ldots, x_n$ *from smallest to largest. Call this new list* $\mathbf{y} = (y_1, y_2, \ldots, y_n)$. *If* n *is odd, we set* $x_{\text{med}} = y_{\frac{n+1}{2}}$. *If* n *is even, we take the median to be the average of the two middle components. That is,* $x_{\text{med}} = \frac{1}{2}\left(y_{n/2} + y_{n/2+1}\right)$. □

The mean and the median are two descriptive statistics that tell us something about the "center" of the data. Unfortunately, neither tell us about the spread of the data. We use the ideas of variance and standard deviation to assist us in analyzing the spread of a data set.

Definition A.3 (Variance and Standard Deviation). *Suppose that a list of numbers* $\mathbf{x} = (x_1, x_2, \ldots, x_n)$ *has mean* $\overline{x}$. *The* variance *of this data set is defined by*

$$\boxed{s^2 = \frac{1}{n-1}\sum_{k=1}^{n}(x_k - \overline{x})^2.} \tag{A.1}$$

The standard deviation *is defined to be the nonnegative square root of the variance.* □

It is interesting to note that the divisor in (A.1) is $n - 1$ rather than n. You will sometimes see the sample variance defined with an n in the divisor, but this produces a *biased estimator*. If you are interested in exploring this topic further, consult an introductory text on probability and statistics such as DeGroot [33]. The following example illustrates the descriptive statistics defined thus far.

Example A.1 (Computing Descriptive Statistics). *Find the mean, median, variance, and standard deviation of* $\mathbf{x} = (6, -3, 17, 0, 5)$, $\mathbf{y} = (12, 19, -6, 0, -2, -11)$.

Solution. *For the list* $\mathbf{x}$, *we have* $\overline{x} = \frac{1}{5}(6 - 3 + 17 + 0 + 5) = 5$. *If we sort this list, we have* $(-3, 0, 5, 6, 17)$ *so that* $x_{\text{med}} = 5$. *Finally, we compute the variance as*

$$s^2 = \frac{1}{4}\left((6-5)^2 + (-3-5)^2 + (17-5)^2 + (0-5)^2 + (5-5)^2\right) = 58.5.$$

Thus, the standard deviation is $s = \sqrt{58.5} \approx 7.64853$.

For the list $\mathbf{y}$, *we have* $\overline{y} = \frac{1}{6}(12 + 19 - 6 + 0 - 2 - 11) = 2$. *If we sort the list, we obtain* $(-11, -6, -2, 0, 12, 19)$, *and since* $n = 6$ *is even, we take* $y_{\text{med}} = \frac{-2+0}{2} = -1$. *For the variance, we compute*

$$\begin{aligned} s^2 &= \frac{1}{5}\left((12-2)^2 + (19-2)^2 + (-6-2)^2 + (0-2)^2\right. \\ &\qquad \left. + (-2-2)^2 + (-11-2)^2\right) \\ &= 128.4. \end{aligned}$$

Thus the standard deviation is $s = \sqrt{128.4} \approx 11.3314$. □

The *median absolute deviation* turns out to be a useful descriptive statistic for estimating the noise level in noisy signals.

Definition A.4 (Median Absolute Deviation). *Let* $\mathbf{v} = (v_1, \ldots, v_N)$ *and let* $xmedev$ *denote the median of* $\mathbf{v}$. *Form the list* $\mathbf{w} = (|v_1 - v_{\text{med}}|, \ldots, |v_N - v_{\text{med}}|)$. *We define the* median absolute deviation *of* $\mathbf{v}$ *as*

$$MAD(\mathbf{v}) = w_{\text{med}}.$$

□

Example A.2 (Computing the MAD). *Consider the list* $\mathbf{y}$ *from Example A.1. To compute* $MAD(\mathbf{y})$, *we recall that* $y_{med} = -1$ *and then form the new vector*

$$\begin{aligned}\mathbf{w} &= (\,|12 - (-1)|, |19 - (-1)|, |-6 - (-1)|,\\ &\quad |0 - (-1)|, |-2 - (-1)|, |-11 - (-1)|\,)\\ &= (13, 20, 5, 1, 1, 10)\,.\end{aligned}$$

We next sort this list to obtain $(1, 1, 5, 10, 13, 20)$. *Since* $n = 6$ *is even, we form the median of this list by averaging five and ten. Hence,* $MAD(\mathbf{y}) = 7.5$. □

PROBLEMS

A.1 Let $\mathbf{x} = (-3, 5, 12, -9, 5, 0, 2)$. Find $\overline{x}$, x_{med}, MAD($\mathbf{x}$), the variance s^2, and the standard deviation s for $\mathbf{x}$.

A.2 The mean of $x_1, x_2, \ldots, x_9$ is $\overline{x} = 7$. Suppose that $x_{10} = 12$. Find $\overline{x}$ for $x_1, x_2, \ldots, x_9, x_{10}$.

A.3 Let $\mathbf{x} = (x_1, \ldots, x_n)$ and let a be any number. Form $\mathbf{y}$ by adding a to each element of $\mathbf{x}$. That is, $\mathbf{y} = (x_1 + a, x_2 + a, \ldots, x_n + a)$. Show that $\overline{y} = \overline{x} + a$ and that the variance of $\mathbf{y}$ is the same as the variance of $\mathbf{x}$.

A.4 Let $\mathbf{x} = (1, 2, \ldots, n)$. Find $\overline{x}$ and the variance of $\mathbf{x}$. *Hint:* See the hint given in Problem 2.1 in Section 2.1.

A.5 Let $\mathbf{x} = (1, 2, \ldots, n)$. Show that $\overline{x} = x_{\text{med}}$.

A.6 Let $\mathbf{x} = (x_1, \ldots, x_n)$ with $a \le x_k \le b$ for each $k = 1, 2, \ldots, n$. Show that $a \le \overline{x} \le b$.

A.7 Let $\mathbf{x} = (x_1, x_2, \ldots, x_n)$. Show that the variance s^2 can be written as

$$s^2 = \frac{1}{n-1}\sum_{k=1}^{n} x_k^2 - \frac{n}{n-1}\overline{x}^2.$$

A.8 Suppose $\mathbf{y} \in \mathbb{R}^N$ has mean $\overline{y}$ and variance s_y^2. For real numbers a and b with $b \ne 0$, define $\mathbf{z} \in \mathbf{R}^N$ componentwise by the rule $z_k = (y_k - a)/b$, $k = 1, \ldots, N$.

(a) Show that $\overline{z} = (\overline{y} - a)/b$.

(b) Show that $s_z^2 = s_y^2/b^2$.

(c) What are $\overline{z}$ and s_z^2 if $a = \overline{y}$ and $b = s_y$?

A.2 Sample Spaces, Probability, and Random Variables

In probability and statistics, we often analyze an experiment or repeated experiments. As a basis for this analysis, we need to identify a *sample space*.

Sample Spaces

Definition A.5 (Sample Space). *The* sample space *for an experiment is the set of all possible outcomes of the experiment.* □

Example A.3 (Identifying Sample Spaces). *Here are some examples of sample spaces.*

(a) The experiment is tossing a penny, nickel, and dime. If we let H denote heads and T denote tails, then the sample space is

$$S = \{HHH, HHT, HTH, HTT, THH, THT, TTH, TTT\}.$$

(b) The experiment is to toss a standard red die and a standard green die and record the numbers that appear on top of the dice. The sample space is the set of ordered pairs

$$S = \{(j, k) \mid j, k = 1, \ldots, 6\}.$$

The previous two sample spaces are examples of finite sample spaces. Many experiments have sample spaces that are not finite.

(c) The experiment of measuring the height of the woman from a certain population. Although there are finitely many women in the population, their heights (in feet) can assume any value in the interval $[0, 9]$.

(d) The experiment of measuring the difference of the true arrival time of a city bus versus the expected arrival time at a certain stop results in an infinite number of possible outcomes.

□

Probability

Let A be any subset of a sample space S. The set A is called an *event*. If $A_1, A_2, \ldots$ is a sequence of events, we say that $A_1, A_2, \ldots$ is a *sequence of disjoint events* if $A_j \cap A_k = \emptyset$ whenever $j \neq k$.

We assign a *probability* to an event A. This value, denoted by $Pr(A)$, indicates the likeliness that A will occur. The range of the numbers is zero (impossible that A occurs) to one (A certainly occurs). Mathematically, we can think of Pr as a function that maps an event to some number in $[0,1]$. The function Pr must satisfy the following three axioms:

(Pr 1) For any event A in sample space S, we have $Pr(A) \geq 0$.

(Pr 2) If S is the sample space, then $Pr(S) = 1$.

(Pr 3) Let $A_1, A_2, \ldots$ be a sequence of disjoint events. Then

$$Pr\left(A_1 \cup A_2 \cup \ldots\right) = \sum_{k=1}^{\infty} Pr(A_k).$$

Given a function Pr that satisfies the probability axioms (Pr 1) – (Pr 3), we can prove several properties obeyed by Pr.

Proposition A.1. *Let S be a sample space and Pr a probability function that satisfies (Pr 1) – (Pr 3). Then*

(a) $Pr(\emptyset) = 0$.

(b) If $A_1, \ldots, A_n$ is a finite list of disjoint events, then

$$Pr\left(A_1 \cup A_2 \cup \cdots \cup A_n\right) = \sum_{k=1}^{n} Pr(A_k).$$

(c) If A^c denotes all the elements in S but not in A, then $Pr(A^c) = 1 - Pr(A)$.

(d) If A and B are events in S with $A \subseteq B$, then $Pr(A) \leq Pr(B)$.

(e) If A and B are any events in S, then

$$Pr(A \cup B) = Pr(A) + Pr(B) - Pr(A \cap B).$$

□

Proof. We will prove (a) and leave the remaining proofs as problems. For (a), let $A_k = \emptyset$ for $k = 1, 2, \ldots$. Then by (Pr 3), we have

$$Pr\left(A_1 \cup A_2 \cup \cdots\right) = \sum_{k=1}^{\infty} Pr\left(A_k\right).$$

Now $A_1 \cup A_2 \cup \cdots = \emptyset$, so the identity above can be written as

$$Pr(\emptyset) = \sum_{k=1}^{\infty} Pr(\emptyset).$$

The only way the infinite series on the right converges is if each term is zero. □

Let's look at an example illustrating these properties.

Example A.4 (Using Probability Properties). *A small remote town has two grocery stores and the citizens only shop for groceries at these two stores. The probability that a citizen shops at Store 1 is* 0.65 *and the probability that a citizen shops at Store 2 is* 0.43.

(a) What is the probability that a citizen does not shop at Store 2?

(b) What is the probability that a citizen shops at both stores?

Solution. *Let* A *be the event that a citizen shops at Store 1 and* B *be the event that a citizen shops at Store 2. For (a), we can use (c) of Proposition A.1. We have* $Pr(B^c) = 1 - Pr(B) = 1 - 0.43 = 0.57$. *For (b), we need to compute* $Pr(A \cap B)$. *We use the last property of Proposition A.1. Since* $A \cup B = S$, *we have*

$$\begin{aligned} 1 = Pr(S) = Pr(A \cup B) &= Pr(A) + Pr(B) - Pr(A \cap B) \\ &= 0.65 + 0.43 - Pr(A \cap B) \\ &= 1.08 - Pr(A \cap B). \end{aligned}$$

We can solve for $Pr(A \cap B)$ *in the above equation to find that* $Pr(A \cap B) = 0.08$. □

Random Variables

We are now ready to define a *random variable.* Random variables are very important in statistical modeling.

Definition A.6 (Random Variable). *Let* S *be a sample space. A real-valued function* $X : S \to \mathbb{R}$ *is called a* random variable. □

Example A.5 (Identifying Random Variables). *Here are some examples of random variables.*

(a) Consider the sample space S *from Example A.3(a). Let's define the random variable* X *to be number of tails recorded when the three coins are tossed. So if the three coins were tossed and the outcome was* HTH, *then* $X = 1$.

(b) Consider the sample space S *from Example A.3(b). We can define a random variable* X *to be the sum of the numbers on the dice. For example, if the roll was* $(2, 5)$, *then* $X = 7$. *We could also define a random variable* X *to be the number of even numbers that appear on the roll. So if the roll was* $(2, 5)$, *then* $X = 1$.

(c) For the sample space S *from Example A.3(c), the height of a woman chosen from the population is a random variable.*

□

PROBLEMS

A.9 Suppose that the experiment we conduct is to record various two-card hands dealt from a deck of cards. Describe the sample space.

A.10 Suppose that the experiment we conduct is to measure the outcome of drawing three balls from a container containing two red ball and five blue balls. Describe the sample space.

A.11 Prove (b) – (e) of Proposition A.1.

A.12 Children at a summer camp have the opportunity to attend either a music class or a science class (scheduled at different times). All children must attend at least one class. The probability that a student attends music class is 0.39 and the probability that a student attends science class is 0.67. What is the probability a student attends both classes?

A.13 Let S be the sample space that consists of the possible sums that occur when two fair dice are rolled. Describe three possible random variables for S.

A.14 Let A and B be two events. Let C be the event that exactly one of the events will occur. Write $Pr(C)$ in terms of probabilities involving A and B.

A.15 Let S be the sample space that consists of the possible temperature measurements at an airport. Describe three possible random variables for S.

A.3 Continuous Distributions

The random variables in parts (a) and (b) of Example A.5 are examples of *discrete random variables* – i.e. random variables can only assume a finite number of values. A *continuous random variable* can assume every value in a given interval. In part (c) of Example A.5, X can theoretically assume any positive value, say between zero and nine feet.

Distributions and Probability Density Functions

In this appendix we are primarily concerned with continuous random variables. We have need to study the *distribution* of a continuous random variable. The following definition makes precise what we mean by the distribution of a continuous random variable.

Definition A.7 (Probability Density Function). *Let X be a random variable. We say that X has a* continuous distribution *if there exists a nonnegative real-valued function $f(t)$ so that for every interval (or finite union of intervals) $A \subseteq \mathbb{R}$, the probability that X assumes a value in A can be written as the integral of f over A.*

If we use the notation Pr to denote probability, we have

$$Pr(X \in A) = \int_A f(t)\,\mathrm{d}t. \tag{A.2}$$

The function $f(t)$ is called a probability density function *and in addition to being nonnegative for $t \in \mathbb{R}$, we have*

$$\int_{-\infty}^{\infty} f(t)\,\mathrm{d}t = 1. \tag{A.3}$$

□

The relation (A.3) follows immediately if we take $A = \mathbb{R}$. The probability that X assumes a real value is one, so we have

$$1 = Pr(X \in \mathbb{R}) = \int_{-\infty}^{\infty} f(t)\,\mathrm{d}t.$$

Note that if $A = [a, b]$, then (A.3) becomes

$$Pr(a \le X \le b) = \int_a^b f(t)\,\mathrm{d}t.$$

A simple example of a distribution is the *uniform distribution*. Any random variable X whose probability density function $f(t)$ is constant on the set of nonzero values of X is called a uniform distribution.

Example A.6 (Uniform Distribution). *Political polling is a complicated process, and the results are often rounded to the nearest percent when they are presented to the public. Suppose it is reported that support for a candidate is $p\%$ of the voters, $p = 0, 1, 2, \ldots, 100$. Then the actual percent X of those who support the candidate could be thought of as following a uniform distribution where the probability density function is*

$$f(t) = \begin{cases} 1, & t \in [p - \frac{1}{2}, p + \frac{1}{2}) \\ 0, & t \notin [p - \frac{1}{2}, p + \frac{1}{2}). \end{cases} \tag{A.4}$$

This probability density function is plotted in Figure A.1.

Certainly, $f(t) \ge 0$ and it is easy to check that $\int_{\mathbb{R}} f(t)\,\mathrm{d}t = 1$. □

Example A.7 (Computing Probabilities). *Suppose X is the random variable that represents the percentage of tax returns that have errors. It is known (see [43, page 119]) that X follows the distribution*

$$f(t) = \begin{cases} 90t(1-t)^8, & 0 < t < 1 \\ 0, & \text{otherwise}. \end{cases}$$

It can easily be verified that $f(t) \ge 0$ and $\int_{\mathbb{R}} f(t)\,\mathrm{d}t = 1$.

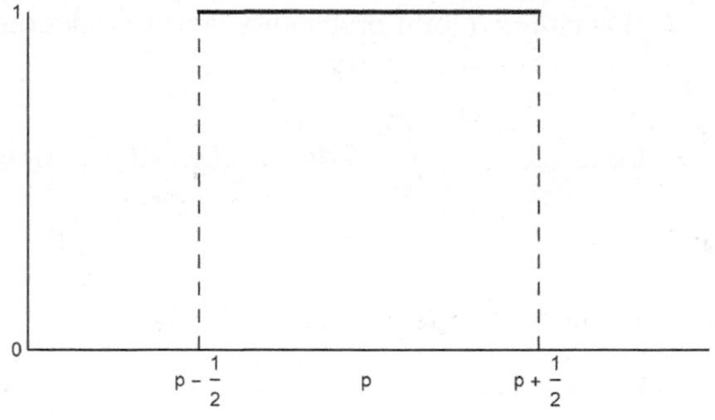

Figure A.1 The uniform distribution for X for Example A.6.

(a) Find the probability that there will be fewer than 10% erroneous tax forms in a given year.

(b) Find the probability that between 10% *and* 25% *of the tax forms for a given year will have errors.*

Solution. *For part (a) we compute*

$$Pr(X < 0.1) = \int_0^{0.1} 90t(1-t)^8 \, \mathrm{d}t \approx 26.39\%$$

and for part (b) we compute

$$Pr(0.1 < X < 0.25) = \int_{0.1}^{0.25} 90t(1-t)^8 \, \mathrm{d}t \approx 49.21\%.$$

□

Distributions Involving Several Random Variables

We can extend the notion of the distribution of a random variable X to higher dimensions. Much of what follows utilizes ideas of multiple integrals from multivariate calculus.

Definition A.8 (Joint Distribution). *If $X_1, \ldots, X_N$ are random variables, we say that $X_1, \ldots, X_N$ have a* continuous joint distribution *if there exists a nonnegative function $f(t_1, \ldots, t_N)$ defined for all points $(t_1, \ldots, t_N) \in \mathbb{R}^N$ so that for every N-dimensional interval (or finite union of intervals) $A \subseteq \mathbb{R}^N$, the probability that $(X_1, \ldots, X_N)$ assumes a value in A can be written as the integral of f over A. That is,*

$$Pr\left((X_1, \ldots, X_N) \in A\right) = \int_A f\left(t_1, \ldots, t_N\right) \mathrm{d}t_1 \cdots \mathrm{d}t_N.$$

The function $f(t_1, \ldots, t_N)$ *is called a* joint probability density function. *This function also satisfies*

$$1 = Pr\left((X_1, \ldots, X_N) \in \mathbb{R}^N\right) = \int_{\mathbb{R}^N} f(t_1, \ldots, t_N)\, \mathrm{d}t_1 \ldots \mathrm{d}t_N = 1.$$

□

Here is an example of a joint probability function in $\mathbb{R}^2$.

Example A.8 (Bivariate Distribution). *Consider the function*

$$f(x,y) = \begin{cases} cx^2y & 0 \le y \le 4 - x^2 \\ 0 & \text{otherwise.} \end{cases}$$

Find the value for c that makes $f(x,y)$ *a joint probability density function on* $\mathbb{R}^2$ *and then find* $P(Y \ge -2X + 4)$.

Solution. *Let* $A = \{(x,y) \mid 0 \le y \le 4 - x^2\}$. *The set A is plotted in Figure A.2(a). For* $f(x,y)$ *to be a joint probability density function on* $\mathbb{R}^2$, *we need* $f(x,y) \ge 0$ *on* $\mathbb{R}^2$ *and*

$$\int_{\mathbb{R}^2} f(x,y)\, \mathrm{d}y\, \mathrm{d}x = 1.$$

Certainly, for $c > 0$, *we have* $f(x,y) \ge 0$ *for all* $(x,y) \in \mathbb{R}^2$. *Since* $f(x,y) = 0$ *for* $(x,y) \notin A$, *we have*

$$\begin{aligned}\int_{\mathbb{R}^2} f(x,y)\, \mathrm{d}y\, \mathrm{d}x &= \int_A f(x,y)\, \mathrm{d}y\, \mathrm{d}x = \int_{x=-2}^{2} \int_{y=0}^{4-x^2} cx^2y\, \mathrm{d}y\, \mathrm{d}x \\ &= c\int_{-2}^{2} x^2 \left(\int_0^{4-x^2} y\, \mathrm{d}y\right) \mathrm{d}x = \frac{c}{2}\int_{-2}^{2} x^2(4-x^2)^2\, \mathrm{d}x \\ &= c\int_0^2 x^2(4-x^2)^2\, \mathrm{d}x = \frac{1024}{105}\, c.\end{aligned}$$

Since we want the integral equal to one, we choose $c = \frac{105}{1024}$.

To find the desired probability, we must integrate over the region where A intersects the set $B = \{(x,y) \mid y \ge -2x + 4\}$. *This region is plotted in Figure A.2(b).*

We can thus compute

$$Pr(Y \ge -2X + 4) = \int_{x=0}^{2} \int_{y=-2x+4}^{4-x^2} \frac{105}{1024}\, x^2y\, \mathrm{d}y\, \mathrm{d}x = \frac{9}{32}.$$

□

In many applications we are interested in identifying the distribution function of a particular random variable when we have started with a joint density distribution.

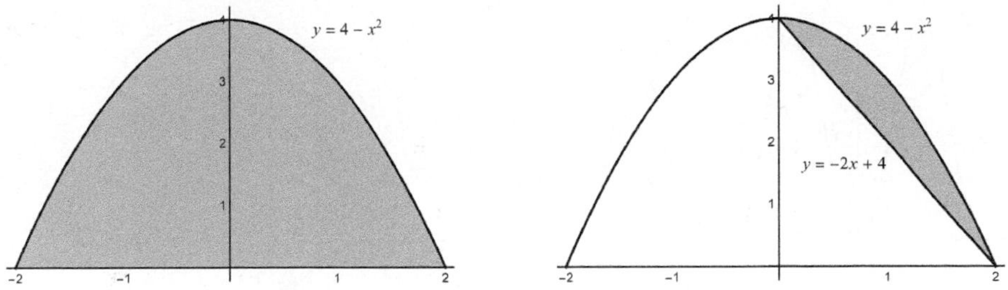

(a) A (b) $A \cap B$

Figure A.2 The sets A and $A \cap B$ from Example A.8.

Definition A.9 (Marginal Distribution). *Suppose that f is the joint probability distribution function of random variables $X_1, X_2, \ldots, X_N$. Then the distribution function f_j of random variable X_j is called the* marginal distribution function *of X_j. In the case where $X_1, X_2, \ldots, X_N$ are continuous random variables, we can write $f_j(x)$ as*

$$f_j(x) = \int_{\mathbb{R}^{N-1}} f(x_1, x_2, \ldots, x_N)\, \mathrm{d}x_1 \cdots \mathrm{d}x_{j-1}\, \mathrm{d}x_{j+1} \cdots \mathrm{d}x_N. \tag{A.5}$$

□

Thus, we see from (A.5) that we form the marginal distribution function $f_j(x)$ by "integrating out" the remaining variables in the joint distribution $f(x_1, \ldots, x_N)$.

Example A.9 (Marginal Distribution). *Let $f(x, y)$ be the joint probability density function from Example A.8. Compute the marginal distributions $f_1(x)$ and $f_2(y)$.*

Solution. *We first note from Figure A.2 that $f_1(x) = 0$ for $x < -2$ and $x > 2$. For $-2 \leq x \leq 2$, we need to integrate $f_1(x)$ over all possible values of y. But for a fixed $x \in [-2, 2]$, y ranges from zero to $4 - x^2$, so we can compute*

$$f_1(x) = \int_{-\infty}^{\infty} f(x, y)\, \mathrm{d}y = \int_0^{4-x^2} \frac{105}{1024} x^2 y\, \mathrm{d}y = \frac{105}{2048} x^2 (4 - x^2)^2.$$

For $f_2(y)$, we again use Figure A.2 to see that $f_2(y) = 0$ for $y < 0$ and $y > 4$. Moreover, when $0 \leq y \leq 4$, we can solve $y = 4 - x^2$ for x and deduce that $f(x, y) = 0$ unless $-\sqrt{4-y} \leq x \leq \sqrt{4-y}$. So the marginal distribution is

$$f_2(y) = \int_{-\infty}^{\infty} f(x, y)\, \mathrm{d}x = \frac{105}{1024}\, y \int_{-\sqrt{4-y}}^{\sqrt{4-y}} x^2\, \mathrm{d}x = \frac{35}{512}\, y\, (4 - y)^{3/2}.$$

The marginal distributions $f_1(x)$ and $f_2(y)$ are plotted in Figure A.3.

□

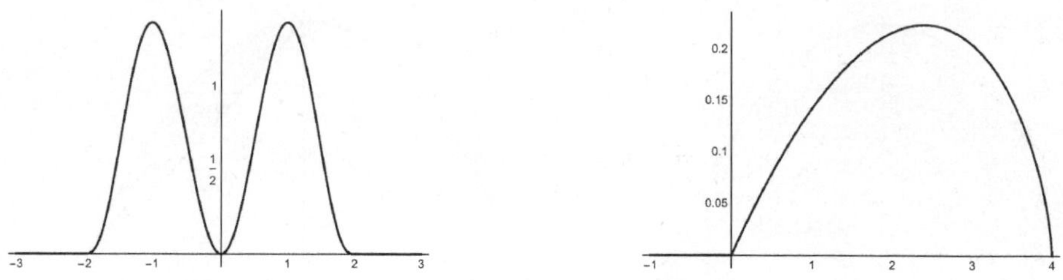

(a) $f_1(x)$ (b) $f_2(y)$

Figure A.3 Marginal distributions.

Independent Random Variables

We are often interested in random variables that are *independent*. For example, suppose that X and Y are random variables. Then X and Y are independent if the value of X has no affect on the value of Y and vice versa. We can write this notion mathematically as follows:

Definition A.10 (Independent Random Variables). *Let X and Y be random variables. Then for any two sets of real numbers A and B, we say X and Y are* independent *if*

$$Pr(X \in A \text{ and } Y \in B) = Pr(X \in A) \cdot Pr(Y \in B).$$

We can easily extend this definition for N random variables. We say that the random variables $X_1, \ldots, X_N$ are independent *if for any N sets of real numbers $A_1, \ldots, A_N$, we have*

$$Pr(X_1 \in A_1, X_2 \in A_2, \ldots, X_N \in A_N) = Pr(X_1 \in A_1) \cdots Pr(X_N \in A_N).$$

□

The next result follows without proof. The proof uses Definition A.10 along with marginal density functions (see [33], pp. 150-151).

Proposition A.2. *Suppose that $X_1, X_2, \ldots, X_N$ are continuous random variables with joint probability density function $f(x_1, \ldots, x_N)$. Further, let $f_j(x_j)$ denote the marginal probability density function of X_j, $j = 1, \ldots, N$. Then $X_1, X_2, \ldots, X_N$ are independent random variables if and only if*

$$f(x_1, x_2, \ldots, x_N) = f_1(x_1) f_2(x_2) \cdots f_N(x_N).$$

□

PROBLEMS

A.16 Suppose that $f(t)$ is the uniform distribution on $[a, b]$ with $a < b$. Find $f(t)$.

A.17 Let $r \in \mathbb{R}$ and set $S = \{r\}$. Using (A.2), compute $Pr(X \in S)$. Why does your answer makes sense for a continuous random variable X?

A.18 Consider the function

$$f(t) = \begin{cases} ce^{-2t}, & t \geq 0 \\ 0, & t < 0. \end{cases}$$

(a) Find the value for c so that $f(t)$ is a probability density function.

(b) Find $Pr(2 < X < 3)$.

(c) Find $Pr(X > 1)$.

A.19 Let c be a constant and suppose that X and Y have a continuous joint distribution

$$f(x, y) = \begin{cases} c, & 0 \leq x \leq 2,\ 0 \leq y \leq 3, \\ 0, & \text{otherwise.} \end{cases}$$

(a) What value must c assume?

(b) Find the marginal densities $f_1(x)$ and $f_2(y)$.

A.20 Suppose that a point (X, Y) is selected at random from the semicircle $0 \leq y \leq \sqrt{4 - x^2}$, $-2 \leq x \leq 2$.

(a) Find the joint probability density function for X and Y. *Hint:* Since the point is chosen at random, the joint probability density function must be constant inside the semicircle and zero outside the semicircle.

(b) Find $Pr(X \geq Y)$.

A.21 Let X and Y be random variables with joint probability density function

$$f(x, y) = \begin{cases} 9xy, & 0 \leq x \leq 2,\ 0 \leq y \leq \frac{1}{3} \\ 0, & \text{otherwise.} \end{cases}$$

Are X and Y independent random variables?

A.22 For the joint density function in Problem A.20, find the marginal densities $f_1(x)$ and $f_2(y)$.

A.4 Expectation

If you have had a calculus class, you may remember covering a section on moments and centers of mass. In particular, the x-coordinate of the center of mass of a plate is given by

$$\overline{x} = \frac{1}{A}\int_a^b x f(x)\,\mathrm{d}x. \tag{A.6}$$

where A is the area of the plate. This concept can serve as motivation for the statistical concept of *expectation* or *expected value*.

Definition A.11 (Expected Value). *The* expected value *of a random variable X with probability density function $f(t)$ is*

$$\boxed{E(X) = \int_{-\infty}^{\infty} t f(t)\,\mathrm{d}t.}$$

□

Suppose that the probability density function $f(t)$ is zero outside an interval $[a, b]$. Then

$$E(X) = \int_a^b t f(t)\,\mathrm{d}t$$

looks a lot like (A.6). As a matter of fact, it is the same since the "plate" we consider is the region bounded by $f(t)$ and the t-axis, and we know from (A.3) that this area is one.

The expected value of X is also called the *mean* of X. We can think of the expected value as an average value of the random variable X.

Example A.10 (Expected Value). *Consider the probability density function given by* (A.4). *We compute the expected value*

$$E(X) = \int_{-\infty}^{\infty} t f(t)\,\mathrm{d}t = \int_{p-1/2}^{p+1/2} t\,\mathrm{d}t = p.$$

What this computation tells us is that p is the average value for X. □

We will have the occasion to compute the expected value of a function $g(x_1, \ldots, x_N)$.

Definition A.12 (Expected Value of a Function). *Suppose that $X_1, \ldots, X_N$ are random variables with joint probability density function $f(x_1, \ldots, x_N)$. Let $g : \mathbb{R}^N \mapsto \mathbb{R}$ be some function. Then we define* the expected value of function $g(x_1, \ldots, x_N)$ *as*

$$E\left(g\left(x_1, \ldots, x_N\right)\right) = \int_{\mathbb{R}^N} g\left(x_1, \ldots, x_N\right) f\left(x_1, \ldots, x_N\right)\,\mathrm{d}x_1 \cdots \mathrm{d}x_N.$$

□

Properties of Expected Values

There are several useful properties for expected values. We have the following result regarding the expected value of linear combinations of random variables.

Proposition A.3 (Addition Properties of Expected Value). *Let a and b be any real numbers and suppose that X and Y are random variables. Then*

(a) $E(a) = a$,

(b) $E(aX + b) = aE(X) + b$,

(c) $E(X + Y) = E(X) + E(Y)$,

More generally, if $X_1, \ldots, X_N$ are random variables, then

(d) $E\left(\sum_{k=1}^{N} X_k\right) = \sum_{k=1}^{N} E(X_k)$.

□

Proof. The proof is left as Problem A.25. □

A useful property of expectation as it pertains to expected values of products of independent random variables is stated below.

Proposition A.4 (Products of Expected Values). *Suppose that $X_1, X_2, \ldots, X_N$ are independent random variables such that $E(X_j)$ exists for $j = 1, 2, \ldots, N$. Suppose that f is the joint probability density function for $X_1, \ldots, X_N$ and f_j is the marginal density function for X_j, $j = 1, \ldots, N$. Then*

$$E(X_1 \cdot X_2 \cdots X_N) = E(X_1) \cdot E(X_2) \cdots E(X_N).$$

□

Proof of Proposition A.4. We start by writing

$$E(X_1 \cdot X_2 \cdots X_N) = \int_{\mathbb{R}^N} x_1 \cdot x_2 \cdots x_N f(x_1, \ldots, x_N)\, \mathrm{d}x_1 \cdots \mathrm{d}x_N.$$

Since $X_1, \ldots, X_N$ are independent we can write this integral as

$$\begin{aligned}
E(X_1 \cdot X_2 \cdots X_N) &= \int_{\mathbb{R}^N} x_1 \cdot x_2 \cdots x_N f(x_1, \ldots, x_N)\, \mathrm{d}x_1 \cdots \mathrm{d}x_N \\
&= \int_{\mathbb{R}^N} x_1 \cdots x_N f_1(x_1) \cdots f_N(x_N)\, \mathrm{d}x_1 \cdots \mathrm{d}x_N \\
&= \int_{\mathbb{R}} \cdots \int_{\mathbb{R}} x_N f_N(x_N) \cdots x_2 f_2(x_2)\, x_1 f_1(x_1)\, \mathrm{d}x_1 \cdots \mathrm{d}x_N.
\end{aligned}$$

Consider the innermost integral. We can factor out $x_N f_N(x_N) \cdots x_2 f_2(x_2)$ so that the innermost integral is simply

$$\int_{\mathbb{R}} x_1 f_1(x_1)\, \mathrm{d}x_1 = E(X_1).$$

Now $E(X_1)$ is a constant and it can be factored out of the integral to leave

$$E(X_1 \cdot X_2 \cdots X_N) = E(X_1) \int_{\mathbb{R}} \cdots \int_{\mathbb{R}} x_N f_N(x_N) \cdots x_2 f_2(x_2)\, \mathrm{d}x_2 \cdots \mathrm{d}x_N.$$

Again, we can factor out all but $x_2 f_2(x_2)$ from the innermost integral. Thus the innermost integral is

$$\int_{\mathbb{R}} x_2 f_2(x_2)\, \mathrm{d}x_2 = E(X_2).$$

Continuing this process gives the desired result. □

Variance

While the expected value is useful for analyzing a distribution, it is restricted in what it can tell us. Consider the probability density functions

$$f_1(x) = \begin{cases} 1, & -\frac{1}{2} \le x \le \frac{1}{2} \\ 0, & \text{otherwise} \end{cases}$$

and

$$f_2(x) = \frac{1}{2} e^{-|x|}.$$

Both have expected value zero but as you can see in Figure A.4, the distributions are quite different.

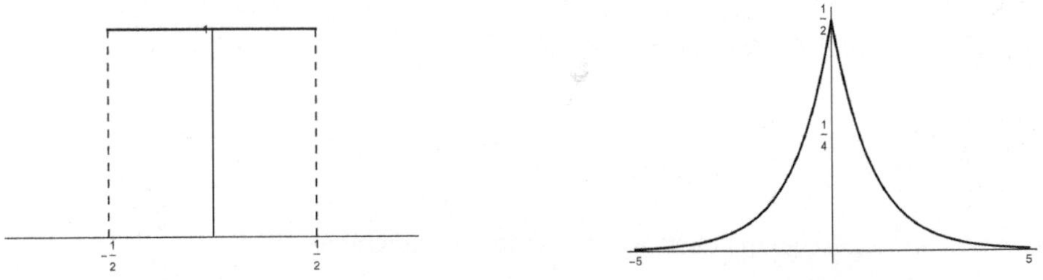

(a) $f_1(x)$ (b) $f_2(x)$

Figure A.4 The probability density functions $f_1(x)$ and $f_2(x)$.

Another measure is needed to help us better analyze distributions. The tool that is often used in this endeavor is the *variance* of a distribution. The variance is a measure that indicates how much the distribution is spread out.

Definition A.13 (Variance). *Let X be a random variable with $\mu = E(X)$. Then the* variance *of X is given by*

$$\boxed{\operatorname{Var}(X) = E\left((X-\mu)^2\right).}$$

The standard deviation *of X is given by the square root of* $\operatorname{Var}(X)$. □

Example A.11 (Computing Variance). *Find the variance for the distributions $f_1(x)$ and $f_2(x)$ plotted in Figure A.4.*

Solution. *For $f_1(x)$, $\mu = 0$, so*

$$\operatorname{Var}(X) = E\left(X^2\right) = \int_{\mathbb{R}} x^2 f_1(x)\,\mathrm{d}x = \int_{-1/2}^{1/2} x^2\,\mathrm{d}x = \frac{1}{12}.$$

Now $\mu = 0$ for $f_2(x)$ as well so

$$\operatorname{Var}(X) = E\left(X^2\right) = \frac{1}{2}\int_{\mathbb{R}} x^2 e^{-|x|}\,\mathrm{d}x = \int_0^\infty x^2 e^{-x}\,\mathrm{d}x.$$

We can use integration by parts twice to see that

$$\int x^2 e^{-x}\,\mathrm{d}x = -\left(x^2+2x+2\right)e^{-x}$$

so that

$$\int_0^\infty x^2 e^{-x}\,\mathrm{d}x = \lim_{L\to\infty} -\left(x^2+2x+2\right)e^{-x}\Big|_0^L = 2.$$

□

Properties of Variance

Just as in the case of expectation, we can list some basic properties obeyed by the variance of a distribution.

Proposition A.5 (Properties of Variance). *Let X be a random variable with a and b any constants and $\mu = E(X)$.*

(a) $\operatorname{Var}(aX+b) = a^2\operatorname{Var}(X)$,

(b) $\operatorname{Var}(X) = E\left(X^2\right) - \mu^2$.

□

Proof. For (a), we first use Proposition A.3(c) to write $E(aX+b) = aE(X)+b = a\mu+b$. We can thus compute

$$\begin{aligned}
\text{Var}(aX+b) &= E\left((aX+b-a\mu-b)^2\right) \\
&= E\left((aX-a\mu)^2\right) \\
&= a^2E\left((X-\mu)^2\right) \\
&= a^2\,\text{Var}(X).
\end{aligned}$$

Note that we have also utilized (b) from Proposition A.3 (with $b=0$) in the argument.

For (b), we again use Proposition A.3 to write

$$\begin{aligned}
\text{Var}(X) &= E\left((X-\mu)^2\right) \\
&= E\left(X^2-2X\mu+\mu^2\right) \\
&= E\left(X^2\right)-2\mu E(X)+E\left(\mu^2\right) \\
&= E\left(X^2\right)-2\mu^2+\mu^2 \\
&= E\left(X^2\right)-\mu^2.
\end{aligned}$$

□

Proposition A.6 (Sums of Variances). *Let $X_1, X_2, \ldots, X_N$ be independent random variables. Then*

$$\text{Var}\left(X_1+X_2+\cdots+X_N\right)=\text{Var}\left(X_1\right)+\text{Var}\left(X_2\right)+\cdots+\text{Var}\left(X_N\right).$$

□

Proof. This proof is by induction. Let $N=2$ and set $\mu_1=E(X_1)$ and $\mu_2=E(X_2)$. From (c) of Proposition A.3 we have

$$E\left(X_1+X_2\right)=E\left(X_1\right)+E\left(X_2\right)=\mu_1+\mu_2.$$

We use this identity to compute

$$\begin{aligned}
\text{Var}\left(X_1+X_2\right) &= E\left(\left(X_1+X_2-\mu_1-\mu_2\right)^2\right) \\
&= E\left(\left(X_1-\mu_1\right)^2+\left(X_2-\mu_2\right)^2+2\left(X_1-\mu_1\right)\left(X_2-\mu_2\right)\right) \\
&= E\left(\left(X_1-\mu_1\right)^2\right)+E\left(\left(X_2-\mu_2\right)^2\right)+2E\left(\left(X_1-\mu_1\right)\left(X_2-\mu_2\right)\right) \\
&= \text{Var}\left(X_1\right)+\text{Var}\left(X_2\right)+2E\left(\left(X_1-\mu_1\right)\left(X_2-\mu_2\right)\right).
\end{aligned}$$

Since X_1 and X_2 are independent, we can use Proposition A.4 to infer that

$$E\left(\left(X_1-\mu_1\right)\left(X_2-\mu_2\right)\right)=E\left(X_1-\mu_1\right)E\left(X_2-\mu_2\right).$$

But $E\left(X_1-\mu_1\right)=E\left(X_1\right)-E\left(\mu_1\right)=\mu_1-\mu_1=0$. In a similar way, $E\left(X_2-\mu_2\right)=0$, so the result holds for $N=2$. You are asked to prove the induction step in Problem A.31. □

PROBLEMS

A.23 Let X be a random variable whose distribution is

$$f(t) = \begin{cases} 3t^2, & 0 < t < 1 \\ 0, & \text{otherwise} \end{cases}$$

Find $E(X)$.

A.24 Let X be a random variable with probability density function f and let g be a function. Consider the statement $E(g(x)) = g(E(x))$. Either prove this statement or find a counterexample that renders it false.

A.25 Prove Proposition A.3.

A.26 Let X be a random variable with probability distribution function $f(t)$.

(a) Suppose that there exists constant a such that $Pr(X \geq a) = 1$. Show that $E(X) \geq a$.

(b) Suppose that there exists constant b such that $Pr(X \leq b) = 1$. Show that $E(X) \leq b$.

(c) Use (a) and (b) to show that if $Pr(a \leq X \leq b) = 1$, then $a \leq E(X) \leq b$.

A.27 Let a be constant and assume that $Pr(X \geq a) = 1$ and $E(X) = a$. Show that

(a) $Pr(X > a) = 0$.

(b) $Pr(X = a) = 1$.

Hint: Problem A.26 will be useful.

A.28 Use Proposition A.3 to show that if $X_1, X_2, \ldots, X_N$ are random variables such that each $E(X_j)$ exists for $j = 1, \ldots, N$ and $a_1, a_2, \ldots, a_N$ and b are any constants, then

$$E(a_1X_1 + a_2X_2 + \cdots + a_NX_N + b) = \sum_{j=1}^{N} a_k E(X_k) + b.$$

A.29 Suppose that $f(t)$ is the uniform distribution on $[0, 3]$. That is,

$$f(t) = \begin{cases} \frac{1}{3}, & 0 \leq t \leq 3 \\ 0, & \text{otherwise} \end{cases}$$

Assume that X_1 and X_2 are random variables. Find $E((X_1 + X_2)(2X_1 - X_2))$. You may assume that $X_1 + X_2$ and $2X_1 - X_2$ are independent random variables.

A.30 Find $\text{Var}(X)$ for the distribution in Problem A.23.

A.31 Complete the proof of Proposition A.6. That is, assume Var $(X_1 + \cdots + X_N) =$ Var $(X_1) + \cdots +$ Var (X_N) and show that

$$\text{Var}\,(X_1 + \cdots + X_N + X_{N+1}) = \text{Var}\,(X_1) + \cdots + \text{Var}\,(X_N) + \text{Var}\,(X_{N+1})\,.$$

A.32 Use Proposition A.6 in conjunction with Proposition A.5 to show that if $X_1, X_2, \ldots, X_N$ are independent random variables and $b, c_1, c_2, \ldots, c_N$ are constants, then

$$\text{Var}\,(c_1 X_1 + \cdots c_N X_N + b) = c_1^2 \text{Var}\,(X_1) + \cdots c_N^2 \text{Var}\,(X_N)\,.$$

A.33 Suppose X_1 and X_2 are independent random variables with $E\,(X_1) = E\,(X_2)$. Assume also that Var (X_1) and Var (X_2) exist. Show that

$$E\Big((X_1 - X_2)^2\Big) = \text{Var}\,(X_1) + \text{Var}\,(X_2)\,.$$

A.5 Two Special Distributions

We make use of two special distributions in this book. The first is the well-known normal distribution and the second is the χ^2 distribution.

The Normal Distribution and the χ^2 Distribution

Definition A.14 (Normal Distribution). *A random variable X has a* normal distribution *if for real numbers μ and σ with $\sigma > 0$, the probability density function is*

$$\boxed{f(t) = \frac{1}{\sqrt{2\pi}\,\sigma} e^{-\frac{(t-\mu)^2}{2\sigma^2}}.} \tag{A.7}$$

$\mu = E(X)$ is the mean and σ^2 is the variance. □

Since $\sigma > 0$, it is clear that $f(t) > 0$. In Problem A.37, you will show that $\int_{\mathbb{R}} f(t)\,\mathrm{d}t = 1$.

The probability density function $f(t)$ from (A.7) with mean $\mu = 0$ is plotted in Figure A.5.

Definition A.15 (χ^2 Distribution). *Let n be a positive integer and X a random variable. Then X has a* χ^2 distribution with n degrees of freedom *if the probability density function is*

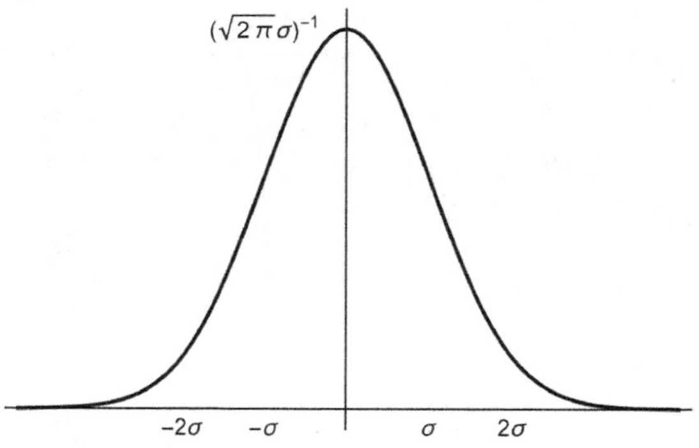

Figure A.5 The normal distribution with mean zero and standard deviation σ.

$$f(t) = \begin{cases} \frac{1}{2^{n/2}\Gamma(n/2)}\, t^{(n/2)-1}\, e^{-t/2}, & t > 0 \\ 0, & \text{otherwise} \end{cases} \tag{A.8}$$

□

The *gamma function* $\Gamma(t)$ (see Problem A.38) is utilized in the definition of the χ^2 distribution. You can think of $\Gamma(t)$ as a generalized factorial function. Indeed, $\Gamma(k) = (k-1)!$ whenever k is a positive integer. In the case where k is an odd positive integer, formulas exist for computing $\Gamma(k/2)$. For more information on the gamma function, see Carlson [19].

It can be shown that $\Gamma(n/2) > 0$ whenever n is a positive integer, so that $f(t) \geq 0$. In Problem A.42 you will show that $\int_{\mathbb{R}} f(t)\, dt = 1$.

In Figure A.6, χ^2 distributions are plotted for $n = 3, 4, 5, 6$. In Problem A.41, you are asked to plot the χ^2 distributions for $n = 1, 2, 7$.

The expected value for X when X has the χ^2 distribution is easy to compute.

Proposition A.7 (Expected Value of χ^2 Distribution). *If X is a χ^2 random variable with n degrees of freedom, then*

$$E(X) = n.$$

□

Proof. The proof is left as Problem A.43. □

The following result appears in [99]. We need it in Chapter 6.

Theorem A.1. *Suppose that X is a normally distributed random variable with mean $E(X) = \mu$ and standard deviation σ. Define the random variable*

$$Z = \frac{X - \mu}{\sigma}.$$

Then Z^2 has a χ^2 distribution with one degree of freedom. □

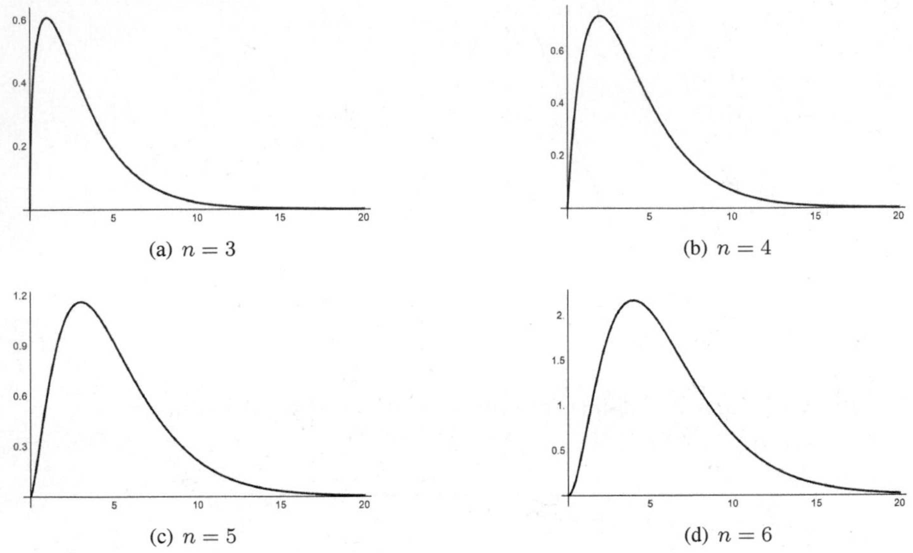

Figure A.6 The χ^2 probability density functions with $n = 3, 4, 5, 6$.

PROBLEMS

A.34 Use a **CAS** to plot the following normal distributions on the same set of axes.

(a) $\mu = 1, \sigma = 2$

(b) $\mu = 1, \sigma = \frac{1}{4}$

(c) $\mu = 1, \sigma = 5$

A.35 Suppose that X is normally distributed with mean zero and $\sigma = 1$. Use a **CAS** (if necessary) to compute the following probabilities.

(a) $Pr(X \geq 0)$

(b) $Pr(X < 1.5)$

(c) $Pr(|X| < 2.5)$

(d) $Pr(0 \leq -3X + 2 \leq 4)$

A.36 Use what you know about the normal distribution with mean zero to compute the following integrals:

(a) $\int_{\mathbb{R}} e^{-t^2/2}\, dt$

(b) $\int\limits_0^\infty e^{-t^2/2}\,dt$

(c) $\int\limits_0^\infty e^{-2t^2}\,dt$

A.37 In this problem you will show that the probability density function (A.7) has a unit integral. This problem requires some knowledge of multi-variable calculus. The following steps will help you organize your work.

(a) First consider the integral $I = \int_{\mathbb{R}} e^{-x^2}\,dx$. Then

$$I^2 = \left(\int_{\mathbb{R}} e^{-x^2}\,dx\right)^2 = \int_{\mathbb{R}} e^{-x^2}\,dx \int_{\mathbb{R}} e^{-y^2}\,dy.$$

Convert the integral above to polar coordinates and show that

$$I^2 = 2\pi \int_0^\infty re^{-r^2}\,dr.$$

(b) Use u-substitution to compute the integral above and thereby show that $I^2 = \pi$, or equivalently,

$$I = \int\limits_{\mathbb{R}} e^{-x^2}\,dx = \sqrt{\pi}. \tag{A.9}$$

(c) Now consider the probability density function (A.7). Make the substitution $u = \frac{t-\mu}{\sqrt{2}\,\sigma}$ in conjunction with (A.9) to show that the normal probability density function has a unit integral.

A.38 In this problem we formally define the gamma function and prove a fundamental property about it. For $a > 0$, define the *gamma function* as

$$\Gamma(a) = \int_0^\infty t^{a-1}e^{-t}\,dt. \tag{A.10}$$

(a) Show that $\Gamma(1) = 1$.

(b) Use integration by parts to show that for $a > 1, \Gamma(a) = (a-1)\Gamma(a-1)$.

(c) Use (a) and (b) to show that if n is a positive integer, then $\Gamma(n) = (n-1)!$.

A.39 In this problem you will show that $\Gamma(1/2) = \sqrt{\pi}$. *Hint:* Use (A.10) with $a = \frac{1}{2}$ to write down the integral form of $\Gamma(1/2)$ and then make the substitution $u = \sqrt{t}$. Use (A.9).

A.40 Use Problem A.39 along with Problem A.38(b) to show that $\Gamma(7/2) = \frac{15\sqrt{\pi}}{8}$.

A.41 Use a **CAS** to sketch the χ^2 distribution for $n = 1, 2, 7$. You will find Problems A.39 and A.40 helpful.

A.42 In this problem, you will show that the probability density function (A.8) has unit integral. *Hint:* Make the substitution $u = t/2$ in (A.8), simplify, and then use (A.10).

A.43 In this problem you will prove Proposition A.7. The following steps will help you organize your work.

(a) Show that $E(X) = \frac{1}{2^{n/2}\Gamma(n/2)} \int\limits_0^\infty t^{n/2}e^{-t/2}\,dt$.

(b) Make an appropriate u-substitution to show that

$$E(X) = \frac{2}{\Gamma(n/2)}\Gamma\left(\frac{n}{2}+1\right).$$

(c) Use Problem A.38(b) on (b) to conclude that $E(X) = n$.

A.44 Let $k = 1, 2, \ldots$ and suppose that X is a random variable that has a χ^2 distribution with n degrees of freedom. In this problem you will compute $E(X^k)$. The following steps will help you organize your work:

(a) Write down the integral that represents $E\left(X^k\right)$ and then make an appropriate u-substitution to write

$$E\left(X^k\right) = \frac{2^k}{\Gamma(n/2)}\int_0^\infty u^{n/2+k-1}e^{-u}\,du = \frac{2^k}{\Gamma(n/2)}\Gamma\left(\frac{n}{2}+k\right).$$

(b) Repeatedly use (b) from Problem A.38 to show that

$$\Gamma\left(\frac{n}{2}+k\right) = \frac{n}{2}\left(\frac{n}{2}+1\right)\cdots\left(\frac{n}{2}+k-1\right)\Gamma(n/2)$$

and thus show that

$$E(X^k) = 2^k\frac{n}{2}\left(\frac{n}{2}+1\right)\cdots\left(\frac{n}{2}+k-1\right).$$

A.45 Use Problem A.44 and Proposition A.7 to show that $\mathrm{Var}(X) = 2n$ when X has the χ^2 distribution with n degrees of freedom.

REFERENCES

1. E. Aboufadel, J. Olsen, and J. Windle. Breaking the Holiday Inn Priority Club CAPTCHA. *The Coll. Math. J.*, 36(2), March 2005.

2. Robert G. Bartle and Donald R. Sherbert. *Introduction to Real Analysis, 3rd ed.* Wiley, Hoboken, NJ, 1999.

3. G. Benke, M. Bozek-Kuzmicki, D. Colella, G. M. Jacyna, and J. J. Benedetto. Wavelet-based analysis of electroencephalogram (EEG) signals for detection and localization of epileptic seizures. In Harold H. Szu, editor, *Proceedings of SPIE: Wavelet Applications II*, volume 2491, pages 760–769, April 1995.

4. J. Berger, R. R. Coifman, and M. J. Goldberg. Removing noise from music using local trigonometric bases and wavelet packets. *J. Audio Eng. Soc.*, 42(10):808–818, 1994.

5. Jonathan Berger. Brahms at the piano. CCRMA, Stanford University, Stanford, CA, 1999. `ccrma.stanford.edu/groups/edison/brahms/brahms.html`.

6. A. Bijaoui, E. Slezak, F. Rue, and E. Lega. Wavelets and the study of the distant universe. *Proc. of the IEEE*, 84(4):670–679, 1996.

7. Albert Boggess and Francis J. Narcowich. *A First Course in Wavelets with Fourier Analysis*. Prentice Hall, Upper Saddle River, NJ, 2001.

8. J. N. Bradley, C. M. Brislawn, and T. Hopper. WSQ gray-scale fingerprint image compression specification. Technical Report Revision 1, Federal Bureau of Investigation, Washington, DC, February 1992.

Discrete Wavelet Transformations: An Elementary Approach With Applications, Second Edition.
Patrick J. Van Fleet.

9. J. N. Bradley, C. M. Brislawn, and T. Hopper. The FBI wavelet/scalar quantization standard for gray-scale fingerprint image compression. In *Proceedings of SPIE: Visual Information Processing*, volume 1961, pages 293–304, Orlando, FL, April 1993.

10. J. N. Bradley, C. M. Brislawn, R. J. Onyshczak, and T. Hopper. The FBI compression standard for digitized fingerprint images. In *Proceedings of SPIE: Applied Digital Image Processing XIX*, volume 2847, pages 344–355, Denver, CO, August 1996.

11. C. M. Brislawn. Fingerprints go digital. *Notices Amer. Math. Soc.*, 42(11):1278–1283, November 1995.

12. Richard A. Brualdi. *Introductory Combinatorics, 4th ed.* Prentice Hall, Upper Saddle River, NJ, 2004.

13. A. Bruce and H.-Y. Gao. Waveshrink with firm shrinkage. *Statistica Sinica*, 7:855–874, 1997.

14. A. G. Bruce and H.-Y. Gao. Understanding waveshrink: Variance and bias estimation. *Biometrika*, 83:727–745, 1996.

15. Richard L. Burden and J. Douglas Faires. *Numerical Analysis, 8th ed.* Brooks/Cole, Pacific Grove, CA, 2005.

16. P. J. Burt and E. H. Adelson. The Laplacian pyramid as a compact image code. *IEEE Trans. Comm.*, 31(4):532–540, April 1983.

17. A. Calderbank, I. Daubechies, W. Sweldens, and B.-L. Yeo. Wavelet transforms that map integers to integers. *Appl. Comp. Harm. Anal.*, 5(3):332–369, 1998.

18. J. Canny. A computational approach to edge detection. *IEEE Trans. Pattern Anal. and Mach. Intell.*, 8:679–714, 1986.

19. Bille C. Carlson. *Special Functions of Applied Mathematics*. Academic Press, San Diego, CA, 1977.

20. C. Christopoulos, A. Skodras, and T. Ebrahimi. The JPEG2000 still image coding system: An overview. *IEEE Trans. on Consumer Electronics*, 46(4):1103–1127, November 2000.

21. Charles K. Chui. *An Introduction to Wavelets, Volume 1, (Wavelet Analysis and Its Applications)*. Academic Press, San Diego, CA, 1992.

22. B. Cipra. Wavelet applications come to the fore. *SIAM News*, 26(7), November 1993.

23. A. Cohen, I. Daubechies, and J.-C. Feauveau. Biorthogonal bases of compactly supported wavelets. *Comm. Pure Appl. Math.*, 45:485–560, 1992.

24. R. Coifman, Y. Meyer, and M. V. Wickerhauser. Size properties of wavelet packets. In *Wavelets and Their Applications*, pages 453–470. Jones and Bartlett, Boston, MA, 1992.

25. R. Coifman, Y. Meyer, and M. V. Wickerhauser. Wavelet analysis and signal processing. In *Wavelets and Their Applications*, pages 153–178. Jones and Bartlett, Boston, MA, 1992.

26. R. Coifman and M. V. Wickerhauser. Entropy-based algorithms for best basis selection. *IEEE Trans. Inform. Theory*, 38:713–718, 1992.

27. Keith Conrad. Probability distributions and maximum entropy. University of Connecticut, 2005. `math.uconn.edu/~kconrad/blurbs/`.

28. Wolfgang Dahmen, Andrew Kurdila, and Peter Oswald. *Multiscale Wavelet Methods for Partial Differential Equations (Wavelet Analysis and Its Applications)*. Academic Press, San Diego, CA, 1997.

29. I. Daubechies. Orthogonal bases of compactly supported wavelets. *Comm. Pure Appl. Math.*, 41:909–996, 1988.

30. I. Daubechies. Orthonormal bases of compactly supported wavelets: II. Variations on a theme. *SIAM J. Math. Anal.*, 24(2):499–519, 1993.

31. I. Daubechies and W. Sweldens. Factoring wavelet transforms into lifting steps. *J. Fourier Anal. Appl.*, 4(3):245–267, 1998.

32. Ingrid Daubechies. *Ten Lectures on Wavelets*. Society for Industrial and Applied Mathematics, Philadelphia, PA, 1992.

33. Morris H. DeGroot and Mark J. Schervish. *Probability and Statistics, 4th ed.* Addison Wesley, Reading, MA, 2012.

34. Inc. DigitalGlobe. Worldview-2 satellite. Digital Globe, Westminster, CO, 2016. `www.digitalglobe.com/`.

35. D. Donoho. Wavelet shrinkage and W.V.D.: A 10-minute tour. In Y. Meyer and S. Rogues, editors, *Progress in Wavelet Analysis and Applications*, Editions Frontiers, pages 109–128, Toulouse, France, 1992.

36. D. Donoho and I. Johnstone. Ideal spatial adaptation via wavelet shrinkage. *Biometrika*, 81:425–455, 1994.

37. D. Donoho and I. Johnstone. Adapting to unknown smoothness via wavelet shrinkage. *J. Amer. Stat. Assoc.*, 90(432):1200–1224, December 1995.

38. David Donoho, Mark Reynold Duncan, Xiaoming Huo, and Ofer Levi-Tsabari. Wavelab 802, August 1999. `statweb.stanford.edu/~wavelab/`.

39. R. Dugad, K. Rataconda, and N. Ahuja. A new wavelet-based scheme for watermarking images. In *Proceedings of the International Conference on Image Processing 1998*, volume 2, pages 419–423, Chicago, IL, October 1998.

40. Paul M. Embree and Bruce Kimble. *C Language Algorithms for Digital Signal Processing*. Prentice Hall, Upper Saddle River, NJ, 1991.

41. Federal Bureau of Investigation. Criminal Justice Information Services (CJIS) Web site. Federal Bureau of Investigation, Washington, DC, 2009. `www.fbi.gov/services/cjis`.

42. Michael W. Frazier. *An Introduction to Wavelets Through Linear Algebra*. Undergraduate Texts in Mathematics. Springer-Verlag, New York, NY, 1999.

43. John E. Freund. *Mathematical Statistics*. Prentice-Hall, Upper Saddle River, NJ, Second edition, 1971.

44. A. Gavlasová, A. Procházka, and M. Mudrová. Wavelet based image segmentation. In *Proceedings of the 14th Annual Conference of Technical Computing*, Prague, 2006.

45. Ramazan Gençay, Faruk Selçuk, and Brandon Whitcher. *An Introduction to Wavelets and Other Filtering Methods in Finance and Economics*. Academic Press, San Diego, CA, 2002.

46. Allen Gersho and Robert M. Gray. *Vector Quantization and Signal Compression*. Kluwer Academic, Norwell, MA, 1992.

47. Rafael C. Gonzalez and Richard E. Woods. *Digital Image Processing*. Addison-Wesley, Boston, MA, 1992.

48. Rafael C. Gonzalez and Richard E. Woods. *Digital Image Processing, 3rd ed.* Pearson Prentice Hall, Upper Saddle River, NJ, 2006.

49. Rafael C. Gonzalez, Richard E. Woods, and Steven L. Eddins. *Digital Image Processing Using Matlab*. Pearson Prentice Hall., Upper Saddle River, NJ, 2004.

50. A. Grossmann and J. Morlet. Decomposition of Hardy functions into square integrable wavelets of constant shape. *SIAM J. Math. Anal.*, 15(4):723–736, July 1984.

51. A. Haar. Zur theorie der orthogonalen funktionensysteme, (erste mitteilung). *Mathematische Annalen*, 69:331–371, 1910.

52. F. Hampel. The influence curve and its role in robust estimation. *J. Amer. Stat. Assoc.*, 69(346):383–393, June 1974.

53. Mark R. Hawthorne. *Fingerprints: Analysis and Understanding*. CRC Press, Boca Raton, FL, 2008.

54. T. He, S. Wang, and A. Kaufman. Wavelet-based volume morphing. In *Proceedings of Visualization 94*, pages 85–92, Washington, DC, October 1994.

55. M. Hennessey, J. Jalkio, C. Greene, and C. Sullivan. Optimal routing of a sailboat in steady winds. Technical report, University of St. Thomas, St. Paul, MN, May 2007.

56. S. G. Henry. Catch the (seismic) wavelet. *AAPG Explorer*, pages 36–38, March 1997.

57. S. G. Henry. Zero phase can aid interpretation. *AAPG Explorer*, pages 66–69, April 1997.

58. Barbara Burke Hubbard. *The World According to Wavelets: The Story of a Mathematical Technique in the Making*. A. K. Peters, Wellesley, MA, Second edition, 1998.

59. D. A. Huffman. A method for the construction of minimum-redundancy codes. *Proc. Inst. Radio Eng.*, 40:1098–1101, September 1952.

60. ISO/IEC JTC 1/SC 29. Information technology – JPEG 2000 image coding system: Core coding system. Technical Report 15444-1:2016, International Organization for Standardization, Geneva, Switzerland, 2016.

61. Arne Jensen and Anders la Cour-Harbo. *Ripples in Mathematics: The Discrete Wavelet Transform*. Springer-Verlag, New York, NY, 2001.

62. David W. Kammler. *A First Course in Fourier Analysis*. Prentice Hall, Upper Saddle River, NJ, 2000.

63. Fritz Keinert. *Wavelets and Multiwavelets*. Chapman & Hall/CRC, Boca Raton, FL, 2004.

64. Thomas W. Körner. *Fourier Analysis*. Cambridge University Press, New York, NY, 1989.

65. David Lay, Steven Lay, and Judi McDonald. *Linear Algebra and Its Applications*. Pearson, London, 5th edition, 2015.

66. D. LeGall and A. Tabatabai. Subband coding of digital images using symmetric short kernel filters and arithmetic coding techniques. In *Proceedings of the ICASSP 1988, IEEE*, pages 761–764, New York, April 1988.

67. Jun Li. A wavelet approach to edge detection. Master's thesis, Sam Houston State University, Huntsville, TX, 2003.

68. S. Mallat. Multiresolution approximations and wavelet orthonormal bases of $l^2(\mathbb{R})$. *Trans. Amer. Math. Soc.*, 315:69–87, September 1989.

69. S. Mallat and W. L. Hwang. Singularity detection and processing with wavelets. *IEEE Trans. Inf. Th.*, 38(2):617–643, March 1992.

70. D. Marr and E. Hildreth. Theory of edge detection. *Proc. R. Soc. London*, 207, 1980.

71. Yves Meyer. *Wavelets and Operators*. Advanced Mathematics. Cambridge University Press, New York, NY, 1992.

72. N. Mitianoudis, Georgios Tzimiropoulos, and Tania Stathaki. Fast wavelet-based pan-sharpening of multi-spectral images. In *Proceedings of the 2010 IEEE International Conference on Imaging Systems and Techniques*, Thessaloniki, July 2010.

73. Steve Oualline. *Practical C++ Programming*. O'Reilly, Sebastopol, CA, 2003.

74. William B. Pennebaker and Joan L. Mitchell. *JPEG Still Image Compression Standard*. Van Nostrand Reinhold, New York, 1992.

75. Charles A. Poynton. *A Technical Introduction to Digital Video*. Wiley, Hoboken, NJ, 1996.

76. K. Ramamohan Rao and Patrick Yip. *Discrete Cosine Transform: Algorithms, Advantages, Applications*. Academic Press, San Diego, CA, 1990.

77. David K. Ruch and Patrick J. Van Fleet. *Wavelet Theory: An Elementary Approach with Applications*. Wiley, Hoboken, NJ, 2009.

78. Walter Rudin. *Principles of Mathematical Analysis, 3rd ed.* McGraw-Hill, New York, NY, 1976.

79. John C. Russ. *The Image Processing Handbook, 4th ed.* CRC Press, Boca Raton, FL, 2002.

80. Khalid Sayood. *Introduction to Data Compression, 2nd ed.* Morgan Kaufmann, San Francisco, CA, 2000.

81. C. E. Shannon. A mathematical theory of communication. *Bell Syst. Tech. J.*, 27(3):379–423, July 1948. Continued in 27(4): 623-656, October 1948.

82. Steven Smith. *Digital Signal Processing: A Practical Guide for Engineers and Scientists*. Newnes, Elsevier Science, Amsterdam, 2002.

83. C. M. Stein. Estimation of the mean of a multivariate normal distribution. *Ann. Stat.*, 9(6):1135–1151, November 1981.

84. James Stewart. *Calculus, 8th ed.* Cengage Learning, Boston, MA, 2015.

85. G. Strang. The discrete cosine transform. *SIAM Rev.*, 41(1):135–147, 1999.

86. Gilbert Strang. *Introduction to Linear Algebra, 3rd ed.* Wellesley Cambridge Press, Wellesley, MA, 2003.

87. Gilbert Strang. *Linear Algebra and Its Applications, 4th ed.* Brooks/Cole, Pacific Grove, CA, 2006.

88. Gilbert Strang and Truong Nguyen. *Wavelets and Filter Banks*. Wellesley Cambridge Press, Wellesley, MA, 1996.

89. United States Geological Survey. Landsat 8. USGS National Center, Reston, VA, 2016. `landsat.usgs.gov/landsat-8`.

90. Web Technology Surveys. Usage of image file formats for websites. Q-Success, 2017. `https://w3techs.com/technologies/overview/image_format/all`.

91. W. Sweldens. Wavelets and the lifting scheme: A 5 minute tour. *Z. Angew. Math. Mech.*, 76 (Suppl. 2):41–44, 1996.

92. W. Sweldens. The lifting scheme: A construction of second generation wavelets. *SIAM J. Math. Anal.*, 29(2):511–546, 1997.

93. C. Taswell. The what, how, and why of wavelet shrinkage denoising. Technical Report CT-1998-09, Computational Toolsmiths, Stanford, CA, January 1999.

94. D. Taubman. High performance scalable image compression with EBCOT. *IEEE Trans. Image Proc.*, 9:1158–1170, July 2000.

95. David Taubman and Michael Marcellin. *JPEG2000: Image Compression Fundamentals, Standards and Practice*. The International Series in Engineering and Computer Science. Kluwer Academic Publishers, Norwell, MA, 2002.

96. M. Unser and T. Blu. Mathematical properties of the JPEG2000 wavelet filters. *IEEE Trans. Image Proc.*, 12(9):1080–1090, 2003.

97. K. Varma and A. Bell. JPG2000 – choices and tradeoffs for encoders. *IEEE Signal Proc. Mag.*, pages 70–75, November 2004.

98. Brani Vidakovic. *Statistical Modeling by Wavelets*. Wiley, Hoboken, NJ, 1999.

99. Dennis D. Wackerly, William Mendenhall III, and Richard L. Scheaffer. *Mathematical Statistics with Applications*. Duxbury, Belmont, CA, Sixth edition, 2002.

100. David F. Walnut. *An Introduction to Wavelet Analysis*. Birkhäuser, Cambridge, MA, 2002.

101. Peter Wayner. *Compression Algorithms for Real Programmers*. Morgan Kaufmann, San Francisco, CA, 2000.

102. Mladen Victor Wickerhauser. *Adapted Wavelet Analysis from Theory to Software*. A. K. Peters, Wellesley, MA, 1994.

103. X.-G. Xia, C. G.Boncelet, and G. R. Arce. A multiresolution watermark for digital images. In *International Conference on Image Processing 1997*, volume 1, pages 548–551, Santa Barbara, CA, October 1997.

104. B. L. Yoon and P. P. Vaidyanathan. Wavelet-based denoising by customized thresholding. In *Proceedings of the 29th International Conference on Acoustics, Speech, and Signal Processing*, Montreal, Canada, May 2004.

Index

Patrick J. Van Fleet.